High Performance Computing in Science and Engineering ’14

Wolfgang E. Nagel • Dietmar H. Kröner •
Michael M. Resch
Editors

High Performance Computing in Science and Engineering '14

Transactions of the High Performance Computing Center, Stuttgart (HLRS) 2014

Editors
Wolfgang E. Nagel
Zentrum für Informationsdienste
und Hochleistungsrechnen (ZIH)
Technische Universität Dresden
Dresden
Germany

Dietmar H. Kröner
Abteilung für Angewandte Mathematik
Universität Freiburg
Freiburg
Germany

Michael M. Resch
Höchstleistungsrechenzentrum
Stuttgart (HLRS)
Universität Stuttgart
Stuttgart
Germany

Front cover figure: Shape of the flame front of a lean propane-air flame which interacts with large-scale turbulent vortexes. The flame front is colored by the heat release rate. At the locations where the flame front is bended towards the burnt gases, the intensity of the reaction rate of the fuel is increased considerably (red color). Details can be found in "Lean Premixed Flames: A Direct Numerical Simulation Study of the Effect of Lewis Number at Large Scale Turbulence", by J.A. Denev, I. Naydenova and H. Bockhorn, Engler-Bunte Institute, Combustion Division, Karlsruhe Institute of Technology, Karlsruhe, Germany, on page 237ff

---

ISBN 978-3-319-10809-4 ISBN 978-3-319-10810-0 (eBook)
DOI 10.1007/978-3-319-10810-0
Springer Cham Heidelberg New York Dordrecht London

Library of Congress Control Number: 2014959094

Mathematics Subject Classification (2010): 65Cxx, 65C99, 68U20

Printed on acid-free paper

Springer is part of Springer Science+Business Media (www.springer.com)

Preface

The new Cray XC40 supercomputer Hornet – completion of installation in November 2014 – is the new flagship computer at HLRS replacing the previous supercomputer Hermit after approximately 3 years in service. With a peak performance of 3.8 PFLOP/s it almost quadruples the performance of its predecessor. The system is based on the new Intel Xeon processors, formerly code-named 'Haswell', and the Cray Aries system interconnect. In its current configuration state, Hornet consists of 21 cabinets hosting 3,944 compute nodes, which sum up to a total of 94,656 compute cores. The system's main memory capacity is 493 terabytes. Users will benefit in particular from the quadrupled storage space the HLRS supercomputing infrastructure provides: 5.4 petabytes of file storage with an input/output speed in the range of 150 gigabyte/s is available to meet the performance challenges of today's most demanding users of HPC (high-performance computing) systems. The Cray XC40 system Hornet ranks as number 16 in the current Top500 list and achieved a Linpack value of 2,763 TFLOP/s. As part of the GCS, HLRS participates in the European project PRACE (Partnership for Advanced Computing in Europe), extending its reach to all European member countries, and has already provided hundreds of millions of cpu hours to the European user community. Additionally, HLRS participates with partners in Germany in two Exascale Software Initiatives at European Level, namely CRESTA (http://cresta.epcc.ed.ac.uk/) and Mont-Blanc 2 (http://www.montblanc-project.eu/), where the challenges on the efficient use of current and future computing systems are investigated.

While the GCS has successfully addressed the high-end computing needs, it was clear from the very beginning that an additional layer of support is required to maintain the longevity of the centre, via a network of competence centres across Germany. This gap is addressed by the Gauß-Allianz (GA), in which regional and local centres teamed up to create the necessary infrastructure, knowledge, and the required methods and tools. The mission of the GA is to coordinate the HPC-related activities of its members. By providing versatile computing architectures and by combining the expertise of the participating centres, the necessary ecosystem for computational science has been created. In October 2014, a new project has been started to strengthen the governance of the GA over the next 3 years.

Most of the projects from the second BMBF HPC-call have ended in mid-2014, while the projects from the third BMBF HPC-call will run for approximately one more year. This call was directed towards proposals that enable and support the execution of petascale applications on more than 100,000 processors. While the projects of the first funding round started in early 2009 and those of the second round in April 2011, the third (and thus far last) call had been delayed again by more than 18 months. Nevertheless, all experts and administration authorities continue to acknowledge the strong need for such a funding program, given that the main issue identified in nearly all applications is that of *scalability*. There is still the strong hope – and need – of follow-up calls over the next 5-years, for projects that develop scalable algorithms, methods, and tools to support massively parallel systems. This can be seen as a very large investment. Nevertheless, in relation to the investment in computing hardware within Germany over this 5 year period, the investment in software is still comparatively very small amounting to less than 10 % of the hardware investment. Furthermore, the investment in software will produce the 'brains' that will be needed to use the newly developed innovative methods and tools, to accomplish technological breakthroughs in scientific as well as industrial fields of applications.

It is widely known that the long-term target is aimed not only at petascale but at exascale systems as well. We do not only need competitive hardware but also excellent software and methods to address – and solve – the most demanding problems in science and engineering. The success of this approach is highly important for our community and will also greatly influence the development of new technologies and industrial products. Beyond being highly significant, the success of this approach will finally determine whether Germany will be an accepted partner alongside the leading technology and research nations worldwide.

With the support of the German Research Foundation (DFG), in January 2013 we have started the additional Priority Program 1648 'Software for Exascale Computing (SPPEXA)' in the field of HPC. As time is passing by so fast, a new round of proposals are already expected to be submitted on January 15, 2015, in response to the second phase of this Priority Program. This present call invites proposals for the second 3-year funding period from 2016 to 2018 and is now opening up to foster international collaboration. The call is intended to support collaborative projects of bilateral or trilateral research teams, bringing together researchers from France (ANR), Germany (DFG), and Japan (JST).

Since 1996, the HLRS has supported the scientific community as part of its official mission. As in the previous years, the major results of the last 12 months were presented at the 16th Annual Results and Review Workshop on High Performance Computing in Science and Engineering, held on September 29–30, 2014, at the Universität Stuttgart. The workshop proceedings contain the written versions of the research work presented. The papers were selected from all projects running at the HLRS and the SSC Karlsruhe during a 1-year period between October 2013 and September 2014. Overall, a number of 45 papers were chosen from Physics; Molecules, Surfaces, and Solids; Reactive Flow; Computational Fluid Dynamics (CFD); Transport and Climate; and Miscellaneous Topics. The largest number of contributions originated from the CFD field, as in many previous years, with 18 papers. Even though such a small collection cannot entirely represent an area this vast, the selected papers demonstrate the state-of-the-art use of high-performance computing in Germany. The authors were encouraged to emphasize the computational techniques used in solving the scientific or engineering problems examined. This is an often forgotten aspect and was the major focus of the workshop. Nevertheless, the importance of the newly computed scientific results for the specific disciplines is impressive.

We gratefully acknowledge the continuing support of the federal state of Baden-Württemberg in promoting and supporting high-performance computing. Grateful acknowledgments are also due to the German Research Foundation (Deutsche Forschungsgemeinschaft (DFG)) and the Germany Ministry for Research and Education (BMBF), as many projects pursued on the HLRS and SSC computing machines could not have been carried out without their support. Also, we thank Springer Verlag for publishing this volume and, thus, for helping to position the national scientific activities in an international framework. We hope that this series of publications contributes to the global promotion of high-performance scientific computing.

Dresden, Germany — Wolfgang E. Nagel
Freiburg, Germany — Dietmar H. Kröner
Stuttgart, Germany — Michael M. Resch
November 2014

Contents

Part I Physics

Part I
Physics

Peter Nielaba

In this section, nine physics projects are described, which achieved important scientific results by using the HPC resources of the HLRS. Fascinating new results are being presented in the following pages for quantum systems (quarks, many body dynamics, atomic and molecular collisions), soft matter systems (colloids), earth science (subsurface structures) and astrophysical systems (solar corona, cosmic ray induced air showers, search for gravitational wave signals).

The studies of the (colloidal) soft matter systems have focused on their dynamical properties, the effect of boundary conditions, the determination of interfacial tensions, and on crystallization phenomena.

F. Schmitz, A. Statt, P. Virnau and K. Binder from the University of Mainz in their project *colloid* have investigated logarithmic finite-size corrections in the determination of the interfacial tension with large scale Monte Carlo simulations, and the effect of applied boundary conditions and the ensemble. By computations on Cray XE6 (HERMIT) at the HLRS, using Monte Carlo simulations, ensemble switch methods and finite size scaling techniques, properties of Lennard-Jones systems and of hard sphere systems have been studied and analyzed as well as of the Ising model, in particular the Laplace pressure of a crystalline nucleus surrounded by a liquid, in comparison to the results of the classical nucleation theory.

J. Zhou, J. Smiatek, E. S. Asmolov, O. I. Vinogradova and F. Schmid from the University of Mainz in their project *CCAC* studied the effect of a tunable partial-slip (Navier) boundary condition with arbitrary slip length in particle-based computer simulations, using Dissipative Particle Dynamics, in particular on the flow past a patterned surface with alternating no-slip/partial-slip stripes, and on the diffusion of a spherical colloidal particle. The simulations have been done on HERMIT.

K. Kratzer, D. Roehm and A. Arnold from the University of Stuttgart studied in their project *HYNUC 2* the heterogeneous and homogeneous crystallization

P. Nielaba (✉)
Fachbereich Physik, Universität Konstanz, 78457 Konstanz, Germany
e-mail: peter.nielaba@uni-konstanz.de

of charged spherical (colloidal) particles by Molecular Dynamics simulations, using the ESPRESSO package, and Forward Flux Sampling techniques, in order to analyze the ordering processes in crystallization phenomena. The later stages of crystallization have been studied for systems confined between two planar walls, with a particular focus on the effect of hydrodynamic interactions on the crystallization dynamics, utilizing a combination of computations on CPUs (for the particle dynamics) and on GPUs (Lattice Boltzmann fluid). In particular, the particle diffusion close to the crystal front and the van Hove correlation function have been analyzed.

By using the Nehalem cluster, it was possible to to study the effect of hydrodynamic interaction on the crystal growth. The NVIDIA Tesla C1060 cards in the Nehalem cluster accelerated the simulations including hydrodynamic interactions by a factor of about 20.

In the last granting period, quantum mechanical properties of quarks have been investigated as well as atomic and molecular collisions and the quantum many body dynamics of trapped bosonic systems.

B. M. McLaughlin, C. P. Ballance, M. S. Pindzola, S. Schippers and A. Müller from the Universities of Belfast (BMM), Auburn (CPB and MSP) and Giessen (SS and AM) investigated in their project *PAMOP* atomic, molecular and optical collisions on petaflop machines. The Schrödinger and Dirac equations have been solved with the R-matrix or R-matrix with pseudo-states approach. Various systems and phenomena have been investigated, ranging from X-ray and inner-shell processes (atomic oxygen and nitrogen ions) to heavy atom systems (Xe ions and tungsten ions), computations have been done on the Cray XE6 at the HLRS.

S. Klaiman, A.U.J. Lode, K. Sakman, O.I. Streltsova, O.E. Alon, L.S. Cederbaum, and A.I. Streltsov from the Universities of Heidelberg (SK, LSC, AIS), Basel (AUJL), Stanford (KS), Dubna (OIS) and Haifa (OEA) studied in their project *MCTDHB* trapped interacting many-boson systems by their method termed multiconfigurational time-dependent Hartree for bosons (MCTDHB). The principal investigators have focused on four applications: (a) universality of the fragmentation dynamics in double wells, (b) novel many-body spectral features in trapped systems, (c) efficient protocol to control the many-particle tunneling dynamics to open space, (d) physics behind the formation of patterns in the ground states of trapped bosonic systems with strong finite- and long-range repulsive interactions and the origin of their dynamical stability. The computations have been done at the HLRS.

S. Krieg and the Wuppertal-Budapest collaboration in their project *HighPQCD* aim at a high precision calculation of the equation of state of Quantum Chromodynamics (QCD) including the (dynamical) effects of the charm quark. The principal investigators (PIs) use a lattice discretized version of QCD, and most of the finite temperature ensembles have been generated on less scalable architectures, such as GPU clusters, the zero temperature renormalization runs, however, require a scalable architecture, and the PIs used HERMIT with up to 256 nodes to generate these essential ensembles.

On different length scales compared to the quantum and soft matter systems described above, the project *AEFWT* has focused on the recovery of subsurface images from seismic data, and the project *Brush* on the treatment of nano flare heating in the solar corona. In addition, the team of the support project *SimLabEA* has collaborated with a scientific community of nuclear, particle and astrophysics, and has focused on cosmic ray induced air showers, stability of atomic isotopes at finite temperatures and high densities, and on the search for gravitational wave signals.

A. Kurzmann, S. Butzer and T. Bohlen from the Karlsruhe Institute of Technology have applied and developed in their project *AEFWT* a seismic inversion and imaging technique that uses the full information content of the seismic recordings. Each echo or reflection from geological discontinuities in the earth is used to reveal the earth structure. For an estimation of subsurface parameters the principal investigators (PIs) exploit the full information contained in the seismic waveforms ("acoustic and elastic full waveform tomography", FWT), and find a subsurface model by iteratively minimizing the misfit between observed and modeled data by using a conjugate gradient method. Three dimensional FWT aims to reconstruct 3D subsurface structures in high resolution. The PIs discuss optimization methods and present model results (boxmodel in transmission geometry, 3D Marmousi model). Computations have been done on HERMIT, the studies require about 8,500 simulations, a typical run using up to 4,096 cores and 65.000 core hours.

S. Bingert and H. Peter from the MPI für Sonnensystemforschung in Göttingen studied in their project *Brush* the nanoflare heating in the solar corona. In this project, a three dimensional magneto-hydrodynamical (3D MHD) model has been used to investigate the statistics of the spatial and temporal distribution of the coronal heating. The model describes the evolution of the solar corona above an observed Active Region. Emerging heating processes in the upper atmosphere, which are in general transient in time and space, have been studied and compared to observed flare statistics. The computations have been done on HERMIT and require for a high resolution study up to 2,024 cores and 8 million core hours.

The team of the support project *SimLabEA* from the Steinbruch Centre for Computing of the Karlsruhe Institute of Technology has collaborated with a scientific community of nuclear, particle and astrophysics, and has focused on cosmic ray induced air showers, stability of atomic isotopes at finite temperatures and high densities, and on the search for gravitational wave signals. Computer time on HERMIT has been used to analyze and optimize the codes CORSIKA (for simulation of cosmic rays/air showers), AtomicClusters (evaluation of in-medium properties of nuclear clusters), PolGrawAllSky (for searching of gravitational waves signals) and THiSMPI (to simulate particle accelerations in supernovae-shock fronts).

Investigation of Finite-Size Effects in the Determination of Interfacial Tensions

Fabian Schmitz, Antonia Statt, Peter Virnau, and Kurt Binder

Abstract The interfacial tension between coexisting phases of a material is an important parameter in the description of many phenomena such as crystallization, and even today its accurate measurement remains difficult. We have studied logarithmic finite-size corrections in the determination of the interfacial tension with large scale Monte Carlo simulations, and have identified several novel contributions which not only depend on the ensemble, but also on the type of the applied boundary conditions. We present results for the Lennard-Jones system and the Ising model, as well as for hard spheres, which are particularly challenging. In the future, these findings will contribute to the understanding and determination of highly accurate interfacial properties with computer simulations, and will be used in the study of nucleation of colloidal crystals. As a first application, we compare the Laplace pressure of a crystalline nucleus surrounded by liquid as obtained from simulations with classical nucleation theory.

1 Introduction

The computation of excess free energy due to interfaces (also called surface tension or interfacial tension) for condensed matter systems is still an outstanding challenge. First of all, on a molecular scale, interfaces are diffuse: thus for a vapor-liquid interface, it is not straightforward to distinguish a local excursion of the interface position from density fluctuations in the coexisting vapor and liquid phases near the interface. In addition, interfaces are mesoscopic objects, and may exhibit fluctuations from the molecular scale to the scale of the simulation box, which are hard to sample exhaustively. A system containing one or more interfaces is necessarily anisotropic, directions parallel and perpendicular to the interface(s) are not equivalent, and boundary conditions matter. Thus, the sampling of the physical

F. Schmitz (✉) • P. Virnau • K. Binder
Institut für Physik, Johannes Gutenberg-Universität, Staudinger Weg 7, D-55099 Mainz, Germany
e-mail: schmifa@uni-mainz.de

A. Statt
Graduate School Materials Science in Mainz, Staudinger Weg 9, D-55099 Mainz, Germany

W.E. Nagel et al. (eds.), *High Performance Computing in Science and Engineering '14*, DOI 10.1007/978-3-319-10810-0_1

effects of interfaces that are present in a system requires huge computational efforts on supercomputers, since the interfacial tension is not a straightforward output variable of a simulation (unlike quantities such as internal energy, pair correlation functions, etc.). One typically needs to compute the difference in free energy between two systems, one system with interfaces, and the other system without. Finding efficient algorithms for this task has been a longstanding challenge. The present paper describes work on a recently developed algorithm that has great potential for this task. As a first step, we focus on the finite size effects, which are anyway ubiquitous in simulations, but particularly harmful here, since interfaces exhibit long wavelength fluctuations of an anisotropic character. In the following section we will first introduce out new computational approach to solve this task, namely the ensemble switch method.

2 The Ensemble Switch Method

The determination of the interfacial tension between coexisting phases is a non-trivial task, especially if a crystalline phase occurs. To compute the interfacial tension using Monte Carlo simulations, we use a method which is based on the "ensemble switch method", which has been used successfully to calculate wall tensions between a phase and various types of walls [4]. Our generalization of this method allows us to calculate interfacial tensions directly [12].

The idea of our approach is to calculate the free energy difference between two systems with Hamiltonians $\mathcal{H}_0$ and $\mathcal{H}_1$ respectively, differing only by the absence or presence of interfaces (cf. Fig. 1). The first system, characterized by $\mathcal{H}_0$, consists of two separate boxes of length $L_z/2$ and width L, each filled with one phase, which are called A and B (A, B = crystal, liquid, vapor, ...). The two boxes have periodic boundary conditions individually and are therefore completely separated. The other

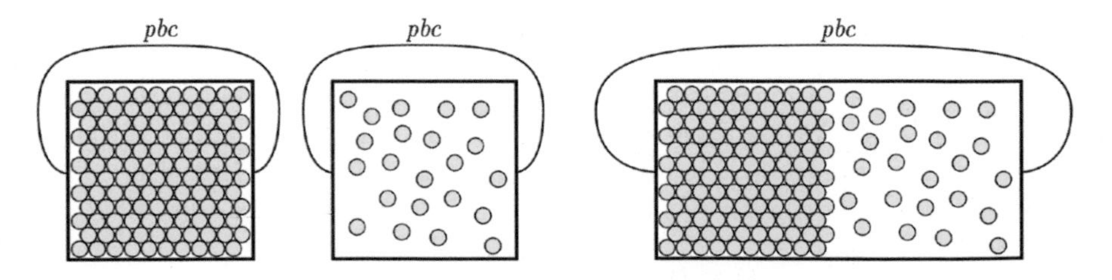

Fig. 1 Visualization of the ensemble switch method. The interfacial tension is computed via a thermodynamic integration from a reference system with Hamiltonian $\mathcal{H}_0$, which does not contain any interfaces, to a system with Hamiltonian $\mathcal{H}_1$, where interfaces are present. The free energy difference between those systems is then given by the interfacial tension times the total area of all interfaces. Here, the initial state consists of two separate boxes, each with periodic boundary conditions in all directions, containing a homogeneous phase, e.g. a crystal, liquid or vapor. Then the two boxes are combined by changing the periodic boundary conditions continuously from one state to the other. The intermediate systems have the Hamiltonian $\mathcal{H}_\kappa = \kappa\,\mathcal{H}_1 + (1-\kappa)\,\mathcal{H}_0$

system has the Hamiltonian $\mathcal{H}_1$ and consists of one large box of length L_z and width L, where the two phases are in direct coexistence and thereby form two interfaces. Apart from the boundary conditions, the two systems are identical. One can do a thermodynamic integration from one system to the other via a reaction coordinate $\kappa \in [0, 1]$. At $\kappa = 0$ or 1, the system is characterized by the Hamiltonian $\mathcal{H}_0$ or $\mathcal{H}_1$, respectively, while the intermediate systems are defined by the Hamiltonian

$$\mathcal{H}(\kappa) = \kappa\,\mathcal{H}_1 + (1-\kappa)\,\mathcal{H}_0 \ . \tag{1}$$

Except for the cases $\kappa = 0, 1$, which correspond to real physical systems, the intermediate systems are unphysical, but nevertheless well-defined in this context. If such an integration over κ is performed, the free energy difference between the two systems of interest is given by

$$\Delta F = (F_{\text{A,bulk}} + F_{\text{B,bulk}} + 2\gamma_{AB}L^2) - (F_{\text{A,bulk}} + F_{\text{B,bulk}}) = 2\gamma_{AB}L^2 \ , \tag{2}$$

where γ_{AB} is the interfacial tension to be measured and L^2 the size of one interface. Because of periodic boundary conditions, one has two interfaces between the two phases. This approach can be combined with Wang-Landau Sampling [15], Successive Umbrella Sampling [13, 14] or similar advanced simulation techniques.

For the Monte Carlo simulation, one only needs two kinds of moves. Apart from canonical moves, i.e. trial moves to translate a single particle, where the energy difference is calculated via the Hamiltonian $H(\kappa)$, one needs to implement the ensemble switch move. Here, the value of κ is changed, changing the system's internal energy according to $U(\kappa) = \kappa U_1 + (1-\kappa)U_0$ where U_0 and U_1 are the internal energies of the systems with two separate boxes and with one combined box, respectively, and $U(\kappa)$ is the internal energy of the system for the current value κ. This move is computationally very cheap if the energies U_0 and U_1 are kept up to date during the simulation.

In the simulation, the interval $[0, 1]$ is subdivided into a discrete set of κ_i, so that the free energy difference can be calculated from κ_i to κ_{i+1}. The simulation can be parallelized by using successive umbrella sampling and assigning one window $[\kappa_i, \kappa_{i+1}]$ to one core. To obtain more accurate results, it is better not to let κ vary linearly from window to window. Since the intermediate steps are non-physical anyway, one can choose an arbitrary set of $\{\kappa_i\}$. We choose functions which vary slowly near 0 and 1, e.g.

$$\kappa_i = \sin^2\left(\frac{\pi}{2}x\right) \ , \text{or} \tag{3a}$$

$$\kappa_i = \begin{cases} \frac{(2x)^a}{2} & x < 0.5 \\ 1 - \frac{(2(1-x))^a}{2} & x \geq 0.5 \end{cases} \ , \tag{3b}$$

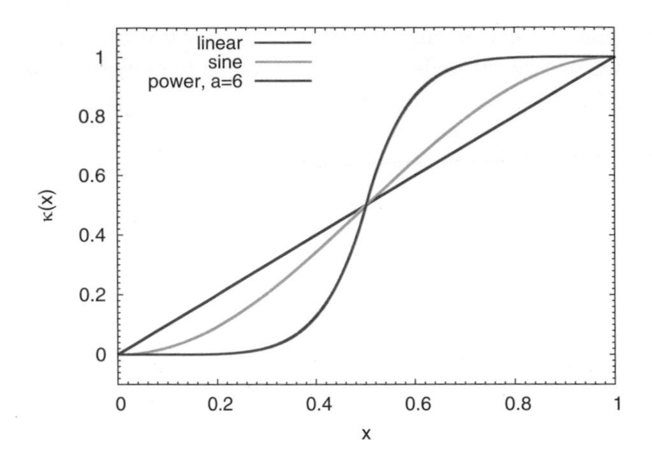

Fig. 2 Some choices for the mapping $i \to \kappa_i$. Instead of a linear mapping, one can make smaller windows where more accuracy is needed. The curve labeled sine shows Eq. (3a) while the curve labeled power shows Eq. (3b) with $a = 6$. Both functions have smaller steps near 0 and 1, while the step size is larger near $x = 0.5$

where $x = i/M$ and M is the number of κ values, which in our case is usually 1,024. The second function has a parameter a to tune the exact shape of the function. Higher numbers result in a small slope at the beginning and a steep slope near $\kappa = 0.5$. Figure 2 shows examples of the mapping $i/M \to \kappa_i$.

Another means to optimize successive umbrella sampling is to introduce a bias [13]. To do this, one equilibrates the system and then does a short pre-production run to estimate the probability ratio within the two states κ_i and κ_{i+1} of the window. Then one uses this ratio as a fixed bias for the production run. This bias makes the two states equally likely and therefore reduces the computational effort.

Figure 3 shows the resulting curves $\beta F(\kappa)$ for various model systems, namely hard spheres (for solid-liquid coexistence) and spheres with a Lennard-Jones (LJ) potential (for vapor-liquid coexistence) in $d = 3$ dimensions. In order to reduce the computational effort, the Lennard-Jones potential is truncated at a distance $r_{\text{cut}} = 2 \cdot 2^{1/6}$ and shifted [$C = 127/16{,}384$] so that the potential is continuous at r_{cut}:

$$U(r) = \begin{cases} 4\varepsilon\left[\left(\frac{\sigma}{r}\right)^{12} - \left(\frac{\sigma}{r}\right)^{6} + C\right] & , r \leq r_{\text{cut}} \\ 0 & , r > r_{\text{cut}} \end{cases} . \tag{4}$$

In this work, the temperature for the Lennard-Jones vapor-liquid coexistence is $k_{\text{B}}T/\varepsilon = 0.78$ throughout. For a Lennard-Jones-like potential, the integration curve (Fig. 3a) is smooth and exhibits a hump, so that the free energy of some intermediate states it higher than the free energy of the final state. The reason behind this is that the interface, which forms during the integration, cannot move at first, but when κ is large enough, it can begin to explore the length L_z of the box. This effect is

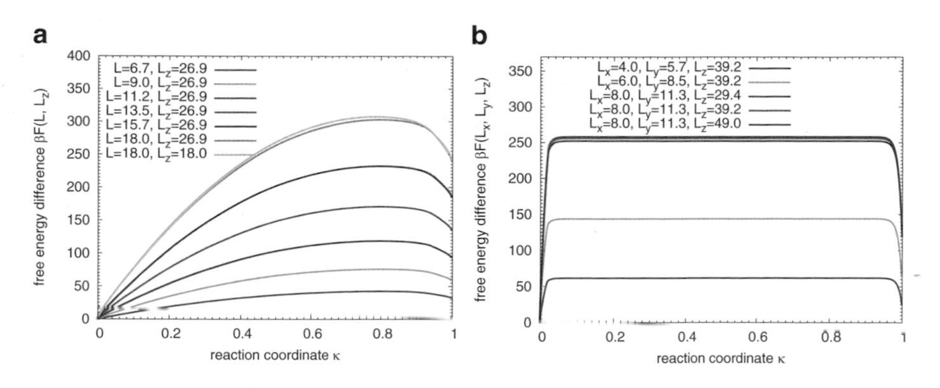

Fig. 3 Free energy difference $\beta F(\kappa)$ versus the integration coordinate κ for various box geometries for a LJ system (*left*) and a hard sphere system (*right*). The difference between the value at $\kappa = 1$ and $\kappa = 0$ is the free energy needed to create the interfaces in the systems and is therefore proportional to the interfacial tension

explained in more detail in Sect. 4. Of course, the difference between $\kappa = 0$ and $\kappa = 1$ is proportional to the interfacial area and therefore scales like L^2. If L_z is changed while L is constant, the free energy difference decreases slightly because of the translational entropy of the interfaces (cf. Sect. 4).

In the hard sphere case (Fig. 3b), the free energy difference $\beta\Delta F(\kappa)$ rises much more rapidly with κ already at small κ, and stays almost constant for a wide range of κ unlike in the case of vapor-liquid interfaces in the LJ-type model. But the variation with L and L_z is qualitatively similar: the free energy difference scales roughly like L^2, and decreases with increasing L_z, due to the interfaces' translational degrees of freedom. The dependence on L and L_z will be discussed in more detail in the next section.

3 Finite Size Scaling of the Interfacial Tension

The interfacial tension γ is defined as the amount of free energy per unit area to create an interface. In the thermodynamic limit, i.e. for infinite volume, the interfacial tension is well-defined by

$$\beta\gamma_\infty = \lim_{V\to\infty} \frac{\beta\Delta F}{A}\,, \tag{5}$$

where ΔF is the free energy to create an interface with area A. Since in computer simulations, the boxes one can consider with an acceptable amount of computational resources are always finite and far away from being large enough to ignore finite-size effects, it is crucial to analyze the limit in Eq. (5) systematically and extrapolate to the thermodynamic limit.

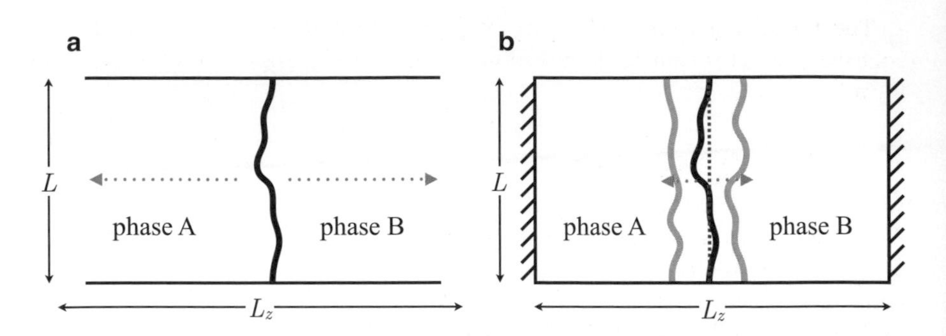

Fig. 4 Sketch of the origin of the logarithmic contributions. *Left*: translational entropy. If the interface position is not fixed, it can explore the whole simulation box. This corresponds to a translational degree of freedom which decreases the cost of free energy to form an interface. *Right*: domain breathing. If the interface position is fixed within the simulation box, the interface can still fluctuate around its average position. One can show that this has an influence on the cost of free energy, depending on the dimensions of the simulation box. In both cases, the interface exhibits capillary waves. The interface is not flat but its shape is the sum of capillary wave modes. The capillary wave spectrum is, however, restricted by the periodic boundary conditions, because only wavelengths compatible with the box dimensions are possible: the entropy connected to capillary waves is reduced and hence the free energy cost is increased

In a cuboid d-dimensional simulation box with linear dimension L except for one direction, in which the box is elongated and has a length $L_z \geq L$, an interface always aligns itself to be perpendicular to the z-direction, in order to minimize its free energy, and hence has an area L^{d-1}. The fact that a finite interface can explore the whole length L_z of the simulation box, as indicated in Fig. 4a, corresponds to a degree of freedom, which gives rise to an entropy contribution

$$\Delta S = k_{\mathrm{B}} \ln\left(\frac{L_z}{l_z}\right) , \tag{6}$$

where l_z is a natural length scale so that L_z/l_z corresponds to the number of possible positions of the interface in the simulation box. The free energy changes according to

$$\Delta F = \Delta U - T\Delta S = -k_{\mathrm{B}}T \ln\left(\frac{L_z}{l_z}\right) . \tag{7}$$

Note that the free energy cost of an interface is decreased. This is a well-known effect in the literature [6, 9]. For example, one can observe that for extremely elongated boxes ($L_z \gg L$), the gain of entropy outweighs the cost of energy so that the system creates new interfaces spontaneously [16]. It is remarkable that the effect of translational entropy on the free energy exhibits a logarithmic dependence on the box length L_z with a prefactor -1 which is independent of the details of the model under consideration.

The translational entropy is one of three effects which lead to logarithmic contributions. They can be described by a general finite-size scaling ansatz of the form

$$\beta\gamma(L, L_z) = \beta\gamma_\infty - x_\perp \frac{\ln L_z}{L^{d-1}} + x_\parallel \frac{\ln L}{L^{d-1}} + \frac{\text{const}}{L^{d-1}} . \tag{8}$$

All length scales in the logarithmic terms (like l_z in Eq. (7)) can be absorbed by the constant in the last term, which is the next-to-leading order contribution. The prefactors $x_\perp$ and $x_\parallel$ are determined by the degrees of freedom of the interface(s). In the following the other two effects and their influence on the prefactors will be motivated.

If considering rough interfaces, one has to deal with the phenomenon of capillary waves. A rough interface cannot be described by a flat plane, but it is a rather fuzzy object because of microscopic fluctuations. From a coarse-grained perspective, the shape of the interface can be decomposed into modes. The frequency (or wavelength) spectrum of an infinite interface is continuous. In a finite simulation box, however, the modes must be compatible with the periodic boundary conditions. This constraint leads to a discretization of the spectrum and hence to a loss of entropy [3] (l_{cw} is a short wavelength cutoff to the capillary wave spectrum)

$$\Delta S = \frac{3-d}{2} \ln\left(\frac{L}{l_{\text{cw}}}\right) , \tag{9}$$

which depends on L only and therefore contributes to the prefactor $x_\parallel$ in Eq. (8). Note that the prefactor depends on the dimensionality and vanishes[1] in $d = 3$.

The third and final effect has been discovered recently [12] and plays a very prominent role because it influences both $x_\perp$ and $x_\parallel$. This effect is called domain breathing and occurs in canonical ensembles. If the particle number is fixed, the average volume fractions of the coexisting phases are fixed by their coexistence densities

$$\rho V = \langle\rho_1 V_1\rangle + \langle\rho_2 V_2\rangle . \tag{10}$$

For simplicity, we consider the case $\langle V_1\rangle = \langle V_2\rangle$, i.e. the particle number is $N = (\langle\rho_1\rangle + \langle\rho_2\rangle)V/2$. It is important to note that although the average position of the interface between the phases is located in the center of the box, the actual interfacial position can fluctuate around its mean position by spontaneous fluctuations in the

[1] Instead of a logarithmic dependence like in Eq. (9), a very weak L-dependence of the form $\Delta S \propto \ln(\ln(L))$ is expected, which is beyond the scope of current computer simulations. Therefore, we use Eq. (9) also for $d = 3$.

Table 1 The universal prefactors $x_\perp$ and $x_\|$ for various choices of boundary conditions (periodic or antiperiodic) and the ensemble (canonical or grandcanonical). Note that $x_\perp$ is independent of the dimensionality of the interface while $x_\|$ depends on d

d	BC	Ensemble	$x_\perp$	$x_\|$
2	Antiperiodic	Grandcanonical	1	1/2
3	Antiperiodic	Grandcanonical	1	0
2	Antiperiodic	Canonical	1/2	1
3	Antiperiodic	Canonical	1/2	1
2	Periodic	Canonical	3/4	3/4
3	Periodic	Canonical	3/4	1/2

bulk of the two phases, as indicateed in Fig. 4b. A straightforward calculation [12] shows that the mean squared displacement is non-zero and corresponds to an entropy contribution

$$\Delta S = -\frac{1}{2}\ln\left(\frac{L_z}{l_{\text{db}}}\right) + \frac{d-1}{2}\ln\left(\frac{L}{l_{\text{db}}}\right). \tag{11}$$

Apparently, this effect contributes to both prefactors $x_\perp$ and $x_\|$. Again, the minimum length scale l_{db} on which these fluctuations contribute to the free energy is not discussed further here.

The exact values of the prefactors $x_\perp$ and $x_\|$ in Eq. (8) depend on the dimensionality d and the choice of boundary conditions and the ensemble (cf. Table 1). For a given choice, the prefactors are universal in the sense that they do not depend on details of the system like the pair potential between the particles or whether it is a Ising-like or an off-lattice model. In contrast, the length scales l_z, l_{cw} and l_{db}, which contribute to the term const $/L^{d-1}$ in Eq. (8) surely will depend on such details.

4 Results

Here, we show results obtained from the ensemble switch method when applied to various model systems. The easiest model system to study phase coexistence is the Ising model, which can be used to study the coexistence of a liquid (spin up rich phase) and a vaporous phase (spin down rich phase). An advantage of the Ising model is that one can apply different combinations of boundary conditions (periodic BC or antiperiodic BC) and ensembles (canonical (c) or grandcanonical (gc)). The two-dimensional variant is also attractive because the interfacial tension is exactly known [10]. Hence, the Ising model is a very good choice to test the finite-size scaling ansatz Eq. (8).

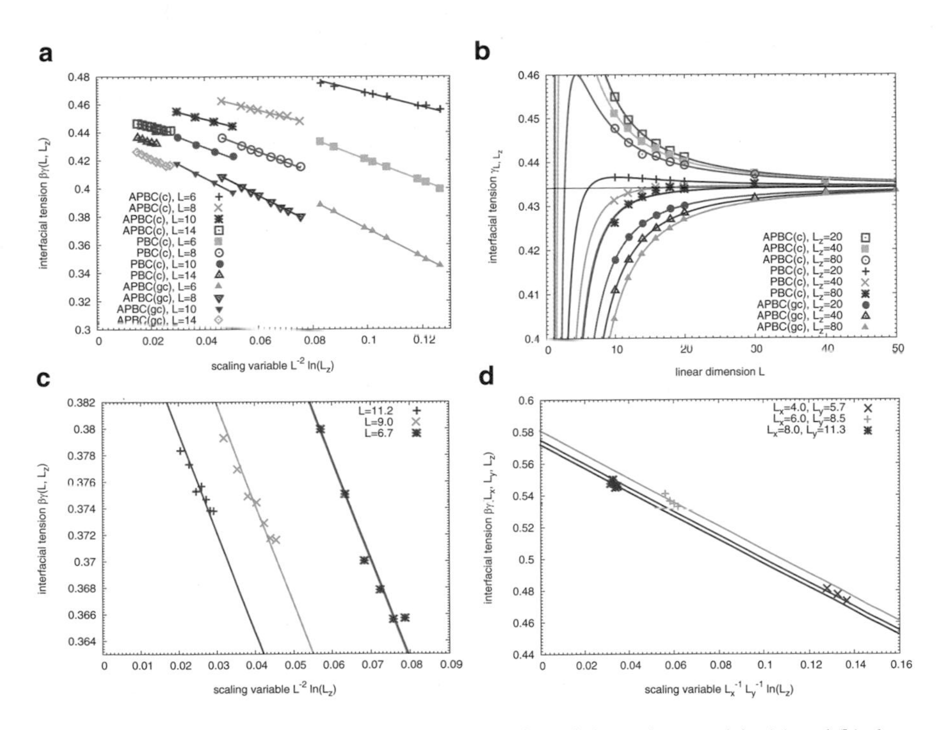

Fig. 5 Finite-size scaling of the interfacial tension $\gamma(L, L_z)$ for various models. (**a**) and (**b**) show data for a 3d Ising model at $k_B T/J = 3.0$ for three combinations of boundary conditions (periodic (PBC) and antiperiodic (APBC)) and ensembles (canonical (c) and grandcanonical (gc)). (**c**) shows data for a Lennard-Jones liquid-vapor coexistence and (**d**) shows results for hard spheres with fcc110 orientation in the crystal. (**a**), (**c**) and (**d**) show data at fixed L, so that according to Eq. (8), the data should be straight lines with slope $-x_\perp$ if plotted against $L^{-2}\ln(L_z)$. (**b**) shows data at fixed L_z where the lines are one-parameter fits. (**b**) is taken from [12]

We focus on the 3d Ising model first. To test Eq. (8), it is useful to test the dependence on L and L_z independently. If L is fixed, and the data for various L_z is plotted against $L^{-2}\ln(L_z)$, the scaling ansatz suggests that the data can be represented by straight lines with slope $-x_\perp$. Figure 5a confirms that the above logarithmic corrections are indeed present. The slopes of the curves agree with the predicted values of $x_\perp$ collected in Table 1 for all combinations of BC and ensembles. Note that the slopes do not depend on L at all.

The L-dependence is more complicated, for the next-to-leading order amplitude in Eq. (8) is unknown and the L-dependence of $\beta\gamma(L, L_z)$ is non-linear. Hence, one has to do a fit where the amplitude is an unknown parameter. Figure 5b shows data at constant L_z for the 3d Ising model. The fit curves describe the data very well for $x_\parallel$ from Table 1. Alternatively, one can take $x_\parallel$ as a second unknown parameter, but this yields results compatible with the expected values.

The next step is to test the method with off-lattice systems. To study the interfacial tension for a vapor-liquid coexistence, we use a model with point particles, where the pair potential $U(r)$ is the truncated and shifted Lennard-Jones model (Eq. (4)). The ensemble switch method is also applicable for hard sphere systems, where coexistence of crystalline and fluid phases can be studied. Here, the interfacial tension is especially hard to compute. For off-lattice models, like simulation of colloids with Lennard-Jones or hard sphere potentials, there is no straightforward method to implement antiperiodic boundary conditions. Hence, it is of great benefit to analyze the Ising model carefully in order to gain insight for the case where one simulates a box with periodic boundary conditions in a canonical ensemble.

As was claimed in Sect. 3, the prefactors $x_\perp$ and $x_\parallel$ do not depend on the model. Figure 5c, d show that the data is compatible with the slope $-x_\perp$ if plotted against $A^{-1}\ln(L_z)$, A being the box cross section, for the Lennard-Jones particles for hard spheres. The quality of the data in Fig. 5c, d is not as good as for the Ising model. Of course, the Ising model is a simpler model, where one can easily simulate comparatively large systems with feasible effort. Furthermore, for the Ising model one can use moves where arbitrarily chosen pairs of spins of opposite signs are exchanged instead of local canonical moves. Conducting the ensemble switch method for off-lattice models takes a significantly larger amount of computational resources until sufficient convergence is achieved. Nevertheless, the data suggests that the finite-size scaling ansatz Eq. (8) remains valid for off-lattice models. The next step is to test the L-dependence for fixed L_z. If the full scaling ansatz is validated, this will enable us to make much better predictions of the interfacial tension γ^∞, which is an important parameter for classical nucleation theory and other fields of research.

5 Future Applications

One potential application of the described method is to test classical nucleation theory (CNT) [7], since it is not fully understood on a quantitative level. In CNT, the barrier of homogeneous nucleation is given by two contributions, the free energy gain of creating a droplet and the free energy loss due to surface tension of the newly created interface. The underlying assumption is, that macroscopic properties of the system can be applied to describe microscopic droplets. We are going to study the coexistence of a crystalline nucleus in a liquid environment. Hence, we require a model, which shows phase separation into a liquid-like and a solid-like phase. Therefore, we use a soft extension of the well-known effective Asakura-Oosawa (AO) model [1], which has the great advantage that one can integrate the degrees of freedom of the polymers out and replace them by an effective attractive interaction between the colloids, if the diameter ratio $q = \sigma_p/\sigma_c$ of polymers and colloids is smaller than 0.154 [5]. It has the following form

$$U(r) = \begin{cases} \infty & , r \leq \sigma_c \\ -\eta_p^r \left(\frac{1+q}{q}\right)^3 \left[1 - \frac{3r}{2\sigma_c(1+q)} + \frac{r^3}{2\sigma_c^3(1+q)^3}\right] & , \sigma_c < r < \sigma_c + \sigma_p \\ 0 & , r \geq \sigma_c + \sigma_p \end{cases} \quad (12)$$

where η_p^r is the polymer reservoir packing fraction. To compute the pressure using a virial expression, the repulsive part of the potential needs to be continuous, which is why we use the following soft repulsive part instead of the hard sphere repulsion in Eq. (12)

$$U(r) = 4\left[\left(\frac{b\sigma}{r-e\sigma}\right)^{12} + \left(\frac{b\sigma}{r-e\sigma}\right)^6 - \left(\frac{b}{1+q-e}\right)^{12} - \left(\frac{b}{1+q-e}\right)^6\right], r \leq \sigma_c \quad (13)$$

where b and e control the strength and the zero crossing point of the repulsive part,[2] respectively. This potential is a continuous fit to the effective Asakura-Oosawa model [5].

Our aim is to measure the pressure p_f in the liquid surrounding a crystalline nucleus. A typical system configuration containing a nucleus is shown in Fig. 6. The value of the pressure is not the same as in pure liquid. It is enhanced due to the existence of a curved interface of the nucleus, the so-called Laplace pressure. The

Fig. 6 Typical snapshot of a crystalline nucleus in liquid. The simulation volume is $21.484318 \times 21.484318 \times 21.484318$ with periodic boundary conditions and contains 10,000 particles, resulting in a packing fraction of $\eta_c = 0.528$. For distinguishing the different phases, we use averaged Steinhardt bond order parameters, as defined in Ref. [8]. *Blue* colored particles are liquid-like, *red* ones are solid-like and the interface is shown in *green*

[2]In our case, the values are chosen in such a way that $U(r = \sigma_c) = 1$, corresponding to $b = 0.01$ and $e = 0.988571$.

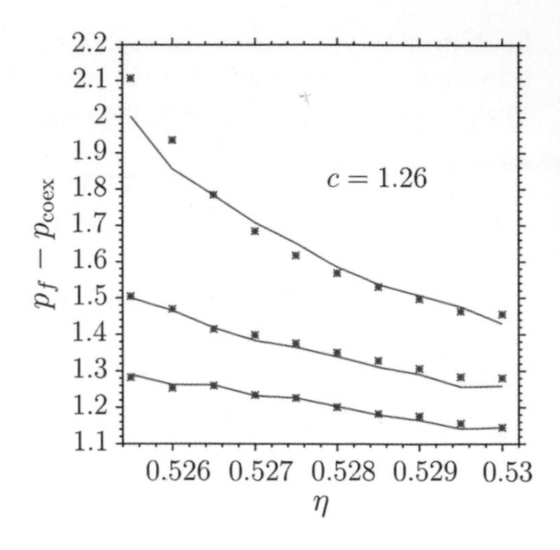

Fig. 7 Pressure in the fluid phase of a system which consists of a crystalline nucleus surrounded by liquid. The three sets of data points correspond to systems with (from top to bottom) $N = 6{,}000$, 8,000 and 10,000 particles. The curves are theoretical estimates from Eq. (16) multiplied by a constant $c = 1.26$

Laplace pressure can be calculated [2] if one assumes a spherical nucleus of radius R^* and an isotropic interfacial tension γ:

$$p_c - p_f = 2\gamma/R^* \,, \tag{14}$$

$$p_c = p_{\mathrm{coex}} + (p_f - p_{\mathrm{coex}})v_f/v_c \,, \tag{15}$$

where v_f/v_c is the volume ratio of the surrounding fluid and the nucleus, p_f is the pressure of the fluid and p_{coex} is the coexistence pressure, which can be determined independently. If we combine the two Eqs. (14) and (15), we also obtain

$$p_f - p_{\mathrm{coex}} = \frac{2\gamma}{R^*}\frac{\eta_f}{\eta_c - \eta_f} \tag{16}$$

with $\eta_{f,c}$ being the respective packing fraction of the fluid or the crystal. To identify the crystal and determine the critical radius R^* within the simulation box (see Fig. 6), we apply averaged Steinhardt order parameters [8].

In Fig. 7, we compare the pressure in the fluid phase with Eq. (16). For γ, we use an estimate for the interfacial stiffness $\tilde{\gamma}$ [17]. If the result on the right-hand side is multiplied by $c = 1.26$, the two curves match. This prefactor can either be attributed to the difference between γ and $\tilde{\gamma}$ or deviations from the spherical shape of the nucleus [11] or both. Therefore, the determination of γ is crucial for the comparison between classical nucleation theory and simulation results. The "ensemble switch" method can be easily applied to the effective Asakura-Oosawa model and the next step will be to calculate the interfacial tension for this model.

6 Conclusion

In this report, we have introduced an extension of the ensemble switch method which allows us to determine interfacial tensions in various combinations of ensembles and boundary conditions. This method was applied to Ising models as well as continuous models with Lennard-Jones or hard sphere interactions to study logarithmic finite-size corrections including a previously unknown correction due to "domain breathing". As a first practical application, we have outlined a procedure to compute the Laplace pressure of a crystalline nucleus surrounded by liquid to compare results with classical nucleation theory.

Acknowledgements We would like to thank the DFG for funding (Vi 237/4-3), and the HLRS for a generous computing grant on HERMIT.

References

1. Asakura, S., Oosawa, F.: Interaction between particles suspended in solutions of macromolecules. J. Polym. Sci. **33**, 183–192 (1958)
2. Block, B., Deb, D., Schmitz, F., Statt, A., Tröster, A., Winkler, A., Zykova-Timan, T., Virnau, P., Binder, K.: Computer simulation of heterogeneous nucleation of colloidal crystals at planar walls. Eur. Phys. J. Spec. Top. **223**, 347 (2014)
3. Brézin, E., Zinn-Justin, J.: Finite size effects in phase transitions. Nucl. Phys. B **257**, 867 (1985)
4. Deb, D., Winkler, A., Yamani, M.H., Oettel, M., Virnau, P., Binder, K.: Hard sphere fluids at a soft repulsive wall: A comparative study using Monte Carlo and density functional methods. J. Chem. Phys. **134**, 214706 (2011)
5. Dijkstra, M., Brader, J., Evans, R.: Phase behaviour and structure of model colloid-polymer mixtures. J. Phys. Condens. Matter **11**, 10079 (1999)
6. Gelfand, M.P., Fisher, M.E.: Finite-size effects in fluid interfaces. Physica A **166**, 1 (1990)
7. Kashchiev, D.: Nucleation: Basic Theory with Applications. Butterworth-Heinemann, Oxford (2000)
8. Lechner, W., Dellago, C.: Accurate determination of crystal structures based on averaged local bond order parameters. J. Chem. Phys. **129**, 114707 (2008)
9. Caselle, M., Hasenbusch, M., Panero, M.: High precision monte carlo simulations of interfaces in the three-dimensional ising model: a comparison with the nambu-goto effective string model. J. High Energy Phys. **0603**, 084 (2006)
10. Onsager, L.: Crystal Statistics. I. A Two-Dimensional Model with an Order-Disorder Transition. Phys. Rev. **65**, 117 (1944)
11. Schmitz, F., Virnau, P., Binder, K.: Logarithmic finite-size effects on interfacial free energies: Phenomenological theory and Monte Carlo studies. Phys. Rev. E **87**, 053302 (2013)
12. Schmitz, F., Virnau, P., Binder, K.: Determination of the Origin and Magnitude of Logarithmic Finite-Size Effects on Interfacial Tension: Role of Interfacial Fluctuations and Domain Breathing. Phys. Rev. Lett. **112**, 125701 (2014)
13. Virnau, P., Müller, M.: Calculation of free energy through successive umbrella sampling. J. Chem. Phys. **120**, 10925 (2004)
14. Virnau, P., Müller, M., MacDowell, L.G., Binder, K.: Phase behavior of n-alkanes in supercritical solution: A Monte Carlo study. J. Chem. Phys. **121**, 2169 (2004)
15. Wang, F., Landau, D.P.: Determining the density of states for classical statistical models: A random walk algorithm to produce a flat histogram. Phys. Rev. E **64**, 056101 (2001)

16. Winkler, A., Wilms, D., Virnau, P., Binder, K.: Capillary condensation in cylindrical pores: Monte Carlo study of the interplay of surface and finite size effects. J. Chem. Phys. **133**, 164702 (2010)
17. Zykova-Timan, T., Horbach, J., Binder, K.: Monte Carlo simulations of the solid-liquid transition in hard spheres and colloid-polymer mixtures. J. Chem. Phys. **133**, 014705 (2010)

Application of Tunable-Slip Boundary Conditions in Particle-Based Simulations

Jiajia Zhou, Jens Smiatek, Evgeny S. Asmolov, Olga I. Vinogradova, and Friederike Schmid

Abstract Compared to macroscopic systems, fluids on the micro- and nanoscales have a larger surface-to-volume ratio, thus the boundary condition becomes crucial in determining the fluid properties. No-slip boundary condition has been applied successfully to wide ranges of macroscopic phenomena, but its validity in microscopic scale is questionable. A more realistic description is that the flow exhibits slippage at the surface, which can be characterized by a Navier slip length. We present a tunable-slip method by implementing Navier boundary condition in particle-based computer simulations (Dissipative Particle Dynamics as an example). To demonstrate the validity and versatility of our method, we have investigated two model systems: (i) the flow past a patterned surface with alternating no-slip/partial-slip stripes and (ii) the diffusion of a spherical colloidal particle.

J. Zhou (✉) • F. Schmid
Institut für Physik, Johannes Gutenberg-Universität Mainz, Staudingerweg 7, D-55099 Mainz, Germany
e-mail: zhou@uni-mainz.de; friederike.schmid@uni-mainz.de

J. Smiatek
Institut für Computerphysik, Universität Stuttgart, Allmandring 3, 70569 Stuttgart, Germany

E.S. Asmolov
A.N. Frumkin Institute of Physical Chemistry and Electrochemistry, Russian Academy of Science, 31 Leninsky Prospect, 119071 Moscow, Russia

Central Aero-Hydrodynamic Institute, 140180 Zhukovsky, Moscow region, Russia

Institute of Mechanics, M.V. Lomonosov Moscow State University, 119991 Moscow, Russia

O.I. Vinogradova
A.N. Frumkin Institute of Physical Chemistry and Electrochemistry, Russian Academy of Science, 31 Leninsky Prospect, 119071 Moscow, Russia

Department of Physics, M.V. Lomonosov Moscow State University, 119991 Moscow, Russia

DWI – Leibniz Institute for Interactive Materials, RWTH Aachen, Forckenbeckstraße 50, 52056 Aachen, Germany

W.E. Nagel et al. (eds.), *High Performance Computing in Science and Engineering '14*, DOI 10.1007/978-3-319-10810-0_2

1 Introduction

Modeling fluids in small length scales from micrometer to nanometer not only is a fundamental problem in fluid mechanics, but also plays a paramount role in modern fluidic devices [1]. These micro- or nanofluidic devices can be found in wide range of applications and in many different fields. Examples include the development of inkjet printheads for xerography, lab-on-a-chip technology, and manipulation and separation of bio-molecules. Compared to macroscopic systems, microscopic fluids have a larger surface-to-volume ratio, therefore boundary conditions become crucial in determining the hydrodynamic properties. The most well-known boundary condition is the no-slip boundary condition, i.e., the fluid velocity vanishes at a fluid/solid interface. No-slip boundary condition has been successfully applied to wide ranges of macroscopic phenomena, but it has no microscopic justification and its validity in small length scales is questionable. A more general boundary condition is the Navier boundary condition, which is described in the following.

Assume that a fluid/solid interface lies in the xOy plane and the fluid occupies $z > 0$ half-space (Fig. 1). Let us consider the case of a laminar Couette flow in the y-direction, with a slip velocity v_s at $z = 0$. The tangential force per unit area (component of the shear stress σ_{yz}), to the first-order approximation, is proportional to the slip velocity v_s,

$$\sigma_{yz} = \zeta_s v_s, \tag{1}$$

where ζ_s is the surface friction coefficient. For Newtonian fluids, the shear stress can also be written as

$$\sigma_{yz} = \eta \frac{\partial v_y}{\partial z}, \tag{2}$$

where η is the shear viscosity of the fluid. Combining Eqs. (1) and (2), one arrives the Navier boundary condition [2]

$$v_s = \frac{\eta}{\zeta_s} \frac{\partial v_y}{\partial z} = b \frac{\partial v_y}{\partial z}. \tag{3}$$

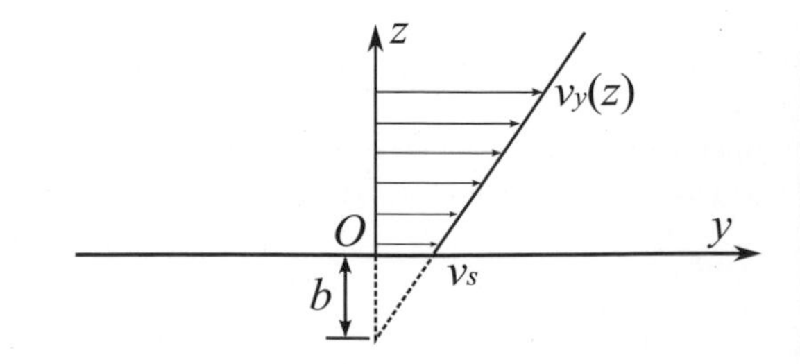

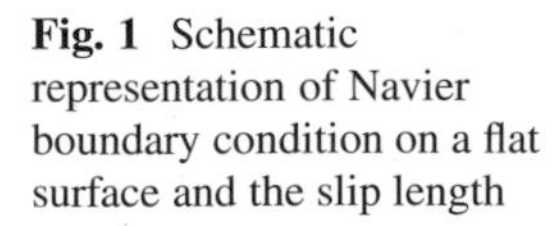
Fig. 1 Schematic representation of Navier boundary condition on a flat surface and the slip length

The degree of slippage at the surface is quantified in terms of the Navier slip length $b = \eta/\zeta_s$. No-slip boundary condition is recovered for $b = 0$, while for ideal frictionless surface we have $b \to \infty$. The physical meaning of the slip length is the distance within the solid at which the flow velocity extrapolates to zero.

2 Implementation

The main question we try to address in this contribution is the implementation of Navier boundary condition with tunable slip length in particle-based computer simulations. Any models of a solid surface require at least two components: (a) an excluded interaction to prevent fluid particles penetrating the hard surface; and (b) an effective interaction due to the surface friction. In the following, we shall use Dissipative Particle Dynamics (DPD) [3–5] to demonstrate the tunable-slip method, although the basic idea is quite general and can be applied to other simulation techniques. In the following we only give a brief description and introduce parameters for late discussions. For a more detailed description, please consult the origin reference [6].

We model the impermeable surface using a pure repulsive interaction of the Weeks-Chandler-Andersen form [7],

$$V(z) = \begin{cases} 4\varepsilon[(\frac{\sigma}{z})^{12} - (\frac{\sigma}{z})^{6} + \frac{1}{4}], & z < \sqrt[6]{2}\sigma \\ 0, & z \geq \sqrt[6]{2}\sigma \end{cases} \tag{4}$$

where z is the distance between the fluid particle and the wall. The WCA parameters also set the unit system (ε for the energy, σ for the length, and m for the particle mass).

The interactions between the surface and fluid can be very complicated and depend on the chemical details. But in the mesoscale, the main effect of the surface is to provide friction. Inspired by the form of Eq. (1), we model the fluid/surface friction by a dissipative force to the fluid, which depends on the relative velocity between the fluid particle and the wall,

$$\mathbf{F}_i^{\mathrm{D}} = -\gamma_{\mathrm{L}}\omega_{\mathrm{L}}(z)(\mathbf{v}_i - \mathbf{v}_{\mathrm{wall}}). \tag{5}$$

The friction constant γ_{L} is an adjustable parameter that characterizes the strength of the wall friction, and it can be used to vary the slip length. The position-dependent function $\omega_{\mathrm{L}}(z)$ is a monotonically decreasing function of the wall-particle separation, and vanishes at certain cutoff z_c to mimic the finite range of the wall interaction. Additionally, a random force obeying the fluctuation-dissipation theorem is required to ensure the correct equilibrium statistics,

$$\mathbf{F}_i^{\mathrm{R}} = \sqrt{2k_B T \gamma_{\mathrm{L}}\omega_{\mathrm{L}}(z)}\boldsymbol{\xi}_i, \tag{6}$$

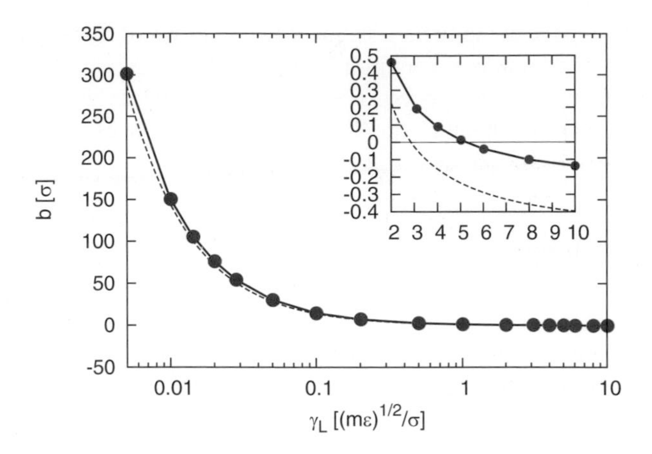

Fig. 2 Relation between the slip length b and the wall friction parameter $\gamma_{\rm L}$. The *inset* shows an enlarged portion of the region where the slip length is zero. No-slip boundary condition can be implemented by using $\gamma_{\rm L} = 5.26\sqrt{m\varepsilon}/\sigma$. *Dashed curves* show analytical prediction Eq. (7)

where k_B is the Boltzmann constant, T the temperature, and $\boldsymbol{\xi}_i$ a vector whose component is a Gaussian distributed random variable with zero mean and unit variance.

The slip length b can be computed analytically as a function of a dimensionless parameter $\alpha = z_c^2 \gamma_{\rm L} \rho / \eta$, where ρ is the fluid number density [6]. For a linear function $\omega_{\rm L}(z) = 1 - z/z_c$, the relation is

$$\frac{b}{z_c} = \frac{2}{\alpha} - \frac{7}{15} - \frac{19}{1800}\alpha + \cdots \tag{7}$$

This formula works quite well for large values of b, but shows deviation near the no-slip region ($b/\sigma \sim 1$). The slip length can also be measured by short simulations of Poiseuille and Couette flow in a thin channel geometry. Figure 2 shows the relation between the slip length b and the wall friction parameter $\gamma_{\rm L}$. One can then choose a suitable value of $\gamma_{\rm L}$ based on the requirement of the slip length.

To demonstrate the validity and versatility of our method, we investigate two model systems: the flow past a patterned surface with alternating no-slip/partial-slip stripes (Sect. 3) and free diffusion of a slipping colloidal particle (Sect. 4).

3 Striped Surfaces

Patterned surfaces play a major role in micro- and nanofluidics. One important example is the superhydrophobic Cassie surfaces, where no-slip and partial-slip area arrange in a striped pattern (see Fig. 3a). The no-slip region consists of fluid/solid

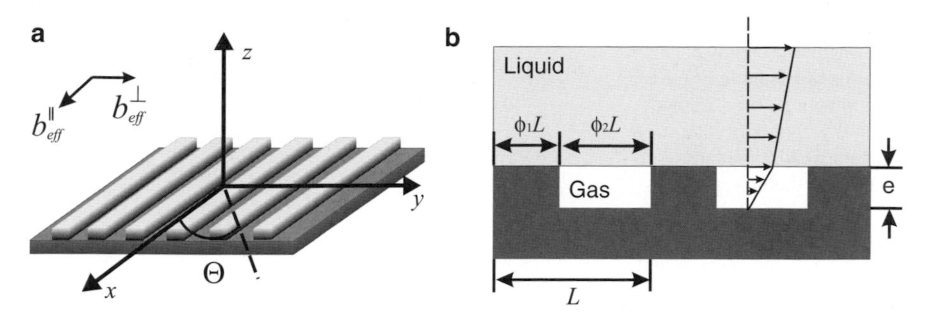

Fig. 3 Sketch of the striped surface: $\Theta = 0$ to longitudinal stripes, $\Theta = \pi/2$ corresponds to transverse stripes (**a**), and of the liquid interface in the Cassie state (**b**)

interface, while the partial-slip region consists of trapped gas which is stabilized by a rough wall texture. These superhydrophobic surfaces exhibit very low friction, and the drag reduction is associated with the large slip length over the fluid/gas interfaces.

We simulate the superhydrophobic Cassie surface using the "gas-cushion model" [8]. Instead of modeling the physically corrugated surface, we use a *flat* surface with alternating no-slip and partial-slip boundary conditions. The value of the slip length b over the gas sector is related to the thickness of the gas region e by

$$b \simeq \frac{\eta}{\eta_g} e, \tag{8}$$

where η and η_g are shear viscosities of a gas and a liquid (see Fig. 3b). We consider a striped pattern with a periodicity L. The area fractions of the solid and gas sectors are ϕ_1 and ϕ_2, respectively. For inhomogeneous surfaces, the effective slip length depends on the flow direction Θ and is in general a tensorial quantity $\mathbf{b}_{\text{eff}}$. For the special case of striped surfaces, the eigenvalues of the slip length tensor correspond to the flow direction parallel to the stripes (maximum slip $b^{\parallel}_{\text{eff}}$) and perpendicular to the stripes (minimum slip $b^{\perp}_{\text{eff}}$).

We compare our simulation results with the numerical solution to the Stokes equations and some analytical formulas. The results for the perfect-slip gas sector ($b \gg L$) are well-known [9, 10]

$$\frac{b^{\parallel}_{\text{eff}}}{L} = \frac{2b^{\perp}_{\text{eff}}}{L} \simeq \frac{1}{\pi} \ln\left[\sec\left(\frac{\pi\phi_2}{2}\right)\right]. \tag{9}$$

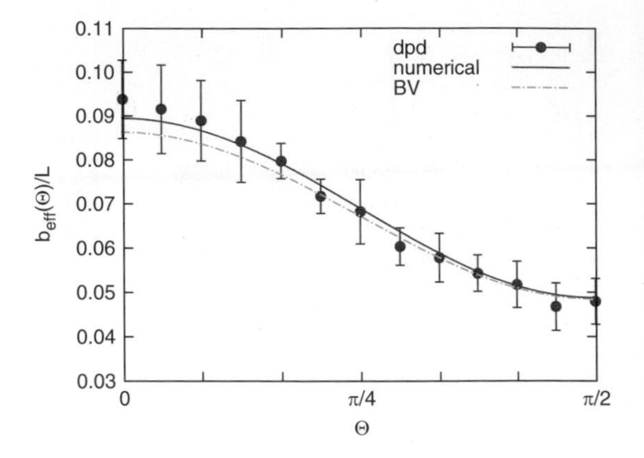

Fig. 4 Effective downstream slip length, $b_{\mathrm{eff}}(\Theta)$ as a function of tilt angle Θ for a pattern with $b/L = 1.0$ and $\phi_2 = 0.5$. Symbols with error bars are simulation data. Curves are theoretical values calculated using Eq. (12) with eigenvalues obtained by a numerical method (*solid*) and by Eqs. (10 and 11) (*dot-dashed*)

For arbitrary value of b, Belyaev and Vinogradova suggested approximate expressions [11],

$$\frac{b_{\mathrm{eff}}^{\parallel}}{L} \simeq \frac{1}{\pi} \frac{\ln\left[\sec\left(\frac{\pi\phi_2}{2}\right)\right]}{1 + \frac{L}{\pi b}\ln\left[\sec\left(\frac{\pi\phi_2}{2}\right) + \tan\left(\frac{\pi\phi_2}{2}\right)\right]}, \tag{10}$$

$$\frac{b_{\mathrm{eff}}^{\perp}}{L} \simeq \frac{1}{2\pi} \frac{\ln\left[\sec\left(\frac{\pi\phi_2}{2}\right)\right]}{1 + \frac{L}{2\pi b}\ln\left[\sec\left(\frac{\pi\phi_2}{2}\right) + \tan\left(\frac{\pi\phi_2}{2}\right)\right]}. \tag{11}$$

These formulas are accurate over wide range of the parameters and recover to Eq. (9) at large b value.

We start with varying Θ for patterns that has equal areas of no-slip and partial-slip regions ($\phi_2 = 0.5$) and the slip length is in the intermediate region $b/L = 1.0$. Figure 4 shows the results for the effective downstream slip lengths, $b_{\mathrm{eff}}(\Theta)$, and the theoretical curve

$$b_{\mathrm{eff}}(\Theta) = b_{\mathrm{eff}}^{\parallel} \cos^2\Theta + b_{\mathrm{eff}}^{\perp} \sin^2\Theta. \tag{12}$$

The simulation data are in good agreement with theoretical predictions, confirming the anisotropy of the flow and the validity of the concept of a tensorial slip for striped surfaces.

Next we examine the effect of varying the fraction of gas/liquid interface, ϕ_2. Figure 5a shows the effective slip lengths $b_{\mathrm{eff}}^{\parallel}$ and $b_{\mathrm{eff}}^{\perp}$ as a function of ϕ_2. The striped pattern again has $b/L = 1.0$. The results clearly demonstrate that the gas fraction ϕ_2 is the main factor in determining the value of effective slip; the slip length increases significantly when the gas fraction increases to unity.

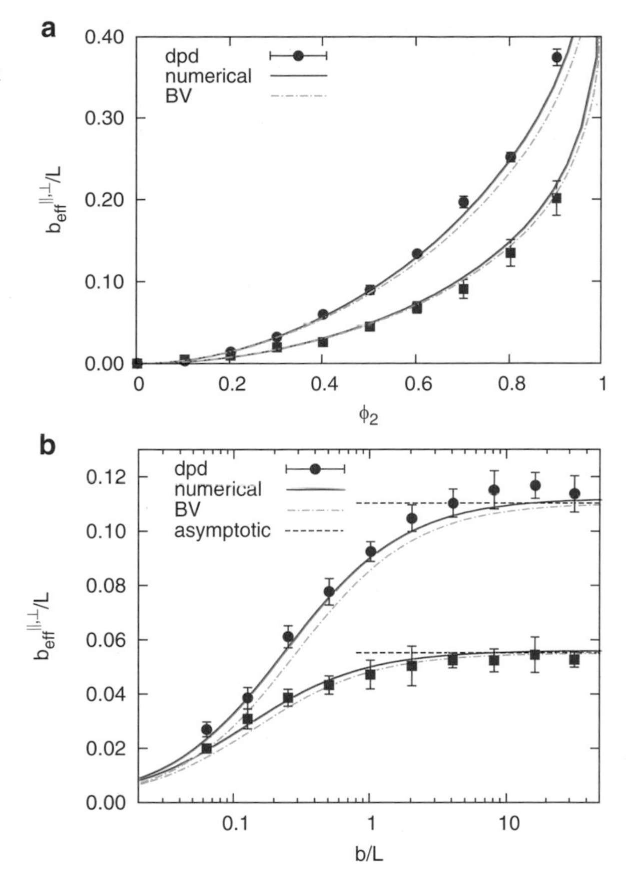

Fig. 5 (**a**) Effective slip lengths (*symbols*) as a function of gas-sector fraction ϕ_2 for $b/L = 1.0$. (**b**) Effective slip lengths as a function of local slip length b for $\phi_2 = 0.5$. The numerical results are shown as *solid curves*. Also shown are analytic expressions Eqs. (10 and 11) (*dot-dashed*), and asymptotic formulas Eq. (9) (*dashed lines*, for (**b**) only)

Finally, we present the effective slip lengths as a function of the local slip length b at gas/liquid interface in Fig. 5b. The pattern has equal fractions of solid and gas, $\phi_1 = \phi_2 = 0.5$. The analytic formulas Eqs. (10) and (11) are shown in dot-dashed lines, which are accurate over wide range of b value. The effective slip lengths reach the asymptotic values predicted by Eq. (9) at large values of b/L. The excellent agreement between the simulation data and theoretic predictions promotes studies of more complex systems and applications, such as thin channels with symmetric stripes [12], anisotropic flow over weakly slipping stripes [13], trapezoidal grooves [14], electroosmotic flows [15], polyelectrolyte electrophoresis [16, 17], and separation of chiral particles [18].

4 Colloidal Particle

Colloidal dispersions have numerous applications in different fields such as chemistry, biology, medicine, and engineering [19, 20]. Better understanding of the dynamics of colloidal particles can provide insights to improve the material properties. Molecular simulations can shed light on the dynamic phenomena of colloidal particles in a well-defined model system. Due to the size difference between the colloids and solvent molecules, it is difficult to simulate the system on the smaller length scale (the solvent), and one has to rely on coarse-graining technique to reach meaningful time scales. The general idea is to couple the solute with a mesoscopic model for Navier-Stokes fluids. One of the examples is the coupling scheme developed by Ahlrichs and Dünweg [21], which combines a Lattice-Boltzmann approach for the fluid and a continuum Molecular Dynamics model for the polymer chains. The method has been extended to colloidal particles [22, 23]. We shall discuss a similar colloid model based on DPD.

The interactions between the colloid and fluid can be separated into two components, similar to the case of flat surface presented in Sect. 2. One component is the hard-core interaction to prevent the fluid entering the colloid. We use a pure repulsive potential of WCA form (Eq. (4)). The other one is the friction between the colloid surface and the fluid, which can be modeled by a pair of dissipative and random forces (cf. Eqs. (5) and (6)). The complication arises for colloidal particles which have a curved surface. There are two possible solutions. The first is to treat the surface as a continuous two-dimensional object and replace z in the friction force pair by the shortest distance between the fluid particle and the curved surface. This can be easily implemented for simple geometry such as spheres, but becomes complicated for colloids with irregular shape. The second solution is to discretize the surface by many interaction sites, whose positions are fixed in the local frame of the colloid. These sites interact with the solvent through the DPD dissipative and stochastic interactions, with a friction coefficient γ_{L} (Eqs. (5) and (6) but with z replaced by bead separation r). Boundary conditions on the colloid surface can be tuned by vary the value of γ_{L}. In this work, we opt to the second option for better extensibility. Figure 6 shows a representative snapshot of a colloidal particle decorated by many surface sites.

The total force exerted on the colloid is given by the sum over all forces on the surface sites, plus the conservative excluded volume interaction,

$$\mathbf{F}_{\mathrm{C}} = \sum_{i=1}^{N} \left[\mathbf{F}_i^{\mathrm{D}}(\mathbf{r}_i) + \mathbf{F}_i^{\mathrm{R}}(\mathbf{r}_i) \right] + \mathbf{F}^{\mathrm{WCA}}. \tag{13}$$

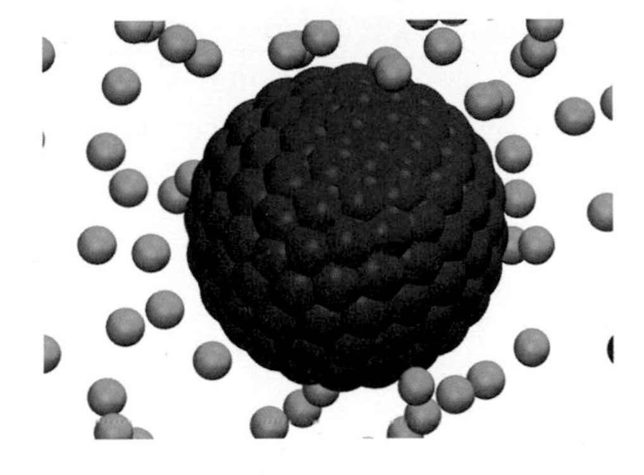

Fig. 6 Snapshot of a colloidal particle in solution. The surface sites are represented by the *dark beads*, and the *light beads* are fluid particles. Only selective solvent beads are shown here for clarity

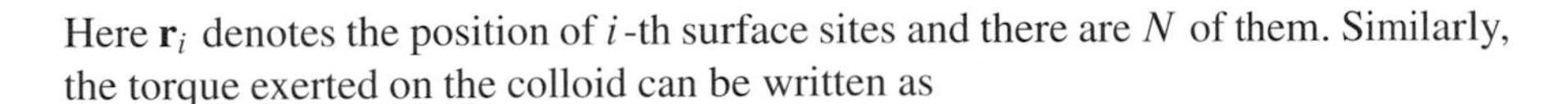

Here $\mathbf{r}_i$ denotes the position of i-th surface sites and there are N of them. Similarly, the torque exerted on the colloid can be written as

$$\mathbf{T}_C = \sum_{i=1}^{N} \left[\mathbf{F}_i^D(\mathbf{r}_i) + \mathbf{F}_i^R(\mathbf{r}_i)\right] \times (\mathbf{r}_i - \mathbf{r}_{cm}), \tag{14}$$

where $\mathbf{r}_{cm}$ is the position of the colloid's center-of-mass. Note that the excluded volume interaction does not contribute to the torque because the associated force points towards the center for spherical colloids, but this is in general not true for irregular shape. The total force and torque are then used to update the position and velocity (both translational and rotational) of the colloid at the next time step.

The diffusion constant of the colloid can be calculated from the velocity autocorrelation function using the Green-Kubo relation or by a linear fit to the mean-square displacement. Due to the periodic boundary condition implemented in simulations, the diffusion constant for a single colloid in a finite simulation box depends on the box size. Figure 7a demonstrates the finite-size effect by plotting the mean-square displacement as a function of time for two different simulation boxes $L = 10\,\sigma$ and $L = 30\,\sigma$.

The diffusion constant increases with increasing box size. For small simulation box, the long-wavelength hydrodynamic modes are suppressed due to the coupling between the colloid and its periodic images. The diffusion constant can be written in terms of an expansion of $1/L$, the reciprocal of the box size [24]

$$D = \frac{k_B T}{6\pi\eta}\left(\frac{1}{R} - \frac{2.837}{L} + \frac{4.19R^2}{L^3} + \cdots\right). \tag{15}$$

In Fig. 7b, simulation results of the diffusion constant are plotted in terms of $1/L$. The simulation results and the hydrodynamic theory show reasonable agreement.

Most of the studies on colloid dynamics have been focus on the no-slip boundary condition. We have investigated the response of charged colloids under alternating electric fields [25–28]. No-slip is valid for large colloids, but may be questionable for nanometer-sized or hydrophobic particles. Furthermore, one can adjust the

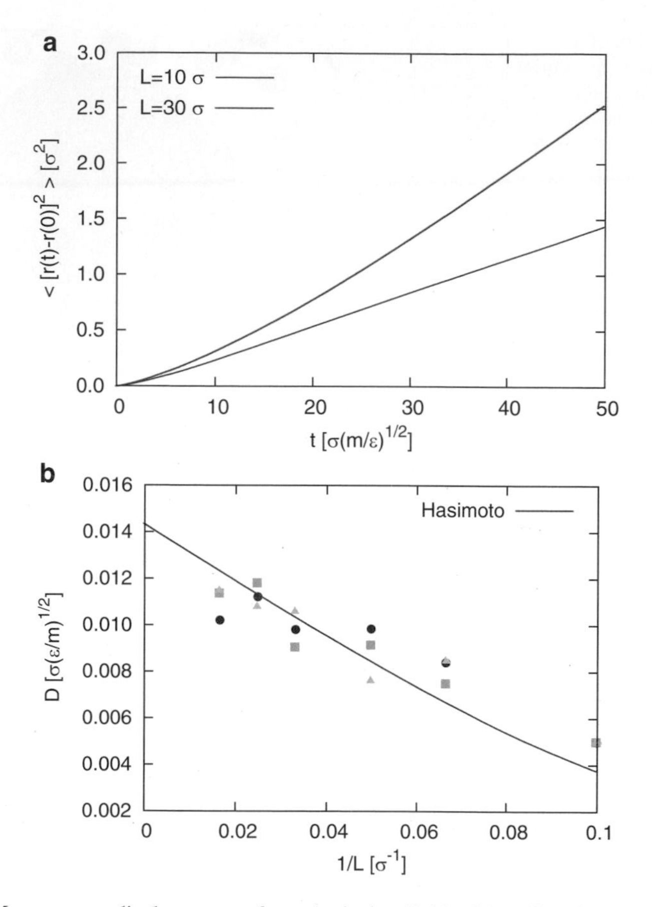

Fig. 7 (**a**) Mean-square displacement of a spherical colloid with radius $R = 3.0\,\sigma$ for two different sizes of simulation box, $L = 10\,\sigma$ and $L = 30\,\sigma$. (**b**) Diffusion constant D for a spherical colloid of radius $R = 3.0\,\sigma$ as a function of the reciprocal of the box size $1/L$. The *curve* is the prediction from Eq. (15). Different symbols correspond to simulation runs with different initialization

boundary condition by modifying the surface properties. One advantage of our colloid model is the ability to adjust the boundary condition from no-slip to full-slip by changing the friction coefficient γ_L. Figure 8 illustrates the change of the diffusion constant by varying γ_L in a simulation box $L = 30\,\sigma$. This freedom provides opportunities to study the effect of hydrodynamic slip on the colloid dynamics [29, 30].

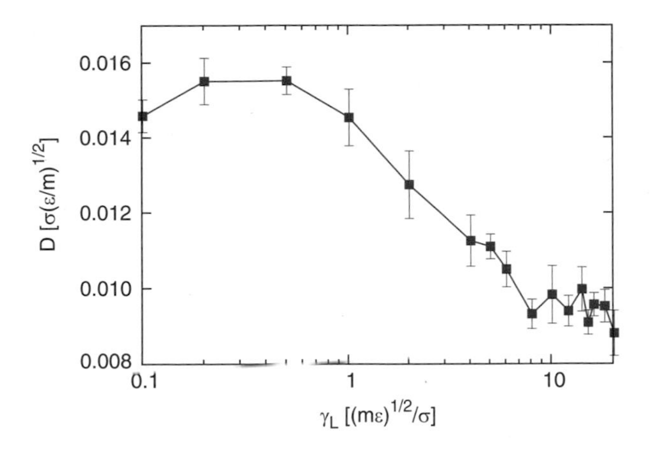

Fig. 8 Diffusion constant D for a spherical colloid of radius $R = 3.0\,\sigma$ as a function of the surface-fluid DPD friction coefficient γ_L. The simulation box has a size of $L = 30\,\sigma$. No-slip is realized for $\gamma_L > 10\,\sqrt{m\varepsilon}/\sigma$

5 Summary

We have presented a coarse-grained method to implement Navier boundary condition with arbitrary slip length in particle-based simulations. We have validated our method by simulating a flat homogeneous surface and proposed an analytical relation between the slip length and simulation parameters. To illustrate the versatility of the method, we extend the method to inhomogeneous surfaces by investigating Newtonian flow over superhydrophobic striped surface, and to curved surfaces by studying the dynamics of single spherical colloid. Our method provides a general tool to exam the slip-dependent phenomena in particle-based simulations and should be suitable to study more complex system, for example, flow over structured surface or channel and dynamics of aspherical colloids.

Acknowledgements We thank the HLRS Stuttgart for a generous grant of computer time on HERMIT. This research was supported by the DFG through the SFB TR6, SFB 985, SFB 1066, and by RAS through its priority program "Assembly and Investigation of Macromolecular Structures of New Generations".

References

1. Squires, T.M., Quake, S.R.: Microfluidics: Fluid physics at the nanoliter scale. Rev. Mod. Phys. **77**, 977 (2005)
2. Bocquet, L., Barrat, J.L.: Flow boundary conditions from nano- to micro-scales. Soft Matter **3**, 685 (2007)
3. Hoogerbrugge, P.J., Koelman, J.M.V.A.: Simulating microscopic hydrodynamic phenomena with dissipative particle dynamics. Europhys. Lett. **19**, 155 (1992)

4. Español, P., Warren, P.B.: Statistical mechanics of dissipative particle dynamics. Europhys. Lett. **30**, 191 (1995)
5. Groot, R.D., Warren, P.B.: Dissipative particle dynamics: bridging the gap between atomistic and mesoscopic simulation. J. Chem. Phys. **107**, 4423 (1997)
6. Smiatek, J., Allen, M., Schmid, F.: Tunable-slip boundaries for coarse-grained simulations of fluid flow. Eur. Phys. J. E **26**, 115 (2008)
7. Weeks, J.D., Chandler, D., Andersen, H.C.: Role of repulsive forces in determining the equilibrium structure of simple liquids. J. Chem. Phys. **54**, 5237 (1971)
8. Vinogradova, O.I.: Drainage of a thin liquid film confined between hydrophobic surfaces. Langmuir **11**, 2213 (1995)
9. Philip, J.R.: Flows satisfying mixed no-slip and no-shear conditions. J. Appl. Math. Phys. **23**, 353 (1972)
10. Lauga, E., Stone, H.A.: Effective slip in pressure-driven stokes flow. J. Fluid Mech. **489**, 55 (2003)
11. Belyaev, A.V., Vinogradova, O.I.: Effective slip in pressure-driven flow past superhydrophobic stripes. J. Fluid Mech. **652**, 489 (2010)
12. Zhou, J., Belyaev, A.V., Schmid, F., Vinogradova, O.I.: Anisotropic flow in striped superhydrophobic channels. J. Chem. Phys. **136**, 194706 (2012)
13. Asmolov, E.S., Zhou, J., Schmid, F., Vinogradova, O.I.: Effective slip-length tensor for a flow over weakly slipping stripes. Phys. Rev. E **88**, 023004 (2013)
14. Zhou, J., Asmolov, E.S., Schmid, F., Vinogradova, O.I.: Effective slippage on superhydrophobic trapezoidal grooves. J. Chem. Phys. **139**, 174708 (2013)
15. Smiatek, J., Sega, M., Holm, C., Schiller, U.D., Schmid, F.: Mesoscopic simulations of the counterion-induced electro-osmotic flow: A comparative study. J. Chem. Phys. **130**, 244702 (2009)
16. Smiatek, J., Schmid, F.: Polyelectrolyte electrophoresis in nanochannels: A dissipative particle dynamics simulation. J. Phys. Chem. B **114**, 6266 (2010)
17. Smiatek, J., Schmid, F.: Mesoscopic simulations of polyelectrolyte electrophoresis in nanochannels. Comput. Phys. Commun. **182**, 1941 (2011)
18. Meinhardt, S., Smiatek, J., Eichhorn, R., Schmid, F.: Separation of chiral particles in micro- or nanofluidic channels. Phys. Rev. Lett. **108**, 214504 (2012)
19. Russel, W.B., Saville, D.A., Schowalter, W.: Colloidal Dispersions. Cambridge University Press, Cambridge (1989)
20. Dhont, J.: An Introduction to Dynamics of Colloids. Elsevier, Amsterdam (1996)
21. Ahlrichs, P., Dünweg, B.: Simulation of a single polymer chain in solution by combining lattice boltzmann and molecular dynamics. J. Chem. Phys. **111**, 8225 (1999)
22. Lobaskin, V., Dünweg, B.: A new model for simulating colloidal dynamics. New J. Phys. **6**, 54 (2004)
23. Chatterji, A., Horbach, J.: Combining molecular dynamics with lattice boltzmann: A hybrid method for the simulation of (charged) colloidal systems. J. Chem. Phys. **122**, 184903 (2005)
24. Hasimoto, H.: On the periodic fundamental solutions of the stokes equations and their application to viscous flow past a cubic array of spheres. J. Fluid Mech. **5**, 317 (1959)
25. Zhou, J., Schmid, F.: Dielectric response of nanoscopic spherical colloids in alternating electric fields: a dissipative particle dynamics simulation. J. Phys. Condens. Matter **24**, 464112 (2012)
26. Zhou, J., Schmid, F.: AC-field-induced polarization for uncharged colloids in salt solution: a dissipative particle dynamics simulation. Eur. Phys. J. E **36**, 33 (2013)
27. Zhou, J., Schmitz, R., Dünweg, B., Schmid, F.: Dynamic and dielectric response of charged colloids in electrolyte solutions to external electric fields. J. Chem. Phys. **139**, 024901 (2013)
28. Zhou, J., Schmid, F.: Eur. Phys. Computer simulations of charged colloids in alternating electric fields. J. Spec. Top. **222**, 2911 (2013)
29. Swan, J.W., Khair, A.S.: On the hydrodynamics of 'slip-slick' spheres. J. Fluid Mech. **606**, 115 (2008)
30. Khair, A.S., Squires, T.M.: The influence of hydrodynamic slip on the electrophoretic mobility of a spherical colloidal particle. Phys. Fluids **21**, 042001 (2009)

Homogeneous and Heterogeneous Crystallization of Charged Colloidal Particles

Kai Kratzer, Dominic Roehm, and Axel Arnold

Abstract Crystallization of macroions happens in many applications like protein purification, photonic crystals or structure determination. However, the mechanism of crystallization in these systems is only poorly understood. We thus study homogeneous nucleation and crystallization of charged spherical particles using molecular dynamics (MD) computer simulations with the software package ESPResSo.

Nucleation, i.e. crystallization from a homogeneous bulk liquid, has a high energy barrier, which is overcome by rare spontaneous fluctuations. Thus nucleation is a rare event, which can only be studied in computer simulations using special sampling techniques. We apply Forward Flux Sampling (FFS) embedded in our Flexible Rare Event Sampling Harness System to run simulations as close as possible to the coexistence line. Here the supersaturation of the system is low and the crystallization process can be investigated step-by-step. Based on investigations of crystallization pathways of our system, we found that the crystallization can follow a mechanism with both a smooth or stepwise ordering process, depending on the position in the phase diagram.

To study the later stage of crystallization, we use a system of particles confined between two planar walls. In this system, we study the influence of hydrodynamic interactions (HIs) on the crystallization dynamics. To model long-range hydrodynamic interactions, we couple the particle dynamics computed on the CPU with a Lattice- Boltzmann (LB) fluid computed on the GPU. Our results show a significant effect of the HIs on the crystallization dynamics. In order to obtain a deeper insight of the underlying effects, we measured the particle diffusion nearby the crystal front and the corresponding van Hove correlation function.

K. Kratzer (✉) • D. Roehm • A. Arnold
Institute for Computational Physics, Allmandring 3, 70569 Stuttgart, Germany
e-mail: kratzer@icp.uni-stuttgart.de; dominic.roehm@icp.uni-stuttgart.de; arnolda@icp.uni-stuttgart.de

W.E. Nagel et al. (eds.), *High Performance Computing in Science and Engineering '14*, DOI 10.1007/978-3-319-10810-0_3

1 Introduction

Crystallization plays a crucial role in many applications, from purification of proteins in chemistry, photonic crystals in physics to protein or drug structure determination in medicine. The optimal amounts of ingredients for the particular applications are mainly determined by trial and error in experiments, due to a lack of systematic understanding of the influence of individual parameters. Using our simulations, we want to devise a systematic theory of nucleation in such systems of charged macromolecules. Rather than studying specific model systems, we concentrate on a generic unique feature of such systems, namely the influence of long range interactions, which are of electrostatic and hydrodynamic nature.

However, investigations in such systems are difficult because the waiting time between nucleation events is much longer than the time of the event itself, meaning that a lot of time is spent in waiting for the interesting process to occur. This can be overcome by rare event sampling techniques, e.g. the Forward Flux Sampling (FFS). To use these techniques efficiently, we introduce the Flexible Rare Event Sampling Harness System (FRESHS) [2], which can be used with different Molecular Dynamics (MD) codes like GROMACS [3], LAMMPS [4] or in our case ESPResSo [5, 6].

The computational demands rise even further if we take hydrodynamic interactions into account because of the number of solvent molecules. A solution is the reduction of the degrees of freedom with the Lattice-Boltzmann (LB) method, which models the solvent as a lattice fluid. This is much faster than explicit modeling of (water-)molecules, but needs still more computation time compared to the simulation of the charged macromolecules with pure Langevin dynamics. To reduce this time significantly, we model the lattice fluid on a GPU. Thereby, the lattice fluid on the GPU is computed at the same time as the computation of the forces of the macromolecules on the CPU. Since the influence of hydrodynamic interactions both on nucleation and the later crystal growth is unknown, we first concentrate on the latter, which can be greatly facilitated by the presence of a wall that lowers the nucleation barrier and makes direct simulations feasible.

In Sects. 2.1 and 2.2 we introduce the simulation model and an important quantity, the Steinhardt order parameter, which allows to determine whether a particle is in the fluid or solid state. In Sect. 2.3 the rare event sampling method is presented and in Sect. 2.4 the multi-relaxation time Lattice-Boltzmann method on the GPU and its attachment to ESPResSo are introduced. In Sect. 3.1 we present the results of the homogeneous crystallization obtained with rare event sampling simulations and in Sect. 3.2 the results of the effects of hydrodynamic correlations on crystallization in the heterogeneous case.

All calculations have been performed on the Nehalem cluster, using the NVIDIA Tesla C1060 GPUs to compute hydrodynamic interactions and the front end Nehalem processors for the particle dynamics.

2 Background

2.1 Model

The particles in our simulations interact via a screened Coulomb potential, also known as Yukawa potential, which is of the form

$$U_{\text{Yukawa}}(r) = \epsilon \frac{\exp(-\kappa(r/\sigma - 1))}{r/\sigma}. \tag{1}$$

This potential is suitable for modeling large charged particles whose electrostatic interactions are screened by surrounding ionic particles. The screening length κ^{-1} describes the range of electrostatic interactions under screening by the surrounding salt, ϵ describes the interaction strength and σ the apparent diameter of the particles.

In our simulation of the homogeneous nucleation, we add a Weeks-Chandler-Andersen (WCA) short range hard core repulsion to the Yukawa interaction, given by

$$U_{\text{WCA}}(r) = k_B T \begin{cases} 4\left(\left(\frac{\sigma}{r}\right)^{12} - \left(\frac{\sigma}{r}\right)^{6} + \frac{1}{4}\right) & r < \sigma^{\frac{1}{6}} \\ 0 & \text{else,} \end{cases} \tag{2}$$

where $k_B T$ denotes the thermal energy, given by the Boltzmann constant k_B and the temperature T. This repulsive WCA potential is used to characterize the excluded volume of the particles, which is important at high pressures, which are in turn necessary for spontaneous homogeneous nucleation.

2.2 Order Parameter

To monitor the progress of crystal growth we identify solid particles using the averaged Steinhardt order parameter [7, 8]:

$$\overline{q}_l(i) = \sqrt{\frac{4\pi}{2l+1} \sum_{m=-l}^{l} |\overline{q}_{lm}(i)|^2}, \tag{3}$$

with

$$\overline{q}_{lm}(i) = \frac{1}{\tilde{N}_b(i)} \sum_{k=0}^{\tilde{N}_b(i)} q_{lm}(k). \tag{4}$$

Table 1 Mean values of $\overline{q}_4$ and $\overline{q}_6$ for BCC, FCC structures and the undercooled liquid from [8]

	$\overline{q}_4$	$\overline{q}_6$
BCC	0.033406	0.408018
FCC	0.158180	0.491385
HCP	0.084052	0.421810
LIQ	0.031246	0.161962

where

$$q_{lm}(k) = \frac{1}{N_b(k)} \sum_{j=1}^{N_b(k)} Y_{lm}(r_{kj}) \tag{5}$$

is a complex vector based on the spherical harmonics Y_{lm} of order l. $N_b(k)$ is the number of nearest neighbors of particle k and r_{kj} the distance vector between particles k and j. For the detection of solid particles we use the parameter with $l = 6$. The literature values for the $\overline{q}_4$ and $\overline{q}_6$ order parameters in the relevant phases are given in Table 1. To detect solid particles, the $\overline{q}_6$ parameter is sufficient. However, the parameters for the different crystal structures are close to each other, therefore we use in addition $\overline{q}_4$ with $l = 4$ when distinguishing between the different structures. In the homogeneous simulations we use a cluster algorithm to find the largest cluster of solid particles in the system, whose size is considered as another order parameter. In the heterogeneous simulations studying the crystallization in front of a wall, we detect the cluster directly from the $\overline{q}_6$ profile perpendicular to the wall.

2.3 *Rare Event Sampling*

We use the Forward Flux Sampling (FFS) technique to compute the transition rate k_{AB} from the initial state A (e.g. liquid) to the final state B (e.g. most of the particles in the largest solid cluster). The initial and final states are characterized by an order parameter λ and the transition region is divided into a set of n non-intersecting interfaces with order parameter values λ_i (see Fig. 1a). Then the rate is

$$k_{AB} = \Phi \prod_{i=0}^{n-1} p_i, \tag{6}$$

where Φ is the escape flux from the initial state, and p_i is the probability to reach interface λ_{i+1} from interface λ_i before returning to A. The advantage of this form is that the probabilities p_i as well as the escape flux Φ are large enough to be computed by standard simulation techniques.

There exist several variants for calculating the escape flux Φ and the transition probabilities p_i [9–11]. In this work, we use the direct FFS algorithm (DFFS) [10–12]. In this algorithm, the escape flux is calculated by an initial run started from a random state in A. Every time when this run crosses the border λ_A in positive direction, a system configuration is stored. From the simulation time t of this run and the number of configurations q on λ_A one obtains the escape flux

$$\Phi = \frac{q}{t}. \tag{7}$$

The next steps are performed in an iterative way: From the set of configurations on λ_i a system state is drawn at random and the simulation continued until either λ_{i+1} or λ_A is reached. If λ_{i+1} is reached, a system snapshot is stored. The transition probability is then given by the fraction of successful runs M_s divided by the number of total runs M_t,

$$p_i = \frac{M_s}{M_t}. \tag{8}$$

At the end, the transition trajectories for studying physical pathways can be extracted from such an FFS simulation by tracking back the successful runs which reached B (see Fig. 1b). For good statistics, it is vital that several *independent* successful trajectories have been generated. Figure 1b thus shows a case of bad sampling: while there are seemingly 22 different transition trajectories, these in fact differ only on the last three interfaces and share most of the transition path.

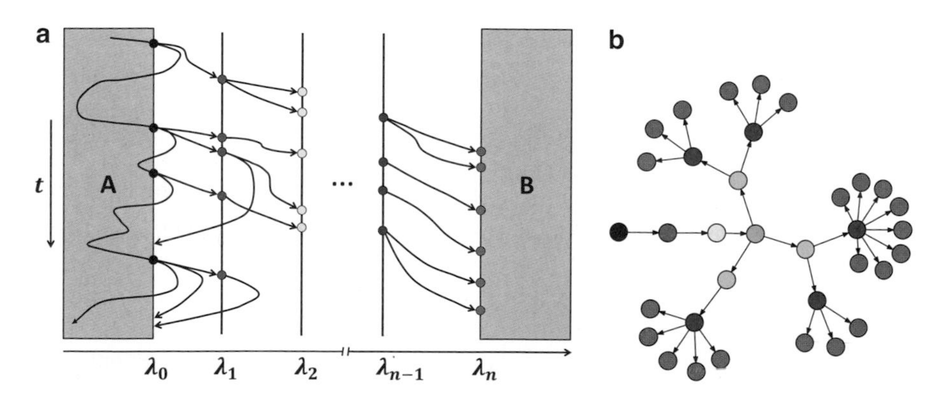

Fig. 1 (**a**) FFS method, schematic. The way from A to B is characterized by an order parameter λ. The intermediate region is subdivided by multiple interfaces at values λ_i. The escape flux from A and the transition probability p_i between the successive interfaces can be calculated by conventional simulations. The *dots* represent system configurations stored at the particular interfaces. (**b**) Successful pathway from a real simulation run, the color coding is per interface, the trajectory starts at λ_0 (*black dot*). The *arrows* point towards B. The flower-like look arises from the fact that in the last step the energy barrier has been overcome and nearly every trial run is successful

2.3.1 Implementation: FRESHS

FFS (and other sampling methods of the 'splitting' family) can be implemented in a generic way: The workflow of the method is controlled by a server which steers multiple calculation clients which simulate the trajectory fragments with the particular simulation tool, driven by so-called harness scripts. Parallelization is obtained in an asynchronous way by connecting multiple clients to the server. This asynchronous parallelization is necessary, because the duration of a simulation run in FFS is not known beforehand, e.g. trajectories are finished if either λ_{i+1} or λ_A is reached, but the number of simulation steps is not limited.

In FFS, one iteration must be completed before drawing configurations for the next iteration. This can lead to situations, where only one client is still calculating on the current set and the other clients would have to wait. We use look-ahead runs (so-called 'ghost runs') to bridge these waiting times by pre-calculations which are stored in a separate database and later transferred if the point of this set is drawn.

Another crucial aspect of an FFS simulation is the placement of the interfaces. To maximize the efficiency, interfaces must be placed at optimized positions λ_i. This can also be obtained in context of this parallel workflow system by estimating the efficiency for a certain interface position by a few number of trial runs. For further details refer to [2, 13].

2.4 Coupling Molecular Dynamics and Lattice Boltzmann Fluid

While hydrodynamic interactions between macromolecules do play an important role, the dynamics of individual solvent particles are irrelevant. Simulating the solvent particles explicitly is therefore not necessary. This allows for the use of a computationally cheaper approach to reproduce the correct hydrodynamics, the Lattice-Boltzmann (LB) method [14, 15]. In this approach, the particles are represented by population counts discretized in space, time and velocity. Because thermal fluctuations are important on this scale, and in fact drive nucleation, they have to be included on the LB fluid level. To this aim, we use the fluctuating LB method [16], which requires more computational effort per grid point than conventional LB, since it utilizes the multiple relaxation time (MRT) scheme [17] and a Gaussian random number per degree of freedom. To implement the confining walls that facilitate crystallization, we use no-slip boundary conditions on individual nodes of the discretized particle populations [18]. The coupling to the embedded macromolecules is achieved by a frictional point-particle coupling [19]. This enables us to use a much coarser LB discretization, and the distances between our particles are large enough due to the long-range electrostatic repulsion that the near-field approximation is sufficiently hidden.

The dynamics of the embedded particles are computed on conventional processors using Message Passing Interface (MPI) parallelization [20]. We use a spatial domain decomposition to distribute the particles among processors and linked cells to speed up the conservative force computation. A velocity Verlet integrator [21] propagates the particles with second order accuracy, which is sufficient for our purposes due to the strong thermal fluctuations of the fluid.

The LB algorithm, and in particular its more costly thermal variant, are ideally suited for SIMD processors like Graphics Processing Units (GPUs), and there are a number of GPU-accelerated LB implementations available [22, 23]. But our code is the first to implement the fluctuating LB equations and frictional particle coupling on the GPU, making it ideal for simulations of colloidal suspensions [24]. We optimized the overall simulation by interweaving conservative force computation on the CPU and fluid computation on the GPU. The code is part of the open-source package ESPResSo [5, 6], available at http://www.espressomd.org.

3 Results

3.1 *Homogeneous Crystallization Using Rare Event Sampling*

3.1.1 Computational Details

We performed MD simulations of 8,192 Yukawa particles using optimized and parallel FFS simulations in the context of FRESHS [2]. Thereby, the physics is calculated using ESPResSo [5,6]. We simulate a cubic simulation box with periodic boundary conditions in the NPT ensemble, where the pressure and temperature are kept constant. The pressure is controlled by an Anderson barostat, the temperature control is performed by a Langevin thermostat which models temperature fluctuations, but suppresses hydrodynamic interactions. Note that we will show in the following that hydrodynamic interactions do play a role for the dynamics of such Yukawa systems. However, the computational effort of Lattice Boltzmann or similar methods is too high to study homogeneous crystallization close to the coexistence line, even when using GPUs, and the dynamics do not change the nucleation pathways, that we are mostly interested in, just the time scale.

The parameters for the FRESHS server are as follows: We use an automatic, optimized interface placement which is based on the exploring scouts method [13] with a target interface transition probability $p = 0.5$. Ghost runs are used for pre-calculation of data points to bridge waiting times on iteration step change and with that to guarantee a maximum utilization load of the processors. During the calculation, up to 100 processors (400 cores) have been used simultaneously. Note, that the absolute number of processors can change dynamically due to the asynchronous parallelization scheme, and new clients can be queued at any time. With such a setup, the simulation advances efficiently through configuration space from A to B.

Because of the very high energy barrier in our simulations, a lot of configuration points must be collected in the beginning to assure accurate statistics [9]. Therefore, we collect $O(10^4)$ points at the first interface λ_A. The further number of points is managed by the server and the automatic interface placement: We collect at least $M = 10^3$ points on the subsequent interfaces, but the maximal decay of the number of different origin points at λ_A when backtracking the simulation from the current interface can be specified. If necessary, the FRESHS server automatically increases the number of runs.

Note that we have to store at least $N = 10^4 + n \times 10^3$ points on disk in total, where n is the number of interfaces. E.g. for one of our simulations, we accumulate a total number of $N = 31{,}864$ points. One system configuration takes about 1.4 MB of storage, which means we need 44.6 GB of storage just for the configuration points of this simulation. The other information about the successful and non-successful runs as well as the order parameter distributions are stored in the database, which contained 3.5×10^6 entries in this example and has a size of about 1 GB. For fast access, this database was located in the main memory (RAM) of the server machine during runtime. In this simulation, we performed 5.6×10^8 integration steps in total. We will now present the results of two simulations of this kind.

3.1.2 Results

Here, we present the results of two points in the phase diagram for an inverse screening length $\kappa = 5$ and pressure $p = 25$ with two different interaction strengths $\epsilon_1 = 2$ and $\epsilon_2 = 20$. These points are both close to the fluid-solid coexistence line in the phase diagram where the transition rate is low due to the small difference in chemical potential μ [25]. At these points, the nucleation mechanism can be investigated step-by-step.

For FFS, the system state A corresponds to the undercooled and therefore metastable liquid phase, where the number of solid particles (determined via $\overline{q}_6 \geq 0.3$) in the largest cluster is less than $\lambda_A = 5$. The system state B is reached if about 90 % of the solid particles are in the largest cluster, $\lambda_B = 7{,}300$. The following results are produced by backtracking a successful pathway (Fig. 2), then reproducing it by simulating again starting with loading the particular configuration point with the appropriate random number seed and storing the crystal cluster information with great detail during the reproduction runs. Then, in the post-processing, a radial analysis concerning the $\overline{q}_4$ and $\overline{q}_6$ parameters of such a cluster is performed (Fig. 3).

At $\epsilon = 2$, we observe that both the $\overline{q}_4$ and $\overline{q}_6$ parameters increase smoothly during the crystallization process (arrows in Fig. 3a). This means that the ordering has no intermediate states where e.g. a shell of another crystal structure like BCC or HCP is formed before finally converting to FCC. Such observations have been reported before [26], but that can be explained by the usual order parameter binning of the order parameter that is used to determine the type of local structure, compare

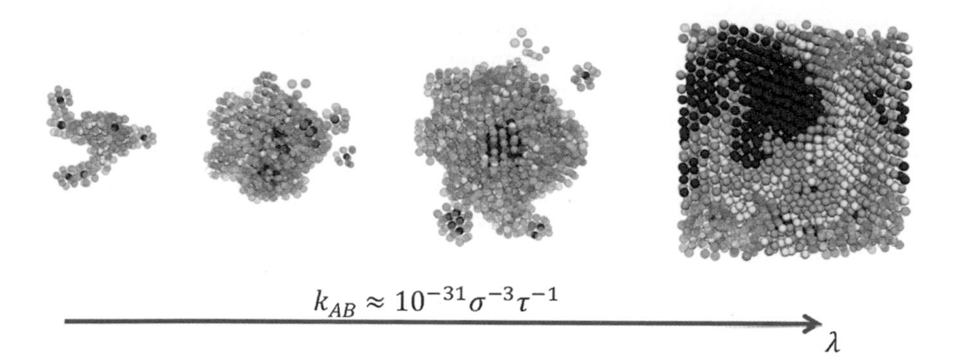

Fig. 2 Snapshots of a successful pathway for $\epsilon = 2$, extracted from the FFS simulation by backtracking from B. The transition rate for this crystallization path is $k_{AB} = 10^{-31}\sigma^{-3}\tau^{-1}$ which is very rare. In a conventional brute-force simulation this investigation would not have been possible. The color indicates different crystal structures: *yellow* is BCC-like, *orange* is HCP-like and *red* is FCC-like. The *blue* particles represent the liquid shell around the solid particles. Note, that this is a classification due to the $\overline{q}_4$ order parameter, but the transition between the structures is smooth at $\epsilon = 2$

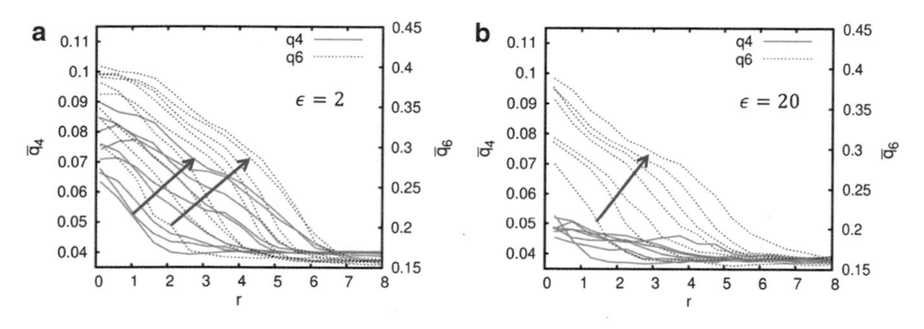

Fig. 3 Radial distribution of the $\overline{q}_4$ and $\overline{q}_6$ order parameters: (**a**) For a contact value $\epsilon = 2$ and (**b**) for $\epsilon = 20$

Fig. 2. In contrast, at $\epsilon = 20$ (Fig. 3b) we see that the $\overline{q}_6$ parameter increases smoothly, while the $\overline{q}_4$ only fluctuates a bit, but does not increase. Note, that also at this phase point the more stable phase is a FCC crystal, while the combination of high $\overline{q}_6$, but low $\overline{q}_4$ rather implies a BCC crystal structure. We interpret this such that we see a transition to a second metastable state, where the particles are more BCC-like. This is possible, since we drive the system only towards formation of a crystal using FFS, but do not enforce specifically the FCC structure. If the metastable phase has a much lower energy barrier than the final phase, it is very likely that first nuclei of this phase are formed and only later transform into the final phase. However, since FFS with our current order parameter does not drive this second transformation, the system just remains in the metastable basin. At present, we therefore run simulations to explicitly investigate this transition from this metastable BCC-like phase to the FCC structure.

Note that without the combination of rare event sampling methods, our framework FRESHS, and high performance computing resources these simulations would not have been possible. At pressure $p = 25$, which is relatively low for the purely repulsive system, the transition rates for such events are $k_{AB} = 10^{-31}\sigma^{-3}\tau^{-1}$ for $\epsilon = 2$ and $10^{-12}\sigma^{-3}\tau^{-1}$ for $\epsilon = 20$. This means that that one would have to compute roughly 10^{29} and 10^{9} time steps, respectively, to observe a single nucleation event during a conventional brute-force computer simulations. Obtaining sufficient statistics is thus impossible without rare event sampling.

3.2 Heterogeneous Crystallization Using Hydrodynamics

3.2.1 Effects of Hydrodynamic Correlations in Crystallization of Yukawa-Type Colloids

The role of hydrodynamic interactions in the growth of colloidal crystals has not been studied so far, since in experiments, it is impossible to control these interactions. Computer simulations, in fact, allow to switch these interactions on or off, but hydrodynamic interactions are expensive to include. Since the time-scale on which hydrodynamic interactions take place are much smaller than the ones of the colloidal dynamics, it is usually argued that the crystallization dynamics are not influenced by hydrodynamic interactions. However, this assumption has not yet been tested.

We use ESPResSo with a GPU-accelerated LB solver in order to study the influence of hydrodynamically induced correlations on the crystallization in a soft sphere system with Yukawa interactions. In order to switch these interactions off, we simply replace the LB fluid with a Langevin thermostat. The Langevin thermostat produces the same long-term diffusion in the bulk with appropriate parameters, but suppresses any hydrodynamic interactions. In order to induce controlled nucleation, we introduce two planar walls. This lowers the nucleation barrier sufficiently and direct simulations in the constant volume isothermal ensemble show crystallization. We can thus prepare our system as an undercooled liquid and let it crystallize spontaneously.

For our simulations, we use a system of 16,384 particles in a simulation box of size $66 \times 16 \times 16$ confined by two planar walls located at $x = 0.5$ and $x = 65.5$ (Fig. 4). In order to track the progress of the crystal front, we record the density profile and the average $\overline{q}_6$ bond-order parameter across the simulation box, which we computed on-the-fly to reduce the amount of output data. Usual production runs use a single eight core node with MPI parallelization and the attached GPU, and take up to six wall time hours on the Nehalem cluster. To sample sufficient statistics dozen successful runs are needed, since under unfortunate conditions, even heterogeneous nucleation can take rather long to set in. Furthermore, runs have to be performed for different hydrodynamic interaction strengths.

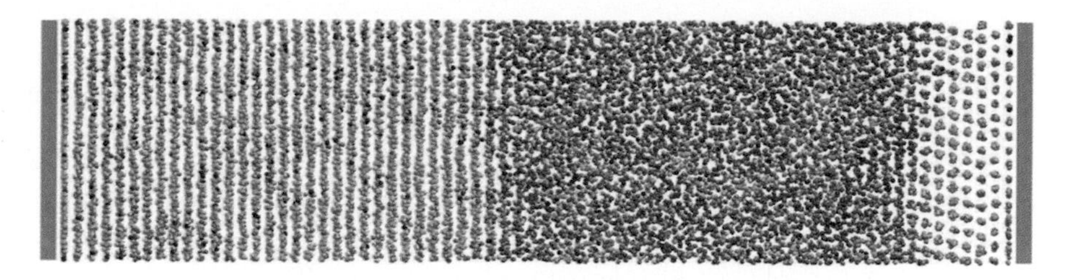

Fig. 4 Snapshot of the colloidal crystal growing from the walls, color refers to the $\overline{q}_6$ parameter (*blue* to *green*: solid, *red*: fluid). Single crystal layers can be seen, starting from both walls

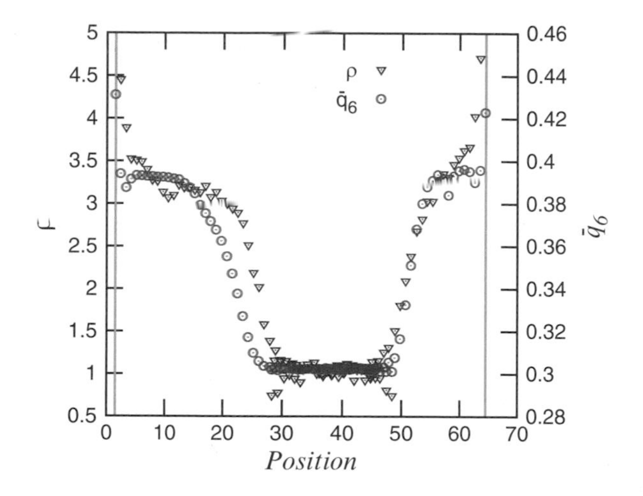

Fig. 5 Values of the $\overline{q}_6$ order parameter and the density in x-direction of the simulation box. The location of the crystal front can be detected more precisely by analyzing the $\overline{q}_6$ profile compared to the density profile. A higher concentration is observed in the vicinity of the growing crystal. Based on the shape of the profile a fitting procedure is used to evaluate the actual position of the crystal front

Our simulation parameters were chosen to form a BCC-solid close to the solid-fluid transition line. The undercooled fluid confined between two planar walls then grows into a crystal with BCC structure starting from an HCP wall layer. In Fig. 5 the peak positions of the measured average $\overline{q}_6$ bond-order parameter and the density are shown. Confined colloidal fluids show a strong layering near walls, where regions of high density interleave with regions of very low density. As the order parameter can only be detected in the regions where particles are, we report only the positions of peaks in the density or the $\overline{q}_6$ profile. As one can clearly see, the average $\overline{q}_6$ gives much smoother results, and it is possible to clearly distinguish the remaining liquid and the growing crystal. Moreover, the crystal growth happens in two steps: first, the density increases and layers are formed, and only then the local order is established. Surprisingly, hydrodynamic correlations do play a noticable role: they slow down the speed with which the front grows by up to a factor of 3 [1].

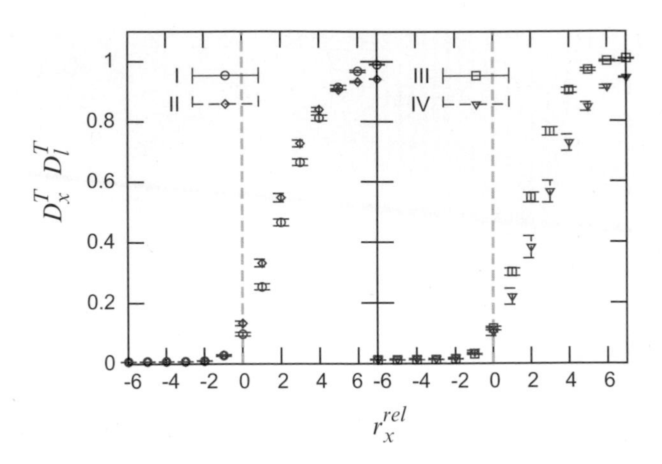

Fig. 6 Long-time diffusion coefficients in the directions of growth D_x^T normalized by the long-time tracer diffusion coefficient D_l^T in the bulk, averaged over a slice of width $2a$ into the direction of growth relative to the position of the front r_x^{rel}. The figure illustrates that there is almost no difference in the long time diffusion coefficients in case of strong HIs compared to weak/no HIs. The *blue rhombuses/triangles* are the results for simulations with HI and the *red circles/squares* are for simulations without HI. On the left hand side the HI is weak and on the right hand side strong

To better resolve the effects of the HIs on diffusive and ordering processes we took a look at several important quantities. First we investigated the tracer diffusion into the direction of growth. Therefore we binned our system along the x-axis and measured the mean square displacement (MSD) of the particles inside the single bins. Afterwards, we used the location of the front, which we estimated from the $\overline{q}_6$ profile in a post-processing step to determine the tracer diffusion in co-moving frames around the actual position of the crystal front. Figure 6 shows the results of the tracer diffusion normalized by the tracer particle diffusion in the bulk. The position of the crystal front is located at 0, while -6 is deep inside the crystal phase and $+6$ is in the bulk. In all cases, the diffusion of the particles into the direction of growth inside the crystal phase is zero, as expected, while it goes up to one in the liquid, where we expect bulk tracer diffusion. On the left hand side of Fig. 6 the diffusion coefficients in case of weak HIs (LV case I and LB case II) show virtually no difference as expected. On the right hand side in the cases of strong HIs (LV case III and LB case IV) one recognizes a small reduction of the diffusion even far away of the crystal front. This simply reflects that the growth of the crystal is hindered by the hydrodynamic interactions. The lower crystal growth speed directly leads to a reduced tracer particle motion.

As a second step, we investigated the van Hove auto-correlation function. The van Hove function is the time resolved particle distribution function and thus gives information on mean field ordering processes. Strictly speaking, it measures the probability density of finding at time t a particle at a given distance from the origin, provided that there was another particle located at the origin at time zero. For $t = 0$

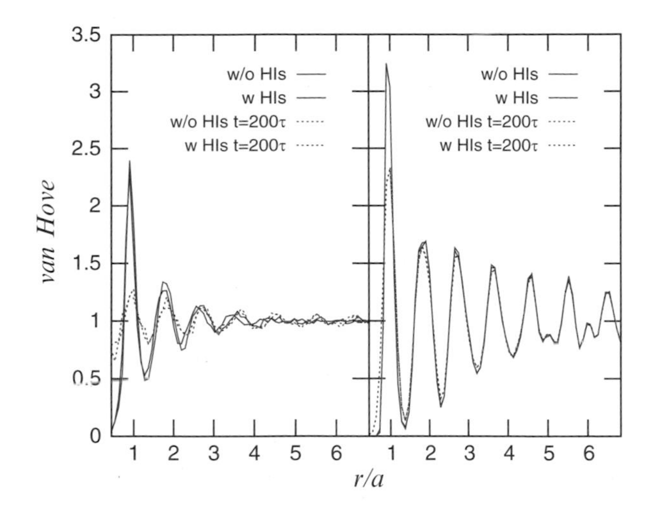

Fig. 7 Van Hove function for time zero (RDF) and 200τ. The *left-hand side* shows results in the liquid phase with and without HIs. For short times the ordering is enhanced without HIs, while for longer times the differences become negligible. The *right-hand side* shows the results within the crystal phase for both cases, with and without HIs, where there even these small differences disappear

the van Hove function is thus equal to the radial distribution function (RDF). For times $t > 0$ the van Hove function measures possible changes of the structure over time.

Using the van Hove function perpendicular to the direction of the front growth allows us to investigate the average ordering process. For that, we again used a binning of our system, where the position and exact width of the bins are such that every bin contains a single crystal layer. Afterwards, we evaluated the two-dimensional van Hove function inside these layers. The left-hand side of Fig. 7 shows the van Hove functions in the liquid phase perpendicular to the direction of growth. In Fig. 7 the van Hove auto-correlation function is shown in the solid phase (right-hand side) for $t = 0$ (RDF) and for $t = 200\tau$. Apparently, the crystal structure and its time dependence are equal with and without HIs and the structure as well as the locations of the particles inside the crystal are absolutely stable. As expected, the static structure of the crystal is not affected by the HIs. Surprisingly, the van Hove function is very crystal-like already in the region where the density is high, but no local order has been established (compare Fig. 5). The establishing of crystal symmetry is thus a purely local effect that cannot be captured by the averaging van Hove function. Further investigations of this pre-ordering region to elucidate its role in crystal growth are presented in Roehm et al. [1].

4 Conclusion

Using two state of the art techniques in computational physics, Forward Flux Sampling and GPU-accelerated Lattice-Boltzmann, we addressed the problems of homogeneous and heterogeneous crystallization of charged macromolecules like proteins or colloidal particles.

Forward Flux Sampling allows to perform rare event sampling using large numbers of processors in a trivially parallel fashion. But this is necessary, since even with this special technique, rare event sampling is computationally expensive. Our efficient FFS implementation in FRESHS allows for the investigation of nucleation of particles with Yukawa interactions close to the coexistence line. It shows that the order is usually gradually established, which should not be mistaken for a surrounding shell of a different phase. Yet, at certain areas of the phase diagram, this system also exhibits a metastable crystal phase, which however needs to undergo a second phase transition rather than just being a precursor to the final phase.

The NVIDIA Tesla C1060 cards in the Nehalem cluster accelerate simulations including hydrodynamic interactions by a factor of $\approx$20. This has enabled us to run several thousand simulations in feasible time, which allowed us to study the effect of hydrodynamic interactions on crystal growth. Only a large GPU cluster such as the Nehalem cluster at HLRS makes such studies possible, where our results show effects of HIs on the local arrangement dynamics as well as a delay of the growth velocity. Further details on the effects of hydrodynamic interactions on the crystallization of soft colloids are presented in Ref. [1].

References

1. Roehm, D., Kesselheim, S., Arnold, A.:Hydrodynamic interactions slow down crystallization of soft colloids. Soft Matter **10**, 5503 (2014). doi:10.1039/C4SM00686K, http://dx.doi.org/10.1039/C4SM00686K
2. Kratzer, K., Berryman, J.T., Taudt, A., Zeman, J., Arnold, A.: The Flexible Rare Event Sampling Harness System (FRESHS). Comput. Phys. Commun. **185**(7), 1875 (2014). doi:10.1016/j.cpc.2014.03.013
3. Hess, B., Kutzner, C., van der Spoel, D., Lindahl, E.: GROMACS 4: Algorithms for highly efficient, load-balanced, and scalable molecular simulation. J. Chem. Theory Comput. **4**(3), 435 (2008)
4. Plimpton, S.J.: Fast parallel algorithms for short-range molecular dynamics. J. Comput. Phys. **117**, 1 (1995)
5. Limbach, H.J., Arnold, A., Mann, B.A., Holm, C.: ESPResSo - an extensible simulation package for research on soft matter systems. Comput. Phys. Comm. **174**(9), 704 (2006). doi:10.1016/j.cpc.2005.10.005
6. Arnold, A., Lenz, O., Kesselheim, S., Weeber, R., Fahrenberger, F., Röhm, D., Košovan, P., Holm, C. In: ESPResSo 3.1 - molecular dynamics software for coarse-grained models. Griebel, M., Schweitzer, M.A. (eds.) Meshfree Methods for Partial Differential Equations VI. Lecture Notes in Computational Science and Engineering, vol. 89, pp. 1–23. Springer, New York (2013). doi:10.1007/978-3-642-32979-1_1, http://www.springer.com/mathematics/computational+science+%26+engineering/book/978-3-642-32978-4

7. Steinhardt, P., Nelson, D., Ronchetti, M.:Bond-orientational order in liquids and glasses. Phys. Rev. B **28**(2), 784 (1983)
8. Lechner, W., Dellago, C.: Accurate determination of crystal structures based on averaged local bond order parameters. J. Chem. Phys. **129**(11), 114707 (2008)
9. Allen, R.J., Frenkel, D., ten Wolde, P.R.: Simulating rare events in equilibrium or nonequilibrium stochastic systems. J. Chem. Phys. **124**, 024102 (2006)
10. Allen, R.J., Valeriani, C., ten Wolde, P.R.: Forward Flux Sampling for rare event simulations. J. Phys. Condens. Matter **21**(46), 463102 (2009)
11. Escobedo, F.A., Borrero, E.E., Araque, J.C.: Transition Path Sampling and Forward Flux Sampling. Applications to biological systems. J. Phys. Condens. Matter **21**(33), 333101 (2009)
12. Allen, R.J., Warren, P.B., ten Wolde, P.R.: Sampling rare switching events in biochemical networks. Phys. Rev. Lett. **94**, 018104 (2005)
13. Kratzer, K., Arnold, A., Allen, R.J.: Automatic, optimized interface placement in Forward Flux Sampling simulations. J. Chem. Phys. **138**(16), 164112 (2013)
14. Succi, S.: The Lattice Boltzmann equation for fluid dynamics and beyond. Oxford University Press, New York (2001)
15. Bhatnagar, P.L., Gross, E.P., Krook, M.: A model for collision processes in gases. I. Small amplitude processes in charged and neutral one-component systems. Phys. Rev. **94**(3), 511 (1954)
16. Dünweg, B., Ladd, A.J.C. Lattice Boltzmann simulations of soft matter systems. In: Holm, C., Kremer, K. (eds.) Advanced Computer Simulation Approaches for Soft Matter Sciences III. Advances in Polymer Science, vol. 221, pp. 89–166. Springer, Berlin (2009). doi:10.1007/12_2008_4
17. d'Humieres, D.: Multiple-relaxation-time Lattice Boltzmann models in three dimensions. Philos. Trans. R. Soc. Lond. Ser. A Math. Phys. Eng. Sci. **360**(1792), 437 (2002)
18. Ladd, A.J.C.: Numerical simulations of particulate suspensions via a discretized Boltzmann equation. Part 1. Theoretical foundation. J. Fluid Mech. **271**, 285 (1994)
19. Ahlrichs, P., Dünweg, B.: Lattice-Boltzmann simulation of polymer-solvent systems. Int. J. Mod. Phys. C **9**(8), 1429 (1998)
20. Consortium, M.: The message passing interface (mpi) standard (2004). http://www.mcs.anl.gov/research/projects/mpi/Homepage
21. Frenkel, D., Smit, B.: Understanding Molecular Simulation, 2nd edn. Academic, San Diego (2002)
22. Li, W., Wei, X., Kaufman, A.: Implementing Lattice Boltzmann computation on graphics hardware. Vis. Comput. **19**(7), 444 (2003)
23. Feichtinger, C., Donath, S., Köstler, H., Götz, J., Rüde, U.: Walberla: HPC software design for computational engineering simulations. J. Comput. Sci. 2(2), 105–112 (2011)
24. Röhm, D., Arnold, A.: Lattice Boltzmann simulations on GPUs with ESPResSo. Eur. Phys. J. Spec. Top. **210**, 89 (2012). http://www.springerlink.com/content/f7702046h324u123/?MUD=MP
25. Azhar, F.E., Baus, M., Ryckaert, J.P., Meijer, E.J.: Line of triple points for the hard-core Yukawa model: A computer simulation study. J. Chem. Phys. **112**(11), 5121 (2000)
26. Auer, S., Frenkel, D.: Crystallization of weakly charged colloidal spheres: a numerical study. J. Phys. Condens. Matter **14**(33), 7667 (2002)

PAMOP: Petascale Atomic, Molecular and Optical Collision Calculations

B.M. McLaughlin, C.P. Ballance, M.S. Pindzola, and A. Müller

Abstract Petaflop architectures are currently being utilized efficiently to perform large scale computations in Atomic, Molecular and Optical Collisions. We solve the Schrödinger or Dirac equation for the appropriate collision problem using the R-matrix or R-matrix with pseudo-states approach. We briefly outline the parallel methodology used and implemented for the current suite of Breit-Pauli and DARC codes. In this report, various examples are shown of our theoretical results compared with experimental results obtained from Synchrotron Radiation facilities where the Cray architecture at HLRS is playing an integral part in our computational projects.

1 Introduction

Our research efforts continue to focus on the development of computational methods to solve the Schrödinger and Dirac equations for atomic and molecular collision processes. Access to leadership-class computers such as the Cray XE6 at HLRS allows us to benchmark our theoretical solutions against dedicated collision experiments at synchrotron facilities such as the Advanced Light Source (ALS), Astrid II, BESSY II, SOLEIL and Petra III and to provide atomic and molecular data for ongoing research in laboratory and astrophysical plasma science. In order to have direct comparisons with experiment, semi-relativistic or fully relativistic

B.M. McLaughlin (✉)
School of Mathematics & Physics, Centre for Theoretical Atomic, Molecular and Optical Physics (CTAMOP), The David Bates Building, Queen's University, 7 College Park, Belfast BT7 1NN, UK
e-mail: b.mclaughlin@qub.ac.uk

C.P. Ballance • M.S. Pindzola
206 Allison Laboratory, Department of Physics, Auburn University, Auburn, AL 36849, USA
e-mail: ballance@physics.auburn.edu; pindzola@physics.auburn.edu

A. Müller
Institut für Atom- und Molekülphysik, Justus-Liebig-Universität Giessen, 35392 Giessen, Germany
e-mail: Alfred.Mueller@iamp.physik.uni-giessen.de

W.E. Nagel et al. (eds.), *High Performance Computing in Science and Engineering '14*, DOI 10.1007/978-3-319-10810-0_4

computations, involving a large number of target-coupled states are required to achieve spectroscopic accuracy. These computations could not be even attempted without access to HPC resources such as those available at leadership computational centers in Europe (HLRS) and the USA (NERSC, NICS and ORNL). We use the R-matrix and R-matrix with pseudo-states (RMPS) methods to solve the Schrödinger and Dirac equations for atomic and molecular collision processes.

Satellites such as *Chandra* and *XMM-Newton* are currently providing a wealth of X-ray spectra on many astronomical objects, but a serious lack of adequate atomic data, particularly in the K-shell energy range, impedes the interpretation of these spectra. Spectroscopy in the soft X-ray region (0.5–4.5 nm), including K-shell transitions of singly and multiply charged ionic forms of atomic elements such as Be, B, C, N, O, Ne, S and Si, as well as L-shell transitions of Fe and Ni, provides a valuable probe of the extreme environments in astrophysical sources such as active galactic nuclei (AGN's), X-ray binary systems, and cataclysmic variables [10, 11, 14]. For example, K-shell photoabsorption cross sections for the carbon isonuclear sequence have been used to model the Chandra X-ray absorption spectrum of the bright blazar Mkn 421 [9].

The motivation for our work is multi-fold; (a) Astrophysical Applications [5, 9], (b) Fusion and plasma modelling, JET, ITER, (c) Fundamental interest and (d) Support of experimental measurements and Satellite observations. In the case of heavy atomic systems [12,13], little atomic data exists and our work provides results for new frontiers on the application of the R-matrix; Breit-Pauli and DARC parallel suite of codes. Our highly efficient R-matrix codes are widely applicable to the support of present experiments being performed at synchrotron radiations facilities such as; ALS, ASTRID II, SOLEIL, PETRA III, BESSY II. Various examples of large scale calculations are presented to illustrate the predictive nature of the method.

The main question asked of any method is, how do we deal with the many body problem? In our case we use first principle methods (ab initio) to solve our dynamical equations of motion. Ab initio methods provide highly accurate, reliable atomic and molecular data (using state-of-the-art techniques) for solving the Schrödinger and Dirac equation. The R-matrix non-perturbative method is used to model accurately a wide variety of atomic, molecular and optical processes such as; electron impact ionization (EII), electron impact excitation (EIE), single and double photoionization and inner-shell X-ray processes. The R-matrix method provides highly accurate cross sections and rates used as input for astrophysical modeling codes such as; CLOUDY, CHIANTI, AtomDB, XSTAR necessary for interpreting experiment/satellite observations of astrophysical objects and fusion and plasma modeling for JET and ITER.

2 Parallel R-matrix Photoionization

The use of massively parallel architectures allows one to attempt calculations which previously could not have been addressed. This approach enables large scale relativistic calculations for trans-iron elements such as; Kr-ions, Xe-ions [12] and W-ions [2,3]. It allows one to provide atomic data in the absence of experiment and takes advantage of the linear algebra libraries available on most architectures. We fill in our sea of ignorance i.e. provide data on atomic elements where none have previously existed. The present approach has the capability to cater for Hamiltonian matrices in excess of 250 × 250 K. Examples are presented for both valence and inner-shell photoionization for systems of prime interest to astrophysics and for complex species necessary for plasma modeling in fusion tokamaks.

The development of the dipole codes benefit from similar modifications and developments made to the existing excitation R-matrix codes. In this case all the eigenvectors from a pair of dipole allowed symmetries are required for bound-free dipole matrix formation. Every dipole matrix pair is carried out concurrently with groups of processors assigned to an individual dipole. The method is applicable to photoionization, dielectronic-recombination or radiation damped excitation and now reduces to the time taken for a single dipole formation. The method so far implemented on various parallel architectures has the capacity to cater for photoionization calculations involving 500–1,000 levels. This dramatically improves (a) the residual ion structure, (b) ionization potential, (c) resonance structure and (d) can deal with in excess of 5,000 close-coupled channels.

3 Scalability

As regards the scalability of our R-matrix codes, we find from experience on a variety of peta-flop architectures that various modules within this suite of codes scale well, upwards to 100,000 cores. In practical calculations for cross sections on various systems it is necessary to perform fine energy resolution of resonance features (10^{-8} (Ry) $\sim$ 1.36 meV) observed in photoionization cross sections. This involves many (6–30 million) incident photon energies, vital when comparing with high precision measurements, like those for Xe^{+} ions made at the Advanced Light Source synchrotron radiation facility in Berkeley, California, USA where energy resolutions of 4 meV FWHM are achieved [12].

The formation of many real symmetric matrices (Hamiltonians), typically 60–150 K, requires anywhere from 10 to 500 GB of storage. The diagonalization of each matrix, from which *every* eigenvalue and *every* eigenvector is required is achieved through use of the ScaLapack package. In particular routines : **pdsyevx** and **pdsyevd**, where preference is given to the latter, as it ensures orthogonality between all eigenvectors. In typical collision calculations, matrices vary in size from 2×2 K to 200×200 K, depending on the complexity of the atomic target. The formation of the continuum-continuum part of the $N + 1$ electron Hamiltonian is the most time consuming. Therefore if there are several thousand scattering channels (nchan) then there are $[nchan \times (nchan + 1)/2]$ matrix blocks. Each block represents a partial wave and each subgroup reads a single Hamiltonian and diagonalizes it in parallel, concurrently with each other. So there is endless scalability. R-matrix close-coupling calculations are therefore reduced to the time required for a single partial wave.

Optimization of the R-matrix codes on a variety of HPC architectures implements several good coding practices. We use code inlining (which reduces overhead from calling subroutines), loop unrolling, modularization, unit strides, vectorization and dynamic array passing. Highly optimized libraries such as; ScaLapack, BLAS and BLACS are used extensively in the codes, which are available at HLRS. On Cray architectures, such as the Cray XE6 at Stuttgart, the R-matrix codes were profiled with performance utilities such as CrayPat which gives the user an indication of their scalability. Standard MPI/FORTRAN 90/95 programming is used on the world's leading-edge peta-flop high performance supercomputers. Serial (legacy) code to parallel R-matrix implementation is carried out using a small subset of basic MPI commands.

In Table 1 we show details of test runs for the outer region module PSTBF0DAMP for K-shell photoionization of B^+ using 249-coupled states with 400 coupled channels and 409,600 energy points and an increasing number of CPU cores. A factor of 4 speed up is achieved by using up to 8,192 cores. The computations were carried out on the Cray XE6 (Hopper) at NERSC, comparable to the Hermit architecture at HLRS, Stuttgart using the CrayPat utility. Note, for actual production runs, timings would be a factor of 10 larger, as one would require a mesh of 4,096,000 energy points to fully resolve the resonances features observed in the spectrum [17]. We present the timings for core sizes varying from 1,024 to 8,192 again for B^+ K-shell photoionization in its ground state. The computations were performed with the outer region module PSTGBF0DAMP for 249-states and 400-coupled channels. The main work horse in our linear algebra code is the

Table 1 B^+, 249-states, 400 coupled channels, 409,600 energy points running on an increasing number of cores. The results are from the module PSTGB0FDAMP for the photoionization cross-section calculations of ground state of the B^+ ion carried out on the Hopper architectures at NERSC comparable to Hermit at HLRS. Results presented illustrate the speed up factor with increasing number of CPU cores and the total number of core hours

CRAY XE6 CPU cores	Hopper (NERSC) absolute timing (s)	Hopper (NERSC) speed up factor	Hopper (NERSC) total core hours
1,024	584.19	1.0000	166.1155
2,048	430.80	1.3584	245.0077
4,096	223.08	2.6183	253.8154
8,192	149.70	3.9018	340.6506

ScaLAPACK libraries. The goals of the ScaLAPACK project are the same as those of LAPACK; Efficiency (to run as fast as possible), Scalability (as the problem size grows so do the numbers of processors grow), Realiability (including error bounds), Flexibility (so users can construct new routines from well-designed parts) and Ease of Use (by making the interface to LAPACK and ScaLAPACK look as similar as possible). Many of these goals, particularly portability are aided by developing and promoting standards, especially for low level communication and computation routines.

Access to the Cray XE6 at HLRS, January 2013–May 2013, was provided for exploratory computations, to test our parallel Breit-Pauli and DARC codes. During this period, Connor Ballance and Brendan McLaughlin, worked closely with Stefan Andersson, the resident Cray Research consultant at HLRS, profiling and benchmarking our codes on this Cray XE6. A variety of different detailed calculations were made on the Cray XE6 HLRS architecture (Hermit) resulting in several publications [1, 8, 15]. A formal proposal was submitted to HLRS in June 2013 for access to this architecture for a project lead by Professor Alfred Müller from the University of Giessen.

4 X-Ray and Inner-Shell Processes

4.1 Atomic Oxygen

An accurate description of the photoionization/photoabsorption of atomic oxygen is important for a number of atmospheric and astrophysical applications. Photo-absorption of atomic oxygen in the energy region below the $1\,s^{-1}$ threshold in X-ray spectroscopy from *Chandra* and *XMM-Newton* is observed in a variety of X-ray binary spectra. Photo-absorption cross sections determined from an R-matrix method with pseudo-states (RMPS) and high precision measurements from the Advanced Light Source (ALS) are presented in Fig. 1. High-resolution spectroscopy with $E/\Delta E \approx 4{,}250 \pm 400$ were obtained for photon energies from 520 to 555 eV at an energy resolution of 124 ± 12 meV FWHM. *K*-shell photoabsorption cross-section measurements were made on atomic oxygen at the ALS. Natural line widths Γ are extracted for the $1s^{-1}2s^2 2p^4(^4P)np\,^3P^\circ$ and $1s^{-1}2s^2 2p^4(^2P)np\,^3P^\circ$ Rydberg resonances series and compared with theoretical predictions. Accurate cross sections and line widths are obtained for applications in X-ray astronomy. Excellent agreement between theory and the ALS measurements is shown which will have profound implications for the modelling of X-ray spectra and spectral diagnostics. Further details can be found in our recent study on this complex [15].

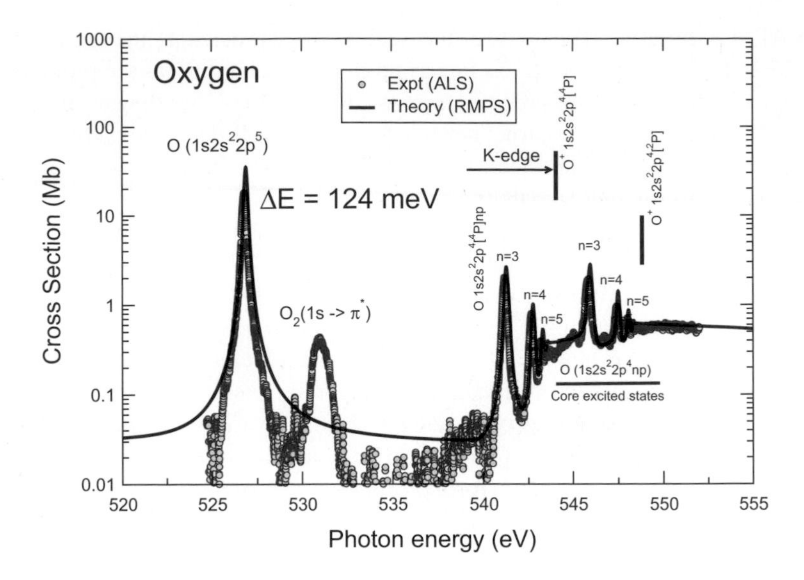

Fig. 1 (Colour online) Atomic oxygen photo-absorption cross sections taken at 124 meV FWHM compared with theoretical estimates. The R-matrix calculations shown are from the R-matrix with pseudo-states method (RMPS: *solid black line*, present results) convoluted with a Gaussian profile of 124 meV FWMH [15]

4.2 Nitrogen Ions

Recent studies on K-shell photoionization of neutral nitrogen and oxygen showed excellent agreement with high resolution measurements made at the Advanced Light Source (ALS) radiation facility [15, 20] as have similar cross section calculations on singly and multiply ionized stages of atomic nitrogen compared with high resolution measurements at the SOLEIL synchrotron facility [1, 7, 8]. The majority of the high-resolution experimental data from third generation light sources show excellent agreement with the state-of-the-art R-matrix method and with other modern theoretical approaches.

The investigations on Li-like, Be-like and B-like atomic nitrogen ions gives accurate values of photoionization cross sections produced by X-rays in the vicinity of the K-edge, where strong $n = 2$ inner-shell resonance states are observed. N^{2+} ions produced in the SOLEIL synchrotron radiation experiments are not purely in their ground state (see Fig. 2). K-shell photoionization contributes to the ionization balance in a more complicated way than outer shell photoionization. In fact K-shell photoionization when followed by Auger decay couples three or more ionization stages instead of two in the usual equations of ionization equilibrium [1, 8].

The R-matrix with pseudo-states method (RMPS) was used to determine all the cross sections (in LS – coupling) with 390 levels of the N^{3+} residual ion included in the close-coupling calculations. Since metastable states are present in the parent ion beam, theoretical PI cross-section calculations are required for both

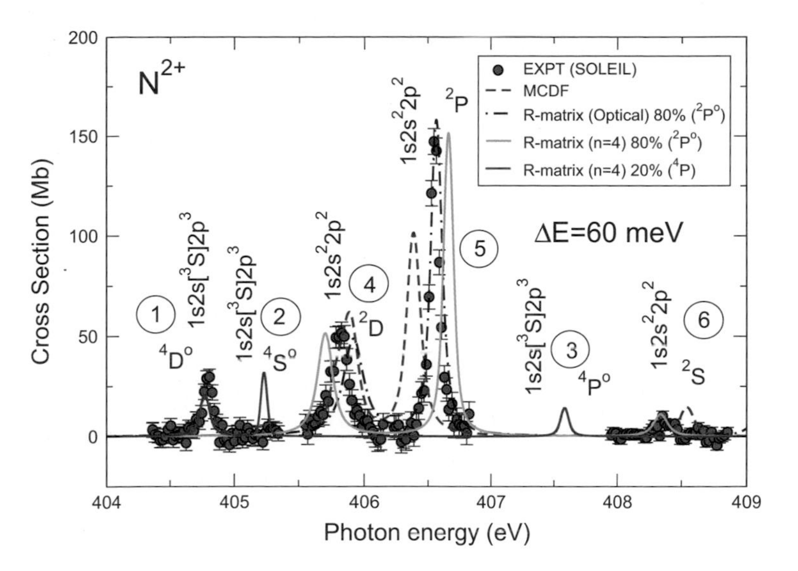

Fig. 2 (Colour online) Photoionization cross sections for N^{2+} ions measured with a 60 meV band pass at SOLEIL. *Solid circles*: total photoionization. The error bars give the statistical uncertainty of the experimental data. The MCDF (*dashed line*), R-matrix RMPS (*solid line*, *red*, ^4P, *green*, $^2P^o$), and the Optical potential (*dash-dot line*, $^2P^o$ only) calculations shown were obtained by convolution with a Gaussian distribution having a profile width at FWHM of 60 meV and a weighting of the ground and metastable states to simulate the measurements [8]

the $1s^2 2s^2 2p\ ^2P^o$ ground state and the $1s^2 2s 2p^2\ ^4$P metastable states of the N^{2+} ion for a proper comparison with experiment. The scattering wavefunctions were generated by allowing two-electron promotions out of selected base configurations of N^{2+}. Scattering calculations were performed with 20 continuum functions and a boundary radius of 9.4 Bohr radii. For the ^{2}P^o ground state and the ^4P metastable states the electron-ion collision problem was solved with a fine energy mesh of 2×10^{-7} Rydbergs ($\approx$2.72 μeV) to delineate all the resonance features in the PI cross sections. Radiation and Auger damping were also included in the cross section calculations.

For a direct comparison with the SOLEIL results, the R-matrix cross section calculations were convoluted with a Gaussian function of appropriate width and an admixture of 80 % ground and 20 % metastable states used to best simulate experiment. The peaks found in the theoretical photoionization cross section spectrum were fitted to Fano profiles for overlapping resonances as opposed to the energy derivative of the eigenphase sum method [1, 8].

For Be-like ions both the initial ^1S ground state and the ^{3}P° metastable states were required (see Fig. 3). Cross section calculations were carried out in *LS*-coupling with 390-levels retained in the close-coupling expansion using the R-matrix with pseudo states method (RMPS). The Hartree-Fock $1s$, $2s$ and $2p$ were used with n = 3 physical and n = 4 pseudo orbitals of the residual N^{4+} ion.

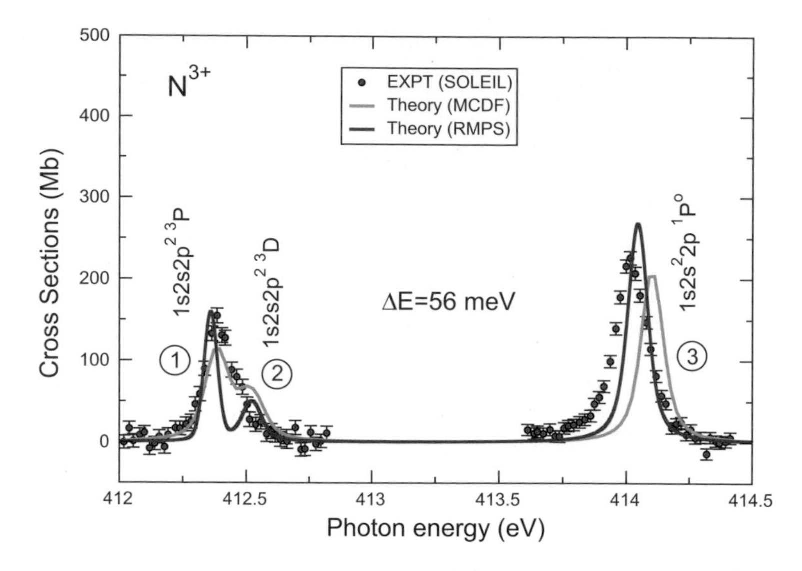

Fig. 3 (Colour online) Photoionization cross sections for Be-like atomic nitrogen (N^{3+}) ions measured with a 56 meV band pass at the SOLEIL radiation facility. *Solid circles*: total photoionization. The error bars give the statistical uncertainty of the experimental data. The MCDF (*solid green line*) and R-matrix (*solid red line*) calculations shown are convoluted with a Gaussian profile of 56 meV FWHM and an appropriate weighting of the ground and metastable states to simulate the measurements. For the metastable $^3P^\circ$ state, the MCDF calculations have been shifted up by +1.46 eV in order to match experiment [1]

The n = 4 pseudo-orbitals were determined by energy optimization on the ground state of the N^{4+} ion, with the atomic structure code CIV3. The N^{4+} residual 390 ion states used multi-configuration interaction target wave functions. The non-relativistic R-matrix method determined the energies of the N^{3+} bound states and all the appropriate cross sections. We determined PI cross sections for the $1s^22s^2$ ^{1}S ground state and the $1s^22s2p$ ^{3}P° metastable state.

For Li-like systems, intermediate-coupling photoionization cross section calculations were performed using the semi-relativistic Breit-Pauli approximation (see Fig. 4). An appropriate number of N^{5+} residual ion states (19 *LS*, 31 *LSJ* levels) were included in our intermediate coupling calculations. The n = 4 basis set of N^{5+} orbitals obtained from the atomic-structure code CIV3 were used to represent the wave-functions. Photoionization cross-section calculations were then performed in intermediate coupling for the $1s^22s$ $^2S_{1/2}$ initial state of the N^{4+} ion in order to incorporate relativistic effects via the semi-relativistic Breit-Pauli approximation.

For cross section calculations, He-like *LS* states were retained: $1s^2$ ^1S, $1sns$ $^{1,3}S$, $1snp$ $^{1,3}P^\circ$, $1snd$ $^{1,3}D$, and $1snf$ $^{1,3}F^\circ$, n $\leq$ 4, of the N^{5+} ion core giving rise to 31 *LSJ* states in the intermediate close-coupling expansions for the $J = 1/2$ initial scattering symmetry of the Li-like N^{4+} ion. The n = 4 pseudo states are included in an attempt to model core relaxation, electron correlations effects and the infinite number of states (bound and continuum) left out by the truncation of

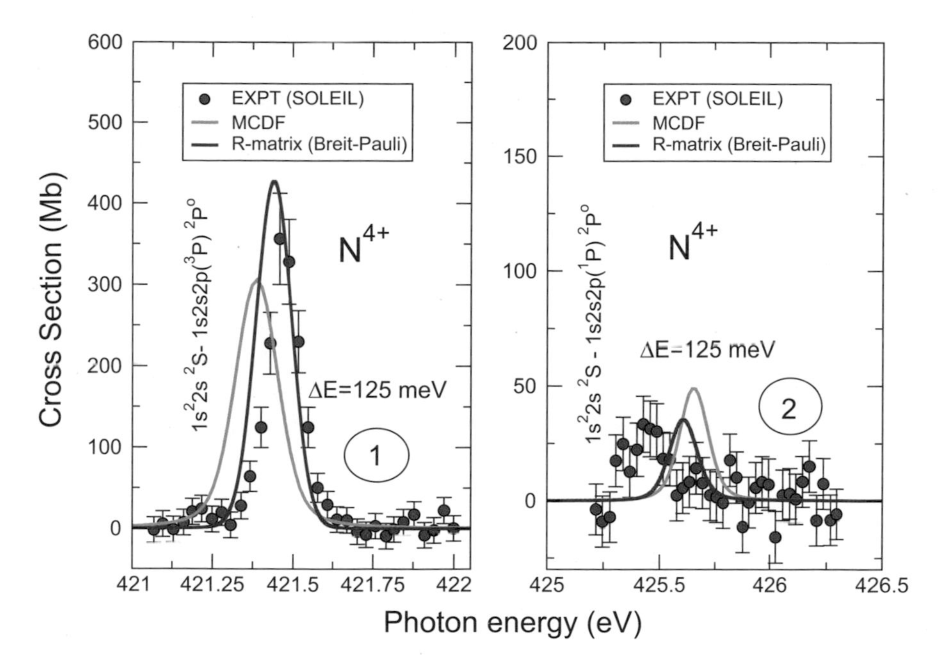

Fig. 4 (Colour online) Photoionization cross sections for Li-like atomic nitrogen (N^{4+}) ions measured with a 125 meV FWHM band pass at the SOLEIL radiation facility. *Solid circles*: absolute total photoionization cross sections. The error bars give the statistical uncertainty of the experimental data. R-matrix (*solid red line*, 31 levels) intermediate coupling, MCDF (*solid green line*), calculations shown are convolution with a Gaussian profile of 125 meV FWHM to simulate the measurements [1]

the close-coupling expansion in our work. For the structure calculations of the residual N^{5+} ion, all n = 3 physical orbitals and n = 4 correlation orbitals were included in the multi-configuration-interaction target wave-functions expansions used to describe the states.

5 Heavy Atomic Systems

5.1 Xe Ions

Photoionization cross sections of heavy atomic elements, in low stages of ionization, are currently of interest both experimentally and theoretically and for applications in astrophysics. The data from such processes have many applications in planetary nebulae, where they are of use in identifying weak emission lines of n-capture elements in NGC 3242.

Xenon ions are of importance in man-made plasmas such as XUV light sources for semiconductor lithography, ion thrusters for space craft propulsion and nuclear fusion plasmas. Xenon ions have also been detected in cosmic objects, e.g., in several planetary nebulae and in the ejected envelopes of low- and intermediate-mass stars [12]. Collision processes with highly charged xenon ions are of interest for UV-radiation generation in plasma discharges, for fusion research and for space craft propulsion. Here we report theoretical and experimental results for the photoionization of Ag-like (Xe^{7+}) xenon ions which were measured at the photon-ion end station of ALS on beamline 10.0.1. Compared with the only previous experimental study of Bizau and co-workers [4] of this reaction,the present cross-sections were obtained at higher energy resolution (38–100 meV versus 200–500 meV) and on an absolute cross-section scale. In the experimental photon energy range of 95–145 eV the cross-section is dominated by resonances associated with $4d \rightarrow 5f$ excitation and subsequent autoionization. The theoretical results were obtained using the Dirac Coulomb R-matrix approximation [6, 12, 13]. The small resonances below the ground-state ionization threshold, located at about 106 eV [19], are due to the presence of metastable $Xe^{7+}(4d^{10}4f\ ^2F^{\circ}_{5/2,7/2})$ ions with an excitation energy of 32.9 eV [19] in the ion beam. In the experimental photon energy range of 95–145 eV the cross-section is dominated by resonances associated with $4d \rightarrow 5f$ excitation and subsequent autoionization. The most prominent feature in the measured spectrum is the giant $4d^9 5s5f\ ^2P^{\circ}$ resonance located at 122.139 ± 0.01 eV, which reaches a peak cross-section of 1.2 GB at 38 meV photon energy spread [21]. The experimental resonance strength of (161 ± 31) Mb eV (corresponding to an absorption oscillator strength of 1.47 ± 0.28), width of 76 ± 3 meV, is in suitable agreement with the present theoretical estimate and with previous investigations [4]. The high-precision cross-section measurements obtained from the Advanced Light source compared with large-scale theoretical calculations obtained from a Dirac Coulomb R-matrix approach over the entire photon energy region investigated show suitable agreement.

The present work on Ag-like ions of Xenon provides a benchmark for future work. In addition to the direct photoionization process, indirect excitation processes occur for the interaction of a photon with the $4d^{10}5s\ ^2S_{1/2}$ ground-state and the metastable $4d^{10}4f\ ^2F^{\circ}_{5/2,7/2}$ levels of the Xe^{7+} ion. These intermediate resonance states can then decay to the ground state or energy accessible excited states.

Photoionization cross-sections on this complex ion were performed for the ground ($4d^{10}5s$) and the excited metastable ($4d^{10}4f$) levels associated with the Ag-like xenon ion to benchmark our theoretical results with the present high resolution experimental measurements made at the Advanced Light Source radiation facility in Berkeley [18]. The atomic structure calculations were carried out using the GRASP code. Initial scattering calculations were performed using 249-levels arising from 12 configurations of the Pd-like (Xe^{8+}) residual ion. Further collision models were investigated where both a 508-level and 526-level approximation were used in the close-coupling calculations. For the $4d^{10}4f\ ^2F_{7/2,5/2}$ metastable states of this ion, both a 249-level and a 528-level model were investigated [18].

Photoionization cross-section calculations were performed in the Dirac Coulomb approximation using the DARC codes [6, 12, 13] for different scattering models for the ground $4d^{10}5s\ ^2S_{1/2}$ state of the Xe^{7+} ion. The R-matrix boundary radius of 12.03 Bohr radii was sufficient to envelop the radial extent of all the n = 6 atomic orbitals of the residual Xe^{7+} ion. A basis of 16 continuum orbitals was sufficient to span the incident experimental photon energy range from threshold up to 150 eV. Dipole selection rules for the ground-state photoionization require only the bound-free dipole matrices, $2J^{\pi} = 1^e \rightarrow 2J^{\pi} = 1^{\circ}, 3^{\circ}$. For the excited metastable states, $2J^{\pi} = 5^o \rightarrow 2J^{\pi} = 3^e, 5^e, 7^e$ and $2J^{\pi} = 7^o \rightarrow 2J^{\pi} = 5^e, 7^e, 9^e$ are necessary.

In the experimental photon energy range of 95–145 eV the cross-section is dominated by resonances associated with $4d \rightarrow 5f$ excitation and subsequent autoionization. We performed 249-level, 508-level and 526-level DARC photoionization cross-section calculations for the $Xe^{7+}(4d^{10}5s\ ^2S_{1/2})$ ground state to check on the convergence of our results. Figure 5, includes the theoretical cross-sections

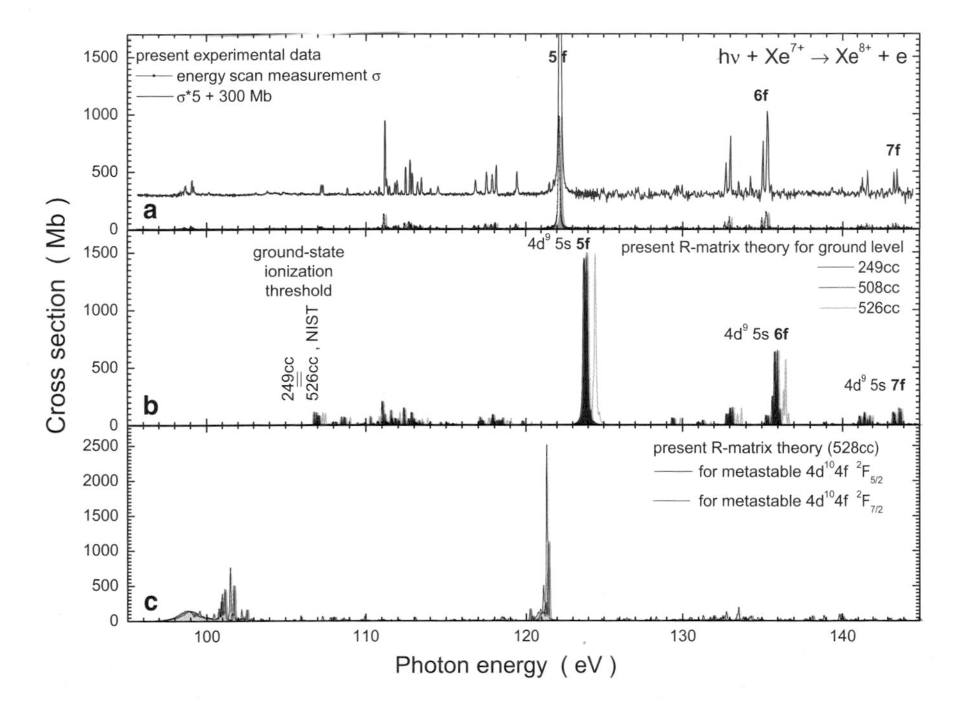

Fig. 5 (Colour online) The experimental and theoretical results of the present study of photoionization of Xe^{7+} ions. The *top* (panel **a**) displays the experimental energy-scan data taken at 65 meV resolution (data points connected by straight lines with light shading). In order to visualize the contributions of resonances other than the dominating $4d^95s5f\ ^2P^o$ peak, the experimental data were multiplied by a factor of 5 and offset by 300 Mb (*solid line, red online*). Panel (**b**) displays the present DARC results for a 249-level, a 508-level and a 526-level calculation, abbreviated as 249cc, 508cc and 526cc, respectively, for photoionization of ground-state Xe^{7+} ions. Panel (**c**) shows 528-level DARC results for the two metastable fine-structure components of the long lived excited $4d^{10}4f\ ^2F^o$ term. The theoretical cross sections were convoluted with a 65 meV FWHM Gaussian in order to be comparable with the experiment

results from the 249-state, 508-state, 526-state and 528-state models, compared to experiment. We find best agreement with experiment from an admixture of 97.6 % ground and 2.4 % metastable state. Given the complexity of this system, satisfactory agreement with experiment is obtained over the photon energy region investigated. We see that the small resonances below the ground-state ionization threshold, occurring at about 106 eV [19], are due to the presence of metastable $Xe^{7+}(4d^{10}4f\ ^2F^o_{5/2,7/2})$ ions with an excitation energy of 32.9 eV [19] in the ion beam.

Recently, the photon-ion merged-beams technique has been employed at the new Photon-Ion spectrometer at PETRA III (PIPE), in Hamburg, Germany, for measuring multiple photoionization of Xe^{q+} (q = 1–5) ions [22]. Prominent ionization features have been observed in the photon-energy range 650–800 eV, which are associated with excitation or ionization of an inner-shell 3d electron. Large-scale DARC calculations are planned for the various charged states of these Xe ions.

5.2 *Tungsten (W) Ions*

The choice of materials for the plasma facing components in fusion experiments is guided by competing desirables: on the one hand the material should have a high thermal conductivity, high threshold for melting and sputtering, and low erosion rate under plasma contact, and on the other hand as a plasma impurity it should not cause excessive radiative energy loss. The default choice of material for present experiments is carbon (or graphite), however tritium is easily trapped in carbon-based walls and for that reason carbon is at present held to be unacceptable for use in a D-T fusion experiment such as ITER. In its place, tungsten (symbol W, atomic number 74) is the favoured material for the wall regions of highest particle and heat load in a fusion reactor vessel.

ITER is scheduled to start operation with a W-Be-C wall for a brief initial campaign before switching to W-Be or W alone for the main D-D and D-T experimental program. The attractiveness of tungsten is due to its high thermal conductivity, its high melting point, and its resistance to sputtering and erosion, and is in spite of a severe negative factor (that as a high-Z plasma impurity tungsten) does not become fully stripped of electrons and radiate copiously, so that the tolerable fraction of tungsten impurity in the plasma is at most 2×10^{-5}.

W ions impurities in a fusion plasma causes critical radiation loss and minuscule concentrations prevent ignition. High resolution experiments are currently available from the ALS on low ionization stages of W ions. We use the Dirac-Atomic-R-matrix-Codes (DARC) to perform large scale calculations for the single photoionization process and compare our results with experiment. These systems are an excellent test bed for the photoionization (PI) process where excellent agreement is achieved between theory and experiment providing a road-map for electron – impact excitation (EIE). For photoionization of the W^{3+} ion of tungsten we use a 379-level scattering model, obtained from 9 configuration state functions

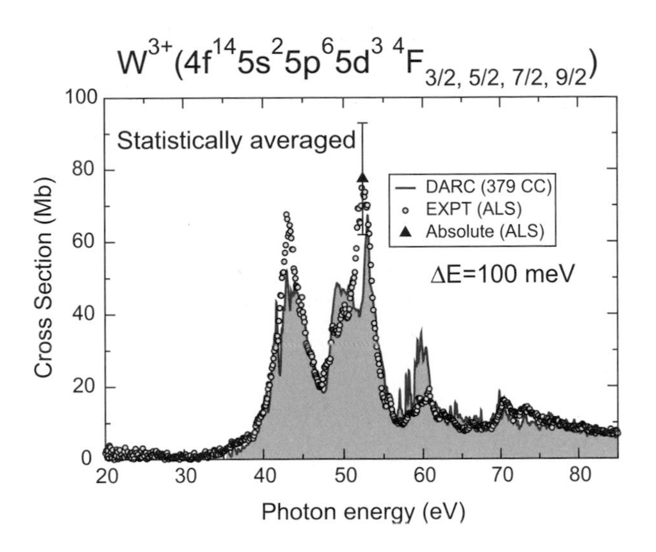

Fig. 6 Photoionization of W^{3+} ions over the photon energy range 20–90 eV. Theoretical work (*solid red line*: DARC) from the 379-level approximation calculations, convoluted with a Gaussian profile of 100 meV FWHM and statistically averaged over the fine structure $J = 3/2$, 5/2, 7/2 and 9/2 levels. The *solid circles* (*yellow*) are from the experimental measurements made at the ALS using a band width of 100 meV [16]. The *solid triangle* is the absolute photoionization cross section accurate to within 20 %

of the residual ion. Figure 6 shows our theoretical results from this 379-level model obtained from the DARC codes compared with measurements made at the Advanced Light source [16]. The comparison made in Fig. 6 illustrates suitable agreement between the theoretical results and the ALS experimental measurements which are accurate to within 20 %.

Acknowledgements A Müller acknowledge support by Deutsche Forschungsgemeinschaft under project number Mu 1068/10 and through NATO Collaborative Linkage grant 976362 as well as by the US Department of Energy (DoE) under contract DE-AC03-76SF-00098 and grant DE-FG02-03ER15424. M S Pindzola and C P Ballance were supported by US Department of Energy (DOE) and US National Science Foundation grants through Auburn University. B M McLaughlin acknowledges support from the US National Science Foundation through a grant to ITAMP at the Harvard-Smithsonian Center for Astrophysics, under the visitor's program, the RTRA network *Triangle de le Physique* and a visiting research fellowship (VRF) from Queen's University Belfast. This research used computational resources at the National Energy Research Scientific Computing Center in Oakland, CA, USA, the Kraken XT5 facility at the National Institute for Computational Science (NICS) in Knoxville, TN, USA and at the High Performance Computing Center Stuttgart (HLRS) of the University of Stuttgart, Stuttgart, Germany. The Kraken XT5 facility is a resource of the Extreme Science and Engineering Discovery Environment (XSEDE), which is supported by National Science Foundation grant number OCI-1053575. The Oak Ridge Leadership Computing Facility at the Oak Ridge National Laboratory, provided additional computational resources, which is supported by the Office of Science of the U.S. Department of Energy under Contract No. DE-AC05-00OR22725. The Advanced Light Source is supported by the Director, Office of Science, Office of Basic Energy Sciences, of the US Department of Energy under Contract No. DE-AC02-05CH11231.

References

1. Al Shorman, M.M., Gharaibeh, M.F., Bizau, J.M., Cubaynes, D., Guilbaud, S., El Hassan, N., Miron, C., Nicolas, C., Robert, E., Sakho, I., Blancard, C., McLaughlin, B.M.: K-Shell photoionization of Be-like and Li-like ions of atomic nitrogen: experiment and theory. J. Phys. B: At. Mol. Opt. Phys. **46**, 195701 (2013)
2. Ballance, C.P., Griffin, D.C.: Relativistic radiatively damped R-matrix calculation of the electron-impact excitation of W^{46+}. J. Phys. B: At. Mol. Opt. Phys. **39**, 3617 (2006)
3. Ballance, C.P., Loch, S.D., Pindzola, M.S., Griffin, D.C.: Electron-impact excitation and ionization of W^{3+} for the determination of tungsten influx in a fusion plasma. J. Phys. B: At. Mol. Opt. Phys. **46**, 055202 (2013)
4. Bizau, J.M., Esteva, J.-M., Cubaynes, D., Wuilleumier, F.J., Blancard, C., Compant La Fontaine, A., Couillaud, C., Lachkar, J., Marmoret, R., Rmond, C., Bruneau, J., Hitz, D., Ludwig, P., Delaunay, M.: Photoionization of highly charged ions using an ECR ion source and undulator radiation. Phys. Rev. Lett. **84**, 435 (2000)
5. Covington, A.M., Aguilar, A., Covington, I.R., Hinojosa, G., Shirley, C.A., Phaneuf, R.A., Álvarez, I., Cisneros, C., Dominguez-Lopez, I., Sant'Anna, M.M., Schlachter, A.S., Ballance, C.P., McLaughlin, B.M.: Valence-shell photoionization of chlorinelike Ar^{+} ions. Phys. Rev. A **84**, 013413 (2011)
6. Fivet, V., Bautista, M.A., Ballance, C.P.: Fine-structure photoionization cross sections of Fe II. J. Phys. B: At. Mol. Opt. Phys. **45**, 035201 (2012)
7. Gharaibeh, M.F., Bizau, J.M., Cubaynes, D., Guilbaud, S., El Hassan, N., Al Shorman, M.M., Miron, C., Nicolas, C., Robert, E., Blancard, C., McLaughlin, B.M.: K-shell photoionization of singly ionized atomic nitrogen: experiment and theory. J. Phys. B: At. Mol. Opt. Phys. **44**, 175208 (2011)
8. Gharaibeh, M.F., El Hassan, N., Al Shorman, M.M., Bizau, J.M., Cubaynes, D., Guilbaud, S., Blancard, C., McLaughlin, B.M.: K-shell photoionization of B-like atomic nitrogen ions: experiment and theory. J. Phys. B: At. Mol. Opt. Phys. **47**, 065201 (2014)
9. Hasoglu, M.F., Abdel Naby, S.A., Gorczyca, T.W., Drake, J.J., McLaughlin, B.M.: K-shell photoabsorption studies of the carbon isonuclear sequence. Astrophys. J. **724**, 1296 (2010)
10. Kallman, T.R.: Challenges of plasma modelling: current status and future plans. Space Sci. Rev. **157**, 177 (2010)
11. McLaughlin, B.M.: Inner-shell photoionization, fluorescence and Auger yields. In: Ferland, G., Savin, D.W. (eds.) Spectroscopic Challenges of Photoionized Plasma, Kentucky. ASP Conference Series, vol. 247, p. 87. Astronomical Society of the Pacific, San Francisco (2001)
12. McLaughlin, B.M., Ballance, C.P.: Photoionization cross section calculations for the halogen-like ions Kr^{+} and Xe^{+}. J. Phys. B: At. Mol. Opt. Phys. **45**, 085701 (2012)
13. McLaughlin, B.M., Ballance, C.P.: Photoionization cross-sections for the trans-iron element Se^{+} from 18 eV to 31 eV. J. Phys. B: At. Mol. Opt. Phys. **45**, 095202 (2012)
14. McLaughlin, B.M., Ballance, C.P.: Photoionization, fluorescence and inner-shell processes. In: McGraw-Hill (ed) McGraw-Hill Yearbook of Science and Technology, p. 281. McGraw Hill, New York (2013)
15. McLaughlin, B.M., Ballance, C.P., Bown, K.P., Gardenghi, D.J., Stolte, W.C.: High precision K-shell photoabsorption cross sectons for atomic oxygen: experiment and theory. Astrophys. J **771**, L8 (2013) & **779**, L31 (2013)
16. McLaughlin, B.M., Ballance, C.P., Kilcoyne, A.L.D., Phaneuf, R.A., Hellhund, J., Schippers, S., Müller, A.: Valence shell photoionization of Hf-like (W^{2+}) and Lu-like (W^{3+}) ions of atomic tungsten: experiment and theory. J. Phys. B: At. Mol. Opt. Phys. (2014, submitted for publication)
17. Müller, A., Schippers, S., Phaneuf, R.A., Scully, S.W.J., Aguilar, A., Cisneros, C., Gharaibeh, M.F., Schlachter, A.S., McLaughlin, B.M.: K-shell photoionization of Be-like Boron (B^{+}) ions: experiment and theory. J. Phys. B: At. Mol. Opt. Phys. **47**, 135201 (2014)

18. Müller, A., Schippers, S., Esteves-Macaluso, D., Habibi, M., Aguilar, A., Kilcoyne, A.L.D., Phaneuf, R.A., Ballance, C.P., McLaughlin, B.M.: High resolution valence shell photoionization of Ag-like (Xe^{7+}) xenon ions: experiment and theory. J. Phys. B: At. Mol. Opt. Phys. **47**, 215202 (2014)
19. Saloman, E.B.: Energy levels and observed spectral lines of Xenon, Xe I through Xe LIV. J. Phys. Chem. Ref. Data **33**, 765 (2004)
20. Sant'Anna, M.M., Schlachter, A.S., Öhrwall, G., Stolte, W.C., Lindle, D.W., McLaughlin, B.M.: K-shell X-ray spectroscopy of atomic nitrogen. Phys. Rev. Lett. **107**, 033001 (2011)
21. Schippers, S., Müller, A., Esteves, D., Habibi, M., Aguilar, A., Kilcoyne, A.L.D.: Experimental absolute cross section for photoionization of Xe^{7+}. J. Phys. Conf. Ser. **194**, 022094 (2009)
22. Schippers, S., Ricz, S., Buhr, T., Borovik Jr, A., Hellhund, J., Holste, K., Huber, K., Schäfer, H.-J., Schury, D., Klumpp, S., Mertens, K., Martins, M., Flesch, R., Ulrich, G., Rühl, E., Jahnke, T., Lower, J., Metz, D., Schmidt, L. P. H., Schöffler, M., Williams, J. B., Glaser, L., Scholz, F., Seltmann, J., Viefhaus, J., Dorn, A., Wolf, A., Ullrich J. and Müller, A.: Absolute cross sections for photoionization of Xe^{q+} ions ($1 \leq q \leq 5$) at the 3d ionization threshold. J. Phys. B: At. Mol. Opt. Phys. **47**, 115602 (2014)

Quantum Many-Body Dynamics of Trapped Bosons with the MCTDHB Package: Towards New Horizons with Novel Physics

Shachar Klaiman, Axel U.J. Lode, Kaspar Sakmann, Oksana I. Streltsova, Ofir E. Alon, Lorenz S. Cederbaum, and Alexej I. Streltsov

Abstract The MCTDHB package has been applied to study the physics of trapped interacting many-boson systems by solving the underlying time-dependent (as well as the time-independent) many-boson Schrödinger equation. Here we report on four studies where novel physical ideas and phenomena have been proposed and discovered: (a) Universality of the fragmentation dynamics in double wells – at long propagation times properties of the evolving system saturate to some asymptotic values; (b) Novel many-body spectral features in trapped systems – the newly-developed linear-response theory on-top of MCTDHB predicts the existence of low-lying excitations not described so far by the standard theory even in harmonic potentials; (c) Efficient protocol to control the many-particle tunneling dynamics to open space, by combining the effects of a threshold potential and inter-particle interaction; (d) Physics behind the formation of patterns in the ground states of trapped bosonic systems with strong finite- and long-range repulsive interactions and the origin of their dynamical stability. From the perspective of the required computational resources and numerical algorithms applied, each of these numerically-demanding studies has challenged different aspects of computational physics and mathematics: Long-time propagation – stability of the numerical methods used to integrate the MCTDHB equations-of-motion; Control of the

S. Klaiman • L.S. Cederbaum • A.I. Streltsov (✉)
Theoretische Chemie, Universität Heidelberg, Im Neuenheimer Feld 229, D-69120 Heidelberg, Germany
e-mail: Alexej.Streltsov@pci.uni-heidelberg.de

A.U.J. Lode
Condensed Matter Theory and Quantum Computing Group, Departement für Physik, Universität Basel, Klingelbergstrasse 82, CH-4056 Basel, Switzerland

K. Sakmann
Department of Physics, Stanford University, Stanford, CA 94305, USA

O.I. Streltsova
Laboratory of Information Technologies, Joint Institute for Nuclear Research, Joliot-Curie 6, Dubna, Russia

O.E. Alon
Department of Physics, University of Haifa at Oranim, Tivon 36006, Israel

W.E. Nagel et al. (eds.), *High Performance Computing in Science and Engineering '14*, DOI 10.1007/978-3-319-10810-0_5

tunneling dynamics – a very detailed study where an interplay of the parameters controlling the decay by tunneling dynamics is accompanied by a long-time propagation on huge spatial grids, which are needed to simulate open systems; Excited states of many-body systems – construction and diagonalization of complex non-hermitian linear-response matrices; Finite- and long-range interactions in 1D, 2D, and 3D setups – efficient methods and techniques for evaluation of involved high-dimensional integrals. Implications and further perspectives and future plans are briefly discussed and addressed.

1 Introduction

A key fascinating feature which distinguishes iltra-cold atomic and molecular systems from the more conventional atomic, molecular, solid-state, or nuclear many-body systems is that the density of the *many-body* wave-function is often a directly measurable quantity, and can be detected and analyzed also in the time-dependent experiments. Hence, recent experimental progress in cooling and trapping of ultra-cold atomic and molecular systems has given a new breath to and stimulates the development of new theoretical methods capable of providing the full many-body wave-function in real space by solving the time-dependent many-body Schrödinger equation. Ultimately, it is the Schrödinger equation which governs the physics of these fascinating ultra-cold quantum systems [12, 28, 30].

The underlying many-particle Hamiltonians of these systems generally do not permit analytical solutions, but, fortunately, recent progress in numerical methods, developments of modern computational technologies [58], and available hardware [8, 11, 26] allow one to solve them numerically at a very accurate level [2, 32, 33, 51, 57].

Having at hand these accurate theoretical tools we can apply them to understand known phenomena at a new theoretical level, as well as to discover and predict novel ones. This defines the mainstream activity of our current MCTDHB project within the framework of the HLRS project.

In the present report we summarize our findings on some new physical phenomena predicted to exist in different ultra-cold bosonic systems. The common features shared by these studies are: (i) All of them are done with the help of the MCTDHB method; (ii) The physics involves many-body states not available at the standard, mean-field level.

The paper is organized as follows. In Sect. 1.1 we briefly overview the MCTDHB method. In Sect. 2 we discuss the universality of fragmentation in double-well traps. Section 3 extends our previous study of tunneling dynamics to open space to include a threshold potential, and thereby demonstrates the control of the many-body tunneling process. The just-developed linear-response method for MCTDHB is discussed in Sect. 4, along with some applications to excitation spectra beyond the standard theory. Section 5 deals with static as well as with dynamical properties of trapped systems with finite- and long-range strong repulsive inter-particle interactions in 1D, 2D, and 3D. Finally, Sect. 6 concludes and summarizes our results, and also briefly discusses future perspective and research plans.

1.1 Hamiltonian and the MCTDHB in Brief

The Multi-Configurational Time-Dependent Hartree method for Bosons (MCTDHB) is used to solve the time-dependent Schrödinger equation:

$$\hat{H}\Psi = i\hbar\frac{\partial \Psi}{\partial t}. \quad (1)$$

$\hat{H}$ is the generic many-body Hamiltonian of N identical bosons trapped in an external trap potential $V(\mathbf{r}, t)$ and interacting via general inter-particle interaction potential which can be time-dependent:

$$\hat{H}(\mathbf{r}_1, \ldots, \mathbf{r}_N, t) = \sum_{j=1}^{N}\left[-\frac{\hbar^2}{2m}\nabla^2_{\mathbf{r}_j} + V(\mathbf{r}_j, t)\right] + \sum_{j<k}^{N}\lambda_0 W(\mathbf{r}_j - \mathbf{r}_k, t). \quad (2)$$

Here, λ_0 defines the strength of the interaction and $W(\mathbf{r} - \mathbf{r}') \equiv W(\mathbf{R})$ its shape. Usually, all the computations are done in dimensionless units, i.e., $\hbar = 1, m = 1$. The dimensionless units are obtained by dividing the Hamiltonian by $\frac{\hbar^2}{mL^2}$, where m is the mass of a boson and L is a length scale.

The MCTDHB *ansatz* for the many-body wave-function $\Psi(t)$ is taken as a linear combination of time-dependent permanents:

$$|\Psi(t)\rangle = \sum_{\mathbf{n}} C_{\mathbf{n}}(t)\,|\mathbf{n}; t\rangle, \quad (3)$$

where the summation runs over all possible configurations whose occupations $\mathbf{n} = (n_1, n_2, n_3, \ldots, n_M)$ preserve the total number of bosons N. The expansion coefficients $\{C_{\mathbf{n}}(t)\}$ and shapes of the orbitals $\{\phi_k(\mathbf{r}, t)\}$ are variational time-dependent parameters of the MCTDHB method, determined by the time-dependent variational principle. From a mathematical perspective, to solve a time-dependent problem means to specify an initial many-body state and to find how the many-body function evolves. The initial MCTDHB many-body state is given by initial expansion coefficients $\{C_{\mathbf{n}}(t = 0)\}$ and by initial shapes of the orbitals $\{\phi_k(\mathbf{r}, t = 0)\}$. To determine their evolution one has to solve the respective governing equations which we list below. The equations-of-motion for the expansion coefficients $\{C_{\mathbf{n}}(t)\}$ read:

$$\mathbf{H}(t)\mathbf{C}(t) = i\frac{\partial \mathbf{C}(t)}{\partial t}, \quad (4)$$

where $\mathbf{H}(t)$ is the Hamiltonian matrix with elements $H_{\mathbf{nn}'}(t) = \left\langle \mathbf{n}; t \left| \hat{H} \right| \mathbf{n}'; t \right\rangle$. The equations-of-motion for the orbitals $\phi_j(\mathbf{r}, t)$, $j = 1, \ldots, M$ are:

$$i \left|\dot{\phi}_j\right\rangle = \hat{\mathbf{P}} \left[\hat{h} \left|\phi_j\right\rangle + \lambda_0 \sum_{k,s,q,l=1}^{M} \{\boldsymbol{\rho}(t)\}_{jk}^{-1} \rho_{ksql} \hat{W}_{sl} \left|\phi_q\right\rangle \right], \tag{5}$$

where $\hat{\mathbf{P}} = 1 - \sum_{j'=1}^{M} \left|\phi_{j'}\right\rangle\left\langle \phi_{j'}\right|$ is a projector and

$$\hat{W}_{sl}(\mathbf{r}, t) = \int \phi_s^*(\mathbf{r}', t) W(\mathbf{r} - \mathbf{r}') \phi_l(\mathbf{r}', t) d\mathbf{r}' \tag{6}$$

are *local* time-dependent potentials which play a crucial role in the dynamics (see Sect. 4). ρ_{kq} and ρ_{ksql} are elements of the reduced one- and two-body density matrices available from the many-body wave-function $\Psi(t)$ at every point in time. The details on the numerical and algorithmical implementation of the MCTDHB method have been reported in peer-review journals, see [2, 32, 33, 51, 57], in books [37, 47], as well as in our previous HLRS report for 2012 [34]. Here, we only mention that the use of M optimized time-dependent orbitals leads to much faster numerical convergence to the full many-body Schrödinger results than an expansion in M time-independent orbitals. Thereby, problems involving large numbers of bosons can be practically solved on the full many-body Schrödinger level.

Now we concentrate on applications of the MCTDHB method to investigate physics. An intuitive picture to study the dynamics means to prepare the system in the ground state of an initial Hamiltonian and then to change some parameters of this Hamiltonian, e.g., to quench the strength of the interaction, to move the origin of the trap, to alter its frequency, or to superimpose another trapping potential, etc. Since the initially-prepared state is no longer an eigenstate of the new Hamiltonian, the system will evolve in time. We would like to monitor the evolution of the many-body wave-function and some of the system's properties which can be computed at every time point.

1.2 Analysis of the Many-Body System

Having the MCTDHB(M) (or, in the static case, the MCHB [56]) many-body wave-function $|\Psi\rangle$ at hand, one can compute quantities of interest and analyze the physics behind. Let us show how to quantify the Bose-Einstein condensation and fragmentation phenomena. The reduced one-body density matrix is directly available:

$$\rho(\mathbf{r}_1, \mathbf{r}_1'; t) = N \int \Psi^*(\mathbf{r}_1', \mathbf{r}_2, \ldots, \mathbf{r}_N; t) \Psi(\mathbf{r}_1, \mathbf{r}_2, \ldots, \mathbf{r}_N; t) d\mathbf{r}_2 d\mathbf{r}_3 \cdots d\mathbf{r}_N.$$

We can diagonalized it and write its eigen-decomposition:

$$\rho(\mathbf{r},\mathbf{r}';t) = \sum_{k,q=1}^{M} \rho_{kq}(t)\phi_k^*(\mathbf{r}_1',t)\phi_q(\mathbf{r}_1,t) = \sum_{k=1}^{M} n_k(t)\phi_k^{*NO}(\mathbf{r}',t)\phi_k^{NO}(\mathbf{r},t).$$

The obtained eigenvalues n_k and eigenvectors ϕ_k^{NO} are called natural occupation numbers and natural orbitals, respectively. This diagonalization procedure is often referred to as natural-orbital analysis. The natural occupation numbers n_k can be considered as average numbers of bosons residing in ϕ_k^{NO}. The natural orbital analysis is used to characterize the system: The system is referred to as condensed [45] when only a single natural orbital has a macroscopic occupation and *fragmented* if several natural orbitals are macroscopically occupied [41, 42]. The fragmentation phenomenon has been observed experimentally [25, 44] and is generally well understood theoretically [39, 41, 42, 54, 56].

Usually, to visualize the evolution of the systems one plots only the diagonal part $\rho(\mathbf{r},t) \equiv \rho(\mathbf{r},\mathbf{r};t)$ which is simply referred to as the density.

Sometimes, it is convenient to quantify fragmentation using the quantity

$$f = \frac{1}{N}\sum_{i>1} n_i, \tag{7}$$

which whenever unambiguous we will refer to as the fragmentation for simplicity. For fully-condensed states, i.e. $n_1 = N$, one finds $f = 0$, whereas $0 \leq f < 1$ in general. At the MCTDHB(M) level, the reduced one-body density matrix $\{\rho_{ij}\}$ is an M–by–M matrix, implying $f \leq \frac{M-1}{M}$. Fragmentation manifests itself in correlation functions and fringe visibilities [50]. As mentioned, several experiments have measured the effects of fragmentation, see, e.g., Refs. [25, 44, 46].

2 Universality of Fragmentation

In this project we investigate a one-dimensional bosonic Josephson junction for initially fully-condensed Bose-Einstein condensates (BECs) made of up to 10,000 bosons, by solving the time-dependent Schrödinger equation for the long-time many-body dynamics. Full report has been published as a research paper entitled “Universality of fragmentation in the Schrödinger dynamics of bosonic Josephson junctions” in Phys. Rev. A **89**, 023602 (2014) [52].

The main results are as follows: We report on the existence of a *universal* many-body fragmentation *dynamics* in bosonic Josephson junctions; We find that systems consisting of different numbers of bosons all fragment to the same value at a fixed mean-field interaction strength, and; We show how this universality of fragmentation manifests itself in observables, i.e., can be detected. The fact that the fragmentation phenomenon occurs even for the largest particle numbers considered in our investigations indicates that there is no weakly-interacting limit where the mean-field (Gross-Pitaevskii) theory becomes valid at long times. This observation has very strong theoretical consequences – there are effects existing for large particle numbers that are not described by the simple and standard Gross-Pitaevskii mean-field theory.

2.1 System, Hamiltonian and Parameters

In this study we choose for the potential $V(x)$ a double-well constructed by connecting two harmonic potentials

$$V_{\pm}(x) = \frac{1}{2}(x \pm 2)^2 \tag{8}$$

with a natural cubic spline at $|x| = 0.5$. This uniquely defines the external potential $V(x)$. The even-symmetry ground state ϕ_g and odd-symmetry excited state ϕ_u of the one-particle double-well Hamiltonian $\hat{h} = -\frac{1}{2}\nabla_x^2 + V(x)$ allow one to define several commonly-used parameters. First, two left- and right-localized Wannier functions $\phi_{L,R} = \frac{1}{\sqrt{2}}(\phi_g \pm \phi_u)$ are constructed. This defines the hopping parameter $J = -\langle\phi_L|h|\phi_R\rangle = 2.2334 \times 10^{-2}$, as well as the interaction parameter $U = \lambda_0 \int dx|\phi_L(x)|^4$ and the single-particle tunneling time-scale $t_{Rabi} = \pi/J = 140.66$.

To quantify the dynamics of Bose-Einstein condensates in double-well potentials we compute the integral over the probability density in the left half of the double-well potential

$$p_L(t) = \frac{1}{N}\int_{-\infty}^{0} \rho(x,t)dx. \tag{9}$$

It gives a measure for the fraction of bosons in the left well. The probability in the right half of space then follows from the normalization, $p_R(t) = 1 - p_L(t)$.

2.2 Dynamical Scenario and Novel Physical Results

We choose fully-condensed initial states which are the mean-field (Gross-Pitaevskii) ground states of the potential $V_+(x)$. Thus, initially the BECs are located in the left well. For each particle number N the interaction strength λ_0 is chosen such that the mean-field interaction strength $\lambda = \lambda_0(N-1)$ is constant. We begin with $\lambda = 0.152$ ($\Lambda = \frac{U(N-1)}{2J} = 1.33$) which corresponds to a regime where the tunneling is prohibited, i.e., we are well below the critical value $\Lambda_c = 2$ for self-trapping [38, 53]. Self-trapping – is a regime of the parameters where tunneling becomes

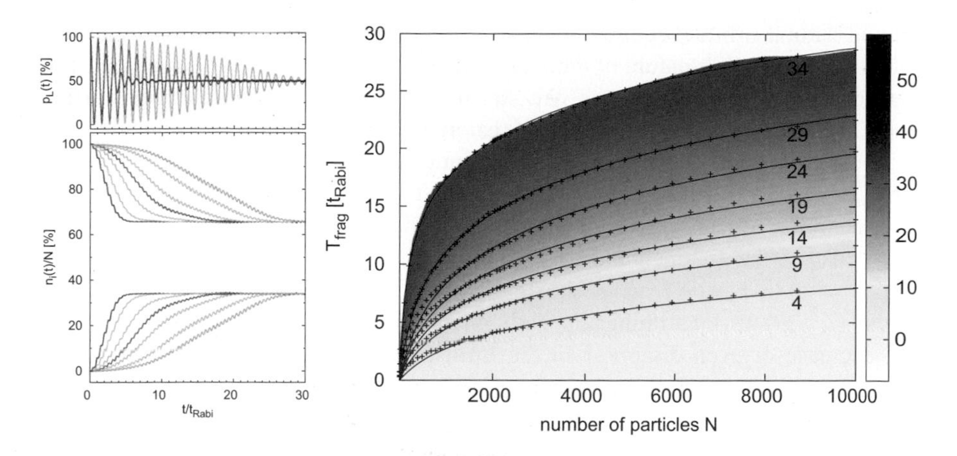

Fig. 1 Universality of the many-body Schrödinger fragmentation dynamics. *Left–top*: Shown is the probability in the left potential well $p_L(t)$ as a function of time for different numbers of particles. For all particle numbers the initial state is fully condensed, all bosons are in the *left* well, $p_L(0) = 1.0$, and the mean-field interaction strength is $\lambda = 0.152$ ($\Lambda = 1.33$). As the BECs tunnel back and forth through the barrier, the density oscillations collapse. Particle numbers: $N = 100$ (*blue*), 1,000 (*magenta*), and 10,000 (*orange*). *Left–bottom*: Shown are the corresponding natural occupations n_i/N as a function of time. The *gray lines* represent particle numbers in between: $N = 200$, 500, 2,000, and 5,000. Only two natural occupations n_1 and n_2 become significantly occupied in the dynamics. Thus, the fragmentation f is given here by $f \approx n_2/N$, see Eq. (7). The fragmentation reaches a plateau at $f = f_{col} = 34\,\%$ after the collapse of the density oscillations for all particle numbers. The nature of the BEC changes from condensed to fragmented. *Right panel*: Many-body Schrödinger dynamics fragmentation times. Shown are the times T_{frag} that are needed to reach a given fragmentation f as a function of the number of bosons for the same parameters as in *left panels*. The fragmentation values (*crosses*) are well described by a logarithmic fit function (*lines*), here shown for the values $f = 4, 9, \ldots, 34\,\%$. This logarithmic dependence of T_{frag} on the number of particles implies the breakdown of Gross-Pitaevskii mean-field theory even in the limit of large particle numbers. See Phys. Rev. A **89**, 023602 (2014), Ref. [52], for more details. All quantities shown are dimensionless

impossible (within two-mode theories). The inhibition of tunneling takes place if $\Lambda \geq \Lambda_c = 2$.

The left–top panel of Fig. 1 shows the probability $p_L(t)$ in the left well as a function of time for $N = 100$–10,000 bosons. The density tunnels back and forth through the potential barrier. However, the amplitude of the oscillations decreases and eventually the oscillations collapse. With increasing particle number the collapse occurs at later times. It is well known that the collapse is present in the many-body dynamics, but not within (two-mode) Gross-Pitaevskii mean-field theory [38, 51]. Here we report on investigations of the many-body nature of the BEC during the collapse.

In the left–bottom panel of Fig. 1 the corresponding natural occupations are shown. Since the many-body wave-function is initially condensed, there is only one natural occupation, $n_1 = N$ at $t = 0$, and the fragmentation f is zero, see Eq. (7). However, as the condensate tunnels back and forth through the barrier a second natural orbital becomes occupied, implying that the systems become two-fold fragmented. The nature of the BEC changes from condensed to fragmented as the density oscillations collapse. We stress that these results represent the many-body Schrödinger dynamics of the full real-space Hamiltonian (2).

It is worthwhile to define the fragmentation time T_{frag} as the first time at which a certain fragmentation f is reached. The right panel of Fig. 1 shows the fragmentation time T_{frag} as a function of N. Clearly, T_{frag} increases with N. For any value of the fragmentation, T_{frag} is well described by a fit to the function $T(N) = a\ln(1+bN)$, as is shown here for the values $f = 4, 9, \ldots, 34\,\%$. Thus, T_{frag} grows logarithmically with N. Consequently, the fragmentation does not decrease or even disappear in the limit of large N. Even for $N = 10{,}000$ bosons the fragmentation rises to about 10 % in less than a dozen t_{Rabi}. As Gross-Pitaevskii theory does not allow BECs to fragment at all, T_{frag} defines a measure for the breakdown of mean-field theory. Thus, we find here a logarithmic dependence of the breakdown of mean-field theory on the number of particles within the time-dependent many-body Schrödinger equation. All in all, these figures confirm and establish the universality of fragmentation of the many-body Schrödinger dynamics.

2.3 Complexity of the Employed Numerics

The presented numerical results solve the time-dependent Schrödinger equation, i.e., propagate it till $t \approx 30t_{Rabi}$, implying that the MCTDHB equations-of-motion have been solved till $t \approx 4{,}200$ in dimensionless units of time. The numerical accuracy of long-time propagations of the MCTDHB equations presents a formidable challenge to numerical integrators. The numerical accuracy of the results was insured by back-propagating the equations-of-motion starting from the state at the final time point.

For the smallest particle number studied, $N = 100$ bosons, $M = 4$ orbitals were used in the MCTDHB computation, and the result was found to be numerically exact. However, only $M = 2$ time-dependent variationally optimized orbitals are needed here: There is only a small quantitative difference between computations with $M = 4$ and $M = 2$ orbitals. Until time $t = 30t_{Rabi}$ the occupations of the third and fourth natural orbitals together stay below 0.1 % and are not shown in the left panels of Fig. 1. Therefore, $M = 2$ orbitals were used in all computations for larger particle numbers.

3 Efficient Control of the Tunneling to Open Space

In this project we generalize our findings on many-boson tunneling to open space previously reported in Proc. Natl. Acad. Sci. USA, **109**, 13521 (2012) [33]. A full report has been published as a research paper entitled "Controlling the velocities and the number of emitted particles in the tunneling to open space dynamics" in Phys. Rev. A **89**, 053620 (2014) [35].

In this study we devise a scheme to control the many-boson tunneling process from a trap through a potential barrier to open space. The threshold of the trapping potential and the inter-particle interaction are found to be the ideal parameters to control the number of ejected particles and their velocities. Employing the MCTDHB method we were able to compute dynamics for $N = 2$, 3, and 101 interacting bosons in one spatial dimension numerically exactly, solving thereby an intricate many-body problem of general interest. We have found that the devised control scheme for the many-boson tunneling process performs very well for the dynamics of the momentum density, the correlations and coherence, as well as for the number of particles remaining in the trap. To interpret the many-body tunneling process, a transparent model assembling the many-body process from single-particle emission processes has been derived. Analyzing the energetics of available decay channels together with the numerical results rule out the situation of two (or more) bosons tunneling together to open space.

3.1 System, Hamiltonian and Parameters

A key idea is to control the many-body tunneling process to open space by employing a one-particle potential $V(x)$ with a threshold, see the left panel of Fig. 2. To investigate the control mechanisms we use the following smooth polynomial

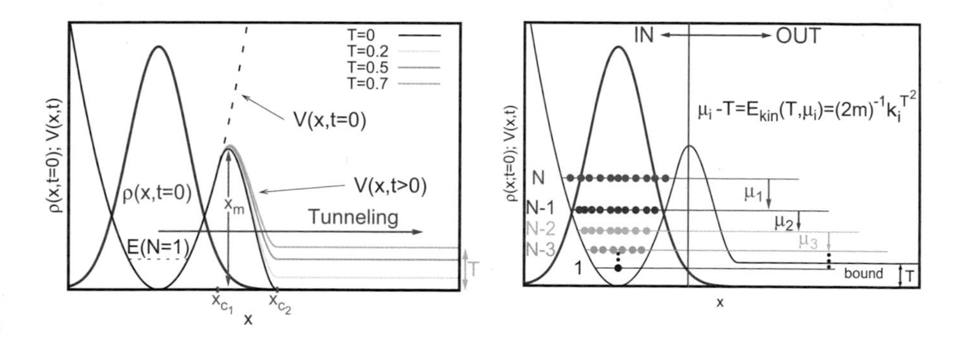

Fig. 2 Controlling the dynamics of many-boson tunneling to open space. *Left panel*: Protocol for the tunneling dynamics with non-zero potential threshold. *Right panel*: Static mean-field scheme to model and explain the tunneling processes with a threshold T. The bosons are tunneling from the interior "IN" to the exterior "OUT" region of space (separated by the *vertical red line*). If the threshold T is large enough, some of the states can become bound (see, e.g., the $N = 1$ state indicated by the *lowest black line*). If the state is not bound, the chemical potential μ_i is used to overcome the threshold T and the remainder is transformed to kinetic energy $E_{kin}(T, \mu_i)$. See Phys. Rev. A **89**, 053620 (2014), Ref. [33, 35], for more details. All quantities shown are dimensionless

continuation of the harmonic trap $V_h(x) = \frac{1}{2}x^2$ from $x_{c1} = 2$ to $x_{c2} = 4$ with a threshold T. Initially $V(x, t = 0) = \frac{1}{2}x^2$, then:

$$V(x, t > 0) = \Theta(x_{c1} - x) \cdot \frac{1}{2}x^2 + \Theta(x - x_{c1})\Theta(x_{c2} - x)P(x) + \Theta(x - x_{c2})T.$$

The details on the polynomial $P(x)$ are given in Ref. [35]. Here, $\Theta(\cdot)$ is the Heaviside step function. In Fig. 2 we depict this potential for various values of T. At $t = 0^+$ the harmonic potential is suddenly changed to the open trap, initiating thereby the tunneling process.

The density $\rho(x, t)$ describes the probability to find a single particle in the many-body system at a certain position x at a given time t. To measure the number of particles remaining inside the parabolic part of the potential (see Fig. 2), it is instructive to define the non-escape probability,

$$P^x_{not}(t, T) = \int_{-\infty}^{x_m} \rho(x, t)dx. \tag{10}$$

x_m is the coordinate where the potential berrier reaches its maximum.. It depends on the threshold T and is marked for the $T = 0$ case by a red vertical line in the right panel of Fig. 2.

3.2 Dynamical Scenario of Controlling Protocol

We propose the following protocol for the tunneling dynamics with non-zero potential threshold. The initial density (blue line in Fig. 2) is prepared as the ground state of the parabolic trap (black dashed $V(x, t = 0)$). Subsequently, the potential is transformed to its open form with a threshold (various solid colored lines, $V(x, t > 0)$). Following this transformation the particles can tunnel to open space. The tunneling process can be controlled by the threshold T as well as by the interaction strength λ_0. The energy of a single, parabolically trapped particle, $E(N = 1)$, is indicated by the horizontal black dashed line in Fig. 2 to guide the eye.

Consider the system as split into an "IN" part, to the left of the maximum of the barrier, and an "OUT" part to the right of it, see the right panel of Fig. 2. The "IN" part is the part of the potential that is classically allowed, i.e., classical particles would be indefinitely confined in the "IN" region.

Suppose a single boson has escaped from the "IN" to the "OUT" region. Following the analysis in Ref. [33], the available energy of this boson must come from the energy difference of the trapped systems with N and with $N - 1$ particles, $E^N - E^{N-1} = \mu_1$ – the chemical potential of the N-particle system. With this in principle available energy, the ejected boson has to overcome the threshold T. Hence, after tunneling it remains with an energy $(\mu_1 - T)$ in the "OUT" part of the potential. As the potential in the "OUT" part is flat and the density can be assumed to be small, the ejected boson will convert its available energy to kinetic energy. Analogously, the other particles which are ejected have their available energies from chemical potentials μ_i of $(N - i)$ particles left in the trap.

The partition of the systems depicted in the right panel of Fig. 2 naturally implies the notation $|N_{IN}, N_{OUT}\rangle$, where N_{IN} counts the number of particles in the "IN" region and N_{OUT} in the "OUT" region. The notation $|N_{IN}, N_{OUT}\rangle$ is referred to as the counting statistics of the system (it should not be confused with the notation of a Fock state). To adjust the energies of the bosons inside the trap one can tune the interaction λ_0, whereas the energies of the bosons outside the trap can be tuned by the threshold T. These control parameters permit manipulation of the counting statistics of a given state, namely, a many-body state with desired occupations of the internal and external parts $|N_{IN}, N_{OUT}\rangle$ can be prepared deterministically.

Having learned the role of the threshold and interaction in controlling the tunneling to open space dynamics, we confirm the above explanation by numerical simulations. We consider a system of an initially-trapped $N = 101$ interacting bosons, and aim at devising a scheme allowing roughly half of the bosons to tunnel out, or, equivalently, $N_{IN} = 51$ bosons staying trapped. We first fix the threshold T and then tune the interaction λ_0 appropriately. It is also possible to choose the reverse strategy (i.e., fixing the interaction and varying the threshold), yet, for brevity, we focus here on the first one.

Figure 3a shows the energetics of the counting statistics states $|41, 60\rangle$, $|51, 50\rangle$, and $|61, 40\rangle$ for the threshold $T = 0.6$ as a function of λ_0 which are relevant for

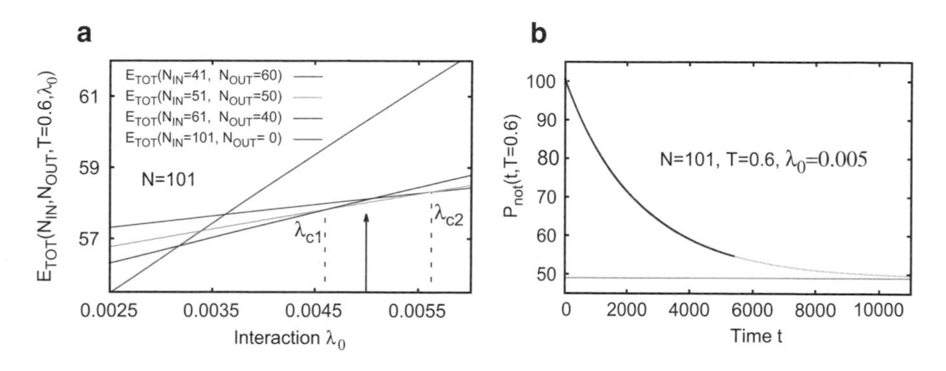

Fig. 3 Controlling the tunneling to open space of an interacting $N = 101$ boson system. (**a**) Energies of selected final states $|41, 60\rangle$, $|51, 50\rangle$, and $|61, 40\rangle$ for the threshold $T = 0.6$ as a function of the interaction λ_0. For reference the energy of the initial state $|101, 0\rangle$ is also depicted. When one tunes λ_0 such that it is in between the critical interactions λ_{c1} and λ_{c2} (marked by the *black dashed vertical lines*), the energy of the $|51, 50\rangle$ state (*green line*) is lower than the energy of the $|41, 60\rangle$ state (*red line*) and the $|61, 40\rangle$ state (*blue line*). An interaction chosen in this interval is expected to sustain the control objective. The *black arrow* shows the interaction $\lambda_0 = 0.005$ chosen for the tunneling process. (**b**) The non-escape probability (*red solid line*). According to a least squares fit (see *green dotted line*), the final state is $|49, 52\rangle$ (corresponding to the *horizontal dotted line*) and hence in the range assessed from the energetics in the *left panel*. See Phys. Rev. A **89**, 053620 (2014), Ref. [35], for more details. All quantities shown are dimensionless

our analysis. We have identified a whole range of interaction strengths where out of these states the counting statistics state $|51, 50\rangle$ is lowest in energy, see Fig. 3a. We expect an interaction chosen in this interval, say $\lambda_0 = 0.005$, to sustain the control objective of roughly half of the bosons tunneling out.

In Fig. 3b the non-escape probability of the $N = 101$ interacting bosons' system tunneling to open space is computed numerically. We stress that this is a demanding time-dependent many-body task. An analysis of the results indicates that the counting statistics converges to the state $|49, 52\rangle$. This is in excellent agreement with our control objective!

3.3 Complexity of the Employed Numerics

In the present work the MCTDHB Software Package [58] was employed to obtain converged solutions of the time-dependent Schrödinger equation. The computations for $N = 2$, 3, and 101 interacting bosons used one-dimensional grids of sizes of up to $[-5; 7{,}465]$ in dimensionless units, represented by up to $2^{16} = 65{,}536$ time-independent basis functions (grid points), with up to $M = 14$ time-adaptive orbitals, and ran up to more than $t = 5{,}000$ in dimensionless units of the propagation time. It noteworthy that the above domain corresponds to a length of more than 7 mm (!) and the above time to more than 6 s (!) in dimension-full units. The long-term propagation of the $N = 101$ system was used to verify the numerical exactness of the recently developed recursive MCTDHB (R-MCTDHB) program, see http://ultracold.org as well as Ref. [31].

4 Excitation Spectrum by Linear Response

In this project we continue to invent and develop new theoretical tools capable of describing interacting many-boson systems. This direction of our work was inspired by the fact that the linear-response of the time-dependent Schrödinger equation with respect to perturbation of typically the ground state gives rise to the exact excitation spectra and corresponding eigenfunctions of the quantum system, see, e.g., Ref. [14, 43]. In the derivation, no assumptions are made on the quantum system's Hamiltonian, except that the system is perturbed by a weak time-periodic field (hence, linear response). Since the MCTDHB method is capable of providing essentially numerically-exact solutions for the ground states [32], we expect that the linear-response (LR) idea applied on-top of MCTDHB will equip us with a powerful LR-MCTDHB method, which would be capable of computing excited states very accurately. The present derivation applies to bound systems.

Full discussions have been reported in two research papers entitled "Excitation spectra of many-body systems by linear response: General theory and applications to trapped condensates" published in Phys. Rev. A **88**, 023606 (2013) [20] and

"Unified view on linear response of interacting identical and distinguishable particles from multiconfigurational time-dependent Hartree methods" published in J. Chem. Phys. **140**, 034108 (2014) [3]. These papers significantly contribute to three different aspects of modern physics. First – to theoretical many-body physics – the detailed derivations of the LR-MCTDHB equations are presented in closed form, implying their generality and applicability also to fermions (LR-MCTDHF) and mixtures. Second – to new understanding of physics, because first applications of the developed theory for bosons have revealed unexpected physical results – novel many-body spectral features (low-lying excitations) even in harmonically-trapped systems. Third – to numerical methods and algorithms, since the LR-MCTDHB theory requires construction and diagonalization of complex non-hermitian linear-response matrices which is a challenging computational problem.

4.1 The LR-MCTDHB Theory in a Glimpse

The key idea was to derive the linear-response theory from the MCTDHB theory by using a small perturbation *around* the static MCHB solution, typically the ground state. The latter is computed by imaginary-time propagation using MCTDHB. Thus, we denote a small, time-dependent and periodic perturbation in the external applied potential as $\hat{h}(\mathbf{r}) \to \hat{h}(\mathbf{r}) + \delta\hat{h}(\mathbf{r},t)$, where

$$\delta\hat{h}(\mathbf{r},t) = f^{+}(\mathbf{r})e^{-i\omega t} + f^{-}(\mathbf{r})e^{i\omega t}. \tag{11}$$

Hereby, the probe frequency is ω and the driving amplitudes $f^{\pm}$ are real. Further, we have the following ansatz for the perturbed wave-function:

$$\begin{aligned} \phi_k(\mathbf{r},t) &\approx \phi_k(\mathbf{r}) + \delta\phi_k(\mathbf{r},t), \quad & \delta\phi_k(\mathbf{r},t) &= u_k(\mathbf{r})e^{-i\omega t} + v_k^*(\mathbf{r})e^{+i\omega t} \\ \mathbf{C}(t) &\approx e^{-i\varepsilon t}[\mathbf{C} + \delta\mathbf{C}(t)], \quad & \delta\mathbf{C}(t) &= \mathbf{C}_u e^{-i\omega t} + \mathbf{C}_v^* e^{+i\omega t}, \end{aligned} \tag{12}$$

$k = 1, \ldots, M$, and ε is the ground-state energy. The coupled responses of the orbitals and coefficients result in the linear-response matrix $\boldsymbol{\mathcal{L}}$ which acts on an $2(M + N_{conf})$-dimensional vector composed of the orbitals' and coefficients' response amplitudes $(\mathbf{u}, \mathbf{v}, \mathbf{C}_u, \mathbf{C}_v)^T$, with $\mathbf{u} = \{|u_q\rangle\}$ and $\mathbf{v} = \{|v_q\rangle\}$, $q = 1, \ldots, M$.

To solve the LR-MCTDHB linear-response system and hence the Schrödinger equation in linear response, we have to diagonalize the linear-response matrix $\boldsymbol{\mathcal{L}}$ and find its eigenvalues – the excitations energies $\{\omega_k\}$ – and eigenvectors

$$\boldsymbol{\mathcal{L}} \begin{pmatrix} \mathbf{u}^k \\ \mathbf{v}^k \\ \mathbf{C}_u^k \\ \mathbf{C}_v^k \end{pmatrix} = w_k \begin{pmatrix} \mathbf{u}^k \\ \mathbf{v}^k \\ \mathbf{C}_u^k \\ \mathbf{C}_v^k \end{pmatrix} \equiv w_k \mathbf{R}^k. \tag{13}$$

In Refs. [3,20] we have given the detailed derivations and explanations of all aspects of the LR-MTDHB theory. Here we only mention that the general structure of the linear-response matrix, i.e., the sizes of its sub-matrices is

$$\mathcal{L} \sim 2 \left(\begin{array}{c||c} M \times M & M \times N_{conf} \\ \hline\hline N_{conf} \times M & N_{conf} \times N_{conf} \end{array} \right). \tag{14}$$

Technically, the linear-response matrix $\mathcal{L}$ and the vector $\mathcal{R}$ which collects the perturbing fields are fully determined and can be constructed from the static MCHB solution at hands.

A reference point. We would like to mention that the MCTDHB(1) theory is fully equivalent to the Gross-Pitaevskii mean-field theory, so the linear response on-top of MCTDHB(1) is equivalent to the linear response of the Gross-Pitaevskii (LR-GP) equation. Another name of the popular LR-GP theory is the Bogoliubov – de Gennes (BdG) equations, see, e.g., Refs. [9, 13, 15, 49].

4.2 System, Hamiltonian and Parameters

The first application of the LR-MCTDHB theory was to the commonly-used contact inter-particle interaction potential $W(x - x') = \delta(x - x')$ of strength λ_0. We have studied the response of the many-boson systems trapped in a one-dimensional harmonic potential and shallow symmetric double-well potential:

$$V(x) = \omega_{ho}^2 x^2/2\,, \quad V(x) = b/2 \cdot \cos\left(\frac{\pi}{3}x\right) + \omega_{ho}^2 x^2/2\,, \tag{15}$$

with frequency $\omega_{ho} = \sqrt{2}$ and $b = 5$.

The first step was to compute by imaginary-time propagation the MCHB ground state of the systems made of $N = 100$ bosons with inter-particle interaction strength $\lambda_0 = 0.01$. The computations reveal that the fragmentation, i.e., the population of higher orbitals, in the ground states of these systems is negligible: 0.001 % for bosons in the harmonic trap and 0.03 % in the double well, respectively. The common belief is that in this case the Gross-Pitaevskii mean-field theory should provide very accurate description of the ground state. Indeed, the comparison of the MCTDHB and Gross-Piteavskii results for the ground states confirms this supposition.

The second common belief, naturally followed from the above one, was that for almost fully-condensed ground states the linear response of the Gross-Pitaevskii mean-field theory, LR-GP, should provide correct and complete description also for excited states, at least for the low-lying excitations. In other words, the popular LR-GP theory was considered to be the main source of information on excited states available in such systems. Let us see how the LR-MCTDHB results change the BdG excitation picture.

4.3 Novel Physical Results

In Fig. 4 we compare the LR-GP (BdG) and LR-MCTDHB(2) excitation spectra of $N = 100$ trapped interacting bosons for the harmonic (left panel) and the double-well (right panel) potentials. The ground-state densities obtained for these systems within the Gross-Pitaevskii and MCTDHB(2) theories are very similar and shown schematically in the insets together with the trapping potentials. The x-axes in Fig. 4 indicate excitation energies in units of the trap frequency ω_{ho}. The height of the lines or, equivalently, the position of the points, indicate response weights. We have chosen linear $f^+ = f^- = x$ and quadratic $f^+ = f^- = x^2$ perturbations to study both odd- and even-parity excitations separately. The solid (red) lines with triangles correspond to the odd excitations, and the dashed (green) lines with squares to even ones.

The left panel of Fig. 4 contrasts the predictions of the standard LR-GP (BdG) and our many-body LR-MCTDHB theories in the situation where the initial state in the harmonic trap is completely condensed, i.e., the GP and LR-GP theories are believed to provide adequate descriptions. As one can see, the many-body LR-MCTDHB theory contains the mean-field excitations, but predicts also additional many-body excited states which are out of the realm of the mean-field linear response. The response of these many-body excited states to the perturbations studied strongly depends on the details of the system, namely on the inter-particle interactions and total number of atoms. In the right panel of Fig. 4 one can see that the excitation spectrum of a condensate in the shallow-double well at the LR-MCTDHB level of theory possesses some additional spectral features not described by the standard LR-GP theory. In particular, the energy of the lowest-in-energy

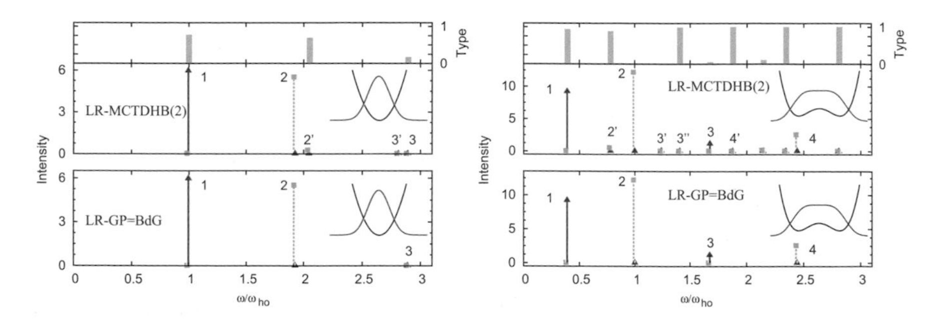

Fig. 4 In trapped systems there are low-lying excitations of many-body nature, not described by the standard linear-response theory for BECs (LR-GP). Comparisons of the standard LR-GP (BdG) and LR-MCTDHB(2) linear-response spectra computed for the systems of $N = 100$ interacting bosons trapped in a harmonic potential (*left panel*) and in a very shallow double-well potential trap (*right panel*). The perturbations are linear and quadratic. The trap potential and ground-state density (*red solid line*) are shown in the insets. The key observation here, when comparing the results in harmonic and double-well traps, is that the number of low-lying excitations not described by the BdG theory in double-well systems is much larger. See Phys. Rev. A **88**, 023606 (2013), Ref. [20], for more details. All quantities shown are dimensionless

excitation of even symmetry is almost 25 % lower than predicted by LR-GP theory. Moreover, this excitation is not of a single-particle nature as predicted by LR-GP theory, but it is rather comprised by a transfer of two bosons from the condensate to the lowest ungerade mode. Generally, in shallow double-well systems the number of low-lying non-mean-field states is larger in comparison with the harmonic case. The existence of low-lying excited states, not-described by the LR-GP theory, can have very important consequences to the quantum dynamics and temperature properties of ultra-cold systems trapped in anharmonic potentials.

Finally, we would like to mention that in Ref. [20] we have explicitly shown that the MCTDHB(M) theory is capable of providing numerically-exact description of the excited states. Thus, we have thereby expanded the prospects and validity of the MCTDHB concept also to describe static excited states (benchmarks for the dynamics have been provided and proven in Refs. [2, 32, 33, 51]). So, by increasing the number M of the self-consistent (time-adaptive) modes used in the computations we increase the quality of description of the ground-state wave-function and thereby the quality of the linear-response-computed excited states.

4.4 Complexity of the Employed Numerics

The linear-response matrix $\mathcal{L}$ acts on the $2(M + N_{conf})$-dimensional vector composed of the orbitals' and coefficients' response amplitudes, $(\mathbf{u}, \mathbf{v}, \mathbf{C}_u, \mathbf{C}_v)^T$, where $N_{conf} = \binom{N+M-1}{N}$ is the number of the involved configurations. Hence, the main complexity comes from the need to diagonalize a large complex non-Hermitian matrix. Moreover, since the spectrum contains positive and negative frequencies there is no efficient diagonalizer capable of providing selected roots. Another complication comes from the fact that usually along with the excitation energies one needs also properties of the excited states, hence we have to compute the eigenvectors $\{\mathbf{R}^k\}$ as well.

5 Dynamical Stability of Strongly Interacting Systems

In this project we open new horizons for the MCTDHB ideology in several directions: We use general inter-particle interactions of finite- and long-range and we investigate systems in one- (1D), two- (2D), and three-dimensional (3D) setups. These investigations have been inspired and motivated by recent experiments with ultra-cold polarized clouds of chromium ^{52}Cr [17, 60], dysprosium ^{164}Dy [36], and erbium ^{168}Er [1] as well as by progress in experimental realization of so-called "Rydberg-dressed" systems [61]. These experiments have ultimately shown that the short-range (contact) inter-particle interaction potential alone cannot describe the observed physics and it should be augmented by an additional finite- and long-range terms.

We have studied statics and non-equilibrium quantum dynamics of trapped bosons interacting by strong repulsive inter-particle interaction potentials of finite-range in one, two, and three spatial dimensions with MCTDHB. Some preliminary results have already been published as research Letters entitled "Quantum systems of ultracold bosons with customized interparticle interactions" in Phys. Rev. A **88**, 041602(R) (2013) [55] where statics has been investigated and "Generic regimes of quantum many-body dynamics of trapped bosonic systems with strong repulsive interactions" in Phys. Rev. A **89**, 061602(R) (2014) [59] where we report on novel dynamical phenomena.

5.1 System, Hamiltonian and Parameters

Generally, strong inter-particle interactions manifest in quantum many-body systems via formations of non-trivial structures in the ground states' densities and developments of strong correlations therein, see, e.g., Refs. [4, 5, 16, 23, 29, 40, 48]. The first two research questions we have addressed were: (i) To investigate the microscopic details of how repulsive inter-particle potentials create the non-trivial features (humps) in the densities of ultra-cold systems confined in simple barrier-less traps, and; (ii) To understand which characteristic of a general inter-particle interaction function $W(\mathbf{R})$ favors these density modulations and thereby controls the accompanying developments of correlations and fragmentation. The next demanding step is to explore how stable these interaction-induced modulations of the ground-states' densities are against external and internal disturbances in bosonic systems confined in 1D, 2D, and 3D setups.

We have considered several one-dimensional barrier-less external traps:

$$V(x) = 0.5x^2, \quad V(x) = 0.5x^6, \quad V(x) = \{-x : x < 0; 3x/4 : x \geq 0\}, \quad (16)$$

namely, a standard harmonic, an anharmonic, and an asymmetric–linear confining potentials. We have examined the following inter-particle interaction functions: Screened Coulomb $1/\sqrt{(|x - x'|/D)^{2n} + 1)}$, exponential $\exp\left[-\frac{1}{2}(|x - x'|/D)^n\right]$, and Sech-shaped $\operatorname{sech}\left[(|x - x'|/D)^n\right]$ of half-width D with $n = 1$, as well as their *sharper* analogues with $n = 2$. The sketches of the trapping and inter-particle potentials are shown in Fig. 5. In the following we use, respectively, the shorthand notations $\mathbf{exp}\left[-\mathbf{R}^n\right]$, $1/\mathbf{R}^n$, and $\mathbf{sech}\left[\mathbf{R}^n\right]$. To study the dynamics in 1D, 2D, and 3D we have used a more "realistic" inter-particle interaction function $W(\mathbf{R}) = 1/((|\mathbf{r} - \mathbf{r}'|/D)^n + 1)$ of half-width $D = 4$ with $n = 4$, because inter-particle interactions of similar shapes naturally appear in the so-called "Rydberg-dressed" ultra-cold systems [10, 22, 27] which are of current experimental interest [61]. Let us see what is the physics found with the help of the MCTDHB method.

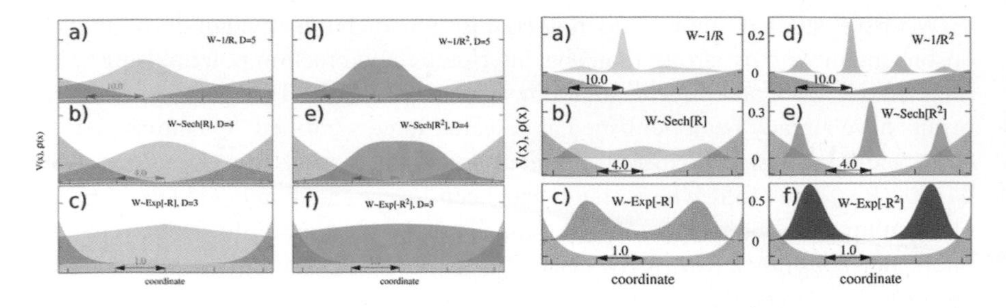

Fig. 5 Schematic plot of the trapping and inter-particle interaction potentials used to verify the diversity and generality of fragmentation phenomena in strongly-repulsive systems. *Left panels*: The depicted one-dimensional trapping potentials $V(\mathbf{r})$ and inter-particle $W(\mathbf{R})$ interaction potential are scaled and shifted for better presentation. *Right panels*: Shown are the densities of one-dimensional systems made of $N = 108$ interacting bosons and the corresponding trapping potentials, scaled and shifted for better visualization. See Phys. Rev. A **88**, 041602(R) (2013), Ref. [55], for more details. All quantities shown are dimensionless

5.2 Novel Physical Results

The main message of the static study was that a strong inter-particle repulsion leads to formation of multi-hump, localized, fragmented structures irrespective to the shapes of the inter-particle $W(\mathbf{R})$ and trapping $V(\mathbf{r})$ potentials used. In the left panels of Fig. 5 we show the variety of one-dimensional trapping potentials $V(\mathbf{r})$ and inter-particle interaction potentials $W(\mathbf{R})$ used. In the right panels of Fig. 5 we plot the corresponding ground-state densities. Specifically, in the right–upper panels: Asymmetric–linear trap and screened Coulomb inter-boson interactions of half-width $D = 5$ with $n = 1$ (a) and $n = 2$ (d); Right–middle panels: Harmonic trap and Sech-shaped interactions with $D = 4$ and $\lambda_0 = 1.0$, and $n = 1$ (b), $n = 2$ (e); Right–lower panels: Anharmonic trap and exponential inter-particle interactions with $D = 3$ and $\lambda_0 = 1.5$, and $n = 1$ (c), $n = 2$ (f). All show the diversity and generality of the fragmentation phenomenon.

These computations predict that, to minimize the strong inter-particle repulsion, the density of trapped repulsive ultra-cold bosons inevitably splits and fragments into multi-hump structures. The physics behind is an interplay between classical "electrostatic" repulsion, which pushes the bosons from the trap center towards its edges, provoking thereby the formation of multi-hump structures in the density, and quantum mechanics, which governs the loss of inter-hump coherence and emergence of fragmentation. Inter-particle interaction potentials with *sharper* edges ($n = 2$) enhance fragmentation, compare, respectively, windows (a, b, c) and (d, e, f) in the right panels of Fig. 5.

In the next step, the stability of these multi-hump states was investigated as a time-dependent process by solving the real-space time-dependent Schrödinger equation for several dynamical scenarios involving manipulations with the strength

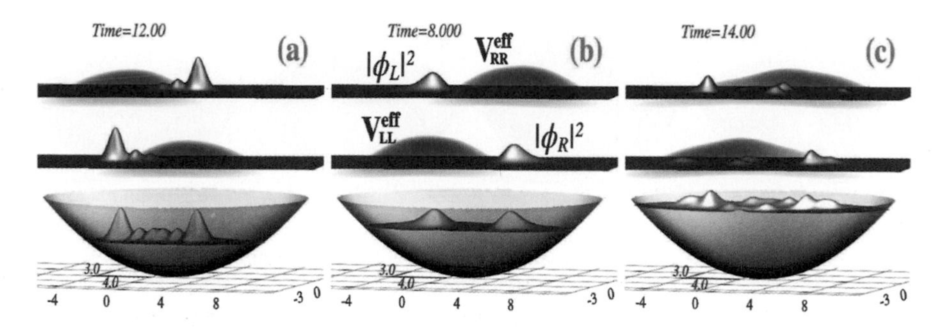

Fig. 6 Dynamics of trapped strongly-interacting repulsive bosons in 2D. Evolutions of a two-fold fragmented initial state induced by a sudden quench of the inter-particle repulsion with a simultaneous displacement of the harmonic trap, $V(x, y) = 0.5x^2 + 1.5y^2 \rightarrow V(x-1.5, y-0.5)$. Two generic dynamical regimes of strongly-interacting trapped bosons can take place. In the first, non-violent regime the overall "topology" of the ground-state density is preserved during the propagation, whereas in the second, highly-non-equilibrium regime "explosive" quantum many-body dynamics emerges. The over-a-barrier dynamics (**a**,**c**) happens when the energy-per-particle of the out-of-equilibrium state is larger than the heights of the induced barriers, otherwise the dynamics is under-a-barrier (**b**). The emerging physics is explained in terms of the densities $|\phi_k(\mathbf{r}, t)|^2$ of the *left* (*right*) fragment and the time-dependent barriers induced by the complimentary *right* (*left*) fragment, $V^{eff}_{jj}(\mathbf{r}, t) = \lambda_0 \frac{N}{2} \int |\phi_j(\mathbf{r}', t)|^2 W(\mathbf{r} - \mathbf{r}') d\mathbf{r}'$. See Phys. Rev. A **89**, 061602(R) (2014), Ref. [59], for more details. All quantities shown are dimensionless

λ_0 of the inter-boson interaction potential $W(\mathbf{r} - \mathbf{r}') \equiv W(\mathbf{R})$ and the external trap $V(\mathbf{r}, t)$. Namely, we use different time-dependent processes to destabilize the systems' ground states: A sudden quench of the strength of the inter-particle repulsion is accompanied by a sudden displacement of the trap. Two qualitatively different but otherwise generic dynamical quantum many-body behaviors are discovered. In the first, the overall "topology" of the ground-state density is preserved, whereas in the second the density totally "explodes". An intuitive many-body time-dependent model was devised to interpret and explain the observations. The model's main constituents are the interaction-induced time-dependent barriers, see Fig. 6 and Ref. [59] for more details. The generality of the discovered scenarios was explicitly confirmed in traps of various shapes and dimensionality, and inter-particle interactions of different forms and ranges.

Summarizing, in the non-violent evolutions, see Fig. 6b in 2D, the superposition of the interaction-induced time-dependent barriers $V^{eff}_{jj}(\mathbf{r}, t)$ and external trap $V(\mathbf{r})$ results in effective potentials which are high enough to confine the fragments also when they are moving. Hence, the physics behind the non-violent dynamics of the trapped strongly-interacting fragmented systems is indeed an under-a-barrier dynamics.

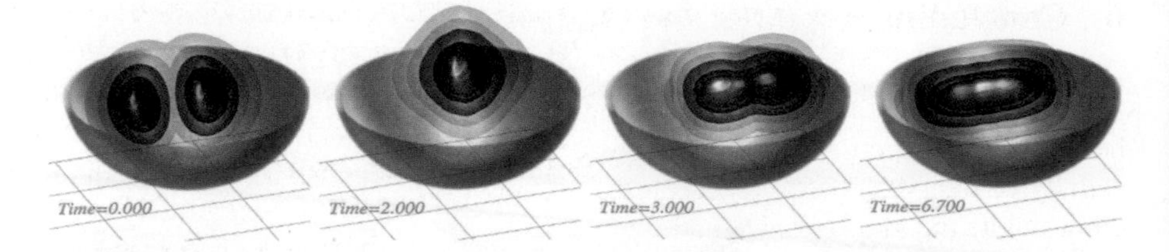

Fig. 7 Over-a-barrier regime in the dynamics of strongly-interacting trapped repulsive bosons in 3D. The evolution of a two-hump two-fold fragmented initial state is induced by a strong, sudden decrease of the repulsion from $\lambda_0 = 0.5 \rightarrow 0.1$ accompanied by a sudden displacement of the trap, $V(x,y,z) = 0.5x^2 + 1.5y^2 + 1.5z^2 \rightarrow V(x-1.5, y-0.5, z-0.5)$. To visualize the 3D functions we plot several iso-surfaces of the density and an equipotential cut of the trap. The snapshots of the density are taken to show initial ($t = 0$), coalescing ($t = 2$), penetrating ($t = 3$), and highly-excited ($t = 6.7$) momentary instances of the violent over-a-barrier dynamics. See Phys. Rev. A **89**, 061602(R) (2014), Ref. [59], for more details. All quantities shown are dimensionless

Complimentary, the violent dynamics appears when the induced barriers are not high enough to keep the sub-clouds apart from each other. This is achieved by the sudden decreases of the inter-particle repulsion from $\lambda_0 = 0.5 \rightarrow 0.1$, depicted in Fig. 6a for 2D and Fig. 7 for 3D. The sub-clouds start to leak out and eventually become delocalized over the entire trap. During the violent evolutions the available energy is redistributed between the excited states which have multi-node structures and can be delocalized over the entire trap. As a result, the density reveals the observed highly violent explosive behavior. So, the physics behind is a complex over-a-barrier dynamics.

5.3 *Complexity of the Employed Numerics*

The non-contact finite- and long-range inter-particle interaction potentials $W(\mathbf{R})$ used in this project imply that at each micro-iteration all involved integrals, e.g., $V_{jj}^{eff}(\mathbf{r},t) = \lambda_0 \frac{N}{2} \int |\phi_j(\mathbf{r}',t)|^2 W(\mathbf{r}-\mathbf{r}') d\mathbf{r}'$ have to be recomputed. This a challenging problem even in the one-dimensional case. In the presented 2D and 3D dynamical studies, similar integrals comprise the main time- and memory-consuming parts. We have examined several different product grids used to represent the involved single-particle functions $\phi_k(\mathbf{r},t)$ and, as a preliminary outcome, found out that the FFT technique seems to be the most efficient, at least for traps of general shapes. However, this is an on-going investigation and numerical performance and different paralellization techniques have to be further tested and compared.

6 Concluding Remarks and Outlook

The performance of the MCTDHB Package tested on Hermit and Laki has already been discussed in our previous HLRS report 2012, see Ref. [34]. The mainstream activity of our MCTDHB project in the present year has been the application of our accurate theoretical tools to understand known phenomena at a new theoretical level, as well as the discovery and prediction of novel physical phenomena. We have also expanded the MCTDHB methodology to include excited states via the derivation and implementation of a linear-response theory on-top of the MCTDHB equations-of-motion. This opens up a broad perspective to access excited states and their properties directly in large interacting quantum systems.

From the perspective of the required computational resources and numerical algorithms applied, each of these numerically-demanding studies has challenged different aspects of computational physics and mathematics: (i) Long-time propagation – stability of the numerical methods used to integrate the MCTDHB equations-of-motion; (ii) Control of the tunneling dynamics – a very detailed study where an interplay of the parameters controlling the decay by tunneling dynamics is accompanied by a long-time propagation on huge spatial grids, which are needed to simulate open systems; (iii) Excited states of many-body systems – construction and diagonalization of complex non-hermitian linear-response matrices; (iv) Finite- and long-range interactions in 1D, 2D, and 3D setups – efficient methods and techniques for evaluation of involved high-dimensional integrals. In turn, these challenges – that we have successfully been coping with in the past year – pave the way to further research and developments to come.

In future we plan to keep a similar line of development and to explore and push the new horizons opened in the presently-reported studies further. The rich physics "hiding" in systems with short-, finite-, and long-range inter-particle interactions, in 1D, 2D and 3D, their excitations, and non-equilibrium dynamics, all beyond standard mean-field and on an accurate many-body level, utilizing MCTDHB and its just-derived linear-response theory on-top, LR-MCTDHB, are high on our agenda. These directions also imply a serious improvement of the numerical methods and algorithms used, including efficient usage of the available graphical accelerators and modern co-processors, as well as the further developments of software packages [31,58]. With the growing proliferation of the MCTDHB method, see, e.g., Refs. [6, 7, 18, 19, 21, 24], we believe many more exciting developments and thrilling physics are ahead of us.

References

1. Aikawa, K., Frisch, A., Mark, M., Baier, S., Rietzler, A., Grimm, R., Ferlaino, F.: Bose-Einstein Condensation of Erbium. Phys. Rev. Lett. **108**, 210401 (2012)
2. Alon, O.E., Streltsov, A.I., Cederbaum, L.S.: Multiconfigurational time-dependent Hartree method for bosons: Many-body dynamics of bosonic systems. Phys. Rev. A **77**, 033613 (2008)

3. Alon, O.E., Streltsov, A.I., Cederbaum, L.S.: Unified view on linear response of interacting identical and distinguishable particles from multiconfigurational time-dependent Hartree methods. J. Chem. Phys. **140**, 034108 (2014)
4. Astrakharchik, G.E., Girardeau, M.D.: Exact ground-state properties of a one-dimensional Coulomb gas. Phys. Rev. B **83**, 153303 (2011)
5. Baranov, M.A.: Theoretical progress in many-body physics with ultracold dipolar gases. Phys. Rep. **464**, 71 (2008)
6. Březinová, I., Lode, A.U.J., Streltsov, A.I., Alon, O.E., Cederbaum, L.S., Burgdörfer, J.: Wave chaos as signature for depletion of a Bose-Einstein condensate. Phys. Rev. A **86**, 013630 (2012)
7. Březinová, I., Lode, A.U.J., Streltsov, A.I., Cederbaum, L.S., Alon, O.E., Collins, L.A., Schneider, B.I., Burgdörfer, J.: Elastic scattering of a Bose-Einstein condensate at a potential landscape. J. Phys. Conf. Ser. **488**, 012032 (2014)
8. bwGRiD, member of the German D-grid initiative, funded by the Ministry for Education and Research (Bundesministerium für Bildung und Forschung) and the Ministry for Science, Research and Arts Baden-Württemberg (Ministerium für Wissenschaft, Forschung und Kunst Baden-Württemberg). http://www.bw-grid.de
9. Castin, Y., Dum, R.: Low-temperature Bose-Einstein condensates in time-dependent traps: Beyond the U(1) symmetry-breaking approach. Phys. Rev. A **57**, 3008 (1998)
10. Comparat, D., Pillet, P.: Dipole blockade in a cold Rydberg atomic sample [Invited]. J. Opt. Soc. Am. B **6**, A208 (2010)
11. Cray XE6 cluster Hermit and NEC Nehalem cluster Luki at the High Performance Computing Center Stuttgart (HLRS). https://www.hlrs.de
12. Dalfovo, F., Giorgini, S., Pitaevskii, L.P., Stringari, S.: Theory of Bose-Einstein condensation in trapped gases. Rev. Mod. Phys. **71**, 463 (1999)
13. Esry, B.D.: Hartree-Fock theory for Bose-Einstein condensates and the inclusion of correlation effects. Phys. Rev. A **55**, 1147 (1997)
14. Fetter, A.L., Walecka, J.D.A.: Quantum Theory of Many-Particle Systems. McGraw-Hill, New York (1971)
15. Gardiner, C.W.:Particle-number-conserving Bogoliubov method which demonstrates the validity of the time-dependent Gross-Pitaevskii equation for a highly condensed Bose gas. Phys. Rev. A **56**, 1414 (1997)
16. Girardeau, M.: Relationship between Systems of Impenetrable Bosons and Fermions in One Dimension. J. Math. Phys. **1**, 516 (1960)
17. Griesmaier, A., Werner, J., Hensler, S., Stuhler, J., Pfau, T.: Bose-Einstein Condensation of Chromium. Phys. Rev. Lett. **94**, 160401 (2005)
18. Grond, J., Schmiedmayer, J., Hohenester, U.: Optimizing number squeezing when splitting a mesoscopic condensate. Phys. Rev. A **79**, 021603(R) (2009)
19. Grond, J., Betz, T., Hohenester, U., Mauser, N.J., Schmiedmayer, J., Schumm, T., New J.: Shapiro effect in atomchip-based bosonic Josephson junctions. Phys. **13**, 065026 (2011)
20. Grond, J., Streltsov, A.I., Lode, A.U.J., Sakmann, K., Cederbaum, L.S., Alon, O.E.: Excitation spectra of many-body systems by linear response: General theory and applications to trapped condensates. Phys. Rev. A **88**, 023606 (2013)
21. Heimsoth, M., Hochstuhl, D., Creffield, C.E., Carr, L.D., Sols, F.: Effective Josephson dynamics in resonantly driven Bose-Einstein condensates. New J. Phys. **15**, 103006 (2013)
22. Henkel, N., Nath, R., Pohl, T.: Three-Dimensional Roton Excitations and Supersolid Formation in Rydberg-Excited Bose-Einstein Condensates. Phys. Rev. Lett. **104**, 195302 (2010)
23. Henkel, N., Cinti, F., Jain, P., Pupillo, G., Pohl, T.: Supersolid Vortex Crystals in Rydberg-Dressed Bose-Einstein Condensates. Phys. Rev. Lett. **108**, 265301 (2012)
24. Hochstuhl, D., Hinz, C.M., Bonitz, M.: Time-dependent multiconfiguration methods for the numerical simulation of photoionization processes of many-electron atoms. Eur. Phys. J. Spec. Top. **223**, 177 (2014)
25. Hofferberth, S., Lesanovsky, I., Fischer, B., Verdu, J., Schmiedmayer, J.: Radiofrequency-dressed-state potentials for neutral atoms. Nat. Phys. **2**, 710 (2006)

26. Hybrid computing complex K100 (Keldysh Institute of Applied Mathematics, RAS). http://www.kiam.ru
27. Johnson, J.E., Rolston, S.L.: Interactions between Rydberg-dressed atoms. Phys. Rev. A **82**, 033412 (2010)
28. Köhler, T., Góral, K., Julienne, P.S.: Production of cold molecules via magnetically tunable Feshbach resonances. Rev. Mod. Phys. **78**, 1311 (2006)
29. Lahaye, T., Menotti, C., Santos, L., Lewenstein, M., Pfau, T.: The physics of dipolar bosonic quantum gases. Rep. Prog. Phys. **72**, 126401 (2009)
30. Leggett, A.J.: Bose-Einstein condensation in the alkali gases: Some fundamental concepts. Rev. Mod. Phys. **73**, 307 (2001)
31. Lode, A.U.J., Tsatsos, M.C.: The recursive multiconfigurational time-dependent Hartree for Bosons package. http://ultracold.org; http://r-mctdhb.org; http://schroedinger.org (2014)
32. Lode, A.U.J., Sakmann, K., Alon, O.E., Cederbaum, L.S., Streltsov, A.I.: Numerically exact quantum dynamics of bosons with time-dependent interactions of harmonic type. Phys. Rev. A **86**, 063606 (2012)
33. Lode, A.U.J., Streltsov, A.I., Sakmann, K., Alon, O.E., Cederbaum, L.S.: How an interacting many-body system tunnels through a potential barrier to open space. Proc. Natl. Acad. Sci. USA **109**, 13521 (2012)
34. Lode, A.U.J., Sakmann, K., Doganov, R.A., Grond, J., Alon, O.E., Streltsov, A.I., Cederbaum, L.S.: Numerically-exact Schrödinger dynamics of closed and open many-boson systems with the MCTDHB package, HLRS report for 2012. In: Nagel, W.E., Kröner, D.H., Resch, M.M. (eds.) High Performance Computing in Science and Engineering '13: Transactions of the High Performance Computing Center, Stuttgart (HLRS) 2013. Springer, Heidelberg (2013)
35. Lode, A.U.J., Klaiman, S., Alon, O.E., Streltsov, A.I., Cederbaum, L.S.: Controlling the velocities and the number of emitted particles in the tunneling to open space dynamics. Phys. Rev. A **89**, 053620 (2014)
36. Lu, M., Burdick, N.Q., Youn, S.H., Lev, B.L.: Strongly Dipolar Bose-Einstein Condensate of Dysprosium. Phys. Rev. Lett. **107**, 190401 (2011)
37. Meyer, H.-D., Gatti, F., Worth, G.A. (eds.): Multidimensional Quantum Dynamics: MCTDH Theory and Applications. Wiley-VCH, Weinheim (2009)
38. Milburn, G.J., Corney, J., Wright, E.M., Walls, D.F.: Quantum dynamics of an atomic Bose-Einstein condensate in a double-well potential. Phys. Rev. A **55**, 4318 (1997)
39. Mueller, E.J., Ho, T.-L., Ueda, M., Baym, G.: Fragmentation of Bose-Einstein condensates. Phys. Rev. A **74**, 033612 (2006)
40. Nest, M., Klamroth, T., Saalfrank, P.: Unified view on multiconfigurational time-propagation for systems consisting of identical particles. J. Chem. Phys. **127**, 154103 (2005)
41. Nozières, P.: Some Comments on Bose-Einstein Condensation. In: Griffin, A., Snoke, D.W., Stringari, S. (eds.) Bose-Einstein Condensation. Cambridge University Press, Cambridge (1996)
42. Nozières, P., Saint James, D.: Particle vs. pair condensation in attractive Bose liquids. J. Phys. (Fr.) **43**, 1133 (1982)
43. Olsen, J., Jørgensen, P.: Linear and nonlinear response functions for an exact state and for an MCSCF state. J. Chem. Phys. **82**, 3235 (1985)
44. Orzel, C., Tuchman, A.K., Fenselau, M.L., Yasuda, M., Kasevich, M.A.: Squeezed States in a Bose-Einstein Condensate. Science **291**, 2386 (2001)
45. Penrose, O., Onsager, L.: Bose-Einstein Condensation and Liquid Helium. Phys. Rev. **104**, 576 (1956)
46. Perrin, A., Chang, H., Krachmalnicoff, V., Schellekens, M., Boiron, D., Aspect, A., Westbrook, C.I.: Observation of Atom Pairs in Spontaneous Four-Wave Mixing of Two Colliding Bose-Einstein Condensates. Phys. Rev. Lett. **99**, 150405 (2007)
47. Proukakis, N.P., Gardiner, S.A., Davis, M.J., Szymanska, M.H. (eds.): Quantum Gases: Finite Temperature and Non-equilibrium Dynamics. Cold Atoms Series, vol. 1. Imperial College Press, London (2013)

48. Pupillo, G., Micheli, A., Boninsegni, M., Lesanovsky, I., Zoller, P.: Strongly Correlated Gases of Rydberg-Dressed Atoms: Quantum and Classical Dynamics. Phys. Rev. Lett. **104**, 223002 (2010)
49. Ruprecht, P.A., Edwards, M., Burnett, K., Clark, C.W.: Probing the linear and nonlinear excitations of Bose-condensed neutral atoms in a trap. Phys. Rev. A **54**, 4178 (1996)
50. Sakmann, K., Streltsov, A.I., Alon, O.E., Cederbaum, L.S.: Reduced density matrices and coherence of trapped interacting bosons. Phys. Rev. A **78**, 023615 (2008)
51. Sakmann, K., Streltsov, A.I., Alon, O.E., Cederbaum, L.S.: Exact Quantum Dynamics of a Bosonic Josephson Junction. Phys. Rev. Lett. **103**, 220601 (2009)
52. Sakmann, K., Streltsov, A.I., Alon, O.E., Cederbaum, L.S.: Universality of fragmentation in the Schrödinger dynamics of bosonic Josephson junctions. Phys. Rev. A **89**, 023602 (2014)
53. Smerzi, A., Fantoni, S., Giovanazzi, S., Shenoy, S.R.: Quantum Coherent Atomic Tunneling between Two Trapped Bose-Einstein Condensates. Phys. Rev. Lett. **79**, 4950 (1997)
54. Spekkens, R.W., Sipe, J.E.: Spatial fragmentation of a Bose-Einstein condensate in a double-well potential. Phys. Rev. A **59**, 3868 (1999)
55. Streltsov, A.I.: Quantum systems of ultracold bosons with customized interparticle interactions. Phys. Rev. A **88**, 041602(R) (2013)
56. Streltsov, A.I., Alon, O.E., Cederbaum, L.S.: General variational many-body theory with complete self-consistency for trapped bosonic systems. Phys. Rev. A **73**, 063626 (2006)
57. Streltsov, A.I., Alon, O.E., Cederbaum, L.S.: Role of Excited States in the Splitting of a Trapped Interacting Bose-Einstein Condensate by a Time-Dependent Barrier. Phys. Rev. Lett. **99**, 030402 (2007)
58. Streltsov, A.I., Sakmann, K., Lode, A.U.J., Alon, O.E., Cederbaum, L.S.: The multiconfigurational time-dependent Hartree for Bosons package, version 2.3, Heidelberg. http://MCTDHB.org (2013)
59. Streltsova, O.I., Alon, O.E., Cederbaum, L.S., Streltsov, A.I.: Generic regimes of quantum many-body dynamics of trapped bosonic systems with strong repulsive interactions. Phys. Rev. A **89**, 061602(R) (2014)
60. Stuhler, J., Griesmaier, A., Koch, T., Fattori, M., Pfau, T., Giovanazzi, S., Pedri, P., Santos, L.: Observation of Dipole-Dipole Interaction in a Degenerate Quantum Gas. Phys. Rev. Lett. **95**, 150406 (2005)
61. Viteau, M., Bason, M.G., Radogostowicz, J., Malossi, N., Ciampini, D., Morsch, O., Arimondo, E.: Rydberg Excitations in Bose-Einstein Condensates in Quasi-One-Dimensional Potentials and Optical Lattices. Phys. Rev. Lett. **107**, 060402 (2011)

Thermodynamics with 2 + 1 + 1 Dynamical Quark Flavors

Stefan Krieg

Abstract We report on our calculation of the equation of state of Quantum Chromodynamics (QCD) from first principles, through simulations of Lattice QCD. We use an improved lattice action and $N_f = 2+1+1$ dynamical quark flavors and physical quark mass parameters.

1 Introduction

The aim of our project is to compute the charmed equation of state for Quantum Chromodynamics (for details, see [1]). We are using the lattice discretized version of Quantum Chromodynamics, called lattice QCD, which allows simulations of the theory through importance sampling methods. Our results are important input quantities for phenomenological calculations and are required to understand experiments aiming to generate a new state of matter, called Quark-Gluon-Plasma, such as the upcoming FAIR at GSI, Darmstadt.

The present status of the field is marked by our papers on the $N_f = 2+1$[1] equation of state [2, 3]. In the time since the publication of the aforementioned works, the hotQCD collaboration have improved the precision of their results. It was found that some discrepancies between our and their results still remain (see e.g. [4]). It is the aim of this work to provide a high precision calculation of the equation of state of QCD including the (dynamical) effects of the charm quark, in order to remedy the above situation.

[1]This refers to dynamical up/down and strange quarks – including a dynamical charm quark is what was proposed here.

S. Krieg (✉)
for the Wuppertal-Budapest collaboration

Bergische Universität Wuppertal, Fachbereich C - Physik, D-42119 Wuppertal, Germany

IAS, Jülich Supercomputing Centre, Forschungszentrum Juelich GmbH, D-52425 Jülich, Germany
e-mail: krieg@uni-wuppertal.de

W.E. Nagel et al. (eds.), *High Performance Computing in Science and Engineering '14*, DOI 10.1007/978-3-319-10810-0_6

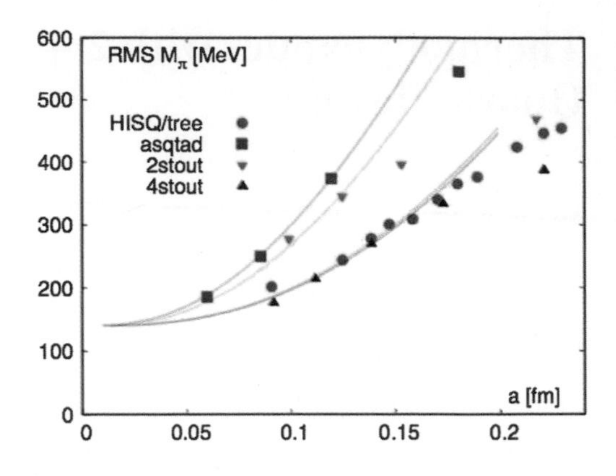

Fig. 1 RMS pion mass for different staggered fermion actions, in the continuum limit

Our simulations are performed using so-called staggered fermions. In the continuum limit, i.e. at vanishing lattice spacing a, one staggered Dirac operator implements four flavors of mass degenerate fermions. At finite lattice spacing, however, discretization effects induce an interaction between these would be flavors lifting the degeneracy. The "flavors" are, consequentially, renamed to "tastes", and the interactions are referred to as "taste-breaking" effects. Even though the tastes are not degenerate, in simulations one takes the fourth root of the staggered fermion determinant to implement a single flavor. This procedure is not proven to be correct – however, practical evidence suggests that is does not induce errors visible with present day statistics.

Taste-breaking is most severely felt at low pion masses and large lattice spacing, as the pion sector is distorted through the taste-breaking artifacts: there is one would-be Goldstone boson, and 15 additional heavier "pions", which results in an RMS pion mass larger than the mass of the would-be Goldstone boson. This effect is depicted in Fig. 1 for different staggered type fermion actions. As can be seen for this figure, the previously used twice stout smeared action ("2stout") has a larger RMS pion mass and thus taste-breaking effects than the HISQ/tree action. If, however, the number of smearing steps is increased to four, with slightly smaller smearing strength ("4stout"), the RMS pion mass measured agrees with that of the HISQ/tree action. In order to have an improved pion sector, we, therefore, opted to switch to this new action and to restart our production runs.

2 Status

The status of our production is summarized in Table 1 and Fig. 2. We generated most finite temperature ensembles on less scalable architectures, such as GPU clusters, available to the collaboration at Wuppertal and Budapest universities. Our (zero temperature) renormalization runs require a scalable architecture, and we, as

Table 1 Production status with the new "4stout" action

Data set	Status
$N_t = 6$	Ready
$N_t = 8$	Ready
$N_t = 10$	Ready
$N_t = 12$	Production

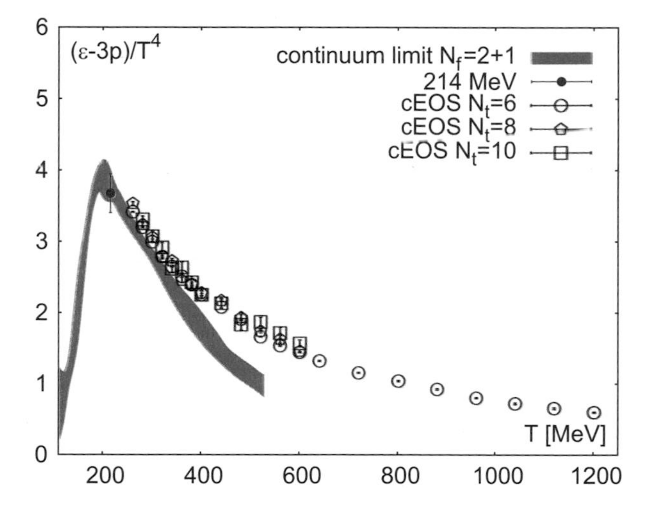

Fig. 2 Production status with the new "4stout" action. Presently available statistics with $N_t = 6, 8, 10$ is shown and compared to our result of [3]. Clearly, the curves start to deviate in a temperature range of $T = 300 \ldots 400$ MeV, as suggested by perturbative calculations [5]

proposed, used HERMIT to generate these essential ensembles. Our present and upcoming finite temperature simulations on finer, thus, larger lattices also benefit from scalability.

2.1 Results

In order to be able to reach very fine lattice spacings, or, equivalently, large temperatures at $N_t = 12$, we had to extend our line-of-constant-physics (LCP) beyond the range of lattice spacings available to us. Our previous strategy was the following ($m_c = 11.85\, m_s$):

1. Simulate a "reasonable" rectangle of up/down and strange quark mass values and to measure M_π/f_π and M_K/f_π.
2. Interpolate to the quark mass values where above ratios take their physical values. This gives m_{ud} and m_s.
3. Extract the lattice spacing by interpolating af_π to the above quark mass point.

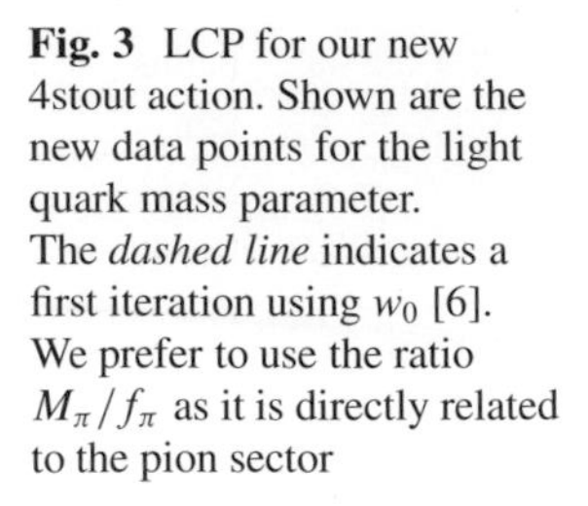

Fig. 3 LCP for our new 4stout action. Shown are the new data points for the light quark mass parameter. The *dashed line* indicates a first iteration using w_0 [6]. We prefer to use the ratio M_π/f_π as it is directly related to the pion sector

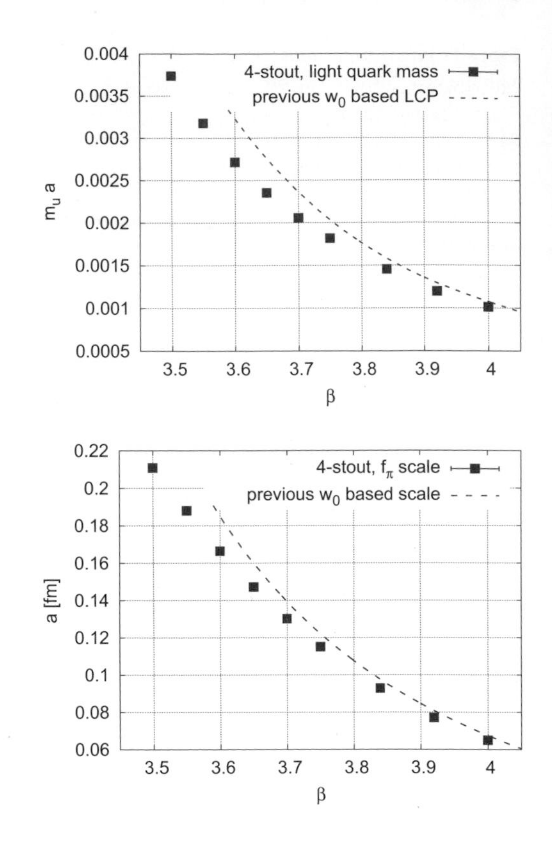

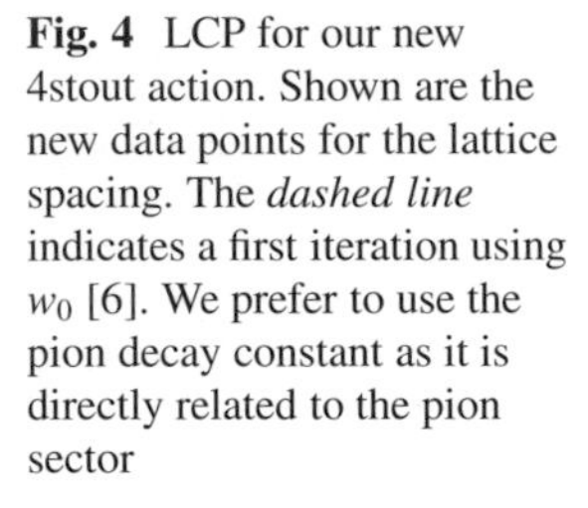

Fig. 4 LCP for our new 4stout action. Shown are the new data points for the lattice spacing. The *dashed line* indicates a first iteration using w_0 [6]. We prefer to use the pion decay constant as it is directly related to the pion sector

The results of this procedure are shown in Figs. 3 and 4, for the data points with $\beta < 3.9$. The remaining points have been generated with the procedure described in the following. At very fine lattice spacing, the zero temperature runs at physical mass parameters would require enormous lattices and would also likely face issues related to the freezing of the topological charge at $a \approx 0.5$ fm. We, therefore, adapted our strategy:

1. For every $\beta < 3.9$: Simulate $N_f = 3{+}1$ flavors in the flavor-symmetric point [7] ($\bar{m} = (2m_{ud,phys} + m_{s,phys})/3$), using the quark mass parameters found above.
2. Measure f_{PS} and M_{PS}/f_{PS}.
3. Perform a continuum extrapolation.

We now can estimate the expected values for f_{PS} and M_{PS}/f_{PS} at the target $\beta > 3.9$, see Figs. 5 and 6. Simulating several $\bar{m}$ values at these β we can find the value where the expected M_{PS}/f_{PS} is reproduced. This gives $\bar{m}(\beta)$, while the lattice spacing will be set through f_{PS}. The procedure is shown for one β in Figs. 7 and 8. It gives the LCP entries at $\beta > 3.9$ as shown above in Figs. 3 and 4.

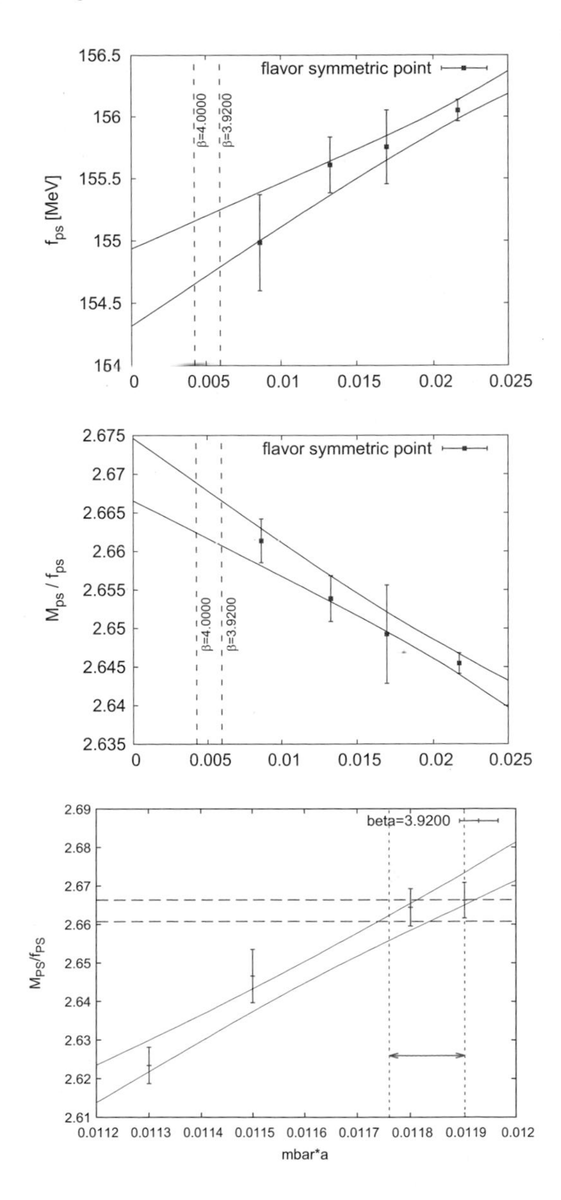

Fig. 5 Continuum extrapolation of the pion decay constant in the flavor-symmetric point (see text) using our new 4stout action. The target beta values are indicated by *dashed lines*. This extrapolation provides expected values for the pion decay constant at these β values

Fig. 6 Continuum extrapolation of the ration M_{PS}/f_{PS} in the flavor-symmetric point (see text) using our new 4stout action. The target beta values are indicated by *dashed lines*. This extrapolation provides expected values for the ratio at these β values

Fig. 7 Calculation of the (1σ region of) bare mass parameter for our new 4stout action. The *blue band marks* the expected value for the ration M_{PS}/f_{PS} at this β value, taken from the continuum extrapolation as shown in Fig. 6

2.2 Additional Results

The reduced taste breaking allowed us to determine a further observable, which is of particular interest of the heavy ion community. While the equation of state is an input to the relativistic hydrodynamic calculations that reproduce the observed

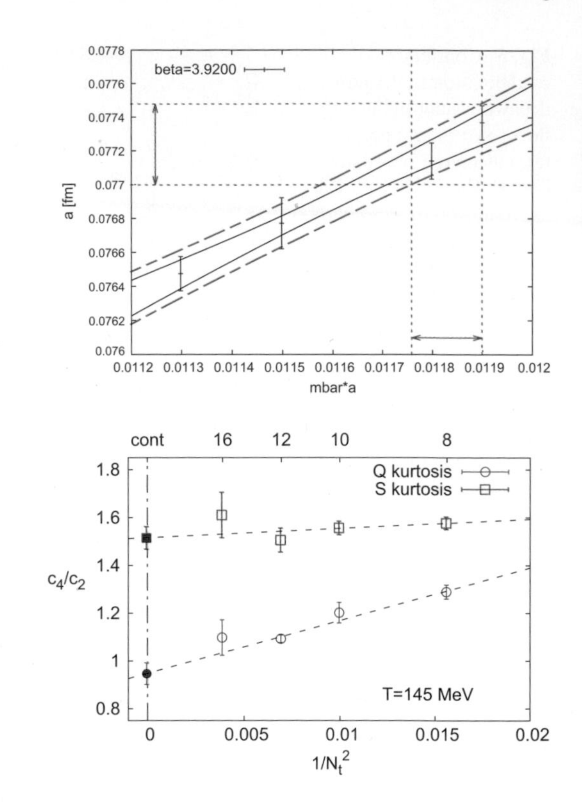

Fig. 8 Calculation of the lattice spacing at $\beta = 3.92$ for our new 4stout action. The correct value for $a\bar{m}$ has been extracted as shown in Fig. 7. The *blue band marks* the region defined by the measured f_π divided by the expected value from the continuum extrapolation (including errors)

Fig. 9 Continuum limit of the c_4/c_2 (kurtosis $\times$ variance) at $T = 145\,\text{MeV}$ temperature for the net charge and strangeness. These data were taken using our new 4stout action

flow characteristics, there is an observable that can and is being directly measured at LHC: the non-gaussianity of the net charge and strangeness fluctuations at freeze-out temperature. We concentrated all our efforts on a single temperature, so far, and show a continuum extrapolation in Fig. 9. This of course will need to be extended to a couple of more temperatures. The temperature dependence could be then used to measure the freeze-out temperature in heavy-ion collisions.

3 Production Specifics and Performance

Most of our production is done using modest partition sizes, as we found these to be most efficient for our implementation.

3.1 Performance

Our code shows nice scaling properties on HERMIT. For our scaling analysis below, we used two lattices ($N_s = 32$ and 48) and several partition sizes up to 256 nodes. We timed the most time consuming part of the code: the fermion matrix multiplication. The results are summarized in the following table:

No of nodes	Gflop/node $N_s = 32$	Gflop/node $N_s = 48$
1	16.3	15.4
2	16.8	16.0
4	16.5	16.2
8	16.3	16.3
16	16.3	16.3
32	16.8	16.0
64	17.1	16.5
128	19.2	16.5
256	16.3	16.0

3.2 *Production*

Given the nice scaling properties of our code, we were able to run at the sweet spot for queue throughput, which we found to be located at a job size of 64 nodes. Larger job sizes proved to have a scheduling probability sufficiently low that benefits in the runtime due to the larger number of cores were compensated and the overall production throughput decreased. We, therefore, opted to stay at jobs sizes with 64 nodes.

4 Outlook

At the present level of ensemble generation, we believe we will be able to publish within 2014. Since we, unfortunately, ran out of quota very quickly, we were not able keep up our earlier estimates. Still, HERMIT has proved to be an essential tool to be able to achieve this goal.

References

1. Borsanyi, S., Endrodi, G., Fodor, Z., Katz, S.D., Krieg, S., et al.: The QCD equation of state and the effects of the charm, PoS Lattice **2011**, 201 (2011)
2. Borsanyi, S., Endrodi, G., Fodor, Z., Jakovac, A., Katz, S.D., et al.: The QCD equation of state with dynamical quarks, JHEP **1011**, 077 (2010). doi:10.1007/JHEP11(2010)077
3. Borsanyi, S., Fodor, Z., Hoelbling, C., Katz, S.D., Krieg, S., et al.: Full result for the QCD equation of state with 2+1 flavors, Phys. Lett. **B730**, 99–104 (2014). doi:10.1016/j.physletb.2014.01.007

4. Petreczky, P.: On trace anomaly in 2+1 flavor QCD, PoS Lattice **2012**, 069 (2012)
5. Laine, M., Schroder, Y.: Quark mass thresholds in QCD thermodynamics, Phys. Rev. **D73**, 085009 (2006). doi:10.1103/PhysRevD.73.085009
6. Borsanyi, S., Durr, S., Fodor, Z., Hoelbling, C., Katz, S.D., et al.: High-precision scale setting in lattice QCD, JHEP **1209**, 010 (2012). doi:10.1007/JHEP09(2012)010
7. Horsley, R., et al.: Isospin breaking in octet baryon mass splittings, Phys. Rev. **D86**, 114511 (2012). doi:10.1103/PhysRevD.86.114511

Acoustic and Elastic Full Waveform Tomography

A. Kurzmann, S. Butzer, and T. Bohlen

Abstract For a better estimation of subsurface parameters we develop imaging methods that can exploit the richness of full seismic waveforms. Full waveform tomography (FWT) is a powerful imaging method and emerges as an important procedure in hydrocarbon exploration and underground construction. It is able to recover high-resolution multi-parameter subsurface images from recorded seismic data. For the reconstruction of 3D subsurface structures we apply large-scale 3D elastic and acoustic FWT, which require extensive optimization of runtime and performance. For an improvement of the gradient based optimization method we apply the inverse of a diagonal Hessian approximation for preconditioning of the gradients. However, its calculation is computationally expensive, as it requires one additional forward simulation for each receiver of the underlying seismic acquisition geometry. Therefore, we calculate it only once for an iteration subset of each stage of the multi-stage workflow considering several frequency ranges. The performance is shown for a simple transmission geometry application. Source and receiver artifacts were removed sufficiently and the inversion was successfully performed. The second application shows the effects of the number of sources – correlating with the number of simulations and, thus, mainly affecting the computational efforts – on the reconstruction of a 3D acoustic model in reflection geometry. To allow a successful inversion of the 3D structures and to avoid artifacts due to spatial aliasing, a reasonable number of sources is required. This essential amount of sources depends on the choice of seismic frequencies. Thus, we recommend to reduce simulations at early FWT stages (low frequencies) and to increase the number of sources with increasing frequencies.

1 Introduction

Living in a time where natural resources are scarce and precious, the number of underground constructions is increasing, and the storage of waste and other material in the earth is necessary, it is important to find accurate ways of mapping geological

---

A. Kurzmann (✉) • S. Butzer • T. Bohlen
Karlsruhe Institute of Technology, Karlsruhe, Germany
e-mail: andre.kurzmann@kit.edu

W.E. Nagel et al. (eds.), *High Performance Computing in Science and Engineering '14*, DOI 10.1007/978-3-319-10810-0_7

structures in the Earth's interior. A considerable amount of money and effort has been spent on the field of reflection seismology, the science of collecting echoes and transforming them into images of the subsurface. Conventional seismic imaging methods applied in reflection seismology utilize a small portion of the information of the echoes we obtain from the subsurface. Most methods analyze the arrival time of the echoes or specific signal amplitudes only.

We further develop full waveform tomography (FWT) – a new seismic inversion and imaging technique that uses the full information content of the seismic recordings. Each echo or reflection from geological discontinuities in the earth is used to unscrample the earth structure. It iteratively retrieves multi-parameter models of the subsurface by solving the full wave equations. It allows for a mapping of structures on spatial scales down to less than the seismic wavelength, hence providing a tremendous improvement of resolution compared to traveltime tomography based on ray-theory.

In this work, we investigate strategies to improve the performance of our 3D elastic and 3D acoustic FWT implementations. That is, in case of the elastic FWT we focus on the development of efficient preconditioning techniques stabilizing the optimization method and allowing the reconstruction of meaningful subsurface models from seismic data. Furthermore, the acoustic FWT is utilized to balance computational efforts and accuracy of the recovered subsurface model. In other words, we estimate a measure which defines the amount of computational efforts to obtain reasonable subsurface models.

2 Method and Implementation

Full waveform tomography aims to find the optimal subsurface model by iteratively minimizing the misfit between observed and modeled data. The implementation is based on the conjugate gradient method as described by [8, 14].

2.1 Seismic Modeling

Seismic modeling is the fundamental part of full waveform tomography and requires nearly all the computation time. In dependence of the field of application, the wave-propagation physics for an underlying subsurface model has to be described by an appropriate wave equation. On the one hand, this comprises the acoustic or elastic wave equations. On the other hand, they have to be solved for two-dimensional or three-dimensional subsurface models (corresponding applications are referred to as 2D or 3D FWT). The numerical implementation of the wave equations consists of a time-domain finite-difference (FD) time-stepping method in cartesian coordinates. In the elastic case the FD-scheme solves the velocity-stress formulation by utilizing particle velocities and stress components of the wavefield. However, the acoustic approximation is limited to the pressure formulation with only one wavefield component. Due to finite model sizes, the wave equations are expanded by perfectly matched layer terms (PML) to avoid artificial boundary reflections.

2.2 Full Waveform Tomography

The solution of the inverse problem comprises several steps shown in Fig. 1. The method is initialized by the choice of a 2D or 3D initial parameter model. Seismic velocities or mass density are assigned to the model m_0 at the first iteration.

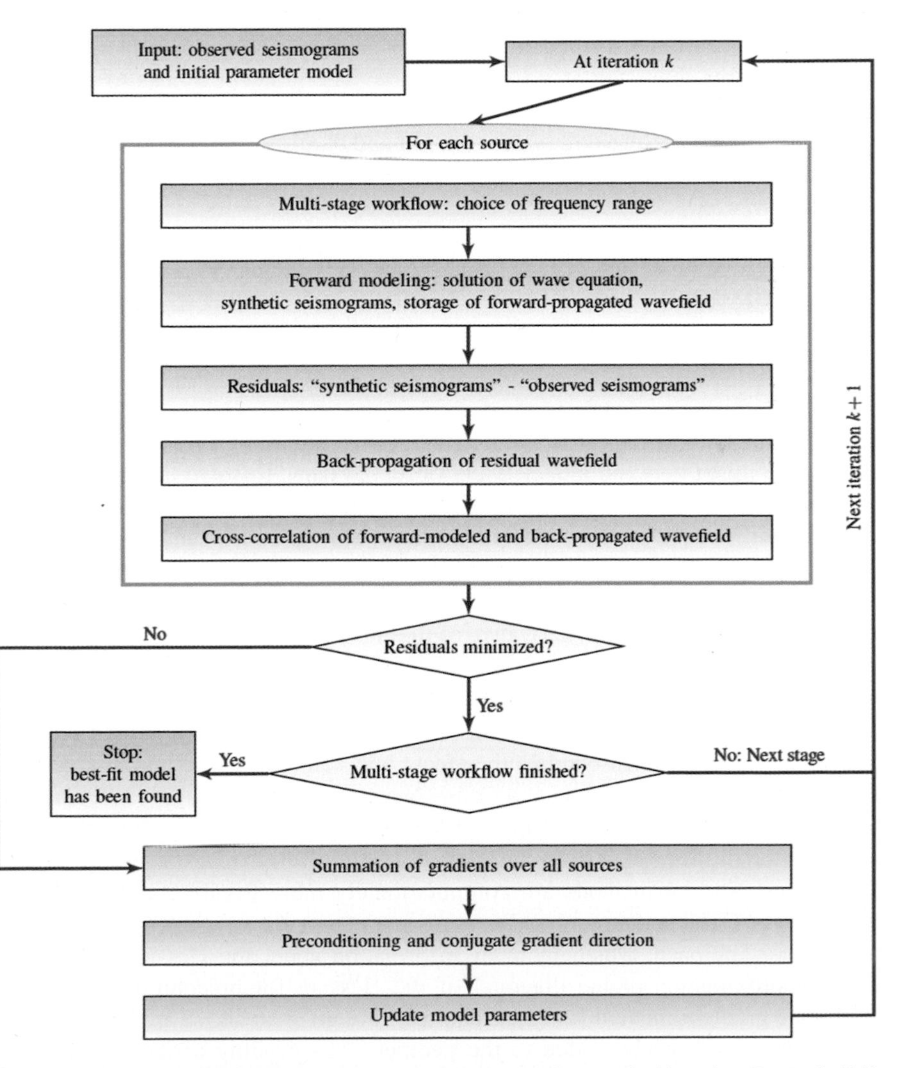

Fig. 1 General time-domain FWT scheme as an iterative gradient method based on Tarantola [14]

The initial model can be either assumed using a priori information or computed by conventional imaging methods. For each source of the acquisition geometry seismic modeling is applied, i.e., the wavefield is emitted by the source and forward-propagates across the medium. This wavefield has to be stored in memory with respect to the whole volume and propagation time. Synthetic seismic data is recorded at the receivers and the difference of observed and synthetic data is calculated – resulting in residuals. For each source the residual wavefield is back-propagated from the receivers to the source position. The cross-correlation of forward- and back-propagated wavefields yields shot-specific steepest-descent gradients δm_n. The computation of the final gradient is given by the summation of all shot-specific gradients. The result is the global gradient of the entire acquisition geometry. An optimized gradient $\widetilde{\delta m}_n$ is computed by subsequent preconditioning and application of the method of conjugate gradient. The update of the model parameter is the final step of a FWT iteration. The gradient d_n has to be scaled by an optimal step length α to get a proper model update at iteration step n. The estimation of α is performed at each iteration and requires additional modelings.

3 3D Elastic FWT: Applying a Diagonal Hessian Approximation for Preconditioning

3.1 Motivation

The FWT aims to resolve structures of the subsurface in high resolution by minimizing the misfit between modeled and observed data. To solve this optimization problem, we use the conjugate gradient approach [8, 14], which uses the gradient of the misfit function to approach its minimum. This approach can be implemented very efficiently with the adjoint method [8] and is thus realizable for large model and dataset applications. Another class of optimization methods, the Newton methods, take into account the second derivative of the misfit function, the so-called Hessian matrix. The use of the Hessian can significantly improve the performance of FWT, by speeding up convergence and improving resolution. The Hessian matrix can account for geometrical amplitude effects in the gradient due to source receiver coverage and for limited-bandwidth effects and can thus focus and sharpen the image [2, 10]. The full-Newton and Gauss-Newton methods explicitly calculate the Hessian or in the latter case the approximate Hessian. However, these methods are computationally very expensive and thus not attractive for realistic problem sizes [10]. Quasi-Newton methods are computationally more feasible, as they do not calculate the Hessian matrix directly.

Another approach, which includes information about the Hessian, is the use of an approximation of the diagonal of the Hessian for preconditioning in the conjugate gradient method [2, 12]. Generally, the gradient shows high amplitudes near sources and receivers due to the geometrical spreading of the forward and adjoint wavefields. Thus, a thorough preconditioning of the gradients is required for a successful inversion. This can be done by Gaussian tapering around sources and receivers. Still, a more physical and sophisticated approach is the use of the diagonal of the Hessian, which can correctly account for the geometrical amplitude effects in the gradient.

3.2 Theory and Implementation

3.2.1 Newton and Conjugate Gradient Methods

FWT aims to minimize the misfit beween observed and modeled data. We use the L_2-norm based misfit function E given as

$$E = \frac{1}{2} \sum_{sources} \int dt \sum_{receivers} \delta u_i(\mathbf{x}_s, \mathbf{x}_r, t) \delta u_i(\mathbf{x}_s, \mathbf{x}_r, t) \tag{1}$$

with the i-th component of the displacement residual $\delta u_i = u_i - u_{i,obs}$ at source position $\mathbf{x}_s$ and receiver position $\mathbf{x}_r$. Different optimization approaches can be used for minimizing the misfit and a good discussion about different Newton methods and gradient methods can be found in [10]. We will give a short overview here.

A Taylor expansion of the misfit function E around the model parameters $\mathbf{m} = (m_1, \ldots, m_n)^T$ up to second order leads to:

$$E(\mathbf{m} + \delta\mathbf{m}) = E(\mathbf{m}) + \delta\mathbf{m}^T \nabla_m E(\mathbf{m}) + \frac{1}{2} \delta\mathbf{m}^T \mathbf{H} \delta\mathbf{m} + O(|\delta\mathbf{m}|^3). \tag{2}$$

Hereby $\mathbf{H}$ is defined as the second derivative of the misfit with respect to the model parameters, i.e.,

$$H_{ij} = \frac{\partial^2 E(\mathbf{m})}{\partial m_i \partial m_j} \quad (i = 1, \ldots n) \;\; (j = 1, \ldots n). \tag{3}$$

The index n is the total number of model parameters. To find the minimum of the misfit function the deviation of Eq. 2 with respect to the model perturbation $\delta\mathbf{m}$ is set zero. This gives us the following model update:

$$\delta\mathbf{m} = -\mathbf{H}^{-1} \nabla_m E. \tag{4}$$

The Newton method uses this model update for an iterative solution and updates the model in iteration $k + 1$ with

$$\mathbf{m}_{k+1} = \mathbf{m}_k - \mathbf{H}_k^{-1} \nabla_m E_k. \tag{5}$$

The use of the Hessian leads to a good convergence. However, the calculation and inversion of the large $n \times n$ Hessian matrix is highly expensive. The gradient method thus uses the following simplified update:

$$\mathbf{m}_{k+1} = \mathbf{m}_k - \alpha_k \mathbf{P} \nabla_m E_k. \tag{6}$$

In this case, the model is updated in gradient direction using an appropriate step length α. In this case, the gradient is not scaled and preconditioned by the inverse Hessian. Consequently, a preconditioning operator $\mathbf{P}$ and a the estimation of α are required for the inversion to succeed. The corresponding convergence is lower. A slight improvement in convergence of the gradient method is gained by using the conjugate gradient $\mathbf{c}$ as search direction [8], which is given by

$$\mathbf{c}_k = \mathbf{P} \nabla_m E_k + \beta_k \mathbf{c}_{k-1}, \tag{7}$$

where the scalar β is calculated as given by [8]. This gives as the following model update:

$$\mathbf{m}_{k+1} = \mathbf{m}_k - \alpha_k \mathbf{c}_k. \tag{8}$$

The preconditioning operator used in the gradient method is the replace of the Hessian operator. Thus, it is natural to use some approximation of the Hessian for preconditioning, which can lead to improvements in the gradient method.

3.2.2 Gradient Calculation

We use the adjoint-state method to calculate the gradients [8–10, 14]. Hereby we use a time-frequency approach [13], where the forward-propagation is done in time domain and the gradients are calculated for few discrete frequencies in frequency domain. To transform the wavefields from time to frequency domain, a discrete Fourier transform is performed on the fly. A detailed introduction of our implementation can be found in [3]. For forward modeling, we use the 3D viscoelastic finite-difference code (SOFI), which is based on a velocity-stress formulation [1]. The following steps are performed to calculate the gradient direction in each iteration k:

1. Forward-propagation of source wavefield across the medium
2. Calculate residual between modeled and observed seismograms
3. Back-propagation of residual wavefield from receivers across the medium with time-reversed residuals acting as source-time function
   in step (1) and (3): discrete Fourier transformation on the fly steps
   (1)–(3): these steps are performed for each source

4. Gradient ($\nabla_m E_k$) calculated as multiplication of forward wavefield and conjugate back-propagated wavefield in frequency domain and summed up over all frequencies and sources

Afterwards, a preconditioning operator is applied to the gradients and the conjugate gradient direction $\mathbf{c}_k$ (Eq. 7) is calculated. To estimate an optimal step length α_k we use the misfit value of zero step length and of two additional test step lengths calculated for a subset of shots. The model can then be updated according to Eq. 8. Most of the computational time in FWT is spent for wavefield modelings. In the gradient method the number of forward modelings is 2 $\times$ (number of shots) + 2 $\times$ (number shots step length calculation).

3.2.3 Calculation of the Diagonal Hessian Approximation

In this section, we will introduce how the diagonal Hessian approximation $\mathbf{H}_D$, that we use for preconditioning, is calculated. Detailed discussions about the calculation of the Hessian matrix can be found in [10, 11]. The Hessian (Eq. 3) can be calculated as

$$\mathbf{H} = \mathrm{Re}(\mathbf{J}^t \mathbf{J}^*) + \mathbf{R}. \tag{9}$$

The second term R is generally small [10], and we only use the first term, known as the approximate Hessian. $\mathbf{J}$ is the Jacobian matrix, which is defined as

$$J_{ij} = \frac{\partial u_i}{\partial m_j} \qquad i = 1, 2, \ldots, w, \quad j = 1, 2, \ldots, n. \tag{10}$$

The index i runs over all wavefield parameters w and the index j runs over all model parameters n. For preconditioning the calculation of $\mathbf{H}$ is restrained to the diagonal elements of the approximate Hessian, with

$$H_{jj} = \sum_i \frac{\partial u_i}{\partial m_j} \frac{\partial u_i^*}{\partial m_j}. \tag{11}$$

The full Hessian is very large with $n \times n$ elements, whereas the diagonal Hessian only consists of n elements. The Jacobian matrix is not explicitely calculated in the gradient method and additional computations are required to calculate it for the diagonal Hessian. The following steps describe, how the Jacobian matrices are constructed and how the diagonal Hessian approximation $\mathbf{H}_D$ is calculated.

1. Forward-propagation from each source into medium; this is already done for gradient computation
2. Back-propagation of delta functions from each receiver into media to find the Green's receiver functions
   in steps (1) and (2): discrete Fourier transforms on the fly

3. Calculation of Jacobian matrices for each source-receiver combination by multiplication of forward wavefield and conjugate receiver Green's functions in frequency domain
4. Calculate diagonal Hessian approximation $\mathbf{H}_D$ as multipliaction of the complex Jacobian matrices with their conjugate, summed up over all frequencies and source-receiver combinations

The calculation of Jacobian matrices for each source-receiver combination is required. Hence, either the Green's receiver functions or the forward-propagated wavefields need to be stored. In our implementation, the Hessian calculation is performed in frequency domain, which means, that these wavefields are stored only for few discrete frequencies.

In the gradient calculation, the back-propagated wavefield is generated at all receivers simultanously. For the calculation of $\mathbf{H}_D$ we need the Green's receiver functions. Thus, one forward-propagation for each receiver is performed additionally to the gradient calculations. Depending on the number of receivers, this can be very time consuming. To compute the full diagonal approximate Hessian, the Green's receiver functions are calculated for each spatial direction, which requires $3\times$ (number receivers) modelings in step 2. We use only the component which dominates the forward wavefield.

3.2.4 Application of Hessian Preconditioning

After calculation of $\mathbf{H}_D$, as described in the last section, we use this operator to precondition the gradients. The preconditioning operator is then given by:

$$\mathbf{P} = (\mathbf{H}_D + \epsilon \mathbf{I})^{-1}. \tag{12}$$

$\mathbf{I}$ is the identity matrix. Hereby a water level ϵ is added to $\mathbf{H}_D$ to stabilize the inversion. At the moment, we estimate this water level empirically. The inversion of $\mathbf{H}_D$ is straightforward, because we use only the diagonal part of the Hessian. The Hessian is calculated only once for each frequency stage, because its calculation might become quite expensive. Then the same $\mathbf{P}$ is applied within this frequency stage.

3.3 Example: Box in Transmission Geometry

3.3.1 Model and Inversion Setup

Even though the implementation of the Hessian preconditioning is of main interest for its application to FWT of complex models, we choose a relatively simple test to show its effects on the gradients and to prove its performance in the inversion. We

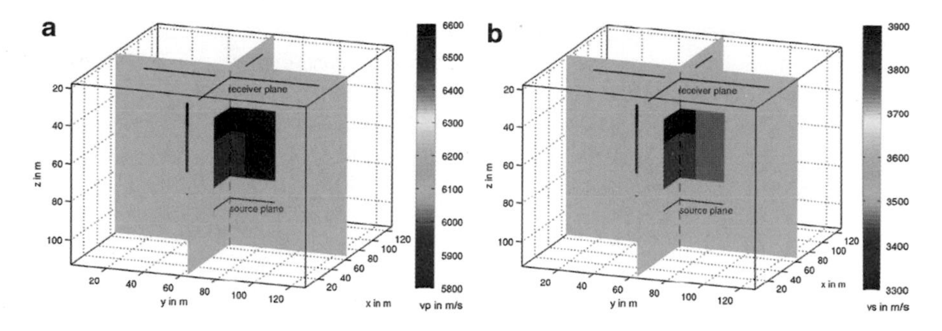

Fig. 2 Real box model of v_p (**a**) and v_s (**b**) with indication of source and receiver plane

consider a transmission geometry test of a box model with the size of $160 \times 160 \times 184$ grid points corresponding to $128 \times 128 \times 147.2$ m in x-, y- and z-direction. The true models for the seismic velocities are shown in Fig. 2, the density model is chosen constant and known with 2,800 kg/m^3. The seismic velocity models contain a box, which is divided into four differently-sized parts with different positive and negative velocity variations. The v_p/v_s ratio is not constant. The model does not contain a free surface. As a starting model, the homogeneous background velocities outside the box are used. Sources and receivers are arranged within x–y-planes, as indicated in Fig. 2. We use $12(3 \times 4)$ sources in 88 m depth and $169(13 \times 13)$ receivers in 24 m depth. The sources are vertically directed point forces with $\sin^3$-wavelets as source time functions and a dominant frequency of 200 Hz.

In total, we performed 75 iterations, divided into 4 different frequency stages ranging from 160 to 290 Hz. In each stage five discrete frequencies in 10 Hz intervals were used for inversion and frequencies increased from stage to stage.

For the general gradient method 30 forward modelings are calculated within each iteration. One hundred and sixty-nine additional forward modelings are required for the calculation of the Green's receiver functions, needed for the diagonal Hessian preconditioning matrix $\mathbf{H}_D$. This matrix was calculated only once for each frequency stage and used for the whole stage. Thus, for the full inversion 667 forward modelings are added to the 2,250 forward modelings of the general gradient approach, which is an increase of about 30 % of runtime in this example.

3.3.2 Hessian Preconditioning

Figure 3 shows the effects of the diagonal Hessian preconditioning on the gradients of v_p and v_s for the first iteration, and thus for the first frequency stage. The gradients before preconditioning, normalized to their maximum are shown in Fig. 3a for v_p and (b) for v_s. The high amplitudes around sources and receivers are clearly visible. Without preconditioning, the model update is only significant within these areas, and the inversion fails. Figure 3c, d shows the logarithm of the normalized diagonal Hessian approximation $\mathbf{H}_D$ for v_p and v_s. The Hessian matrix covers

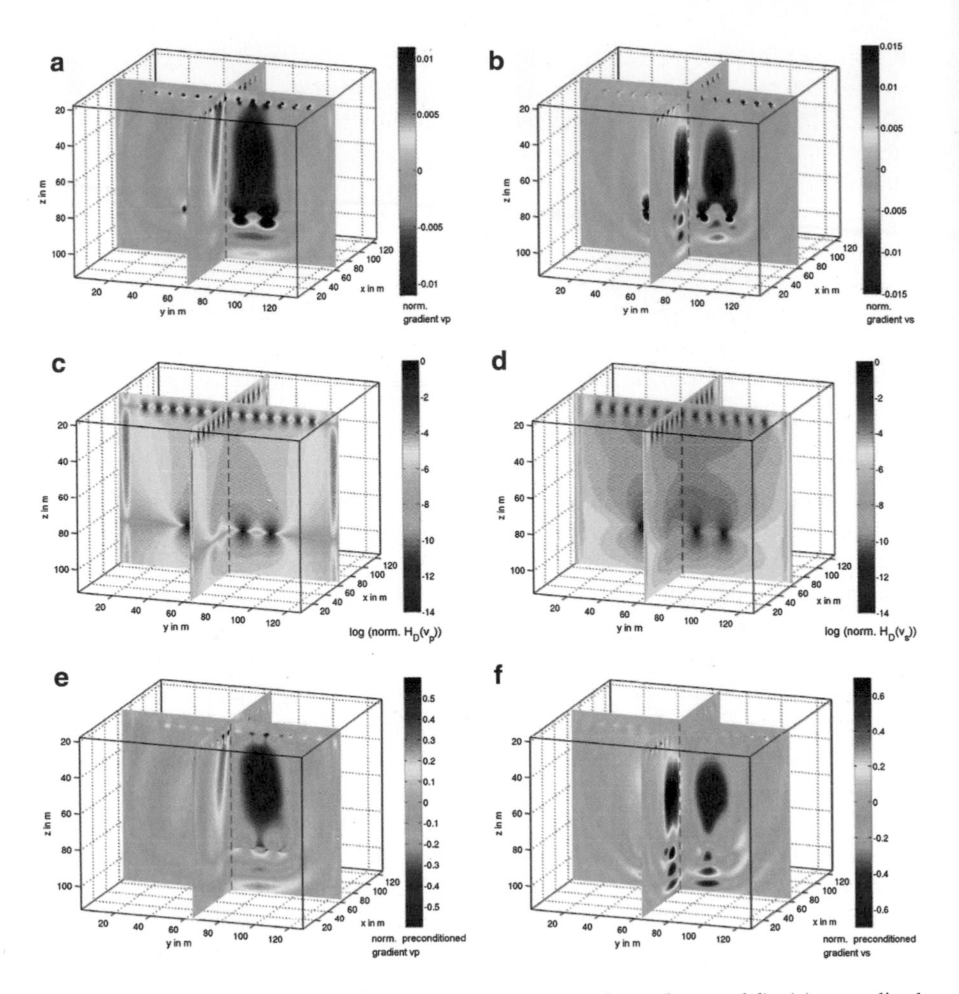

Fig. 3 Effects of Hessian preconditioning on v_p and v_s gradients (box model): (**a**) normalized gradient v_p, (**b**) normalized gradient v_s, (**c**) logarithm of normalized $\mathbf{H}_D(v_p)$, (**d**) logarithm of normalized $\mathbf{H}_D(v_s)$, (**e**) normalized preconditioned gradient v_p and (**f**) normalized preconditioned gradient v_s

several orders of magnitude and, like the gradient, it shows extremly high values at source and receiver positions. The influence of the geometric amplitude decay of the wavefield is clearly visible. Areas with no or very low wavefield coverage show very low values. This is for example visible in the blue areas of the v_p Hessian. The application of the inverse Hessian in such areas would thus lead to an enormous enhancement of the gradient, even though we have no or very little information in our data. To avoid this, the water level is added to the Hessian, which, at the moment, we determine manually. $\mathbf{H}_D$ is used as preconditioner according to Eq. 12 and applied to the gradient. The normalized preconditioned gradient is shown in Fig. 3e, f for v_p and v_s, respectively. The high amplitudes at the sources are corrected. Most of the high amplitudes around the receivers also vanished, however, some receiver artifacts are still visible. The reason for this is probably the approximation we do by calculating only Green's receiver functions from delta functions applied as vertical forces. Hence, some additional damping at receiver positions might be required. The highest amplitude in the gradient now concentrates on the box area. Some effects of the preconditiong are also visible within the box area, where amplitudes of the preconditioned gradient are higher in the middle part of the box. However, for transmission geometry, these effects are low. Unfortunately, some artifact below the source plane are increased after preconditioning in the gradient of v_s. However, the inversion results show, that these artifacts in the gradients have no significant impact on the final inversion result.

3.3.3 Inversion Results

Figure 4 shows the data fit of observed, starting and final inverted data for some representative traces for the vertical component. The data is lowpass-filtered with a corner frequency of 290 Hz, which is the maximum frequency used for the inversion. It is visible, that the seismograms of the homogeneous starting model are already relatively close to the data. The final inverted data and the observed data show a nearly perfect fit. The final inverted models of v_p and v_s are shown for a horizontal slice in Fig. 5. For comparison, the real models are plotted. Overall, the box could be successfully reconstructed by the FWT. The final inverted models in Fig. 5b, d the three sub-boxes are successfully resolved. Due to the smaller wavelengths of the shear wave, the shape of the sub-boxes is clearer in v_s compared to v_p.

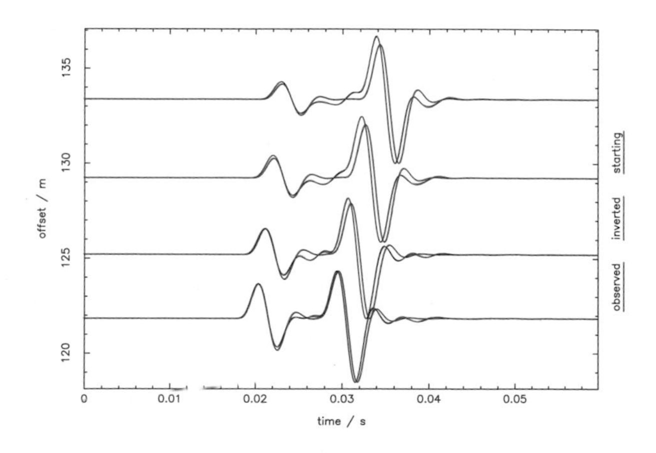

Fig. 4 Seismograms of vertical component for observed, starting and final inverted data for few representative traces

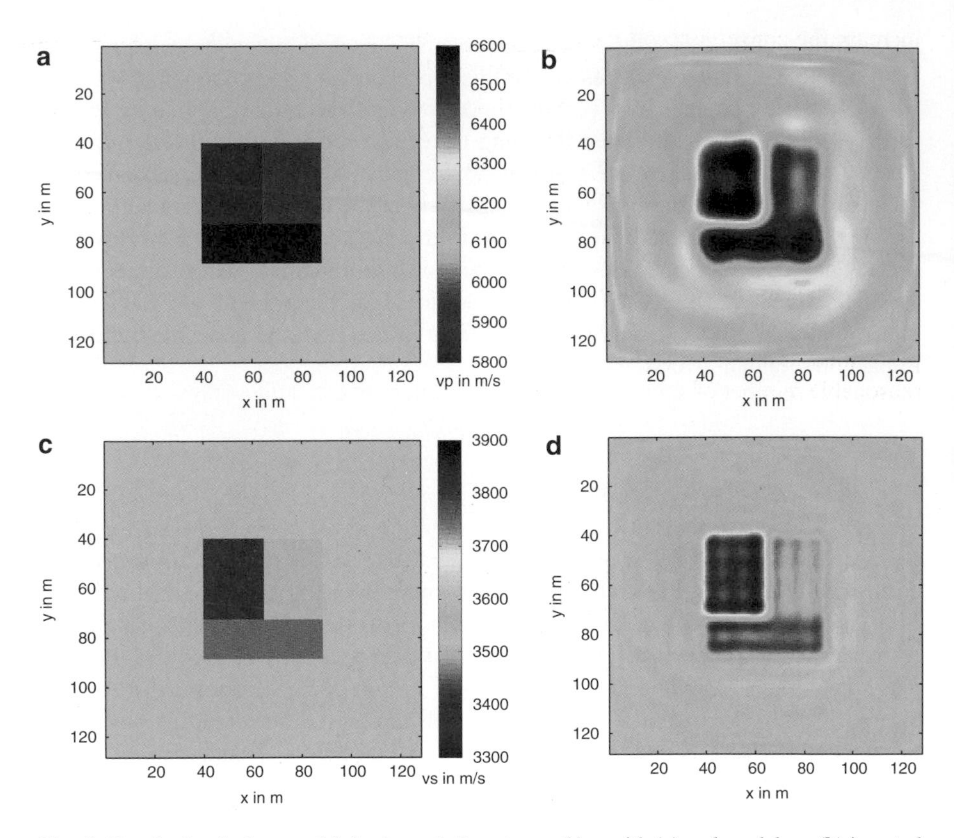

Fig. 5 Results for the box model: horizontal slice at $z = 64$ m with (**a**) real model v_p, (**b**) inverted model v_p, (**c**) real model v_s and (**d**) inverted model v_s

4 3D Acoustic FWT: Requirements and Feasibility Study

4.1 Motivation

It is well-known, that the analysis of marine seismic data in presence of complex subsurface structures might require the application of 3D imaging techniques, such as 3D full waveform tomography (FWT). Using the example of our acoustic time-domain implementation this work is an overview of computational efforts of 3D FWT. Although this implementation is limited to the description of the acoustic wavefield, the computational times are significantly high, e.g., between two and three orders of magnitude higher compared to a similar 2D acoustic FWT. In order to allow a reasonable performance of more accurate seismic modeling, such as the consideration of attenuation which multiplies the efforts by a factor of at least two, the exploitation of methodological and numerical strategies is indispensable. An automatic step-length estimation has proved to be a very efficient technique to increase the convergence of the conjugate gradient method (e.g., [4, 5]). On the one hand, it requires expensive seismic modeling, but, on the other hand, the number of necessary iterations is reduced tremendously. In addition, a massively parallelized forward modeling – involving the Message Passing Interface (MPI) – combines domain decomposition [1] and simultaneous simulations of wavefield propagations (cp. Sect. 2) [4, 6]. It minimizes the dependency of hardware bottlenecks and maximizes the exploitation of computational resources.

However, most of the computation time is consumed by forward modeling. Therefore, apart from the previous optimization strategies, a numerical study using a 3D expansion of the Marmousi model and different acquisition geometries was performed. That is, this investigation covers the crucial point of choosing an reasonable number of sources to allow the recovery of a meaningful velocity model.

4.2 3D Marmousi Model and Acquisition Geometry

The model used in this FWT experiment is a 3D expansion of the 2D Marmousi model (Fig. 6) described by [7, 15] with a physical size of $9.6 \times 4.8 \times 3.52$ km. The initial model consists of a simple 1D gradient assumption and a known water layer. The acquisition geometry is located at the sea surface and comprises 9 receiver lines (green lines) as well as between 4 and 60 sources (red dots). In total, 1,002 receivers were used. The recording length of seismic data amounts to 6 s.

4.3 Results of 3D Marmousi Experiment

We estimated a measure to distinguish reasonable and artificial results caused by spatial aliasing and, thus, an insufficient illumination. It is given by the number of dominant wavelengths per source spacing. The results of the feasibility study are composed of exemplary cross-sections (Fig. 7) of the 3D model, which are located at y = 2.4 km (compare x–z-slice shown in the 3D model). The columns of the figure array represent the intermediate results within a multi-stage FWT workflow

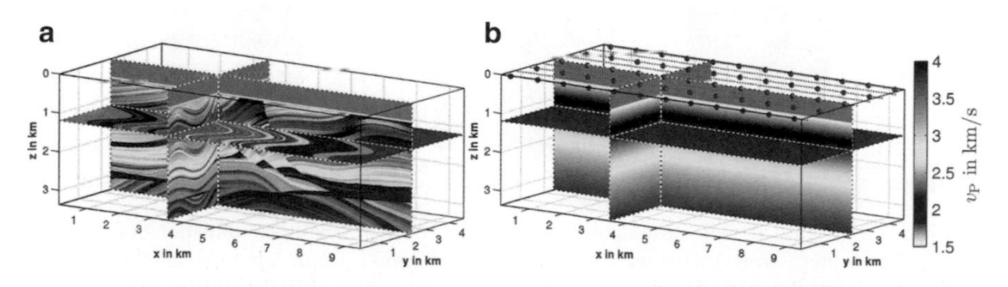

Fig. 6 3D Marmousi model: (**a**) true velocity model, (**b**) initial model for FWT and acquisition geometry with receivers (*green*) and sources (*red*, maximum number of 60 used in this study)

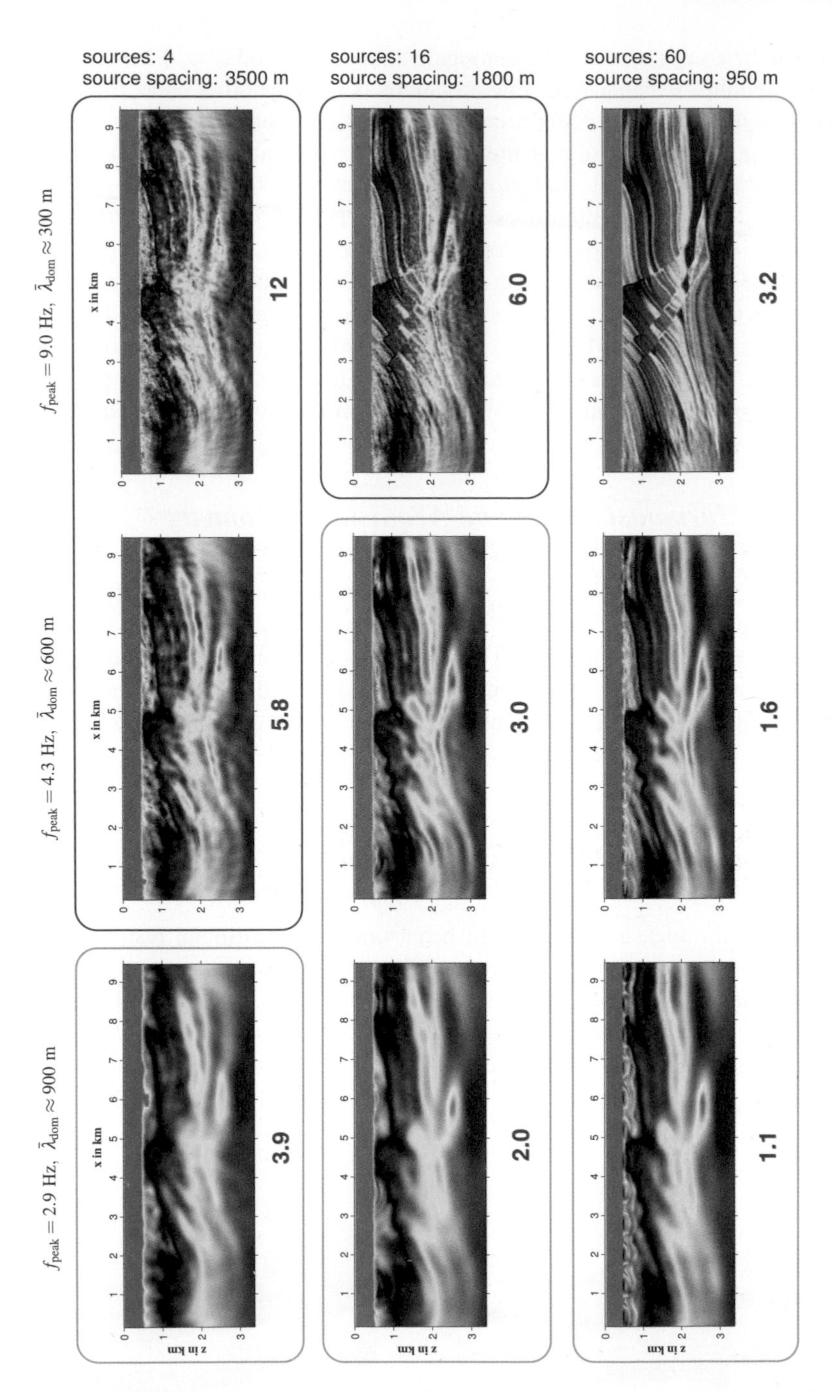

Fig. 7 3D Marmousi model: Results of 3D feasibility study involving combinations of different numbers of sources (*rows*) and frequency contents (*columns*). From *left* to *right*, the columns represent the consideration of increasing frequency content within a FWT workflow (final result in right column). The ratio of average source spacing and dominant wavelength is given by the *blue number*. *Green*/*red boxes* highlight the reconstruction of reasonable/artificial velocity models

considering several frequency contents of the recorded data (peak frequency) and corresponding dominant wavelengths. The rows belong to the FWT experiments using a different number of sources. The accuracy measure mentioned above is highlighted in blue color. Reasonable results with respect to a certain frequency content are highlighted by green boxes, while red boxes emphasize the occurrence of spatial aliasing and artificial model reconstructions. Yellow boxes indicate problems in the determination of the critical accuracy measure. Here, we used the average wavelengths of the adjacent stages of frequency filtering.

On average, a critical value between three and four dominant wavelengths $\bar{\lambda}_{\text{dom}}$ per source spacing d_s should not be exceeded. For example, regarding the experiment with 60 sources, an according average source spacing of 950 m and a dominant wavelength of 300 m, $d_s/\bar{\lambda}_{\text{dom}}$ amounts to 3.2. In contrast, significant artifacts can be observed in results of experiments with 4 sources ($d_s/\bar{\lambda}_{\text{dom}} = 12$) or 16 sources ($d_s/\bar{\lambda}_{\text{dom}} = 6$) considering the inversion of the full frequency content. In case of 4 sources, a reliable v_P model can be obtained at the inversion of frequencies up to 2.9 Hz. The involvement of higher frequencies leads to spatial aliasing. In contrast, the usage of 60 sources allows a satisfactory inversion of the full frequency content. In conclusion, in case of this 3D experiment, the inversion for several frequency contents might be connected to a variable amount of sources. This would lead to a significant reduction of computational efforts of the 3D time-domain FWT.

5 Summary

3D FWT aims to reconstruct 3D subsurface structures in high resolution. Still it is computationally expensive and needs to be optimized regarding runtime and performance.

Table 1 Computational efforts of exemplary 3D FWTs on the supercomputer HERMIT

	3D acoustic	3D elastic
Discretization and efforts of finite-difference modeling		
No. of grid points	$162 \cdot 10^6$	$4.7 \cdot 10^6$
Grid size in grid points	$960 \times 480 \times 352$	$160 \times 160 \times 184$
No. of time steps	4,600	1,200
No. of sources	16	12
Total no. of simulations	8,480	2,917
Resource consumption and computational performance		
No. of cores	4,096	512
Overall memory usage	5.4 TB	100 GB
No. of FWT iterations	106	75
Total computation time	15.8 h	14 h
Total core hours	65,000	7,168

To optimize the 3D elastic FWT performance, we apply a preconditioning operator to the gradients, which is based on the inverse diagonal Hessian approximation. The calculation of this preconditioning matrix is performed only once for each frequency stage and requires one additional forward simulation for each receiver. We showed its application for a box model in transmission geometry. High amplitudes at sources and receivers could be well corrected and a successfull inversion was performed. In future tests, we will test the performance of Hessian preconditioning for surface acquisition geometries. Additionally, we will implement and test the performance of the L-BFGS method which, combined with Hessian preconditioning, should lead to an even better performance.

Different numbers of sources were tested for the 3D acoustic FWT of the 3D Marmousi model. In order to allow a satisfactory reconstruction of the P-wave velocity model, i.e., to avoid an artificial recovery due to spatial aliasing, a reasonable number of sources was found. That is, the average source spacing should not exceed a value between three and four dominant wavelengths. Consequently, the feasibility of 3D FWT benefits from a combination of methodological improvements and an adequate choice of acquisition geometry.

Regarding computational efforts, that is, memory consumption and computation time (Table 1), our examples of 3D elastic and 3D acoustic FWT represent applications at a manageable size. More realistic problems could easily exceed resources available on current computers.

References

1. Bohlen, T.: Parallel 3-D viscoelastic finite difference seismic modeling. Comput. Geosci. **28**, 887–899 (2002)
2. Brossier, R., Operto, S., Virieux, J.: Seismic imaging of complex onshore structures by 2D elastic frequency-domain full-waveform inversion. Geophysics **74**(6), WCC105–WCC118 (2009)
3. Butzer, S., Kurzmann, A., Bohlen, T.: 3D elastic full-waveform inversion of small-scale heterogeneities in transmission geometry. Geophys. Prospect. **61**(6), 1238–1251 (2013)
4. Kurzmann, A.: Applications of 2D and 3D full waveform tomography in acoustic and viscoacoustic complex media. Dissertation, Karlsruhe Institute of Technology (2012)
5. Kurzmann, A., Köhn, D., Bohlen, T.: Comparison of acoustic full waveform tomography in the time- and frequency-domain. In: 70th EAGE Conference and Technical Exhibition, Rome (2008)
6. Kurzmann, A., Köhn, D., Przebindowska, A., Nguyen, N., Bohlen, T.: 2D acoustic full waveform tomography: performance and optimization. In: 71st EAGE Conference and Technical Exhibition, Amsterdam (2009)
7. Martin, G.S., Wiley, R., Marfurt, K.J.: Marmousi2 – an elastic upgrade for Marmousi. Lead. Edge **25**, 156–166 (2006)
8. Mora, P.: Nonlinear two-dimensional elastic inversion of multioffset seismic data. Geophys. **52**, 1211–1228 (1987)
9. Plessix, R.-E.: A review of the adjoint-state method for computing the gradient of a functional with geophysical applications. Geophys. J. Int. **167**, 495–503 (2006)
10. Pratt, R., Chin, C., Hicks, G.: Gauss-Newton and full Newton methods in frequency-space seismic waveform inversion. Geophys. J. Int. **133**, 341–362 (1998)

11. Sheen, D.-H., Tunkay, K., Baag, C.-E., Ortoleva, P.: Time domain Gauss-Newton seismic waveform inversion in elastic media. Geophys. J. Int. **167**, 1373–1384 (2006)
12. Shin, C., Jang, S., Min, D.-J.: Improved amplitude preservation for prestack depth migration by inverse scattering theory. Geophys. Prospect. **49**, 592–606 (2001)
13. Sirgue, L., Etgen, J.T., Albertin, U.: 3D frequency domain waveform inversion using time domain finite difference methods. In: 70th EAGE Conference and Technical Exhibition, Rome (2008)
14. Tarantola, A.: Inversion of seismic reflection data in the acoustic approximation. Geophysics **49**, 1259–1266 (1984)
15. Versteeg, R.: The marmousi experience: velocity model determination on a synthetic complex data set. Lead. Edge **13**, 927–936 (1994)

Nanoflare Heating in the Solar Corona

Sven Bingert and Hardi Peter

Abstract We employ a three dimensional magneto-hydrodynamical (3D MHD) model to investigate the statistics of the spatial and temporal distribution of the coronal heating. The model describes the evolution of the solar corona above an observed Active Region. This model is additionally compared to coronal models where the underlying photospheric magnetic field consists of simplified magnetic configurations. In all models random like photospheric motions braid the magnetic field. This induces currents in the upper atmosphere eventually leading to coronal heating by their dissipation. We analyze the heating process which is in general transient in time and space. We compare the results to observed flare statistics.

1 Introduction

The coronal heating problem is one of the major topics in astrophysics. The question is how the million degree hot plasma in the solar corona is heated and how these high temperatures are maintained. With actual observations it is nowadays common knowledge that the coronal heating process is closely related to the magnetic fields. These fields are generated in the solar interior and expand outwards into the solar atmosphere. Granular motions in the photosphere shuffle around the magnetic field lines. This generates a Poynting flux that transports the energy into the corona in a non-thermal mechanism. The relevant question is then how the magnetic energy is actually dissipated. The term of the magnetic resistivity in the induction equation is responsible for the dissipation. In our magneto-hydrodynamical approach we can only test parametrizations of the resistivity that is in general a function the local particle properties of the gas.

S. Bingert (✉) • H. Peter
Max-Planck-Institut für Sonnensystemforschung, Göttingen, Germany
e-mail: bingert@mps.mpg.de; peter@mps.mpg.de

W.E. Nagel et al. (eds.), *High Performance Computing in Science and Engineering '14*, DOI 10.1007/978-3-319-10810-0_8

2 Coronal Models

We employ different kind of 3D numerical models for the investigation of the coronal heating. They range from very simplified low-resolution models to realistic high-resolution large-scale coronal simulations. This allows us to combine parameter studies with more realistic models. In the most complex model we use the full set of MHD equations for a single fluid. These equations are the conservation of mass, the equation of motion, the energy equation, and the induction equation. To complete we need the equation of state for an ideal gas. To account for the solar coronal environment we have to include important physical processes besides the standard MHD equations for a viscous, resistive, and compressional gas. These are the field-aligned (non-isotropic) Spitzer heat conduction [10] and the optical thin radiative loss [5] of the coronal plasma. These important effects and their different dependencies on temperatures and plasma densities are important to obtain the correct scale of coronal pressure in the simulations as it is at the real sun. The full set of equations and all parameters used in complex model can be found in [1].

2.1 Large Scale High Resolution Model

The biggest coronal model is the high resolution model with the most extensive set of physical parameters. For this numerical experiment we used $512 \times 512 \times 256$ grid points. The underlying magnetogram is based on an actual coronal observation (cf. Fig. 1 left). This magnetogram is the key ingredient for the energy balance and transport. We use observed magnetic fields in order to allow us the comparison of our results with solar observations. By using time dependent observations we not only determine the spatial but also the temporal scales for the energy input into the model.

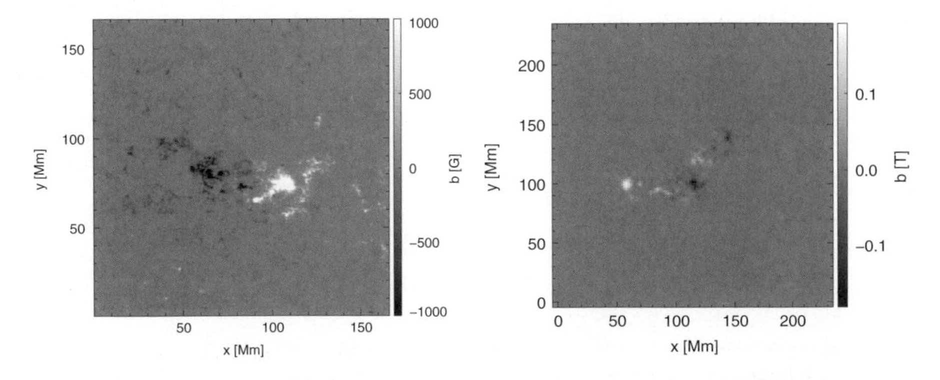

Fig. 1 *Left*: The first snapshot of the time series of the Active Region AR11102 observed by SDO/HMI [9]. *Right*: The first snapshot of the time series of AR11158 also observed by SDO/HMI

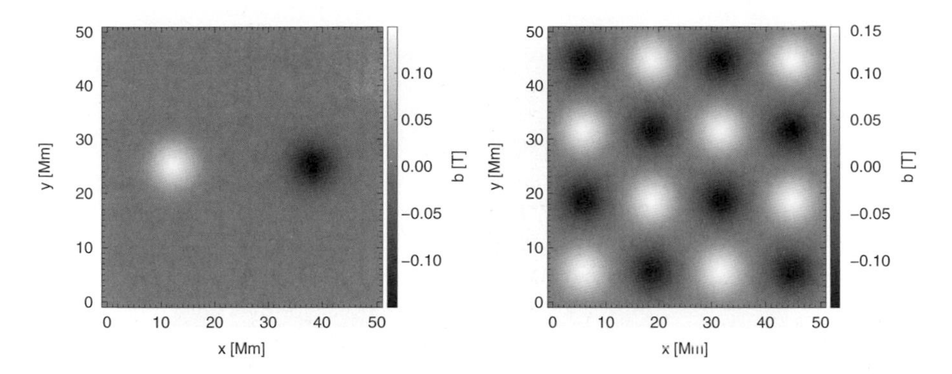

Fig. 2 *Left*: Initial magnetogram for the dipol field. *Right*: Initial magnetogram for the multipol field

2.2 Low Resolution Models

We also run series of smaller simulations consisting of 128^3 grid points. Together with using simplified physical modules the completion could be sped up. Here we list three representative models used for the investigation.

First we conducted a low resolution model covering the same physics as the high resolution model. This model uses the observed Active Region AR11158 (cf. Fig. 1 right). This AR produced a well observed strong X-class flare and is investigate in several publications of different groups (e.g. [4]). The other two models used simplified magnetic lower boundaries as depicted in Fig. 2. In the time dependent model these fields are also shuffled around by granular motions. As the granular motions are suppressed at strong magnetic fields the overall structure does not change much during the evolution of the model.

3 Results

We used the forward model approach to investigate the coronal heating. In this approach a numerical model with certain physical features is conducted. Based on the results of the model synthetic emissions are computed which eventually can be compared to real observations. Therefore a first measure of the model success is the visual comparison to the real Sun. Figure 3 depicts the synthesized emission of an optical thin emission line. The same line is observed by the space based solar telescope SDO/AIA [6]. The figure shows several thin loops which connect the main polarities of the underlying magnetogram. Not only the visual impression fits to the observation but also the absolute emission compares very well. This reflects that the coronal pressure is in the same order of magnitude as on the real sun.

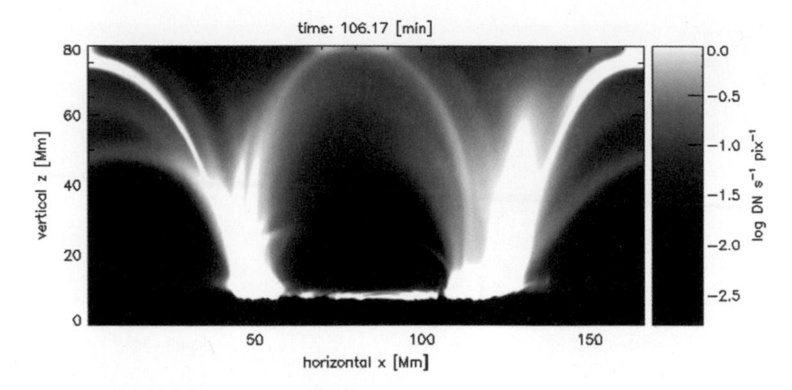

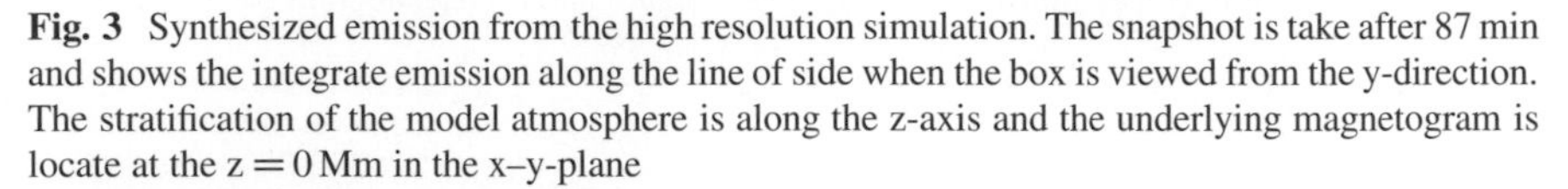
Fig. 3 Synthesized emission from the high resolution simulation. The snapshot is take after 87 min and shows the integrate emission along the line of side when the box is viewed from the y-direction. The stratification of the model atmosphere is along the z-axis and the underlying magnetogram is locate at the $z = 0$ Mm in the x–y-plane

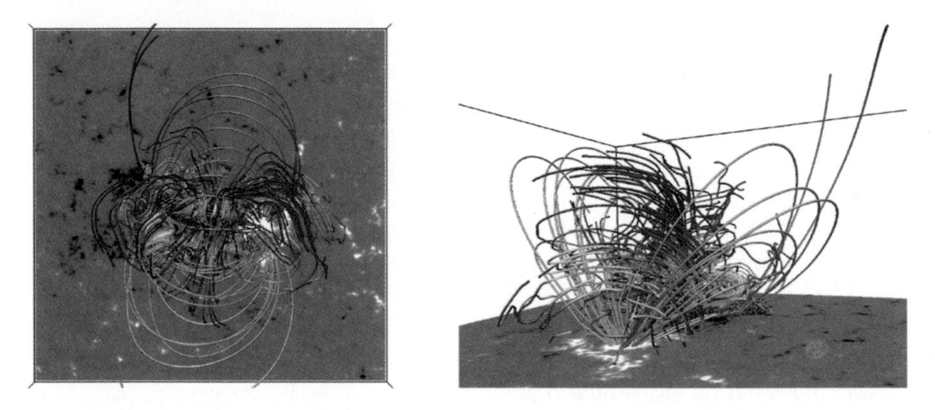

Fig. 4 The figures depict magnetic field lines (*grey*) flow lines of the Poynting flux (color) and the underlying magnetgram (*greyscale*). *Left*: View from *top*. *Right*: View from an arbitrary angle. As the Poynting flux has to be perpendicular to the magnetic field lines the figures show regions where the Poynting flux is arranged in vortices around the magnetic field lines. At these areas the spatial averaged flux does point in the direction of the magnetic field

In order to understand the heating it is also important to investigate the energy flux into the upper atmosphere. The energy is transported via the Poynting flux which is generated at the lower boundary of the numerical domain. Looking at coronal observations the energy seems to be distributed in loop like structures. But the Poynting flux $\mathbf{S} = \mathbf{E} \times \mathbf{B}$, where $\mathbf{E}$ is the electric field and $\mathbf{B}$ is the magnetic field, is perpendicular to the magnetic field. Figure 4 depicts the Poynting flux together with the magnetic field lines. This figure shows that the Poynting flux indeed is directed in the direction of the magnetic field lines, e.g. loops, when averaged over a certain area. The averaging has to cover the vortex like structures of the

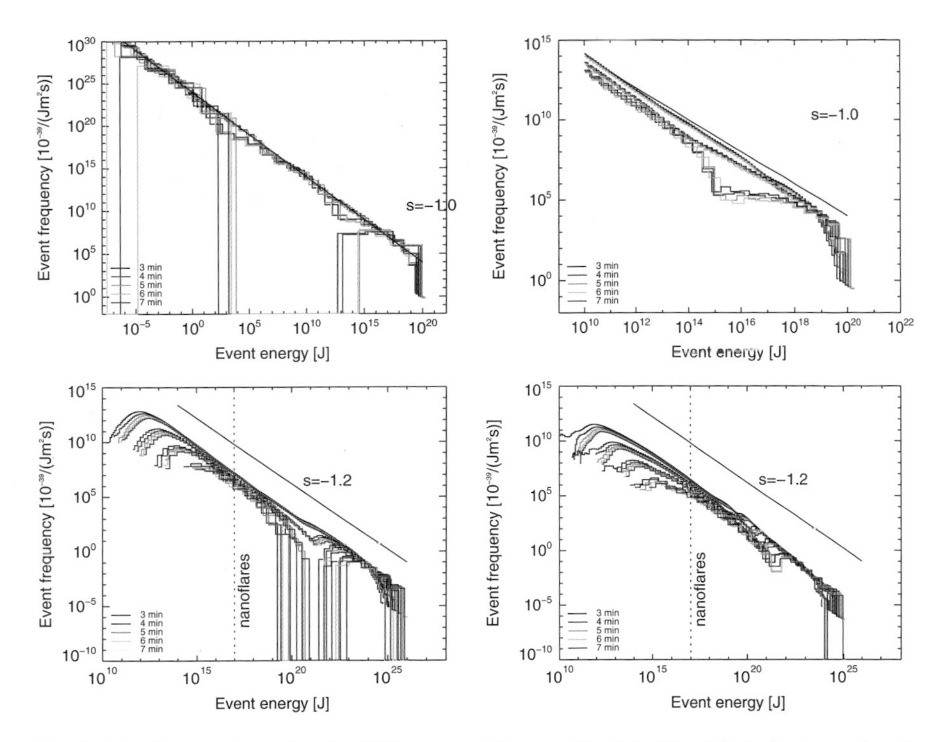

Fig. 5 Nanoflare statistics for the different model types. *Top left*: Simplified physics and only hyper-resistivity. *Top right*: Simplified physics and hyper-resistivity combined with constant background resistivity. *Bottom left*: Full physics and constant resistivity. *Bottom right*: Full physics and a Alfven speed dependent resistivity

Poynting flux. Thus magnetic energy can flow into loop like structures leading to enhancement of energy available to heat the plasma.

The heating process is the dissipation of magnetic energy. This is due to the finite resistivity of the fluid. In our numerical approach the resistivity can be parameterized in different ways. The analysis of the models with full physics showed that the heating is transient in space in time. So we compared these small scale events with real observations of solar flares [2], which are several orders of magnitude larger. But the distribution of the heating events in the model show the same power law behavior as the distribution of solar flares (Fig. 5 bottom left). Even the slope of the power laws are comparable. In this investigation we also found that dominant heating event has a typical energy in the order of a nanoflare [7]. This model used a constant resistivity and the question is how the distribution is altered when changing the parametrization of the resistivity or the underlying energy input, e.g. the structure of the magnetogram. Therefore we conducted two models with simplified magnetic configurations (cf. Fig. 2). Because of well described symmetry they are also used to compare the results with theoretical studies, which will be published. In these models we additionally varied the magnetic resistivity. In Fig. 5

four different kind of resistivity parametrization are shown. The first is a pure hyper diffusion which is mainly used to stabilize the numerical evolution. In our case this diffusion is implemented in a way that it conserves the total energy and is thus physical somewhat meaningful. The hyper diffusion acts only on small scales and helps to resolve the smallest scales of the numerical experiment. The second example used a hyper diffusion combined with constant background resistivity. The third model used a resistivity which scales with the Alfven velocity of the fluid. This type of resistivity is based on theoretical considerations and is highly depending on space and time. The last example shown in Fig. 5 is the model with a constant resistivity only. This was used in many previous experiments and proved to give good results in a numerical stable simulation. But the resistivity has to be large and thus small scales are not properly resolved.

All four experiments produced a heating rate sufficient to heat the solar corona. But the statistical properties of the heating rate are quiet different. The largest difference is between the hyper-diffusion dominate models and the more physical based resistivities. The pure hyper-diffusion model shows a power law with a slope of one (cf. Fig. 5 top left). Also the average slope of the hyper diffusion mixed with constant resistivity (cf. Fig. 5 top right) shows same the power law. This slope is different from the properties of the observed flare statistics. But the energy range is also quiet different so that these statistics could be combined in one picture. This is ongoing work and will be presented in the next publication.

The models with a constant resistivity and the Alfven velocity dependent resistivity show roughly the same slope at the same energy range. This can be explained by the available energy. The energy input of the model only weakly depends on the resistivity parametrization. Thus the amount of energy which has to be dissipated is roughly the same. If the models dissipate this energy in the same energy range the slope has to be roughly the same. A more detailed analysis of the investigation will be presented in the next publication.

With the complex time dependent models of the solar corona one has a large set of data at hand. Thus the possibilities to analyze are diverse. The results discussed above are only a small selection of the ongoing work. We also investigate the dependent behavior of the visible loop structures. Here topics like the constant cross section of the visible loops [8] is still an open question.

4 Numerical Model

We used the *pgi* compiler instead of the *cray* compiler due to problems with the Fortran namelists used in our code. The performance of the code was only minor affected by the choice of the compiler. We conduct a benchmark to find the optimal number of cores to be used in parallel. The code scaled very well until a certain threshold which was then used.

To improve the speed and to optimize the usage of the Cray system we used the *Cray Performance Analysis Tools*. The tools were easy to apply and we found at least one bottleneck of the code when using large number of cores.

4.1 Pencil-Code

Throughout our experiments we used the Pencil-Code [3]. It is a high-order finite-difference code for astrophysical fluid dynamics. It is written in Fortran90 with many helper scripts in C/Perl/Python to easily apply the code on different systems. MPI is used for the parallelization. The code project started 2001 and is still actively evolving. A group of 15 core developers and many users make the code a vivid project. It is a modular code and highly optimized to run on big system with many processes in parallel. The code was tested and run on almost all major computing centers in Europe.

4.2 Technical Model Setup

As described in Sect. 2 we conducted mainly two different kind of models. The high resolution model was most expensive in CPU time. Here we list the key data for this model

- Platform: CRAY XE6 (Hermit)
- Job runtime: 24 h max wall time of the queuing system, several submissions
- Two thousand and twenty-four cores used in parallel
- Number of grid points: $512 \times 512 \times 256 \approx 67$ million
- Number of time steps is about 14 million
- Average time step length: below 1 ms
- Total cpu hours used: eight million

The low resolution models were less CPU time consuming, still the demand was so high that only on large HPC system these models could be conducted. Here we give the list of the key data

- Number of models: six
- Platform: CRAY XE6 (Hermit)
- Job run time: 24 h max wall time of the queuing system, several submissions
- Two-hundred and fifty-six cores used in parallel
- Average time step length: below 3 ms
- Total CPU hours used: one million

Acknowledgements All numerical experiments were conducted at HLRS within the project *Brush* (acid 12870). Special thanks to the HLRS support team.

References

1. Bingert, S., Peter, H.: Intermittent heating in the solar corona employing a 3D MHD model. A&A **530**, A112 (2011). doi:10.1051/0004-6361/201016019
2. Bingert, S., Peter, H.: Nanoflare statistics in an active region 3D MHD coronal model. A&A **550**, A30 (2013). doi:10.1051/0004-6361/201220469
3. Brandenburg, A., Dobler, W.: Hydromagnetic turbulence in computer simulations. Comput. Phys. Commun. **147**, 471–475 (2002)
4. Cheung, M.C.M., DeRosa, M.L.: A method for data-driven simulations of evolving solar active regions. APJ **757**, 147 (2012). doi:10.1088/0004-637X/757/2/147
5. Cook, J.W., Cheng, C.C., Jacobs, V.L., Antiochos, S.K.: Effect of coronal elemental abundances on the radiative loss function. APJ **338**, 1176–1183 (1989). doi:10.1086/167268
6. Lemen, J.R., Title, A.M., Akin, D.J., Boerner, P.F., Chou, C., Drake, J.F., Duncan, D.W., Edwards, C.G., Friedlaender, F.M., Heyman, G.F., Hurlburt, N.E., Katz, N.L., Kushner, G.D., Levay, M., Lindgren, R.W., Mathur, D.P., McFeaters, E.L., Mitchell, S., Rehse, R.A., Schrijver, C.J., Springer, L.A., Stern, R.A., Tarbell, T.D., Wuelser, J.P., Wolfson, C.J., Yanari, C., Bookbinder, J.A., Cheimets, P.N., Caldwell, D., Deluca, E.E., Gates, R., Golub, L., Park, S., Podgorski, W.A., Bush, R.I., Scherrer, P.H., Gummin, M.A., Smith, P., Auker, G., Jerram, P., Pool, P., Soufli, R., Windt, D.L., Beardsley, S., Clapp, M., Lang, J., Waltham, N.: The atmospheric imaging assembly (AIA) on the solar dynamics observatory (SDO). Sol. Phys. **275**, 17–40 (2012). doi:10.1007/s11207-011-9776-8
7. Parker, E.N.: Nanoflares and the solar X-ray corona. APJ **330**, 474–479 (1988). doi:10.1086/166485
8. Peter, H., Bingert, S.: Constant cross section of loops in the solar corona. A&A **548**, A1 (2012). doi:10.1051/0004-6361/201219473
9. Schou, J., Scherrer, P.H., Bush, R.I., Wachter, R., Couvidat, S., Rabello-Soares, M.C., Bogart, R.S., Hoeksema, J.T., Liu, Y., Duvall, T.L., Akin, D.J., Allard, B.A., Miles, J.W., Rairden, R., Shine, R.A., Tarbell, T.D., Title, A.M., Wolfson, C.J., Elmore, D.F., Norton, A.A., Tomczyk, S.: Design and ground calibration of the helioseismic and magnetic imager (HMI) instrument on the solar dynamics observatory (SDO). Sol. Phys. **275**, 229–259 (2012). doi:10.1007/s11207-011-9842-2
10. Spitzer, L.: Physics of Fully Ionized Gases. Interscience Tracts on Physics and Astronomy, vol. 3. Interscience Publishers, New York (1962)

High-Level Support Activities of Simulation Laboratory E&A Particles

G. Poghosyan, S. Sharma, A. Kaur, V. Jindal, P. Bisht, A. Streit, M. Bejger, A. Królak, T. Klaehn, S. Typel, J. Oehlschläger, T. Pierog, and R. Engel

Abstract The Simulation Laboratory Elementary Particle and Astropartice Physics (SimLab E&A Particle) is one of the new support instruments recently established in the Steinbuch Centre for Computing (SCC) and Jülich Supercomputing Centre (JSC) providing high-level support to supercomputer users. Simulation Laboratory (SimLab) is a community-oriented research and support team of computational scientists with a broad spectrum of competencies from computing to domain-specific knowledge. In collaboration with scientific communities the SimLabs team solves the grand challenge computing problems requiring more than the pure usage of standard services made available by HPC, Grid and Cloud infrastructure providers. The access of SimLab E&A Particles to massively parallel system Cray XE6 Hermit at High Performance Computing Centre HLRS in Stuttgart in framework of Project ACID 12863, is used to work on parallelisation, performance analysis of scalability, efficiency and estimation of potential consumption of CPU time for the codes *CORSIKA* (for simulation of cosmic rays/air showers), *AtomicClusters* (evaluation of in-medium properties of nuclear clusters), *PolGrawAllSky* (for searching of

G. Poghosyan (✉) • A. Streit
Steinbuch Centre for Computing, Karlsruhe Institute of Technology, D-76131 Karlsruhe, Germany
e-mail: poghosyan@kit.edu

S. Sharma • A. Kaur • V. Jindal • P. Bisht
Steinbuch Centre for Computing, Karlsruhe Institute of Technology, D-76131 Karlsruhe, Germany

PEC University of Technology, Chandigarh, India

M. Bejger
N. Copernicus Astronomical Center of the Polish Academy of Sciences, Warsaw, Poland

A. Królak
Institute of Mathematics of the Polish Academy of Sciences, Warszawa, Poland

T. Klaehn
Instytut Fizyki Teoretycznej, Uniwersytet Wroclawski, Wrocław, Poland

S. Typel
GSI Helmholtzzentrum für Schwerionenforschung GmbH, Darmstadt, Germany

J. Oehlschläger • T. Pierog • R. Engel
Institute of Nuclear Physics, Karlsruhe Institute of Technology, Karlsruhe, Germany

W.E. Nagel et al. (eds.), *High Performance Computing in Science and Engineering '14*, DOI 10.1007/978-3-319-10810-0_9

gravitational waves signals). The tests and optimisation of selected simulation codes, allowed us to localise the scientific domains showing strong needs and potential for supercomputing, which will help easily constitute the claim of CPU hours when applying for large scale computation time at national and European supercomputer centres.

1 Introduction

Motivated by the problem that applications developed by scientific communities during last decades are lagging behind progress of High Performance Computing (HPC) hardware, which are now at Petaflop scale and will reach approximately in 2021 the Exascale, a new innovative research and support structure, so-called Simulation Laboratories, are currently being established at different sites in Germany and Europe. SimLabs are addressed to the complementary question of software development and productive usage of current and future high-end systems. They are a joint endeavour of supercomputing centres and the scientific and engineering research communities going well beyond the traditional support desk favoured by many contemporary computer centres [3].

SimLabs are aiming to support or even fully take over the part of research activities of scientific groups related to computation in different scientific domains. Currently five SimLabs at JSC and four at SCC are actively engaged with user groups from different communities jointly researching in disciplines like Energy, Plasma Physics, Biology, Molecular Chemistry, Nanophysics, Climate, Environment, Civil Security, Particle- and Astro-Physics. For the last field the SimLab E&A Particles is active in initiating joint research and development (R&D) projects with user groups from homonymous scientific field, through various workshops, informal cooperation and third-party projects. This allows to initiate the join R&Ds with scientific groups to master often initially unplanned *SPORADIC* changes in scientific simulation codes with goal to reach effective, qualitative and quantitative usage of current and future HPC systems. Short as well as long term co-operations were initiated in last years, where personnel for common research activities are partially provided by the SimLabs and the community itself [9].

Under the concept *SPORADIC* changes of simulation codes (Standardisation, Parallelising, Optimization, Release, Adaptation, Data Intensive Computing) we understand:

- Modernisation of old simulation codes, making code object oriented, standardisation of I/O data formats, usage of modern mathematical and scientific libraries – **S**tandardisation
- Implementation of trivial and advanced algorithms for parallel runs of code, speed-up to solve complex problems in acceptable time scale on multi and many core systems- **P**arallelisation
- Performance-analysis, acceleration of simulations reducing usage of resources for same problem – **O**ptimisation

- Simulation code as a software: licensing, documentation, implementing versioning system and providing repository for storage of source code and simulation results – **R**elease
- Porting of codes and access to high performance super- and distributed computing systems, developing algorithms for use of code at HPC, grid and cloud resources – **A**daptation
- **D**ata **I**ntensive **C**omputing
  - Storing and managing huge outputs of simulation and experimental data
  - Visualisation of results, data intensive statistical analysis
  - Development of user friendly work-flow and bookkeeping interfaces, resource orchestration systems

Standardisation is crucial for modernising the standards used in old codes developed in last decades by scientific communities. By separating different objects, functions, subroutines of code, deploying object oriented architecture or setting standards for I/O data formats, one can improve scalability, interoperation of scientific simulation codes. Promoting appropriate standards leads to higher sustainability for scientific simulation codes, e.g. bringing them into form that it could be used as library from newer simulation software. Even if fundamentals of mathematics, physics, biology or chemistry are not changing, use of well temporised and standard numerical libraries when running scientific simulation codes on modern massively parallel supercomputers (like matrix operations, data sorting and searching or solution of linear differential equations) is of crucial importance for reaching higher performance, efficiency and accuracy of calculations.

Many scientific groups are facing challenge of **Parallelisation** of sequential codes as of it is unavoidable step when mastering more complex simulations by using new HPC systems. Using high-throughput computing infrastructures like Grid or Cloud to run many simultaneous sequential versions of codes in parallel, results in only a very limited gain of performance, as of possibilities to run big amount of decoupled single computers is limited with speed of communication network established between them. Moreover, organising the management system for massively distributing separate runs, collecting results and organising workflows between different coupled simulations is the inherently sequential part of the whole parallelised application limiting the performance (Amdahl's Law [10]).

Current trends in computer hardware are dictating a gradual shift towards the use of clusters, multi-core workstations or massively parallel processing machines [13]. To exploit the available potential of multi-core systems and parallel processor cores, software developers must design and implement applications with a high degree of concurrency. The development of parallelised applications is error prone and time consuming [8]. Software architects and developers are thus confronted with the question of when the additional effort for introducing concurrency into their application pays off [4] and SimLabs helping or even overtaking this time consuming activities being able to master it faster based on previously collected experience in parallelisation of different scientific codes.

Often the parallelised code, would be not automatic lead to high efficiency often able to reach only moderate sizes for parallel computation with less than 1,000 parallel processes. Simply parallelised codes where most of non-scaling sequential elements essentially eliminated, unavoidably have a maximal number of tasks up to which it speeds-up: reduction of execution time when using more parallel tasks. Thus **Optimisation** must be made by identifying bottlenecks and implementing effective domain decomposition, task farming or bookkeeping mechanisms for code running system. As well having a parallel code not yet means efficient use of computation time provided from supercomputing centres. Last could be ensured through performance analysis and optimisation and is crucial to be gained before applying and using large scale access on modern HPC systems. One major hurdle in porting scientific codes both to massively parallel single-site and distributed multi-site network computing platforms is the difficulty encountered in partitioning the problem so that the computation-to-communication ratio for each compute node (process) is maximized and the idle time during which one node waits for other nodes to transfer data is minimized [13].

As **Release** we generally name activities SimLabs provides to communities based on his experience in defining the license standards for scientific codes, especially when it is used by broader community. Here making easy-to-use and user friendly documentations/portals/repositories for download of new versions of codes, supporting in use and future bug fixing is part of release activity.

Adaptation: During the last decades of formation of computing technologies, many scientific simulation codes have been used to solve particular problems on available HPC infrastructures, that-why have been developed without significant necessity to use them on other systems. Therefore an adaptation of the old scientific software to new computational infrastructures builds often an inevitable barrier for scientific community, that SimLabs actively overtaking. This is another important joint activities of SimLab by conducting necessary tests and productive runs for quality checks-proofs of correctness of scientific results, when new improved scientific codes are available or simply code is running on new platforms.

For **Data Intensive Computation** when analysing big data amount for input and managing production of usually larger size of output data, SimLab is improving scientific simulation codes by implementing into codes the usage of modern parallel file systems interfaces like MPI I/O, HDF or NetCDF. Usage of advanced technologies for I/O activities when using HPC storage systems plays important role for reaching optimal scalability. Also construction of graphical visualizations from huge amount of simulated or experimental data is itself a complicated investigation. Here we use own data management tools developed for analysis of data produced already during simulations, as well for post-processing the standard tools are used for long term big data intensive computations. The last but not least, keeping data accurately and reliably reproducible is another important challenge in solution of which is SimLabs are involved.

2 *SPORADIC* Changed Simulation Codes

During last years Simlab E&A Particles at the Karlsruhe Institute of Technology were involved in different joint R&D project for *SPORADIC* changing the community scientific simulation codes.

2.1 Air Shower Induced by Ultra High Energy Cosmic Rays

The study of atmospheric air showers induced by cosmic rays the most energetic and rarest of particles in the universe is one challenges of modern astroparticle physics. Especially cosmic rays with extremely high energies remain mysterious and the Pierre Auger Observatory is working on solving these mysteries. To interpret results of the Pierre Auger Observatory simulations of Extensive Air Showers (EAS) initiated by cosmic rays is necessary. Their fate is determined by the flight path between their entry in the Earth atmosphere and their interaction with air component. The nature and energy of EAS determines the number, types, energies and directions of the secondary particles generated in the interaction which will continue the particle cascade. These stochastic processes are well modelled by Monte Carlo (MC) methods to get correctly not only the mean values (which might be given eventually by analytical approximations) but also the fluctuations and correlations of the measurable shower observables as e.g. lateral distribution of the particles at ground or energy spectra of those particles at different distances to the shower axis.

During the past 20 years the code package CORSIKA [5] for the simulation of EAS has been developed at KIT and became a worldwide standard instrument for scientists, studying the complex simulation of air-showers induced by cosmic rays in the atmosphere. CORSIKA is based on the MC technique and allows the realistic simulation of interaction, propagation, and decay of particles in extensive particle cascades initiated by high-energy cosmic rays from our Galaxy or extragalactic objects. CORSIKA has become the worldwide standard tool for EAS simulations. More than 750 users apply this code in about 40 experiments worldwide.

SimLab E&A Particles in joint research project with AUGER group of Institute of Nuclear Physics of KIT have developed the parallelised version of scientific application CORSIKA that could be used on massively parallel or distributed computing systems [6]. The aim of project is to perform many air shower simulations at ultra-high energies, where one of the problems for simulations comes from the huge number of 10^{11} particles, which are created within the avalanche by the consecutive interactions with air. For that, a new method for book keeping of 10,000 of parallel running sub-shower simulations and produced data is developed and the algorithms of sharing the running processes and collecting the simulated data on multi-core systems is implemented for parallel running of CORSIKA. Maximal scalability of parallelised CORSIKA code is currently 2,500 computing cores which is indispensable for mastering at least partial analysis when ca. 1,000 of proton,

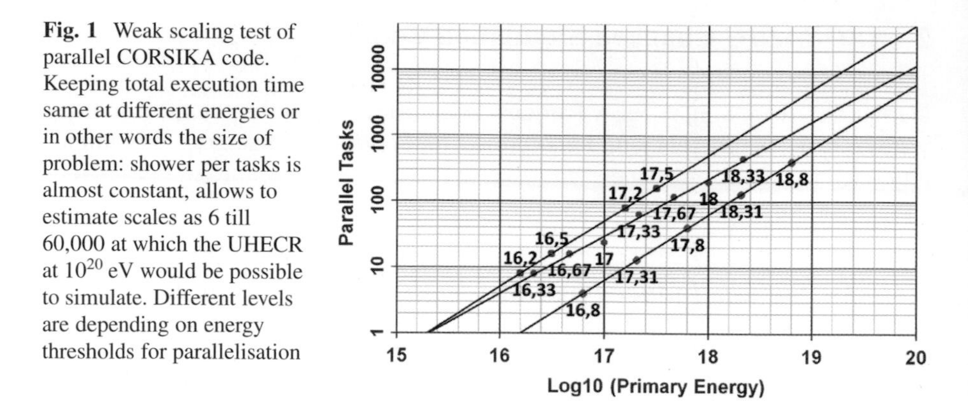

Fig. 1 Weak scaling test of parallel CORSIKA code. Keeping total execution time same at different energies or in other words the size of problem: shower per tasks is almost constant, allows to estimate scales as 6 till 60,000 at which the UHECR at 10^{20} eV would be possible to simulate. Different levels are depending on energy thresholds for parallelisation

Iron, Helium, Carbon and Silicium induced showers at 10^{19} eV Energies consuming 2,000–3,000 CPU hours each will be simulated in productive runs.

Using current parallel code, some full simulations of ultra-high energy cosmic rays (UHECR) at energies of 100 Exa eV ($= 100 \times 10^{18}$ eV) detected by the AUGER experiment were carried out last year for the first time with an acceptable time-to-solution: equivalent simulations with a sequential version of CORSIKA would last decades. Performing a weak scaling tests of parallelised CORSIKA code, when amount of sub-showers per task stays same, we were able to estimate the necessary scalability that code must reach for the simulation of UHECR with energies as high as 10^{20} eV when optimally using supercomputing system, see Fig. 1. But more improvements are necessary to reach higher scalability especially when bottleneck problems of I/O activities of current code CORSIKA would be fully localized and eliminated.

2.2 *Stability of Atomic Isotopes at Finite Temperatures and Extremely High Densities*

The equation of state (EoS) of extremely dense matter is an essential ingredient in astrophysical model simulations. It determines the dynamical evolution of core-collapse supernovae, the static properties of neutron stars, and the conditions for nucleosynthesis. The chemical composition of stellar matter is of particular interest. At densities below nuclear saturation density ($\varrho_{\mathrm{sat}} \approx 2.5 \cdot 10^{14}\,\mathrm{g/cm^3}$) a mixture of nucleons, nuclei and electrons is expected. Their abundancies depend strongly on the thermodynamic conditions such as density ϱ, temperature T, and electron fraction Y_e. Since nuclei, which are composite objects, are immersed in a dense medium, their properties change significantly with the parameters ϱ, T and Y_e due to the nuclear and electromagnetic interaction with the surrounding particles and the action of the Pauli exclusion principle. The latter is the main cause for the dissolution of nuclei with increasing density (Mott effect).

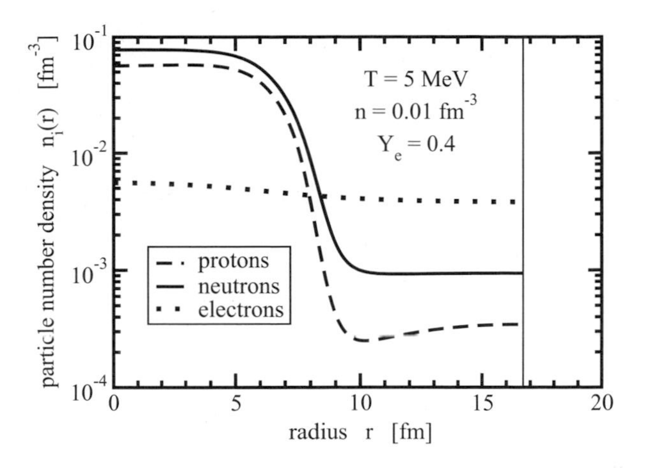

Fig. 2 Radial distribution of proton, neutron, and electron number densities in a spherical Wigner-Seitz cell of radius $R = 16.7$ fm for given temperature T, total baryon number density $n = \varrho/m_{\text{nuc}}$ and electron fraction Y_e obtained from a self-consistent solution of the coupled field equations

In a phenomenological generalized relativistic density functional (gRDF) approach to the EoS of stellar matter [11, 12], the medium-dependent binding energies of nuclear clusters enter in a parametrized form. For light nuclei (^{2}H, ^{3}H, ^{3}He, ^{4}He) they can be obtained by solving few-body in-medium Schrödinger equations with realistic nuclear potentials. For heavier clusters, other (more efficient and applicable) theoretical approaches have to be employed in order to cover essentially the whole nuclear chart. In the present work, binding energy shifts of nuclear clusters are extracted from spherical Wigner-Seitz (WS) cell calculations in Thomas-Fermi (TF) approximation with nucleons and electrons using the same relativistic mean-field model with density-dependent meson-nucleon couplings that underlies the gRDF approach for the EoS. The parametrization of the effective nuclear interaction was adjusted for the change from the original fit in Hartree approximation to the actual TF approximation by an appropriate rescaling of the mass and coupling of the σ meson without affecting the properties of uniform nuclear matter. A typical result of such a calculation is depicted in Fig. 2. The density distributions of protons, neutrons and electrons within the spherical WS cell are found by a self-consistent solution of the coupled set of field equations for given parameters ϱ, T, and Y_e and cell radius R. The formation of a heavy cluster in the centre of the WS cell can be observed clearly.

Because several thousand clusters have to be considered for various values of the thermodynamic parameters temperature, baryon density, asymmetry as well atomic mass and charge number, the whole range of parameters is almost 500 million variations. Herewith, the massive parallel calculations are crucial to complete full analyses. A parallel *AtomicClusters* code for the evaluation of in-medium properties of all possible nuclear clusters was developed in a joint R&D project with the Nuclear Theory group of GSI and is used to model the behaviour of composite particles at zero and finite temperatures. Matter at this conditions can be found in

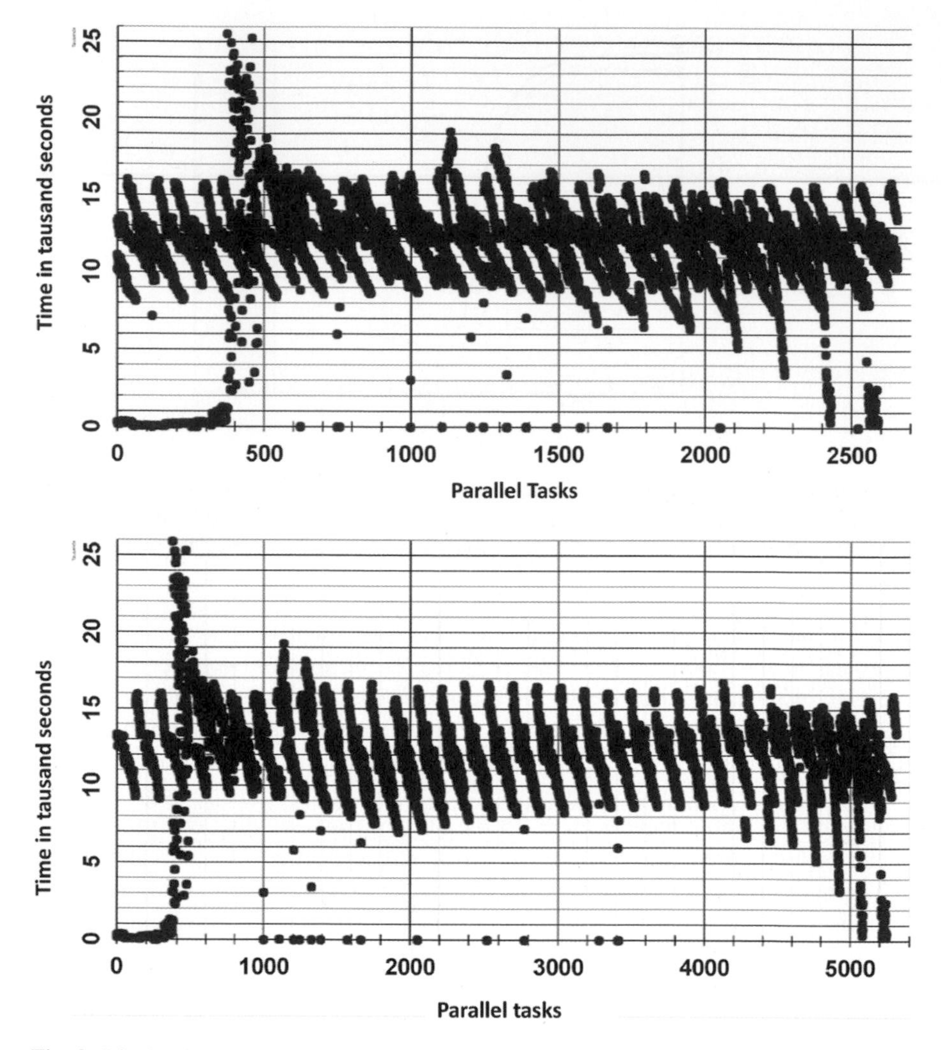

Fig. 3 Distribution of times spent per computation of clusters for given baryon number at temperature $T = 5\,\text{MeV}$. Each point represents a variation in different possible asymmetries. Upper graph is computation with 2,656 tasks/cores and lower part the same simulation using 5,312 parallel tasks

heavy-ion collisions in terrestrial accelerator experiments or in astrophysical environments such as supernovae or compact stars, while a zero temperature is in almost perfect approximation met in neutron stars. The performance of the parallelised code for computation of almost 10,000 variations per given temperature, was tested on CRAY XE6 Hermit system with using up to 10,000 CPU cores in parallel. The load balance of different parallel simulations of atomic isotopes is not optimal as could be seen on Fig. 3. There is an almost 50 % difference in computation time at different values of thermodynamic parameters in particular when varying the baryon density or asymmetry. This leads to the necessity not only to implement an advanced domain decomposition mechanism based not directly on the distribution of parallel computations for a given set of parameters onto available tasks, but also to group non intensive computations into one. The improvement of the load-balance by grouping analysis at different temperatures into one run is the only way to reach scalability for the parallelised *AtomicClusters* code higher than 10,000 parallel tasks. To escape possible I/O bottlenecks due to the writing procedures of separate data generated for each parameter combination computed on many parallel tasks, the MPI I/O system would be implemented and writing would be canalising into less files.

2.3 Search for Gravitational Wave Signals

Detection of a gravitational waves (GWs) will constitute a very precise test of Einstein's theory of relativity and open a new field – gravitational wave astronomy. Despite of the solid theoretical foundation and very strong indirect evidences their existence has never been confirmed directly by any experiment. The primary goal of the modern GW detectors, e.g., LIGO in the USA and Italian-French Virgo is the direct observation of such a gravitational wave. These detectors are state-of-the-art Michelson laser interferometers with two orthogonal arms, each many kilometers long. The effect of an incident gravitational wave on the detector is a measurable strain (length change) of its arms. By comparing the fluctuations of the arm lengths with the theoretical predictions, one can search for gravitational wave signal in the noisy detector output.

As the GW signals are extremely weak, their detection constitutes a major challenge in data analysis and computing. The arithmetic density of the algorithms used in the data analysis varies between several orders of magnitudes up to the practically almost impossible ranges. Depending on the type of search, one has to match the data with several tens of thousands of theoretical waveform template, look for temporal power excesses in the detector output, or analyse the data of several detectors and find correlations. Furthermore when searching for signals emitted by rotating neutron stars one has to perform very high resolution many-dimensional parameter scan in long baseline data stretches. In this case the available computing power directly translates to the search sensitivity, and as such it is of an utmost importance from the first discovery point of view.

The *Polgraw-Virgo* team, working within the **LIGO scientific Collaboration** (LSC)[1] and the **Virgo Collaboration**[2] has developed algorithms and a pipeline called *PolGrawAllSky* to search for GW signals from spinning, non-axisymmetric neutron stars [2]. The algorithm is based on the matched filter technique and the likelihood estimation using the $\mathcal{F}$-statistic [7]. By using the $\mathcal{F}$-statistic we do not need to search for the polarization, amplitude, and phase of the signal, but are

[1] http://www.ligo.org

[2] https://wwwcascina.virgo.infn.it

left with a 4-dimensional space parametrised by frequency, frequency derivative (spindown) and the two angles determining the location of the source in the sky.

Up to now the pipeline was applied to the analysis of data from the Virgo detector first science run using sequential version of the *PolGrawAllSky* code. The analysis involved 5 million CPU hours and took almost 3 years to complete on distributed network computing systems [1]. The sequential code was able to use only one processor core and was run on number of computer clusters with standard queuing systems – such performance is not satisfactory for current and future requirements of the GW data analysis. We estimate, that in order to analyse all the data collected by the current Virgo detector, 250 million CPU hours are required. The advanced detectors are expected by the year 2018 to collect much more data and their analysis will require four times more CPU hours.

In framework of joint R&D project of SimLab with *Polgraw-Virgo* group we have parallelised the *PolGrawAllSky* code implementing the usage of the Message Passing Interface (MPI) library when distributing the searches of gravitational waves at different sky positions. This parallelised version is tested on high performance computers with up to 50,000 of parallel tasks case.

To estimate the scalability we made representative tests with the Gaussian noise data at different band frequencies illustrated on Fig. 4. The tests are made for two different frequencies: 500 and 900 Hz. As of at low frequencies, the scalability

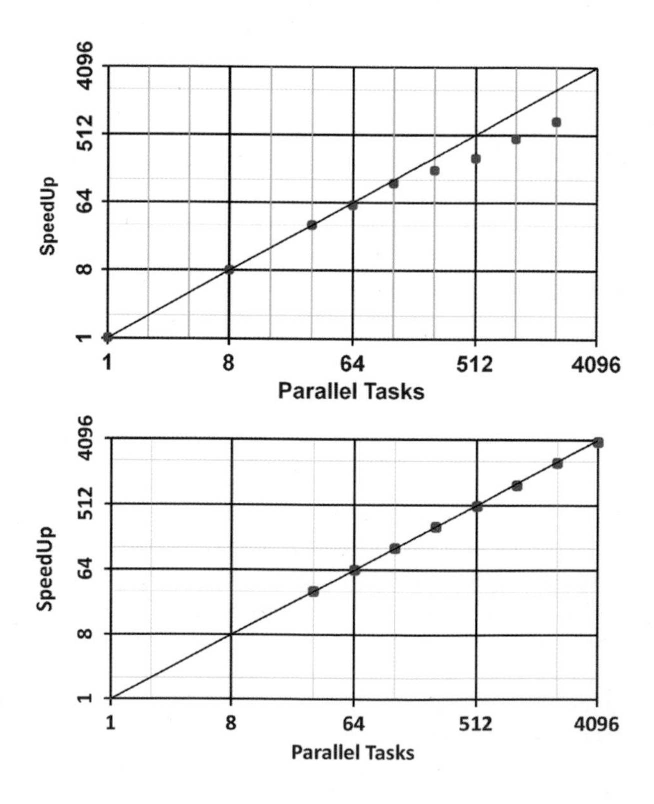

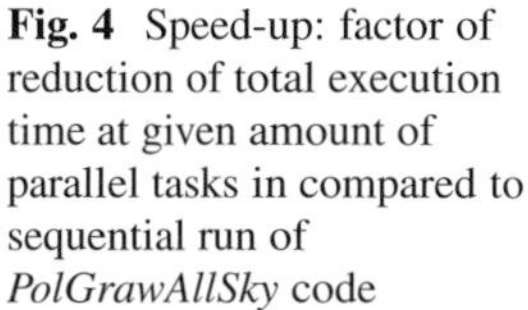
Fig. 4 Speed-up: factor of reduction of total execution time at given amount of parallel tasks in compared to sequential run of *PolGrawAllSky* code

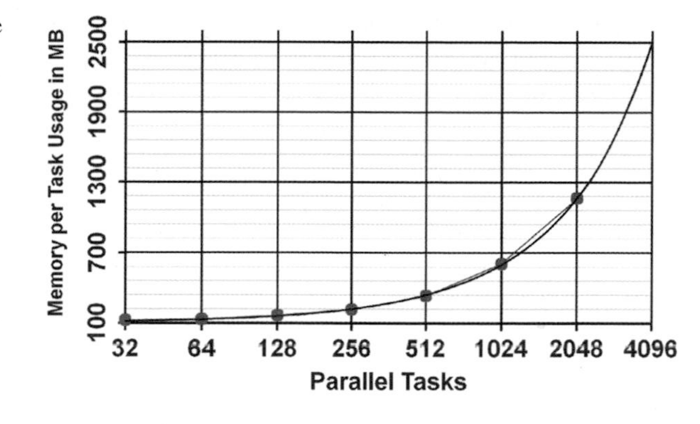

Fig. 5 Total memory usage per tasks of trivially parallelised *PolGrawAllSky* code in Megabytes as a function of number of total parallel tasks

is limited due to lack of complexity of problem, at higher frequencies, when the computation is more intensive, the scalability as high as 8,000 tasks in parallel could be reached.

After initially achieving the acceptable scalability, scaling up encountered problems due to the memory "leaks" at massively parallel runs. Initial parallelised code implemented a non-optimal memory management, due to the fact that the algorithms were designed to use in a sequential ways only. The estimations showed that there is a memory insufficiency for higher frequencies runs when using more parallel tasks,see Fig. 5

We have re-engineered non-optimal memory management algorithms of *PolGrawAllSky* by rearranging internal loops and decrease/optimize the usage of memory per loop, as well as I/O activities. Herewith a stable usage of no more than 100 MB per tasks/CPU were reached and higher scalability was possible. Subsequently, using the estimations for maximal scalability per given frequency band, we grouped different frequencies with varying intensity of computation into one run, thus reaching enough complexity of problem to use as many as 50,727 parallel tasks at once (tested on CRAY HERMIT system).

3 Summary and Outlook

We have parallelised and tested at different level of parallelism the simulation codes *CORSIKA* , *AtomicClusters* and *PolGrawAllSky* . The scaling performance, load balance and memory usage analysis with tools available on Cray XE6 Hermit allowed as to develop the codes increasing their scalability and finding out strategy how to reach effective usage of supercomputing systems, when one will productively run it for particular research or experimental data analysis.

During tests we were able to run advanced *PolGrawAllSky* with up to 50,000 parallel CPU cores, *AtomicClusters* up to 10,000 and *CORSIKA* up to 3,000.

Bottlenecks in I/O activities "rising up" when serial code is simply parallelised, which were possible to localise by debugging the code with more than 1,000 parallel tasks. This was mainly neutralised in *PolGrawAllSky* code by implementing a parallel I/O of MPI and will be made in new versions of *CORSIKA* and *AtomicClusters* for enabling massively parallel runs on HPC clusters using highly efficient interconnects like InfiniBand.

We have identified memory "leaks" and re-engineered non-optimal memory management algorithms of all codes to fit the needs of memory limitations per node on HPC systems.

The performance analysis allowed us to organise the implementation of *MPI_Groups*, MPI I/O and non-blocking global communications into new versions of all mentioned codes.

In order to keep up with the huge size of the parameter space for analysing the future data of cosmic ray, GW detectors and atomic clusters, a hybrid parallelisation of codes would be necessary, when computation could be partially driven on co-processors like Graphical Processor Units (GPU) available on same computational nodes of supercomputing systems. Hybrid parallelisation would be crucial when one want to use optimally future massively parallel Exascale computing systems with more than one million CPUs. Using optimally processors and co-processors in different many-core architectures on can easily escape communication overheads in larger scale simulations.

Acknowledgements The developments of codes and massivly parallel simulations, were possible by using Cray XE6 Hermit at High Performance Computing Center HLRS in Stuttgart.[3] The work of G.Poghosyan and A.Streit were supported by the Helmholtz "Supercomputing" program. S.Sharma, A. Kaur, V.Jindal, P.Bisht thank the Karlsruhe Institute of Technology and PEC University of Technolody for support and hospitability during their internship visits to SimLab E&A Particles. The work of M.Bejger and A.Krolak was supported in part by the Polish Ministry of Science and Higher Education grant DPN/N176/VIRGO/2009 and National Science Center grants UMO-2013/01/ASPERA/ST9/00001 and 2012/07/B/ST9/04420. The work of S. Typel was was supported by the Helmholtz Association (HGF) through the Nuclear Astrophysics Virtual Institute (VH-VI-417). T.Klaehn acknowledges support from the National Science Center of Poland under grant number 2013/09/B/ST2/01560.

References

1. Aasi, J., Abadie, J., Abbott, B.P., Abbott, R., Abbott, T.D., Abernathy, M., Accadia, T., Acernese, F., Adams, C., Adams, T., et al.: Einstein@Home all-sky search for periodic gravitational waves in LIGO S5 data. Phys. Rev. D **87**, 042001 (2013)
2. Astone, P., Borkowski, K.M., Jaranowski, P., Pietka, M., Królak, A.: Data analysis of gravitational-wave signals from spinning neutron stars. V.A narrow-band all-sky search. Phys. Rev. D **82**, 022005 (2010)
3. Attig, N., Esser, R., Gibbon, P.: Simulation laboratories: an innovative community-oriented research and support structure. In: Bubak, M., Turala, M., Wiatr, K. (eds.) Proceedings of the Cracow Grid Workshop (CGW'07), Krakow (2008)

[3]Project ACID 12863

4. Happe, J.: Predicting Software Performance in Symmetric Multi-core and Multiprocessor Environments. The Karlsruhe Series on Software Design and Quality. Volume 3, Universittsverlag Karlsruhe, Karlsruhe (2009). http://dx.doi.org/10.5445/KSP/1000011806
5. Heck, D., Knapp, J., Capdevielle, J.N., Schatz, G., Thouw T.: CORSIKA: a Monte Carlo code to simulate extensive air showers, 90 pages, 10 figures: Forschungszentrum Karlsruhe Report FZKA 6019 (1998)
6. Heck, D., Pierog, T., Engel, R., Thiara, G., Karla, S., Singla, H., Poghosyan, G., Seldner, D., Schmitz, F.: Computational Methods in Science and Engineering, pp. 87–90. KIT Scientific Publishing (2011). http://dx.doi.org/10.5445/KSP/1000023323
7. Jaranowski, P., Królak, A., Schutz, B.F.: Data analysis of gravitational-wave signals from spinning neutron stars: the signal and its detection. Phys. Rev. D **58**, 063001 (1998)
8. Lee, E.A.: The problem with threads. IEEE Comput. **39**(5): 33–42 (2006)
9. Poghosyan, G., Seldner, D., Schmitz, F.: Simulation laboratory astro- and elementary particle physics. In: Computational Methods in Science and Engineering, Proceedings Workshop SimLabs@KIT, Karlsruhe, Nov 2011, pp. 77–84. KIT Scientific Publishing. http://dx.doi.org/10.5445/KSP/1000023323 (2011)
10. Rodgers, D.P.: Improvements in multiprocessor system design. SIGARCH Comput. Archit. News **13**, 225–231 (1985)
11. Typel, S., Röpke, G., Klähn, T., Blaschke, D., Wolter, H.H.: Composition and thermodynamics of nuclear matter with light clusters. Phys. Rev. C **81**, 015803 (2010)
12. Typel, S., Wolter, H.H., Röpke, G., Blaschke, D.: Effects of the liquid-gas phase transition and cluster formation on the symmetry energy. Eur. Phys. J. A **50**, 17 (2014)
13. Vatsa, V.N.: Parallelization of a multiblock flow code: an engineering implementation. Comput. Fluids **38**(4–5), 603–614 (1999)

Part II
Molecules, Surfaces, and Solids

Holger Fehske and Christoph van Wüllen

Our need for technological advances generates a constant demand for materials with improved properties such as strength and biocompatibility, or new materials to be used in electronic devices, for energy conversion or in aerospace engines, just to name a few examples. It is obvious that there is a connection between the microscopic (atomistic) structure of a material and its properties, and first-principle simulations on atomistic scales have emerged as a fruitful approach not only to understand this connection, but also to use this understanding in designing materials with taylored properties. The following contributions demonstrate how research in solid state physics and chemistry has profited from the Cray XE6 "Hermit" technology available at the High Performance Computing Center Stuttgart.

The Stuttgart University and DLR collaboration, composed of D. J. Förster, J. Roth and S. Scharring, H.-A. Eckel, respectively, used a molecular dynamics (IMD program package based) modeling approach of laser-ablation micro-propulsion in order to calculate important quantities for aerospace engineering (microthruster) applications, such as impulse coupling coefficients or specific impulse and angle-distributions of neutral and charged particles. Here the amount of ablated material is of vital importance for the operation-time mission parameter. Since it is impossible to directly include quantum effects in the MD simulations, a phenomenological (two-temperature) model was implemented for the heat transport in the metallic samples and solved by finite difference schemes. The results were compared with a hydrodynamic (virtual laser lab) model for the laser-matter interaction. Taking

H. Fehske (✉)
Institut für Physik, Lehrstuhl Komplexe Quantensysteme, Ernst-Moritz-Arndt-Universität Greifswald, Felix-Hausdorff-Str. 6, 17489 Greifswald, Germany
e-mail: fehske@physik.uni-greifswald.de

C. van Wüllen
Fachbereich Chemie, TU Kaiserslautern, Erwin-Schrödinger-Str. 52, 67663 Kaiserslautern, Germany
e-mail: christoph.vanwullen@chemie.uni-kl.de

the limitations of the simulation approaches into account, the estimated values of the aerospace parameters and experimental values are in satisfactory accord. From a computational resources point of view, the necessity to simulate ns-pulses for satellite propulsion results in much longer runtimes, i.e., a decreasing the performance.

M. Hummel, A.-P. Prskalo, P. Binkele and S. Schmauder from the University Stuttgart (IMWF) performed MD simulations of single- and poly-crystalline structures with up to six million atoms. For the binary aluminum copper alloy the mutually different bond lengths lead to coherent tension and internal stresses. To quantify the residual stresses Gunier-Preston zones were introduced. The interaction of dislocations with the obstacle fields represented by the copper atoms is analyzed, particularly with regard to shearing effects and the influence on the tensile strength. For the simulated polycrystals the transition of the inverse to the normal Hall-Petch effect was examined. Furthermore, the authors use their IMD simulation code to study the cyclic deformation of notched iron and the ultimate tensile strength of nanoporose iron at various temperatures. It was shown that the porose structure of such an iron foam is weakened as the temperature increases; concomitantly the porosity is decreased by the enhanced density of the structure.

The third contribution by Sanna et al. from the University Paderborn deals with the surface charge of ferroelectric materials. In particular, the charge associated to various terminations of the the polar lithium niobate Z-cut surface is estimated from self-consistent density-functional theory (DFT) calculations, using the Vienna Ab initio Simulation Package (VASP) for a periodically slab geometry. For the VASP implementation of DFT on the HLRS CRAY each orbital is treated with a number of cores (corresponding either to the square root of available cores or the number of cores per node). Then, with a combination of parallelisation over bands and momentum-points, an almost linear scaling up to 2,048 cores could be achieved. Using such a kind optimized code, the surface charge of a lithium niobate slab—terminated according to the stable surface structure—has been determined and shown to amount about 16 % of a ideal ferroelectric bulk cut. Demonstrating that (any thermodynamically stable) surface reconstruction lowers the polarisation charge, the authors have the surface charge compensation pegged as a major driving force for the morphologic transformations on lithium niobate surfaces.

Several applications depend on the question what happens at the interface of a material with the environment, and the fluorite-water interface is an important example. This has been investigated experimentally, but numerical simulations are needed for a molecular interpretation of experimental results. The fourth contribution by Khatib and Sulpizi reports numerical studies on what happens with the water structure in the vicinity of a fluorite surface, using molecular dynamics simulations based on density functional theory. Of particular interest is an assignment of signatures seen in vibrational spectroscopy experiments to calculated molecular motions that are close to the fluorite surface.

The advantage of simulations based on a quantum mechanical description of the electrons lies in the possiblity to describe the breaking and formation of chemical bonds. An important chemical reaction is the splitting of water into hydrogen and oxygen on a titania (TiO_2) surface. In this case, solar light can provide the required energy which is then stored as chemical energy for later use. Absorption of a solar photon induces a transition to an electronically excited state where the ball starts rolling. Atomistic simulations of such photochemical processes require highly sophisticated wave function based methods, and the contribution of Mitschker and Klüner reports first steps into a simulation of the H_2O/TiO_2 system.

The last contribution by Stegmüller and Tonner models the growth of semiconductor materials in a chemical vapor deposition process. Here, volatile (gas phase) substances containing gallium or phosphorus decompose on a silicon surface, and finally form GaP films. Besides knowing how the surfaces actually look like, it is important to estimate the mobility of atoms on the surface, since this is a prerequisite for the production of an ordered film. In their work, the authors calculated such "hopping" pathways on pure silicon or on gallium- or phosphorus-terminated GaP surfaces.

All these simulations would be unthinkable without access to leading edge supercomputers, and the projects we have reported on demonstrate the need for such facilities in this research area. But having access to supercomputers also shapes our way of thinking about scientific problems: simulations that we could not even dream of yesterday can be considered "just around the corner" today, and this triggers the development of more and more realistic simulation methods that tomorrow contribute to scientific and technological progress. It is therefore of vital importance that the national supercomputer infrastructure is kept up-to-date.

Molecular Dynamics Simulations of Laser Induced Ablation for Micro Propulsion

Daniel J. Förster, Stefan Scharring, Johannes Roth, and Hans-Albert Eckel

Abstract A new concept of micro propulsion based upon laser ablation MICRO-LAS was introduced by the Institute of Technical Physics (ITP) of DLR Stuttgart. Pulsed lasers are used for material removal of a target. The amount of removed material should be variable due to the tunability of input laser energy and repetition rate, resulting in well defined impulse bits and low small thrusts down to the sub-μN scale. We present a modeling approach of laser ablation in order to calculate important figures of merit in aerospace engineering. The program applied is IMD (http://imd.itap.uni-stuttgart.de), an open source molecular dynamics package of the Institute of Functional Materials Quantum Technologies (FMQ). Results are compared with a hydrodynamic code, VLL (http://vll.ihed.ras.ru), as well as with experimental investigations.

1 Introduction

Very precise position control systems are needed in order to fulfill requirements of projects in earth and space exploration planned in past years. Existing thruster concepts have problems of endurable precise regulation of thrusts down to a few μN, needed for accurate compensation of forces acting on satellites. Examples are chemical engines like hydrazine boosters, pulsed plasma thrusters or arcjets. An alternative to existing concepts in nN-μN regime like nano-FEEPs, MEMS-ion thrusters or HEMP thrusters was introduced by the ITP [26]. It is based upon laser ablation of a (solid) propellant material and in the future shall guarantee position control of pretentiously missions like LISA Pathfinder [1] or MICROSCOPE [3].

D.J. Förster • J. Roth (✉)
Institut für Funktionelle Materialien und Quantentechnologien, Universität Stuttgart, Stuttgart, Germany
e-mail: foerster@itap.physik.uni-stuttgart.de; johannes@itap.physik.uni-stuttgart.de

S. Scharring • H.-A. Eckel
Institut für technische Physik, Deutsches Zentrum für Luft- und Raumfahrttechnik, Stuttgart, Germany
e-mail: stefan.scharring@dlr.de; hans-albert.eckel@dlr.de

W.E. Nagel et al. (eds.), *High Performance Computing in Science and Engineering '14*, DOI 10.1007/978-3-319-10810-0_10

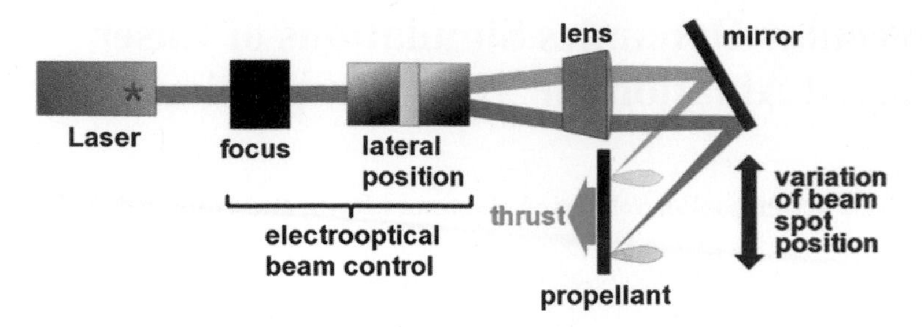

Fig. 1 Concept sketch of MICROLAS propulsion (Adapted from [9])

At the FMQ ultrashort pulses in the fs-regime have been investigated through simulations. An existing Molecular Dynamics code of the institute, IMD, was under further development to enable the simulation of pulsed laser ablation [25]. This report investigates on the validation of IMD for calculations in the aerospace field. At the DLR facility in Stuttgart, laser ablation experiments are develop. Partially, the measurement of important quantities such as impulse coupling coefficient, specific impulse and angle distributions of neutral and charged particles were performed. Data of these will be compared with simulations.

The general concept of laser-ablation micropropulsion was given by Phipps who introduced a thruster concept using ms-laser pulses [14]. Later developments used ns-laser pulses [14]. MICROLAS is a new propulsion concept of a fine tuneable and controllable thruster in the μN-regime. The intended structure is sketched in Fig. 1. Main and probably heaviest component is a laser source which beam propagates through a system of lenses and mirrors without any moving mechanical parts. This new concept is called inertia-free since the only impulse change comes from the ablation process itself. This occurs when the laser beam hits the propellant plate. In experiments at DLR mainly the laser-matter interaction has been analyzed until now in the laboratory.

Main interests in simulations on atomistic scales, on one hand, come from the ablation process itself. The question is what processes in the microscopic world have a direct influence on macroscopic values. How are domains on the surface and within the material reorganized when a laser beam of high energy hits the target? Where does the energy go to? Does overlap of a beam spot with an already dug hole has an influence? What are the ratios of matter states and does ionization play a big role? In the long run, the evolution of the velocity and mass distribution in time and space is of big interest. The amount of ablated material and its potential of contaminating the propulsion system itself plays a big role when looking at main mission parameters like lifetime and possible time of operation. But the relapse of the material is also important in missions where the MICROLAS propulsion is designed for. Dirty lenses or current conducting shells may induce a variety of problems.

2 Modeling

2.1 *Molecular Dynamics and the IMD Code*

In Molecular Dynamics (MD) simulations particles interact with each other under the influence of a potential $V(r)$. The interactions used in this investigation were the **E**mbedded **A**tom **M**ethod-potentials of Ercolessi and Adams for aluminum [4] which are well established for metals. Newton's equations of motion are solved by integrators, in IMD a leapfrog integrator is used [25]. At each simulation step $t \rightarrow t + \Delta t$ the forces are calculated from the interaction potentials by the negative gradient

$$\boldsymbol{F} = -\nabla\, V(\boldsymbol{r}) = -\nabla\, V(r), \tag{1}$$

where the second part comes from the assumption that the interactions just exhibit a proportionality to the distance. With the forces the coordinates are updated to new positions.

2.2 *Laser-Matter Interaction*

When energy is coupled into a sample by laser photons, the electrons react by gaining energy. The energy transport over the sample is no longer just driven by collisions of atoms as in equilibrium but by quantum mechanical effects. Since it is not directly possible in MD simulations to consider quantum effects yet, a macroscopic model was implemented to describe the heat transport within metallic samples properly. The `two temperature model` (TTM) describes the evolution of heat with three measurable main parameters in a set of nonlinear differential equations [25]:

$$C_e(T_e)\frac{\partial T_e}{\partial t} = \nabla[K_e \nabla T_e] - G(T_e - T_l) \tag{2}$$

$$C_l(T_l)\frac{\partial T_l}{\partial t} = \nabla[K_l \nabla T_l] + G(T_e - T_l) \tag{3}$$

The indices e and l here stand for *electrons* and *ions*. The heat transfer from the electron subsystem to the lattice is described by the term $(T_e - T_l)$, with G being the electron-phonon coupling constant. $C_{e,l}$ are heat capacities of the subsystems, $K_{e,l}$ are heat conductivities. The laser source term can be coupled to the heat equations via the electronic system via a laser power density term $S(\boldsymbol{r}, t)$. The phenomenological TTM equations describe heat transport in continuum, so usually they are solved by `finite difference schemes` (FD). Since we simulate on an atomistic level, just the electronic heat transport is described in continuum as in

Eq. (2). The ionic movement is of course influenced by the behavior of the electrons, but their positions directly refer to atomic coordinates in the MD simulation. A hybrid simulation model was implemented in IMD in the following way [7]:

$$C_e(T_e)\frac{\partial T_e}{\partial t} = \nabla[K_e \nabla T_e] - G(T_e - T_l) + S(\boldsymbol{r}, t) \qquad \text{FD} \tag{4}$$

$$m_i \frac{\partial^2 \boldsymbol{r}_i}{\partial t^2} = \boldsymbol{F}_i + \xi\, m_i\, \boldsymbol{v}_i^T \qquad \text{MD} \tag{5}$$

$$\xi = \frac{\frac{1}{n}\sum_{k=1}^{n} G\, V_N\, (T_{e,k} - T_l)}{\sum_j m_j\, (v_j^T)^2} \qquad \text{coupling} \tag{6}$$

m_i, $\boldsymbol{r}_i$ and $\boldsymbol{F}_i$ correspond to the MD system, the index i stands for an atom, the forces $\boldsymbol{F}_i$ come from the interatomic interactions. The ions and electrons are calculated within two different systems, one belonging to MD space, the other to FD space, in which the electron heat transport is calculated. Here, the so called thermal velocity $v_i^T = v_i - v_{\text{com}}$ is introduced, where v_{com} is the center of mass velocity of atoms in a FD cell. The two systems are coupled via ξ. For further details of the model see [23, 25].

The simulation results obtained with the interactions implemented in IMD show good agreements with experimental results in the fs-laser pulse regime [23, 25]. However, there are limitations. A particle based algorithm that takes quantum effects directly into account would perfectly describe the atomistic world. Currently, however, all material parameters needed for the model have to be extracted from experiments. Most parameters are defined as constant at present, although in principle all are at least temperature dependent:

Electron-phonon coupling parameter	$G(T) := G$
Heat capacity	$C(T) := \gamma \cdot T$
Thermal conductivity	$K(T) := K$
Reflectivity	$R(T, \lambda, P) := R$

First principle calculations show that even the behavior of simple metals differs from these assumptions under non-equilibrium conditions [11]. Deviations occur for aluminum, copper and gold in heat conductivity C and the electron-phonon coupling factor G. For C, a simple relation is already implemented, assuming a linear behavior. Here deviations lie in the range of decades, so the linear approximation covers the effect partially although the shape dependence is not completely correct. K and G deviate for aluminum at the same magnitude as the values themselves, so the constant approximation represents measured effects quite well. Reflectivity, however, in general is dependent of at least three physical values being temperature T, wavelength λ and polarization P. The general laser source term applied exhibits the temporal shape of a Gaussian pulse:

$$S(x,y,z,t) = S_0 \; e^{-\frac{1}{2}\frac{(t-t_0)^2}{\sigma_t^2}} \; I(y,z) \, e^{-\mu x}. \tag{7}$$

Here σ_t is the time at which $S(\boldsymbol{r},t)$ decreased to a value of $1/\sqrt{e}\; S_0$, t_0 is the time of the peak maximum. In x-direction the applied laser energy is decaying via the `Lambert-Beer's law` with μ being the inverse absorption length, $[\mu] = m^{-1}$, at which the intensity in x-direction decreases to $1/e\; S_0$. The parameters used in simulations can be connected directly with common laser parameters used in experiments. By integration over time and space the `total laser energy` is provided via

$$E = \frac{(2\pi)^{3/2}\; S_{3D}\; \sigma_t\; \omega_0^2}{(1-R)\;\mu}. \tag{8}$$

By integration over time and just the x-coordinate the `fluence`

$$\Phi = \frac{\sqrt{2\pi}\; S_{1D}\; \sigma_t}{(1-R)\;\mu} \tag{9}$$

can be derived. The reflectivity R is a unitless percentual constant that is fixed and is introduced during integration via

$$E \cdot (1-R) = \int S(\boldsymbol{r},t). \tag{10}$$

The reflectivity R is treated as conversion parameter via

$$\Phi = \frac{\sigma_e}{(1-R)} \tag{11}$$

since in IMD the quantity used to define the applied energy is the fluence parameter σ_e.

2.3 *Calculation of Aerospace Parameters*

Most microthruster systems for space applications are characterized by two main parameters, the so called `specific impulse` I_{sp} and `impulse coupling coefficient` C_m. The specific impulse of classic propulsion systems is defined by [12]

$$I_{sp} = \frac{F}{\dot{m}\; g} = \frac{p}{m\; g} = \frac{\langle v\rangle}{g} \qquad [I_{sp}] = s, \tag{12}$$

where F is the thrust in Newton of the engine, $\dot{m}$ the mass flow rate of the propellant in kg/s and g the earth's standard acceleration on the surface measured in $\mathrm{m/s^2}$. p is the impulse of the exhaust jet, in our case the ablated material and $\langle v \rangle$ the mass weighted velocity of N expelled particles,

$$\langle v \rangle = \frac{\sum_{i=1}^{N} m_i \ v_i}{\sum_{i=1}^{N} m_i}. \tag{13}$$

The impulse coupling coefficient c_m is defined by [14]

$$c_m = \frac{\Delta p}{E_{\text{laser}}} = \frac{p_{\text{plume}}}{E_{\text{laser}}} \qquad [c_m] = \frac{\text{Ns}}{\text{J}} = \frac{\text{N}}{\text{W}}. \tag{14}$$

The ratio of the impulse change Δp to the laser pulse energy E is an important figure of merit of a propulsion system. Under the assumption of a non moving sample before ablation, the second part of the formula holds, which will mainly be used in the following.

2.4 Hydrodynamic Model: Virtual Laser Lab

At the ITP, initial estimation parameters for MICROLAS propulsion were obtained by using `Virtual Laser Lab` (VLL). VLL is an online tool for one-dimensional calculations of laser matter interactions, e.g. the behavior of thin metal foils has been simulated properly [16]. It is solving Maxwell's equations and a hydrodynamic continuum model in one dimension. Equations of state cover two main models, a Drude-like model valid for temperatures below T_{Fermi} and a plasma model for the excited state above T_{Fermi}, both connected by a smooth transition. Also TTM equations are solved and connected to a single fluid material model. More details are given in [16].

3 Results

Simulations were adjusted according to experimental values available in DLR laser laboratory. Main parameters of the system are given in Table 1. For pulse width it follows that $\sigma_t = (2 \cdot \sqrt{2 \cdot ln2})^{-1} \cdot t_{Pulse} = 211$ ps, so a width of 210 ps was chosen.

Table 1 Laser parameters of a recent laser ablation experiment at the DLR using a microchip laser

Maximum energy	Wavelength	Spot diameter at the target	Fluence	Pulse duration
E_{max} (μJ)	λ (μm)	ω_0 (μm)	Φ (J/cm^2)	t_{pulse} (ps)
80	1.064	30	3–10	495

According to the well known scaling law in laser ablation propulsion [13]

$$C_m \sim \left(I\,\lambda\,\sqrt{t_{\text{Pulse}}}\right)^{-1/4} \tag{15}$$

with $\lambda = const.$ and $I = \sigma_e / t_{Pulse}$, σ_e being the fluence, a law for same power input can be derived. For known $\sigma_{e,1}$ and $t_{\text{Pulse},1}$ the fluence for another pulse duration $t_{\text{Pulse},2}$ follows from

$$\sigma_{e,2} = \frac{\sigma_{e,1}}{\sqrt{t_{\text{Pulse},1}}} \cdot \sqrt{t_{\text{Pulse},2}} = \frac{\sigma_{e,1}}{\sigma_{t,1}} \cdot \sqrt{\sigma_{t,2}}, \tag{16}$$

and thus for the chosen laser parameters

$$\sigma_e = \frac{3\,\text{J/cm}^2}{\sqrt{210\,\text{ps}}} \cdot \sqrt{\sigma_t}. \tag{17}$$

One main parameter set (σ_e, σ_t) was simulated, i.e. 9,490 J/m^2, 21 ps. A simulation of the laser system of 30,000 J/m^2, 210 ps as given in laboratory was out of reach due to the limits in computation time. In current implementation it is not possible to restart an interrupted TTM simulation from the last checkpoint. An interruption leads to configurations that are not in thermal equilibrium and thus will not give proper results when the simulation is restarted. The temperature of the ionic system would be the new starting temperature for both electronic and ionic subsystem, resulting in an artificial energy loss.

In Table 2 the parameters used for TTM simulations are given. The Lambert-Beer penetration depth was set to $\mu = 8$ nm [22]. The simulated sample size in z-direction was 1.5 μm, in x- and y- direction 20 nm and included $37.5 \cdot 10^6$ aluminum particles. Crystal structures are fcc with a lattice constant of 4.0513 Å. As a first output, the density distribution over time is given in Fig. 2. The initial surface of aluminum bulk material lies at distance 0 nm, the material is directed along the positive distance axis, negative values describe positions in vacuum in front of the aluminum sheet. On the time axis 0 ps defines the time of maximum intensity, at $3 \cdot \sigma_t = 63$ ps most energy is coupled in. Material begins to melt at −55 ps, evaporation starts at −25 ps and continues over the whole simulation time.

Table 2 Material parameters for heat conduction in TTM-model

Material	Electron heat capacity coefficient γ ($\frac{\text{J}}{\text{m}^3\text{K}^2}$)	Electron thermal conductivity K ($\frac{\text{J}}{\text{K ms}}$)	Electron-phonon-coupling constant G ($\frac{\text{J}}{\text{s m}^3\text{K}}$)
Aluminum	135[a]	235[b]	$5.69 \cdot 10^{17}$ [c]

[a] [11]
[b] [2, p. 759]
[c] [5]

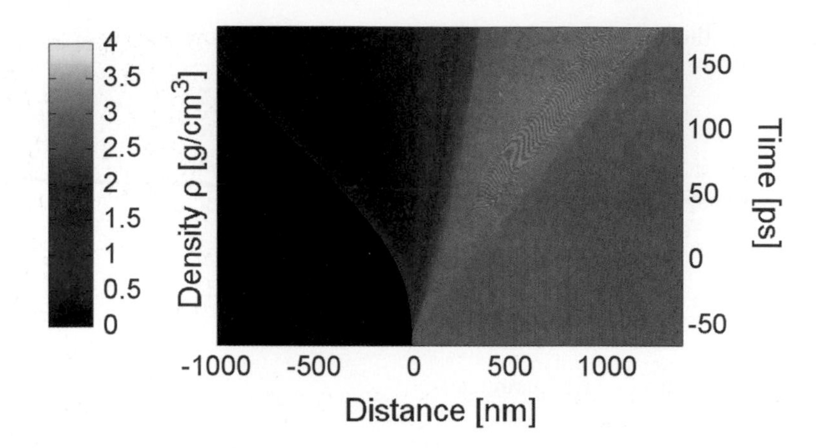

Fig. 2 Density plot for $\sigma_t = 21$ ps, $\sigma_e = 9{,}490\,\mathrm{J/m^2}$

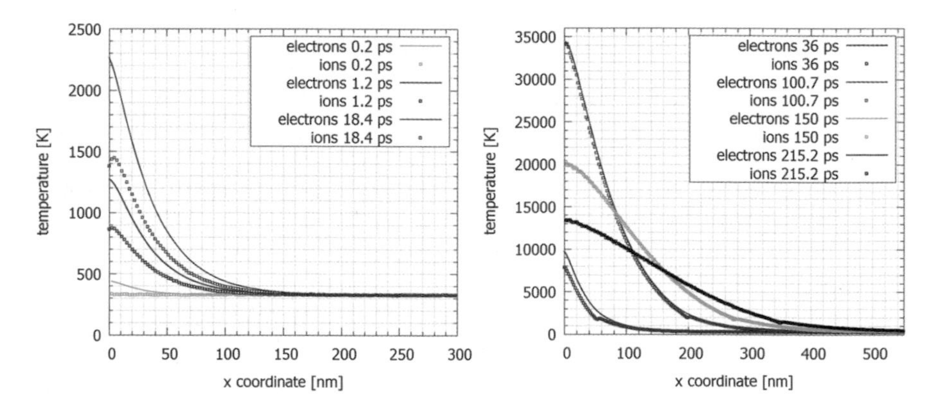

Fig. 3 Spatial distribution of electron and ion temperature. *Left*: at the beginning of the simulation. *Right*: towards the end of the simulation

A pressure wave travels into the material with velocity

$$\frac{1{,}000\,\mathrm{nm}}{170\,\mathrm{ps}} \approx 5.88\,\frac{\mathrm{m}}{\mathrm{s}},$$

which lies in the order of the velocity of sound waves in aluminum ($\nu_{\mathrm{sound}} \approx 6.42$ km/s at 300 K [10, pp. 14–36]). This region of compressed material is followed by a region of undistorted aluminum with normal solid density from $t = 100$ ps on. At the surface of the material on the left a liquid layer develops. A vapor region follows, whereas a fluid mixture of both liquid and fluid separates both less dense regions. Temperature distributions of the electronic and ionic system vs. depth with respect to the initial surface at 0 nm are given in Fig. 3. The time is measured from the beginning of the simulation. In the first 200 fs after simulation begins the lattice is still at 320 K, while the interaction with the light field leads to an increase of electron temperature. It rises within the next 18 ps, the ionic system following due

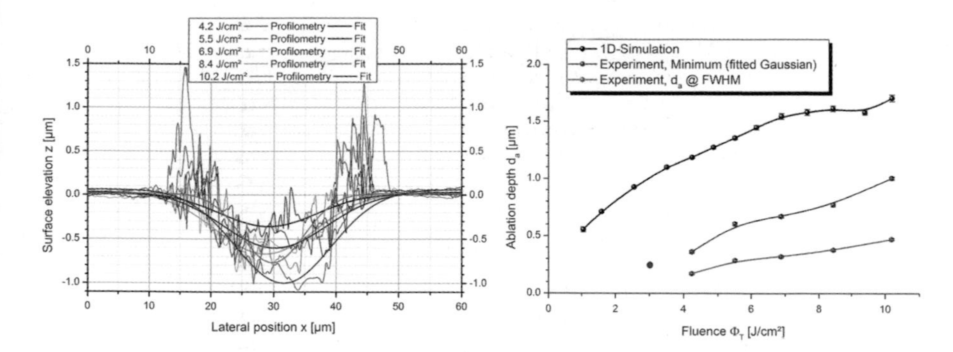

Fig. 4 An estimation of crater depths by Gaussian fits of experimental data (*left*) and in combination with VLL simulation output. The *blue* point refers to the IMD simulation of σ_t = 21 ps (*right*) [8, 21]

to electron-lattice relaxation. In Fig. 3 further time snapshots are given. Around 100 ps after starting the simulation, the current incoupling amounts to ~17 % of the maximum value, and the highest electronic temperature is achieved. Lattice and electronic system differ only by 15 % at maximum at that time, with a decreasing difference to less than 4 % after 215 ps simulation time. This configuration leads to runtimes which can be handled smoothly on multiple nodes at the High Performance Computing Center Stuttgart, so samples of the size used in this work can be simulated properly with irradiation of few ten picoseconds. Longer pulse durations results in non-equilibrated samples at the end of the simulation.

Crater depths of ablation experiments can be compared to simulated depths, here namely melting depths. They are determined by finding the border position between solid and liquid phase in the density plots. White light interferometry of modified aluminum surfaces after laser pulse irradiation allows experimental access to crater depths. Typical experimental data of 210 ps ablation experiments in vacuum is given in Fig. 4 (left). A fit with Gaussian functions to the data points leads to full width half maximum depths as well as minimum depths. In Fig. 4 (right) average values over multiple craters are given. Additionally, liquid-solid interface position in the simulation of parameter set (9,490 J/m^2, 21 ps) are inserted for comparison. Under the pre-described assumptions they fit in quite well.

The checkpoints files given by `IMD` include particle masses, vectorial positions and velocities. The aerospace parameters specific impulse I_{sp} and impulse coupling coefficient C_m were calculated from these data via simple scripts. Figure 5 shows their temporal behavior for the parameter pair (9,490 J/m^2, 21 ps). The way of choosing particles belonging to the ablated plume is quite naive, simply all particles with a spatial coordinate beyond a certain threshold are counted. In this case, all particles being 400 nm away from the initial surface were included. During simulation no recondensation processes could be depicted due to short simulation times since these processes happen on timescales of microseconds, which are not in reach at present. As shown in to Fig. 5 the parameters were least-squares fitted with

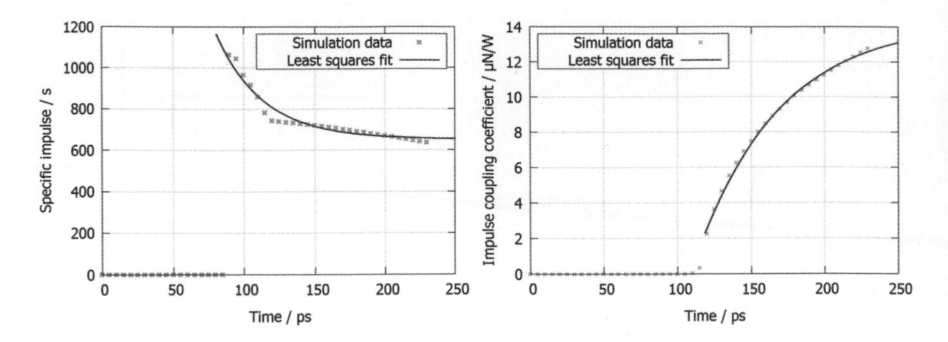

Fig. 5 Time evolutions. *Left*: of specific impulse. *Right*: of impulse coupling coefficient

Table 3 Estimated values of aerospace parameters and experimental values; 130 fs data from [15]

Simulation	C_m (µN/W)	Standard deviation $\sigma(C_m)$(µN/W)	I_{sp} (s)	Standard deviation $\sigma(I_{sp})$ (s)
21 ps	14.4	0.4	651.8	7.6
21 ps, VLL	25.0	0.1	435	3
210 ps, Phipps exper.	24.98	–	–	–
130 fs, Phipps model	36	–	–	–
130 fs, Phipps exper.	18	–	2,245	–

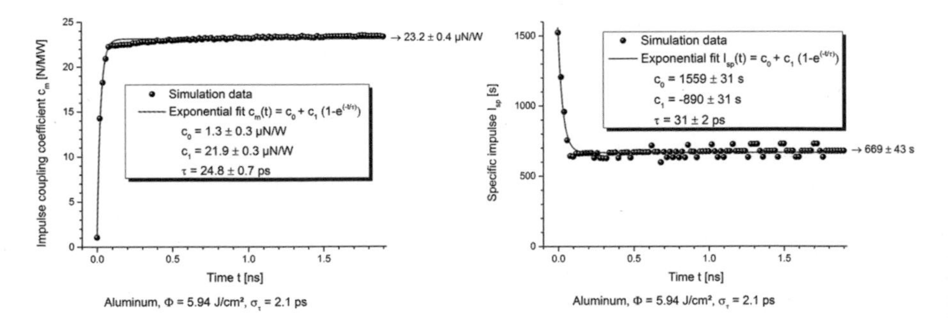

Fig. 6 Impulse coupling coefficient and specific impulse from VLL simulations, $\sigma_t = 2.1$ ps, [21]

two functions of the form

$$I_{sp} = I_1 + I_2 \cdot e^{-a \cdot t}$$
$$C_m = C_1 + C_2 \cdot \left(1 - e^{-b \cdot t}\right),$$

resulting in long time values for $t \to \infty$ given in Table 3. The same was done for hydrodynamic one-dimensional simulations with the code VLL [16], with simulation times in the nanosecond range (cf. Fig. 6). In the table the values are given for the second `IMD` simulation together with the data for 210 ps estimated with the experimental formula derived by Phipps et al.

In experiments with polymer and metal targets the Phipps formula was proven to estimate the aerospace parameters properly down to a lower limit of $t_{pulse} = 100\,\text{ps}$ ($\sigma_t \approx 42.47\,\text{ps}$) [6].
The trend function formula for aluminum leads to an estimated value of [14]:

$$
\begin{aligned}
c_m\,[\mu\text{N/W}] &= \frac{55.6}{(\Phi\ \lambda\ \sqrt{t_{\text{Pulse}}})^{0.301}} \\
&= \frac{55.6}{(2.998\,\text{J/cm}^2 \cdot 1.064\,\mu\text{m} \cdot \sqrt{500\,\text{ps}})^{0.301}} \\
\Rightarrow c_m &\approx 25\,\mu\text{N/W}
\end{aligned}
$$

VLL data seems to fit well although one order of magnitude lies between simulated and experimental pulse duration, and also between VLL and the IMD simulations performed. The results from the latter differ by a factor of 1.5–1.7 compared to VLL values. However, both aerospace parameters computed by IMD were not saturating within computation time and are likely to approach VLL findings. Phipps et al. and other groups did not yet publish findings about pulse durations around the simulated one, i.e. 21 ps. However, they investigated the few 100 fs regime [15] for different materials. They found a slight decrease in C_m compared to the 100 ps regime and state an optimal fluence for which they published C_m and I_{sp}. Although optimal fluence is not directly comparable, the order of magnitude for both aerospace parameters remains the same as for longer pulses. The author's inference is that both will not differ much in the regime in between [15]. Experimental and modeling values are given in Table 3. Since no ultrashort pulse effects are included, a deviation in a factor of 2 from experimental findings is observed. Open questions arise from the simulations with IMD. Longer short pulse simulations are within reach and the future will show a more detailed behavior in this regime. Additionally, the estimation of the aerospace parameters has to be improved.

4 Performance and Benchmarks

The general performance of IMD has been addressed in the basic papers on the implementation of IMD [17,24], where benchmarks are also given. The performance on more recent supercomputer architectures and the performance in simulations of laser ablation especially has been addressed in previous HLRS reports [18–20].

The simulation module for laser ablation was originally intended for fs-pulses only. The simulation of satellite propulsion uses pulses up to ns which leads to much longer simulation runs involving inhomogeneities generated by expansion of the probe and the ablated material. The result is a degradation of the performance.

Rough benchmark behaviours can be seen in Fig. 7. The simulations were carried out on an Intel Xeon Processor X5670 (Nehalem-EP) with 12 threads (6 physical), running at 2.93 GHz and one or several nodes on HLRS Hermit CRAY XE6 system, respectively. Each node has 32 threads (16 physical), being a Dual Socket AMD

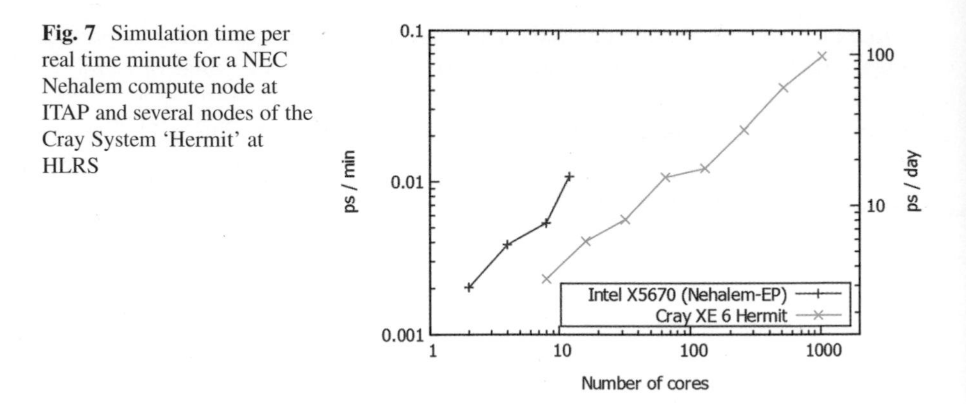

Fig. 7 Simulation time per real time minute for a NEC Nehalem compute node at ITAP and several nodes of the Cray System 'Hermit' at HLRS

Interlagos CPU at 2.3 GHz. All simulations were carried out with a MD system of 12.5 million atoms. A linear scaling behaviour of cores to simulated time over real time in ps/min can be inferred. Although the scaling behaviour and total simulation time on the Nehalem CPU seems to surpass the multi-core system in first place, it reaches its maximum for 12 threads and is only expandable through interconnection of several nodes. Here the optimized system of HLRS plays off the standard simulation node. Still, there is a factor of 6 in total node number to achieve the same ratio of simulation to real time as Nehalem CPU. However, with increasing node number of course the absolute ratio also increases, leading to total simulation times of several 100 ps per day in the range of 1,000 cores.

A general dynamical load balancing scheme has been implemented by Begau and Sutmann (High Performance Computing Center for Materials Science in Bochum) in IMD. In the case of a sphere colliding with a plate a performance gain of a factor of about two could be achieved for 64–256 CPUs. The data indicate that this gain will not decrease for higher CPU numbers. Since the colliding sphere setup is similar to inverse ablation it is likely that the load balancing can be applied to ablation also. Unfortunately, no benchmark data are currently available. The dynamical load balancing scheme of Begau and Sutmann will be adapted expecially to the simulations of laser ablation in the next period of the laser ablation subproject B.5 of SFB 716.

5 Summary

The fully parallelized open source package IMD was evaluated with respect to a new application in the aerospace field. Although underlying limitations and not direct comparability with experiments was given, it is suitable for calculations regarding MICROLAS propulsion. In near future investigations will be focused in first place on ultrashort laser pulses in the femtosecond regime to ensure direct comparability. For longer laser pulses absorptivity will play a main role, changing over time during irradiation. Long term planning will allow for three dimensional spatially resolved crater forming.

Acknowledgements Financial funding from the German Science Foundation DFG for the Collaborative Research Center SFB 716 "Dynamic simulations of systems with large particle numbers" in subproject B.5 "Laser ablation: from simple metals to complex materials" is greatly acknowledged.

References

1. Armano, M., Benedetti, M., Bogenstahl, J., Bortoluzzi, D.: LISA Pathfinder: the experiment and the route to LISA. Class. Quant. Grav. **26**(9), 094001 (2009)
2. Bäuerle, D.: Laser Processing and Chemistry. Springer, Berlin/Heidelberg (2011)
3. Dittus, H., van Zoest, T.: Applications of microthrusters for satellite missions and formation flights scenarios. AIP Conf. Proc. **1402**, 367–373 (2011)
4. Ercolessi, F., Adams, J.B.: Interatomic potentials from first-principles calculations: the force-matching method. Europhys. Lett. **26**, 583 (1994)
5. Hüttner, B., Rohr, G.: On the theory of ps and sub-ps laser pulse interaction with metals i. surface temperature. Appl. Surf. Sci. **103**, 269–274 (1996)
6. Ihlemann, J., Beinhorn, F., Schmidt, H., Luther, K., Troe, J.: Plasma and plume effects on UV laser ablation of polymers. In: Phipps, C.R., et al. (eds.) High-Power Laser Ablation V. Proceedings of SPIE, vol. 5448, pp. 572–580. SPIE, Bellingham (2004)
7. Ivanov, D.S., Zhigilei, L.V.: Combined atomistic-continuum modeling of short-pulse laser melting and disintegration of metal films. Phys. Rev. B **68**, 064114 (2003)
8. Karg, S., Fedotov, V.: Investigation of laser-ablative micropropulsion as an alternative thruster concept for precise satellite attitude and orbit control. In: ONERA-DLR Aerospace Symposium 2013, Palaiseau, 27–29 May 2013
9. Karg, S., Scharring, S., Eckel, H.-A.: Microthruster research activities at dlr stuttgart – status and perspective. AIP Conf. Proc. **1402**, 374–382 (2011)
10. Lide, D.R. (Ed.): CRC Handbook of Chemistry and Physics, 78th edn. 1997–1998. CRC Press, Inc., Boca Raton, New York, London (1997)
11. Lin, Z., Zhigilei, L.V., Celli, V.: Electron-phonon coupling and electron heat capacity of metals under conditions of strong electron-phonon nonequilibrium. Phys. Rev. B **77**(7), 075133 (2008)
12. Messerschmid, E., Fasoulas, S.: Raumfahrtsysteme, 4th edn. Springer, Berlin/Heidelberg (2011)
13. Phipps, C.R., Birkan, M., Bohn, W., Eckel, H.-A., Horisawa, H., Lippert, T., Michaelis, M., et al.: Review: laser-ablation propulsion. J. Prop. Power **26**, 609–637 (2010)
14. Phipps, C.R., Luke, J.R.: Advantages of a ns-pulse micro-laser plasma thruster. AIP Conf. Proc. **664**, 230–239 (2003)
15. Phipps, C.R., Luke, J.R., Funk, D.J., Moore, D.S., Glownia, J., Lippert, T.: Measurements of laser impulse coupling at 130 fs. In: Phipps, C.R., et al. (eds.) High-Power Laser Ablation V. Proceedings of SPIE, vol. 5448, pp. 1201–1209. SPIE, Bellingham (2004)
16. Povarnitsyn, M.E., Andreev, N.E., Levashov, P.R., Khishchenko, K.V., Rosmej, O.N.: Dynamics of thin metal foils irradiated by moderate-contrast high-intensity laser beams. Phys. Plasmas **19**(2), 023110 (2012)
17. Roth, J., Gähler, F., Trebin, H.-R.: A molecular dynamics run with 5.180.116.000 particles. Int. J. Mod. Phys. C **11**, 317–322 (2000)
18. Roth, J., Karlin, J., Sartison, M., Krauß, A., Trebin,H.-R.: Molecular dynamics simulations of laser ablation in metals: parameter dependence, extended models and double pulses. In: Nagel, W.E., Kröner, D.B., Resch, M.M. (eds.) High Performance Computing in Science and Engineering '12, pp. 105–117. Springer, Heidelberg (2012)

19. Roth, J., Karlin, J., Ulrich, C., Trebin, H.-R.: Laser ablation of aluminium: drops and voids. In: Nagel, W.E., Kröner, D.B., Resch, M.M. (eds.) High Performance Computing in Science and Engineering '11, pp. 93–104. Springer, Heidelberg (2011)
20. Roth, J., Trichet, C., Trebin, H.-R., Sonntag, S.: Laser ablation of metals. In: Nagel, W.E., Kröner, D.B., Resch, M.M. (eds.) High Performance Computing in Science and Engineering '10, pp. 159–168. Springer, Heidelberg (2010)
21. Scharring, S.: Post-processing and visualization of simulation data which have been obtained through the use of virtual laser laboratory (2013). http://vll.ihed.ras.ru, unpublished results
22. Sonntag, S.: Computer simulations of laser ablation: from simple metals to complex metallic alloys. PhD thesis, Universität Stuttgart (2011)
23. Sonntag, S., Roth, J., Trebin, H.-R.: Molecular dynamics simulations of laser induced surface melting in orthorhombic $Al_{13}Co_4$. Appl. Phys. A **101**, 77–80 (2010)
24. Stadler, J., Mikulla, R., Trebin, H.-R.: IMD: a software package for molecular dynamics studies on parallel computers. Int. J. Mod. Phys. C **8**, 1131–1140 (1997)
25. Ulrich, C.: Simulation der Laserablation an Metallen. Diploma thesis, Universität Stuttgart (2007)
26. Workshop Mikroantriebe, 16./17.4.2013, Inst. f. tech. Physik, DLR Stuttgart. http://www.dlr.de/tp/Portaldata/39/Resources//Agenda.pdf. Visited at 19 June 2013

MD-Simulations on Metallic Alloys

Martin Hummel, Alen-Pilip Prskalo, Peter Binkele, and Siegfried Schmauder

Abstract Strengthening effects in solid materials depend on different mechanisms, which have been analyzed empirically since very long time. The here presented work shows the investigation on atomistic length scale of precipitate hardening, with detailed look on dislocation obstacle interactions, and grain-boundary-strengthening using Molecular Dynamics to perform numerical simulations on polycrystalline aluminum systems including copper as alloying element. Work hardening effects has been simulated for a notched iron specimen, where periodic loading and compression has been applied. Temperature and density influences are shown for nanoporose iron models.

1 Introduction

Taking a detailed look on various mechanisms lead to the necessity of atomistic length scale simulation methods. Therefore the molecular dynamics (MD) simulation code `IMD` [1] is used for the here presented investigations. It was developed at the Institute of Theoretical and Applied Physics (ITAP) belonging to the University of Stuttgart and is already extensively used on the super computer cluster HERMIT by different researchers.

2 Usage and Performance

2.1 Benchmark

Due to the fact, that there have already been several simulations running on the computation systems of the HLRS, the benchmark and the scalability was discussed

---

M. Hummel (✉) • A.-P. Prskalo • P. Binkele • S. Schmauder
Institut für Materialprüfung, Werkstoffkunde und Festigkeitslehre, Universität Stuttgart, Stuttgart, Germany
e-mail: Martin.Hummel@imwf.uni-stuttgart.de; Alen-Pilip.Prskalo@imwf.uni-stuttgart.de; Peter.Binkele@imwf.uni-stuttgart.de; Siegfried.Schmauder@imwf.uni-stuttgart.de

W.E. Nagel et al. (eds.), *High Performance Computing in Science and Engineering '14*, DOI 10.1007/978-3-319-10810-0_11

in different publications, e.g. by Stadler et al. [1] and other HLRS reports like [2]. Nevertheless one very important point which should be mentioned here is the almost linear scaling of the code with the number of computational nodes. That makes it a perfect tool to be used on massiv parallel computational machines like the CRAY XE6.

2.2 Usage

Traditionally MD simulations are restricted to strongly limited time and length scales. To overcome these boundaries the combination of a simulation code with very good scaling properties, which is given for the here used simulation package `IMD`, and the access to a high number of physical computational nodes is needed. Both traditional boundaries are intended to be left behind with the following presented projects. With up to six million atoms, the systems which are simulated are computationally very demanding and the intention for the year of 2014 is to simulate even larger systems with 50 million atoms or even more. These large systems are necessary to observe the transition from the inverse Hall-Petch effect to the Hall-Petch effect. A usual simulation of the switching from the inverse Hall-Petch effect to the regular Hall-Petch effect is carried out on 2,048 physical CPUs for 24 h. The simulated time is a very important factor for cyclic deformations as well. There many cycles have to be simulated, which is not occupying a very high number of processors, but they are needed for longer time. To obtain 50 cycles a simulation has to be restarted for about 5 times if 512 CPUs are used.

3 Model

The Model we were using is in most cases a polycrystalline system. This configuration consists of a predefined number of grains, divided by a Voronoi tessellation using the program QHull [3], in which the atoms are positioned by our tool NanoSculpt. Depending on the used size of the simulation box and the number of grains, it is possible to construct polycrystalline structures with different mean diameters. In Fig. 1 two structures containing about six million atoms are displayed. The left structure is made of 10 grains, whereas the right system consists of 100 grains, resulting in a diameter of 27 and 12.5 nm. Higher diameters are achievable by this way of construction, though this leeds to higher computational effort to carry out the simulation. The inter atomic potential used for the here presented simulations are of EAM-type (Embedded Atom Method) [4]. For the aluminum-copper systems the potential by Sheng et al. [5] and for Fe systems the ternary potential for Fe-Cu-Ni by Bonny et al. [6] have been applied.

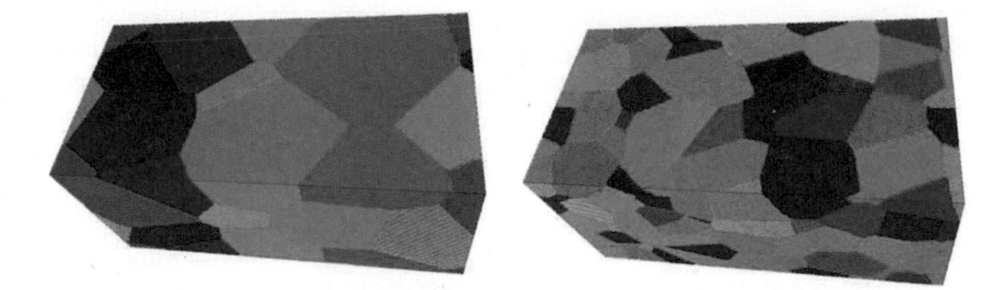

Fig. 1 Two polycrystals with 10 grains on the *left* and 100 grains on the *right side*. The simulation box contains approximately six million aluminum atoms in a volume of 74 · 37 · 37 nm. The distribution is obtained via Voronoi tessellation using the program QHull [3]

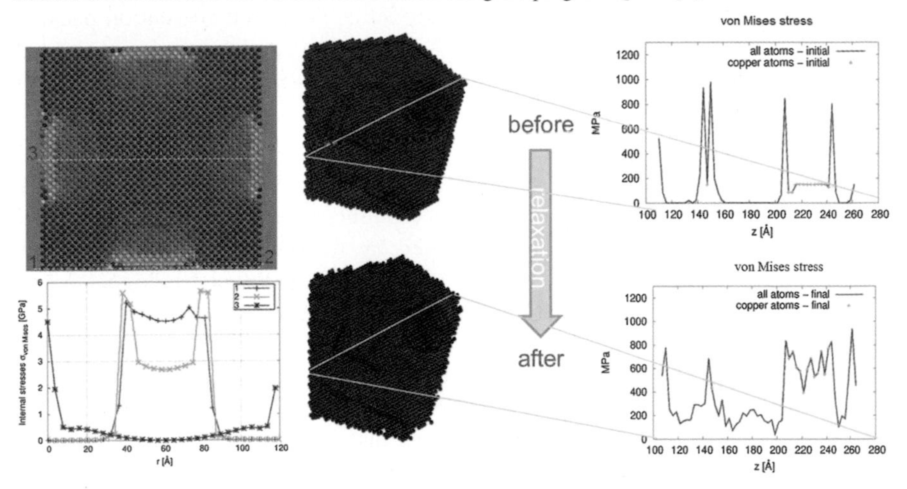

Fig. 2 Cross section of a single crystal with two GP-1 zones under periodic boundary conditions on the *left* and on the *right* a grain of a polycrystal with differently oriented GP-1 zones. The analyzed tension is the von Mises equivalent stress, which was used for coloring

4 Results

4.1 Coherence and Internal Stresses

The binary system of an aluminum copper alloy, where Gunier-Preston (GP) zones are formed as precipitates in higher aging stages, has been observed. The mutually different bond lengths of aluminum and copper atoms in the sample structure lead to coherent tensions. To quantify these residual stresses, the von Mises equivalent stress was used. It was evaluated for the GP-1-zone in the single crystal as well as in multi-grain system. In the crystal, two GP-1-zones were introduced and considered, the stress profile is shown along the in Fig. 2 marked paths. Von Mises equivalent stresses of about 5 GPa are formed at the aluminum-copper interface. The second path provides for the aluminum atoms, which are positioned parallel to the copper plate, a stress value, which is reduced to approximately half of the value of the copper atoms. The range of the stress field along the GP-zone is limited to a few atomic layers. In a direction normal to the disc, however, the tension is noticeable after ten lattice constants. In the grain of the polycrystal in Fig. 2 a similar picture emerges at the beginning of simulated relaxation as in the single crystal. The edges of the GP zones cause high tensions in the configuration. In addition to the previously analyzed effects also the grain boundary areas are under tension. After the molecular dynamics relaxation it is shown that the areas of high residual stresses are expanded. Over the entire region of the cross section there is a non-zero stress. It is further noted that the value of the residual stress at the edges of GP-zones is approximately equal to the values calculated for the atoms at the grain boundaries.

4.2 Dislocation Obstacle Interactions

For the isolated analysis of the interaction of dislocations with the obstacle fields represented by the copper atoms, a monocrystalline system is used. In this single crystal, first, an edge dislocation is inserted by removal of two half-layers of aluminum atoms, afterwards aluminum atoms are substituted by copper atoms according to the desired precipitation. On the aluminum copper system produced in this way shear simulations were performed. In Fig. 3a different stages of the cutting process of a dislocation through a GP-1-zone are exemplarily depicted. Shown are atoms which have a different number of nearest neighbors compared to a perfect fcc-lattice, stacking faults between the partial dislocations and copper atoms, for distinctive marks in the course of the shear stress over the shear. After the impact

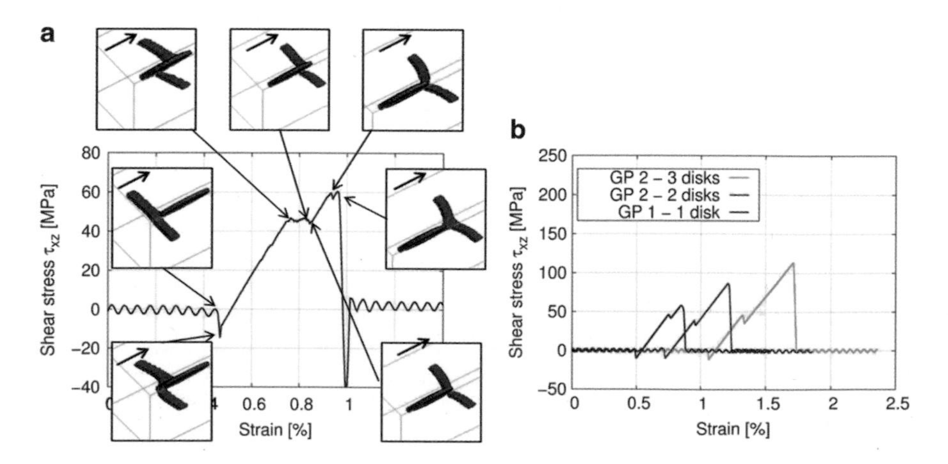

Fig. 3 Different phases of cutting a GP-1-zone in the (001)-orientation by a dislocation (**a**). Transfer of the CRSS during the cutting of the GP-zone with a plurality of disks (**b**)

of the dislocation on the GP-1-zone, one half of the dislocation moves leading along the copper disk. With continued shearing, the second half of the displacement is following up and it involves a further increase of the shear force necessary to establish the depinning. The depinning stresses (CRSS) for the different copper concentrations have been identified. For the GP zones, the value of the CRSS at the same copper concentration increases significantly compared to the value of dissolved copper. The increase of the CRSS in the transition from GP-1- to GP-2-zone is not of that extend. For the (100)- and the (010)-orientation of the GP zone an additional second copper disk increases the CRSS by about 20 % from 165 to 200 GPa, more discs have no additional effect. In the (001)-oriented GP-zone, the CRSS increases linearly with the number of disks, but the increase per slice is less than the CRSS of a GP-1-zone (see Fig. 3b).

4.3 *Influence of Copper on the Tensile Strength*

Molecular Dynamics Simulations have shown that the substitution of aluminum with copper atoms leads to a change in the maximum stress in tensile tests. However, contrary to the experimental experience the maximum stress of the system is reduced by the impurity atoms, as shown in Fig. 4. No dislocations are initially present in the simulations. Therefore the maximum value of the stress is the one which is necessary to generate a dislocation. Here, the inserted copper atoms act as lattice defects, which facilitate the formation of dislocations, so a network of dislocations is build up in the course of the simulation. Comparing the different precipitations according to their aging stages, it is observed that the same number of copper atoms lead to a larger reduction when they are constituted in the discs of GP-zones compared to solid solution crystal.

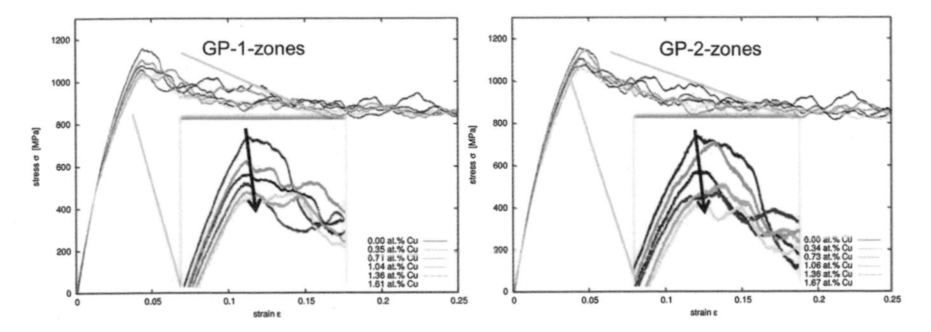

Fig. 4 Decreasing tensile strength with increasing copper concentration in the form of GP-1 *left* and GP-2 zones *right*

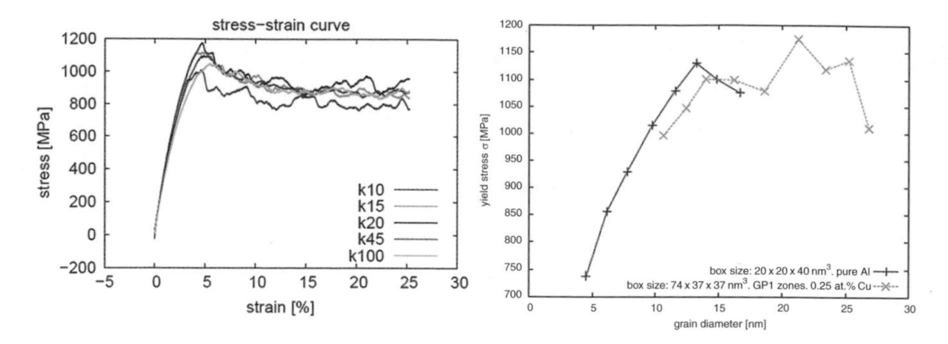

Fig. 5 On the *left* the shape of the stress-strain curves for five selected grain numbers k10, k15, k20, k45 and k100 which correspond to grain diameters from 26.8 to 12.4 nm. On the *right* the maximum stress of the tensile tests for pure polycrystalline aluminum and added GP-1-zones as a function of the grain diameter is shown

4.4 Inverse Hall-Petch Effect

For examining the transition of the inverse to the normal Hall-Petch effect, poly crystals are produced in the computer, having the same volume of the simulation cell but differ in the number of grains. It results in a variation of the average grain diameter of 5–27 nm as exemplarily illustrated in Fig. 1. In Fig. 5 five stress-strain curves are shown. They consist of an elastic part, followed by a peak of the stress. Afterwards the stress decreases to a plateau, at which the values fluctuate slightly. As the relevant variable for the strength the maximum stress of the stress-strain curve was used. The influence of the average grain diameter on the strength was evaluated. In the simulations presented here, a change of the Hall-Petch to the inverse Hall-Petch effect was observed with an average particle diameter of 13.5 nm for pure aluminum. The addition of copper in the appearance of GP-1-zones is resulting in a displacement of the particle diameter – dependent maximum stress to a larger average diameter of 21.25 nm.

4.5 Cyclic Deformation of Notched Iron

The system we investigated contained about half a million iron atoms. They form a cuboid of the size $28.6 \cdot 14.3 \cdot 14.3\,\text{nm}^3$, where a notch $9 \cdot 14.3 \cdot 1.5\,\text{nm}^3$ with triangular notch tip was inserted along the (110) plane. Cyclic deformation of the simulation box was applied in [001]-direction. Therefore the z-component of the simulation cell was elongated with a constant rate of $5 \cdot 10^{-7}$ each time step. After reaching a distension of 7 % we applied pressure at the same rate until we reach 7 % of compression. This procedure was repeated continuously. Periodic boundary

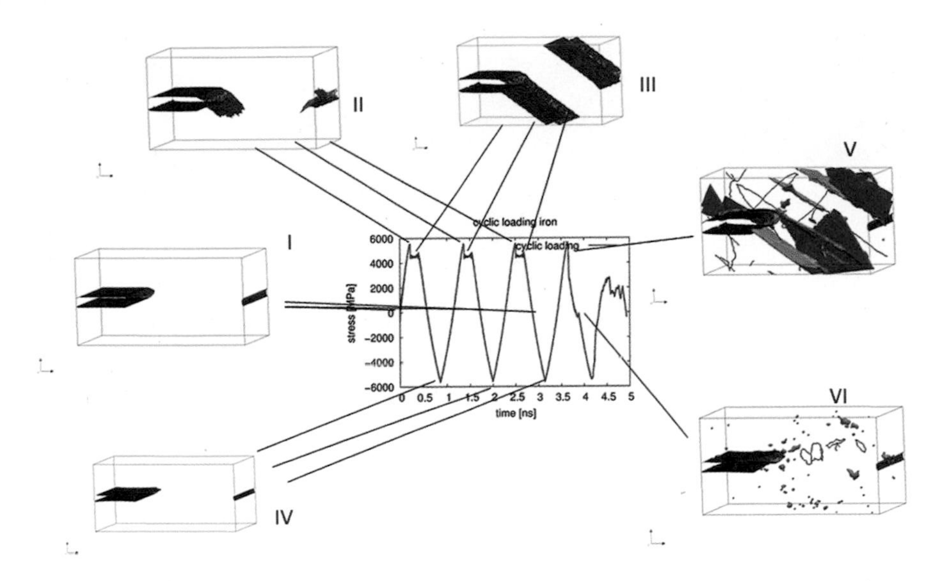

Fig. 6 Stress (MPa) in z-direction in terms of the time (ns). System configurations at different times were depicted. In *blue* are according to DXA [7] "defect surfaces", in *red* are dislocations. The view is from lower left

conditions were used in every direction. Along the cyclic loading simulation, different stages of the systems appeared, which are depicted in Fig. 6.

- Stage I: Configuration under no pressure.
- Stage II: Initiation of reversible local restructuring under tensile loading.
- Stage III: Formation of one continuous plane with non bcc structure.
- Stage IV: Compression leads to a bending of the notch surfaces.
- Stage V: During the fourth loading cycle dislocations are initiated. The Dislocation Extraction Algorithm (DXA) [3] detects "defect surfaces". "The defect surface consists of those parts of the interface mesh, which have not been swept by elastic Burgers circuits." (DXA manual)
- Stage VI: Dislocations remain in the system still when there is no pressure on the system.

For more detailed information the published article [8] is recommended.

4.6 *Nanoporose Iron*

A representative cut-out of the class of advanced materials is investigated with the MD simulations on nanoporose α-iron. Investigations on this system have been carried out in the bachelor thesis of C. Böhm [9]. A polycrystal has been constructed, where the number of grains is chosen according to the number of closed

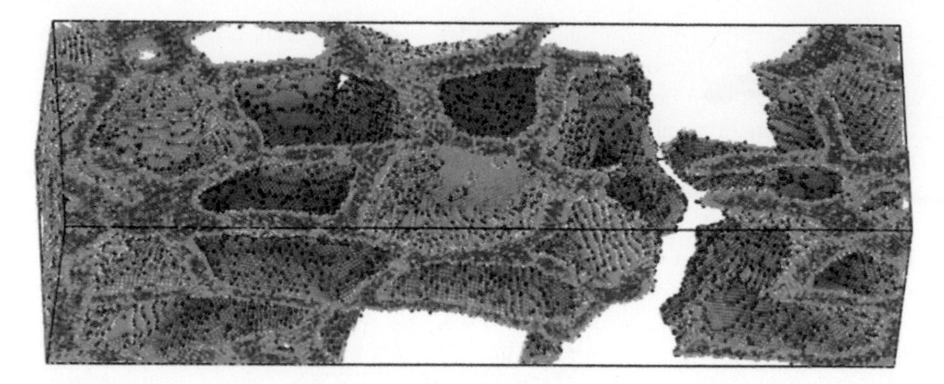

Fig. 7 Iron foam with 20 pores consisting of about 600,000 atoms in a simulation box with dimensions $20 \times 20 \times 40\,\mathrm{nm}^3$ after 60 % elongation. The atoms are colored via their potential energy. Periodic boundary conditions are applied in all three dimensions

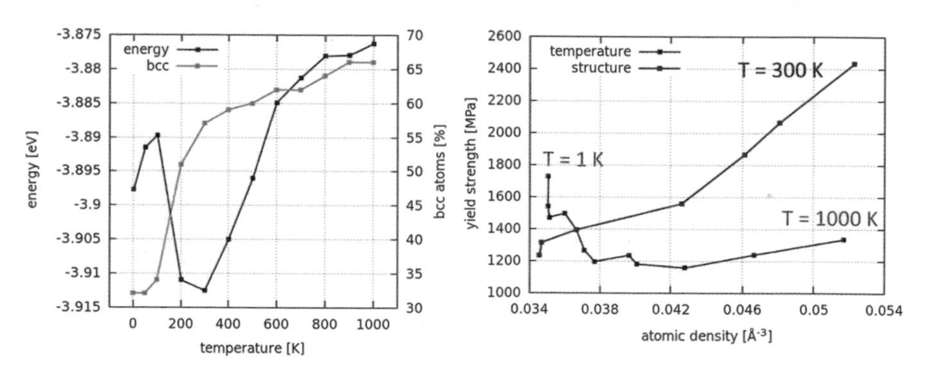

Fig. 8 On the *left side* the fraction of bcc atoms in depending on the applied temperature. It is to remark that surface atoms are detected as non bcc atoms, even if they are positioned on a perfect bcc lattice point. Therefore, the actual fraction is even higher. On the *right* sight is the ultimate tensile strength over the density depicted. The *blue curve* are the configurations which were constructed with different porosities and the *brown curve* is the one structure which was heat treated at different temperatures, where the curves on the left side are from

cells which were desired. Afterwards the bulk material of each grain is removed, that only the grain boundaries remain. During applied thermal treatment the atoms of the former grain boundaries are restructuring and form to a high percentage a perfect bcc structure. Some defects remain, especially at junctions. In dependence of the radius, up to which atoms are taken into account as grain boundary atoms, different porosities have been obtained. As an example a configuration after 60 % loading is shown in Fig. 7.

With the in this way produced initial configurations tensile tests have been performed until failure and therefore material parameters determined. To present one representative of the observed properties the tensile strength is shown in Fig. 8. The necessity of thermal energy for the reordering of the iron atoms can be observed here on the left side. Below 100 K the potential energy increases, whereas the fraction of bcc atoms remains nearly constant. Further increase of the temperature leads to a rapid increase of the quota of bcc atoms and a related decrease of potential energy. This reformation is also visible on the brown curve on the right side, where the ultimate tensile strength is increasing even though the temperature is increased. For even higher temperatures the reordering rate slows down, the potential energy increases due to thermal activation. On the right side in the brown curve there are the densities of the structures at the corresponding temperatures given with their ultimate tensile strengths. For comparison the blue line is added, where the different density is obtained by a different cutoff radius during the constructing process and the simulations have been carried out at 300 K. It is observed that the porose structure is weakened through the higher temperature, but on the other hand the density is increased and therefore the porosity is decreased.

5 Summary

Various simulations have been performed on single and polycrystalline structures. To obtain more realistic results and better comparability with experiments the system sizes of the models should further increase. Especially the investigations on the Hall-Petch effect could lead to the connection between often relatively high ultimate tensile stresses in MD and the lower tensile stresses obtained in experiments by directly showing the influence of the used grain size. For cyclic processes the main focus is in expanding the number of cycles for which the load is applied to get better insight into fatigue processes. Therefore, the further use of the super cluster system HERMIT is irreplaceable.

References

1. Stadler, J., Mikulla, R., Trebin, H.-R.: IMD: a software package for molecular dynamics studies on parallel computers. Int. J. Mod. Phys. C **8**, 1131–1140 (1997)
2. Roth, J., Trichet, C., Trebin, H.-R., Sonntag, S.: Laser ablation of metals. In: Nagel, W.E., Kröner, D.B., Resch, M.M. (eds.) High Performance Computing in Science and Engineering '10, pp. 159–168. Springer, Heidelberg (2011)
3. Barber, C.B., Dobkin, D.P., Huhdanpaa, H.: The Quickhull algorithm for convex hulls. ACM Trans. Math. Softw. **22**(4), 469–483 (1996)
4. Daw, M.S., Baskes, M.I.: Embedded-atom method: derivation and application to impurities, surfaces, and other defects in metals. Phys. Rev. B **29**(12), 6443–6453 (1984)
5. Cheng, Y.Q., Ma, E., Sheng, H.W.: Atomic level structure in multicomponent bulk metallic glass. Phys. Rev. Lett. **102**, 245501-1–245501-4 (2009)
6. Bonny, G., Pasianot, R.C., Castin, N., Malerba, L.: Ternary Fe-Cu-Ni many-body potential to model reactor pressure vessel steels: first validation by simulated thermal annealing. Philos. Mag. **89**, 3531–3546 (2009)
7. Stukowski, A., Bulatov, V.V., Arsenlis, A.: Automated identification and indexing of dislocations in crystal interfaces. Model. Simul. Mater. Sci. Eng. **20**, 085007–085023 (2012)

8. Božić, Ž., Schmauder, S., Mlikota, M., Hummel, M.: Multiscale fatigue crack growth modelling for welded stiffened panels. Fatigue Fract. Engng. Mater. Struct. **37**(9), 1043-1054 (2014)
9. Böhm, C.: Molekulardynamik-Simulationen an nanopor¨sem Eisen zur Evaluierung des Einflusses der Porosität auf die mechanische Belastbarkeit, Bachelor thesis, 2013

Surface Charge of Clean $LiNbO_3$ Z-Cut Surfaces

S. Sanna, U. Gerstmann, E. Rauls, Y. Li, M. Landmann, A. Riefer, M. Rohrmüller, N.J. Vollmers, M. Witte, R. Hölscher, A. Lücke, C. Braun, S. Neufeld, K. Holtgrewe, and W.G. Schmidt

Abstract The geometry of the polar $LiNbO_3$ (0001) surface is strongly temperature dependent. In this work the surface charge associated to various surface terminations is estimated from *first-principles* calculations. All stable terminations are found to lower the polarization charge, showing that the surface charge compensation is a major driving force for surface reconstruction.

1 Introduction

Surfaces of ferroelectric materials, in particular the lithium niobate (0001) surface (LN Z-cut), are widely exploited, e.g., for surface acoustic wave devices [1] or to grow group III nitrides with spatially varied polarity control [2]. Also artificial photosynthesis [3], photocatalytic dye decolorization [4] as well as activation of charged biomolecules [5] were demonstrated. The high surface electric fields were found to efficiently pole electro-optic polymers [6] and to lead to the reversible fragmentation and self-assembling of nematic liquid crystals [7]. Of particular interest is the possibility to switch the surface polarization, and thus the surface chemistry [8].

Given these exciting applications, surprisingly little is known about the microscopic LN surface structure. Only very recently atomically resolved Atomic Force Microscopy (AFM) images of LN Z-cut could be obtained [9]. While no surface reconstruction was found for samples annealed at 1,270 K, a series of structural transformations, including a $\sqrt{7} \times \sqrt{7}$ R19° surface reconstruction [10] were found at samples annealed at lower temperatures. This was interpreted as the result of different charge compensation mechanisms: At low temperatures foreign adsorbates compensate the surface polarization charge. If the temperature is sufficiently high to drive off the adsorbates, surface reconstructions form that lower the electrostatic energy. Finally, at temperatures close to the Curie temperature of LN, the sponta-

S. Sanna • U. Gerstmann • E. Rauls • Y. Li • M. Landmann • A. Riefer • M. Rohrmüller • N.J. Vollmers • M. Witte • R. Hölscher • A. Lücke • C. Braun • S. Neufeld • K. Holtgrewe • W.G. Schmidt (✉)
Lehrstuhl für Theoretische Physik, Universität Paderborn, 33095 Paderborn, Germany
e-mail: W.G.Schmidt@upb.de

W.E. Nagel et al. (eds.), *High Performance Computing in Science and Engineering '14*, DOI 10.1007/978-3-319-10810-0_12

neous polarization, and thus the surface charge, are strongly reduced and the surface reconstructions are quenched [10].

In the present work an approximate method to estimate the surface charge within the density-functional theory (DFT) calculations using periodic boundary conditions (PBCs) is proposed and used in order rationalize the structural models proposed in Ref. [10] in terms of surface charge compensation.

2 Methodology

As in our previous studies on $LiNbO_3$ surfaces [10–13], we perform self-consistent DFT calculations using the Vienna Ab initio Simulation Package (VASP) [14]. All-electron projector-augmented wave potentials within the PW91 formulation of the generalized gradient approximation (GGA) [15] are used. A plane-wave basis set including waves up to an energy of 400 eV is used to expand the electronic orbitals. The atomic positions have been determined minimizing the Hellmann-Feynman forces acting on the single atoms to be below 0.02 eV/Å. A $4 \times 4 \times 1$ Γ-centered Monkhorst-Pack [16] k-point mesh was used to carry out the integration in the Brillouin zone for the simulation of truncated bulk, unreconstructed and $\sqrt{3} \times \sqrt{3}$ reconstructed surfaces. A $2 \times 2 \times 1$ k-point mesh was chosen to sample the much smaller Brillouin zone of the slabs modeling $\sqrt{7} \times \sqrt{7}$ reconstructed surfaces.

VASP offers highly customizable parallelization schemes, which allow for performance optimization on different computational architectures. In particular, parallelization (and data distribution) over bands, parallelization (and data distribution) over plane wave coefficients, as well as parallelization over k-points (no data distribution) are implemented. All of them have to be used at the same time on massively parallel systems or modern multi-core machines, in order to obtain high efficiency. The holds obviously for the HLRS CRAY XE6, the main computational resource used for this project. The performance of the CRAY XE6 in combination with the available parallelization routines has been tested and optimized. Thereby a $LiNbO_3$ slab with C_1 symmetry containing 361 atoms for a total of 2,174 electrons distributed over 1,334 orbitals serves as a realistic test system. All tests have been performed using the VASP Version 5.3.2 and employ a $4 \times 4 \times 4$ k-point mesh, corresponding to 36 irreducible k-points in the Brillouin zone.

The first step of the parallelization procedure is the distribution of the workload related to each orbital on a certain number of cores. Several choices are possible. The simplest option (and default strategy) is to treat one orbital by one core, implying distribution over plane wave coefficients only: all cores will work on every individual band, by distributing the plane wave coefficients over all cores. This is the optimal setting for machines with a single core per node and small communication bandwidth but performs rather inefficiently on the HRLS CRAY. This mode is characterized by heavy memory requirements, as the non-local projector functions must be stored entirely on each core. Furthermore, substantial

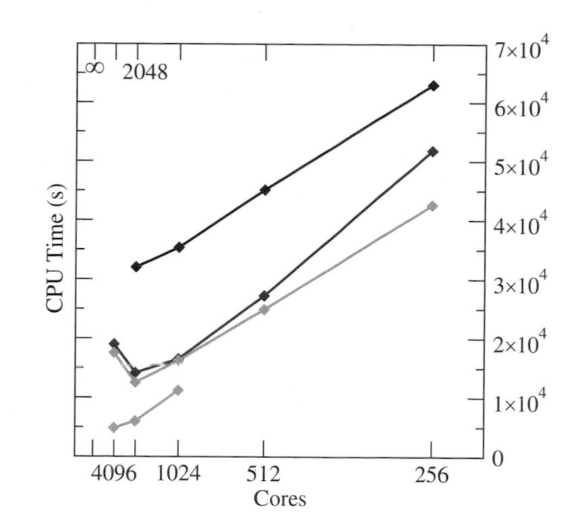

Fig. 1 CPU time on the HRLS CRAY XE6 for the self consistent calculation of the electronic structure of a $LiNbO_3$ slab within different parallelization schemes. See text for details

all-to-all communications are necessary to orthogonalize the band. This setting is usually very slow and should thus be avoided on massively parallel systems.

A better choice for system such as the HRLS CRAY is to treat each orbital with a number of cores corresponding either to the square root of all available cores (at least approximately) or to the number of cores per compute node, e.g. 32. In addition, it also significantly improves the stability of the code due to reduced memory requirements. Therefore, all tests are performed treating each orbital with a number of cores corresponding roughly to the square root of all available cores. The results of the tests on this configuration are reported in Fig. 1 (red line). This setup results in a roughly linear scaling up to 2,048 cores. Beyond this value no speedup is observed anymore. Within the described approach for parallel computing, it is possible to steer the data distribution mode. In particular, the plane wise data distribution in real space can be activated. In this way the communication related to the FFTs can be strongly reduced. Unfortunately, the resulting load balancing is worsened, so that the opportunity of switching on the plane wise data distribution must be tested on the particular computational architecture and in dependence of the number of processors. The results of the corresponding tests are shown in Fig. 1 (orange line). As expected, the advantages of the plane wise data distribution are visible for smaller numbers of processors and vanish for 1,024 cores where load-balancing problems prevail.

The VASP implementation of DFT makes heavy use of mathematical libraries, such as the (scalable) linear algebra package (Sca)LAPACK. The effect of using ScaLAPACK vs. LAPACK libraries is shown in Fig. 1 (black vs. red line). Obviously, using ScaLAPACK reduces the computational time considerably for the system studied here. Starting from the VASP Version 5.3.2 it is possible to switch on an additional parallelization over k points. Thereby the data is not distributed additionally over the k-points. In particular, it is possible to specify the number of k points that are to be treated in parallel, which has been chosen to be 8 out of

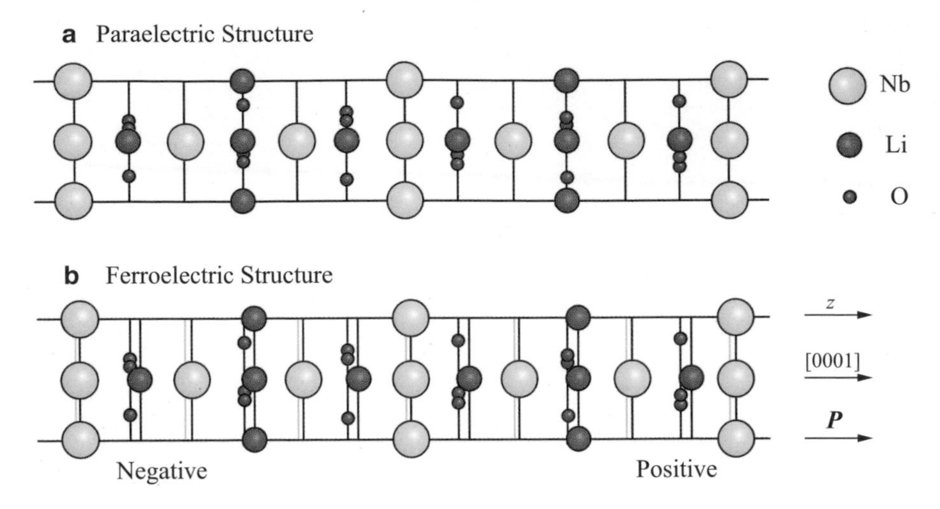

Fig. 2 Schematic representation of the paraelectric (**a**) and ferroelectric (**b**) phases of $LiNbO_3$ along the [0001] crystallographic direction. In the ferroelectric phase Li and Nb atomic layers are shifted with respect to the oxygen sublattice along the crystal *z*-axis. The *gray lines* in the ferroelectric phase represent the positions of the corresponding atomic layers in the paraelectric phase

36 in our test. Then, the set of k points is distributed over 8 groups of compute cores, in a round-robin fashion. This means that N = total-cores/8 compute cores work together on an individual k point. Of course, the number of k points that are treated in parallel must be an integer divisor of the total number of cores. Within this group of N cores that share the work on an individual k point, the usual parallelism over bands and/or plane wave coefficients applies. The results of the corresponding tests are shown in Fig. 1 (blue line). The computational time saving with respect to calculations without k point parallelization amounts to a factor of about two, thereby the scaling is roughly linear up to 4,096 cores. Summarizing, the computational time can be reduced by roughly an order of magnitude with an appropriate choice of the parallelization routines. With a combination of parallelization over bands and k-points a linear scaling at least up to 2,048 cores can be achieved.

Generally, the LN(0001) and (000$\overline{1}$) surfaces are defined as positive and negative Z-cut. The cations in the paraelectric phase are either within (Li^{+1}), or exactly between oxygen layers (Nb^{+5}), as depicted in Fig. 2. In the ferroelectric phase, the cations are shifted along the [0001] crystallographic direction with respect to the oxygen planes, causing a permanent spontaneous polarization parallel to the cationic displacement. Ferroelectric $LiNbO_3$ is thus a stacking of -Nb-O_3-Li- planes along the [0001] direction, and it is not centrosymmetric. This is an issue within the supercell approach, as the two slab surface terminations cannot be made equivalent and only relative energetic comparisons are possible. In this work we model the LN(0001) surface and all its reconstructions with 12 LN trilayers (36 atomic layers) plus surface terminations, resulting in large supercells containing from 60 atoms (for

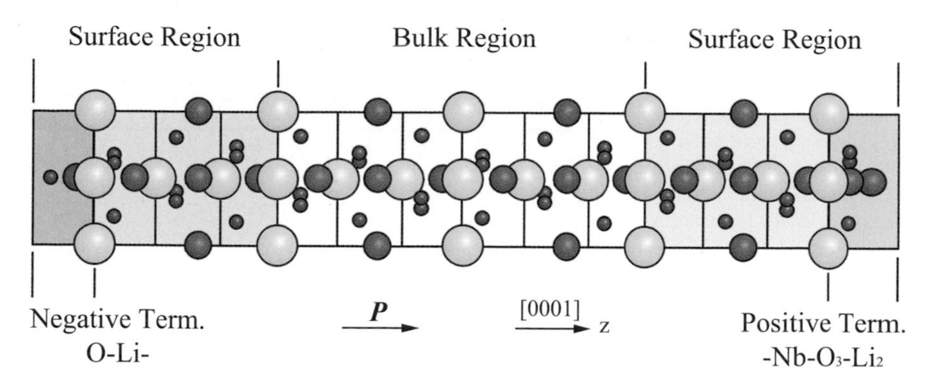

Fig. 3 Slab used to model the thermodynamically stable 1 × 1 (0001) surfaces. The surface termination and the underlying three trilayer (surface region) are free to relax, while the six central trilayer are kept frozen at their bulk positions (bulk region). The positive surface -LN(0001)- is at the right hand side, the negative surface -LN(000$\bar{1}$)- at the left hand side. Atomic color coding as in Fig. 2

the simulation of the truncated bulk) up to 460 atoms for the simulation of $\sqrt{7} \times \sqrt{7}$ reconstructed surfaces. The central part of our slabs (18 atomic layers) are frozen into their ferroelectric positions, while the remaining 18 layers and the surface terminations are free to relax (see Fig. 3). This has two important advantages: On one hand, it saves some computational time. On the other hand, it defines the central part of our slab as a true bulk region, which is not affected by surface effects. A vacuum region of at least 19 Å separates the slab from its periodic images. As we shall see in the following, cells of this size are thick enough to allow for the definition of macroscopic quantities such as inner – $\boldsymbol{E}_D$ – and outer electric field – $\boldsymbol{E}_{EXT}$ – from unit cell averages of electrostatic potential $V_H(\mathbf{t})$ and charge densities $\rho(\mathbf{r})$.

The supercell approach typically makes use of periodic boundary conditions (PBCs), resulting in a vanishing macroscopic internal field $\boldsymbol{E}$ even for ferroelectrics with a spontaneous bulk polarization $\boldsymbol{P}_S^{\text{bulk}}$ (see Fig. 4a). A slab wit a net dipole moment perpendicular to the surface will give rise to an electrostatic potential as illustrated in Fig. 4b. The net dipole moment will in general have contributions from both the bulk polarization and the surface termination. Indeed, a paraelectric material with two non-equivalent surface terminations will give rise to a dipole moment too. This configuration models a surface in the artificial field caused by its neighboring periodic images. Thereby, the electrostatic potential is a continuous function. Neither internal nor external fields have direct physical interpretation as they depend on the cell geometry. Indeed, increasing the thickness of the slab or of the vacuum region the artificial fields become smaller, vanishing in the limit of infinite supercells. In order to correct for the error introduced by the artificial field in finite slabs, dipole corrections can be applied [17, 18]. The correction is introduced by adding an external dipole layer in the vacuum region of the supercell, and leads to the situation depicted in Fig. 4c, with vanishing external electric field.

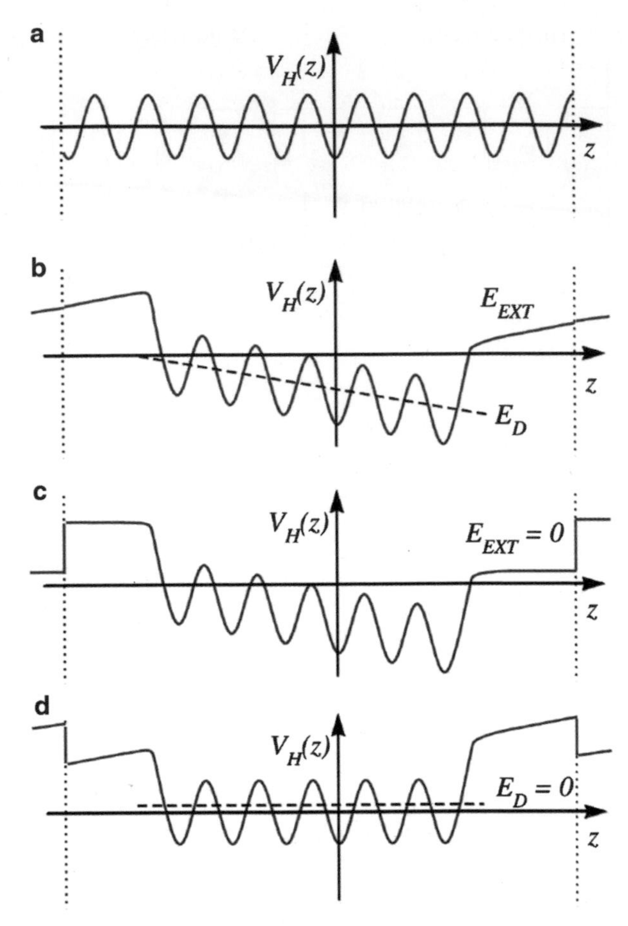

Fig. 4 Planar averaged electrostatic (Hartree) potential of ferroelectric $LiNbO_3$. (**a**) bulk supercell, (**b**) slab of truncated bulk, (**c**) with vanishing external field, and with vanishing internal electric field (**d**). Internal (or depolarization) field $\boldsymbol{E}_D$, external field $\boldsymbol{E}_{EXT}$ and slab/supercell boundaries are indicated

The electrostatic potential is now discontinuous. However, the discontinuity is localized in the vacuum region, where it is supposed not to affect the calculation. The polarization $\boldsymbol{P}$ gives rise to surface polarization charges

$$\sigma_S = \boldsymbol{P} \cdot \hat{\boldsymbol{n}}, \tag{1}$$

where $\hat{\boldsymbol{n}}$ is the surface normal.

Meyer and Vanderbilt [19] pointed out that the depolarization field $\boldsymbol{E}_D$ originating from the surface charges might be large enough to destabilize the ferroelectric state: relaxing a polarized slab under periodic boundary conditions corresponding to vanishing external electric field would result in a paraelectric structure. To overcome this problem they proposed to model polarized slabs with periodic boundary conditions corresponding to vanishing internal field, as illustrated in Fig. 4d. This boundary condition is equivalent to placing the slabs between the grounded plates of a capacitor, and is appropriate to model very thin film geometries, where ferroelectric and dielectric properties are strongly influenced by surface effects [19].

It must be pointed out that none of the described PBCs for the electrostatic potential is intrinsically correct and universally applicable. While the vanishing internal field boundaries are the proper conditions far from the surface [19] and thus ideal for the simulation of the "bulk" of thin films, vanishing external field conditions are more appropriate to model the surface itself. Thus, the choice of the appropriate periodic boundary conditions discussed above depends on the system properties one focuses on. Being interested in surface charges, we adopt vanishing external field boundaries. To prevent the slab to relax into the paraelectric state, the central part of our slabs (18 atomic layers) are frozen into their ferroelectric positions, while the remaining 18 layers and the surface terminations are free to relax. This choice is also justified by the results of Levchenko et al. [20], who pointed out that the structural relaxation in the ferroelectric LN(0001) surfaces is limited to a few outmost atomic layers. The choice of the PBCs affects to some extent the calculated the surface charge. The difference in the surface charge calculated with – Fig. 4c – and without – Fig. 4b – dipole corrections, however, amounts to less than 0.009 C/m^2, which is two orders of magnitude lower than the charge of a LN truncated bulk. For a quantitative estimate of the surface charge, a proper treatment of the pyroelectric effect – currently neglected – will also be necessary.

As seen from Eq. 1, the polarization charge directly depends on the slab polarization. The polarization component perpendicular to the surface is a well defined quantity, and fully determines the surface charge, provided the polarization is homogeneous. The bulk macroscopic polarization is not a well defined quantity within the supercell approach, though. This is due to the fact that the choice of the supercell is not unique: different supercells characterized by different dipole moments may be chosen to model a given crystal. However, this does not affect the calculation of the surface charge related to a given termination. The macroscopic polarization can be correctly calculated as a Berry phase of the electronic orbitals within the modern theory of polarization [21, 22] with respect to a reference phase, usually the corresponding paraelectric phase:

$$\boldsymbol{P} = \int_0^1 \boldsymbol{P}'(\lambda)d\lambda = \boldsymbol{P}(1) - \boldsymbol{P}(0), \tag{2}$$

where $\lambda \in (0, 1)$ is a reaction coordinate which brings the system from the ferroelectric $\boldsymbol{P}(1)$ to the reference (paraelectric) phase $\boldsymbol{P}(0)$. Using the Berry phase approach, we have calculated the spontaneous bulk polarization of stoichiometric LN to be $0.82\,C/m^2$, in good agreement with the commonly accepted value of $0.71\,C/m^2$ [23], and with the value of $0.86\,C/m^2$ calculated in a similar manner by Levchenko et al. [20].

The surface charge σ_S of the positive and negative LN Z-cut is estimated by integrating the planar averaged polarization charge, defined as

$$\rho(z) = \frac{1}{A} \int \int_A \rho(\mathbf{r}) dx dy, \tag{3}$$

from the vacuum center (0) to the cell center (z_{cut}) in both directions

$$\sigma_S = \int_0^{z_{\mathrm{cut}}} \rho(z) dz. \tag{4}$$

Here A is the area of the surface unit cell and $\rho(\mathbf{r})$ is the charge redistribution due to the surface formation (i.e. the difference with respect to the bulk), so that the slab is charge neutral and the integral over the whole supercell is exactly zero.

Furthermore, for the estimation of the surface charge of an ideal bulk cut, we proceed in the spirit of the Berry approach, and calculate the macroscopic surface charge with respect to a reference phase

$$\sigma_S^{cut} = \boldsymbol{P} \cdot \hat{\boldsymbol{n}} = (\boldsymbol{P}_F - \boldsymbol{P}_P) \cdot \hat{\boldsymbol{n}} = \sigma_F^{cut} - \sigma_P^{cut} \tag{5}$$

This procedure yields a surface charge which does not depend on the choice of the slab representing the ideal truncated bulk cut, provided the reference phase is chosen carefully. In our case, the reference structure is a corresponding slab of LN in the paraelectric phase. This method is applied here only for the calculation of an ideal LN bulk cut.

The integration limits of Eq. 4 need to be determined carefully: the charge density oscillations within the slab are larger than the electronic charge accumulation/depletion we want to estimate, and the value of σ_S is extremely sensitive with respect to z_{cut}. We circumvent the choice of some geometrical point z_{cut} to perform "the cut" by dividing the slab into atomic layers. In an small slab consisting of two Nb-O_3-Li tri-layers stapled along the z-direction, the lower trilayer builds the lower (negative) surface, while the upper trilayer builds the upper (positive) surface. Then, the charges of the upper (or lower) slab are summed up to calculate the surface charge. To assign a volume – and the corresponding included charge – to an atomic layer, we use the definition of Bader atomic volumes and Bader charges [24]. This procedure turns out to be very robust and only little influenced by computational parameters. Convergence tests have shown that the calculated surface charge converges quickly with respect to number of grid points in real space. Performing the calculation with or without the core electrons charge does not modify the results within the numerical accuracy.

In summary, we proceed as follows: (i) The total electronic charge of the slab representing a given termination is calculated self-consistently and then partitioned by Bader volumes into upper and lower slab half. (ii) The charge redistribution $\rho(\mathbf{r})$ of Eq. 4 is obtained by subtracting the Bader charges of the corresponding atoms in the bulk phase. (iii) Finally, the charge redistribution of the Bader volumes within the upper and within the lower slab half are summed up and yield the surface charge σ_S.

3 Results

We probe the applicability of the scheme proposed above for the truncated bulk surfaces of LN. Thereby the atomic positions are kept fixed in their bulk positions but the electron distribution is calculated self-consistently. In this case the bulk polarization represents an upper bound of the calculated surface charge since the latter may be partially compensated by the electron transfer through the slab. Considering ferroelectric LN as a stacking of Nb-O_3-Li atomic layers along the [0001] crystallographic direction, three non equivalent atomic cuts can be discriminated. Cutting the crystal above a Li layer (cut 1 in Fig. 5), the slab representing the crystal will be a succession of Nb-O_3-Li trilayers. Cutting the crystal above a Nb layer, the slab will be a succession of O_3-Li-Nb trilayers (cut 2 in Fig. 5). Finally, cutting the crystal above an oxygen layer, the slab will be a succession of Li-Nb-O_3 trilayers (cut 3 in Fig. 5). Twelve trilayers are used for the calculations. For the sake of simplicity, we only consider slabs containing an integer number of LN formula units.

The slabs built to model these cuts are characterized by very different electric dipole moments, both in magnitude and direction, resulting in three different surface charge values σ_F^{cut}. The surface charge of the different cuts has been calculated as described by Eq. 4, and the result are reported in Table 1. However, the total surface

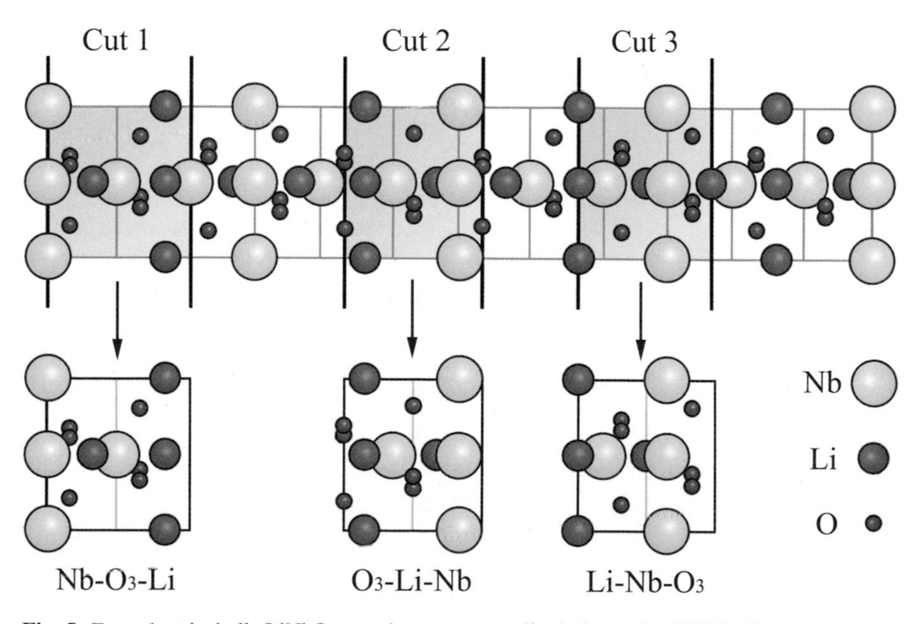

Fig. 5 Ferroelectric bulk $LiNbO_3$ can be cut perpendicularly to the [0001] direction at three non-equivalent planes, resulting in a Li, O_3, and Nb termination (Cut 1, Cut 2 and Cut 3 in the picture). Different cuts are modeled by slabs with different electric dipole moment, both in size and direction. Only slabs containing an integer number of LN formula units are considered

Table 1 Calculated surface charge ($e/(1 \times 1)$ unit cell) of the positive Z-cut for a ferroelectric slab (σ_F^{cut}), for the corresponding reference (paraelectric) phase (σ_P^{cut}) and the value σ_S^{cut} calculated with Eqs. 4 and 5. See Fig. 5 for cut definitions

	Cut 1	Cut 2	Cut 3
σ_F^{cut}	−0.66	1.95	−1.64
σ_P^{cut}	−1.31	1.31	−2.30
σ_S^{cut}	0.65	0.64	0.66

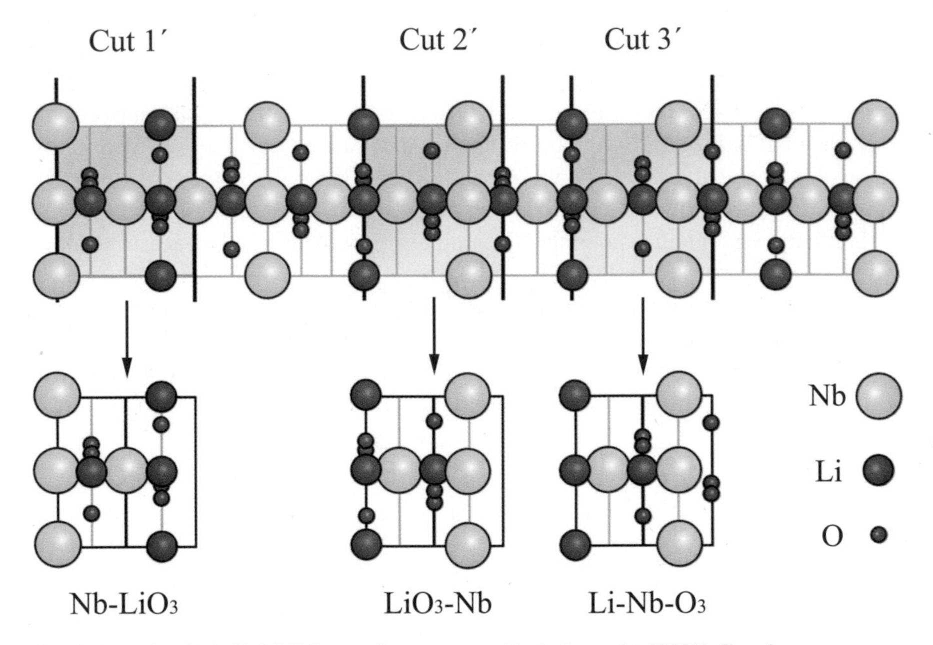

Fig. 6 Paraelectric bulk $LiNbO_3$ can be cut perpendicularly to the [0001] direction at two non-equivalent planes, resulting in a LiO_3 and Nb termination (Cut 1′ and Cut 2′ in the picture). Cut 1′ and Cut 2′ represent the paraelectric reference phase to Cut 1 and Cut 2 of the ferroelectric phase in Fig. 5, Cut 3′ corresponds to the ferroelectric Cut 3

charge σ_S^{cut}has to be calculated with respect to a reference phase, as described by Eq. 5. Thereby, just as in case of the calculation of the volume spontaneous polarization via Berry phase, the correct choice of the (paraelectric) reference phase is crucial. In the case of a LN truncated bulk, it is relatively easy to find the cuts of the paraelectric phase corresponding to those of the ferroelectric phase. These are illustrated in Fig. 6 and denoted cut 1′, cut 2′ and cut 3′. Calculating the corresponding surface charge σ_P^{cut} and inserting it in Eq. 5, one obtains roughly the same value of the surface charge for all the three cuts, namely 0.65 ± 0.01 $e/(1\times1)$-surface unit cell, see Table 1.

Next, σ_S^{cut} has been calculated with slabs containing a different number n of atomic layers. The dependence of σ_S^{cut} on the number of atomic layers is only marginal, see Table 2. However, the value of σ_S^{cut} for cells of infinite size can be

Table 2 Surface charge calculated at the positive LN(0001) for a Cut 1 termination with slabs containing a different number of atomic layers. σ_F^{cut}, σ_P^{cut} and σ_S^{cut} have the same meaning as in Table 1. All values in $e/(1 \times 1)$-surface unit cell

Layer number	18	24	30	36
σ_F^{cut}	−0.62	−0.64	−0.65	−0.66
σ_P^{cut}	−1.28	−1.30	−1.31	−1.31
σ_S^{cut}	0.66	0.66	0.66	0.65

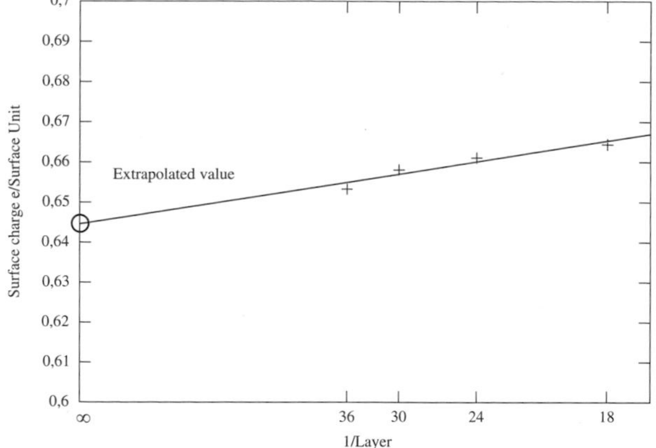

Fig. 7 Surface-charge density σ_S^{cut} as a function of the number n of atomic layer in the slab. The *solid line* is the linear interpolation of the calculated points, the extrapolated value for ideal infinite slabs is indicated

extrapolated plotting its dependence on $1/n$, as shown in Fig. 7. The extrapolated value of the surface charge $\sigma_S^{cut} = 0.64e/(1 \times 1)$ surface unit cell, corresponding to $0.46\,\text{C/m}^2$ is somewhat lower, yet in reasonable agreement with the value expected from the bulk polarization. The deviation from the expected value might be traced back to the uncertainties introduced by the supercell approach as well as to a partial compensation by the electron transfer through the band gap in the calculation, and by the choice of partitioning the cell by Bader volumes. Thus, the calculated value does not allow for a quantitative prediction of the surface charge, however it will suffice for a qualitative discussion.

As a third step we calculated the surface charge for slabs of different periodicity. The calculated values are reported in Table 3. As expected, the charge σ_S^{cut} does not depend on the area of the unit cell used for the calculation.

Our method represents an important step towards the reconciliation of macroscopic and microscopic picture. The surface charge calculated with the three microscopic models is the same, and represents the polarization charge of an ideal bulk cut in our calculation.

After validating the method by application it to an ideal truncated bulk, we proceed with the calculation of the surface charge of realistic surfaces, which are

Table 3 Calculate surface charge ($e/(1\times1)$ unit cell) for the positive Cut 1 termination with slabs of different periodicity. σ_F^{cut}, σ_P^{cut} and σ_S^{cut} have the same meaning as in Table 1

Periodicity	1×1	$\sqrt{3}\times\sqrt{3}$	$\sqrt{7}\times\sqrt{7}$
σ_F^{cut}	−0.62	−0.65	−0.62
σ_P^{cut}	−1.28	−1.31	−1.28
σ_S^{cut}	0.66	0.66	0.66

characterized by relaxation and reconstructions. Levchenko et al. [20] predicted the phase diagram of the LN(0001) from ab initio thermodynamics, finding the stable terminations at the two surfaces to be rather temperature independent. In their work, however, only surfaces with 1×1 periodicity had been considered. Later, AFM measurements and DFT models including cells of different periodicity showed that different reconstructions might be formed at different temperature regimes [10]. Some of them might be inhibited by the presence of adsorbates and can only occur in vacuum. The corresponding phase diagram (now including reconstructed surfaces) has been calculated as a function of the chemical potentials of lithium and oxygen. As the goal of this investigation is to search for a correlation between surface charge and temperature, we first convert the phase diagram of the LN(0001) of Ref. [10] in a function of temperature and Li partial pressure. This can be done within the ideal gas approximation [25], as explained, e.g., in Ref. [13].

The result of this transformation is shown in Fig. 8 for the positive and negative LN Z-cut. A $\sqrt{3}\times\sqrt{3}$ reconstruction built by adding 2 Li atoms to the stable termination is formed at low temperatures at the positive face. With growing temperature, reconstructions containing less Li are formed, till after 1,000 K the 1×1 stable reconstruction predicted by Levchenko et al. [20] is restored. At the negative surface, a $\sqrt{7}\times\sqrt{7}$ reconstruction built adding 2 Li and 2 O atoms to the stable termination is formed at lower temperatures. With increasing temperatures, a further a $\sqrt{7}\times\sqrt{7}$ reconstruction built by adding 2 Li and a single O atoms and a $\sqrt{3}\times\sqrt{3}$ reconstruction obtained by adding a single oxygen atom to the stable termination are formed.

The succession of the surface reconstructions at different temperatures is shown in Table 4. The nominal charge added by the reconstruction at the positive side – also listed in the table – is a decreasing function of the temperature. However, the effective surface charge σ_S is not necessarily proportional to the nominal charge brought by the adsorbed species. Therefore, we calculate σ_S with the method previously illustrated.

At first, the relaxed 1×1 surface is considered. Differently from the truncated bulk termination, where the expected polarization charge is in principle known, there is no reference value for the clean, relaxed LN(0001) surfaces in vacuum. The only measurement of the surface charge of LN(0001) [23] we are aware of has been performed in air. The measured charge refers therefore to an adsorbate-covered surface. A relatively small surface charge of $140\,\mu C/m^2$ is measured for congruently melt, nominally undoped $LiNbO_3$ crystals. This is about $2\cdot10^{-4}$ smaller

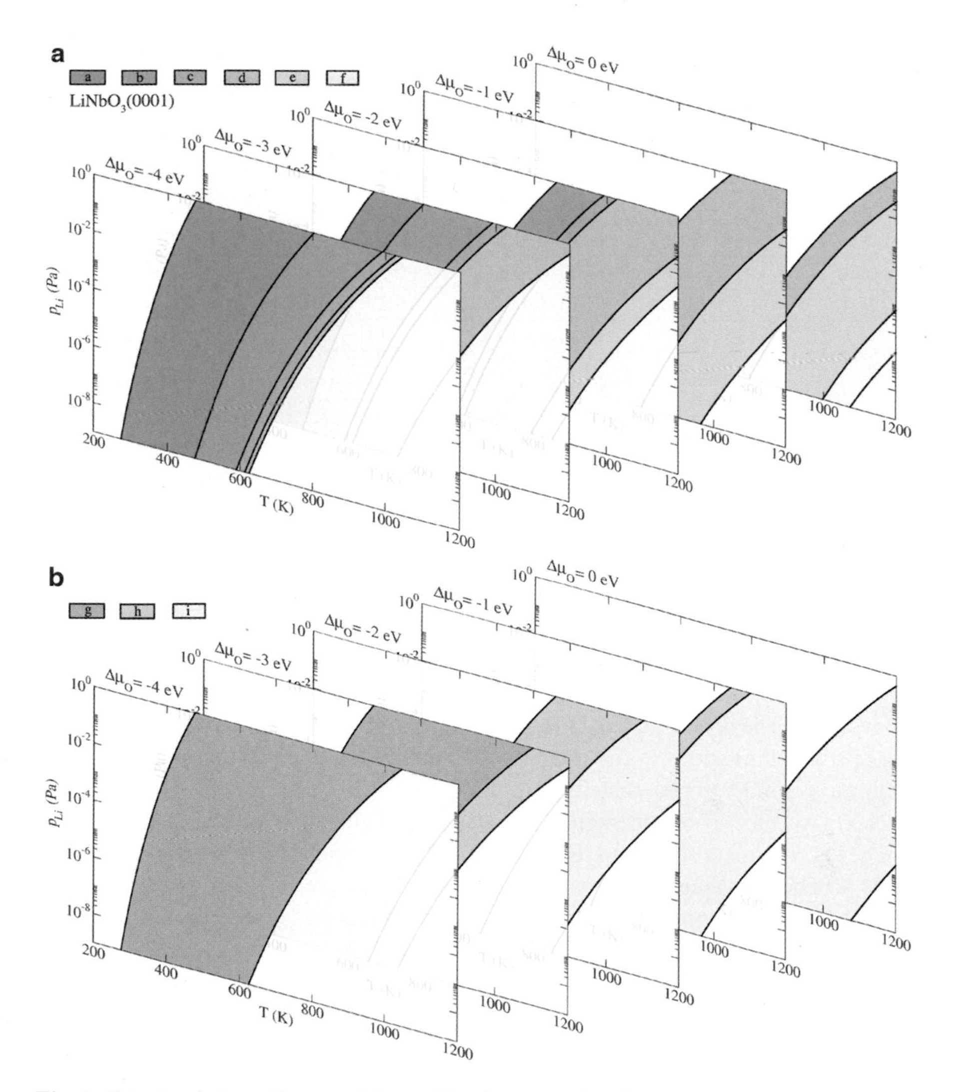

Fig. 8 Calculated phase diagram of the positive (*upper part*) and negative (*lower part*) LN(0001) surface as a function of temperature and pressure. The terminations (a)–(i) are explained in Table 4

than the spontaneous polarization of $0.71\,\mathrm{C/m^2}$, most likely due to the charge compensation by adsorbates.

The calculated surface charge of a $LiNbO_3$ slab terminated according to the stable surface structure (see Fig. 3) amounts to $0.104\ e/(1 \times 1)$ unit cell. This corresponds to about 16 % of the surface charge of the ideal ferroelectric bulk cut. This is in agreement with the predictions of Levchenko and Rappe [20]. In their work, they have proposed a model based on formal oxidation states to explain the stability of the preferred terminations of positive and negative surfaces by

Table 4 Calculated surface charge ($e/(1 \times 1)$ unit cell) for various LN Z-cut surfaces. The temperature range is assigned assuming a Li partial pressure of 10^{-8} Pa. Note that not all the surface reconstructions are allowed for arbitrary values of μ_O, as shown in Fig. 8. The data for positive and negative surfaces are given in the upper and lower part of the table, respectively

Periodicity	Termination	Stability range	Nominal charge	Surface charge σ_S
$\sqrt{3} \times \sqrt{3}$	+2Li (a)	250–450 K	0.66	+0.077
$\sqrt{7} \times \sqrt{7}$	+3Li (b)	450–600 K	0.43	−0.019
$\sqrt{3} \times \sqrt{3}$	+Li (c)	600–620 K	0.33	−0.052
$\sqrt{7} \times \sqrt{7}$	+2Li (d)	620–820 K	0.29	−0.041
$\sqrt{7} \times \sqrt{7}$	+Li (e)	820–1,050 K	0.14	−0.099
1×1	−(f)	1,050–1,150 K	0.00	−0.104
$\sqrt{3} \times \sqrt{3}$	+O (i)	680–1,150 K	0.66	+0.004
$\sqrt{7} \times \sqrt{7}$	+Li+O (h)	620–680 K	0.14	−0.071
$\sqrt{7} \times \sqrt{7}$	+2Li+O (g)	250–620 K	0.00	−0.087

passivation of surface charges with ions. According to this model, the surface charge of a truncated bulk LN(0001) surface is reduced by 80 % by the formation of the stable terminations, a value very close to the reduction by about 84 % calculated here. The experiment [23] shows that the surface charge must be compensated almost completely, because maintaining a non-zero surface charge is energetically unfavorable. The non-complete charge compensation resulting from the calculations is a result of limitations in the models, such as limited unit cell sizes, and most of all limited number of adsorbate types.

Next, we turn to reconstructed surfaces. The calculated values of the surface charge σ_S for reconstructions formed at different temperatures are complied in Table 4. The σ_S values for reconstruction at the positive face are calculated with the stable 1×1 termination at the negative side and vice versa. It is thus possible to estimate the efficiency of the single reconstructions in charge compensation. In all the investigated cases, the formation of an adatom-induced reconstruction yields a polarization charge reduction with respect to the stable 1×1 termination, thus stabilizing the surface. We find the lowering of the surface charge to be roughly proportional to the nominal charge of the adatoms. The $\sqrt{3} \times \sqrt{3}$ reconstruction formed by the addition of two Li-atoms at the positive side and the $\sqrt{3} \times \sqrt{3}$ reconstruction formed by the addition of one O-atom at the negative side (nominal additional charge of ± 0.66 electrons per 1×1 unit cell) lead to an overcompensation of the surface charge.

With increasing temperature surface terminations which only slightly reduce the polarization charge are favored. However, this does not mean that with increasing temperature a larger surface charge will be measured. Indeed, the surface charge in real samples will be a decreasing function of the temperature, as with increasing temperature the cations move closer to their paraelectric configuration, thereby reducing the net polarization. Thus, reconstructions with a minor impact on the surface charge will be sufficient for compensation. At temperatures above 1,200 K, the spontaneous polarization will be low enough that no surface reconstruction is needed and the 1×1 periodicity is recovered. However, word of caution is in order here: The present calculations do not include thermal effects on the bulk polarization. Temperature effects enter exclusively, instead, via the chemical potentials of the surface constituents. This clearly affects the numerical accuracy of the calculated phase diagrams. Nevertheless, the present calculations strongly suggest the surface charge compensation as major driving force for the morphologic transformations observed on LN Z-cut surfaces.

4 Conclusions

A method for the calculation of the macroscopic surface charge within the periodically repeated slab approach has been proposed and applied to ferroelectric LN surfaces. The surface charge of an ideal truncated bulk is calculated to be $\sigma_S^{cut} = 0.46\,\mathrm{C/m^2}$. This value is reduced by about 84 % upon formation of the stable 1×1 surface structure found in Refs. [11, 20]. The surface reconstructions recently revealed by AFM techniques [10] further reduce the polarization charge, thus stabilizing the surface. Indeed, all the thermodynamically stable surface reconstructions lower the polarization charge. This strongly suggests that the surface charge compensation is the major driving force for the observed morphologic transformations. We want to remark that the present models represent clean surfaces of stoichiometric $LiNbO_3$. In congruent material the outgassing of LiO already at moderate temperatures ($\approx$600 K) severely deteriorates the crystalline quality and affects the surface composition. In addition, in real surfaces the presence of external adsorbates, surface defects and step edges will further reduce the surface charge.

Acknowledgements The calculations were done using grants of computer time from the Höchstleistungs-Rechenzentrum Stuttgart (HLRS) and the Paderborn Center for Parallel Computing (PC^2). The Deutsche Forschungsgemeinschaft is acknowledged for financial support.

References

1. Lee, T.C., Lee, J.T., Robert, M.A., Wang, S., Rabson, T.A.: Rabson, surface acoustic wave applications of lithium niobate thin films. Appl. Phys. Lett. **82**(2), 191 (2003)
2. Namkoong, G., Lee, K.K., Madison, S.M., Henderson, W., Ralph, S.E., Doolittle, W.A.: III-nitride integration on ferroelectric materials of lithium niobate by molecular beam epitaxy. Appl. Phys. Lett. **87**(17), 171107 (2005)
3. Stock, M., Dunn, S.: $LiNbO_3$ - A new material for artificial photosynthesis. IEEE Trans. Ultrason. Ferroelectr. Freq. Control **58**(9), 1988 (2011)
4. Stock, M., Dunn, S.: Influence of the ferroelectric nature of lithium niobate to drive photocatalytic dye decolorization under artificial solar light. J. Phys. Chem. C **116**(39), 20854 (2012)
5. Ferris, R., Yellen, B., Zauscher, S.: Ferroelectric Thin Films in Fluidic Environments: A new interface for sensing and manipulation of matter. Small **8**(1), 28 (2011)

6. Huang, S., Luo, J., Yip, H.L., Ayazi, A., Zhou, X.H., Gould, M., Chen, A., Baehr-Jones, T., Hochberg, M., Jen, A.K.Y.: Efficient poling of electro-optic polymers in thin films and silicon slot waveguides by detachable pyroelectric crystals. Adv. Mater. **24**(10), OP42 (2011)
7. Merola, F., Grilli, S., Coppola, S., Vespini, V., De Nicola, S., Maddalena, P., Čarfagna, C., Ferraro, P.: Reversible fragmentation and self-assembling of nematic liquid crystal droplets on functionalized pyroelectric substrates. Adv. Funct. Mater. **22**(15), 3267 (2012)
8. Li, D., Zhao, M.H., Garra, J., Kolpak, A.M., Rappe, A.M., Bonnell, D.A., Vohs, J.M.: Direct in situ determination of the polarization dependence of physisorption on ferroelectric surfaces. Nature Materials **7**(6), 473 (2008)
9. Rode, S., Hölscher, R., Sanna, S., Klassen, S., Kobayashi, K., Yamada, H., Schmidt, W.G., Kühnle, A.: Atomic-resolution imaging of the polar (0001) surface of $LiNbO_3$ in aqueous solution by frequency modulation atomic force microscopy. Phys. Rev. B **86**, 075468 (2012)
10. Sanna, S., Rode, S., Hölscher, R., Klassen, S., Marutschke, C., Kobayashi, K., Yamada, H., Schmidt, W.G., Kühnle, A.: Charge compensation by long-period reconstruction in strongly polar lithium niobate surfaces. Phys. Rev. B **88**, 115422 (2013)
11. Sanna, S., Schmidt, W.G.: Lithium niobate X-cut, Y-cut, and Z-cut surfaces from ab initio theory. Phys. Rev. B **81**, 214116 (2010)
12. Sanna, S., Gavrilenko, A.V., Schmidt, W.G.: Ab initio investigation of the $LiNbO_3$(0001) surface. Phys. Stat. Sol. (c) **7**(2), 145 (2010)
13. Sanna, S., Hölscher, R., Schmidt, W.G.: Polarization-dependent water adsorption on the $LiNbO_3$(0001) surface. Phys. Rev. B **86**, 205407 (2012)
14. Kresse, G., Furthmüller, J.: Efficient iterative schemes for ab initio total-energy calculations using a plane-wave basis set. Phys. Rev. B **54**(16), 11169 (1996)
15. Perdew, J.P., Yue, W.: Accurate and simple density functional for the electronic exchange energy: Generalized gradient approximation. Phys. Rev. B **33**(12), 8800 (1986)
16. Monkhorst, H.J., Pack, J.D.: Special points for Brillouin-zone integrations. Phys. Rev. B **13**, 5188 (1976)
17. Bengtsson, L.: Dipole correction for surface supercell calculations. Phys. Rev. B **59**, 12301 (1999)
18. Neugebauer, J., Scheffler, M.: Adsorbate-substrate and adsorbate-adsorbate interactions of Na and K adlayers on Al(111). Phys. Rev. B **46**, 16067 (1992)
19. Meyer, B., Vanderbilt, D.: Ab initio study of $BaTiO_3$ and $PbTiO_3$ surfaces in external electric fields. Phys. Rev. B **63**(20), 205426 (2001)
20. Levchenko, S.V., Rappe, A.M.: Influence of Ferroelectric Polarization on the Equilibrium Stoichiometry of Lithium Niobate (0001) Surfaces. Phys. Rev. Lett. **100**, 256101 (2008)
21. Vanderbilt, D., King-Smith, R.D.: Electric polarization as a bulk quantity and its relation to surface charge. Phys. Rev. B **48**, 4442 (1993)
22. Resta, R.: Modern theory of polarization in ferroelectrics. Ferroelectrics **151**(1), 49 (1994)
23. Johann, F., Soergel, E.: Quantitative measurement of the surface charge density. Quantitative measurement of the surface charge density, Appl. Phys. Lett. **95**, 232906 (2009)
24. Bader, R.F.W.: Atoms in Molecules – A Quantum Theory. Oxford University Press, Oxford (1990)
25. Landau, L.D., Lifshitz, E.M.: Statistical Physics, Part I, 3rd edn. Butterworth-Heinemann, Oxford (1981)

The Fluorite/Water Interfaces: Structure and Spectroscopy from First Principles Simulations

Rémi Khatib and Marialore Sulpizi

Abstract Despite its relevance to industrial, environmental and medical application, the fluorite/water interface still lacks a microscopic/atomistic characterization. In this contribution we provide the first atomistic description of such interface using *first principles* molecular dynamics simulations. Our models, which explore a wide range of pH, are able to provide a rational of the recent vibrational spectroscopy experiments. In particular we find that at neutral pH the water at the interface is disordered, in agreement with the experimental data, and explaining why no Vibrational Sum Frequency Generation (VSFG) signal is recorded. At high pH, OH groups which localize at the interface are responsible for the "free OH signal" recorded in the vibrational spectroscopy experiments. Finally we propose one possible model for the low pH condition where the F^- vacancies induce a strong layering of the interfacial water.

1 Introduction

Crystal growth and dissolution of fluorite (CaF_2) interfaces have recently attracted much attention thanks to their relevance to industrial, environmental and medical application. In particular, new interest in the material has been promoted by the proposal of using CaF_2 as an analogous of UO_2 in dissolution experiments to understand the long term dissolution behavior of spent nuclear fuel. Atomic scale characterization of fluorite/water interface is still missing and preliminary results are still debated.

The first Vibrational Sum Frequency Generation (VSFG) experiment on water structure at the CaF_2/water interface [1,2] showed dramatic changes in the hydrogen bonding environment with changing the pH of the aqueous phase. In particular at low pH the VSFG experiments suggest that positive charge develops on the surface and facilitates the orientation of water molecules into highly ordered tetrahedrally coordinated states.

R. Khatib • M. Sulpizi (✉)
Johannes Gutenberg University Mainz, Staudinger Weg 7, 55099 Mainz, Germany
e-mail: sulpizi@uni-mainz.de

W.E. Nagel et al. (eds.), *High Performance Computing in Science and Engineering '14*, DOI 10.1007/978-3-319-10810-0_13

However still open questions are: how far does the order extend? Are the water molecules tetrahedrally ordered? How is this order influenced by the surface charge?

At neutral pH the VSFG signal is lost and this has been interpreted as the result of a more random orientation of the interfacial water. Is the water completely disordered here?

Finally in the basic pH regime dissociative absorption is supposed to take place on the solid surface resulting in the formation of Ca-OH species. The questions here are: how do these OH groups contribute to the VSFG spectrum? What type of order in the interfacial water is established?

Very recent Frequency Modulation Atomic Force Microscopy (FM-AFM) experiments [3] have tried to provide new information on the atomic scale behavior analyzing the fluorite/water interface not only as function of the pH but also as function of the concentration of ions in the solution and addressing fluorite/water interfaces with saturated and supersaturated solutions. At low pH atomic scale disorder is found which could be attributed to both partial dissolution of the topmost layer with the creation of F^- vacancies, or to proton adsorption at the interface. Still experiments seem not to be able to distinguish between the two possible scenarios.

From the theoretical point of view there have been to date a number of models to describe mainly bulk and surface properties of CaF_2. For example, bulk properties have been analyzed by both Hartree-Fock methods [4] and Density Functional Theory-based (DFT) electronic structure methods with different functionals [4, 5]. Surface energy calculations showed that the (111) surface is the most stable surface, in agreement with the experiments [4].

A few studies have also addressed the adsorption of water on different surfaces. In particular pioneering work in the group of G. Kresse and N. H. De Leeuw have shown water molecules physisorption on CaF_2 (111) surface [6, 7]. More recently also water adsorption on defects and stepped surfaces have been investigated [8]. All electronic structure investigations have been limited so far to a static picture, where the liquid is actually modelled as a monolayer.

The effect of temperature and dynamics on the CaF_2 surface hydration have been addressed by force field based molecular dynamics simulations, for example in study of the competitive adsorption of water and methanoic acid [7].

In this contribution we would like to shed light on the atomistic structure of the fluorite water interface at different pH. In particular three different models are discussed as representative of low, intermediate and high pH and their structural and vibrational properties are investigated.

Anticipating our results we find that at neutral pH the water at the interface is disordered, in agreement with the experimental data, and explaining why no VSFG signal is recorded. At high pH stable OH groups can localize at the interface which are responsible for the "free OH signal" recorded in the VSFG experiments. Finally we propose one possible model for the low pH condition where the F^- vacancies induce a strong order of the interfacial water.

2 Methodology

2.1 Simulation Setup

We use DFT-based Molecular Dynamic (MD) simulations in order to unravel the microscopic structure of the fluorite/water interface and to provide a molecular interpretation of the VSFG experiments. Different models are used to reproduce the properties of the interfaces for a wide range of pH. The reference system – an interface between CaF_2 (111) and water at neutral pH – is composed of 88 water molecules and 60 f.u. of CaF_2 contained in a $11.59 \times 13.38 \times 34.0\,\text{Å}^3$ cell periodically repeated in the (x, y, z) directions. All the others models have close compositions and size to allow inter-system comparisons. Note that for each system, the thickness of water slabs is around 20 Å along the z-axis, which is thick enough to have bulk properties deep inside these slabs. Simulations are carried out with the package CP2K/Quickstep [9], consisting in Born-Oppenheimer MD (BOMD), BLYP [10, 11] electronic representation including Grimme (D3) correction for dispersion [12], GTH pseudopotentials [13, 14], a combined Plane-Wave (280 Ry density cutoff) and TZV2P basis sets. All the BOMD are performed using the NVT ensemble. The Nosé-Hoover thermostat is used to control the average temperature at 330 K. Trajectories are accumulated for at least 50 ps with a time step of 0.5 fs.

2.2 Vibrational Spectroscopy from Simulations

VSFG is a powerful spectroscopic technique used to probe surfaces and interfaces. This method – first developed by Shen and co-workers [15, 16] – is based on non-linear optics. It involves two input beams of frequency ω_1 and ω_2 which in a non-centrosymmetric medium can induce a response beam of frequency $\omega_{VSFG} = \omega_1 + \omega_2$. Usually, one of the input beams lies in the visible while the other lies in the IR. The intensity of the VSFG beam depends on the second order susceptibility tensor:

$$I_{VSFG} \propto \left|\chi^{(2)}_{pqr}\right|^2, \tag{1}$$

where p, q, r stand for the beams polarization of the VSFG, the visible and the IR beam respectively. The third-rank tensor $\chi^{(2)}_{pqr}$ is frequency dependent and is expressed by the dipole-polarizability time correlation function:

$$\chi^{(2)}_{pqr}(\omega_{\text{VSFG}}, \omega_{\text{vis}}, \omega_{\text{IR}}) = \frac{i\,\omega_{\text{IR}}}{k_{\text{B}}T} \int_0^{\infty} \mathrm{d}t \, \exp(i\,\omega_{\text{IR}}t)\langle \mathscr{A}_{pq}(t)\mathscr{M}_r(0)\rangle, \tag{2}$$

where $\mathscr{A}_{pq}$ is the polarizability tensor and $\mathscr{M}_r$ is the dipole vector of the whole system. Equation 2 shows that a centrosymmetric medium has $\chi^{(2)}_{pqr} = 0$ and therefore no VSFG signal.

The direct calculation of the VSFG signal from Eq. 2 is very expensive since it requires the calculation of the the dipole moment and polarizability of the all system for each step of the trajectory.

Vibrational Density Of States (VDOS) calculations of the mid-IR absorbance spectra are used to calculate the stretching frequencies of the water molecules at the interfaces. These VDOS are based on the Fourier transform of certain autocorrelation function (γ).

$$f(\omega) = F\left[\gamma(t)\right] \tag{3}$$

The Velocity-Velocity Autocorrelation Function (VVAF) has been already successfully used for interpreting IR spectra [17, 18] and to describe water properties at mineral/water interfaces [19, 20]. However, the VVAF is not surface specific, therefore it cannot be directly used to simulate the VSFG spectra. Some new formalisms for surface specific VVAF (ssVVAF) can be introduced and here we focus the following definition:

VVAF:

$$\gamma = \sum_{j=1}^{N} \left\langle \mathbf{v_j}(t).\mathbf{v_j}(0) \right\rangle \tag{4}$$

ssVVAF:

$$\gamma' = \sum_{j \in region} \left\langle v_{z,j}(t) \times v_{z,j}(0) \right\rangle \tag{5}$$

When going from Eqs. 4 to 5, two new elements are introduced, namely:

1. In Eq. 5 only the sum is extended to the atoms which belong to a specific region. This allow us to specifically selected a given thickness layer of water close to the surface or to address a subset of atoms close to the interface. The thickness of the layer can be adjusted either to fit with the experiments or to observe the different contributions when depth increases.
2. In Eq. 5 only the projection of the velocities along the normal axis (z) is included and this select the IR-beam polarization which is commonly used in the experiments and which excites the atoms only along this axis direction.

Finally, to get the VDOS, a numerical Fourier transform is evaluated over a finite interval (1 ps) and a Gaussian apodization – producing a Gaussian broadening of the resulting spectra – is used [21].

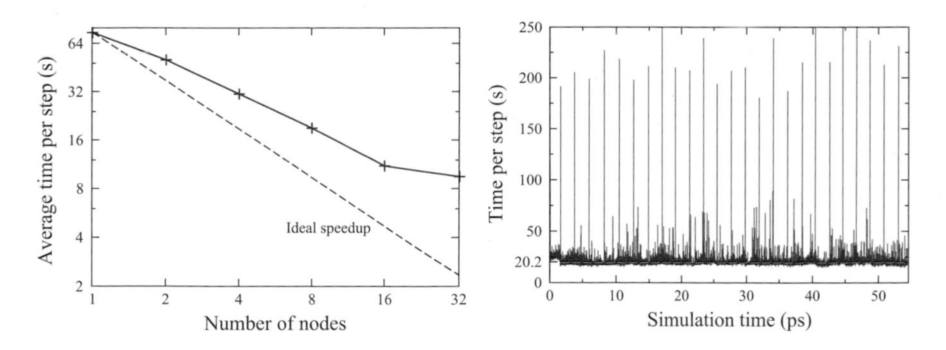

Fig. 1 (*Left*) Average user time required to calculate one step of dynamic for a standard model according to the number of nodes (32 cores) used. The sampling has been done over dynamics of 500 steps. (*Right*) Calculation time needed to compute one step of the dynamic for a standard model. Each peak over 150 s represents the restart of our calculations due to a random guess for the wavefunction initialization. The *black dashed line* represents the average time over the whole dynamic required to calculate one step

2.3 Computational Cost

The reference system described in Sect. 2.1 has been used as a scaling benchmark for standard BOMD. The results obtained for a trajectory of 500 steps are provided in Fig. 1. It is possible to see that up to 16 nodes there is a scaling close to be linear but it is not equals to the ideal speedup. With 32 nodes the calculation efficiency dramatically decreases. To have an acceptable efficiency with an affordable calculation time, most of the calculations are done with eight nodes. With this timing it is possible to obtain a trajectory of 50 ps in 156,000 h (CPU time) on 256 cores. Two picoseconds of trajectories are produced in a single day.

3 Results and Discussions

In order to understand the experimental interfaces, several systems have been simulated with BOMD. In the present contribution, we will discuss in details three of them (one for each class of pH). These three models are – today and according to us – those which have the best agreements with the experimental results. Moreover, their simplicity also allow us to highlight the strengths and the weakness of the ssVVAF that we have introduced in Eq. 5.

3.1 Neutral pH

The system used to simulate the interfacial behaviour of water has been already described in Sect. 2.1 and can be seen on Fig. 2. After an equilibration time of

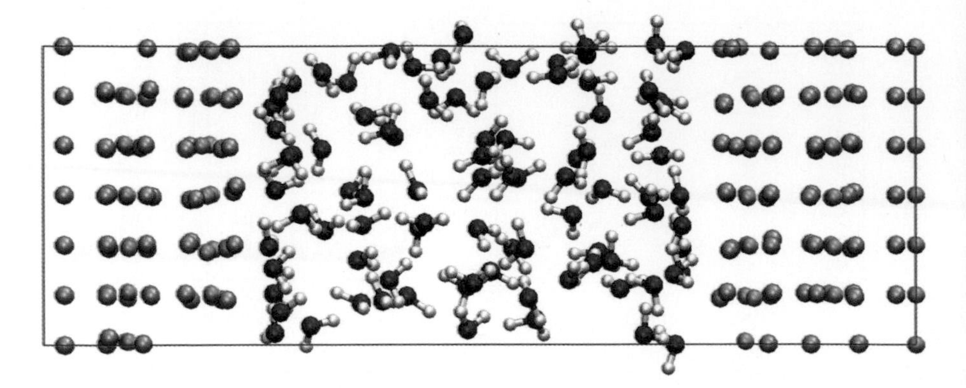

Fig. 2 Random snapshot of the system used to describe the CaF_2/H_2O interface for a neutral pH

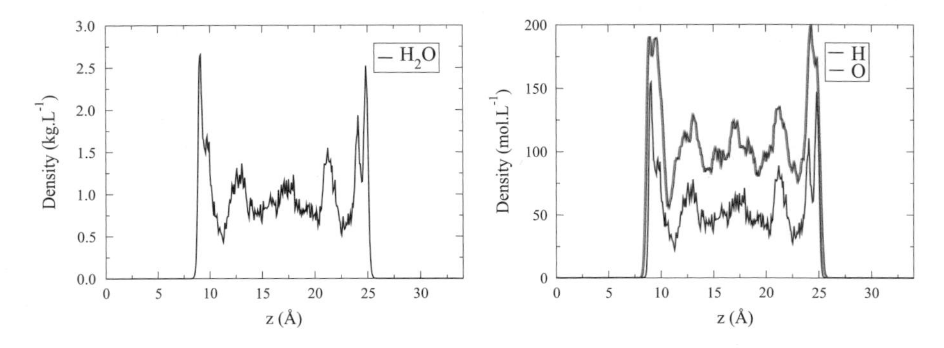

Fig. 3 The effects of interfaces on water density. (*Left*) The volumetric mass density according to the *z*-axis. The water slab is centered at 17 Å. (*Right*) Molar density for O and H according to the *z*-axis

7 ps, the density of the system is calculated averaging over 40 ps (Fig. 3). A strong layering appears at the interface. In this region, we observe that the density is around 1.5 $kg.L^{-1}$. However, the layering tends to be attenuated in the inner part of the slab of water ($z \approx 17$ Å). A density around 1.0 $kg.L^{-1}$ is reached, which is in good agreement with the water density under normal conditions. When the contributions to the density of the H and O atoms are separated, the O and H peaks appears to be in the same position, which means that no strong net dipole is expected at the interface. In this case, the VSFG response should be indeed weak because it depends on the dipole moment (Eq. 2). Experimentally, no response is observed [1].

In Fig. 4 the VDOS associated with the ssVVAF of the hydrogen having a given distance from the fluorite surface are reported. If we analyze the VDOS as function of the increasing thickness of the water layer we can see that all the profiles show a broad band centered around 3,400 cm^{-1}, which suggests that the interfacial water is actually “normal” liquid water.

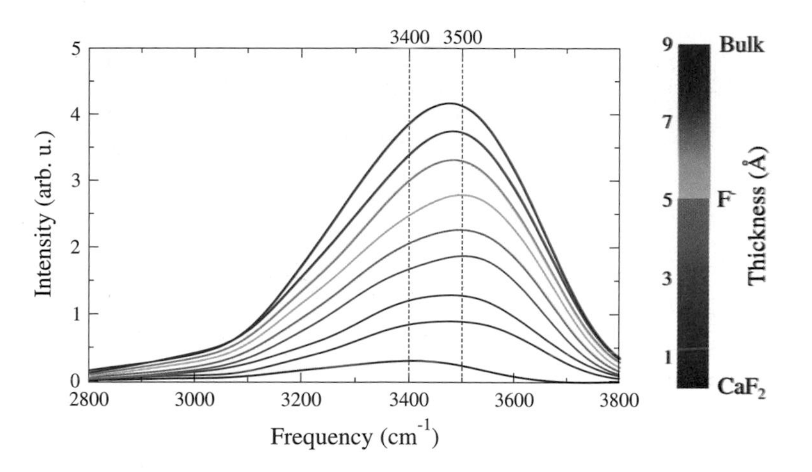

Fig. 4 VDOS associated with the ssVVAF of the hydrogen having a distance to the interface lower than the slab thickness (cf. color code)

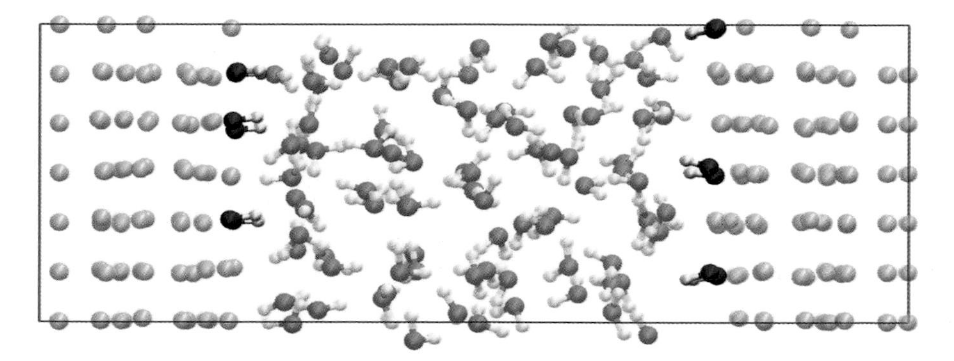

Fig. 5 Random snapshot of the system used to describe the CaF_2/H_2O interface for a high pH. Except for the hydroxide ions, the atoms are transparent in order to highlight the differences with the neutral system

3.2 Basic pH

At high pH, the excess of hydroxide ion is expected to react with the CaF_2 surface. In this case, a anion substitution occurs according to:

$$(CaF_2)_{surf} + HO^-_{aq} \rightarrow (CaFOH)_{surf} + F^-_{aq} \tag{6}$$

In Fig. 5 our high pH model of the interface is presented. In comparison with the neutral system previously discussed, a partial substitution of the topmost fluorite layer has been done: 12 F^- over the 24 available (per box) were replaced by HO^-. We would like to comment that in our model two equivalent surfaces are considered. This choice permit to avoid the creation of an unbalanced electric filed across the

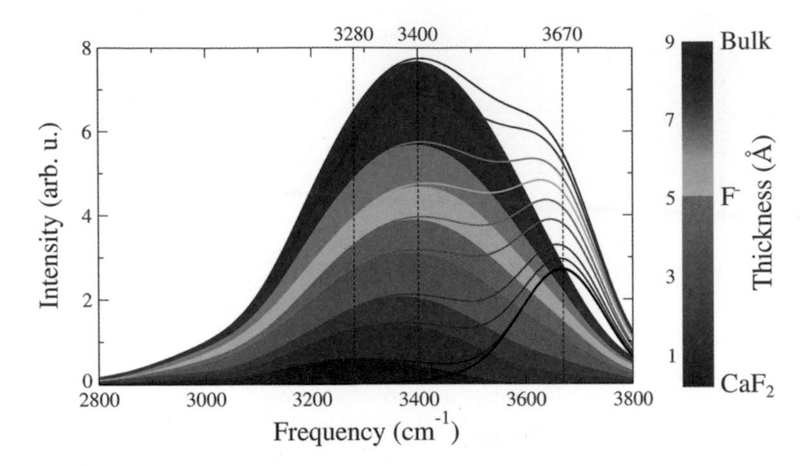

Fig. 6 The *unfilled curves* represent the VDOS associated with the ssVVAF of all the hydrogen having a distance to the interface lower than the slab thickness (cf. color code). These curves can be decomposed in two contributions: the H of water molecules (*filled curves* with the same color code) and the H of the hydroxyl groups (*black line*, constant whatever the slab thickness)

sample and at the same time will permit us to average properties over the two equivalent interfaces.

The results from the VDOS analysis are presented in Fig. 6. for all the values of the interface slab thickness, two peaks can be observed: the first is between 3,280 and 3,400 cm^{-1}, the second is between 3,600 and 3,700 cm^{-1}. Because of the frequencies, one can propose that:

1. The first peak is associated with hydrogen bonded water molecules close to those that we should have in a bulk of water – since the IR-spectrum of water exhibits a peak at 3,404 cm^{-1} [22]. More precisely, the frequency of this peak is less and less red-shifted, so we conclude that the water molecules at the interface make strong H-bond and this effect tends to disappear for water molecules deep inside the slab of water.
2. The second is only associated with the OH groups, because of the strong blue-shift which is characteristic of "free-OH groups".

To prove our assertions, the contributions of H of water molecules and H of hydroxyl groups are split. On Fig. 6, it is possible to see that the water molecules do not contribute at all to the peak at 3,670 cm^{-1}, only the OH groups are responsible of it.

In order to estimate the thickness of the water layer probed by VSFG measurements, we compared our theoretical results with the experimental one obtained by Becraft [1]. In the experiments two peaks are observed: one red-shifted (wide and small), the other is blue-shifted (narrow and intense). The ratio between these two experimental peaks is around 1/10. If we compare this ratio with those obtained from our simulations, two possible interpretations can be proposed: either the probed layer is very thin (≤ 1 Å), or the substitution ratio of F by OH is higher than those we used (1 over 2).

3.3 Acidic pH

Because of the excess of hydronium ion in acidic solution, the expected reaction at the interface is:

$$(CaF_2)_{surf} + H^+_{aq} \rightarrow (CaF^+)_{surf} + HF_{aq}. \quad (7)$$

In order to reproduce the properties of the CaF_2/H_2O interface at low pH, a dissolution of F^- at the interface is simulated (Fig. 7).

Since we cannot directly simulate the dissolution process on the few tens of ps timescale, we assume as starting model a system where partial dissolution is already occurred. In particular we introduce on both fluorite surfaces 4 F^- vacancies, which create an overall 4 positive charges per surface. The positive charge at the surface is counterbalanced by the presence of 4 F^- ions in the bulk solution. In this model a possible excess proton at the interface is not considered.

Moreover, it has to be noticed that not only the chemistry of the system can have an impact on the VSFG simulations, but also the electrical field due to the positive surface and the negative ions in the solutions. The overall electric filed across the interface is strong enough to drive the water orientation over 5 Å, i.e. around two layers of water (Fig. 8).

It is possible to observe the effect of the electrical field on the vibrational properties of water thanks to Fig. 9. The VDOS spectra simulated contains two or three peaks:

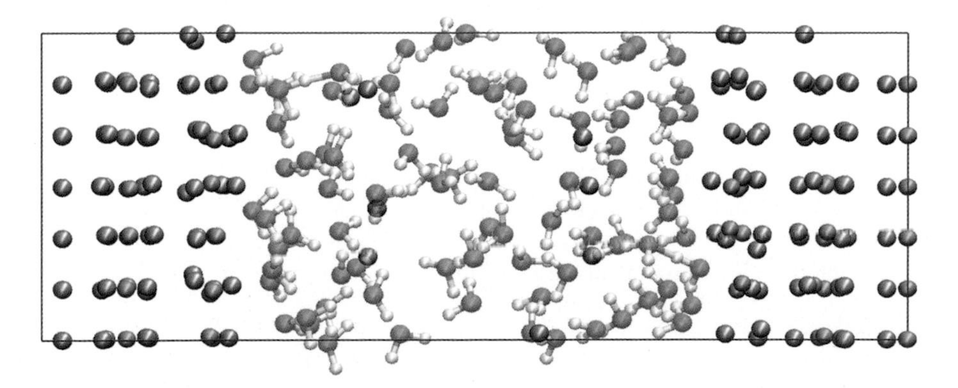

Fig. 7 Random snapshot of the system used to describe the CaF_2/H_2O interface for a low pH. Water molecules are transparent in order to highlight the vacancies and the ions

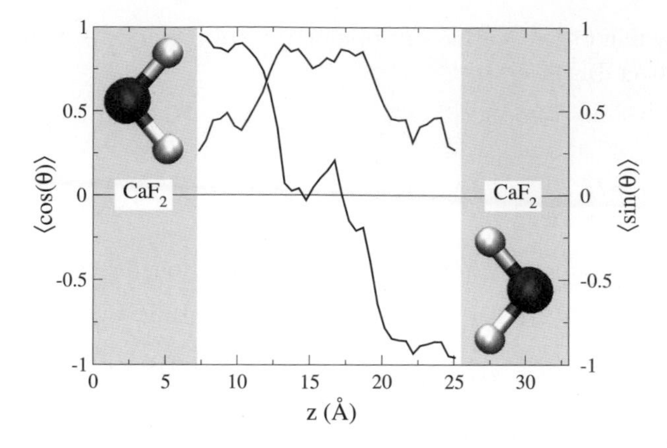

Fig. 8 The effects of interfaces on water molecules orientation. θ is the angle done by the dipole moment of the water molecule and the normal axis (z). Two water molecules are represented to illustrate the main water orientation at the interface

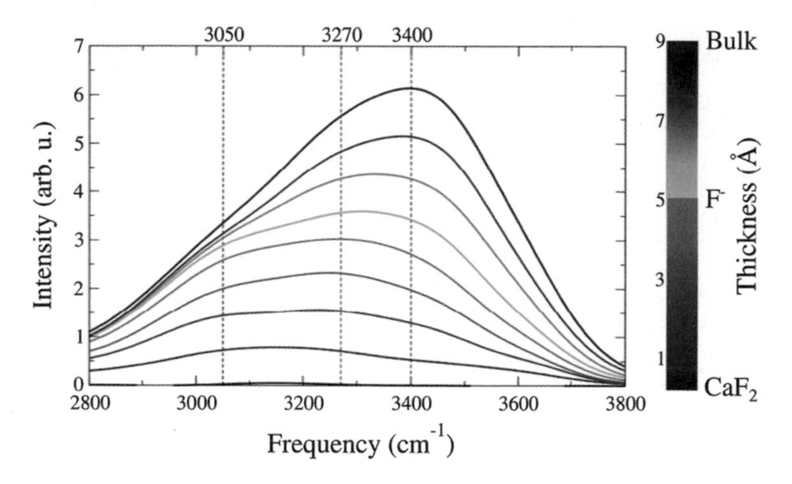

Fig. 9 VDOS associated with the ssVVAF of the hydrogen having a distance to the interface lower than the slab thickness (cf. color code). The average position of F^-_{aq} is around 5 Å

1. The first peak is around $3{,}050\,\text{cm}^{-1}$. When the slab thickness increases, this strongly red-shifted peak increases until saturation around 5 Å. It means that the nature of the water molecule is different before and after 5 Å. This distance to the interface is approximately those of the F^-_{aq} and one can also see on Fig. 8 that the electrical field becomes weaker.
2. One or two other peaks are between 3,270 and $3{,}400\,\text{cm}^{-1}$. It is actually difficult to distinguish if there is only one peak which is shifted or if there are two peaks with a relative intensity which evolves. However, it is interesting to notice that when all the H are taken into account to build the VDOS (thickness $\approx$ 9 Å), the

frequency obtained is around 3,400 cm^{-1} which is the frequency expected for IR spectroscopy of a bulk of water.

It is quite clear from our analysis that the first adsorbed layer of water is responsible for a red-shifted band, while moving away from the surface a conclusive analysis is more difficult. Indeed the presence of two peaks in the water band could be explained either as the results of symmetric and the antisymmetric modes have different frequencies or as the result of two kind of water molecules which are experiencing a different hydrogen bind environment. Further investigation will be required in order to isolate the effect of the explicit counter-ions and allow a direct comparison to the experiments.

4 Conclusions

In this contribution we present a detailed analysis of the fluorite/water interface at different pH conditions. First principles molecular dynamics simulations are used to equilibrate the system and to investigate the water structure at the interface, while surface specific vibrational density of states are used to calculate vibrational properties and to provide a molecular interpretation of the VSFG experiments. At neutral pH a fluorine terminated surface is considered and the water arrangement at the interface is analysed. Although the first water layer strongly adsorbs on the fluorite surface, no special water characterizes the water interfaces providing a rational for the absence of the VSFG signal. The high pH condition is simulated with a partial substitution of the topmost F atoms with OH groups. These OH groups do not form hydrogen bonds with the interfacial water and are responsible for a free OH signal in the VSFG spectrum. Moreover the adjacent water layer shows standard liquid vibrational properties. Finally for the low pH regime we assume a model with partial dissolution of the surface topmost F layer. As a consequence positive charge accumulates at the surface while dissolved F^{-} ions can be found in the bulk solution. The charging of the interface produce a quite strong electric field which is responsible for ordering at least two layers of water molecules. The topmost adsorbed water layer exhibits strongly redshifted vibrational frequency.

References

1. Becraft, K.A., Richmond, G.L.: In situ vibrational spectroscopic studies of the caf2/h2o interface. Langmuir **17**(25), 7721 (2001)
2. Becraft, K.A., Moore, F.G., Richmond, G.L.: Charge reversal behavior at the caf2/h2o/sds interface as studied by vibrational sum frequency spectroscopy. J. Phys. Chem. B **107**(16), 3675 (2003)
3. Kobayashi, N., Itakura, S., Asakawa, H., Fukuma, T.: Atomic-scale processes at the fluorite-water interface visualized by frequency modulation atomic force microscopy. J. Phys. Chem. C **117**(46), 24388–24396 (2013)
4. Shi, H., Eglitis, R.I., Borstel, G.: Ab initio calculations of the caf2 electronic structure and f centers. Phys. Rev. B **72**, 045109 (2005)

5. Verstraete, M., Gonze, X.: First-principles calculation of the electronic, dielectric, and dynamical properties of caf2. Phys. Rev. B **68**, 195123 (2003)
6. de Leeuw, N.H., Cooper, T.G.: A computational study of the surface structure and reactivity of calcium fluoride. J. Mater. Chem. **13**, 93 (2003)
7. de Leeuw, N., Purton, J., Parker, S., Watson, G., Kresse, G.: Density functional theory calculations of adsorption of water at calcium oxide and calcium fluoride surfaces. Surf. Sci. **452**(13), 9 (2000)
8. Foster, A.S., Trevethan, T., Shluger, A.L.: Structure and diffusion of intrinsic defects, adsorbed hydrogen, and water molecules at the surface of alkali-earth fluorides calculated using density functional theory. Phys. Rev. B **80**, 115421 (2009)
9. VandeVondele, J., Krack, M., Mohamed, F., Parrinello, M., Chassaing, T., Hutter, J.: Quickstep: Fast and accurate density functional calculations using a mixed gaussian and plane waves approach. Comput. Phys. Commun. **167**(2), 103 (2005)
10. Becke, A.D.: Density-functional exchange-energy approximation with correct asymptotic behavior. Phys. Rev. A **38**, 3098 (1988)
11. Lee, C., Yang, W., Parr, R.G.: Development of the colle-salvetti correlation-energy formula into a functional of the electron density. Phys. Rev. B **37**, 785 (1988)
12. Grimme, S., Antony, J., Ehrlich, S., Krieg, H.: A consistent and accurate ab initio parametrization of density functional dispersion correction (dft-d) for the 94 elements h-pu. J. Chem. Phys. **132**(15), 154104 (2010)
13. Goedecker, S., Teter, M., Hutter, J.: Separable dual-space gaussian pseudopotentials. Phys. Rev. B **54**, 1703 (1996)
14. Hartwigsen, C., Goedecker, S., Hutter, J.: Relativistic separable dual-space gaussian pseudopotentials from h to rn. Phys. Rev. B **58**, 3641 (1998)
15. Hunt, J., Guyot-Sionnest, P., Shen, Y.: Observation of c-h stretch vibrations of monolayers of molecules optical sum-frequency generation. Chem. Phys. Lett. **133**(3), 189 (1987)
16. Zhu, X.D., Suhr, H., Shen, Y.R.: Surface vibrational spectroscopy by infrared-visible sum frequency generation. Phys. Rev. B **35**, 3047 (1987)
17. Laasonen, K., Sprik, M., Parrinello, M., Car, R.: ab initio liquid water. J. Chem. Phys. **99**(11), 9080 (1993)
18. Frigato, T., VandeVondele, J., Schmidt, B., Schtte, C., Jungwirth, P.: Ab initio molecular dynamics simulation of a medium-sized water cluster anion: From an interior to a surface-located excess electron via a delocalized state. J. Phys. Chem. A **112**(27), 6125 (2008)
19. Sulpizi, M., Gaigeot, M.P., Sprik, M.: The silica-water interface: How the silanols determine the surface acidity and modulate the water properties. J. Chem. Theory Comput. **8**(3), 1037 (2012)
20. Gaigeot, M.P., Sprik, M., Sulpizi, M.: Oxide/water interfaces: how the surface chemistry modifies interfacial water properties. J. Phys.: Condens. Matter **24**(12), 124106 (2012)
21. Gonçalves, S., Bonadeo, H.: Molecular-dynamics calculation of the vibrational densities of states and infrared absorption of crystalline rare-gas mixtures. Phys. Rev. B **46**, 10738 (1992)
22. Max, J.J., Chapados, C.: Isotope effects in liquid water by infrared spectroscopy. iii. h2o and d2o spectra from 6000to0cm1. J. Chem. Phys. **131**(18), 184505 (2009)

Adsorption and Electronic Excitation of Water on TiO_2 (110): Calculation of High-Dimensional Potential Energy Surfaces

Jan Mitschker and Thorsten Klüner

Abstract By combining quantumchemical and quantumdynamical calculations, we aim to understand photochemistry on surfaces from first principles. In this project we investigate the case of water on a titanium dioxide surface. The substrate in its most stable form rutile (110) can act as a photocatalyst for water splitting. Highly accurate potential energy surfaces for the water molecule on this surface were calculated for the electronic ground and a selected excited state. A five-dimensional potential energy surface could be obtained and was fitted with the help of artificial neural networks.

1 Introduction

Investigating any kind of chemical process on a surface is still a great challenge for experimental and theoretical studies. From a theoretical point of view, the size of the system is one limiting factor. Many different approaches have been developed in the last years to overcome these problems. Density functional theory (DFT) in combination with periodic boundary conditions became one of the work-horses. In contrast to simple Hartree-Fock calculations the problem of electron correlation can be dealt with to at least some extent. Nevertheless, some problems and uncertainties remain which become apparent when calculating electronically excited states. For most of these cases DFT cannot be used. But these excited states are very important for photochemical reactions, e.g. reactions that are driven by incoming light. We have studied one of them, the photodesorption of diatomic molecules from metal oxide surfaces, extensively in the last years [1–20]. This reaction can be considered as a prototype of non-adiabatic surface reactions and can help to understand basic mechanistic features of photochemistry on surfaces. The size of the adsorbate has been limited by the computational power available. Now, grown resources give us the possibility to enlarge the system and to study three-atomic adsorbates. Here,

J. Mitschker • T. Klüner (✉)
Institut für Chemie, Theoretische Chemie, Carl von Ossietzky Universität Oldenburg, 26129 Oldenburg, Germany
e-mail: thorsten.kluener@uni-oldenburg.de

W.E. Nagel et al. (eds.), *High Performance Computing in Science and Engineering '14*, DOI 10.1007/978-3-319-10810-0_14

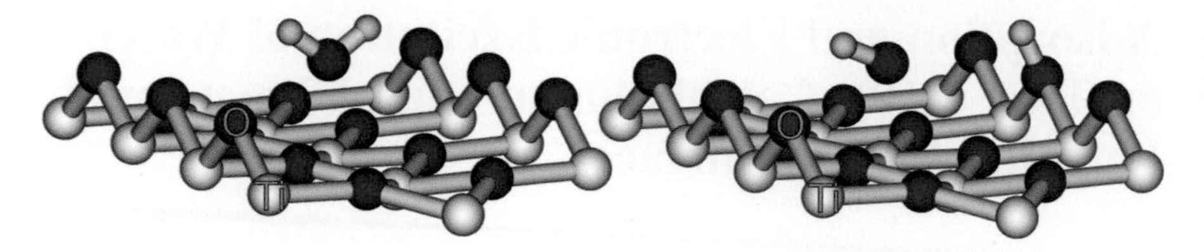

Fig. 1 Adsorption of water on rutile (110): Molecular form (*left*) and dissociated form (*right*)

the water molecule is of great interest because it is omnipresent and some very remarkable reactions can occur. In our study we focus on the water splitting forming hydrogen and oxygen, a reaction that can be observed on some metal oxide surfaces like titanium dioxide (TiO_2). This reaction offers one way to hydrogen production using sun-light as a renewable energy source and is therefore of great ecological and economical interest.

Many further interesting properties have been found in the last decades that are—in contrast to the water splitting—in wide use nowadays. Most prominent examples are self-cleaning surfaces where glass is coated with a thin film of titanium dioxide, or the superhydrophilicity effect [21, 22]. All these effects are connected to the adsorption behaviour of water on titanium dioxide.

The problems that we want to address are very complex. The first question is how the water binds to the surface and how the potential energy surfaces for different electronic states look like. This is a very challenging task since calculations for excited states are computational demanding and the states are not accessible with black-box tools. We will focus on the splitting of *one* O-H bond and the movement of one hydrogen. For this case, two adsorption forms are well-known in literature (see Fig. 1).

2 Computational Details

We used the Molcas program package [23] to perform calculations for the electronic ground state and several electronically excited states. Both problems can not be treated within single-reference methods like Hartree-Fock: in the ground state, stretching one O–H bond and finally breaking it results in a multi-reference situation, because this is not a heterolytic breaking process in all cases. Similarly, the electronic excitation leads to partial occupation of some orbitals which is a multi-reference problem, too. Although there are some extensions of the popular DFT like time-dependent DFT (TD-DFT), it is necessary for our studies to give well-defined states and to preserve the definition for all possible geometries of the adsorbate above the surface. Therefore, we used the Complete Active Space Self-Consistent Field (CASSCF) approach with an active space that can deal with both, dissociation and electronic excitation.

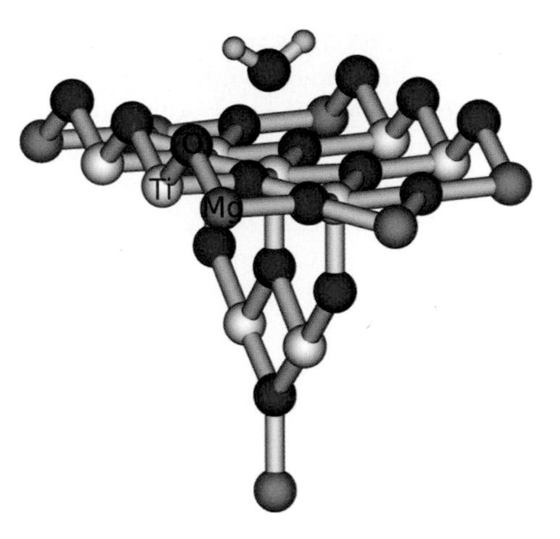

Fig. 2 $Ti_9\,O_{18}Mg_7^{14+}$ cluster used as a model for the rutile (110) surface. For clarity, the point charge field is not shown

For our studies the embedded cluster approach is more favourable than periodic calculations [24]. There are two dominant reasons: first, all modern and highly accurate methods in computational chemistry can be applied to these cluster models. Second, the definition of excited states is less difficult for these finite systems.

During the last years, a special rutile cluster has been developed in our group and it has been used successfully for adsorption and photodesorption phenomena [16,19, 20]. Constraints in the choice of the cluster are on the one hand the computational resources available and on the other hand the accuracy of the simulations. Keeping this in mind, a cluster of 9 titanium and 18 oxygen atoms is a good compromise. The cluster is depicted in Fig. 2. Further magnesium atoms were added at the corners of the cluster in order to saturate the dangling oxygen bonds and to prevent an artificial electron transfer. Note that these atoms are present only for technical reasons and do not represent any dopants.

The cluster is embedded in a finite point charge field of 4,421 point charges. This field is much larger than the cluster and accounts for the long-range coulomb interaction. The point charges were situated at the positions of the titanium and oxygen atoms in the bulk phase. The charges are $+2$ and -1 for titanium and oxygen, respectively. This value is consistent with charges obtained for the atoms in previous calculations and shows that titanium dioxide is not a purely ionic system [25].

The coordinates of the atoms include the relaxation induced by the surface and the adsorbed water. They were determined in a periodic calculation by B. Meyer et al. [26].

The basis set for the cluster has been published elsewhere [16, 19]. For the water system we have only changed the basis for the central bridging oxygens to be identical to the one of the water oxygen. This was motivated by the formation of the O-H bond in the dissociative adsorption form (cf. Fig. 1).

The size of the active space and the basis set for the water molecule have been checked thoroughly. The dissociation and the excitation energy to the first excited state were used as a criterion. We found that an active space of eight electrons in five orbitals (CAS(8,5)) is sufficient. The five orbitals are the occupied orbitals of water except the first and one virtual orbital which corresponds to the anti-bonding orbital of the breaking O-H bond. Although the second occupied orbital is energetically separated from the rest, we found that this orbital sometimes rotates in the active space. Therefore, we have included this orbital. Extending the size to a CAS(8,6), i.e. a full-valence CAS with all anti-bonding orbitals included, does not change the dissociation and excitation energy.

The influence of the basis set is most pronounced for the excited state. Inclusion of diffuse functions is indispensable for correct excitation energies. Although the aug-cc-pVTZ basis set gives very good results, we tried to reduce the number of basis functions and found that our smaller basis set for oxygen (used in [16, 19]) in combination with a cc-pVTZ basis plus a diffuse s-function from the corresponding augmented basis set is totally sufficient. The calculated excitation energy is 7.63 eV on CASPT2 (Complete Active Space Perturbation Theory) level of theory, and the experimental spectrum gives energies in the range of 7.25 up to 7.6 eV [27]. The dissociation (5.23 eV) is higher than the experimental value of 5.10 eV [28, 29] but the missing vibrational zero-point correction shifts the value to lower energies.

For the water adsorbed on the cluster the active space was extended by two further electrons and one orbital. They belong to the bridging oxygen involved in the dissociative adsorption. Therefore, for this system a CAS(10,6) was used.

In the Molcas program package we used the Cholesky decomposition of the two-electron integrals. This option decreases the amount of disk-space required by almost two orders of magnitude and therefore increases the speed dramatically. However, for CASPT2 calculations there is still a large need (some hundred GB) and therefore we have only used the CASSCF method.

The adsorption of a three-atomic molecule like water is (under the assumption of a fixed surface) a nine-dimensional problem. Calculating a pointwise potential energy surface is despite current computational power still unfeasible on a high level of theory. Fortunately, the size can be reduced by exploiting symmetry and further constraints. In Fig. 3 the definition of Jacobi coordinates used in our study is sketched. The great advantage of this coordinate system is that some of the degrees of freedom have a chemical meaning. The distance d corresponds to a bond and the variable R describes the dissociation coordinate. (Note that R is defined as

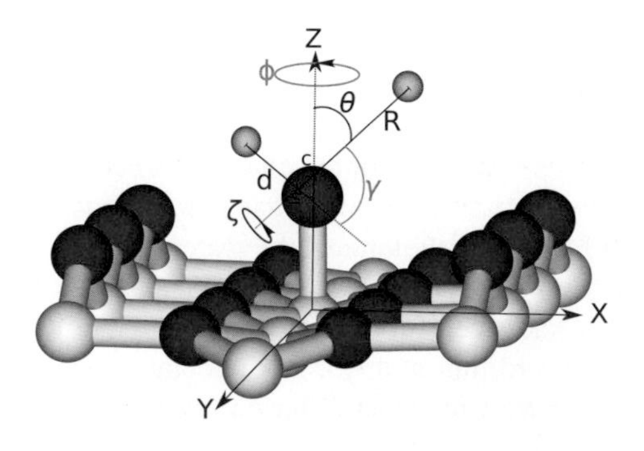

Fig. 3 Definition of the coordinate system for water adsorption on rutile (110)

the distance between one hydrogen and the center of mass of the other two atoms. Because of the difference in atomic masses this is approximately the distance to the oxygen atom.)

We assume that the whole process takes place in the xz plane thus the parameters Y, γ and ζ do not change. Furthermore, the distance d is kept fixed. As a result, the potential energy surface is now five-dimensional. The first assumption introduces a plane of symmetry which can be exploited in the calculation of each data point and (as a side effect) makes the definition of electronically excited states easier.

3 Results and Discussions

3.1 *Electronic Ground State*

As outlined in the sections before, even the electronic ground state is complicated due to the multi-reference character of the wave-function during bond breaking. Simulating this breaking in the gas phase is straight forward but the presence of the cluster introduces a new degree of complexity. The origin can be traced back to the virtual orbitals of the titanium atoms. These orbitals are lower lying than the one of the water molecule. Therefore, excitations into d-orbitals of the titanium atoms are a common problem, especially when the water is close to its equilibrium geometry. In this case the anti-bonding orbital is unnecessary for a proper description of the bonding situation in the water molecule and inclusion of cluster excitations lowers the energy. Various efforts to overcome this problem (e.g. by using the RASSCF approach) were not successful. Finally, we used the following workaround: In a first calculation the one-determinant Hartree-Fock solution was obtained. In a second step all occupied orbitals that are not part of the active space were frozen and one anti-bonding orbital was added to the active space. This procedure prevents excitations of cluster electrons to any active orbital. This is of course only an approximation to the real wave-function but it performs very well for geometries near the equilibrium geometry and the error is low even for larger bond distances. A potential energy curve is shown in Fig. 4.

With this approach we have calculated about 160,000 data points in total for a five-dimensional potential energy surface. The data points were distributed randomly but more dense near the two minima.

The whole potential energy surface is quite complicated and can only be visualized by one and two-dimensional projections for fixed combinations of the three remaining variables. In Fig. 5 two curves for different values of the coordinate Z, which describes the distance of the water molecule (i.e. its center of mass) from the substrate surface, are shown. For this special cut (due to a favourable combination of θ, γ and X) a bond between hydrogen and a bridging oxygen atom can be formed. Normally, enlarging the distance R moves the hydrogen somewhere above the cluster, resulting in a Morse-like potential in Z and R. But here, a second minimum can occur if the bond is formed, as long as the distance is not too short and nuclear repulsion overwhelms this attraction.

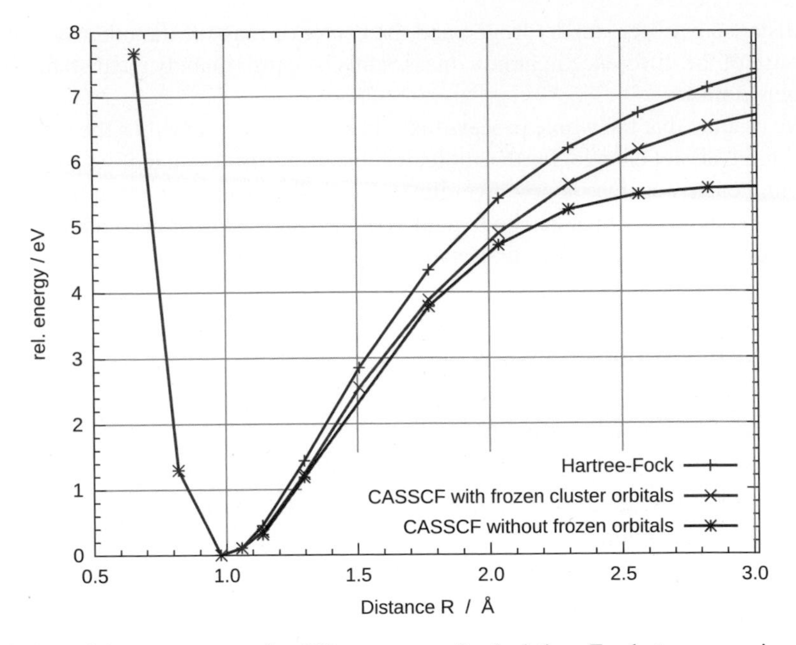

Fig. 4 Potential energy curves for different types of calculation. For better comparison, the energies are given relative to each minimum energy

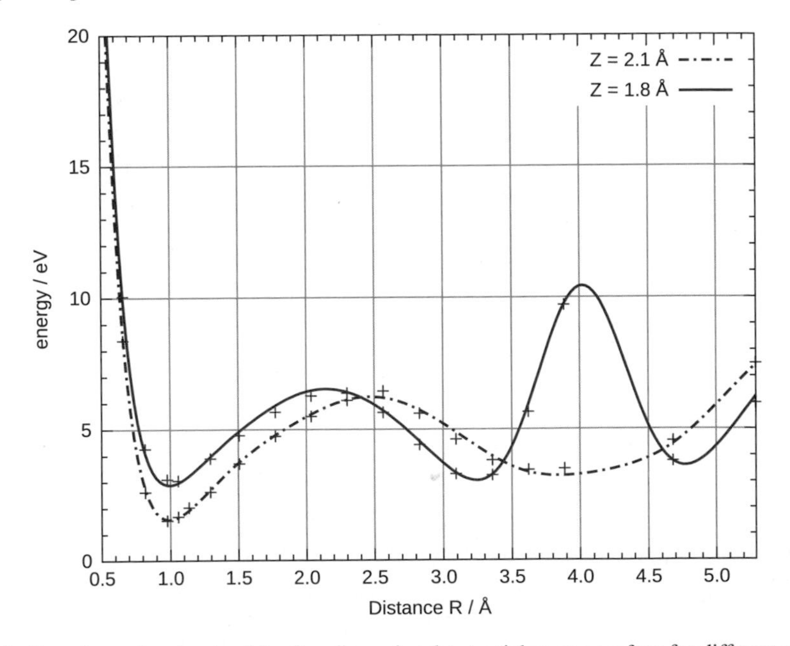

Fig. 5 One-dimensional cuts of the five-dimensional potential energy surface for different values of Z

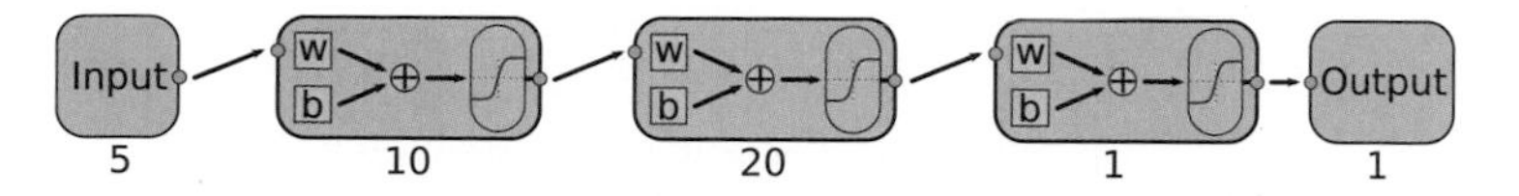

Fig. 6 Structure of the neural network used for data fitting. The numbers correspond to the number of nodes and w and b denote the weight and bias, respectively

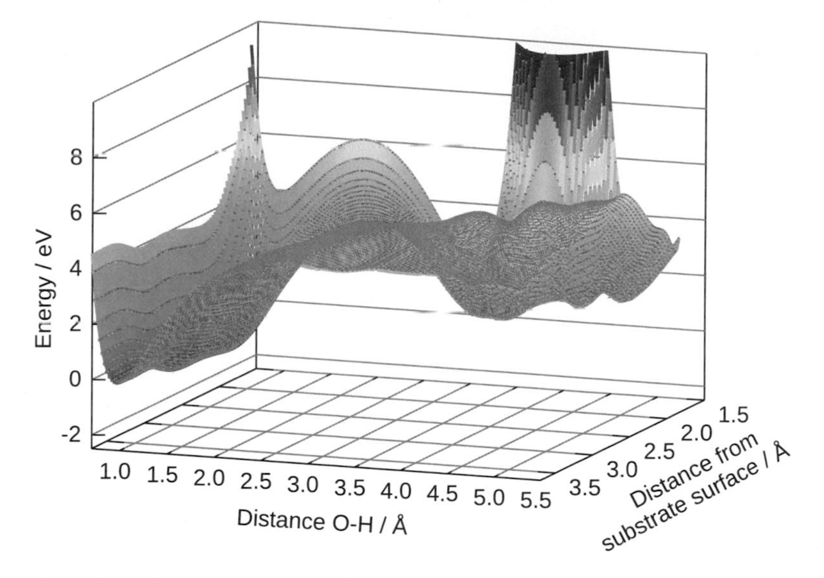

Fig. 7 Two-dimensional cut of the five-dimensional surface for fixed values of θ, X, γ

Similarly, changing the angle γ is less pronounced for small values of R, but results in a strong motion of hydrogen if R is large. Therefore, the topology of two-dimensional cuts strongly depends on the choice of the remaining coordinates. This makes it very complicated to find an analytical expression for the whole potential energy surface. Nevertheless, a way of interpolation is necessary for later quantum dynamical studies, which need a more dense and regular grid of data points. Recently, the idea of using neural-networks for data fitting became very popular in chemistry (For a detailed overview, see e.g. [30–32]). This tool has the great advantage that no analytical expression is necessary. However, a systematic improvement is hard to realize because of the black-box character of this approach. We tried different sizes of neural networks and found that a neural network consisting of two hidden layers works well. The structure of the network is displayed in Fig. 6. For the activation functions sigmoidal functions are used. The whole fitting was performed with the MATLAB program [33]. Although more demanding, we used the learning algorithm with Bayesian regularization because the generalization of the results was better compared to the simple Levenberg-Marquardt procedure.

Extending Fig. 5 to two dimensions results in a potential energy surface as depicted in Fig. 7. Two regions of minima are clearly visible: the molecular

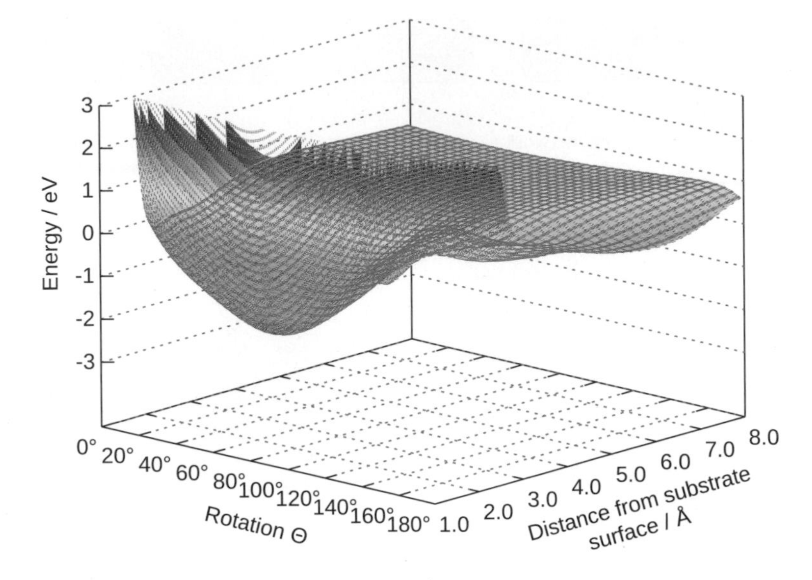

Fig. 8 Adsorption energy as a function of Z and θ

adsorption for values of $R = 0.98$ Å and the dissociative form near $R = 2.55$ Å. Both minima are separated by an energy barrier.

Optimizing each adsorption form gives energies of 2.1 eV for the molecular and 0.9 eV for the dissociative adsorption. This agrees with the literature which emphasizes that for a perfect surface molecular adsorption is the only possible form [34]. However, this does not imply that no dissociation occurs if light is involved because in the excited state this barrier may vanish.

The next cut in Fig. 8 presents the energy as a function of Z and θ for a constant R with molecular adsorption. This picture is comparable to those of other two-dimensional studies with small molecules on this surface [16, 19]. The influence of θ is not unexpected. The water molecule is preferentially adsorbed symmetrically with the oxygen towards the surface. This is the molecular adsorption mentioned above. Rotation around θ increases the energy.

At the beginning of this section, we mentioned that our potential energy surface is five-dimensional and the translation coordinate X is included. However, the importance of this coordinate is not obvious when comparing the two adsorption forms in Fig. 1 where the center of mass (e.g. the OH-fragment) does actually not move. Nevertheless, this coordinate can become important if the way between these two forms is examined. To show its importance, Fig. 9 is a projection of the five-dimensional surface to a two-dimensional one with all other variables optimized ("constraint optimisation"). The graph gives the energy as a function of R and the translation X. Small values of R are for the molecular form and large values for the dissociative one.

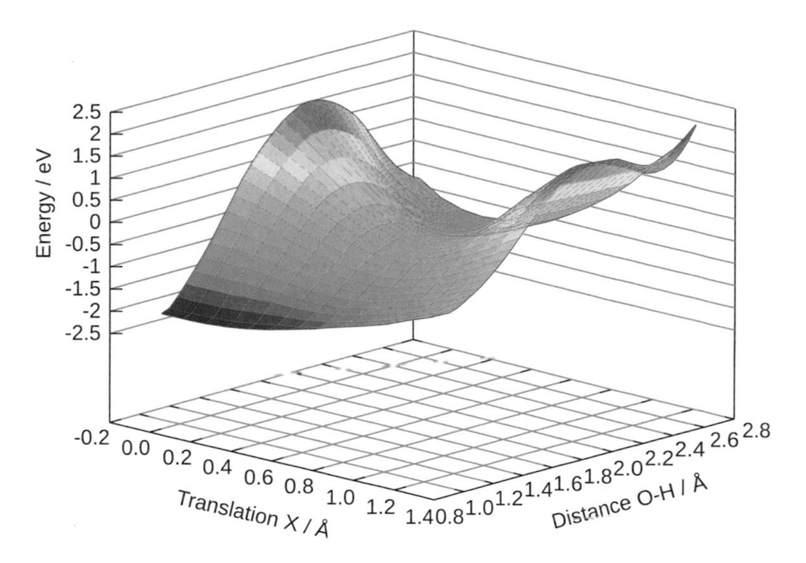

Fig. 9 Two dimensional projection of the potential energy surface with values of θ, γ and Z optimized for each combination of X and R

Without the translation the barrier is very high but moving the molecule in X direction decreases the barrier. This behaviour can easily be understood: for larger X the breaking of one and the forming of a new O-H bond happen simultaneously and no "free" hydrogen atom occurs. Thus, if the dissociation would happen even in the ground state, the whole water molecule would first move towards the bridging oxygen forming a new bond and then only the OH fragment would move back towards the center of the cluster.

3.2 Electronically Excited States

The technical problems in the ground state have been discussed in detail above. For the excited state they are even more pronounced. We therefore choose a very simple approach for the excited state and just removed one electron of the water molecule. This would correspond to a situation when an hole attacks the adsorbate but the electron remains somewhere in the bulk phase and does not interact at the surface.

Calculations on this state are still in progress and we can only present some preliminary results here. Figure 10 consists of the potential energy surface for the ground and the selected excited state. In the latter state the interaction is mostly repulsive even in the area of the ground states minimum. This encourages the idea that after the electronical excitation the dynamics may lead to water splitting.

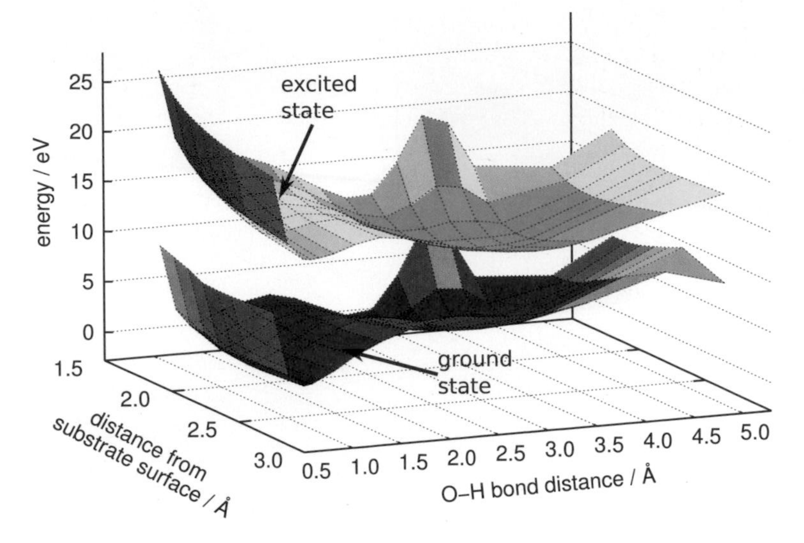

Fig. 10 Potential energy surface for the electronic ground and excited state

4 Conclusion and Outlook

In the contribution we have shown in detail how the water molecule adsorbs on rutile (110) in the electronic ground state. Furthermore, preliminary results for an electronically excited state were presented. A very large number of *ab initio* data points were calculated and fitted with an artificial neural network. This procedure proofed to be very helpful for interpolating the potential energy surfaces. The subsequent construction of a very dense grid of data points is the prerequisite for the treatment of the photodissociation with our program for quantum dynamics (Dyn5d) after some minor changes will have been implemented. With this quantum dynamical study we hope to reveal the dynamics after the excitation with light (e.g. a femtosecond laser pulse) and to describe the process on an atomic and femtosecond time-scale. Our program can account for all the quantum effects because the molecule is described as a wave packet. Therefore, tunneling or isotopic effects can be unveiled.

5 Computational Resources

The great advantage of potential energy surfaces is that each data point is independent of the other ones. Therefore, several points can be calculated simultaneously without any parallelization overhead. Since our CASPT2 calculations perform excessive IO on large files, one bottleneck of our calculations consists of the IO-performance of the file system.

Explorative studies were performed on our local resources, and we used the HLRS for the calculation of ground and excited state potential energy surfaces, respectively. About 18,000 data points (for both states) were calculated on the HLRS systems, each taking about 7 h of CPU time.

Acknowledgements We like to thank Prof. Dr. Bernd Meyer from Erlangen for providing the geometry of the relaxed surfaces. Some of the simulations were performed at the HPC Cluster HERO, located at the University of Oldenburg and funded by the DFG through its Major Research Instrumentation Programme (INST 184/108-1 FUGG) and the Ministry of Science and Culture of the Lower Saxony State.

References

1. Klüner, T., Freund, H.J., Freitag, J., Staemmler, V.: Laser-induced desorption of NO from NiO(100): ab initio calculations of potential surfaces for intermediate excited states. J. Chem. Phys. **104**(24), 10030 (1996). doi.10.1063/1.471747
2. Kluner, T., Freund, H.J., Staemmler, V., Kosloff, R.: Theoretical investigation of laser induced desorption of small molecules from oxide surfaces: a first principles study. Phys. Rev. Lett. **80**(23), 5208 (1998). http://link.aps.org/doi/10.1103/PhysRevLett.80.5208
3. Thiel, S., Pykavy, M., Klüner, T., Freund, H.J., Kosloff, R., Staemmler, V.: Three-dimensional ab initio quantum dynamics of the photodesorption of CO from Cr_2O_3(0001): Stereodynamic Effects. Phys. Rev. Lett. **87**(7), 077601 (2001). http://link.aps.org/doi/10.1103/PhysRevLett.87.077601
4. Pykavy, M., Thiel, S., Klüner, T.: Laser-induced desorption of CO from Cr_2O_3 (0001): ab initio calculation of the four-dimensional potential energy surface for an intermediate excited state. J. Phys. Chem. B **106**(48), 12556 (2002). doi:10.1021/jp026597h
5. Thiel, S., Pykavy, M., Klüner, T., Freund, H.J., Kosloff, R., Staemmler, V.: Rotational alignment in the photodesorption of CO from Cr_2O_3 (0001): a systematic three-dimensional ab initio study. J. Chem. Phys. **116**(2), 762 (2002). http://dx.doi.org/10.1063/1.1425383
6. Borowski, S., Klüner, T., Freund, H.J.: Complete analysis of the angular momentum distribution of molecules desorbing from a surface. J. Chem. Phys. **119**(19), 10367 (2003). http://link.aip.org/link/?JCP/119/10367/1
7. Borowski, S., Klüner, T., Freund, H.J., Klinkmann, I., Al-Shamery, K., Pykavy, M., Staemmler, V.: Lateral velocity distributions in laser-induced desorption of CO from Cr_2O_3 (0001): experiment and theory. Appl. Phys. A: Mater. Sci. Process. **78**(2), 223 (2004). doi:10.1007/s00339-003-2306-2
8. Kröner, D., Mehdaoui, I., Freund, H.J., Klüner, T.: Three-dimensional ab initio simulation of laser-induced desorption of NO from NiO(100). Chem. Phys. Lett. **415**(1–3), 150 (2005). http://www.sciencedirect.com/science/article/B6TFN-4H5MYSJ-6/2/3fc8feb13e07dcdc4a5cac98b7afc2ae
9. Dittrich, S., Klüner, T.: The role of laser pulse duration in the photodesorption of NO/NiO(100). Chem. Phys. Lett. **430**(4–6), 443 (2006). http://www.sciencedirect.com/science/article/B6TFN-4KW5FGW-2/2/13913091185fccbccbd6478fe6bb66d5
10. Mehdaoui, I., Kröner, D., Pykavy, M., Freund, H.J., Klüner, T.: Photo-induced desorption of NO from NiO(100): calculation of the four-dimensional potential energy surfaces and systematic wave packet studies. Phys. Chem. Chem. Phys. **8**, 1584 (2006). doi:10.1039/B512778E
11. Dittrich, S., Klüner, T.: Calculation of thermal effects in the photodesorption of NO from NiO(100). Appl. Phys. A: Mater. Sci. Process. **88**, 571 (2007). doi:10.1007/s00339-007-4055-0

12. Mehdaoui, I., Klüner, T.: Bonding of CO and NO to NiO(100): a strategy for obtaining accurate adsorption energies. J. Phys. Chem. A **111**(50), 13233 (2007). doi:10.1021/jp075703i
13. Mehdaoui, I., Klüner, T.: Understanding surface photochemistry from first principles: the Case of CO-NiO(100). Phys. Rev. Lett. **98**(3), 037601 (2007). http://link.aps.org/doi/10.1103/PhysRevLett.98.037601
14. Mehdaoui, I., Klüner, T.: New mechanistic insight into electronically excited CO-NiO(100): a quantum dynamical analysis. Phys. Chem. Chem. Phys. **10**(31), 4559 (2008). doi:10.1039/b805597a
15. Klüner, T.: Photodesorption of diatomic molecules from surfaces: a theoretical approach based on first principles. Prog. Surf. Sci. **85**(5–8), 279 (2010). http://www.sciencedirect.com/science/article/B6TJF-50RV8BY-1/2/4c11c05444ae24d2c97d877df68052d5
16. Mehring, M., Klüner, T.: Understanding surface photochemistry from first principles: the case of CO-TiO_2(110). Chem. Phys. Lett. **513**(4–6), 212 (2011). http://www.sciencedirect.com/science/article/pii/S0009261411009444
17. Mitschker, J., Klüner, T.: Adsorption and photodesorption of CO from single C60 molecules studied from first principles. Chem. Phys. Lett. **514**(1–3), 83 (2011). doi:10.1016/j.cplett.2011.08.013
18. Mitschker, J., Klüner, T.: New insight into CO photodesorption from C60. J. Phys. Chem. A **116**(46), 11211 (2012). doi:10.1021/jp305133z
19. Mehring, M., Klüner, T.: Calculation of two-dimensional potential energy surfaces of CO on a rutile(110) surface: ground and excited states. Mol. Phys. **111**, 1612 (2013). doi:10.1080/00268976.2013.780106
20. Arndt, M., Murali, S., Klüner, T.: Interaction of NO with the TiO_2(110) surface: a quantum chemical study. Chem. Phys. Lett. **556**, 98 (2013). doi:10.1016/j.cplett.2012.11.057
21. Fujishima, A., Rao, T.N., Tryk, D.A.: Titanium dioxide photocatalysis. J. Photochem. Photobiol. C: Photochem. Rev. **1**(1), 1 (2000). doi:10.1016/S1389-5567(00)00002-2
22. Fujishima, A., Zhang, X., Tryk, D.A.: TiO_2 photocatalysis and related surface phenomena. Surf. Sci. Rep. **63**(12), 515 (2008). doi:10.1016/j.surfrep.2008.10.001
23. Karlström, G., Lindh, R., Malmqvist, P.Å., Roos, B.O., Ryde, U., Veryazov, V., Widmark, P.O., Cossi, M., Schimmelpfennig, B., Neogrády, P., Seijo, L.: MOLCAS: a program package for computational chemistry. Comput. Mater. Sci. **28**, 222 (2003). http://www.sciencedirect.com/science/article/pii/S0927025603001095
24. Sousa, C., Tosoni, S., Illas, F.: Theoretical approaches to excited-State-related phenomena in oxide surfaces. Chem. Rev. **113**(6), 4456 (2013). doi:10.1021/cr300228z
25. Sousa, C., Illas, F.: Ionic-covalent transition in titanium oxides. Phys. Rev. B **50**, 13974 (1994). doi:10.1103/PhysRevB.50.13974
26. Kowalski, P.M., Meyer, B., Marx, D.: Composition, structure, and stability of the rutile TiO_2(110) surface: oxygen depletion, hydroxylation, hydrogen migration, and water adsorption. Phys. Rev. B **79**(11), 115410 (2009). http://link.aps.org/doi/10.1103/PhysRevB.79.115410
27. Mota, R., Parafita, R., Giuliani, A., Hubin-Franskin, M.J., Lourenço, J., Garcia, G., Hoffmann, S., Mason, N., Ribeiro, P., Raposo, M., Limão-Vieira, P.: Water VUV electronic state spectroscopy by synchrotron radiation. Chem. Phys. Lett. **416**(1–3), 152 (2005). doi:10.1016/j.cplett.2005.09.073
28. Harich, S.A., Hwang, D.W.H., Yang, X., Lin, J.J., Yang, X., Dixon, R.N.: Photodissociation of H_2O at 121.6 nm: a state-to-state dynamical picture. J. Chem. Phys. **113**(22), 10073 (2000). doi:http://dx.doi.org/10.1063/1.1322059
29. Maksyutenko, P., Rizzo, T.R., Boyarkin, O.V.: A direct measurement of the dissociation energy of water. J. Chem. Phys. **125**(18), 181101 (2006). doi:http://dx.doi.org/10.1063/1.2387163
30. Bukas, V.J., Meyer, J., Alducin, M., Reuter, K.: Ready, set and no action: a static perspective on potential energy surfaces commonly used in gas-surface dynamics. Z. Phys. Chem. **227**, 1523 (2013). http://www.degruyter.com/view/j/zpch.2013.227.issue-9-11/zpch-2013-0410/zpch-2013-0410.xml

31. Behler, J.: Neural network potential-energy surfaces in chemistry: a tool for large-scale simulations. Phys. Chem. Chem. Phys. **13**, 17930 (2011). doi:10.1039/C1CP21668F
32. Behler, J.: Chemical Modelling: Applications and Theory, vol. 7, pp. 1–41. The Royal Society of Chemistry (2010). doi:10.1039/9781849730884-00001
33. MATLAB, version 8.2.0.701 (R2013b). The MathWorks Inc., Natick (2013)
34. Sun, C., Liu, L.M., Selloni, A., Lu, G.Q.M., Smith, S.C.: Titania-water interactions: a review of theoretical studies. J. Mater. Chem. **20**(46), 10319 (2010). http://dx.doi.org/10.1039/C0JM01491E

GaP/Si: Studying Semiconductor Growth Characteristics with Realistic Quantum-Chemical Models

Andreas Stegmüller and Ralf Tonner

Abstract The understanding of microscopic processes and properties is crucial for the development and efficient production of inorganic III/V semiconductor materials. Those materials are grown in chemical vapour deposition procedures where elementary steps have not yet been thoroughly understood. Ab initio calculations are capable to investigate those atomic and electronic properties. Modern implementations of Density Functional Theory were applied to study layered bulk structures, periodic surface properties and adatom transport on Si(001) and GaP-Si(001) materials. By increasing cell sizes and number of atoms to scales that only supercomputing facilities can handle, a realistic chemical environment can be modeled with increased structural degrees of freedom. Bulk supercells were constructed in order to model realistic interfaces between two thin films in the nanometer scale. Supercell models in slab geometry were set up and converged with respect to the volume of vacuum and number of relaxed atoms for an accurate description of slab surfaces. These studies enable a direct comparison to experimental studies on these materials.

1 Introduction

The growth of semiconductor materials from groups 13 and 15 of the periodic system of the elements, aka groups III and V, onto silicon substrates is studied in this project by means of quantum-chemical calculations. The long-term goals of this demanding study are (a) to generate explicit knowledge of elementary chemical and physical processes occurring during the epitaxial growth of those new materials on silicon and (b) to use this data to predict the formation of high-quality structures suitable for implementation in functional opto-electronic devices in realistic environments.

A. Stegmüller • R. Tonner (✉)
Fachbereich Chemie, Philipps-Universität Marburg, Hans-Meerwein-Straße, 35032 Marburg, Germany
e-mail: tonner@chemie.uni-marburg.de

W.E. Nagel et al. (eds.), *High Performance Computing in Science and Engineering '14*, DOI 10.1007/978-3-319-10810-0_15

As a consequence, not only each individual calculation is demanding in computational capacity, the applied methodologies and models require meticulous validation of their suitability for the posed scientific questions. Hence, this report is initiated by a section on the construction and convergence study of model cells and their electronic energy. For these tasks, high-performance computing facilities were essential as to perform energy calculations of systems with large numbers of atoms and electrons.

As part of a large research cooperative this project contributes to the development of new III/V semiconductor materials with direct band gaps [1–3]. In contrast to conventional silicon-based structures, the ternary and quaternary, mixed materials have enabled optical functionality [4, 5]. This makes the integration into modern chips and devices interesting as those become sensitive to light and have the potential to emit light at defined wavelengths and intensities – features that are commonly summarized as *Silicon Photonics* [6–10].

The deposition of the functionalized materials onto silicon (Si) substrates requires an initial layer of galliumphosphide (GaP) that buffers the lattice mismatch between Si and III/V elements [6, 11]. The material layers are grown in a metal-organic vapour phase epitaxy (MOVPE) procedure from the precursors triethylgallane and *tert*-butylphosphine [11–13]. The substrate is heated and the precursors are separately admitted into the reaction chamber in hydrogen gas streams. Upon molecular decomposition, fragment adsorption, further surface-assisted reactivity, nucleation and growth the resulting thin layers form almost perfect crystal structures with low defect and impurity concentrations[7]. This project addresses those elementary processes occurring during the growth and aims to understand their chemical and/or physical nature. It is the authors' intention to contribute to the optimization of the GaP/Si growth process and predict its equivalent for more complex materials.

Periodic density functional theory (DFT) [14,15] calculations as implemented in the Vienna Ab Initio SImulation Package (VASP) [16,17] represent the predominant choice for energy calculations of the Si surface and related systems [18–20]. In a former study it was shown that DFT on the level of the Generalized Gradient Approximation (GGA) together with a recent correction for dispersion interactions and the Perdew-Becke-Enzerhof functional (PBE-D3) [21–23] are in good agreement with high-level wavefunction-based methods like MP2 and CCSD(T) for the above mentioned precursor systems [24]. Extending this experience with previous studies on extended semiconductor systems[20, 25–27] the applied approximations in this study's computations are suited to produce results on a high-accuracy level. Electronic structure calculations were performed based on DFT and the projector-augmented plane-wave formalism (PAW) [28,29] in a Bloch-periodic[30] description of crystalline surface and bulk structures. Gained energies and structures are utilized to describe chemical reactions, atomic transport during growth and relative crystal stabilities.

In this report's latter section the kinetics of adatom hopping processes (Si atom on different substrates) are presented. The hopping barriers were determined by accurate DFT and applied in a kinetic Monte Carlo simulation[31–33] in order to understand parts of the growth behaviour of GaP on Si(001), where atomic transport at the surface is an essential and determining physical process. In this report, preliminary and preparational work is presented, which was performed in 2013 and is necessary to produce well-converged calculations with reliable results.

2 Construction of Model Cells and Convergence

In order to study the chemical behaviour of precursor fragments close to and attached to a surface it is necessary to construct model cells that fulfill certain requirements. Those are

1. Realistic mapping of the physical situation with respect to chemical configuration and orientation, surface coverage and surface reconstruction,
2. Consistent behaviour of the applied energy methods and resulting structures, and
3. Absence of artefacts.

For many systems the constructed supercells have to be very large, contain many atoms and exhibit multiples of the diamond cubic unit as cell vectors. It is shown in the following section which supercell types were used and how those were constructed from Si's experimental diamond cell with lattice vector a = 5.431 Å [30].

2.1 *Atomic Structure in Periodic Supercells*

The supercells were constructed from the diamond cubic unit cell which contains $Z = 8$ atoms in the conventional unit and is shown in different orientations in the left column of Fig. 1. Its structure is distinctly determined by the lattice constant a, which was kept constant in this report's calculations.

Bulk or slab supercells can be constructed in manyfold ways. On the one hand, Fig. 1 shows slab supercells that were elongated along the z-direction (i.e. <001>in this case) to $3 \times a$ (central column) and $4 \times a$ (right column), respectively. This elongation leads to the formation of two surfaces – at the top of the supercell and at about the center of the structure slab. The region in between is filled with vacuum containing only unoccupied plane wave functions. Due to its periodic character, the structure resembles infinite thin films separated by vacuum in x/y-direction, which are infinitely stacked in z-direction.

In bulk cells (e.g. the single cubic unit cell, left column), on the other hand, any region is filled with atoms and the periodic structure constitutes a bulk phase with the given cell as its translational unit.

Examples for the constructions of bulk supercells are introduced in the following subsection while slab supercells were applied in the subsection below and in the study of adatom hopping on Si(001) and GaP/Si(001) substrates in Sect. 3.

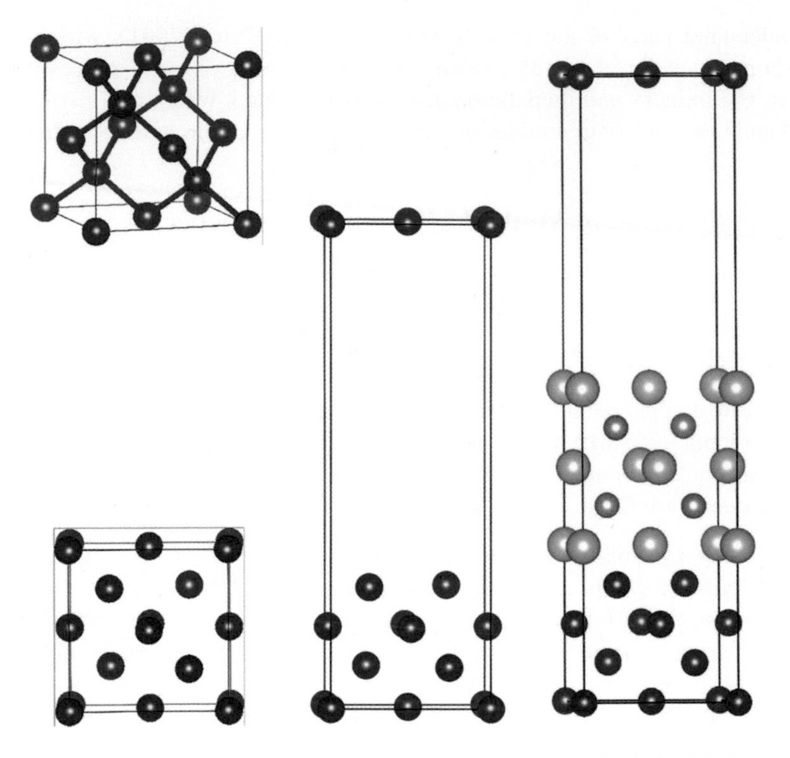

Fig. 1 Diamond cubic unit cells of the silicon crystal with experimentally derived lattice constant $a = 5.431\,\text{Å}$ (Si(001)(1 × 1), *left column*). Supercell slab models were constructed from the diamond cubic unit cell by elongation to 3 × *a* (*central column*) and 4 × *a* (*right column*) in z-direction. The region above the atoms was filled with vacuum and with four layers of Si and GaP, respectively, generating the GaP/Si(001)(1 × 1) slab supercell (*right column*)

2.1.1 Convergence of Layer Thickness in Bulk Models

Layered bulk structures of alternating thin films of Si and GaP are studied by the interface formation energy from the corresponding pure bulk materials. The constructed Si and GaP films have equal thicknesses, respectively, i.e. 8.6, 13.8 and 21.7 Å shown in the three structures in Fig. 2. Accordingly, the number of atoms within one supercell are equal for the two assembled materials Si and GaP, respectively. The supercells were, again, constructed from the diamond cubic unit cell and elongated to $3 \times a$, $5 \times a$ and $8 \times a$ in z-direction and filled with atoms on diamond lattice positions.

The formation of those stacked thin film structures containing two materials was studied for contact facets orthogonal to the <001>direction. The convergence behaviour of the interface formation energies is shown in Fig. 3.

The supercells with an elongation of $8 \times a$ in Z-direction exhibit material film thicknesses of 21.7 Å. The interface formation energies can be regarded as

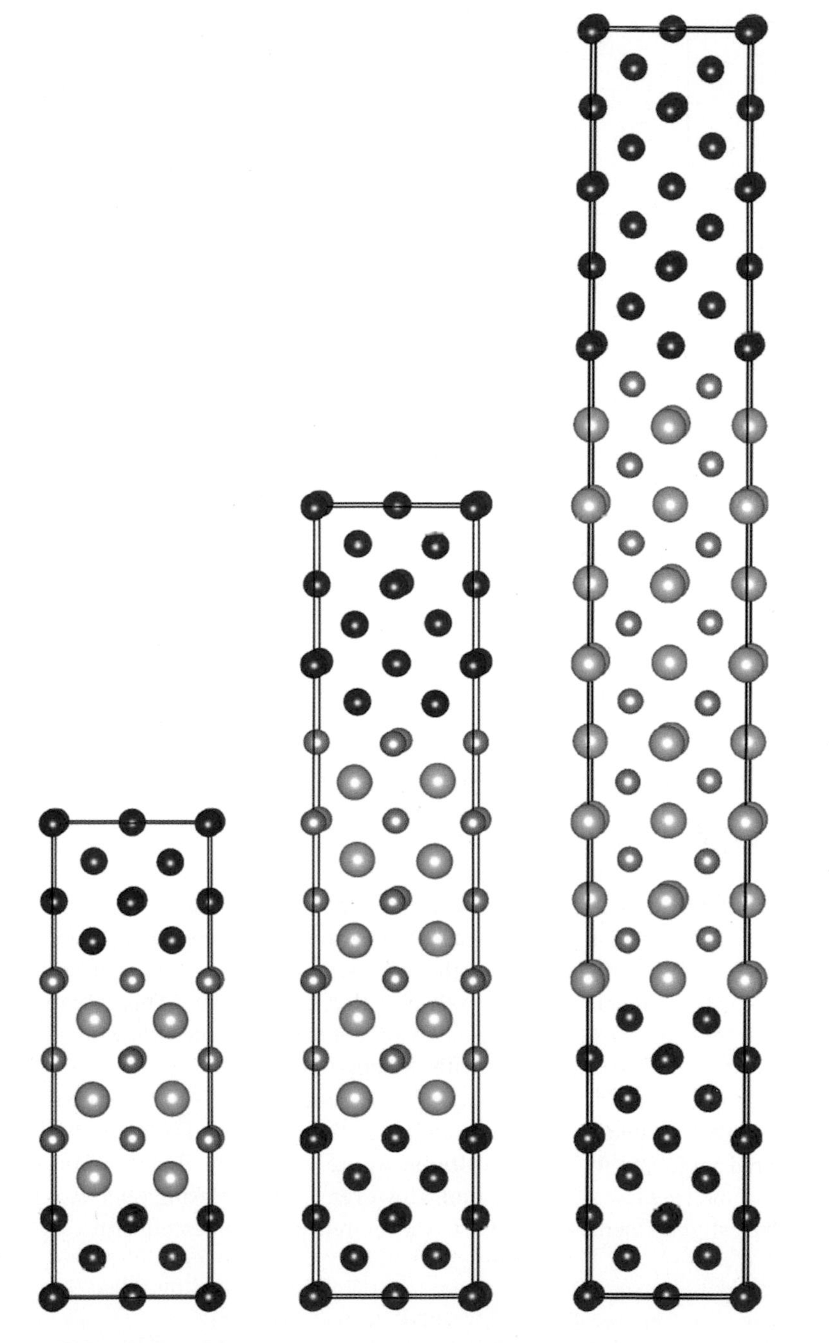

Fig. 2 Bulk supercell models with an elongation factor of $3 \times a$ (*left*), $5 \times a$ (*center*) and $8 \times a$ (*right*) in z-direction (i.e. <001>) relative to the diamond cubic unit cell

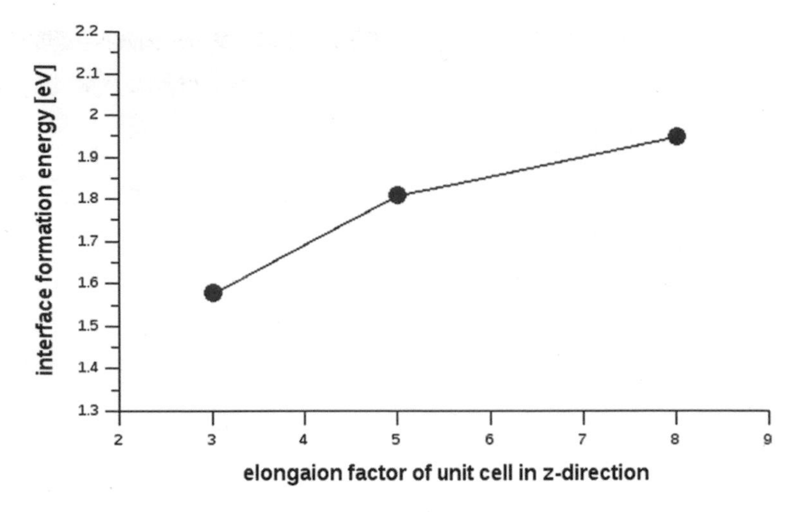

Fig. 3 Convergence behaviour of the interface formation energy between Si and GaP in <001>with respect to the layers' thicknesses in bulk model cells

converged for this supercell size. The calculations[1] of such supercells can efficiently be parallelized (by k-points and electronic bands) and were carried out on 128 or $512cpu$s within a few hours.

2.1.2 Convergence of Vacuum Region in Slab Models

Slab supercells were constructed for the study of large thin films of material-pure Si or GaP structures. Figure 4 show examples of such film cells with a vacuum region of 11.9 Å. The slabs were constructed from the diamond cubic unit cell with an elongation of $8 \times a$ along Z-direction exclusive the vacuum region. In order to relax the electronic structure correctly the distance between the slabs must be sufficiently sized so that no interaction acts from one slab to the other. This is particularly important for those systems where an adsorbate molecule interacts with one of the surfaces. Typically, an interaction with the opposite surface (the neighbouring slab in Z-direction) is not intended, but rather a comparison to unperturbed, relaxed surfaces. The vacuum regions presented in the graph of Fig. 4 are commonly considered large and the energy differences are very small (below 17 meV) with respect to the smallest vacuum region. However, with increasing interslab distance the energy slightly increases relative to the reference slab supercell with a separation of 11.9 Å.

[1] Accurate energy convergence criteria (10^{-6} eV), kinetic energy cutoff for the basis set at 400 eV, 32 k-points, Gaussian smearing ($\sigma = 0.05$).

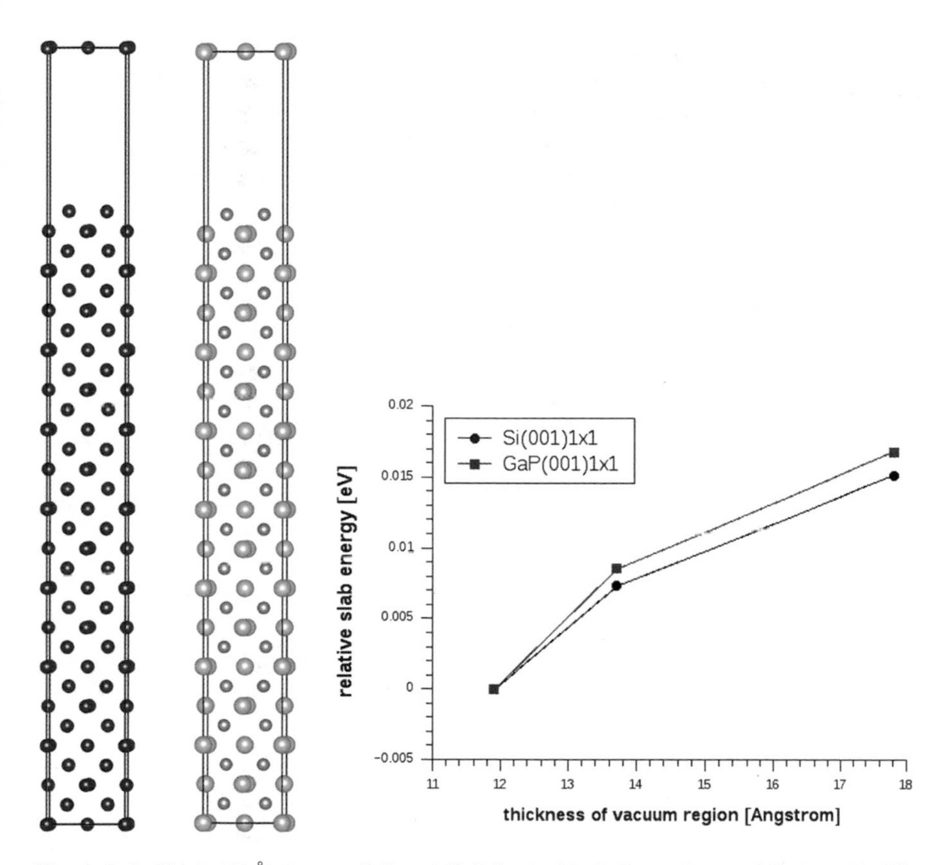

Fig. 4 *Left*: Thick (42 Å) layers of Si and GaP in the ideal diamond crystal lattice with Si's experimental lattice constant ($a = 5.431$ Å). *Right*: The slab electronic energies are presented relative to the illustrated inter-slab separation of 11.9 Å in dependence of an increasing vacuum height

2.2 Surface Reconstruction of Si(001)c(4 × 2) in an Accurate Slab Model

Due to unpaired valence electrons at the top-most Si atoms the Si(001) surface in the ideal diamond lattice conformation is energetically not favourable. It naturally relaxes to surface reconstructions that reorganize atoms so that unpaired electrons, so-called *dangling bonds*, transform to exclusively paired configurations. One predominant configuration is the Si(001)c(4 × 2) reconstruction[34], where a unit cell contains four buckled Si-Si dimers at the facet orthogonal to the <001>-direction (see Fig. 5). Each dimer consists of a covalent Si-Si bond in addition to two bonds per Si atom to the subjacent layer. One of the Si atoms carries paired electrons and is lifted out of the surface while the other is shifted towards the substrate slab

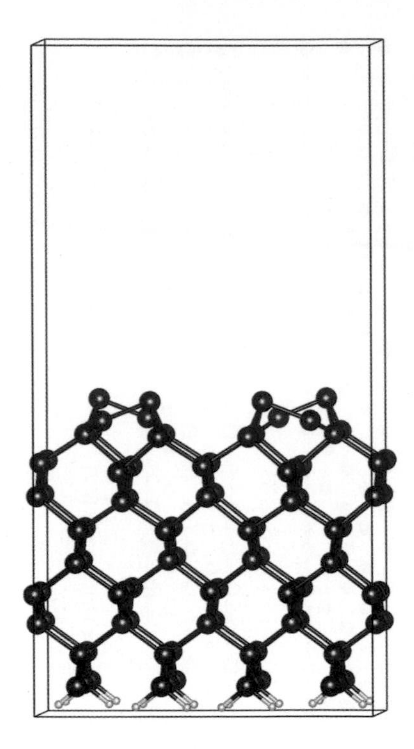

Fig. 5 Large slab model cell of the Si(001)c(4 × 2) surface reconstruction passivated with hydrogen at the *bottom* surface. The structure shown was fully relaxed with only the *bottom* two Si layers and hydrogens constrained to the ideal diamond lattice and pre-relaxed positions, respectively

and exhibits an empty orbital. In the 4 × 2 reconstruction, a mirror plane is found in between neighbouring rows of dimers (comp. Fig. 5 in the direction of view).

A large slab model cell of this reconstruction was constructed and structurally relaxed with high accuracy standards[2] in order to produce a standard configuration for future studies on the Si(001)c(4 × 2) surface reconstruction. The structure model contains ten atomic layers of Si and is saturated at the slab bottom with hydrogen[3] atoms (instead of another Si surface reconstruction). The separation between slabs is 16.2 Å (sufficiently large). The structure was subsequently relaxed including, firstly, the top two layers and, then, stepwise down to the bottom layers. The initial atom positions originated from the ideal diamond lattice.

The energetic differences of an increasing degree of relaxed atomic layers within the Si(001)c(4 × 2) slab is plotted in Fig. 6. The bottom two layers as well as the H atoms are always kept in ideal lattice and pre-relaxed positions, respectively.

Some structural details are summarized in Table 1. There is a significant tilt angle of the Si-Si dimer bond relative to the cell's base plane. This is due to shifts of electron density to the upper lying Si atoms, respectively.

[2]Energy converged to 10^{-7} eV and forces to 10^{-4} eV/ Å at a kinetic energy cutoff of 400 eV with Gaussian-type smearing ($\sigma = 0.05$).

[3]The H atoms were positioned in accordance to silane's, SiH_4, molecular geometry and relaxed in a preparational calculation.

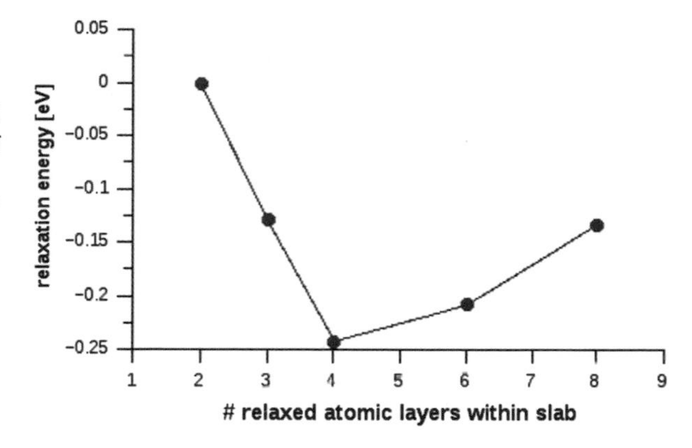

Fig. 6 Stepwise relaxation energies of Si atom layers in the Si(001)c(4 × 2) surface reconstruction slab presented in Fig. 5. The top eight out of ten layers were subsequently relaxed starting from the two top-most, reconstructed (dimer) layers. The initial atom positions corresponded to the ideal diamond lattice positions

Table 1 Heights of the top four Si layers relative to the cell base in the Si(001)c(4 × 2) surface reconstruction (as presented in Fig. 5). Distances are given in Å and angles in ° as obtained by a stepwise relaxation of the top eight of ten atomic layers of the slab

z-height of Si layer	*Top*	*Top* − 1	*Top* − 2	*Top* − 3
Si_{up}/beneath dimer	14.021	12.502	10.948	9.572
Si_{down}/beneath trench	13.169	–	11.215	9.797
Dimer bond length	2.348			
Dimer bond tilt angle	21.3			

The dimerization of the top-layer Si atoms causes the emergence of a trench in between the rows of parallel dimers (direction of view in Fig. 5). It was found that the substrate reacts on this by slightly shifting the third and fourth row atoms (top-2, top-3) upwards.

3 Adatom Hopping in Early GaP/Si Growth

The quality of novel III/V materials is essentially determined by their growth behaviour. In chemical vapour deposition procedures such as MOVPE, the atomic structure is critically determined by the cleanliness of precursor decompositions and uniform distribution of the constituting atoms. Specifically, the nucleation phase of the growth of GaP on Si(001) is determined by the mobility of material fragments adsorbed onto the substrate, which needs to be sufficient to uniformly distribute the atoms over the first atomic layers of the material.

In the following model study it was presumed that the decomposition of precursors (by gas-phase and surface-assisted pathways) is completed before the material nucleation. Hence only atomic species are considered for surface diffusion.

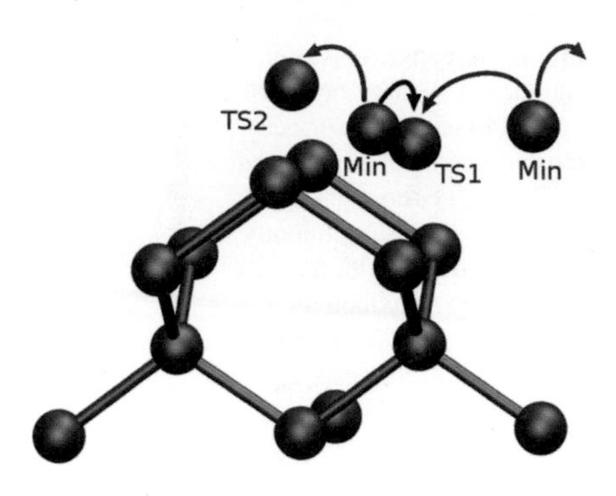

Fig. 7 Minimum (Min) and transition state (TS) positions of a Si adatom for hopping steps along the <110>direction on an ideal Si(001)(1 × 1) surface. The *arrows* indicate the direction of translation from Min to TS; all steps necessary to reach the translationally equivalent Min position are shown for the respective adatom type

Hopping barriers of Si atoms were determined by DFT calculations with high accuracy.[4]

Model slab supercells of the substrates were constructed from Si's diamond cubic unit cell with constrained atom positions. Pure Si(001) as well as Ga- and P-terminated GaP/Si(001) surfaces were considered as diffusion kinetics of the initial GaP monolayer's adatoms (on pure Si) and the following GaP layers' (on GaP/Si) are equally important. The <001>-direction is a typical growth direction of MOVPE experiments. The thermal diffusion barriers for adatoms on the three different substrate surfaces are presented in the following subsections.

3.1 Adatom Hopping on the Si(001) Surface

Slab supercells of Si(001)(1 × 1) as introduced in Fig. 1 (central column) were applied as substrate for the calculation of adatom hopping. Due to the symmetric lattice the configuration al space was limited and minimum positions (Min) of the atomic species adsorbed onto the chemically active (unreconstructed) surface were efficiently determined. The transition states (TS) of the hopping steps, however, were slightly more challenging and computationally demanding.

A hopping step was considered complete as two translationally equivalent Min positions were reached via a (or two subsequent) TS position(s). Both saddle point types of the energy surfaces were characterized by analytical calculations of the adatoms' Hessian matrices: Min positions do not show any and TS positions have one imaginary mode connecting two Min positions along the diffusion pathway.

Figure 7 illustrates Min and TS positions of a Si adatom on a Si(001)(1 × 1) diamond cubic cell slab and indicates the direction of the diffusion hops. Two Min positions with two corresponding TS positions (TS1, TS2) were found within the

---

[4]Analytical Hessian matrices of the adatoms at Min and TS were computed to identify first order saddle points indicating the hopping transition state with the imaginary mode indicating the direction of translation.

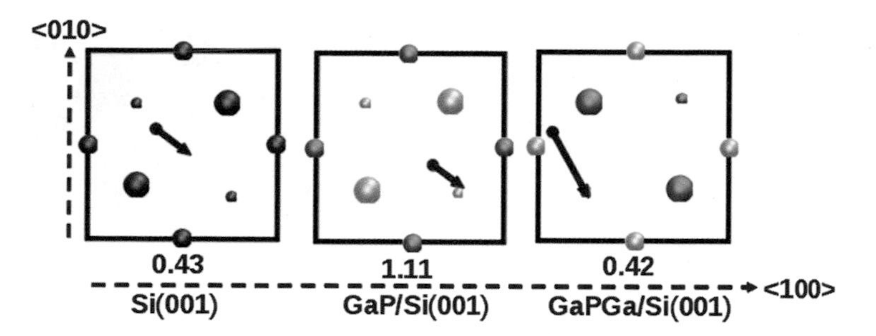

Fig. 8 Minimum (origin of *red arrows*) and transition state (head of *red arrows*) positions of Si adatoms indicating the direction and step width of hopping on a diamond lattice viewed from atop (i.e. <00 − 1>). The underlying substrate surfaces are Si(001)(1 × 1) (*left*), P-terminated GaP/Si(001)(1 × 1) (*center*) and Ga-terminated GaP/Si(001)(1 × 1) (*right*). Hopping barriers corresponding to the dominant diffusion steps for Si adatoms on each substrate are provided below the cell representations in eV

slab cell. The barrier of TS2 was, however, relatively low so that the diffusion rate will be mainly determine by only one TS per adatom hop.

The dominant Min and TS positions on all substrate types considered are summarized in Fig. 8. The direction of the minimum diffusion pathways for the adatom can be observed corresponding to the TSs' imaginary modes. The hopping barriers unveil that the Si adatom has the largest barrier on the P-terminated GaP/Si(001)(1 × 1) surface while the barriers on pure Si and Ga-terminated GaP/Si are less pronounced.

3.2 *Adatom Hopping on the P-Terminated GaP/Si(001) Surface*

The P-terminated GaP/Si(001) substrate slab cell was constructed according to the procedure describe above. It consists of four atomic layers of GaP on top of four layers of Si (see Fig. 1, right column). Figure 9 illustrates the slab and the TS position of the Si adatom. As can be retraced from Fig. 8 the Min and TS positions are shifted with respect to the corresponding positions on the pure Si(001) substrate surface. This might be due to an increased potential for the adatoms to interact chemically with the top-most substrate atoms in the P-terminated substrate as opposed to the Si-termination. Further studies will investigate this behaviour in more detail. The differences in hopping barriers determined in this model study are presumably dominantly determined by effects of different numbers of valence-shell electrons in the substrate species. A quantum-chemical description of these features will benefit the understanding of diffusion mobility differences of the involved species which are expected to determine important material properties.

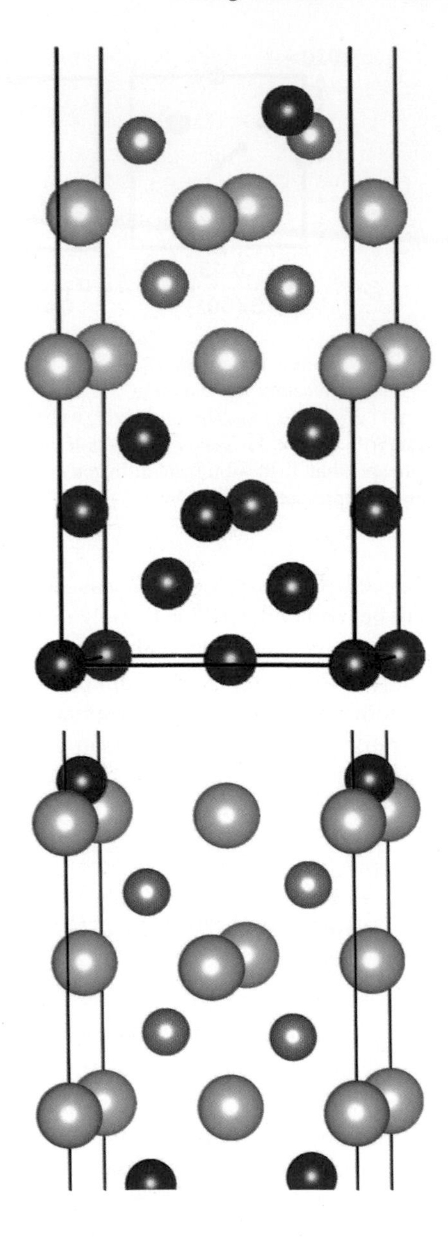

Fig. 9 Transition state (TS) positions of the Si adatom for hopping steps along the <110>direction on an ideal P-terminated GaP/Si(001)(1 × 1) surface

Fig. 10 Transition state (TS) position of a Si adatom for hopping steps along the <110>direction on an ideal P-terminated GaP/Si(001)(1 × 1) surface

3.3 Adatom Hopping on the Ga-Terminated GaP/Si(001) Surface

The slab supercells of Ga-terminated GaP/Si(001) were constructed in accordance to the P-terminated GaP/Si(001)(1 × 1) cell introduced previously. An additional atomic layer of Ga atoms was attached as top layer of the substrate representing ideal Ga-termination.

Figure 10 illustrates the GaP region of the GaP/Si(001) slab together with the Si adatom at its TS position. The additional layer of Ga to the substrate rotates the preferred direction of diffusion steps by 90°. The corresponding barrier is, however, in good agreement with the Si substrate's results indicating similar chemical interaction at Min and TS positions on both substrates.

Further diffusion-related computations will be performed in the future in order to examine the rate- and structure-determining processes during growth of GaP/Si(001). Those will necessarily include relevant fragments of the MOVPE precursors.

Acknowledgements The authors acknowledge the collaborative research training group (Graduiertenkolleg, DFG) 1782 "Functionalization of Semiconductors" as well as the Beilstein Institut, Frankfurt am Main, for financial and further support.

References

1. Jandieri, K., Kunert, B., Liebich, S., Zimprich, M., Volz, K., Stolz, W., Gebhard, F., Baranovski, S.D., Koukourakis, N., Gerhardt, N.C., Hofmann, M.R.: Phys. Rev. B **87**(3), 035303 (2013). doi:10.1103/PhysRevB.87.035303
2. Lange, C., Chatterjee, S., Kunert, B., Volz, K., Stolz, W., Rühle, W.W., Gerhardt, N.C., Hofmann, M.R.: Gain characteristics and lasing of Ga(NAsP) multi-quantum well structures. Phys. Status Solidi (C) **6**(2), 576 (2009). doi:10.1002/pssc.200880360
3. Kunert, B., Volz, K., Koch, J., Stolz, W.: Appl. Phys. Lett. **88**(18), 182108 (2006). doi:10.1063/1.2200758
4. Kunert, B., Volz, K., Stolz, W.: Phys. Status Solidi (B) **244**(8), 2730 (2007). doi:10.1002/pssb.200675609
5. Liebich, S., Zimprich, M., Beyer, A., Lange, C., Franzbach, D.J., Chatterjee, S., Hossain, N., Sweeney, S.J., Volz, K., Kunert, B., Stolz, W.: Appl. Phys. Lett. **99**(7), 071109 (2011). doi:10.1063/1.3624927
6. Németh, I., Kunert, B., Stolz, W., Volz, K.: J. Cryst. Growth **310**(7–9), 1595 (2008). doi:10.1016/j.jcrysgro.2007.11.127
7. Volz, K., Beyer, A., Witte, W., Ohlmann, J., Németh, I., Kunert, B., Stolz, W.: J. Cryst. Growth **315**(1), 37 (2011). doi:10.1016/j.jcrysgro.2010.10.036
8. Liang, D., Bowers, J.E.: Nat. Photonics **4**(8), 511 (2010). doi:10.1038/nphoton.2010.167
9. Dürr, M., Biedermann, A., Hu, Z., Höfer, U., Heinz, T.F.: Science **296**(5574), 1838 (2002). doi:10.1126/science.1070859
10. Kunert, B., Zinnkann, S., Volz, K., Stolz, W.: J. Cryst. Growth **310**(23), 4776 (2008). doi:10.1016/j.jcrysgro.2008.07.097
11. Beyer, A., Ohlmann, J., Liebich, S., Heim, H., Witte, G., Stolz, W., Volz, K.: J. Appl. Phys. **111**(8), 083534 (2012). doi:10.1063/1.4706573
12. Saxler, A., Walker, D., Kung, P., Zhang, X., Razeghi, M., Solomon, J., Mitchel, W.C., Vydyanath, H.R.: Appl. Phys. Lett. **71**(22), 3272 (1997). doi:10.1063/1.120310
13. Fukuda, Y., Kobayashi, T., Mochizuki, S.: Appl. Surf. Sci. **176**, 218 (2001)
14. Hohenberg, P., Kohn, W.: Phys. Rev. **136**(3B), B864 (1964)
15. Kohn, W., Sham, L.J., Others: Phys. Rev. **140**(4A), A1133 (1965)
16. Kresse, G., Furthmüller, J.: Phys. Rev. B Condens. Matter **54**(16), 11169 (1996)

17. Kresse, G., Furthmüller, J.: Comput. Mater. Sci. **6**(15) p. 15 (1996)
18. Schmidt, W.G., Bernholc, J., Bechstedt, F.: Appl. Surf. Sci. **166**, 179 (2000)
19. Hafner, J.: J. Comput. Chem. **29**(13), 2044 (2008). doi:10.1002/jcc
20. Kempisty, P., Krukowski, S., Strak, P., Sakowski, K.: J. Appl. Phys. **106**(5), 054901 (2009). doi:10.1063/1.3204965
21. Perdew, J., Burke, K., Ernzerhof, M.: Phys. Rev. Lett. **77**(18), 3865 (1996)
22. Grimme, S.: Wiley Interdiscip. Rev.: Comput. Mol. Sci. **1**(2), 211 (2011). doi:10.1002/wcms.30
23. Grimme, S., Ehrlich, S., Goerigk, L.: J. Comput. Chem. **32**, 1456 (2011). doi:10.1002/jcc
24. Stegmüller, A., Rosenow, P., Tonner, R.: Phys. Chem. Chem. Phys. **16**, 17018 (2014)
25. Mattsson, A.E., Schultz, P.A., Desjarlais, M.P., Mattsson, T.R., Leung, K.: Model. Simul. Mater. Sci. Eng. **13**(1), R1 (2005). doi:10.1088/0965-0393/13/1/R01
26. Krukowski, S., Kempisty, P., Strak, P.: Cryst. Res. Technol. **44**(10), 1038 (2009). doi:10.1002/crat.200900510
27. Hashemifar, S., Kratzer, P., Scheffler, M.: Phys. Rev. B **82**(21), 1 (2010). doi:10.1103/PhysRevB.82.214417
28. Blöchl, P.: Phys. Rev. B **50**(24), 17953 (1994)
29. Kresse, G., Joubert, D.: Phys. Rev. B **59**(3), 11 (1999)
30. Kittel, C.: Introduction to Solid State Physics. Wiley, New York (1996)
31. Reuter, K.: First-Principles Kinetic Monte Carlo Simulations for Heterogeneous Catalysis: Concepts, Status, and Frontiers. In: Deutschmann, O. (ed.) Modeling Heterogeneous Catalytic Reactions: From the Molecular Process to the Technical System, Chap. 3, p. 71ff. Wiley-VCH, Weinberg (2009)
32. Stampfl, C.: Catal. Today **105**(1), 17 (2005). doi:10.1016/j.cattod.2005.04.015
33. Reuter, K., Frenkel, D., Scheffler, M.: Phys. Rev. Lett. **93**(11), 1 (2004). doi:10.1103/PhysRevLett.93.116105
34. Fritsch, J., Pavone, P.: Surf. Sci. **344**(1–2), 159 (1995). doi:10.1016/0039-6028(95)00802-0

Part III
Reactive Flows

Dietmar Heinrich Kröner

The new contributions of the projects of this section concern the following issues. For all projects challenging real world problems in three space dimensions are considered, two of them are based on the efficient coupling of existing codes. Two projects use direct numerical simulations and one of them the Reynolds-Averaged Navier-Stokes equations. All projects are funded by the DFG.

Feichi Zhang, Henning Bonart, Thorsten Zirwes, Peter Habisreuther, Henning Bockhorn and Nikolaos Zarzalis consider in their project "Direct Numerical Simulation of Chemically Reacting Flows with the Public Domain Code OpenFOAM" complex flow patterns in combustion, characterized by high Reynolds numbers and turbulence. The smallest turbulent length scales are in the order of the typical flame thickness, implying that more involved models have to be used. For the mathematical model they use the compressible reactive flow equations.

The aim of this project is to introduce a conceptually developed solver based on the widespread open-source program OpenFOAM for the flow and Cantera for the reactive part. The discretization is then based on an operator-splitting approach which allows to compute solutions of the flow and chemistry with time scales, that differ by orders of magnitude. The grid is unstructured and consists of up to 144 million cells. As both OpenFOAM and Cantera are freely available, the solver provides a simple access for researchers who are working with modeling of reacting flows.

The simulation has been performed parallel on 8192 processors from the HERMIT cluster. The solver has first been applied to a one dimensional premixed flame for the validation of the implemented algorithms for the calculation of the reaction rates and molecular transports. The calculated burning velocity agrees well with the experimental data. The code has proved to speed-up efficiently. Finally the

D.H. Kröner (✉)
Abteilung für Angewandte Mathematik, Universität Freiburg, Hermann-Herder-Str. 10, 79104 Freiburg, Germany
e-mail: dietmar.kroener@mathematik.uni-freiburg.de

code is applied to simulate the flame propagation in an explosion vessel in three space dimensions of laboratory-scale. The main new result in this project is to prove that the combination of freely available software OpenFOAM and Cantera can be used efficiently on large scale parallel computers.

A similar problem with a different focus has been considered in the contribution of Jordan A. Denev, Iliyana Naydenova and Henning Bockhorn about "Lean Premixed Flames: A Direct Numerical Simulation Study of the Effect of Lewis Number at Large Scale Turbulence". Here four different lean flames – a hydrogen, two methane and a propane flame – having Lewis numbers which are smaller, nearly equal and larger than unity, are considered for their reaction with the turbulent flow field induced by one and the same set of vortices.

Furthermore the effect of fuel type on the flame response to turbulence is considered. Fuels have different transport properties which are due to their Lewis numbers, which are known to respond differently to flame front curvature and aerodynamic strain. The code used in the present work is PARCOMB-3D and it is specialized for the study of non-stationary combustion phenomena. It solves the coupled system of fully compressible Navier-Stokes combined with the reactive transport equations using reduced chemical models. The discretization is based on a three-dimensional DNS with a 4-th order Runge-Kutta explicit time stepping scheme and 6-th order central differencing Cartesian numerical grid with variable grid spacings in space. Also a new method for the generation of turbulent vortices directly inside the computational domain, and not at its boundaries, is developed.

The third contribution of this session concerns the "Parallelization and Performance Analysis of an Implicit Compressible Combustion Code for Aerospace Applications" and is submitted by Roman Keller, Markus Lempke, Yann Hendrik Simsont, Peter Gerlinger and Manfred Aigner. They consider the three dimensional supersonic reactive flows with a detailed chemical reaction model for spacecraft configuration, like scramjets and liquid rocket combustors. The flow is assumed to be a multiphase flow and the liquid phase to be a dilute spray of discrete, non-interacting droplets. For the simulation the in-house code TASCOM3D is coupled with the Lagrangian particle tracking code SPRAYSIM. In the standard setup approximately 13,000 droplet parcels are used to represent the actual spray. The spray initial conditions are derived from measurements. The mesh consists of 850,000 volumes that are distributed over 64 CPUs.

The underlying reactive flow equations are solved on a block structured grid with a finite volume scheme. For the computation of the cell-interface values a reconstruction with up to 6-th order combined with certain limiters, in order to prevent oscillations, is established. Several test configurations are taken into account, also for validation and some of the test cases have been used to assess the scaling performance of the employed numerical tools.

Direct Numerical Simulation of Chemically Reacting Flows with the Public Domain Code OpenFOAM

Feichi Zhang, Henning Bonart, Thorsten Zirwes, Peter Habisreuther, Henning Bockhorn, and Nikolaos Zarzalis

Abstract A new solver for direct numerical simulation (DNS) of chemically reacting flow is introduced, which is developed within the framework of the open-source program OpenFOAM. The code is capable of solving numerically the compressible reactive flow equations employing unstructured grids. Therewith a detailed description of the chemistry, e.g. the reaction rates, and transport, e.g. the diffusion coefficients, has been accomplished by coupling the free chemical kinetics program Cantera. The solver implies a fully implicit scheme of second order for the time derivative and a fourth order interpolation scheme for the discretization of the convective term. An operator-split approach is used by the solver which allows solutions of the flow and chemistry with time scales that differ by orders of magnitude, leading to a significantly improved performance. In addition, the solver has proved to exhibit a good parallel scalability. The implementation of the code has first been validated by means of one-dimensional premixed flames, where the calculated flame profiles are compared with results from the commercially Chemkin code. To demonstrate the applicability of the code for three-dimensional problems, it has been applied to simulate the flame propagation in an explosion vessel of laboratory-scale. A computational grid with 144 million finite volumes has been used for this case. The simulation has been performed parallel on 8192 processors from the HERMIT cluster of HLRS. The calculated burning velocity agrees well with the experimental data.

1 Introduction

Traditional methods for modeling combustion related flows are known as the Reynolds-averaged Navier-Stokes (RANS) approach and large eddy simulation (LES). These are based on the solution of the governing reactive flow equations with

F. Zhang (✉) • H. Bonart • T. Zirwes • P. Habisreuther • H. Bockhorn • N. Zarzalis
Engler-Bunte-Institute, Division of Combustion Technology, Karlsruhe Institute of Technology, Engler-Bunte-Ring 1, 76131 Karlsruhe, Germany
e-mail: feichi.zhang@kit.edu

W.E. Nagel et al. (eds.), *High Performance Computing in Science and Engineering '14*, DOI 10.1007/978-3-319-10810-0_16

help of statistical methods, i.e. the conserved flow variables are decomposed either in a time- or volume-mean part and a fluctuation or residual part, which are then solved by the flow simulation and considered as accurate enough for description of the turbulent process. By doing so, the chemical source terms, i.e. the reaction rates, need to be approximated while averaging or filtering the equations due to their non-linear behaviors. This formulates the main task of the numerous existing concepts for turbulent combustion modeling [1, 2]. These models usually underly different physical restrictions, for example, most of them are valid when the chemical reaction proceeds much faster than the mixing, so that the intrinsic structures of the flame are assumed to be undisturbed by the flow (flamelet assumption). For most industrial applications however, combustion occurs in complex flow patterns characterized by high Reynolds numbers (Re) and turbulence intensities. The smallest turbulent length scales are in the order of the typical flame thickness, suggesting that the flamelet assumption is in most cases violated. Proper and more comprehensive modeling concepts therefore yield importance and are still under development.

The direct numerical simulation (DNS) on the other hand relies on the exact governing equations without averaging or filtering. The full range of time and length scales of both the turbulence and the chemistry is resolved in this case. The resolution of the interactive behavior between the turbulent transport and the complex chemistry in DNS provides greater realism and quantitative predictability, complementing experimental measurements and subordinate numerical modeling (for example, in the framework of RANS or LES) in providing information at critical conditions of future combustion systems. The problem with DNS is however that it is very demanding in computing resources. Terabytes of memory and hard-disk space for raw data as well as millions of cpu-hours are typically required for DNS of even moderately complex cases containing hundreds of millions grid points and dozens of species [3]. This is the main reason why most of the previous combustion DNS have been limited to fundamental research cases, for example, low Re flows with simulation domains of centimeter range or only two dimensions [2, 4, 5].

The rapid growth in high performance computing (HPC) technique in the past decades has presented significant opportunities for DNS of complex reactive flows, such as turbulent combustion [3, 6]. Nevertheless, an efficient implementation of the code still plays a decisive role for the successful completion of high-fidelity simulations. For instance, most of the available DNS codes use a time-explicit finite difference method together with a structured mesh, which are suited only for simple shaped computational domains such as boxes or squares [4–6]. Furthermore, the DNS codes must scale well in the weak sense to be able to run on the full extent of the peta-scale supercomputers.

The aim of the present paper is to introduce a conceptually developed solver based on the widespread open-source programs OpenFOAM [7] and Cantera [8]. It can be used for DNS of compressible reacting flows on unstructured grids and arbitrarily shaped computational domains. The solver has first been applied to a one-dimensional premixed flame for the validation of the implemented algorithms for the calculation of the reaction rates and molecular transports. Thereafter, an experimental device for measuring the burning velocity in an explosion vessel has been simulated to demonstrate the capability of the code for three-dimensional applications. In addition, the code has proved to speed-up efficiently on the computing platform Cray XE6 (HERMIT) [9].

2 Mathematical Description of Chemically Reacting Flows

To perform DNS of compressible reacting flows with N species, the conservation equations for the total mass, the momentum, the species masses and the energy have to be solved numerically together with the equation of state [2]:

$$\frac{\partial \rho}{\partial t} = -\nabla \cdot (\rho \mathbf{v}) \tag{1}$$

$$\frac{\partial}{\partial t}(\rho \mathbf{v}) = -\nabla \cdot (\rho \mathbf{v}\mathbf{v}) - \nabla p + \nabla \cdot \boldsymbol{\tau} + \rho \mathbf{g} \tag{2}$$

$$\frac{\partial}{\partial t}(\rho Y_k) = -\nabla \cdot (\rho Y_k \mathbf{v}) - \nabla \cdot \mathbf{j}_k + \dot{r}_k, \quad k = 1 \cdots N \tag{3}$$

$$\frac{\partial}{\partial t}(\rho h_s) = -\nabla \cdot (\rho h_s \mathbf{v}) - \nabla \cdot \dot{\mathbf{q}} + \frac{\mathrm{D}p}{\mathrm{D}t} + \dot{q}_r \tag{4}$$

$$p = \rho R T \tag{5}$$

where ρ and $\mathbf{v}$ are the density and velocity vector, p and T denote the static pressure and temperature and h_s the sensible enthalpy. Y_k and $\dot{r}_k$ indicate mass fraction and reaction rate of the species k. R is the specific gas constant. The gravitational force $\rho \mathbf{g}$ acts as an external force on the cell volume. The heat source from viscous dissipation and radiation are neglected here.

The mixture-averaged model is applied to evaluate the diffusion flux $\mathbf{j}_k$, the viscous stress flux $\boldsymbol{\tau}$ for a Newtonian fluid and the diffusive heat flux $\dot{\mathbf{q}}$ [10]:

$$\boldsymbol{\tau} = -\mu[\nabla \mathbf{v} + (\nabla \mathbf{v})^T] + \frac{2}{3}\mu(\nabla \cdot \mathbf{v})\mathbf{I} \tag{6}$$

$$\mathbf{j}_k = -\rho D_k \nabla Y_k \tag{7}$$

$$\dot{\mathbf{q}} = -\lambda \nabla T + \sum_{k=1}^{N} \mathbf{j}_k h_k \tag{8}$$

Here D_k is the mixture-averaged diffusion coefficient between the k-th species and the rest mixture, μ and λ are the mixture-averaged dynamic viscosity and thermal conductivity, h_k is the specific enthalpy of the k-th species [10]. The reaction rate $\dot{r}_k$ in the species equation (3) is calculated by the rate law from reaction kinetics together with the extended Arrhenius law [10]. The heat release caused by chemical reactions leads to a source term in the sensible enthalpy equation (4):

$$\dot{q}_r = -\sum \dot{r}_k h_k^0 \tag{9}$$

where h_k^0 is the chemical enthalpy of the k-th species. Equations (1)–(5) constitute a system of interconnected partial differential equations which are highly nonlinear and have to be solved numerically, for example, with the finite volume method. It is noteworthy that a major part of the total computing time (40–60 %) is consumed by the calculation of the reaction source terms $\dot{r}_k$ and the thermophysical transport properties, i.e. μ, D_k and λ, due to the mathematically expensive operations.

3 Development of the DNS Solver

3.1 *Structure of the Solver*

The developed DNS solver is represented by a coupling library implemented in OpenFOAM [7] which connects with the chemical program Cantera [8]. OpenFOAM therewith is used for solving the basic transport equations, whereas Cantera provides the required reaction rates and transport properties.

- OpenFOAM [7] is an open-source toolbox, which has a wide range of available tools for solving general CFD problems, e.g. iterative solvers, discretization schemes, boundary conditions and pre-/post-processing tools. It is capable of handling unstructured grids and parallel simulation with a large number of processors. Moreover, OpenFOAM provides an expressive and versatile syntax for user-own implementations. Due to its advantages, OpenFOAM has found widespread use both in scientific and engineering applications and thereby has been extensively validated.
- Cantera [8] is a free software toolkit for solving problems encountered with chemical kinetics, thermodynamics and transport processes, which can be easily accessed from other environments, e.g. C++, Python or Matlab. The implemented algorithms rigorously follow the kinetic theory of gases [10, 11], ensuring a high level of accuracy. Due to the mathematically expensive operations by solution of reactor networks, extraneous specialized softwares, for example SUNDIALS [12], are used by Cantera to enhance numerical stability and efficiency. In addition, the Chemkin file import mechanism provided by Cantera allows to commit CHEMKIN format files to include reaction mechanisms, transport and thermal coefficients in a simple way.

Figure 1 shows the schematic structure of the coupling library. OpenFOAM has the task to solve the basic conservation equations. The state parameters in terms of pressure, temperature and mixture composition are used as input for the Cantera algorithms which are dynamically called by the coupling interface in order to calculate the thermophysical properties and reaction rates. The implementation of this coupling routine strictly follows the existing structure of the

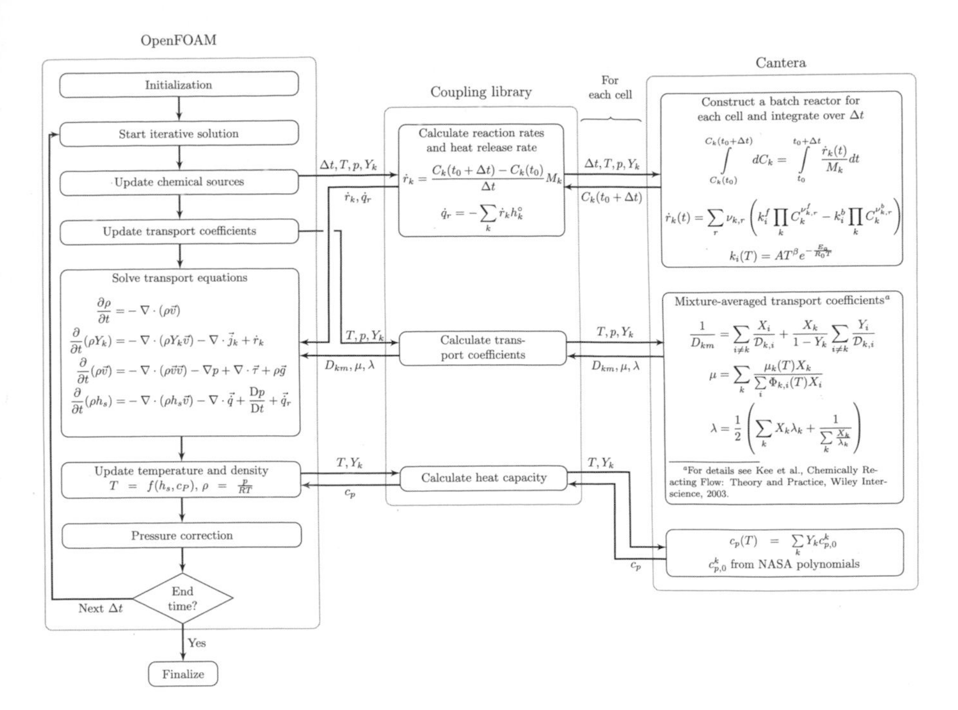

Fig. 1 Simplified flow chart: coupling of OpenFOAM with the chemical library Cantera

standard thermo-physical classes in OpenFOAM and employs templates as well as inheritance to work directly with existent program branches through a run-time selection mechanism, i.e. the user can select required libraries for the solution during run-time. For instance, the coupling allows to use sensible or absolute enthalpies and multicomponent or mixture-averaged transport models [13]. Due to the advantageous features provided by OpenFOAM, the code can easily be extended with other physio-chemical models, for example, thermal radiation or cross diffusion. Because both OpenFOAM and Cantera are freely available, large scale simulations by running the solver in parallel on thousands of processors are feasible and straightforward (see Sect. 4.3).

3.2 Evaluation of Reaction Rates

As a general reaction system consists of many elementary reactions and the individual rate coefficients depend strongly on temperature in a non-linear way, applying the kinetics law for the reaction rates leads to a very stiff, non-linear system of differential equations, which has to be solved iteratively with time steps of the order of $O(10^{-9}–10^{-8}\text{ s})$. On the other hand, the smallest time scale of the flow and transport process is typically of the order of $O(10^{-5}\text{ s})$, so that a disparity exists between the characteristic time scales of the flow and the chemical reaction [14]. For this reason, a general operator splitting technique has been used to calculate the reaction system uncoupled from the solution of the flow equations. By doing so, Cantera creates a batch reactor of zero-dimension for each individual cell and numerically integrates the resulting kinetics equations over the DNS time step Δt, thereby resolving the smallest time scales of the reaction system. The reaction rates $\dot{r}_k$ used for the species equations in OpenFOAM are then evaluated from the change of the molar concentrations C_k via:

$$\dot{r}_k = M_k \frac{\mathrm{d}C_k}{\mathrm{d}t} \approx M_k \frac{C_k(t_0 + \Delta t) - C_k(t_0)}{\Delta t} \tag{10}$$

Eq. (10) reflects a weighted mean rate over the time Δt. The major advantage by using Eq. (10) is attributed to the fact that a considerably larger time step Δt can be used for solution of the reactive flow equations in OpenFOAM than it is necessary for calculation of the chemistry. This significantly reduces the overall computing time as well as the memory overhead bearing an acceptable loss in accuracy [3, 15, 16].

3.3 *Porting of Cantera Code*

The Cantera program [8] contains a large number of thermo-chemical libraries, for example, for the calculation of different reactors or reactor networks, e.g., BR, CSTR or PFR; it also includes reactions occurring within different phases, e.g. gas, liquid, or interface, and different transport models, e.g. the mixture-averaged or multi-component approach. However, only the evaluation of the thermo-physical properties and reaction rates of an ideal gas is used by the current DNS solver. Consequently, the performance of the code is limited by calling functions through the whole program structure of Cantera. Another inconvenience while coupling Cantera is that it has to be installed before running the solver. This makes maintenance of the programs more difficult, because Cantera has to be recompiled each time on the target machines. In addition, the influence of different compilers or compiler settings on the code performance needs to be evaluated before starting productive simulations.

To overcome these shortcomings, the relevant routines in Cantera for calculating the transport coefficients and reaction rates have been extracted and recast to an OpenFOAM library, which works independently from the Cantera program. As the proposed DNS solver applies only the ideal gas phase, most of the Cantera source codes can be discarded. However, because Cantera constitutes of a complex class and file system hierarchy [8], the extraction of the relevant code parts as well as their further adaption for the OpenFOAM program structure represents a difficult task. In this case, the required classes and their source codes have to first be identified and then have to be modified so that they can be run independently from the rest of the codes, i.e. completely decoupled from Cantera. In this way, the solver gets direct access to the internal data structures of Cantera which are not part of a public interface. The extracted code has been constructed as an additional library in OpenFOAM, which connects with the main solver via the coupling interface by replacing the previous, unnecessarily complex Cantera program (see Fig. 1).

It has been noticed by running the DNS code that a major part of the total computing time (40–60 %) is consumed by the evaluation of the reaction source terms and transport coefficients due to the very expensive mathematical operations. The direct implementation of the Cantera routines also enables to apply customized specializations and optimizations of the algorithms for better performance. For example, the interface between OpenFOAM and Cantera has been reconstructed more efficiently without any overhead of abstract communication layer and extra error handling from Cantera. In addition, the extracted code has been further simplified to reduce redundant calculations. For instance, the current solver applies only the batch reactor type of zero-dimension, the original Cantera program however implies a wide range of reactor networks with inlet, outlet or other boundary conditions, which have to be unnecessarily checked. These have been removed by porting the required Cantera code into OpenFOAM. In this way, the total performance of the solver has been improved by up to 20 % through progressive optimization of the extracted code compared to coupling Cantera directly. A detailed description of porting the Cantera code can be found in [17].

3.4 Parallel Performance

The method for parallel computing used by OpenFOAM is known as domain decomposition, where the computational grid associated with internal cell volumes and boundary elements are partitioned to sub-domains and allocated to a number of processors. Therefore, the scale-up factor depends mostly on the number of allocated cells to each single processor. While OpenFOAM is shipped with the OpenMPI library [18], any MPI implementation, such as those optimized for particular hardware platforms (e.g., IBM or CRAY), can be used with OpenFOAM by plugging it in through an interface.

In order to demonstrate the parallel efficiency of the proposed DNS solver on high performance computing platforms, a scale-up study has been performed for the Cray XE6 (HERMIT) cluster [9] maintained by the High Performance Computing Center Stuttgart (HLRS). Figure 2 shows the parallel scalability while running the DNS code on HERMIT for a computational grid with 144 million cells. A detailed description of the setups for this case can be found in Sect. 4.3. The scaling factor is very good, as different threads of the solver consume the most time for calculation rather than for communication and synchronization between the CPUs. Even a super-linear behavior can be detected, indicating that the code is able to exploit the full capacity of the machine. Furthermore, it has been shown from previous studies [19], that the solver is able to speed-up efficiently while a sufficient large

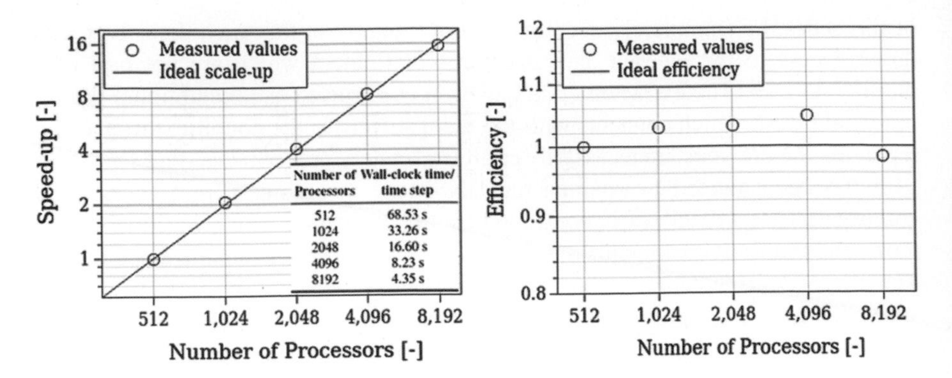

Fig. 2 Parallel scalability while running the DNS solver on the HERMIT cluster [9]: speedup factor (*left*) and efficiency (*right*) normalized to 512 processors

number of cells are allocated to each processor, for example, more than 2,000 cells per core. Similar scale-up behavior has been found for the JUQUEEN cluster with the IBM Blue Gene/Q architecture from the Jülich Supercomputing Centre (JSC) [17, 20]. Therefore, it is evidenced that the solver is suited for massively parallel running on peta-scale supercomputers.

4 Validation

4.1 *Basic Setups*

The current chapter aims to validate the in Sect. 3 presented solver. A one-dimensional premixed flame has first been used to validate especially the calculation of the reaction rates and transport properties. The second case is given by an experimental device for measuring characteristic flame speeds in an explosion vessel, which mainly aims to justify the capability of the code for three-dimensional applications. Both methane and hydrogen have been used as fuel, which are premixed with air and then ignited leading to a flame propagation. The reaction mechanism for the methane/air combustion includes 18 species and 69 fundamental reactions, which are based on [21] and additionally the reaction chain of the short-lived OH^* radical [22]. Li's mechanism containing 9 species and 19 elementary reactions [23] has been used for the hydrogen/air flame taking into account the pressure dependency of the reaction rates. The diffusion coefficients are calculated with the mixture-averaged approach [10] (see Fig. 1).

With help of OpenFOAM's standard libraries [7] the basic equations have been solved employing the finite volume method with a cell-centered storage arrangement. A fully implicit compressible formulation was applied together with the pressure-implicit split-operator (PISO) algorithm with three correction loops [24].

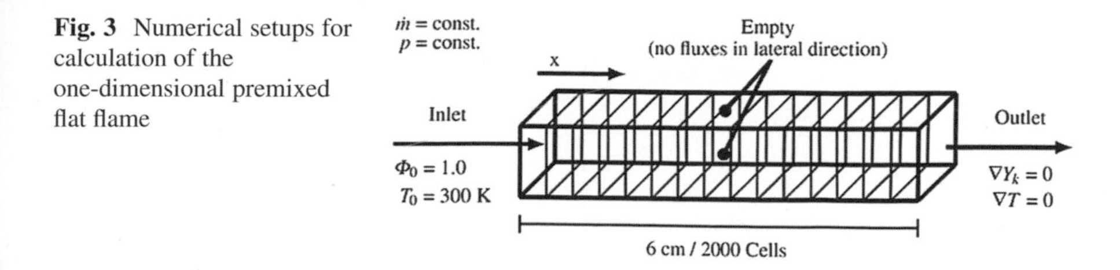

Fig. 3 Numerical setups for calculation of the one-dimensional premixed flat flame

Time derivatives are discretized with a second-order backward scheme. Discretization of convective fluxes was based on a fourth order interpolation scheme. All other diffusion terms are discretized with a second-order central differencing scheme.

4.2 *One-Dimensional Planar Flame*

A freely propagating flat flame of one-dimension is studied at first in order to assess the implementation of detailed chemistry and transport. As shown in Fig. 3, stoichiometric methane/air mixture (equivalence ratio $\Phi_0 = 1.0$) with the temperature $T_0 = 300$ K enters from the left end of the domain at given velocity. Inside the domain the fuel is consumed by the combustion reaction leading to a flame propagation against the fresh gas. The inlet velocity is adjusted to be equal to the burning velocity such that the flame front becomes stationary inside the domain. The static pressure has been kept constant over the whole domain $p = 1$ atm $=$ const. As the current case has only one dimension, the velocity only changes with a variable density due to continuity. Therefore, the solution of the momentum equation can be omitted. The computational domain has a total length of $L = 6$ cm, which is equidistantly discretized by 2,000 elements, i.e. with a resolution of $\Delta = 30\,\mu$m $=$ const. (see Fig. 3).

In Fig. 4, the calculated species mass fractions compare well with those from the commercially program Chemkin [25] employing the same boundary conditions. In Table 1, the evaluated laminar burning velocity $s_{L,0}$ and flame thickness $\delta_{L,0}$ show a quantitatively good agreement with the Chemkin solution as well. The minor deviations are mainly attributed to the different computational grids used for both codes. The CHEMKIN based program package applies a grid which is adaptively refined at the flame front, whereas a static grid is used by the proposed DNS solver in OpenFOAM. Similar results of one-dimensional flame calculations have also been obtained for other equivalence ratios and hydrogen/air mixture [13, 17, 26]. These confirm the validity of the implemented reaction-diffusion algorithms in OpenFOAM, allowing further use of the solver for more complex, three-dimensional cases.

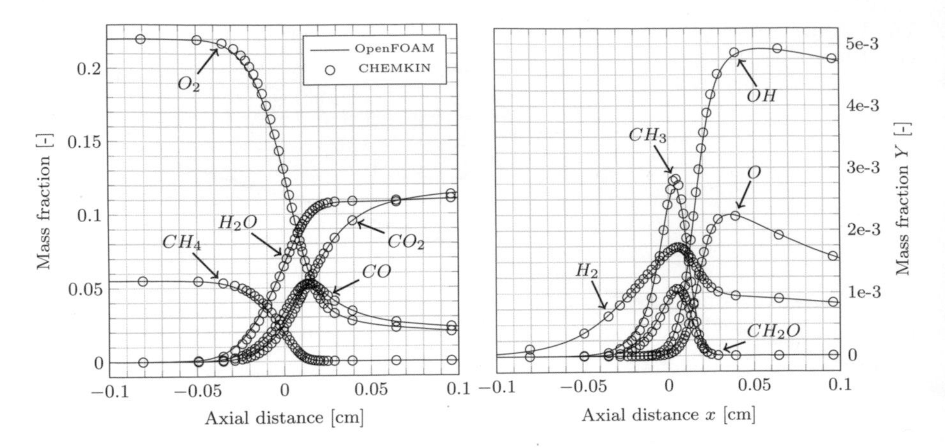

Fig. 4 Comparison of species mass fractions obtained from OpenFOAM and CHEMKIN: main species (*left*) and second order intermediate species (*right*)

Table 1 Laminar burning velocity and flame thickness calculated from the CHEMKIN code and the proposed solver in OpenFOAM: methane/air combustion at $\Phi_0 = 1.0$, $T_0 = 300$ K and $p_0 = 1.0$ atm

	CHEMKIN	OpenFOAM		CHEMKIN	OpenFOAM
$s_{L,0}$	0.374 m/s	0.370 m/s	$\delta_{L,0}$	0.463 mm	0.485 mm

4.3 Three-Dimensional Flame Propagation in an Explosion Vessel

A second case is given by an experimental device for measuring burning velocities, which has been investigated by Weiß [27] at the Engler-Bunte-Institute of KIT. The setup describes a solid enclosed spherical container which is filled with homogeneously mixed fuel and oxidizer gases. A propagating flame is then initiated by a central electrical spark. The non-stationary spherical motion of the flame front may be optically detected. This set-up is designed to measure laminar and turbulent burning velocities [27]. For the current DNS, a laminar hydrogen/air flame with $\Phi_0 = 0.33$ and the initial pressure and temperature by $p_0 = 1$ bar and $T_0 = 300$ K has been considered. The buoyant force caused by large density gradients through the flame front has been included in the DNS, too.

As the flame front propagates with a spherical shape, it is essentially two-dimensional. The flame surface becomes stretched as its radius increases with time, which leads to thermal-diffusive instabilities and breakups of the flame front [2]. Therefore, the computational grid has been built with three dimensions. By use of the axial symmetry, the simulation domain is taken to be 1/4 of the vessel used in the experiment, i.e. 1/4 of a sphere with the radius of $r_0 = 8$ cm, as shown in Fig. 5. A surrounding O-grid is used to generate the mesh from a cube-shaped core region in a structured way (see Fig. 5). The central 1/4 cubic region (5 cm side

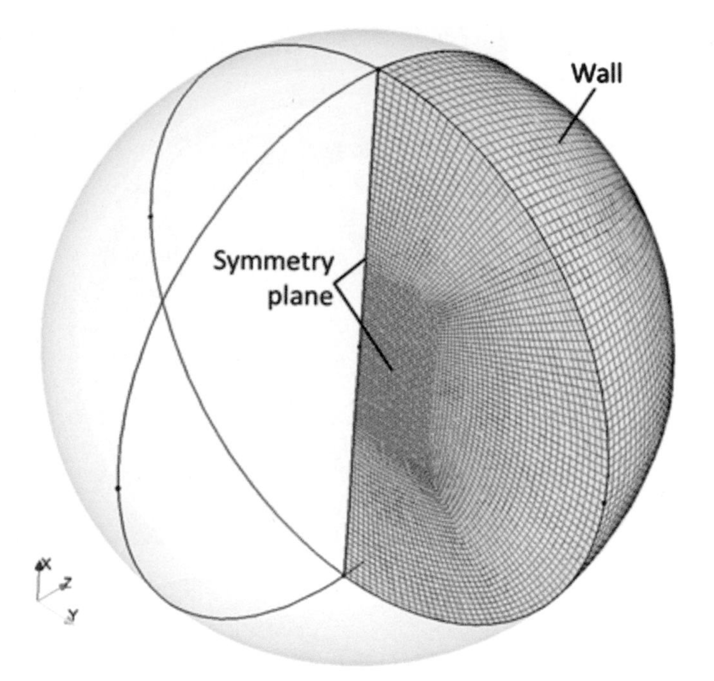

Fig. 5 Computational grid (reduced) and boundary conditions used for DNS of the flame propagation in an explosion vessel

length) is equidistantly discretized by $\Delta_x = \Delta_y = \Delta_z = 0.1$ mm, from which the grid expands with a factor of 1.004 to the wall boundary. Totally, the computational mesh has 144 million cells, ensuring resolution of the flame front with more than 10 grid points for sufficient accuracy. The time step was set to $\Delta t = 0.4\,\mu$s and the simulation has been run for 60 ms (150,000 time steps) by consuming approximately 2.5 million core hours. The DNS has been carried out in parallel by using 8192 processors from the CRAY XE6 cluster in HLRS [9].

Figure 6 shows a couple of contour-plots of the heat release rate, which demonstrate the temporal evolution of the flame front in the bomb vessel. Directly after the spark ignition the flame retains a spherical shape ($t < 20$ ms), whose radius continuously increases due to thermal expansion and consuming fuel from the surrounding fresh mixture. In other words, the flame front is promoted both by the flow motion and the burning velocity. As the flame surface expands towards the vessel wall, the flow is forced to slow down till it becomes reflected by the wall leading to a backflow against the centre. The flow is additionally overlaid with the lift force due to buoyancy ($t > 20$ ms), which interferes with the flame front. In this case, the flame becomes unstable and cannot sustain its spherical propagation, which then breaks up progressively into pieces. Figure 7 compares two snapshots from the high-speed camera measurement and the calculated H_2 mass fraction from

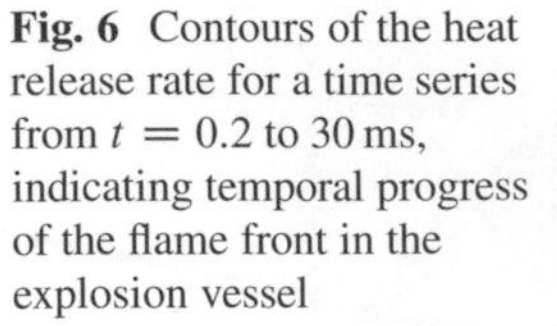

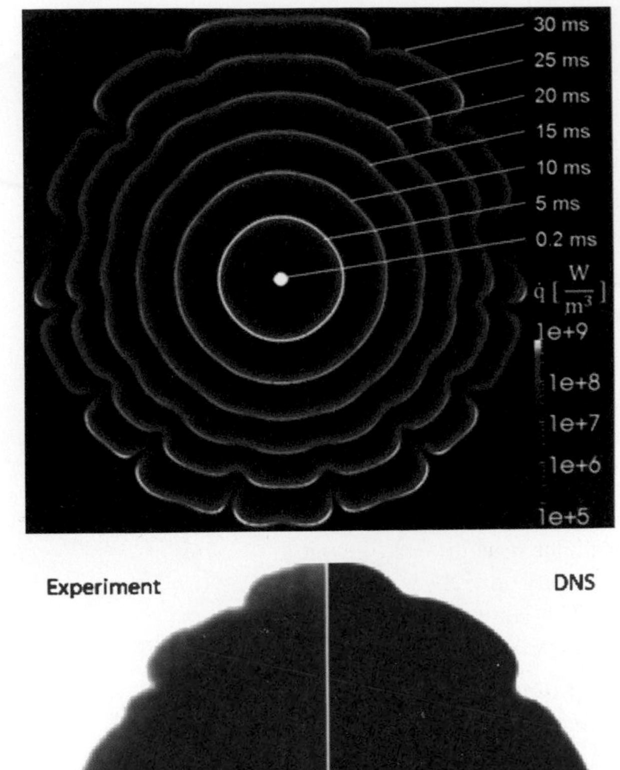

Fig. 6 Contours of the heat release rate for a time series from $t = 0.2$ to 30 ms, indicating temporal progress of the flame front in the explosion vessel

Fig. 7 Comparison of flame geometries obtained from experiment and DNS computation [28]

the DNS, when the flame surface starts to become unstable. Obviously, the shape of the simulated flame front qualitatively agrees well with that of the experiment.

In Fig. 8 the three-dimensional flame front from a later time $t = 0.035$ s is given by the $T = 1{,}100$ K isotherm, where the data from the 1/4 spheric domain have been mirrored over the planes of symmetry to render the whole flame surface.

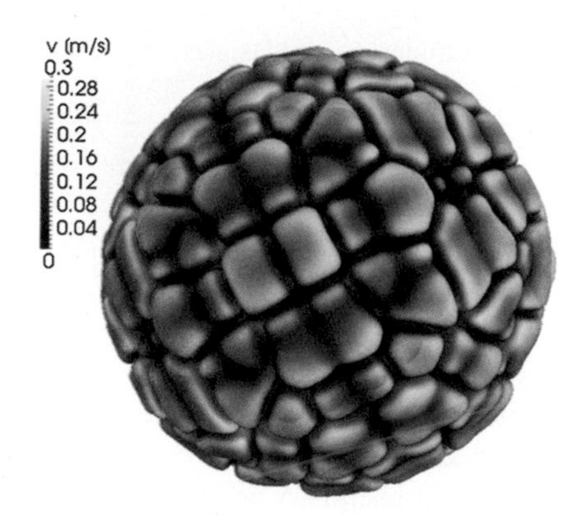

Fig. 8 Collapse of the spherical flame surface leading to formation of cellular structures

The breakup process of the flame can be clearly observed. Its further growth is attributed to an effect called thermodiffusive instability, which is particularly distinctive by lean-premixed hydrogen/air combustion [2]. It leads to formation of cellular structures causing more wrinkling and enlargement of the flame surface in comparison to the pure spheric flame.

The laminar burning velocity $s_{L,0}$ can be evaluated by the growth rate of the radius of the spheric flame front r_{fl}:

$$s_{L,0} = \frac{\rho_{\mathrm{Eq}}}{\rho_0} \frac{\mathrm{d}r_{\mathrm{fl}}}{\mathrm{d}t} \tag{11}$$

where $\rho_{\mathrm{Eq}}/\rho_0 = 0.25$ is the ratio of the gas densities in the burnt and unburnt state. The sphere radius r_{fl} can either be calculated by optically tracking the expansion of the flame, or using the following Eq. (12), which has been derived from thermodynamic relations [27]:

$$r_{fl} = r_0 \left[1 - \left(\frac{p_0}{p} \right)^{1/\kappa} \frac{p_{\mathrm{Eq}} - p}{p_{\mathrm{Eq}} - p_0} \right]^{1/3} \tag{12}$$

where $\kappa = 1.4$ is the isentropic exponent and $p_{\mathrm{Eq}} = 4.85$ bar indicates the equilibrium pressure by an adiabatic, isochoric change of state. Equation (12) reproduces the flame radius in dependence of the pressure progress over time $p(t)$. Table 2 compares the laminar burning speed derived from both experiment by using Eqs. (11) and (12) and DNS calculation, which show a good agreement. The small difference may be attributted to the measurement uncertainties [27]. Therefore, the current DNS could provide concordant results with experiment data, which justifies the capability of the developed DNS code for three-dimensional applications.

Table 2 Comparison of laminar burning velocity obtained from the measurement and DNS

	Experiment	DNS
$s_{L,0}$	0.18 m/s	0.19 m/s

5 Conclusion

The paper introduces a new solver for direct numerical simulation of compressible reactive flows. It is based on the software OpenFOAM for the solution of the governing equations and connects with the chemical code Cantera for the calculation of the thermo-chemistry and transport properties. In order to validate the implemented code, it has been applied to simulate a one-dimensional freely propagating flame and an experimental device for measuring burning velocity of three-dimensions. The calculated results showed a very good agreement with data obtained from the commercially Chemkin program and the measurement. The parallel scalability of the code has been confirmed to be very good while running it in parallel with thousands of processors from the HERMIT cluster of HLRS.

As both OpenFOAM and Cantera are freely available, the solver provides a simple access for researcher communities who are working with modeling of reacting flows, for example, turbulent combustion. Furthermore, it is planned to make the solver available to third parties. It is expect that this will generate significant interest, for example, in the DNS combustion modeling community. Studies of turbulent flames and applications to flow systems are expected to follow from this work. The DNS of simple turbulent combustion cases enables a deeper understanding of the interplay between the underlying flow, chemistry and transport processes. The fine-grained results may complement subordinate modeling concepts and measurement, or even partially replace elaborate experiments, which can then be incorporated in engineering CFD. Consequently, the timescale and cost required for development of future combustion device may be reduced.

Acknowledgements This work was supported by the German Research Council (DFG) through the Research Unit DFG-BO693/27 "Combustion Noise". This research used resources of the High Performance Computing Center Stuttgart (HLRS) at the University of Stuttgart, Germany. The authors gratefully acknowledge assistance from these Communities.

References

1. Peters, N.: Turbulent Combustion. Cambridge Monographs on Mechanics. Cambridge University Press, Cambridge (2000)
2. Poinsot, T., Veynante, D.: Theoretical and Numerical Combustion. Edwards, Philadelphia (2005)
3. Chen, J.H.: Petascale direct numerical simulation of turbulent combustion. Proc. Combust. Inst. **33**, 99–123 (2011)

4. Thévenin, D., Behrendt, F., Maas, U., Przywara, B., Warnatz, J.: Development of a parallel direct simulation code to investigate reactive flows. Comput. Fluids **25**(5), 485–496 (1996)
5. Babkovskaia, N., Haugen, N., Brandenburg, A.: A high-order public domain code for direct numerical simulations of turbulent combustion. J. Comput. Phys. **230**(1), 1–12 (2011)
6. Chen, J.H., Choudhary, A., de Supinski, B., DeVries, M., Hawkes, E.R., Klasky, S., Liao, W.K., Ma, K.L., Mellor-Crummey, J., Podhorszki, N., Sankaran, R., Shende, S., Yoo, C.S.: Terascale direct numerical simulations of turbulent combustion using S3D. Comput. Sci. Discov. **2**(1), 015001 (2009)
7. OpenFOAM Foundation: OpenFOAM – The Open Source CFD Toolbox, Programmer's Guide, Bracknell (2013)
8. Goodwin, D.G.: Cantera C++ User's Guide. California Institute of Technology (2002)
9. Cray Inc.: CrayXE6--HERMIT. http://www.hlrs.de/systems/platforms/cray-xe6-hermit/
10. Kee, R.J., Coltrin, M.E., Glarborg, P.: Chemically Reacting Flow: Theory and Practice. Wiley Interscience, Hoboken (2003)
11. Dixon-Lewis, G.: Flame structure and flame reaction kinetics. II. Transport phenomena in multicomponent systems. Proc. R. Soc. Lond. Ser. A Math. Phys. Sci. **307**, 111–135 (1986)
12. Hindmarsh, A.C., Brown, P.N., Grant, K.E., Lee, S.L., Serban, R., Shumaker, D.E., Woodward, C.S.: Sundials: suite of nonlinear and differential/algebraic equation solvers. ACM Trans. Math. Softw. **31**(3), 363–396 (2005)
13. Bonart, H.: Implementation and validation of a solver for direct numerical simulations of turbulent reacting flows in OpenFOAM. Bachelor's thesis, Karlsruhe Institute of Technology (2012). http://digbib.ubka.uni-karlsruhe.de/volltexte/1000037446
14. Maas, U., Pope, S.: Simplifying chemical kinetics: intrinsic low-dimensional manifolds in composition space. Combust. Flame **88**(34), 239–264 (1992)
15. Day, M., Bell, J., Bremer, P.-T., Pascucci, V., Beckner, V., Lijewski, M.: Turbulence effects on cellular burning structures in lean premixed hydrogen flames. Combust. Flame **156**(5), 1035–1045 (2009)
16. Kanney, J.F., Miller, C.T., Kelley, C.: Convergence of iterative split-operator approaches for approximating nonlinear reactive transport problems. Adv. Water Resour. **26**(3), 247–261 (2003)
17. Zirwes, T.: Weiterentwicklung und Optimierung eines auf OpenFOAM basierten DNS Lösers zur Verbesserung der Effizienz und Handhabung. Bachelor's thesis, Karlsruhe Institute of Technology (2013). http://digbib.ubka.uni-karlsruhe.de/volltexte/1000037538
18. Gabriel, E., Fagg, G.E., Bosilca, G., Angskun, T., Dongarra, J.J., Squyres, J.M., Sahay, V., Kambadur, P., Barrett, B., Lumsdaine, A., Castain, R.H., Daniel, D.J., Graham, R.L., Woodall, T.S.: Open MPI: Goals, concept, and design of a next generation MPI implementation. In: Proceedings, 11th European PVM/MPI Users' Group Meeting, Budapest, pp. 97–104 (2004)
19. Zhang, F., Bonart, H., Habisreuther, P., Bockhorn, H.: Impact of grid refinement on turbulent combustion and combustion noise modeling with large eddy simulation. In: Nagel, W.E., Kröner, D.H., Resch, M.M. (eds.) High Performance Computing in Science and Engineering'13, Stuttgart, pp. 259–274. Springer, Berlin/Heidelberg (2013)
20. IBM: IBM Blue Gene/Q – JUQUEEN. http://www.fz-juelich.de/ias/jsc/EN/Expertise/Supercomputers/JUQUEEN/
21. Kee, R.J., Grcar, J.F., Smooke, M.D., Miller, J.A.: Tech. Rep. SAND85-8240, Sandia National Laboratories Report (1985)
22. Kathrotia, T., Riedel, U., Seipel, A., Moshammer, K., Brockhinke, A.: Experimental and numerical study of chemiluminescent species in low-pressure flames. Appl. Phys. B **107**(3), 571–584 (2012)
23. Li, J., Zhao, Z., Kazakov, A., Dryer, F.L.: An updated comprehensive kinetic model of hydrogen combustion. Int. J. Chem. Kinet. **36**(10), 566–575 (2004)
24. Ferziger, J.H., Perić, M.: Computational Methods for Fluid Dynamics. Springer, Berlin (2010)
25. Kee, R., Rupley, F., Meeks, E., Miller, J.: CHEMKIN-III: a fortran chemical kinetics package for the analysis of gas-phase chemical and plasma kinetics. Tech. Rep. SAND96-8216, Sandia National Laboratories Report (1996)

26. Zhang, F., Bonart, H., Habisreuther, P., Bockhorn, H.: Direct numerical simulations of turbulent combustion with OpenFOAM. In: Proceedings of the 26. Deutscher Flammentag, Duisburg, pp. 867–872 (2013)
27. Weiß, M., Zarzalis, N., Suntz, R.: Experimental study of markstein number effects on laminar flamelet velocity in turbulent premixed flames. Int. J. Heat Mass Transf. **154**, 671–691 (2008)
28. Baust, T.: Druck- und Turbulenzeinfluss auf den Marksteinzahl-Effekt einer instationonren sphrischen Flammenfront. Master's thesis, Karlsruhe Institute of Technology (2013)

Lean Premixed Flames: A Direct Numerical Simulation Study of the Effect of Lewis Number at Large Scale Turbulence

Jordan A. Denev, Iliyana Naydenova, and Henning Bockhorn

Abstract The paper presents results from three-dimensional Direct Numerical Simulations of turbulent lean premixed hydrogen, propane and methane flames. The three fuels studied have Lewis numbers which range from small values (hydrogen) through near-unity values (methane) up to values larger than unity (propane). The computations make use of reduced chemical mechanisms with species number ranging from 9 to 28. It has been found that for the present time-explicit numerical method, the CPU-time scales non-linearly – with the power of 1.6 – with the number of chemical species.

A new method to produce turbulent vortexes directly in the computational domain is presented. The vortexes are produced through localized forcing in physical space. This allows the full control of the length-scale of the vortexes as well as of their spatial and time distribution. The parallelization of the method requires no programming efforts, i.e. the parallelization follows automatically. The set of vortexes created for this particular investigation is identical for all flames studied: two methane, one hydrogen and one propane flames. Despite the identical turbulent vortex set, the flames of the different fuels react quite differently to it, exhibiting qualitatively different combustion regimes as well as different growth of the flame front area and of the consumption flame speed with time.

The computations are carried out on the CRAY XE6 HERMIT supercomputer at the High Performance Computing Center Stuttgart (HLRS). Numerical issues and performance results for the three chemical mechanisms, corresponding to the three fuels used, are presented and discussed.

J.A. Denev (✉) • H. Bockhorn
Engler-Bunte Institute, Combustion Division, Karlsruhe Institute of Technology, Engler-Bunte Ring 1, D-76131 Karlsruhe, Germany
e-mail: jordan.denev@kit.edu; henning.bockhorn@kit.edu

I. Naydenova
College of Energy and Electronics, Technical University of Sofia, Kl. Ochridski Str. 8, 1000 Sofia, Bulgaria
e-mail: inaydenova@tu-sofia.bg

W.E. Nagel et al. (eds.), *High Performance Computing in Science and Engineering '14*, DOI 10.1007/978-3-319-10810-0_17

1 Introduction

It is known that turbulent premixed flames can be controlled by the length scale as well as by the intensity of the turbulent motion [10, 16]. This irregular turbulent motion – with its various time- and length scales – interacts with the time scales of the complex chemical mechanisms governing the conversion of the fuel and the oxidizer into combustion products. In order to understand the complex interaction between the turbulent motion and the chemical processes, Direct Numerical Simulations (DNS) carried out whit chemical mechanisms that are as detailed as possible, become the most powerful tool for understanding and controlling premixed turbulent combustion. However, such DNS require basically high-performance computing on parallel machines. Even then, in order to make such studies feasible, some compromises need to be made with regard, e.g. to the dimensions of the problem, to the size of the computational domain, or to the complexity of the chemical kinetic model.

The present paper aims at the study of three-dimensional turbulent premixed flames and their response to large-scale turbulent motion, i.e. a motion which length scales are larger than the thickness of the flame front. A second target is to study of the effect of fuel type on the flame response to turbulence. Fuels have different transport properties which are expressed by their Lewis numbers, defined as the ratio of species diffusion to the diffusion of heat. The fuels selected for the present study have Lewis numbers ranging from $\mathrm{Le} < 1$ (hydrogen/air flame), through $\mathrm{Le} \approx 1$ (methane/air flames) up to $\mathrm{Le} > 1$ (propane/air flames) which are known to respond differently to flame front curvature and aerodynamic strain, [3, 5, 6].

In order to achieve the above targets, two modelling requirements need to be satisfied. First of all, sufficiently detailed chemical kinetic mechanisms need to be used so that the important reaction pathways and chemical reaction times are adequately presented. As chemical models can be very large, there is a clear need for a trade-off between the modelling complexity and the required CPU-time for the chemical kinetic mechanisms. Therefore, the chemical kinetic models used in the present study are described in detail and their numerical performance is assessed. Second, there is a need to develop three-dimensional turbulent fields, which have a length scale that is prescribed in advance. Furthermore, it is desirable to have a method that develops both large- and small-scale turbulence. As small-scale vortexes decay considerably between the place of their generation – usually at the boundary of the computational domain – and the flame front, a new method that generates turbulence with desired length scales anywhere in the computational domain is required. Therefore, an algorithm for the new method introduced in this paper which possesses the desired features is presented in detail. Its benefits in terms of flexibility, algorithmic implementation and parallelization are discussed.

The last part of the paper shows results from three-dimensional DNS for four turbulent premixed lean flames: hydrogen flame with equivalence ratio $\Phi = 0.33$, propane flame with $\Phi = 0.71$ and two methane flames with $\Phi = 0.55$ and $\Phi = 0.67$. The four flames interact with a turbulent field with identical forcing in physical space. The particular response of each flame to that turbulence is interpreted in order to identify the combustion regime established in each case. Finally, results regarding the time evolution of parameters such as the rms-values of turbulence, the consumption speed and the flame surface are presented and the differences for the four flames are shown and commented. For each flame simulated the computational resources are presented and discussed.

2 Description of the Numerical Method, Initial and Boundary Conditions

The code used in the present work is PARCOMB-3D [19] and it is specialized for the study of non-stationary combustion phenomena. It solves the coupled system of fully compressible Navier-Stokes, species- and energy conservation equations. Temperature is obtained with Newton iterations from the total chemical energy simulated and pressure is obtained from the ideal gas equation of state. The species transport data are simulated in detail – the so-called "complete multicomponent diffusion treatment" is applied. The Sorét effect (molecular diffusion of species due to the presence of temperature gradients) is taken into account for the simulation of the hydrogen flame. Radiative heat transfer is not considered in the present simulations.

Explicit time stepping is used with a 4th order Runge-Kutta scheme. The time-step is variable and it is controlled through the Courant-Friedrichs-Lewy (CFL) number. The CFL-value is prescribed by the user and, as discussed later, depends on the stiffness of the chemical reaction mechanism utilized. Common values for the CFL number for the hydrogen mechanism are close to unity (we used a value of 0.95) and the physical time step varies around the value of 4.3E−08.

The simulation code PARCOMB-3D uses 6th order central differencing in space, along with 3rd order differencing on the domain boundaries. The finite-difference code is able to handle Cartesian numerical grids with variable grid spacing, but in the present work equidistant spacing in all three spatial directions is used. The vortexes that interact with the flame front are several times larger than the flame front itself, and the physical time simulated is quite short, thus preventing the vortexes from splitting and becoming too small in size. In this case the decisive factor for the grid spacing remains the adequate resolution of the steep gradients within the flame front. The flame front, as a rule of thumb, should be resolved with at least 12–15 grid points. Lean flames – as those investigated in the present work – have relatively thick flame fronts which reduces the requirements regarding the grid density. Bearing all above considerations in mind, the present simulations have grid spacing which is equidistant in all three directions with a step of 33 [μm].

The computational domain spans $8 \times 5 \times 5$ [mm]. Along the x-axis inflow and outflow Navier-Stokes characteristic boundary conditions [2, 15] are used; the implementation takes into account also the detailed chemistry equations. All other boundaries have periodic boundary conditions. The code is written in FORTRAN90 and the parallelization is based on MPI 2.0. The domain decomposition is performed from specialized routines at the time of simulations. All simulations are carried out on the CRAY XE6 at the High Performance Computing Center Stuttgart (HLRS).

3 Momentum Sources: A Method for Generating Non-decaying Vortexes in Physical Space with Controlled Length Scale

The interaction of the flame front with turbulent eddies can be studied either statistically, with a thorough simulation of the whole turbulent spectrum, or with a reduced set of well-controlled vortexes. In the present work the second approach has been chosen for it ensures well-defined conditions for the flame-vortex interactions which in turn allow that all four flames are subject to nearly identical conditions.

The new method proposed here is easy to understand and implement and is cheap in a numerical sense. The turbulence generated is not necessarily of a decaying type, although it could be, depending on the specified time window for the action of the momentum sources. This way the method resembles naturally the physics of flows where simultaneously part of the vortexes emerge and part of them decay. The method simply generates additional momentum sources (a kind of the well-known forcing used in the study of channel flows, but being different at each location of the flow) in the Navier-Stokes equations. These sources describe the additional momentum transport responsible for the creation of turbulent vortexes inside the computational domain. Using this method the vortexes are generated directly within the computational domain, not at its boundaries. Numerically the momentum sources are treated the same way the code treats e.g. the buoyant terms (e.g. $\rho.g$, where ρ is the density and g is the gravity vector) in the Navier-Stokes equations and hence they can also be interpreted as spots of localized buoyant forces having a different (for each finite volume or for each group of finite volumes) gravity constant. The x-direction momentum source (MS) at a particular grid point C is calculated from the following equation:

$$MS = C_{MS} \times \frac{\overline{AB}}{\overline{AC} + \overline{BC}} \times cos(\theta_x) \tag{1}$$

Here $\overline{AB}$, $\overline{AC}$ and $\overline{BC}$ are the distances between the grid-point C and the centers A and B of two spheres (each one with a radius $\overline{AB}$), see Fig. 1, C_{MS} is a constant between 10,000 and 100,000 and (θ_x) is the angle between the direction of the momentum source and the x-axis. Prior to the simulation the coordinates of the points A and B, the three angles (θ_x), (θ_y) and (θ_z), C_{MS} as well as the life cycle (in terms of time steps) of the MS are prescribed in a file. Through the choice of the constant C_{MS} and the physical time the momentum source acts, the strength of the

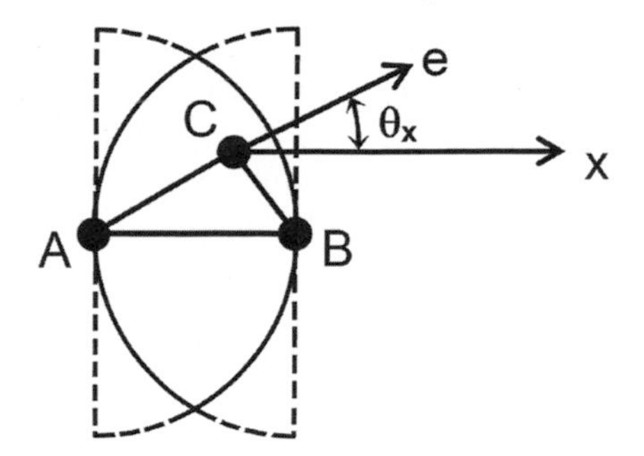

Fig. 1 Scheme, describing the algorithm for vortex-generation using momentum sources. The example is for a momentum source acting in the x-direction and corresponding to grid point C (finite volume C)

velocity induced in the area between the two spheres can be controlled explicitly. The flow induced by one single momentum source resembles the structure of a small jet or of a potential-flow dipole. For the grid points outside the spheres the MS is zero. This creates an abrupt change (jump) in space. This jump in turn is responsible to the appearance of flow shear and instability of the Kelvin-Helmholtz kind thus creating flow vortexes.

The area located between points A and B is the area where a certain MS is active; at the same time this distance prescribes a length-scale for the MS. The momentum source is then added to the other terms of the Navier-Stokes equations for each numerical point that is located in the space between the two spheres in Fig. 1. As the Navier-Stokes equations are discretized in a separate routine that is independent of the parallelization of the code PARCOMB3D, there is no need in any algorithmic effort for the parallelization of the momentum sources which presents a big advantage of the method.

If a large number of vortexes is required (e.g. for the case when small-scale turbulence is studied) the angles and the strength of the individual vortexes (C_{MS}) as well as their life cycle can be generated by random numbers. The example given in Eq. 1 is for the x-momentum equation, for directions y and z the corresponding angle should be used.

The method allows generating turbulence with desired features, e.g. generating only large-scale or only small-scale turbulence which in turn allows studying the impact of small and large eddies on the flame front separately. Also a desired Karlovitz number of a given scale r, defined through the equation [16]:

$$Ka(r) = \frac{u'(r)/r}{\delta/S_l} \tag{2}$$

where S_l is the unstretched flame speed and δ is the flame thickness can be set up through appropriate choice of the C_{MS} constant for that particular vortex size r. In other words, desired ratios of the characteristic chemical time scale and the turbulent time scale can be prescribed using the present vortex generation method.

Two conditions should be satisfied in order to create turbulence with a desired single length scale. The first one is that the momentum sources should have a distance $\overline{AB}$ which is equal to the desired length scale. The second condition is useful when turbulence with only one length scale needs to be generated – in this case additionally the average distance between the different momentum sources should be set equal to the desired length scale. This second condition leads to the requirement that a certain number of momentum sources per unit volume have to be prescribed.

The proposed shape of the MS – between two spheres – is not strictly necessary, many other shapes like ellipsoids or even single spheres or cubes could be used too. Overlapping of the spheres in space (so that a given grid point is associated to more than one MS) is allowed. This corresponds to the real situation where one fluid particle participates simultaneously in the fluid motion of different vortexes, which in the general case can have quite different length scales.

In case of a large amount of vortexes – e.g. many thousands – the vortex database (the particular momentum sources) can be prepared in the preprocessing stage and stored in a file. In the present implementation the vortex database is available for all processors, so that there is no need in special treatment for parallel computations – one more important benefit from the current vortex-generation approach.

4 Description of the Chemical Mechanisms

All numerical calculations were performed using reduced chemical models adopted from various literature sources, described in the following. The chemical kinetic schemes used in this work are briefly summarized in Table 1.

The hydrogen/air flame was modelled applying appropriately modified version of the H_2/O_2 sub-model described in [14]. The current version of the model was adopted from [13]. The thermochemical properties of the involved species are taken from [12].

The methane/air flame calculations were carried out with a reduced methane air chemical kinetic mechanism thoroughly described in [4, 13, 14], using the above described thermochemical data set [12].

The propane/air flame was modelled with a slightly modified version of the kinetic scheme called M5 mechanism adopted from Haworth et al. [8, 11]. The following model modifications were considered in this work: the NO_x chemistry presented in [11] was neglected at this stage of the simulations, because it was absent also in the previous flame models (H_2/air and CH_4/air); the rate coefficients of three sensitive reactions were found to be pressure dependent in [1], therefore,

Table 1 Chemical kinetic models

No	Chemical kinetic mechanism	Number of species	Number of reactions	Reference
1	H_2/air	9	38	[13, 14]
2	CH_4/air	16	50	[4, 17, 18]
3	C_3H_8/air	28	143	[8, 11]

these coefficients were represented as the best fit of their experimentally measured values for pressure of 1 bar. Due to the lack of data on the complete sets of thermochemical and transport properties of the chosen kinetic scheme, relevant data set was adopted from [9]. This choice was made on the fact, that the original C_3H_8 oxidation mechanism [8] is based on an earlier version of the chemical kinetic model [9], even though significantly reduced.

5 Description of Cases Studied and CPU-Time for the Parallel Computations

Four computations of three-dimensional turbulent flames have been carried out as detailed in Table 2.

The time step, listed in the table, is taken at the end of the simulations; it can vary due to the CFL-controlling of the time step. All computations reached physical time around 1 ms.

It can be seen in the table that the methane cases require approx. 2.5 times more CPU-time than the hydrogen flame while for the propane flame the increase is 6.5 times. The increase in CPU-time scales with the ratio of the species number to the power of 1.6, i.e. non-linearly. This non-linearity is a consequence of the additional computational efforts required to calculate the detailed properties of the gas mixture from the properties of each individual species.

The number of grid points per processor for the present simulations lies between 15,000 and 18,000. Reduced to a single processor performance, only 1.67 time-steps in case of propane can be calculated for one CPU-hour. In the case of hydrogen – 5.46 time-steps and for the methane – 2.75 to 3.0 time-steps can be computed for one CPU-hour.

For each simulated case 18 vortexes with identical geometry and identical flow direction have been generated following the algorithm, described in Sect. 3. The average length scale of the momentum sources is 1.01 [mm].

The inflow velocity for the fresh gases is constant and set to 0.105, 0.195, 0.095 and 0.200 [m/s] for the hydrogen, the two methane and the propane flame, correspondingly.

The average velocity and its rms-value close to the flame front are shown in Fig. 2. At the beginning of the simulations the flow is laminar within the

Table 2 Description of cases and required resources for the computations

No	Fuel	Equiv. rat.	Processors	WCT (h)	Iterat.	Time (ms)	Time step (s)	CFL
1	H_2	0.33	300	13.56	22,200	0.947	4.04E−08	0.95
2	CH_4	0.67	360	24.50	26,700	1.021	3.84E−08	0.95
3	CH_4	0.55	360	29.03	28,600	1.090	3.83E−08	0.90
4	C_3H_8	0.71	360	73.01	43,900	1.042	2.38E−08	0.60

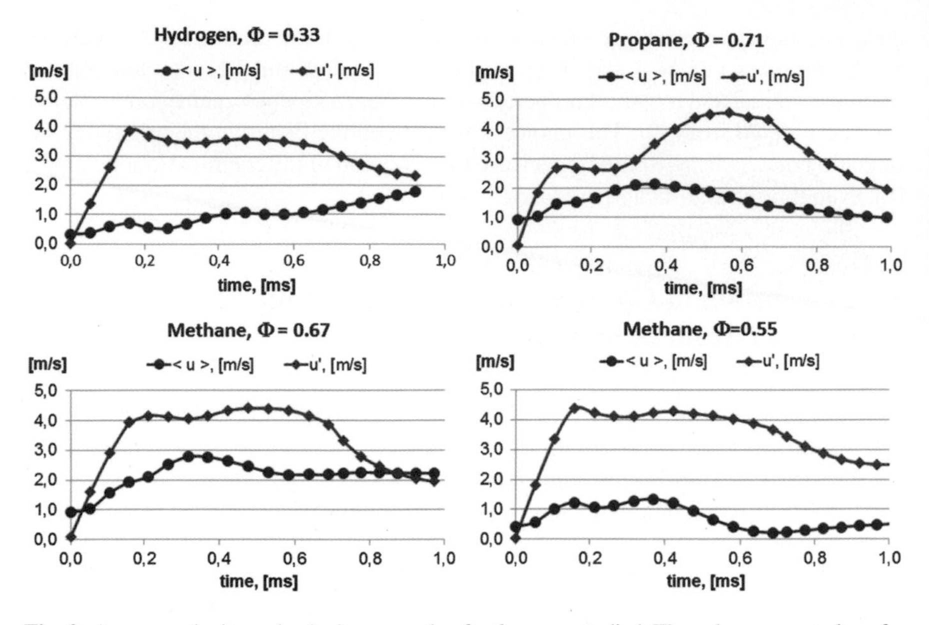

Fig. 2 Average velocity and velocity-rms-value for the cases studied. The values presented are for a narrow region around the flame front as explained in the text

computational domain and therefore the rms-value is almost equal to zero. At a time instance slightly less than 0.2 [ms], the velocity rms-value increases considerably and becomes larger than the average velocity near the flame front. Here should be noted that both average velocity and its rms-value depend on the volume of integration. If the values are averaged throughout the whole computational domain, the real situation at the flame front would be obscured, see also [16]. To avoid this, the domain was first filtered to extract a small area around the flame front – only the region, where the heat release rate is higher than 20 % of the maximum heat release rate of the unstretched planar flame, is taken into consideration in Fig. 2.

One can see that despite of the efforts to keep the vortex set as similar as possible, the velocity fluctuations around the flame front differ. The main reason is in the different flame front shape established during the simulation and in the different flame speed of the different parts of the flame front which drives the flame front towards different parts of the similar vortex set.

6 Combustion Regimes of the Flames Studied

Combustion diagrams for premixed flames allow defining zones of different burning regimes depending on the relationship between characteristics of the flame and of the turbulent motion. The precise quantification of the characteristics of turbulence

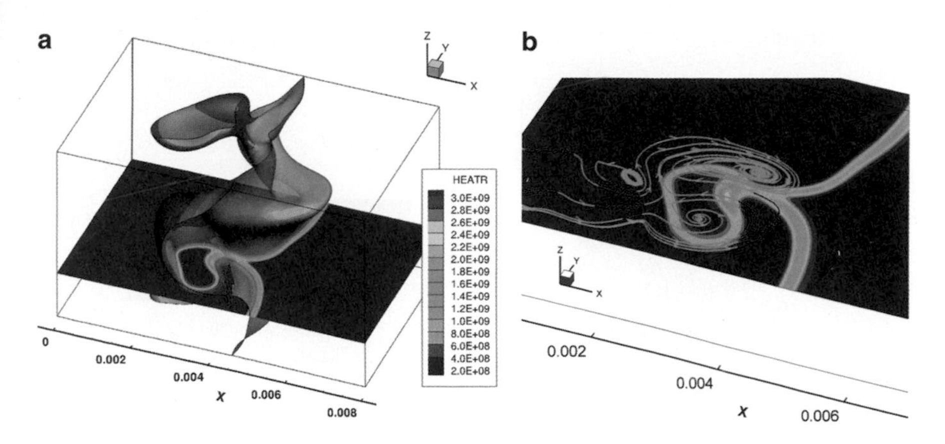

Fig. 3 Flame front wrinkling of the propane flame at time t = 0.00104 [s]. (**a**) The flame front isosurface is colored by the heat release rate, [W/m^3]. The *red arrow* points to the location where the flame front touches itself. (**b**) The same view presented only in the plane z = 0.0013 m with streamlines colored by the vertical velocity component (*red color* means upward and *green* – downward fluid movement). Fresh gases enter the domain from left

(length scales and velocities) faces principal difficulties; for a detailed discussion see [16]. Due to these principal difficulties, the diagrams are based on dimensional analysis and magnitude estimations which in turn hinders the direct comparison with DNS results. In view of the above, the present section aims to analyze and classify the combustion regime arising from the interaction of the flame front and the vortexes, created according to the method, described in Chap. 3 and quantified in Chap. 5. For this purpose, first the wrinkled surface of the flame front of the propane flame (defined as the location where 10 % of the fresh gas fuel mass fraction remains unburned), shown in Fig. 3a is analyzed. As a result of the geometrically complex flame wrinkling and an additional consequent self-touching of the flame front, a ring-like (island-like) structure appears. This ring-like structure is shown in more detail in Fig. 3b, where the surrounding streamlines are also plotted. It contains only burned gases and the temperature in the middle of it is quite high, T = 1,740 [K] (the temperature of burnt gases reaches up to 1,904 [K]). Its existence, together with the fact that the integral flame front thickness remains almost not changed in time, indicates that the present propane flame can be classified to the island formation regime (flamelet regime) as defined in [10]. This regime is typical for the case when the time scale of the chemical reactions remains shorter than the characteristic time of the Kolmogorov-scale turbulence.

Examination of the two methane cases shows that their behavior – regarding the appearance of ring-like structures and the constant in time integral flame front thickness – is qualitatively close to that of the propane flame. Therefore, they also are attributed to the island formation regime (called also thin flame regime with pockets, corrugated flamelet regime), or to the case when the chemical time scale is smaller than any turbulent time scale (Karlovitz number Ka < 1) and at the same

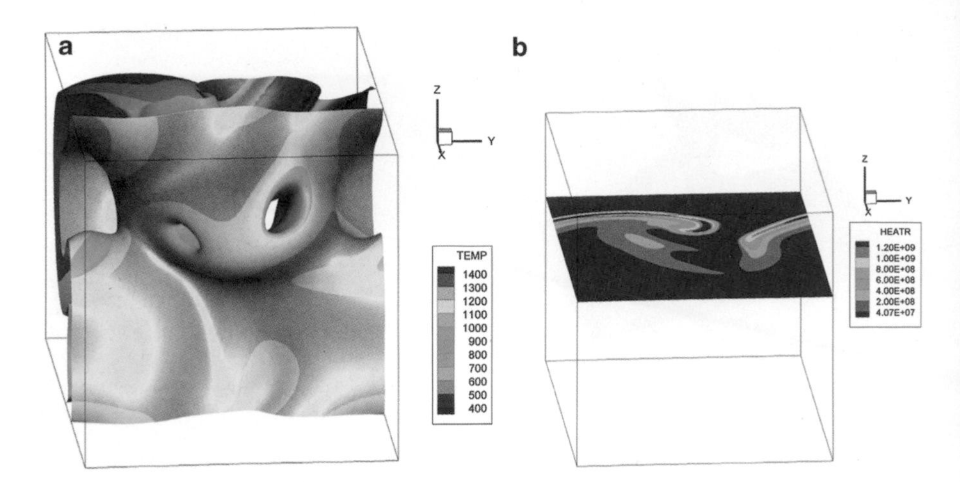

Fig. 4 Flame front of the hydrogen flame with local extinguishing at time t = 0.00078 [s]. (**a**) The flame front is presented by the isosurface of heat release rate 4.07E+7 [W/m^3] colored by the temperature. The two "holes" in the front of the isosurface show the local extinction. (**b**) The same view presenting the heat release rate in the plane z = 0.0033 m. Fresh gases enter the domain from behind

time with a turbulent motion which is faster than the laminar flame speed, see also [16]. Here it should be noted that the island formation in all three cases analyzed so far is a pure effect of the convective transportation and the consequent wrinkling of the flame front. The islands are observed when plane cuts through the highly folded three-dimensional flame surface are drawn (Fig. 3b) which situation appears also in experiments with laser light sheets.

The flame front of the hydrogen flame is shown in Fig. 4a, b. This flame exhibits a qualitatively different behavior – local extinction occurs and hence the flame front is no more continuous. The isosurface shown has a value for the heat release rate of 4.07E+7 [W/m^3]. This value is equal to 20 % of the maximum heat release rate of the unstretched planar laminar flame, which corresponds to our definition of flame front extinguishing. The discontinuous flame front is shown once again on the plane cut in Fig. 4b. According to [10] this is the "distributed reaction zone" called also "torn flame fronts zone".

It should be noted that the hydrogen flame is exposed to turbulent conditions which are quite similar to the other flames, but it finally exhibits a different qualitative behavior. The reason is in the unstable nature of the hydrogen flame – for this flame with $Le \ll 1$ the laminar Markstein number is negative, [7]. Therefore, the response of flames to turbulence cannot be attributed only to the turbulence characteristics itself, neither to the different characteristics of the planar flames (laminar flame speed and thickness), which for the present hydrogen and propane flames differ by no more than 20–30 %.

7 Consumption Speed and Flame Area

Simple modelling approaches assume that the flame speed is linearly proportional to the flame front area, [16]. In the present chapter the time evolution of the consumption flame speed as well as the flame front area are studied. The consumption speed is obtained from the space-averaging of the instantaneous fuel reaction rate throughout the computational domain:

$$Sc = \frac{1}{[A_L(\rho Y_{fuel})]_{init}} \int_V \dot{\omega}_{fuel} dV \tag{3}$$

where A_L is the initial (laminar) flame front area, ρ – the mixture density and Y_{fuel} – the mass fraction of the fuel in the fresh gases. All these values refer to the initial state of the computations. $\dot{\omega}_{fuel}$ is the reaction rate of the fuel in $[\mathrm{kg}_{fuel}/(\mathrm{m}^3.\mathrm{s})]$ and V is the volume of the computational domain.

The results for the scaled flame speed and the scaled flame front area are presented in Fig. 5. It can be seen in the figure, that the connection between the two parameters is different for the different flames. The scaled values of the two parameters are almost equal for the methane flame with equivalence ratio $\Phi = 0.55$ and are quite close to each other for the methane flame with $\Phi = 0.67$. The largest differences are observed for the hydrogen flame, for which the consumption speed grows much quicker with the area growth than in the case of the other two fuels. However, for this hydrogen flame, the growth of the flame front area is still, to a

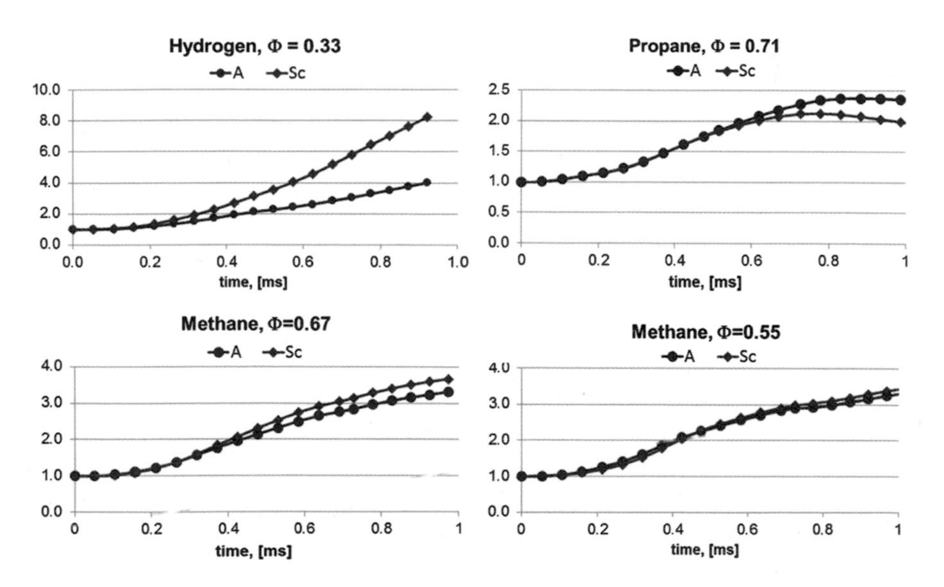

Fig. 5 The area A and the consumption speed Sc for the four flames studied. For each flame the values are scaled with the values of the laminar flame

large extent, linearly proportional to the flame speed growth. The propane flame shows equal growth of the two parameters up to a time of 0.65 [ms]. At later times, the flame front area has a hardly noticeable decrease while the consumption speed decreases quicker. The linear relationship is therefore no more present for the later time instances of these propane flame. However, for the flames studied here and under the conditions of the particular turbulent field created, the linear relationship between the flame front and the flame sped is holds true to a large extent.

8 Conclusions

The paper studies three-dimensional flames interacting with a set of large-scale vortexes. Four different lean flames – a hydrogen, two methane and a propane flame – having Lewis numbers which are smaller, nearly equal and larger than unity, are investigated for their reaction with the turbulent flow field induced by one and the same set of vortexes.

The main outcomes of the paper are as follows.

A new method for generation of turbulent vortexes directly inside the computational domain, and not at its boundaries, is developed and presented in detail. The method exhibits a large degree of flexibility as the vortexes could be decaying, or constantly forced, or both, and their length scale could be either a single one, or cover the whole turbulent spectrum. The method is called the "momentum source" method and has the following important features:

- It is discretized numerically like e.g. the buoyant terms in the Navier-Stokes equations which is straightforward and computationally cheap
- It is easy to implement it for parallel computing into any existing code based on domain decomposition and MPI. No efforts for parallelization of the implementation are required as the momentum sources are treated in exactly the same way as all other regular terms in the system of partial differential equations
- Vortexes can be generated in predefined areas of the domain, e.g. only around the flame front or only in the preheat zone
- It allows a full control of the length-scale of the generated vortexes and therefore a control of the flame stretch (Karlovitz number)

The CPU-time for each of the flames simulated has been compared. It is found that the increase in CPU-time scales with the ratio of the species number to the power of 1.6, i.e. non-linearly. In absolute numbers, the methane flames require 2.5 more CPU-time than the hydrogen flame. The propane flame requires 6.5 more CPU-time compared to the hydrogen flame.

Despite that the same set of vortexes has been created to interact with the four flames and that the flame front thickness and the flame speed for the flames are relatively close to each other, different flame regimes are observed. The two methane flames and the propane flame are attributed to the "island formation regime", while the hydrogen flame is attributed to the "thorn flame fronts zone" as defined in [10].

For the present flames under the conditions of the present turbulent field, the growth of the consumption speed with the flame speed area basically follows a linear relationship. For the methane flame the scaled flame speed and the scaled flame front area change almost equally with time, reaching an increase of 3.5 times after one millisecond. The growth of the turbulent propane flame speed does not exceed 2.2 times of the laminar flame speed. The turbulent hydrogen flame reaches up to 8 times higher consumption speed than the laminar one. For this flame the growth of consumption speed is much larger than the growth of the flame front area.

Acknowledgements The present work was enabled through the funding of the DFG collaborative research center SFB606 "Non-stationary combustion: transport phenomena, chemical reactions and technical systems", sub-project $B8$. The simulations were performed on the national super computer CRAY XE6 at the High Performance Computing Center Stuttgart (HLRS) under the grant with acronym "DNSPREM" which is highly appreciated.

The authors would like to thank their colleagues Prof. Ulrich Maas and Dr.-Ing. Robert Schießl for their valuable help during implementation of different reaction mechanisms and to Georg Bläsinger for the fruitful discussions.

References

1. Baulch, D.L., Bowman, C.T., Cobos, C.J., Cox, R.A., Just, Th., Kerr, J.A., Pilling, M.J., Stocker, D., Troe, J., Tsang, W., Walker, R.W., Warnatz, J.: Evaluated kinetic data for combustion modelling: supplement II. J. Phys. Chem. Ref. Data **34**, 757–1397 (2005)
2. Baum, M., Poinsot, T., Thevenin, D.: Accurate boundary conditions for multicomponent reactive flows. J. Comput. Phys. **116**, 247–261 (1994)
3. Bell, J.P., Cheng, R.K., Day, M.S., Sheperd, I.G.: Numerical simulations of Lewis number effects on lean premixed turbulent flames. Proc. Combust. Inst. **31**, 1309–1317 (2007)
4. Bilger, R.W., Esler, M.B., Starner, S.H.: On reduced mechanisms for methane-air combustion. In: Smooke, M.D. (ed.) Reduced Kinetic Mechanisms and Asymptotic Approximations for Methane-Air Flames. Lecture Notes in Physics, vol. 384, pp. 86–110. Springer, Berlin/Heidelberg (1991)
5. Denev, J.A., Bockhorn, H.: A local look into the unsteady premixed combustion phenomena. In: 26. Deutscher Flammentag, Verbrennung und Feuerungen, VDI-Berichte 2161, VDI, Duisburg-Essen, pp. 601–612, 11–12 Sept 2013
6. Denev, J.A., Bockhorn, H.: Turbulent burning velocity in premixed combustion. In: The Spark Ignition Engine of the Future: International Conference and Exhibition, Strasbourg, 7pp, 4 and 5 Dec 2013
7. Denev, J.A., Vukadinovic, V., Naydenova, I., Zarzalis, N., Bockhorn, H.: Experimental and numerical characterization of H2/air spherically expanding laminar flame at lean conditions. In: Proceedings of the European Combustion Meeting, Paper P1-69, Lund, vol. 12, 25–28 June 2013
8. Haworth, D.C., Blint, R.J., Cuenot, B., Poinsot, T.J.: Numerical simulation of turbulent propane-air combustion with nonhomogeneous reactants. Combust. Flame **121**(4), 395–417 (2000)
9. Heghes, C.I.: C1-C4 Hydrocarbon oxidation mechanism. Phd thesis, Ruprecht-Karls-Heidelberg University (2006)
10. Maas, U., Warnatz, J., Dibble, R.W.: Combustion, 3rd edn. Springer, Berlin (2006)
11. Jimenez, C., Cuenot, B., Poinsot, T., Haworth, D.: Numerical simulations and modelling for lean stratified propane-air flames. Combust. Flame **128**(1–2), 1–21 (2002)

12. Kee, J., Rupley, F., Miller, J.: Chemkin-II: a fortran chemical kinetics package for the analysis of gas-phase chemical kinetics. Technical report, Report No. SAND89-8009B, Sandia National Laboratories (1989)
13. Lecanu, M., Mehravaran, K., Fröhlich, J., Bockhorn, H., Thévenin, D.: Computations of premixed turbulent flames. In: Nagel, W.E., Jaeger, W., Resch, M. (eds.) High Performance Computing in Science and Engineering'07. Transactions of the High Performance Computing Center Stuttgart, pp. 229–239. Springer, Berlin (2008)
14. Miller, J.A., Mitchell, R.E., Smooke, M.D., Kee, R.J.: Toward a comprehensive chemical kinetic mechanism for the oxidation of acetylene: comparison of model predictions with results from flame and shock tube experiments. Proc. Combust. Inst. **19**(4), 181–196 (1982)
15. Poinsot, T., Lele, S.: Boundary conditions for direct simulations of compressible viscous flows. J. Comput. Phys. **101**, 104–129 (1992)
16. Poinsot, T., Veynante, D.: Theoretical and Numerical Combustion, 2nd edn. Edwards, Philadelphia (2005)
17. Smooke, M.D., Giovangigli, V.: Formulation of the premixed and nonpremixed test problems. In: Smooke, M.D. (ed.) Reduced Kinetic Mechanisms and Asymptotic Approximations for Methane-Air Flames. Lecture Notes in Physics, vol. 384, pp. 1–28. Springer, Berlin/Heidelberg (1991)
18. Smooke, M.D., Giovangigli, V.: Premixed and nonpremixed test problem results. In: Smooke, M.D. (ed.) Reduced Kinetic Mechanisms and Asymptotic Approximations for Methane-Air Flames. Lecture Notes in Physics, vol. 384, pp. 29–47. Springer, Berlin/Heidelberg (1991)
19. Thévenin, D., Behrendt, F., Maas, U., Przywara, B., Warnatz, J.: Development of a parallel direct simulation code to investigate reactive flows. Comput. Fluids **25**, 485–496 (1996)

Parallelization and Performance Analysis of an Implicit Compressible Combustion Code for Aerospace Applications

Roman Keller, Markus Lempke, Yann Hendrik Simsont, Peter Gerlinger, and Manfred Aigner

Abstract The compressible, implicit combustion code TASCOM3D is used with and without the spray module SPRAYSIM for different aerospace applications. A number of such cases and analysis of the performance of the code on massively parallel systems will be given. These include supersonic combustion simulations of a complete scramjet model, a model rocket combustor fueled with gaseous oxygen and hydrogen as well as two multiphase simulations. The evaporation of kerosene in a preheated, pressurized channel and the spray combustion in a LOX/GH_2 rocket combustor require an additional numerical tool to account for the liquid phase. Droplet propagation and evaporation is computed by the research code SPRAYSIM. Furthermore investigations with respect to the performance of the employed numerical codes are addressed. With respect to TASCOM3D the influence of block sizing on the performance is investigated intensively both in terms of weak and strong scaling. The strong scaling performance of SPRAYSIM is investigated for both multiphase simulations. It will be shown that both codes show a nearly ideal behavior.

1 Introduction

The in-house code TASCOM3D is used for numerical simulation of combustion processes for spacecraft configurations. The two major research fields are scramjet and liquid rocket combustors that are both challenging due to the extreme conditions prevailing in these devices. The supersonic flow through a scramjet necessitates the compressible flow solver to be able to accurately capture shocks and predict shock/shock as well as shock/boundary-layer interactions. A detailed chemical reaction model and the consideration of turbulence chemistry interaction is crucial

R. Keller (✉) • M. Lempke • Y. Hendrik Simsont • P. Gerlinger • M. Aigner
Institut für Verbrennungstechnik der Luft- und Raumfahrt, Pfaffenwaldring 38-40, 70569 Stuttgart, Germany
e-mail: roman.keller@dlr.de

W.E. Nagel et al. (eds.), *High Performance Computing in Science and Engineering '14*, DOI 10.1007/978-3-319-10810-0_18

to accurately predict the supersonic combustion process. Consequently a stiff system of differential equations has to be solved using an implicit method.

Due to the complexity of the employed models and the resulting high computing times the utilization of high performance computing systems is inevitable. Consequently it is crucial to examine the performance of the simulation codes on the used platform, in this case the Cray XE6 at the High Performance Computing Center Stuttgart (HLRS). Therefore a considerable part of this report is dedicated to this aspect. The liquid droplet tracing code SPRAYSIM, developed at the Institute of Combustion Technology of the German Aerospace Center (DLR), is investigated in terms of strong scaling only. The analysis of the turbulent combustion flow solver TASCOM3D, developed at the Institute of Combustion Technology for Aerospace Engineering (IVLR, University of Stuttgart), considers both strong and weak scaling. For both codes hardware specific aspects are discussed.

A brief introduction of the codes is given in Sect. 2. Wherever possible practical setups are used for the performance assessments. Therefore an overview on a variety of current investigations follows in Sect. 3. Next performance results are presented in Sect. 4 and conclusions for code performance improvements are drawn at the end of the article.

2 Details of the Numerical Codes

2.1 The Gas Phase Code TASCOM3D

The scientific in-house code TASCOM3D (Turbulent All Speed Combustion Multigrid Solver 3D) has been used successfully during the last two decades to simulate reacting and non-reacting super- and subsonic flows. Reacting flows are described by solving the full compressible Navier-Stokes, turbulence and species transport equations. Additionally an assumed PDF (probability density function) approach is available to take turbulence chemistry interaction into account. The three-dimensional conservative form of the Reynolds-Averaged Navier-Stokes (RANS) equations is given by

$$\frac{\partial \mathbf{Q}}{\partial t} + \frac{\partial (\mathbf{F} - \mathbf{F}_v)}{\partial x} + \frac{\partial (\mathbf{G} - \mathbf{G}_v)}{\partial y} + \frac{\partial (\mathbf{H} - \mathbf{H}_v)}{\partial z} = \mathbf{S}, \tag{1}$$

where

$$\mathbf{Q} = [\rho, \rho u, \rho v, \rho w, \rho E, \rho q, \rho \omega, \rho Y_i]^T \,, \; i = 1, 2, \ldots, N_k - 1. \tag{2}$$

The conservative variable vector $\mathbf{Q}$ consists of the density ρ, the velocity components u, v and w, the total specific energy E, the turbulence variables q and ω and the species mass fractions Y_i $(i = 1, 2, \ldots, N_k - 1)$. N_k is the total number of species present in the described gas composition. $\mathbf{F}, \mathbf{G}$, and $\mathbf{H}$ are the vectors specifying the inviscid fluxes in the x-, y- and z-direction, $\mathbf{F}_v$, $\mathbf{G}_v$, and $\mathbf{H}_v$ are the viscous fluxes, respectively. The source vector $\mathbf{S}$ includes terms from turbulence, chemistry and the liquid phase and is given by

$$\mathbf{S} = \left[0, 0, 0, 0, S_E, S_q, S_\omega, S_{Y_i}\right]^T, \; i = 1, 2, \ldots, N_k - 1, \tag{3}$$

where S_E is the energy source term from the liquid phase, S_q and S_ω are the source terms of the turbulence variables and S_{Y_i} the source terms of the species mass fraction. For the turbulence closure the q-ω model of Coakley [1] is used. Additionally some other two-equation models are available like the k-ω model of Wilcox [2] and a DDES approach [3].

The spacial discretization is performed on block structured grids with a finite volume scheme. For the reconstruction of the interface values, MLP^{ld}(Multidimensional Limiting Process – low diffusion) [4] with up to sixth order is used to prevent oscillations at contact discontinuities, frequently appearing in supersonic flows. MLP uses diagonal values to improve the TVD (Total Variation Diminishing) limiters [5]. Using these interface values the AUSM$^+$-up flux vector splitting [6] is used to calculate the inviscid fluxes. The unsteady set of equations (1) is solved with an implicit Lower-Upper-Symmetric Gauss-Seidel (LU-SGS) [7, 8] algorithm. Furthermore finite-rate chemistry is treated fully coupled with the fluid motion. The code is parallelized with MPI (Message Passing Interface). More details concerning TASCOM3D may be found in Refs. [7–10].

2.2 *The Spray Code SPRAYSIM*

For multiphase simulations the flow solver TASCOM3D is coupled with the Lagrangian particle tracking code SPRAYSIM. The liquid phase is assumed to be a dilute spray of discrete, non-interacting droplets. In order to limit the computational effort characteristic groups of droplets are represented by computational parcels. These parcels are tracked through the computational domain and the change in mass, momentum and temperature of the droplets is computed based on the gas phase data along the trajectories. Accordingly the set of ordinary differential equations (ODE)

$$\frac{\mathrm{d}\mathbf{Q}_d}{\mathrm{d}t} = \mathbf{H}_d \tag{4}$$

has to be solved for each group of droplets, where the variable vector $\mathbf{Q}_d$ is given by

$$\mathbf{Q}_d = \left[D_d, u_d, v_d, w_d, T_d\right]^T . \tag{5}$$

The corresponding source term vector $\mathbf{H}_d$ accounts for droplet evaporation and particle drag in the gas field. For more details the interested reader is referred to Lempke et al. [11].

In addition to the provision of the required gas phase values to the Lagrange solver, the spray feedback is considered by regarding the accumulated spray source terms in the mass, momentum, turbulence and species equations (1) of the gas phase solver (two-way coupling) according to

$$\mathbf{S}_{tot} = \mathbf{S} + \mathbf{S}_{spray} \,. \tag{6}$$

Details concerning the calculation of the spray source terms can be found in [12].

The coupling process is iterated until convergence is achieved. At the moment the exchange of gas field values and source terms between the two codes is realized by file I/O. For typically stationary simulations the overhead is negligable compared to the computing time of the gas phase and the spray. However, for unsteady simulations the time for the coupling can account for up to 20 %. Thus it seems worth to switch the code communication to MPI on a long term.

3 Research Topics

3.1 Complete Scramjet Model

Within the collaborative research program GRK 1095 simulations of scramjet combustion chambers and complete scramjet models [13] are performed. Based on the simulations crucial processes in the combustor like mixing, ignition and flame stabilization are analyzed. Figure 1 shows the axial velocity distribution at the symmetry plane of a complete scramjet demonstrator model for flight conditions at Mach 8 at an altitude of 30 km. The flow inside the model is strongly non-uniform and three-dimensional due to the used 3D intake. A passive suction slot at the bottom wall of the isolator captures the cowl shock in order to prevent boundary layer separation. For hydrogen injection a lobed strut injector, which is characterized by high mixing and combustion efficiencies [14, 15], is mounted centrally in the model combustor. Two ignition wedges are placed at the top and bottom wall of the

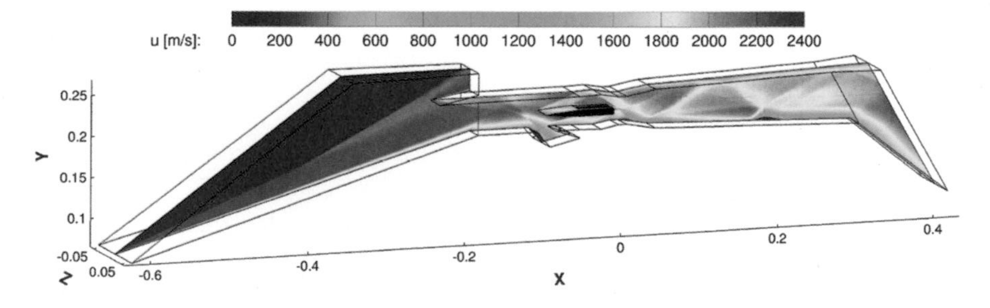

Fig. 1 Axial velocity distribution at the symmetry plane of a complete scramjet model

combustion chamber. They initiate a shock wave system, which locally raises the temperature and consequently ensures ignition and a stable combustion.

3.2 PennState Preburner Combustor

Within the DFG project SFB TRR-40 details of rocket combustion are investigated. One test case here is the PennState Preburner Combustor. Marshall et al. [16] reported the experimental setup that is aimed at the characterization of the chamber wall heat flux in a rocket combustor. Due to the utilization of preburners both oxidizer and fuel are injected into the combustion chamber in gaseous state. The setup was suggested as validation test case at the *3rd International Workshop on Rocket Combustion Modeling* [17] and has since been investigated by numerous research groups (see e.g. Refs. [18–20]).

Two-dimensional axisymmetric URANS calculations with TASCOM3D have been presented by the authors [21] for this setup. Currently three-dimensional simulations of the PennState Preburner Combustor test case on a grid with 16.3 million volumes are performed on the Cray XE6 super computer at HLRS. Although the time-averaging process to obtain a meaningful comparison with the experimental data is still ongoing, Fig. 2 provides an instantaneous visualization of the flame. The iso-surfaces mark a temperature range from 3,100 K (blue) to 3,400 K (red) and reveal the strong corrugation of the flame near the injector. As expected the flame is anchored at the oxidizer post and stretches approximately half-way into the combustion chamber.

Due to the high computational cost of this unsteady 3D-simulation with detailed chemical kinetics (9 species, 21 reactions) it is used in Sect. 4.1 to evaluate the dependence of the code performance on the block sizing under practical conditions.

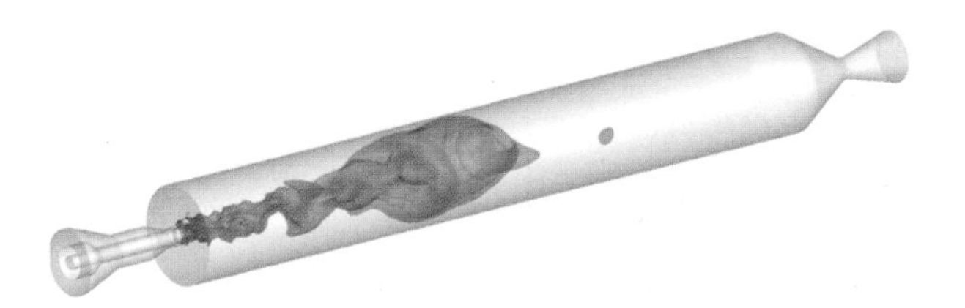

Fig. 2 Visualization of the instantaneous flame structure inside the PennState Preburner Combustor by means of temperature iso-surfaces

3.3 LPPduct Spray Test Case

The test case LPPduct (Lean Premixed Prevaporized) is one of the standard test cases that is used for verification and validation of spray codes. It was also used to test the coupling between TASCOM3D and SPRAYSIM [22]. The subject of investigation is the evaporation of a kerosene spray in a preheated, high pressure duct that is typical for gas turbine applications. Because detailed experiments for model rocket combustors are not available this experiment has been used. The kerosene spray is produced by a planar prefilming airblast atomizer mounted in the center of the channel. Details about the experimental setup can be found in [23].

In the simulation only half of the rectangular channel is considered by employing a symmetry boundary condition for the vertical center plane. In the standard setup approximately 13,000 droplet parcels are used to represent the actual spray. The spray initial conditions are derived from measurements as explained in Ref. [23]. The CFD mesh consists of 850,000 volumes that are distributed over 64 CPUs.

Figure 3 (left) shows the computational domain with spray registration planes in the second half of the chamber. Channel walls are depicted in gray. The actual atomizer is not incorporated in the simulations and its effect on the flow field therefore neglected. Instead the spray parcels are injected along a line in the center of the channel that corresponds to the position of the atomizer edge. To give an impression of the spray evolution Fig. 3 (left) depicts the ratio of the local mass flux to the maximum mass flux in the registration plane.

The spray evaporation process and its influence on the gas field temperature in axial direction are shown in Fig. 3 (right). The liquid mass flow rate normalized by the injection mass flow rate is denoted by the black line. Half of the initial droplet mass already evaporates within the first 20 mm downstream the atomizer. Only around 3 % of the injected droplet mass leaves the computational domain

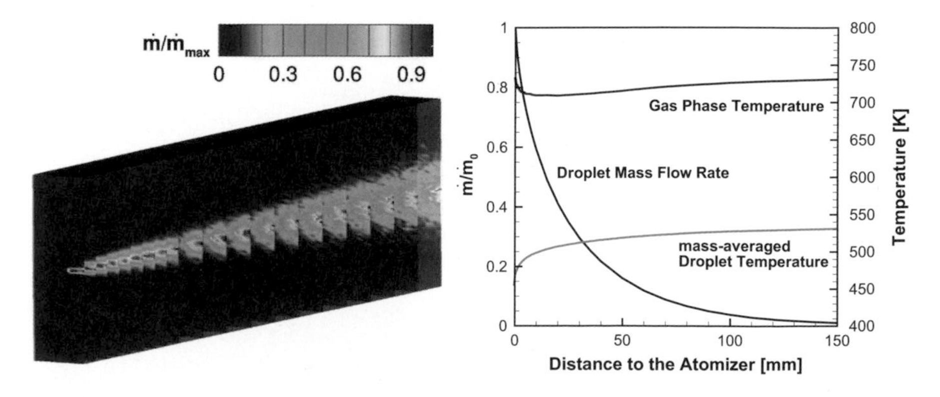

Fig. 3 Visualization of the liquid mass flux normalized by the maximum value in the registration plane (*left*). Axial evolution of droplet evaporation rate (based on the mass flux) and droplet and gas phase temperatures (*right*)

without evaporating. Furthermore the drop in the gas phase temperature due to the heat transfer to the droplets can be observed (blue curve). Accordingly a steep rise in the mass-averaged droplet temperature (green curve) is evident directly behind the atomizer edge.

3.4 Mascotte A-10

Just like the PennState Preburner Combustor the Mascotte A-10 configuration [24] is a model rocket combustor with a shear coaxial injector supplying hydrogen and oxygen as fule. However, in this experiment the oxygen is injected in liquid form at 85 K necessitating a multiphase simulation approach. Results of such simulations, coupling SPRAYSIM and TASCOM3D, have been previously reported by Lempke et al. [25]. Furthermore the test case has been under constant investigation and some adjustments have been performed.

A plot of a simulation result with the revised injector geometry can be seen in Fig. 4. To allow a detailed view of the spray originating from the liquid oxygen core only the first quater of the chamber is shown. It has to be mentioned that for clarity only 1,000 arbitrarily chosen trajectories are presented. All droplets are injected in the computational domain with a velocity of 10 m/s but varying injection angle. Droplets that encounter the fast annular hydrogen coflow experience a strong acceleration in axial direction whereas those traveling near the chamber middle axis show only a moderate rise in velocity. Although Fig. 4 gives the impression that a large amount of droplets travels far into the combustion chamber or hit the channel wall it has to be mentioned that most of the liquid oxygen already evaporates close to

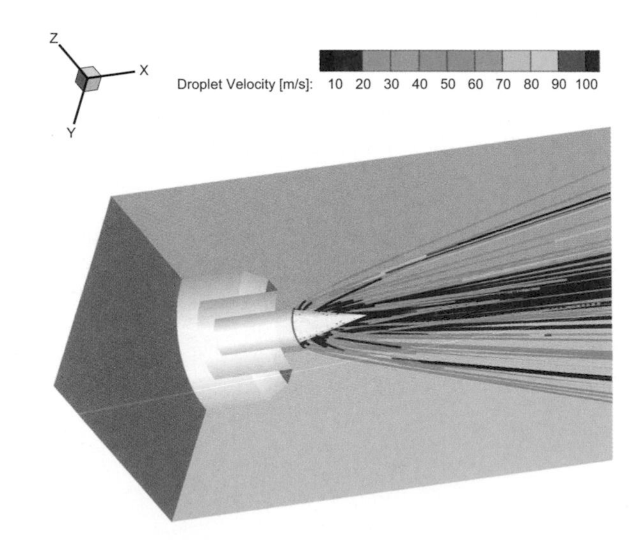

Fig. 4 Droplet trajectories originating from the liquid core in the near injector region. The trajectories are colored according to the droplet velocity

the injector and that the liquid mass flux is strongly concentrated around the injector rotational axis.

4 Performance Analysis

4.1 PennState Preburner Combustor

First, strong scaling is investigated for TASCOM3D using the Pennstate Preburner Combustor test case described in Sect. 3.2. The used standard grid has 16.3 million volumes and 1,000 iterations are calculated. Because of the complex setup of this case, only three different workloads per CPU have been tested. Table 1 summarizes the performance data. The number of volumes for each CPU is reduced from 8192 to 4096 and 2048 while the number of CPUs has been doubled and quadrupled, respectively. With twice the number of CPUs the runtime decreases by a factor of 2.2 with an efficiency of 1.1. For four times the number of CPUs the speedup is 4.8 yielding an efficiency of 1.2. This can be explained by a better calculation efficiency on each CPU due to the smaller local problem size. The MPI communication time decreases as well but the relative time spend in MPI routines increases. This investigation shows an excellent efficiency even if only a doubling and quadrupling of the number of CPUs has been tested.

4.2 3D Supersonic Channel Flow

To examine the performance of the Code more intensively, a special performance test case has been set up: a three-dimensional supersonic channel flow which is geometrically simple. Thus, the number of volumes, the number of CPUs and the block size on each CPU can be changed easily. In the following subsections different core distributions, strong scaling and weak scaling are investigated.

Table 1 Performance comparison for the Pennstate Preburner Combustor

Blocksize	N_{CPU}	Runtime [s]	Speedup [-]	MPI time [-]
$32 \times 16 \times 16$	1,996	2,775.3	–	7.6 %
$16 \times 16 \times 16$	3,992	1,248.7	2.2	9.2 %
$8 \times 16 \times 16$	7,984	577.9	4.8	17.9 %

Table 2 Performance comparison for different core placements

Distance	Runtime [s]	Speedup [-]	Efficiency [-]
1	11,062.8	–	–
2	5,551.5	1.99	1.00
4	4,108.9	2.69	0.67

4.2.1 Core Placement

Each node on the Hermit cluster consists of 32 cores. 16 cores are built into one processor which consists of two dies. Two cores share a floating point unit (FPU) and L2 cache and four cores share a L3 cache with each other. Thus it is not expected that two neighboring cores can work without impacting each other. The performance gain by using different core placements is briefly investigated in this section. Therefore the simulation is devided into eight processes, each having 48^3 cells are run on one node using three different core distributions:

- All eight cores are next to each other.
- Every second core is used, therefore not sharing the FPU and L2 cache.
- Every fourth core is used, therefore not sharing the FPU, L2 and L3 cache.

In Table 2 the achieved runtimes, speedups and efficiencies are shown. The efficiency is calculated by dividing the speedup with the distance between the used cores. Using every second core shows a speedup of almost two and an almost perfect efficiency. This is expected since most of the operations performed are floating point calculations. Thus using every second core only can speed up the calculation by a factor of two without any additional cost. Using every fourth core achieves a speedup of 2.69 and an efficiency of 67 % only.

4.2.2 Strong Scaling

For the strong scaling investigation, the problem size is kept constant. The load is distributed between an increasing number of CPUs while the workload for each CPU is decreasing correspondingly. The used grid consists of 96^3 volumes. The solution is calculated by 1,000 iterations for each setup. For a meaningful strong scaling, every node should have the same number of jobs running on it, otherwise shared cache and memory bandwidth have an effect on the scaling as shown in Sect. 4.2.1. However, the 48^3 blocks exceed the available memory of one single node with 32 GB RAM as can be seen in Fig. 6. Thus, the 48^3 blocks were run with eight processes on one node but using the first eight CPUs only. This is assumed to be almost identical to a node with all cores occupied.

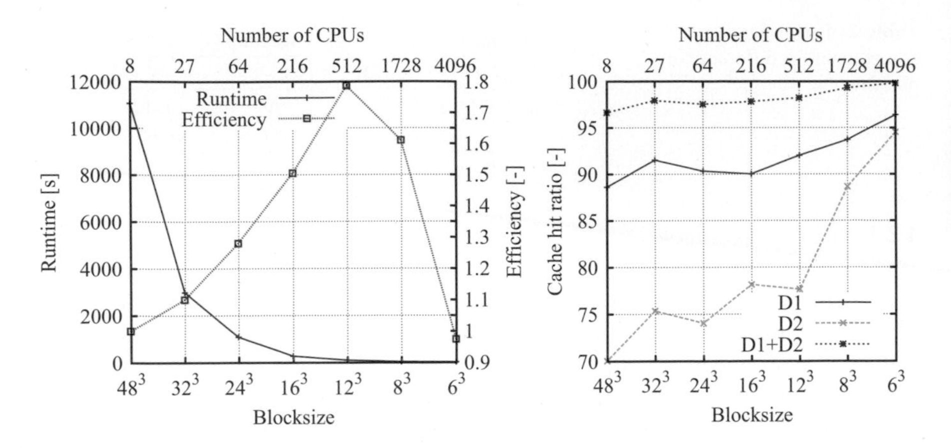

Fig. 5 Strong Scaling runtime, efficiency and cache hit ratio

Figure 5 shows the runtime and the parallelization efficiency. The efficiency is calculated by

$$\eta = \frac{T_{CPU}^{ref}}{T_{CPU}} = \frac{T_R^{ref} \cdot N_{CPU}^{ref}}{T_R \cdot N_{CPU}} \tag{7}$$

where T_{CPU} is the computational cost, T_R is the runtime and N_{CPU} is the number of CPUs used. T_{CPU}^{ref}, T_R^{ref}, and N_{CPU}^{ref} refer to the reference values, in this case the eight CPUs with a block size of 48^3.

In the following study the number of CPUs is varied from eight with a grid size of 48^3 per CPU to 4096 with a grid size of 6^3 per CPU. The calculation time decreases with an increasing number of CPUs from 11,063 s to 22 s (Fig. 5, left figure, red line). This implies a speedup of about 502.9. The efficiency of the parallelization (Fig. 5, left figure, blue line) first increases until a block size of 12^3 is reached. This initial increase in efficiency can be explained by a better local calculation efficiency due to the smaller problem size on each CPU resulting in better cache hit ratios (see Fig. 5, right figure). The decrease in efficiency for grids smaller then 12^3 volumes is due to increasing MPI communication between the processes. The smaller the problem size becomes on each CPU more (relative) time is required for MPI communication. As shown in Fig. 6, the absolute MPI communication for one CPU time decreases with decreasing block size while the relative time spend in MPI increases constantly, except for a peak at a block size of 32^3. The reason for this peak remains unclear to the authors. Figure 6 (right side) visualizes the memory requirements for one CPU.

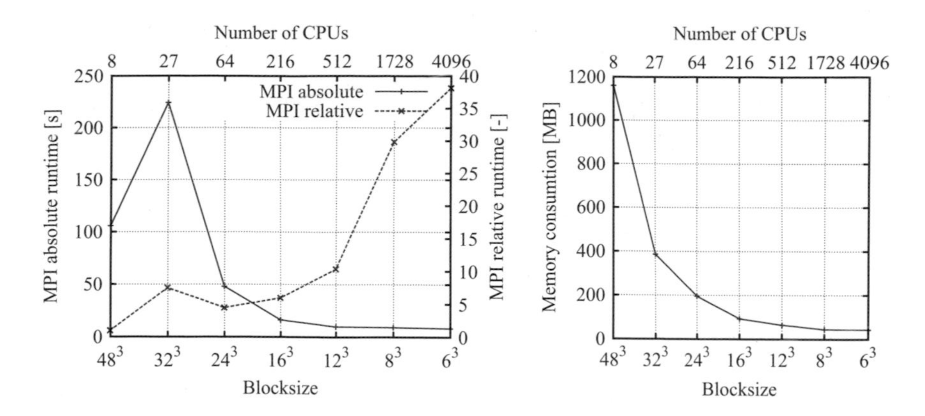

Fig. 6 MPI durations and memory requirement for strong scaling

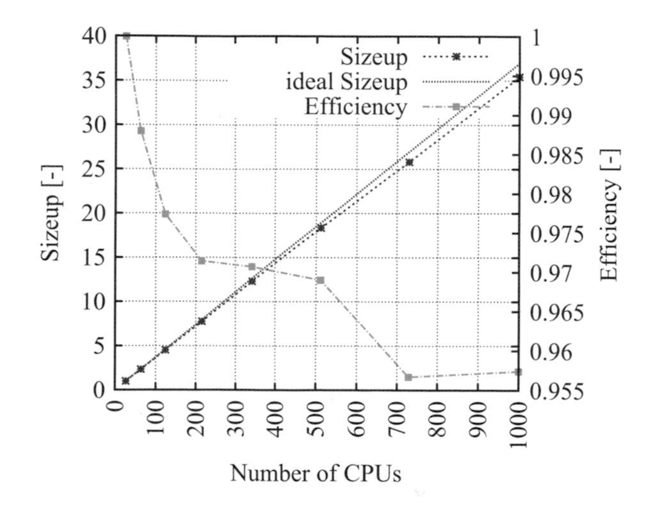

Fig. 7 Weak scaling size up and efficiency

4.2.3 Weak Scaling

To investigate the weak scaling of the code, the workload on each CPU is kept constant and the total problem size is increased together with the number of CPUs employed. Again, for a meaningful comparison, each node in use should be occupied with the same number of processes. Thus the weak scaling study starts with 27 processes which are increased up to 1,000 processes, each having 16^3 cells. Figure 7 shows the size up and efficiency for this test case as well as the ideal speedup. The speedup of the code (TASCOM3D) is almost ideal and for 1,000 CPUs the efficiency is approximately 96 % (please note that the axis for efficiency is shown for the range [0.955,1] only). The losses in the size up can be explained with MPI overhead.

4.3 Lagrangian Spray Simulation (SPRAYSIM)

Two test cases are used to asses the strong scaling properties of the Lagrangian combustion code SPRAYSIM, namely the LPPduct and the Mascotte A-10 multiphase rocket combustor. For this purpose the discretization of the gas field is kept constant while the particle tracking is performed using a varying number of processors.

4.3.1 LPPduct

First results for the LPPduct case (see Sect. 3.3) are given. In this simulation a moderate number of 13,000 parcels are tracked only. Due to the low runtime requirements a large number of tests could be performed. The number of processes was varied between 1 and 32 in two different test series. First all processes were placed on a single node (ideal placement), whereas in a second series all processes were distributed over different nodes. Figure 8 shows the required times and speedup for both cases versus the number of processes used. To account for runtime variations, each test was repeated three times. However, only few severe spikes were encountered, for example for five processes on one node (Fig. 8, left side). It can be stated that the strong scaling performance of SPRAYSIM as far as the actual particle tracing is concerned, is close to ideal. The overall speedup shows a less ideal behaviour due to the rising communication effort. A severe drop in the speedup is observed when the number of processes per node exceeds 16. This can be attributed to the hardware architecture of the AMD Interlagos processors. Up to 16 processes can be distributed such that only one of the two cores that share a FPU (floating point unit) is in use. Moreover this leads to a size doubling of the L2 cache. For more than 16 processes this advantage is lost. The most extreme case, where all processes are distributed on different nodes, has been examined as well

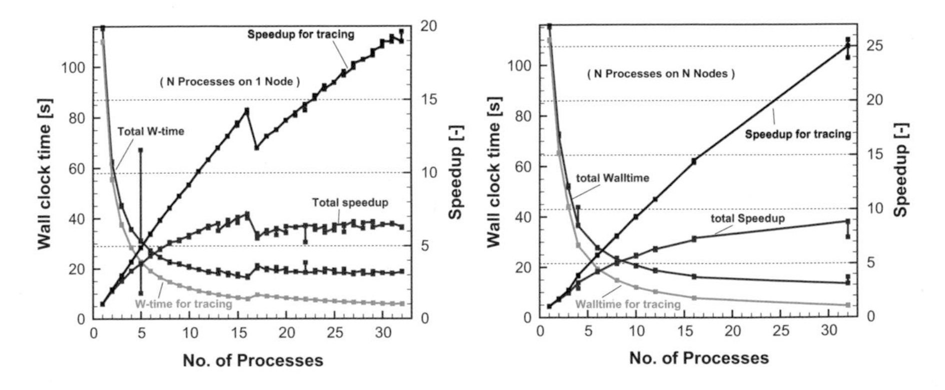

Fig. 8 Strong scaling speedup for the LPPduct case. All processes clustered on one node (*left*) and on different nodes (*right*)

(Fig. 8, right side) despite the fact that in terms of costs this approach is not feasible for practical simulations. Nonetheless it can be observed that the tracing speedup for 32 processes is significantly higher for the distributed case (approximately 25) than for the clustered one (approximately 19). This fact also underlines the excellent performance of the Cray Gemini node-node interconnection. A feasible compromise in this case would be to distribute the 32 processes between two nodes, to take benefit from the processor architecture. However, the discrepancy between the overall run time and the time solely used for particle tracing suggests further optimization of the initialization and the process communication.

4.3.2 Mascotte A-10 Combustor

Due to the much higher computational effort for tracing 200,000 spray parcels in the Mascotte A-10 simulation the aforementioned discrepancy is not as pronounced as can be seen from Fig. 9 where corresponding performance parameters are plotted for the second spray test case. Both the speedup for tracing the spray parcels as well as the total speedup show a close-to-ideal linear scaling. However, this test series reveals another issue that needs to be addressed. Due to the large number of parcels to be tracked each process requires a significant amount of RAM that exceeds even the capacity of the 64 GB RAM nodes available on the XE6 installation. On the regular nodes a maximum of eight Processes could be placed together on a node.

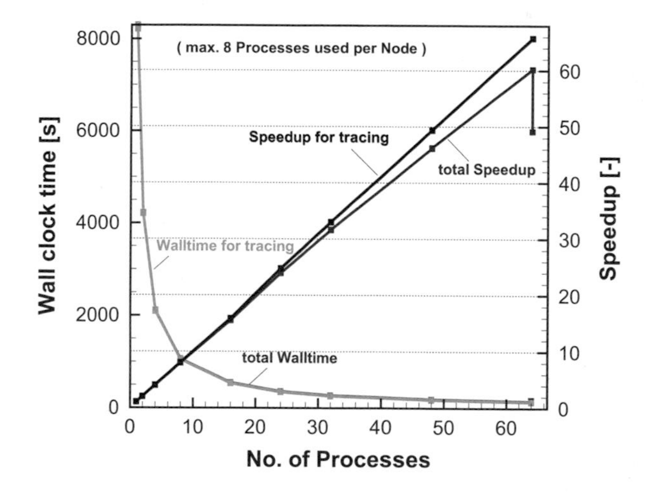

Fig. 9 Strong scaling speedup for the Mascotte A-10 configuration

5 Conclusion

Some major research activities associated with the code TASCOM3D have been outlined. These include the investigations of scramjet configurations as well as multiphase simulations of liquid rocket combustors. Some of the test cases have been used to assess the scaling performance of the employed numerical tools, namely TASCOM3D and SPRAYSIM, under certain criteria.

TASCOM3D shows nearly ideal weak scaling performance, which is only limited by increasing MPI overhead. The strong scaling shows superideal performances. This is due to the increased efficiency for smaller workloads on each CPU. Therefore the main focus for code performance improvements should be a better scalar performance, especially for larger domains. The decreasing efficiency is an indicator that there is a large potential for improvement by better matching the algorithm implementation on the target architecture. First investigations using CrayPat confirm this claim. For isolated routines and loops the performance could be improved and the size dependency reduced. However, so far no overall improvement was achieved and the process is still ongoing.

The scaling performance of the Lagrangian particle tracking code SPRAYSIM, that was recently parallelized using MPI, is very satisfactory. However, the test series revealed certain aspects that need further consideration. First, it is important to consider the specifics of the AMD Interlagos processor architecture whenever possible and economically feasible. Secondly, the first implementation of the parallelization in SPRAYSIM needs to be modified under certain aspects. For simulations with a small amount of parcels the overhead induced by the code initialization and the processor communication can not be neglected. The strategy to obtain all necessary information on one process and subsequently communicate this information has proven to be not efficient enough. First tests with a revised strategy where all processes read the information simultaneously yielded promising results in this aspect. Furthermore the RAM requirements for large test cases (high number of cell volumes as well as parcels) can be a limiting factor. Therefore shared memory strategies are currently under investigation.

Acknowledgements The presented work was performed within the framework of the SFBTR 40 and the GRK 1095/2 funded by the Deutsche Forschungsgemeinschaft (DFG). This support is greatly appreciated. All simulations were performed on the Cray XE6 'HERMIT' Cluster at the High Performance Computing Center Stuttgart (HLRS) under the grant number scrcomb. The authors wish to thank for the computing time and the technical support, especially Aniello Esposito from Cray.

References

1. Coakley, T.J., Huang, P.G.: Turbulence modeling for high speed flows. In: AIAA-92-0436, Reno, Nevada (1992)
2. Wilcox, D.C.: Formulation of the k-ω turbulence model revisited. AIAA J. **46**, 2823–2838 (2008)
3. Spalart, P.R., Deck, S., Shur, M.L., Squires, K.D. Strelets, M., Travin, A.: A new version of detached-eddy simulation, resistant to ambiguous grid densities. Theor. Comput. Fluid Dyn. **20**, 181–195 (2006)
4. Gerlinger, P.: Multi-dimensional limiting for high-order schemes including turbulence and combustion. J. Comput. Phys. **231**, 2199–2228 (2012)
5. Yoon, S.-H., Kim, C., Kim, K.-H.: Multi-dimensional limiting process for three-dimensional flow physics analyses. J. Comput. Phys. **227**, 6001–6043 (2008)
6. Liou, M.-S.: A sequel to AUSM, Part II: AUSM^{+}-up for all speeds. J. Comput. Phys. **214**, 137–170 (2006)
7. Gerlinger, P., Brüggemann, D.: An implicit multigrid scheme for the compressible Navier-Stokes equations with low-Reynolds-number turbulence closure. J. Fluids Eng. **120**, 257–262 (1998)
8. Gerlinger, P., Möbus, H., Brüggemann, D.: An implicit multigrid method for turbulent combustion. J. Comput. Phys. **167**, 247–276 (2001)
9. Gerlinger, P.: Investigations of an assumed PDF approach for finite-rate chemistry. Combust. Sci. Technol. **175**, 841–872 (2003)
10. Stoll, P., Gerlinger, P., Brüggemann, D.: Domain decomposition for an implicit LU-SGS scheme using overlapping grids. In: AIAA-97–1896, Snowmass Village, Colorado, USA (1997)
11. Lempke, M., Gerlinger, P., Rachner, M., Aigner, M.: Euler-Lagrange simulation of a LOX/H2 model combustor with single shear coaxial injector. In: Nagel, W.E., Kröner, D.B., Resch, M.M. (eds.) High Performance Computing in Science and Engineering'10, Stuttgart, Germany, pp. 203–215. Springer, Heidelberg (2010)
12. Le Clercq, P., Doué, N., Rachner, M., Aigner, M.: Validation of a multicomponent-fuel model for spray computations. In: AIAA 2009-1188, Proceedings of the 47th Aerospace Sciences Meeting, Orlando, 5–8 Jan 2009
13. Simsont, Y.H., Gerlinger, P.: Numerical investigation of a complete scramjet model with central strut injection. In: ISABE Paper 2013-1619, Busan, Korea (2013)
14. Gerlinger, P., Stoll, P., Kindler, M., Schneider, F., Aigner, M.: Numerical investigation of mixing and combustion enhancement in supersonic combustors by strut induced streamwise vorticity. Aerosp. Sci. Technol. **12**, 159–168 (2008)
15. Kindler, M., Gerlinger, P., Aigner, M.: Numerical investigation of mixing enhancement by Lobed Strut injectors in turbulent reactive supersonic flow. In: ISABE Paper 2007-1314, Beijng, China (2007)
16. Marshall, W., Pal, S., Woodward, R., Santoro, R.: Benchmark wall heat flux measurements in a uni-element GO_2/GH_2 single element combustor. In: AIAA 2005-3572. Tucson, Arizona, USA (2005)
17. Pal, S., Marshall, W., Woodward, R., Santoro, R.: Wall heat flux measurements in a uni-element GO_2/GH_2 shear coaxial injector. In: 3rd International Workshop on Rocket Combustion Modeling, Vernon (2006)
18. Tucker, K., Menon, S., Merkle, C.L., Oefelein, J.C., Yang, V.: Validation of high-fidelity CFD simulations for rocket injector design. In: AIAA-2008-5226, Hartford, Connecticut (2008)
19. Lian, C., Merkle, C.L.: Contrast between steady and time-averaged unsteady combustion simulations. In: AIAA-2010-371, Orlando, Florida (2010)
20. Ivancic, B., Riedmann, H., Frey, M., Knab, O., Karl, S., Hannemann, K.: Investigation of different modeling approaches for CFD simulation of high pressure rocket combustors. In: Proceedings of the 5th European Conference for Aeronautics and Space Sciences (EUCASS), Munich, Germany (2013)

21. Lempke, M., Gerlinger, P., Kindler, M., Rachner, M., Aigner, M.: Assumed and transported PDF simulations of model rocket combustors. In: Sonderforschungsbereich/Transregio 40 – Annual Report. Universität München, München, Germany (2012)
22. Lempke, M., Gerlinger, P., Rachner, M., Aigner, M.: Numerical simulation of subcritical model rocket combustors with single shear coaxial injectors. In: Sonderforschungsbereich/Transregio 40 – Annual Report. Universität München, München, Germany (2010)
23. Rachner, M., Brandt, M., Eickhoff, H., et al.: A numerical and experimental study of fuel evaporation and mixing for lean premixed combustion at high pressure. In: Twenty-Sixth Symposium (International) on Combustion/The Combustion Institute, pp. 2741–2748. Naples, Italy (1996)
24. Vingert, L., Habiballah, M.: Test case RCM2: cryogenic spray combustion at 10 bar at MASCOTTE. In: 2nd International Workshop on Rocket Combustion Modeling, Lampoldshausen (2001)
25. Lempke, M., Gerlinger, P., Rachner, M., Aigner, M.: Steady and unsteady RANS simulations of cryogenic rocket combustors. In: 49th AIAA Aerospace Sciences Meeting, AIAA-2011-101, Orlando, Florida (2011)

Part IV
Computational Fluid Dynamics

Ewald Krämer

High Performance Computing (HPC) is an absolutely indispensable means for leading edge research in many areas of fluid dynamics. This is, among others, emphasized by the large number of projects that were run on the supercomputers of the HLRS during the reporting period. Thirty eight annual reports were submitted and underwent a peer review process. Due to limited space, only 18 projects could be selected for publication in the present book, which means that a number of high-qualified contributions could not be admitted.

The spectrum of projects ranges from Direct Numerical Simulations (DNS) over Large Eddy Simulations (LES) and hybrid methods (DES) to Reynolds Averaged Navier Stokes (RANS, URANS) simulations on the one hand, and from code development over fundamental research up to application oriented research on the other hand. In-house codes were used as well as commercial codes. All CFD-projects presented in this book were run on the Cray XE6 HERMIT, which offers a peak performance of more than 1 PFLOP/s.

The first three projects were granted computer resources within the frame of the biannual "Call for Large-Scale Projects" of the Gauss Centre for Supercomputing (GCS). Projects considered in these calls require more than 35 million core hours per year.

Wilke and Sesterhenn of the Technical University Berlin, Fachgebiet Numerische Fluiddynamik, have investigated via DNS the heat transfer of a confined round jet impinging on a flat plate. In order to properly resolve the Kolmogorov length scales, a grid with 135 million points was used. Their self-developed code had successfully been installed on HERMIT and has typically been run with 8,192 parallel cores showing a nearly linear scaling behaviour for the given problem. Results are shown in terms of distributions of radial velocity, temperature, and Nusselt number. The

E. Krämer (✉)
Institut für Aerodynamik und Gasdynamik, Universität Stuttgart, Pfaffenwaldring 21, 70550 Stuttgart, Germany
e-mail: kraemer@iag.uni-stuttgart.de

typical vortical structures, primary and secondary ring vortices, are well predicted. Preliminary comparisons with experiments (performed at higher Reynolds numbers) show good qualitative agreement.

An impressive research work has been done over the last years at the Institute of Aerodynamics of the RWTH Aachen (AIA) in the development of numerical simulations methods for various applications in fluid dynamics. The contribution of Gadeschi, Siewert, Lintermann, Meinke, and Schröder describes numerical methods based on hierarchical Cartesian grids for the simulation of multi-scale particle-laden flows. These include Eulerian-Lagrangian approaches for fully resolved moving particles with conjugated heat transfer as well as a one-way coupled Lagrangian particle model for large-scale particle simulations. For both approaches exemplary results are shown. For the latter one, simulations were performed on HERMIT demonstrating the good scaling behaviour of the parallelisation strategy. For an efficient generation of the required very large grids, an automatic hierarchical Cartesian mesh generator was established employing a Hilbert space-filling curve for optimal domain decomposition. The mesh generation was also done on a supercomputer. An example is shown for a mesh with approx. 10^{10} cells on the IBM Blue-Gene/Q (JUQUEEN) of the Jülich Supercomputer Centre (JSC), where nearly linear strong scaling was achieved up to 32,768 cores. The hierarchical structure of the grid allows solving multiple phases by different methods independently.

The near-wake behaviour of discrete roughness elements located on an unswept (2-d) and on a swept (3-d) aircraft wing as an array in spanwise direction has been investigated by Kurz and Kloker of the Institute of Aerodynamics and Gas Dynamics (IAG) at the University of Stuttgart. Applying spanwise periodic boundary conditions, roughness elements with different heights and slope angles were introduced into a laminar 2-d and 3-d boundary layer, respectively, with strongly negative pressure gradients. In a first step, the steady base flows were generated using the finite-volume RANS-code TAU developed by the German Aerospace Center (DLR) and then transferred to a finer grid for the subsequent DNS computations with the compressible in-house code NS3D, employing both explicit and compact difference schemes of 8th and 6th order, respectively. The authors have done a very careful analysis of the local flows both for sub-effective elements, where they have found that in the 3-d case the element triggers steady crossflow vortices that become secondary unstable and, thus, shift the transition location upstream, and for super-effective elements, which directly lead to flow tripping. The authors have also done a comprehensive performance analysis of the code in general and for the present numerical set-up. They show that the compact difference scheme performs poorly, while the explicit scheme and the newly developed sub-domain compact scheme reveal a very good parallel efficiency. Additionally, a comparison was made between the CRAY XE6 (HERMIT) and the recently installed CRAY XC40 (HORNET) in terms of computational time per grid point and time step. It turned out that HORNET was 3–4 times faster (without any specific code optimization done for the XC40).

DNS has also been applied by Rauschenberger, Birkefeld, Reitzle, Meister, and Weigand from the Institute of Aerospace Thermodynamics, University of

Stuttgart, who present a new parallelized method for the simulation of rigid particles in multiphase flows implemented in their incompressible in-house code FS3D. Usually, rigid bodies are treated as Lagrangian particles, but the present scheme is formulated in an Eulerian framework with the code being based on the Volume-of-Fluid (VoF) method. The capability of the code is shown for the known case of two spheres in a Newtonian fluid placed one above the other and then being dropped. Comparisons were performed with numerical results from literature. Deviations were found and possible explanations are given, indicating that the FS3D results are more realistic than the results obtained with other solvers. The performance analysis on HERMIT reveals some drawbacks of the code in terms of parallel efficiency.

Ukai, Kronenburg, and Stein from the Institute for Combustion Technology, University of Stuttgart, present a new CMC-model, a so-called two-conditional moment approach. CMC- (conditional moment closure) methods are used to model the interaction between turbulence, fuel evaporation, and chemistry. They use their model in a two-phase LES context for the simulation of a spray flame. Different model strategies and the governing equations are described in detail, and results are shown in comparison to experiments. The two-conditional moment approach improves the agreement significantly, which is explained by a better prediction of the temperature distribution. With their contribution, the authors have clearly advanced the scientific knowledge in their field, and their findings are important for other researchers using these types of models. The strong scaling behaviour of their in-house code BOFFIN on HERMIT is satisfactory only up to 250 cores, but a restructuring of the code is contemplated aiming at an efficiency of above 70 % for 1,000 cores.

Also the next two contributions deal with the application of LES. Both come from the Institute of Aerodynamics and Fluid Mechanics of the Technical University Munich. Eberhardt and Hickel scrutinized own new as well as previously published LES results for the very challenging problem of a sonic jet injected into a supersonic crossflow. It is interesting to see that all LES results are in excellent agreement to each other, but that there are still considerable deviations from the measurements. Moreover, even the sensitivity study on the influence of uncertain flow parameters has not been able to explain these discrepancies. So, the basic question can be posed, whether the deviations are due to uncertainties in the experiments or to fundamental errors in the LES. The own results were achieved using the in-house implicit LES-code INCA with the Adaptive Local Deconvolution Method (ALDM) as subgrid-scale model. Five hundred and twelve parallel cores on HERMIT were usually employed.

In the second contribution of the institute, Egerer, Hickel, Schmidt, and Adams report on LES results of temporarily cavitating shear layers. Again, the INCA code with ALDM has been used with the cavitation being modelled by a homogeneous equilibrium mixture model. Results are compared with experimental data available in literature. Questions related to the interaction of turbulence and cavitation, the resulting modulation of turbulence due to phase change, and its influence on the properties of the shear layer are addressed. Compressibility effects have been taken into account. In general, the numerical results compare well with the experimental data, with slight differences in the temporal variations. An almost linear speed-up has been achieved with the INCA code on HERMIT with approx. 30,000 grid cells per core for the investigated sample problem.

The objective of the work done by Statnikov, Sayadi, Meinke, Schmid, and Schröder from the AIA was to study the turbulent wakes of generic space launcher configurations in order to better understand the base-flow dynamics and its controllability. The unsteady flow fields for a transonic and a hypersonic freestream Mach number and for configurations with sting support and nozzle jet, respectively, were generated by the in-house zonal RANS/LES finite-volume flow solver, with the LES formulation being based on the monotone integrated LES (MILES) approach modelling the impact of the subgrid scales by numerical dissipation. For the spectral analysis of the wake, Dynamic Mode Decomposition (DMD) together with classical statistical analysis and temporal filtering methods has been applied to the flow field. The use of DMD is this context is new and promising. The authors can impressively show that DMD, especially when applied to different flow quantities, e.g. the velocity and pressure signals or the three velocity components (composite DMD), allows a precise identification of significant coherent flow structures by providing a clear picture of the 3-dimensional modal shapes and the respective frequencies. This is very beneficial for an improved understanding of the complex base-flow dynamics. Depending on the investigated flow case, 512–4,096 cores were used, leading to 44,000–18,500 grid points per CPU. However, for the DMD analysis of the unsteady flow fields, the virtual memory demand is the driving parameter.

Rotating flows are the subject of the next three contributions. Weihing, Schulz, Lutz, and Krämer from the IAG in Stuttgart have performed CFD simulations of wind turbines under realistic operating conditions. Two cases are presented in detail. Case 1 represents two offshore wind turbines located 6.7 rotor diameters behind each other with a lateral offset of half a rotor diameter. The aim is to estimate the impact of the wake of the front turbine on the load distribution of the partly shadowed downstream turbine. At the inlet of the computational domain, a turbulent atmospheric boundary layer is prescribed, which had been provided by partners from ForWind, Oldenburg. In case 2, a wind turbine is located in complex terrain on the top of a hill. Again, a turbulent atmospheric boundary layer is prescribed at the inlet. In both cases, the simulations were performed using the DLR's FLOWer code in DES mode in order to adequately propagate the atmospheric turbulence through the domain. Structured meshes for the full turbines, i.e. including rotor blades, hub, nacelle, and tower, where generated by means of the Chimera technique. The overall mesh sizes were 110 million cells for case 1 and 68 million cells for case 2. The computations were performed on HERMIT using 1,700 and 1,024 cores, respectively. Results of the unsteady flow simulations are shown in terms of wake development, turbulence intensity, and power output.

The commercial flow solvers ANSYS-CFX and OpenFOAM have been used by Krappel, Ruprecht, and Riedelbauch from the Institute of Fluid Mechanics and Hydraulic Machinery at the University of Stuttgart in order to simulate the flow in a Francis turbine and in a Francis pump under part load conditions. Two different kinds of hybrid RANS-LES approaches have been used, a Scale Adaptive

Simulation (SAS) and an Improved Delayed Detached Eddy Simulation (IDDES), and compared to RANS computations. Operation under part load conditions causes high fluctuations and dynamic loads in the runners and the downstream tubes. A characteristic vortex rope establishes behind the hub of the runner due to a rotating low pressure zone that propagates downstream. For the Francis turbine, for which CFX with the SAS turbulence model was applied, a much better resolution of the flow structures could be achieved on a grid with 40 million elements compared to RANS-SST simulations. The same is true for the Francis pump, using OpenFOAM in IDDES mode. In the latter case, comparisons with PIV and LDA measurements could be performed, which again show the superiority of the hybrid approach. It is worth mentioning that the used version 14.5 of ANSYS-CFX shows a significant degradation in the parallel efficiency when using more than 128 cores. However, as the CRAY-MPI is supported in the new version 15.0, better scaling should be possible in future.

The helicopter group of the IAG in Stuttgart has been engaged for some time in the implementation of higher order schemes into the FLOWer code. The lower numerical dissipation of these schemes leads to a much better preservation of the rotor wake. In their contribution, Stanger, Kutz, Kowarsch, Busch, Keßler, and Krämer show the benefits of the high-order approach for two applications, an isolated trimmed, structure-coupled helicopter rotor in free flight and a model rotor operating in ground effect. Since the initially implemented 5th order WENO scheme had shown significant improvements but also some drawbacks, a CRWENO scheme was introduced, which merges the advantages of a compact state reconstruction with the robust WENO scheme. In the first test case, the benefits of the new high order scheme compared to the original WENO and the standard 2nd order JST scheme have impressively been demonstrated for the isolated rotor employing a relatively coarse grid, which is representative for optimization and performance prediction tasks. In case details of the flow field have to be studied, like for the rotor in ground effect, finer meshes are necessary. In this application, a one-way coupling had been applied between the highly resolved flow field and a particle model developed by a team of the University of Magdeburg, to enable simulation of the so-called brown-out phenomenon. Additionally, the aerodynamic and aero-acoustic influence of different blade numbers on a Contra-Rotating Open Rotor (CROR) has been investigated in this project. Simulations were run on 800 cores using 50 million grid cells. For the case shown, the 2nd order scheme had been employed, but WENO simulations have already been started. It will be interesting to see the differences especially in the aero-acoustic results provided through a better representation of the wake-rotor interferences.

So-called shock-control bumps (SCB) are currently subject of research due to their potential to reduce the wave drag of an aircraft operating at transonic speed. However, such bumps also have an impact on the buffet behaviour of the wing. Therefore, investigations at the IAG have been performed to study these impacts. Bogdanski, Gansel, Lutz, and Krämer present the results of their study on transonic airfoils for given SCBs optimized purely under performance aspects as well as for bumps designed for a higher lift coefficient and a lower Mach number close to the buffet boundary. The basic finding is that bumps in general have a detrimental effect on the buffet behaviour, but might provide, if designed properly, a higher usable lift. Furthermore, while SCBs are beneficial (in terms of drag) at the design point, they generally reduce the aerodynamic efficiency under off-design conditions. The DLR's TAU code that was used for the investigations shows a nearly linear speed-up up to 16,384 cores (given the mesh is large enough). Since the investigations were performed for small spanwise sections only (quasi 2-d), a significantly smaller amount of parallel cores was used for this specific study. However, an interesting comparison has been made showing the influence of the employed partitioning tools (Tau native, GNU Chaco) and compilers (Cray, GNU, Intel, PGI) on the parallel efficiency.

The contribution of Volkmer and Vogt from the Institute of Thermal Turbomachinery (ITSM) of the University of Stuttgart presents a comparison of URANS results for two different turbulence models (SAS and SST) for a flow in a gas turbine exhaust diffuser that shows intricate separation phenomena. The commercial ANSYS-CFX code (version 14.0) has been employed for this study. Experimental data obtained in the institute's own test rig have been used for comparison. The differences in the flow fields generated with both turbulence models are elaborated in very detail, and it can be shown that the results obtained with the more time intensive SAS approach are, for the given application, generally not better than with the SST model. While downstream of the internal hub, the SAS results match the measured velocity profiles better than the SST results, due to too a large predicted separation in the annular section, the pressure recovery along the diffuser is lower with the SAS model compared to SST and the experiments, respectively.

Zhang and Laurien from the Institute of Nuclear Technology and Energy Systems (IKE), University of Stuttgart, did numerical simulations of flow with volume condensation in presence of non-condensable gases inside a pressurized water reactor (PWR). This scenario might occur inside a nuclear power plant after an accident, i.e. a leak in the primary circuit, which results in hydrogen and steam injection into the reactor containment. Besides the wall condensation, also the volume condensation phenomenon is of interest for safety considerations. Whereas a wall condensation model is available in the applied ANSYS-CFX code, a new phase exchange model for volume condensation model is presented that takes into account more than one non-condensable gas in a two-phase flow. The new model was validated with a condensation experiment performed in the German THAI facility. Thereafter, the flow and the phase change in a generic containment have been analysed, and results for the time-dependent hydrogen and temperature distributions as well as for the wall condensation rate inside the containment are shown.

The next contribution deals with the Discontinuous Galerkin (DG) method, which may be considered as a combination of finite-element schemes (with a continuous higher order polynomial in each grid cell) and a finite-volume scheme (allowing for discontinuities at the cell faces), and which is supposed to have the potential to some day replace the well-established finite volume production codes, at least for certain applications. The working group of Munz at the IAG in Stuttgart, has built-up a DG-based high order simulation framework, which has been tailored for high performance computing aiming at solving large scale problems. In the present report, Atak, Beck, Flad, Frank, Hindenlang, and Munz show results from their latest code, FLEXI, that uses a very efficient variant of a DG formulation, the so-called spectral element method (DG-SEM). It has been built from ground up for high performance computing and solves three-dimensional advection dominated problems on unstructured meshes very efficiently and with high accuracy. DNS and LES results of compressible unsteady and turbulent flows are presented, which comprise aero-acoustic sound generation on an airfoil, supersonic turbulent boundary layer flow, and decaying turbulence for LES modelling development. The runs were performed on 5,000–10,000 cores on HERMIT exploiting the high parallel efficiency of the code.

The last three contributions in the CFD section of this book are based on the Lattice Boltzmann approach. Krafczyk, Kucher, Wang, and Geier from the Institute for Computational Modeling in Civil Engineering of the TU Braunschweig report about their DNS and LES studies of turbulent flows around a flat plate, in a porous channel, and over a porous medium with different Reynolds roughness numbers. Comparisons with experiments show the validity of their approach. A newly developed cumulant Lattice Boltzmann scheme, implemented in their in-house code VirtualFluids, has been used. It distinguishes from previous LBM schemes by an increased stability and low numerical dispersion and shows a favourable behaviour with respect to parallelization efficiency and accuracy. The code exhibits a good weak scaling behaviour with a parallel efficiency of 83 % for more than 60,000 cores.

Masilamani, Zudrop, Klimach, and Roller from the Chair for Simulation Techniques and Scientific Computing at the University of Siegen, and Johannink from the AVT research group (Aachen Chemical Engineering) at the RWTH Aachen present the results of their simulations of electro-membrane processes for desalination of sea water. They have used a multi-species Lattice Boltzmann Method to study the hydrodynamics and ionic mass transport of sea water flowing through a channel with a complex spacer structure that separates the ion exchange membrane. The code had been developed with a strong focus on a high computational performance. A very good strong and weak scaling is demonstrated up to 1,024 nodes (i.e. 32,768 cores) on HERMIT. The most detailed simulations were done with 520 million cells, but for engineering design purposes coarser grids were used, which still provide sufficient resolution to approximate the pressure for the investigated Reynolds numbers. One aim of the current project has been a systematic investigation on the influence of the spacer geometry on the performance of electrodialysis processes, which shall help develop an engineering model for the relevant design parameters. Results of a multi-species LBM simulation are shown for the transport of ionic species (sodium-chloride) in an aqueous solvent through a spacer structure under a fixed external electrical field.

The contribution of Krüger, Frijters, Günther, Kaoui, and Harting is about mesoscale simulations of fluid-fluid interfaces conducted in cooperation between the Department of Applied Physics of the Eindhoven University of Technology, the Institute for Materials and Processes of the University of Edinburgh, the Chair of Theoretical Physics I of the University of Bayreuth, and the Institute for Computational Physics of the University of Stuttgart. Applying the LBM, they focus on deformable capsules and vesicles on the one hand, which are handled with the immersed boundary method, and on nanoparticle stabilized emulsions on the other hand, using the Shan-Chen pseudopotential approach in combination with a molecular dynamics algorithm. The respective methods are explained in detail, and new and interesting results obtained with their self-developed LB3D code for the above mentioned applications are shown. In previous reports, the excellent performance and scaling behaviour on the supercomputers of the HLRS and the SCC had already been demonstrated.

In summary, again a large variety of ambitious project has been performed at the HLRS in the field of Computational Fluid Dynamics. Most of them were of very high quality and none of them had been realizable without an access to HPC facilities. This demonstrates the high value as well as the indispensability of supercomputing in this area. Thanks have to be given to the staff of the HLRS and CRAY for their valuable support of the individual projects.

Numerical Simulation of Impinging Jets

Robert Wilke and Jörn Sesterhenn

Abstract This report concentrates on the investigation of heat transfer of a confined round impinging jet. A direct numerical simulation was performed at a Reynolds number of $Re = 3{,}300$ using a grid size of $512 \times 512 \times 512$ points. It is shown that the dissipative scales are well resolved. This enables the examination of the impact of the jet's turbulent flow field on the heat transfer of the impinged plate. In this study the distribution of the local Nusselt number is presented and related to the instantaneous flow field of the jet. First results of turbulent statistics are shown.

1 Introduction

Impinging jets provide an effective cooling method for various applications such as the cooling of turbine blades of aircraft. An increase of efficiency not only reduces the required cooling air mass flow and consequently the fuel consumption, but also enables new combustion concepts with even higher cooling demands to be applied in the future.

Heat transfer due to forced convection of a jet impinging on a flat plate has been studied for decades. General information including schematic illustrations of the flow fields as well as distributions of local Nusselt numbers for plenty of different geometrical configurations and Reynolds numbers *Re* can be found in several reviews, such as [7, 8, 12, 13] based on experimental and numerical results. Since experiments cannot provide all quantities of the entire flow domain spatially and temporally well resolved, the understanding of the turbulent flow field requires simulations. Existing publications of numerical nature use either turbulence modelling for the closure of the Reynolds-averaged Navier-Stokes (RANS) equations, e.g. [15], or large eddy simulation (LES), e.g. [4]. Almost all available direct numerical simulations (DNS) are either two-dimensional, e.g. [3], or do not exhibit an appropriate spatial resolution in the three-dimensional case,

R. Wilke (✉) • J. Sesterhenn
Fachgebiet Numerische Fluiddynamik, Technische Universität Berlin, Müller-Breslau-Str. 11, 10623 Berlin, Germany
e-mail: robert.wilke@tnt.tu-berlin.de; joern.sesterhenn@tu-berlin.de;
http://www.cfd.tu-berlin.de/

W.E. Nagel et al. (eds.), *High Performance Computing in Science and Engineering '14*, DOI 10.1007/978-3-319-10810-0_19

e.g. [6]. Recent investigations come from Dairay et al [5]. He conducted a DNS of a round impinging jet. Since those computations require tremendous amounts of computing time, their presented results are preliminary and consider the Nusselt number and the friction coefficient only.

This study deals with a direct numerical simulation of a turbulent round impinging jet. According to Hrycak [7], for free jets as well as for impinging jets four different states dependent on the Reynolds number exist. Those states characterise the jet with regard to turbulence, ranging from dissipated laminar ($Re < 300$) to fully turbulent ($Re > 3{,}000$). To ensure simulating a fully turbulent impinging jet, a Reynolds number of $Re = 3{,}300$ is chosen.

2 Numerical Method

The governing Navier-Stokes equations are formulated in a characteristic pressure-velocity-entropy-formulation, as described by Sesterhenn [11] and solved directly numerically. This formulation has advantages in the fields of boundary conditions, parallelization and space discretisation. No turbulence modelling is required since the smallest scales of turbulent motion are resolved. The spatial discretisation uses 6th order compact central schemes of Lele [9] for the diffusive terms and compact 5th order upwind finite differences of Adams et al. [1] for the convective terms. To advance in time a 4th order Runge-Kutta scheme is applied.

The present simulation is conducted on a numerical grid of size $512 \times 512 \times 512$ points for the computational domain sized $L_x \times L_y \times L_z = 12D \times 5D \times 12D$, where D is the inlet diameter, see Fig. 1. A confined impinging jet is characterised

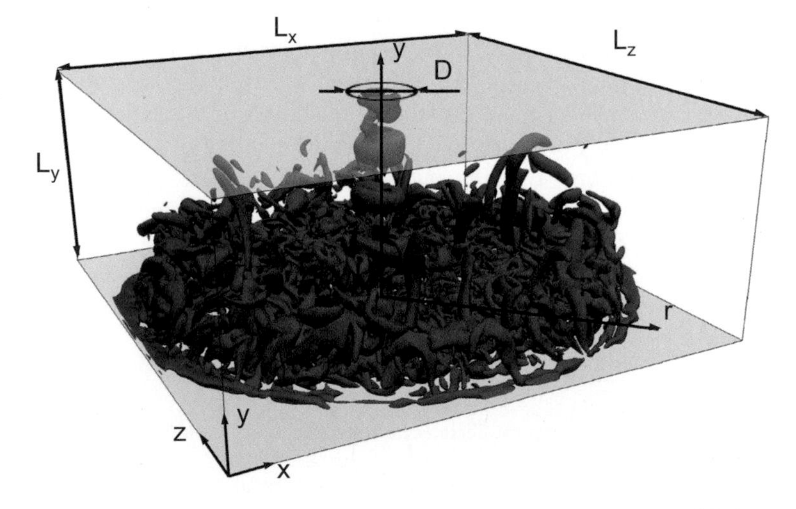

Fig. 1 Computational domain with instantaneous iso-surface of Q-criterion $Q = \frac{1}{2}(P^2 + \omega_{ij}\omega_{ij} - S_{ij}S_{ij})$, $P = -\frac{\partial u_i}{\partial x_i}$, $\omega_{ij} = \frac{1}{2}\left(\frac{\partial u_i}{\partial x_j} - \frac{\partial u_j}{\partial x_i}\right)$, $S_{ij} = \frac{1}{2}\left(\frac{\partial u_i}{\partial x_j} + \frac{\partial u_j}{\partial x_i}\right)$ at $Q = 1{,}000\,\mathrm{m^2 s^{-4}}$

Table 1 Physical parameters of the simulation. $p_o, p_\infty, T_o, T_W, Re, Pr, \kappa, R$ denote the total pressure, ambient pressure, total temperature, wall temperature, Reynolds number, Prandtl number, ratio of specific heats and the specific gas constant

p_o/p	p_∞	T_o	T_W	Re	Pr	κ	R
1.5	10^5 Pa	293.15 K	373.15 K	3,300	0.71	1.4	287 J/(kg K)

by the presence of two walls, the impinging plate and the orifice plate. The grid is refined in those wall-adjacent regions in order to ascertain a maximum value in time and space of the dimensionless wall distance y^+ of the closest grid point to the wall smaller than one for both plates. For the x- and z-direction a slight symmetric grid stretching is applied which refines the shear layer of the jet. The refinements use hyperbolic tangent respectively hyperbolic sin functions and lead to minimal and maximal spacings of $0.017D \leq \Delta x = \Delta z \leq 0.039D$ and $0.0017D \leq \Delta y \leq 0.016D$. The maximum change of the mesh spacing is 0.5 % respectively 1 %. For the circumferential averaging an equidistant grid of size $n_\theta \times n_r = 1{,}024 \times 512$ is applied for each slice at constant height. Table 1 shows the physical parameters of the simulation.

The computational domain is delimited by four non-reflecting boundary conditions, one isothermal wall which is the impinging plate and one boundary consisting of an isothermal wall and the inlet. The walls are fully acoustically reflective. The location of the nozzle is defined using a hyperbolic tangent profile with a disturbed thin laminar annular shear layer as described in [14].

A sponge region is applied for the outlet area $r/D > 5$, that smoothly forces the values of pressure, velocity and entropy to reference values. This destroys vortices before leaving the computational domain. The reference values at the outlet were obtained by a preliminary large eddy simulation of a greater domain.

3 Results and Discussion

3.1 Kolmogorov Scales

The scales of turbulent motion span a huge range from the size of the domain to the smallest energy dissipating ones. Since the turbulent kinetic energy is transferred downwards to smaller and smaller scales, the smallest ones have to be resolved by the numerical grid in order to obtain a reliable solution of the turbulent flow. They are given with the kinematic viscosity ν and dissipation rate ϵ by

$$l_\eta \approx \left(\frac{\nu^3}{\epsilon}\right)^{\frac{1}{4}} \qquad (1)$$

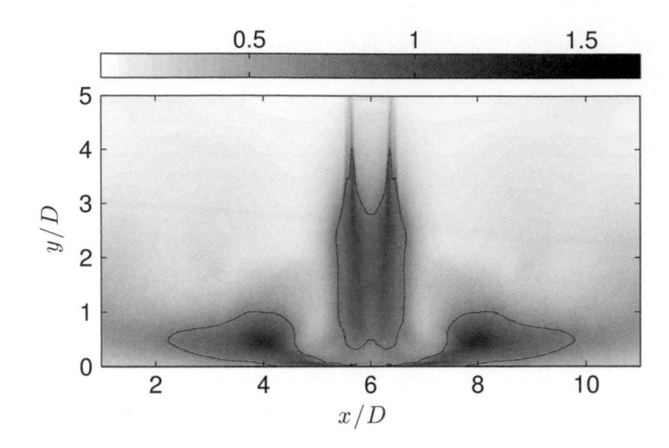

Fig. 2 Grid spacing related to Kolmogorov length scale $(h_x h_y h_z)^{1/3} / l_\eta$. The *solid line* represents the value 1

and are valid for isotropic turbulence, that occurs at sufficiently high Reynolds numbers. The Reynolds number of 3,300 can be considered as low or *not sufficiently high* to obtain isotopic turbulence. Therefore the Kolmogorov microscales provide a conservative clue. The ratio of the mesh width to the Kolmogorov length scale $(h_x h_y h_z)^{1/3} / l_\eta$ is shown in Fig. 2 and reaches a maximum value of 1.6 at the lower wall. For supersonic turbulent boundary layers, Pirozzoli et al. [10] showed that the typical size of small-scale eddies is about $5..6 l_\eta$. A strongly different behaviour for the present boundary layer of high subsonic Mach number is not expected. The maximum ratio in the area of the free jet is $(h_x h_y h_z)^{1/3} / l_\eta = 1.3$ and of the wall jet is $(h_x h_y h_z)^{1/3} / l_\eta = 1.5$.

3.2 Mean and Instantaneous Flow Field

In addition to the criterion due to turbulent motion of the jet, the boundary layer including the viscous sub-layer also has to be resolved appropriately in order to achieve reliable results of the heat transfer at the impinging plate. The maximum dimensionless wall distance

$$y^+ = \frac{u_\tau y}{\nu} \tag{2}$$

of the present simulation occurs at $r/D = 0.46$ and reaches a value of $y^+ = 0.64$. The minimum number of points in the viscous sub-layer $y^+ \leq 5$ is 7 for the entire domain. Figure 3 shows the velocity- and temperature boundary layer profile for different distances from the stagnation point. u^+, u_τ and τ_W are the dimensionless radial velocity, the friction velocity and the wall shear stress:

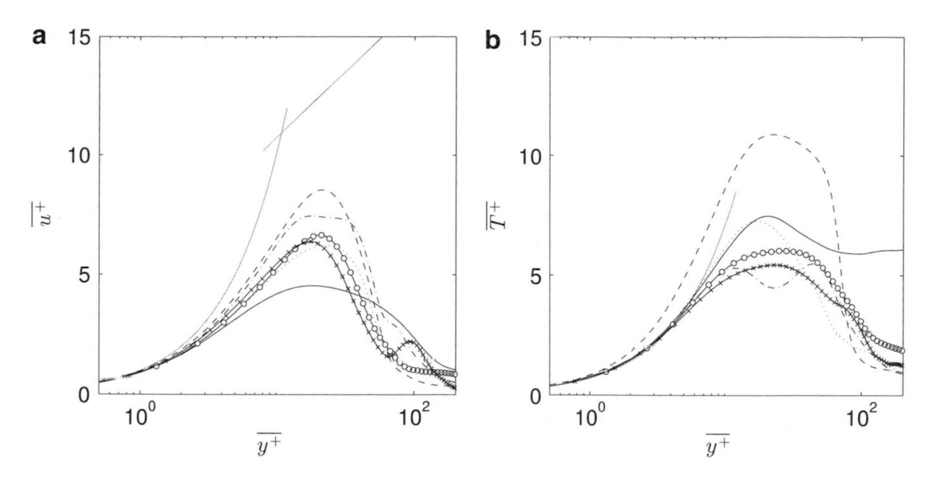

Fig. 3 Mean boundary layer profiles for different radial positions. —— : $r/D = 0.5$, - - - : $r/D = 1.0$; ...... : $r/D = 1.5$; -·-· : $r/D = 2.0$; o-o-o-o : $r/D = 3.0$; ×-×-× : $r/D = 4.5$; —— : $u^+ = y^+$, $u^+ = \ln(y^+)/0.41 + 5.1$ respectively $T^+ = Pr\, y^+$. (**a**) Dimensionless velocity. (**b**) Dimensionless temperature

$$u^+ = \frac{u}{u_\tau} \quad , \quad u_\tau = \sqrt{\frac{\tau_W}{\rho}} \quad , \quad \tau_W = \eta \left(\frac{\partial u}{\partial y} \right)_W \quad . \tag{3}$$

The dimensionless temperature is given by

$$T^+ = \frac{T_W - T}{T_\tau} \quad , T_\tau = \frac{q_W}{\rho c_p u_\tau} \quad , \tag{4}$$

where T_τ is the friction temperature and q_W the wall heat flux.

The main flow characteristics can be compared with experimental results from literature. Figure 4 shows the mean distributions of the radial velocity and temperature for different radial positions and Fig. 5 for different distances from the wall. The maximal mean radial velocity increases from the stagnation point, reaches its maximum value at $r/D = 0.8$ and then decreases. The maximal temperature difference $T - T_o$ increases monotonously with greater radii as a consequence of turbulent mixing of the cold impinging jet and the hot environment. For the same reason, the profile at $r/D = 0.5$ which is located in the shear layer of the jet is at lower temperatures than the other ones.

The hcat transfer at the impinging plate is strongly related to the vortical structures of the turbulent flow field. In the shear layer of the jet (primary) ring vortices develop and grow until they collide with the wall and then stretch and move in radial direction. As soon as the primary toroidal vortex passes the deceleration area of the wall jet $r/D > 0.8$ the flow separates and forms a new secondary counter-rotating ring vortex that enhances the local heat transfer, directly followed by a likewise annular area of poor heat transfer due to separation.

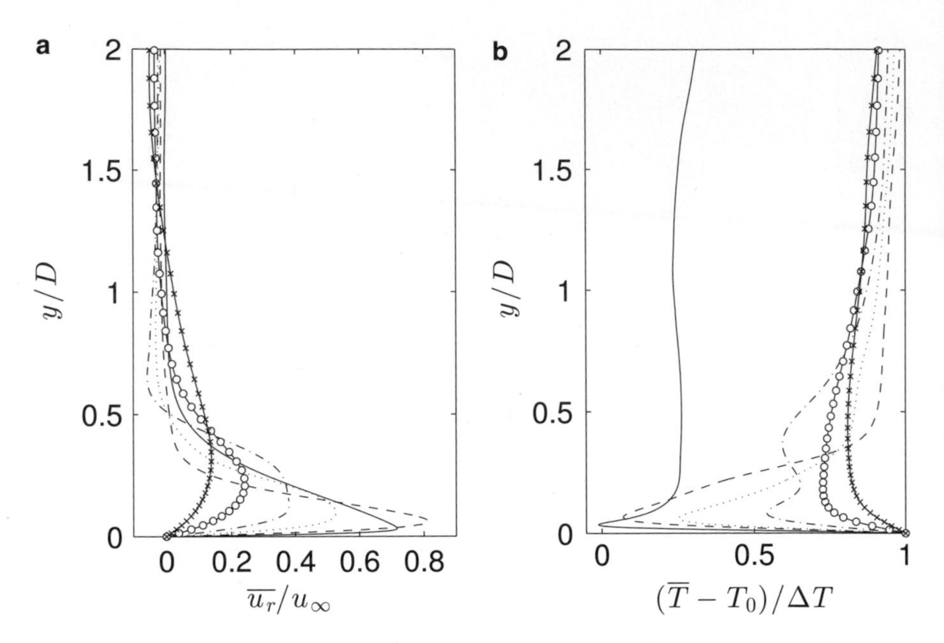

Fig. 4 Mean distributions for different radial positions r/D ___ : $r/D = 0.5$, _ _ _ : $r/D = 1.0$; ...... : $r/D = 1.5$; _._._ : $r/D = 2.0$; o-o-o-o : $r/D = 3.0$; x-x-x-x : $r/D = 4.5$. (**a**) Radial velocity. (**b**) Temperature difference $T - T_o$ relative to $\Delta T = T_W - T_o$

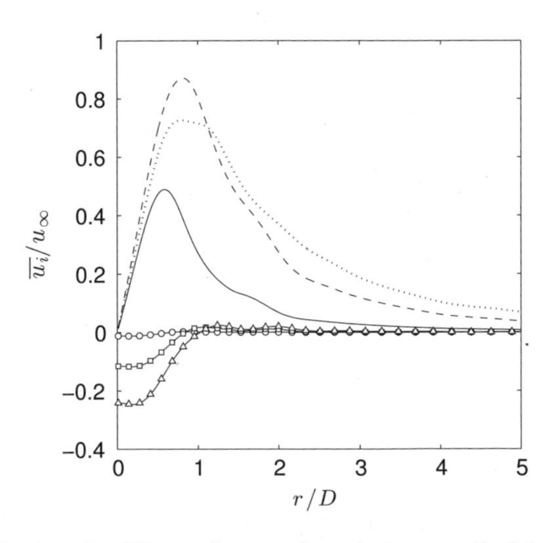

Fig. 5 Mean distributions for different distances from the lower wall of the radial u_r and wall normal velocity component v. ___ : u_r at $y/D = 0.01$, _ _ _ : u_r at $y/D = 0.05$; ...... : u_r at $y/D = 0.1$; _._._ : $r/D = 2.0$; o-o-o-o : $\overline{v}$ at $y/D = 0.01$; □-□-□-□ : $\overline{v}$ at $y/D = 0.05$; △-△-△-△ : $\overline{v}$ at $y/D = 0.10$

Travelling downstream the vortex pair becomes unstable and breaks down into smaller structures that rise. As a consequence the two rings at the wall of very high and low heat transfer vanish and the cycle restarts.

Figures 6 and 7 show the birth of those vortex rings on a x-y plane through the center of the jet. The background pictures the Mach number. Thereon contour lines of same values of the Q-criterion are drawn. The two vertical lines show the position of the maximal radial velocity. In addition, the Nusselt number is represented on the x-z plane at the wall. The first picture of the series is taken at a time step where the primary ring vortex reaches the deceleration area of the wall jet. Beside the stagnation area no strong peak in heat transfer is present. In the next three pictures, the secondary counter-rotating ring vortex is born and travels downstream. A strong annular peak in the Nusselt number can be observed which travels together with the vortex pair until they collapse and the strong heat transfer disappears.

The investigated life cycle leads to a secondary area located at $r/D = 1..1.4$ of high values of the radial Nusselt number

$$Nu(r) = -\frac{D}{\Delta T}\frac{\partial T}{\partial y}(r)\bigg|_W \tag{5}$$

that is shown in Fig. 8a. The primary maximum exists at $r/D = 0.18$.

Buchlin[2] reported the inner or primary maximum at $r/D = 0.7$ and the outer or secondary at $r/D = 2.4$ for experimental data at $Re = 60{,}000, h/D = 1$ with the remark, that the strength of both maxima decrease and the inner one moves to the stagnation point with increased nozzle to plate distances h/D and decreased Reynolds number. Since direct numerical simulations presently cannot reach high Reynolds numbers like $Re = 60{,}000$ of the experiments, no direct comparison between the numeric values can be performed. Nevertheless Buchelins conclusions concerning the phenomenon of secondary vortex rings, the movement of the secondary maximum towards the stagnation point and the loss of strength of both maxima match the results of the present simulation well.

3.3 Turbulent Statistics

Figures 9 and 10 show first results of the Reynolds stresses as well as the turbulent heat fluxes. Thosc results are not completely statistically converged and therefore will be related to the heat transfer and flow phenomena in the final paper.

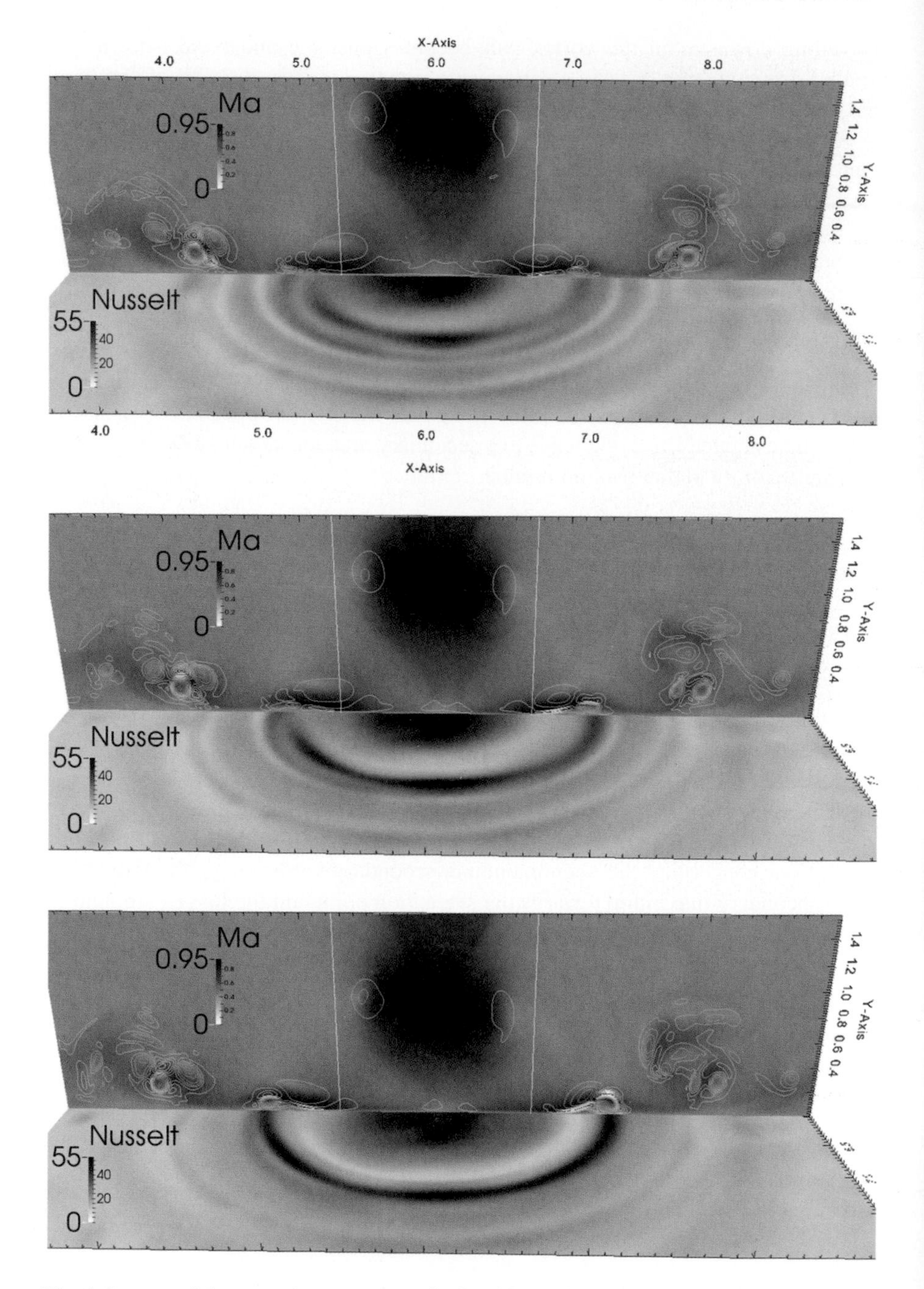

Fig. 6 Impact of the secondary vortex on the local Nusselt number illustrated by contours of equal values of Q above the Mach number Ma on a x-y-plane through the center of the jet and the Nusselt number at the impinging plate (1)

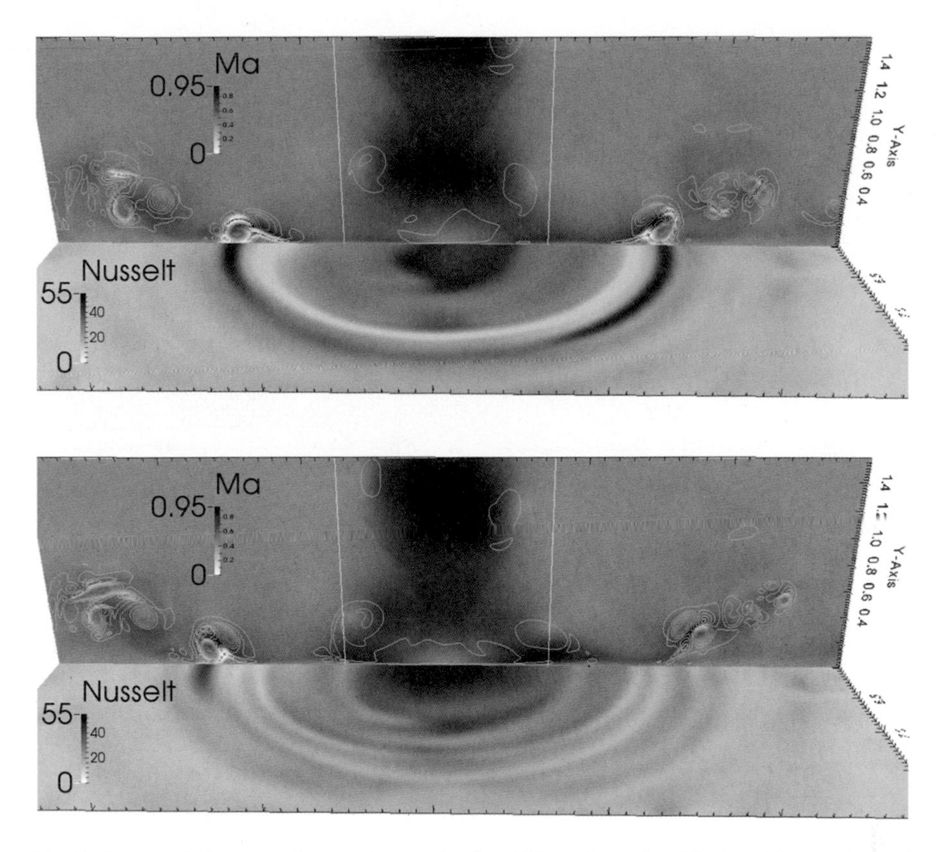

Fig. 7 Impact of the secondary vortex on the local Nusselt number illustrated by contours of equal values of Q above the Mach number Ma on a x-y-plane through the center of the jet and the Nusselt number at the impinging plate (2)

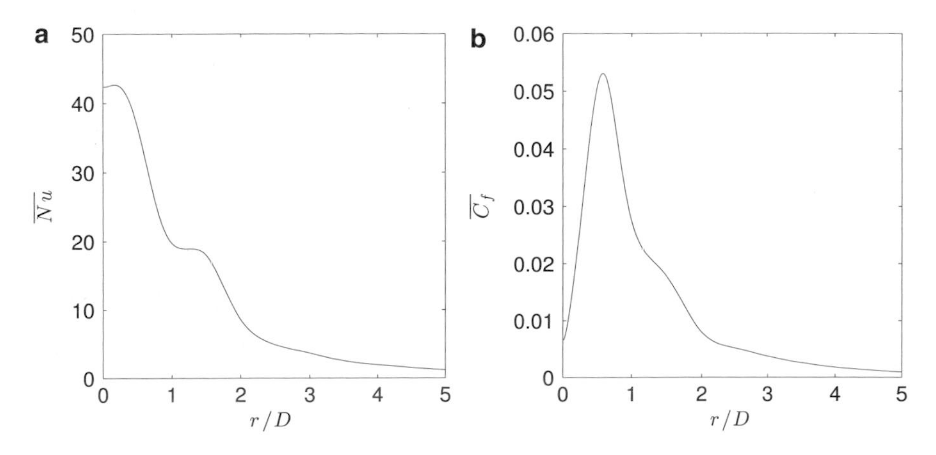

Fig. 8 Mean radial distributions for the local Nusselt number Nu and the skin friction coefficient $C_f = \frac{2\tau_W}{\rho u_\infty^2}$. (**a**) Nusselt number. (**b**) Skin friction coefficient

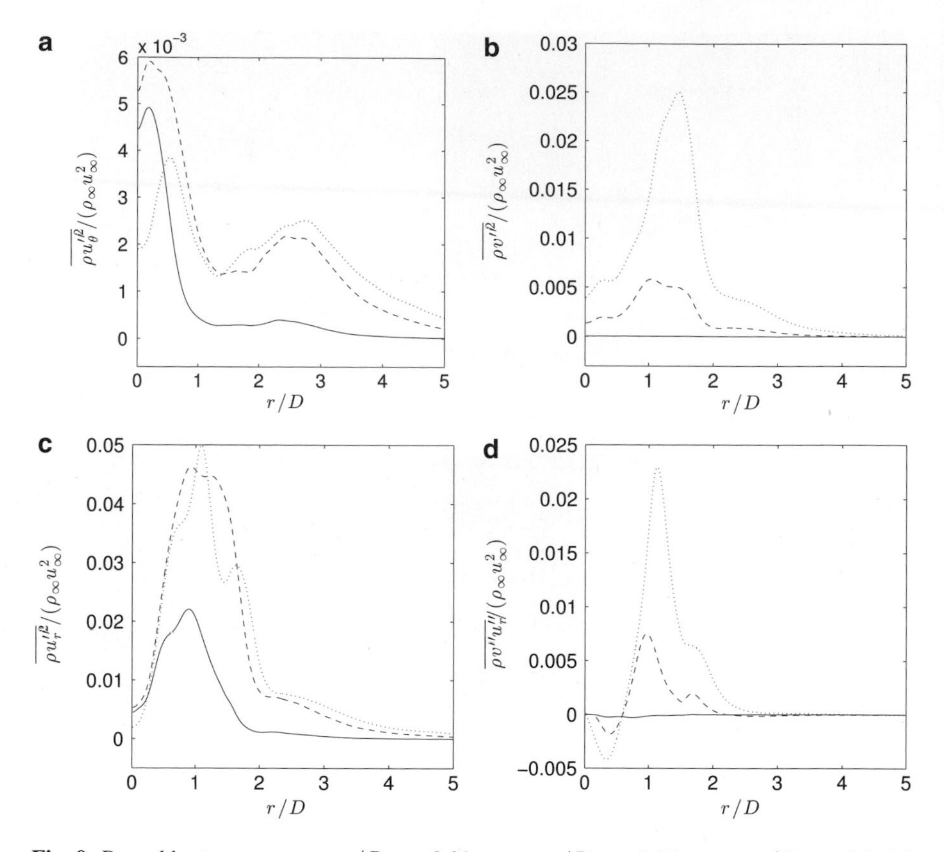

Fig. 9 Reynolds stresses. ——— : $y/D = 0.01$, - - - : $y/D = 0.05$; ...... : $y/D = 0.1$. (**a**) Reynolds stress tensor $RST_{\theta\theta}$. (**b**) Reynolds stress tensor RST_{yy}. (**c**) Reynolds stress tensor RST_{rr}. (**d**) Reynolds stress tensor RST_{yr}

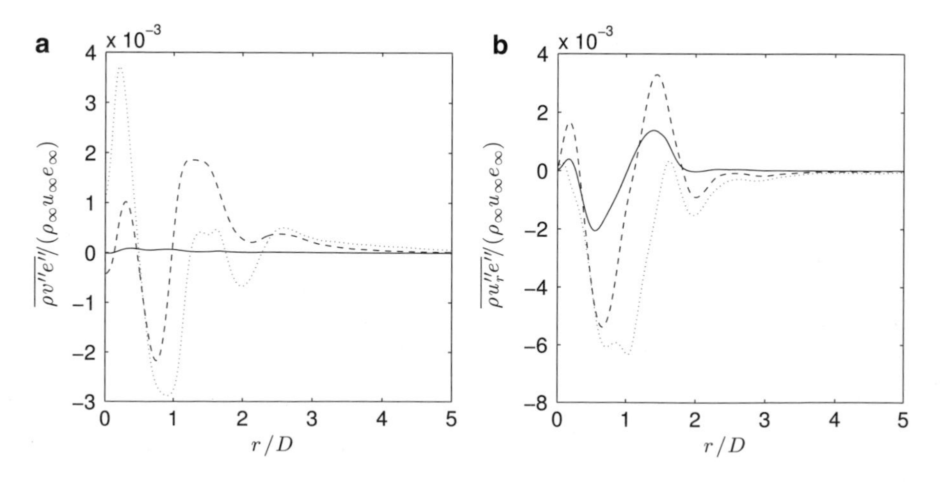

Fig. 10 Turbulent heat fluxes. ——— : $y/D = 0.01$, - - - : $y/D = 0.05$; ...... : $y/D = 0.1$. (**a**) Turbulent heat flux normal to the wall. (**b**) Turbulent heat flux in radial direction

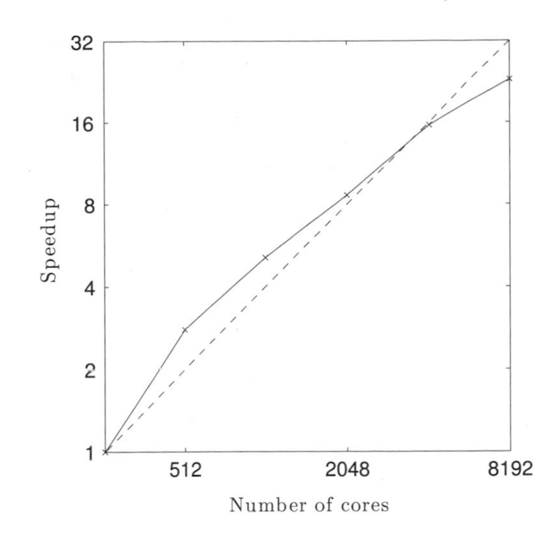

Fig. 11 Scalability of the code; simulations run with 512^3 grid points on CRAY XE6 (Hermit)

4 High Performance Computing

Investigating physics by means of direct numerical simulation require huge computing capacity, which can only be provided by the most powerful high performance computers that are available nowadays. The Kolomgorov's scales that need to be resolved lead to those capacities in the order of much more than a million core hours per computation.

The code is parallelized with MPI libraries. It has been successfully ported to CRAY XE6 (Hermit) and is typically used with $32 \times 16 \times 16 = 8{,}192$ cores. The typical run times of the simulation are 24 h whenever available. The code uses auto-vectorisation. The scalability plot is shown in Fig. 11 and shows nearly perfect linear scaling for the given problem.

5 Conclusion

This paper presents a direct numerical simulation of a subsonic confined impinging jet. The chosen grid size of 512^3 points matches the requirements for reliable results on turbulent heat transfer. The Nusselt number distribution is presented and features a week primary and secondary maximum which is caused by the occurrence of secondary vortex rings that locally increase the heat transfer in an annular shape. Those first results agree with experimental data of [2]. In future work a DNS at $Re = 8{,}000$ will be performed and directly compared to experimental results of our partner project within the Collaborative Research Centre 1029.

Acknowledgements The simulations were performed on the national supercomputer Cray XE6 at the High Performance Computing Center Stuttgart (HLRS) under the grant number GCS-NOIJ/12993.

The authors gratefully acknowledge support by the Deutsche Forschungsgemeinschaft (DFG) as part of collaborative research center SFB 1029 "Substantial efficiency increase in gas turbines through direct use of coupled unsteady combustion and flow dynamics".

References

1. Adams, N.: A high-resolution hybrid compact-ENO scheme for shock-turbulence interaction problems. J. Comput. Phys. **127**, 27–51 (1996). http://dx.doi.org/10.1006/jcph.1996.0156. doi:10.1006/jcph.1996.0156
2. Buchlin, J.: Convective heat transfer in impinging-gas-jet arrangements. J. Appl. Fluid Mech. **4**, 3 (2011)
3. Chung, Y.M., Luo, K.H.: Unsteady heat transfer analysis of an impinging jet. J. Heat Trans. **124**, 12, (6), 1039–1048 (2002). http://dx.doi.org/10.1115/1.1469522. ISBN 0022–1481
4. Cziesla, T., Biswas, G., Chattopadhyay, H., Mitra, N.: Large-eddy simulation of flow and heat transfer in an impinging slot jet. Int. J. Heat Fluid Flow **22**(5), 500–508 (2001). http://dx.doi.org/http://dx.doi.org/10.1016/S0142-727X(01)00105-9. doi:http://dx.doi.org/10.1016/S0142--727X(01)00105--9. ISSN 0142–727X
5. Dairay, T., Fortune, V., Lamballais, E., Brizzi, L.E.: Direct numerical simulation of the heat transfer of an impinging jet. In: 14th European Turbulence Conference, Lyon, Sept 2013. Department of Fluid Flow, Heat Transfer and Combustion, Institute PRIME. CNRS – Universite de Poitiers ENSMA
6. Hattori, H., Nagano, Y.: Direct numerical simulation of turbulent heat transfer in plane impinging jet. Int. J. Heat Fluid Flow **25**(5), 749–758 (2004). http://dx.doi.org/http://dx.doi.org/10.1016/j.ijheatfluidflow.2004.05.004. doi:http://dx.doi.org/10.1016/j.ijheatfluidflow.2004.05.004. ISSN 0142–727X. Selected papers from the 4th International Symposium on Turbulence Heat and Mass Transfer
7. Hrycak, P.: Heat transfer from impinging jets. A literature review. New Jersey Institute of Technology, 1981. Forschungsbericht
8. Jambunathan, K., Lai, E., Moss, M., Button, B.: A review of heat transfer data for single circular jet impingement. Int. J. Heat Fluid Flow **13**(2), 106–115 (1992). http://dx.doi.org/http://dx.doi.org/10.1016/0142-727X(92)90017-4. doi:http://dx.doi.org/10.1016/0142-727X(92)90017-4. ISSN 0142–727X
9. Lele, S.K.: Compact finite difference schemes with spectral-like resolution. J. Comput. Phys. **103**(1), 16–42 (1992). http://dx.doi.org/10.1016/0021-9991(92)90324-R. doi:10.1016/0021–9991(92)90324–R
10. Pirozzoli, S., Bernardini, M., Grasso, F.: Characterization of coherent vortical structures in a supersonic turbulent boundary layer. J. Fluid Mech. **613**(10), 205–231 (2008). http://dx.doi.org/10.1017/S0022112008003005. doi:10.1017/S0022112008003005. ISSN 1469–7645
11. Sesterhenn, J.: A characteristic-type formulation of the Navier-Stokes equations for high order upwind schemes. Comput. Fluids **30**(1), 37–67 (2001). http://dx.doi.org/DOI:10.1016/S0045-7930(00)00002-5. doi:10.1016/S0045–7930(00)00002–5. ISSN 0045–7930
12. Viskanta, R.: Heat transfer to impinging isothermal gas and flame jets. Exp. Therm. Fluid Sci. **6**(2), 111–134 (1993). http://dx.doi.org/http://dx.doi.org/10.1016/0894-1777(93)90022-B. doi:http://dx.doi.org/10.1016/0894-1777(93)90022-B. ISSN 0894–1777
13. Weigand, B., Spring, S.: Multiple jet impingement – a review. Heat Trans. Res. **42**(2), 101–142 (2011). ISSN 1064–2285

14. Wilke, R., Sesterhenn, J.: Direct numerical simulation of heat transfer of a round subsonic impinging jet. In: Active Flow and Combustion Control 2014. Notes on Numerical Fluid Mechanics and Multidisciplinary Design Conference. Springer, Cham (2014)
15. Zuckerman, N., Lior, N.: Impingement heat transfer: correlations and numerical modeling. J. Heat Trans. **127**(5), 544–552 (2005). http://dx.doi.org/10.1115/1.1861921. ISBN 0022–1481

Near-Wake Behavior of Discrete-Roughness Arrays in 2-d and 3-d Laminar Boundary Layers

Holger B.E. Kurz and Markus J. Kloker

Abstract A spanwise row of medium-sized cylindrical roughness elements is introduced to two different base flows, a 2-d boundary layer and a respective 3-d boundary layer, as can be found on unswept and swept wings, respectively, both with strongly negative streamwise pressure gradient. For both cases, sub-effective and super-effective, i.e. directly flow tripping, elements are investigated. Similarities and differences in the flow fields around the roughness elements are highlighted. It turns out that, when comparing the results in a coordinate system aligned with the local flow direction at the boundary layer edge, the qualitative differences in the near wakes are rather subtle. Nevertheless, the elements in the 3-d boundary layer trigger steady crossflow vortices that will ultimately become secondarily unstable, and therefore lead at first to a continuously upstream-shifting transition location with increasing roughness height. Performance data for our direct numerical simulation code *NS3D* are given for the CRAY XE6 and the current test and development system CRAY XC40.

1 Introduction

First investigations of roughness elements in 2-d boundary layers date back to the middle of the past century (see, inter alia, [3] and [9]), where the capability of elements with a height sufficient to cause transition to turbulence directly downstream has been identified. However, the vortex dynamics involved are complex, and little is known about their sensitivity to the properties of the surrounding boundary layer. Especially favourable pressure gradients, as present in the front region on real wing profiles, stabilize the laminar boundary layer substantially for non-swept wings. In case of a linearly stable boundary layer one may assume that a larger roughness is needed to directly trip turbulent flow as in a swept-wing boundary layer, being unstable with respect to steady disturbances (crossflow vortex modes)

H.B.E. Kurz (✉) • M.J. Kloker
Institute of Aerodynamics and Gas Dynamics, University of Stuttgart, Pfaffenwaldring 21, D-70550 Stuttgart, Germany
e-mail: kurz@iag.uni-stuttgart.de; kloker@iag.uni-stuttgart.de

W.E. Nagel et al. (eds.), *High Performance Computing in Science and Engineering '14*, DOI 10.1007/978-3-319-10810-0_20

induced by surface non-uniformities. The present investigation focuses on near-effective roughness elements in a 2-d and a 3-d boundary layer, both on a wing profile in the region of negative pressure gradient.

2 Computational Setup

2.1 Direct Numerical Simulations

We solve the unsteady compressible, three-dimensional Navier-Stokes equations in non-dimensional conservative formulation

$$\frac{\partial Q}{\partial t} + \frac{\partial F}{\partial x} + \frac{\partial G}{\partial y} + \frac{\partial H}{\partial z} = 0 \quad , \tag{1}$$

with the solution vector $Q = (\rho, \rho u, \rho v, \rho w, E)^T$ and the flux vectors F, G and H. A calorically perfect gas is assumed, with constant specific heat ratio $\kappa = 1.4$ and Prandtl number $Pr = 0.71$; the dynamic viscosity is implemented as a function of the local temperature following Sutherland's law. For further details and the description of boundary conditions, see [1]. Direct numerical simulations (DNS) are performed with our high-order finite difference (FD) inhouse code *NS3D*, employing both explicit and compact difference schemes of 8th and 6th order, respectively. For enhanced performance on massively parallel computers, a subdomain compact scheme combining the accuracy of compact schemes in the interior of each subdomain with the computationally efficient explicit formulation at (overlapping) sub-domain boundaries is implemented, reducing idle time compared to the fully compact scheme (see [7]). Derivative computation takes place in a uniformly spaced cartesian computational space (σ, θ, ζ) for optimal performance. The actual grid geometry is considered via a general grid transformation, converting the internal derivatives $(\frac{\partial \Phi}{\partial \sigma}, \frac{\partial \Phi}{\partial \theta}, \frac{\partial \Phi}{\partial \zeta})$ to physical values, i.e.

$$\frac{\partial \Phi}{\partial x} = f_1\left(\frac{\partial \Phi}{\partial \sigma}, \frac{\partial \Phi}{\partial \theta}, \frac{\partial \Phi}{\partial \zeta}\right) \tag{2}$$

$$\frac{\partial \Phi}{\partial y} = f_2\left(\frac{\partial \Phi}{\partial \sigma}, \frac{\partial \Phi}{\partial \theta}, \frac{\partial \Phi}{\partial \zeta}\right) \tag{3}$$

$$\frac{\partial \Phi}{\partial z} = f_3\left(\frac{\partial \Phi}{\partial \sigma}, \frac{\partial \Phi}{\partial \theta}, \frac{\partial \Phi}{\partial \zeta}\right) \tag{4}$$

Higher derivatives are treated analogously. For a detailed description of the method and its accuracy on non-uniform grids, see [2]. Evaluation of (2)–(4) is rather costly for the general full 3-d case, therefore a simplification of the relations for specific geometries can reduce the computational effort for the transformation without sacrificing the accuracy of the scheme, see Sect. 4.2. The temporal integration is accomplished by the classical 4th-order Runge-Kutta scheme.

2.2 *Base Flow*

The flow past an infinite span NACA 671-215 profile is investigated for two different sweep angles $\beta = 0.0°$ and $\beta = 35.0°$, where details on the latter case can also be found in [8]. The chord length $\bar{L} = 2.0$m and the angle of attack $\alpha = -6.1°$ are fixed in both cases, as is the free-stream Reynolds number $Re_L = 6.9 \cdot 10^6$ based on the chord length and the velocitiy component paralled to the chord, U_∞. The velocity component in spanwise direction is W, and the combination of U and W yields the total free stream velocity Q, cf. Fig. 1. Dimensional values are denoted by an overbar. $\bar{L}$ is used as reference length for the global cartesian coordinates x, y and z, as well as for the tangential and wall-normal coordinates ξ and η. For all other lengths, e.g. the roughness properties, the displacement thickness of the 3-d base flow at the position of the roughness $x_R = 0.02$ is used, $\delta_{1,s,R} = 8.5 \cdot 10^{-5}$. All other quantities are non-dimensionalized using the respective free-stream values. Additional base flow properties for both setups are given in Table 1.

The steady base flows are computed with the Reynolds-Averaged Navier-Stokes (RANS) finite-volume solver TAU, using a hybrid grid. The boundary layer is resolved by a structured grid block on the profile surface, while the circular farfield

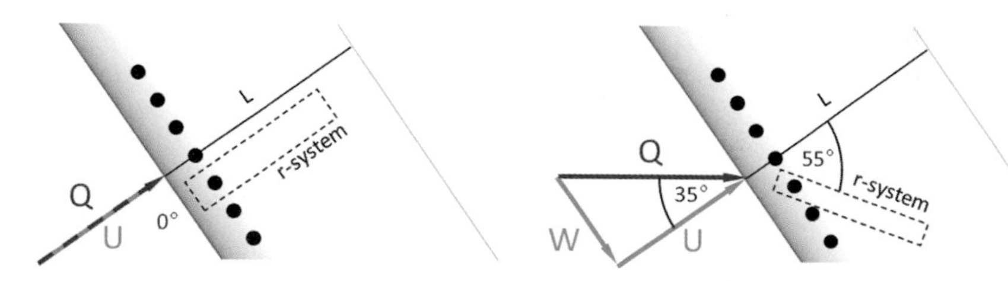

Fig. 1 Top-down views with free-stream velocities of the 2-d and the 3-d case. The former results from the latter by ommiting the spanwise component W. *Black circles* indicate periodic roughness arrays. *Dashed boxes* illustrate position and orientation of the r-system, used for flow visualizations of the near wakes

Table 1 Free-stream properties of the 2-d and the 3-d setup

	2-d setup	3-d setup
β	0.0°	35.0°
Ma_U	0.53	0.53
Ma_Q	0.53	0.65
Re_L	$6.9 \cdot 10^6$	$6.9 \cdot 10^6$
Q	157 m/s	192 m/s
U	157 m/s	157 m/s
W	0 m/s	110 m/s

domain is discretized by unstructured prisms and has a radius of 100 chord lengths. The same grid is used for the 2-d and the 3-d setup, since the boundary-layer thicknesses differ not to much. Transition is prescribed at $x = 0.2$ on both upper and lower side of the profile, while the e^N-transition prediction tool gives $x = 0.1$ on the upper and $x = 0.2$ on the lower profile side in case of the swept-wing base flow. For the subsequent DNS, only the laminar boundary layer region on the upper side is of interest. Small oscillations at the wall, primarily in the pressure, in the converged solution are then removed by a 10th-order filter (see [5]), before the flow field is transferred to a finer grid for the DNS using a 6th-order Lagrangian interpolation method, exploiting the ordered data structure in the boundary-layer block from the RANS simulation. The highly-resolved DNS grid is confined to $0.015 \leq x \leq 0.0325$ and has a spanwise extent of $\lambda_z = 0.002$, using periodic boundary conditions in the spanwise direction. The total resolution for the 2-d base flow is $N_x \times N_y \times N_z = 987 \times 188 \times 128 = 24 \cdot 10^6$ and $1{,}258 \times 188 \times 128 = 30 \cdot 10^6$ grid points for the 3-d case.

A perspective view of the base-flow configuration is depicted in Fig. 2, including the c_p-distribution and selected streamlines.

2.3 *Perturbed Flow: Roughness Elements*

The flow past localized roughness elements is simulated with the DNS code *NS3D*. For a careful verification study for 3-d boundary layers, see [4]. The roughness

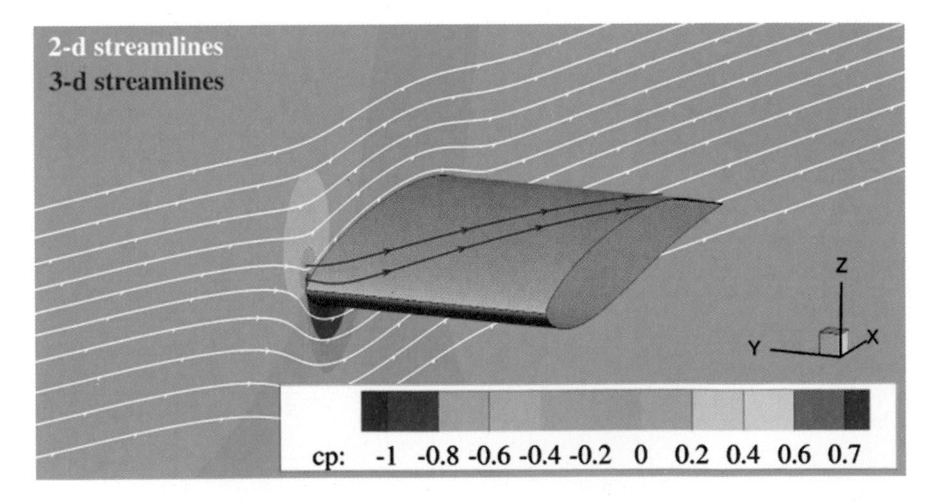

Fig. 2 Perspective view of the infinite-span wing profile in the 3-d setup. The *contour* shows the distribution of the pressure coefficient c_p in a chordwise plane and corresponding in-plane streamlines in *white* (W set to zero). 3-d streamlines near the boundary layer edge are shown in *blue*

Table 2 Shape parameters of investigated roughness elements

k	d	r_a	S_R	Slope angle
1.3750	5.882	0.2338	2.0	45.0°
2.0000	5.882	0.3400	2.0	45.0°
2.5000	5.882	0.4250	4.0	63.4°
3.0000	5.882	0.5100	4.0	63.4°

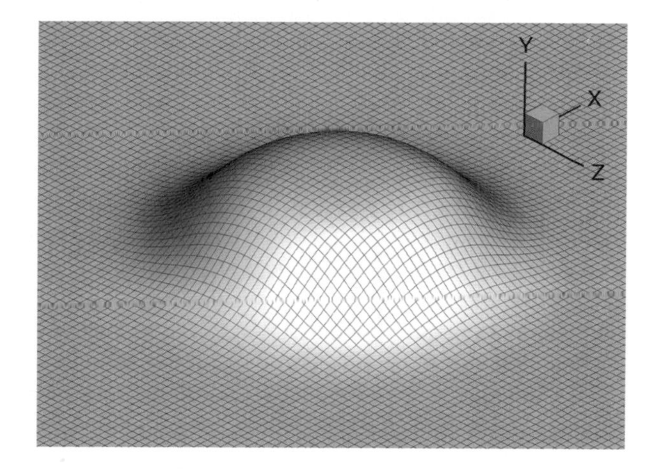

Fig. 3 Meshed roughness element, obtained by a smooth elevation of the gridded surface. The shown element has a height of $k = 2.0$, corresponding to $Re_{kk} = 881$ in the 3-d boundary layer

element constituting the spanwise array under consideration is defined by the smooth shape function

$$\eta_R(r,\phi) = \frac{k}{2}\left(\tanh\left(\frac{S_R}{k}\left(\frac{d}{2} - r\right)\right) + 1\right), \qquad r \geq 0, \quad 0 \leq \phi < 2\pi \quad , \tag{5}$$

with height k and nominal diameter d; S_R/k controls the slope angle of the element. The subscript R refers to values at the roughness. The shape parameters for the elements investigated in the present study are summarized in Table 2, including the aspect ratio $r_a = k/d$. The geometry is then transferred to the curved surface of the wing profile and placed at the chordwise position $x_R = 0.02$. A meshed roughness element is shown in Fig. 3.

The effect of the roughness elements on the boundary layer are characterized by the Reynolds number

$$Re_{kk} = \frac{\bar{k} \cdot |\mathbf{V}(k)| \cdot \bar{\rho}(\bar{k})}{\bar{\mu}(\bar{k})} \quad , \tag{6}$$

with the velocity vector $\bar{\mathbf{V}}$, the density $\bar{\rho}$ and the dynamic viscosity $\bar{\mu}$, both taken from the unperturbed base flow at the height of the roughness $\bar{k}$. This Reynolds number is taken as a measure for the roughness being sub- or super-effective. The present study is designated to determine whether a unique value $Re_{kk,eff}$ for direct flow tripping exists that is valid for both 2-d and 3-d base flows.

Table 2 shows shape parameters for roughness elements introduced to the 2-d and 3-d base flow. For the former, larger heights k are needed to obtain Re_{kk} values comparable to the elements for the 3-d base flow. This is due to the smaller velocity magnitude at the edge of the 2-d boundary layer, resulting from the lack of a spanwise velocity component, cf. Fig. 1 and Eq. (6).

3 Results

We now investigate roughness elements in the effective range and compare results for the 2-d and the 3-d base flow. We start with sub-effective heights, where the roughness elements induce steady vortices without amplifying strongly unsteady background disturbances such that they grow large enough to trigger transition. At larger element heights, the near-wake behavior changes from convective to a global instability, where unsteady disturbances emerge directly downstream of the roughness elements due to a timewise growth in this region up to saturation.

Note that below the effective roughness height, a second characteristic height is referred to as *critical* height. It constitutes the boundary between completely ineffective and transition-influencing height. For swept boundary layers, the critial height can be substantially smaller than the effective height.

For visualization purposes, the flow fields are transformed to the r-system, depicted in Fig. 1 for both investigated base flows. The origin in these systems is defined in the center of the included roughness element, the respective downstream coordinate is denoted ξ_r. In the 3-d case, the r-system is rotated about the arbitrary angle 55° to align approximately with the local flow direction.

3.1 *Sub-effective Roughness Elements*

A pulse-like unsteady disturbance background is forced by means of a disturbance strip at the wall, some roughness diameters upstream of the element, to have controlled unsteady "background" disturbances. The wall-normal mass flux ρv is prescribed, and the pulse-like behavior is achieved by superposition of a fundamental frequency ω_0 with 49 higher harmonics, cf. inset in Fig. 4. The fundamental frequency corresponds to one of the most amplified travelling crossflow modes in the 3-d base flow. In the 2-d base flow, the disturbances are constant along the spanwise coordinate z, in the 3-d boundary layer the distribution is modulated with one spanwise harmonic.

Figure 4 shows the downstream amplitude development of steady and unsteady streamwise velocity disturbances in the investigated boundary layers, obtained from a temporal Fourier analysis of one fundamental period. The Fourier analysis is performed for every grid point, then the maximum value from every η-z-plane is taken and plotted against the streamwise coordinate. The vertical lines indicate

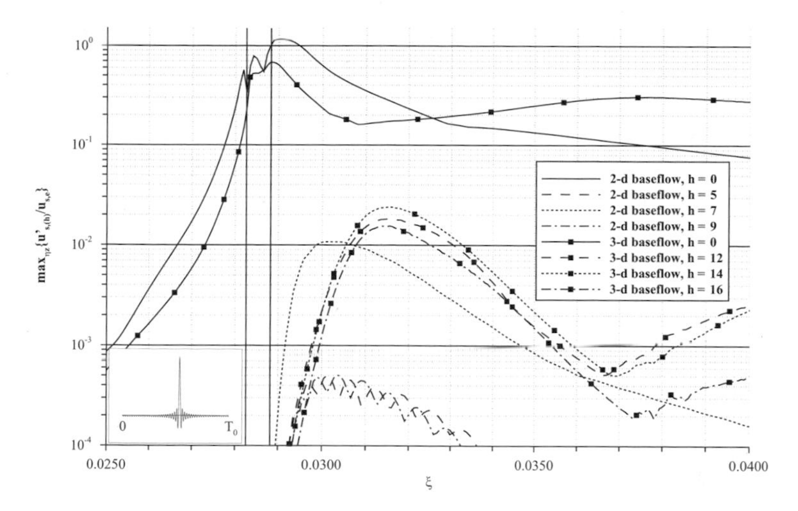

Fig. 4 Downstream development of selected steady and unsteady modal amplitudes for sub-effective roughness elements in a 2-d and a 3-d boundary layer, plotted against the tangential coordinate ξ. 2-d results without symbols, 3-d results with squares. The *inset* shows a time signal of the pulsing amplitude

the roughness-element position. In the 2-d boundary layer, the steady disturbance, represented by $h = \omega/\omega_0 = 0$, reaches a maximum over the roughness element, but then decays; the vortices in the near wake relax into so-called (u-) streaks that decay less strong and may be even transiently amplified. In the 3-d case, a similar shoot-up is observed at the roughness. However, shortly downstream, the amplitude starts to grow and settles at a level of about 30 %. This observation is in accordance with results from linear stability theory (LST), predicting exponential amplification of zero-frequency (crossflow vortex) modes in the 3-d boundary layer, while no such amplification is present in the 2-d case. In the latter, only a transient, algebraic growth of high- and low-velocity streaks can appear.

Selected unsteady disturbance amplitudes are also given in Fig. 4. For the 2-d base flow, one distinct frequency $h = 7$ grows to a level of $1 \cdot 10^{-2}$ right behind the element, but then undergoes a decay again. The amplitudes of all other frequencies show a similar behavior, however at a lower level. Again, the results for the 3-d boundary layer seem to resemble these findings at first glance. However, the growth rate is somewhat lower, but since the region of amplification is longer, the maximum amplitude is higher than in the 2-d boundary layer. At $\xi = 0.0370$, unsteady disturbances are amplified again, a consequence of a secondary instability induced by the large amplitude steady disturbances.

Altogether, the near wake of the investigated roughness elements gives rise to an amplification of unsteady background disturbances in both base flows under consideration. In the observed cases with sub-effective roughness elements, the amplitudes gained are too small to influence the near wake and the vortex structures observed.

A closer inspection shows that any unsteady disturbance amplitudes shoot up *behind* the roughness element – no amplification is observed at the actual roughness position.

The flow field behind sub-effective roughness elements is illustrated by streamlines in Fig. 5. The *right column* shows an element with $k = 2.500$, $Re_{kk} = 537$, in the 2-d base flow. A set of 10 streamlines is introduced upstream of the element at the heights $0.10k$, $0.45k$ and $0.60k$, where k is the height of the roughness. At the lowest height, the roughness displaces the streamlines to the left and the right of the element. The displacement is preserved further downstream. At $0.45k$, the formation of a horseshoe vortex in front of the element is indicated by a roll-up of the flow. The inner streamlines turn inwards behind the element, and roll up in a recirculation zone. When starting at $0.60k$, the flow does not enter the recirculation zone anymore. All streamlines are perfectly symmetric with respect to the chordwise symmetry plane through the roughness center.

In the right column, a roughness with $k = 1.375$, $Re_{kk} = 487$ is introduced in the 3-d boundary layer. Note the basic crossflow is from the lower right to the upper left corner of the plot. It appears that the lower regions of the recirculation zone are fed by fluid coming from the right of the element, while fluid from the left enters the zone only at a larger wall distance. However, the roll up in front of the element and the formation of a recirculation zone can be observed again. We conclude that the 3-d base flow leads to a distinct asymmetry in the near wake of the element.

The formation of vortices is visualized in Fig. 6 by isosurfaces of $\lambda_2 = -20{,}000$. The isosurfaces are colored with the streamwise vorticity, where light gray color indicates a clockwise rotation sense when looking downstreams; black structures rotate counter-clockwise. In both cases, the formation of a horseshoe vortex right upstream of the element can be observed.

Behind the element in the 2-d base flow, two dominant pairs of counter rotating vortices appear downstream of the complex vortex system surrounding the roughness. The outer pair represents the extended legs of the horseshoe vortex, while the inner pair is a result of the roll-up of streamlines behind the element, cf. Fig. 5. The outer pair pushes high-momentum fluid from the upper region of the boundary layer towards the wall in between, whereas the inner pair lifts slow fluid away from the wall due to its reverse rotation sense. As an effect, a pronounced low-speed streak appears in the downstream center line of the roughness, as can be seen from the velocity contours plotted in crosscuts in Fig. 6. Such low speed streaks are known to be prone to instability and may be the reason for the amplification of unsteady disturbances in the near wake, in addition to the recirculation zone behind the element.

In the 3-d boundary layer, the vortex "pairs" clearly are asymmetric, meaning that vortices with a rotation sense corresponding to the basic crossflow direction near the wall are supported, while the other vortex legs are suppressed. Here, the deformation of the underlying boundary layer is not as distinct. Recall the lower growth rate for unsteady disturbances in the 3-d boundary layer in the near wake. Note also that the recirculation region topology is virtually the same for the 2-d and the 3-d base flow (not shown).

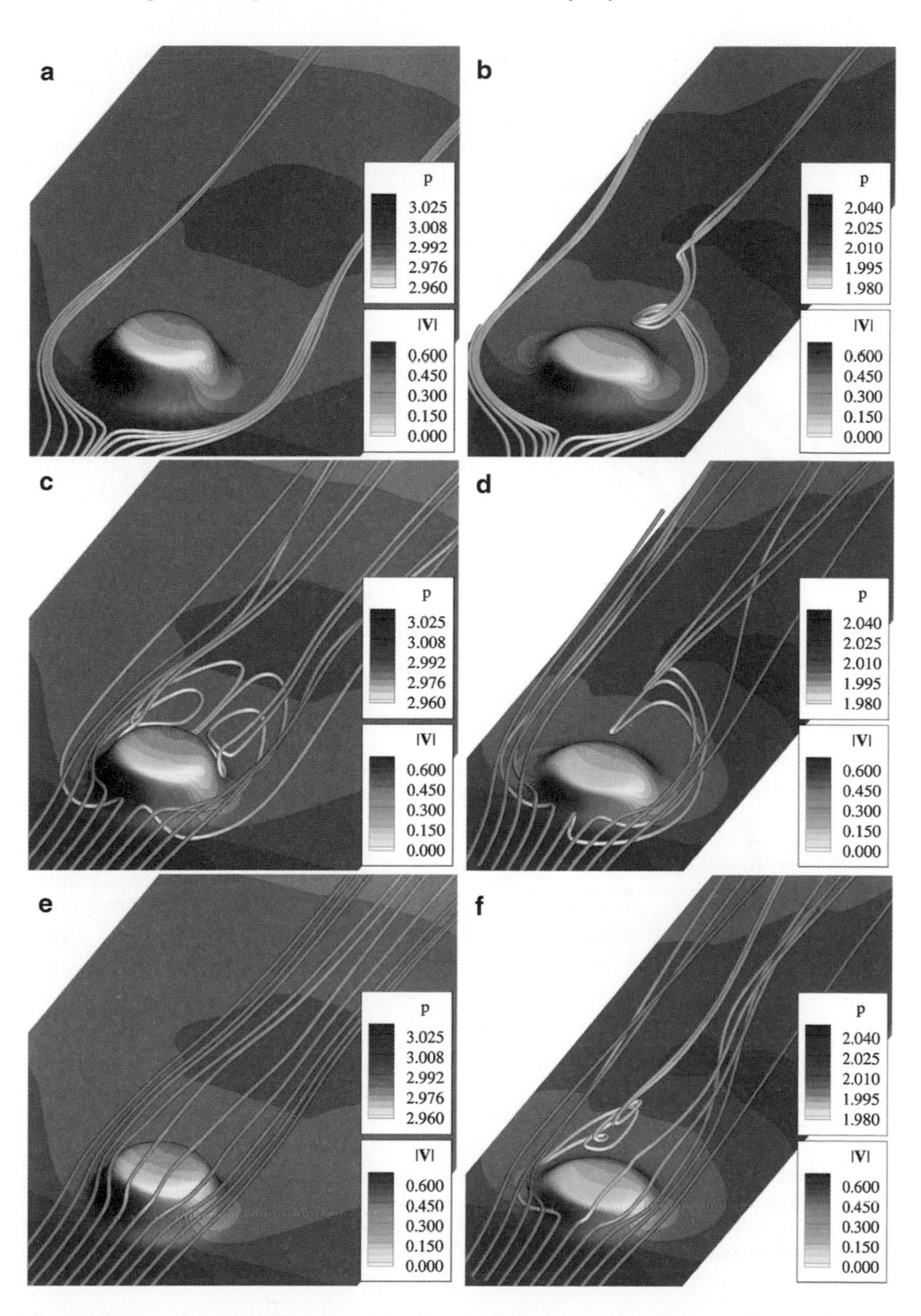

Fig. 5 Wall pressure and streamlines introduced at different heights of the roughness elements. Streamlines are colored with the local velocity magnitude. *Left column*: 2-d base flow, $k = 2.500$, $Re_{kk} = 537$; right column: 3-d base flow, $k = 1.375$, $Re_{kk} = 487$. *First row*: Streamlines at $0.10k$; *second row*: streamlines at $0.45k$; *third row*: streamlines at $0.60k$

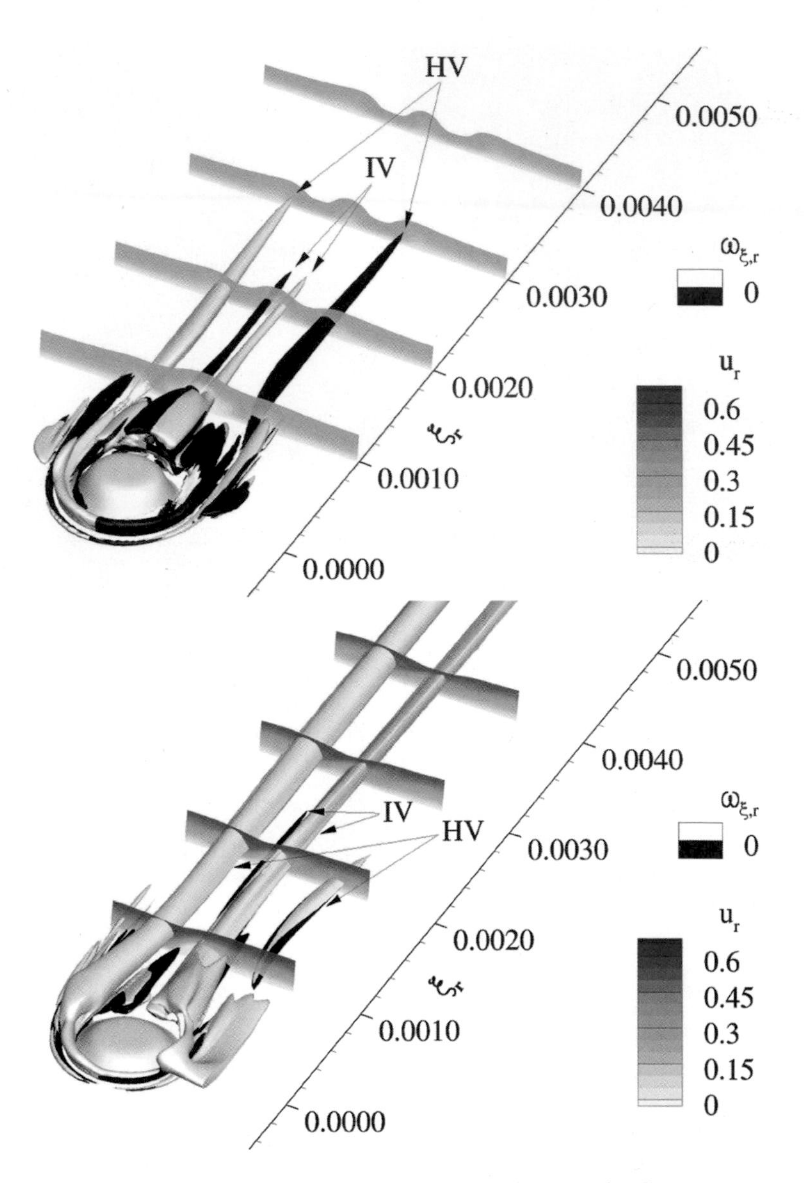

Fig. 6 Vortex systems generated by sub-effective roughness elements, visualized by isosurfaces at $\lambda_2 = -20{,}000$. *IV* and *HV* denote inner vortex pair and horseshoe vortex legs, respectively. *White* color indicates clockwise rotation sense, *black* color counter-clockwise rotation. The *slices* show contours of the streamwise veloctiy component u_r. *Upper*: 2-d base flow, $k = 2.500$, $Re_{kk} = 537$. *Lower*:3-d base flow, $k = 1.375$, $Re_{kk} = 487$

In order to connect the above findings with the results of the temporal Fourier analysis, isosurfaces of the modal amplitudes corresponding to the most amplified frequencies are illustrated in Fig. 7. For the 2-d and 3-d boundary layer, we pick $h = 7$ and $h = 14$, respectively. For the visualization, the temporal amplitude

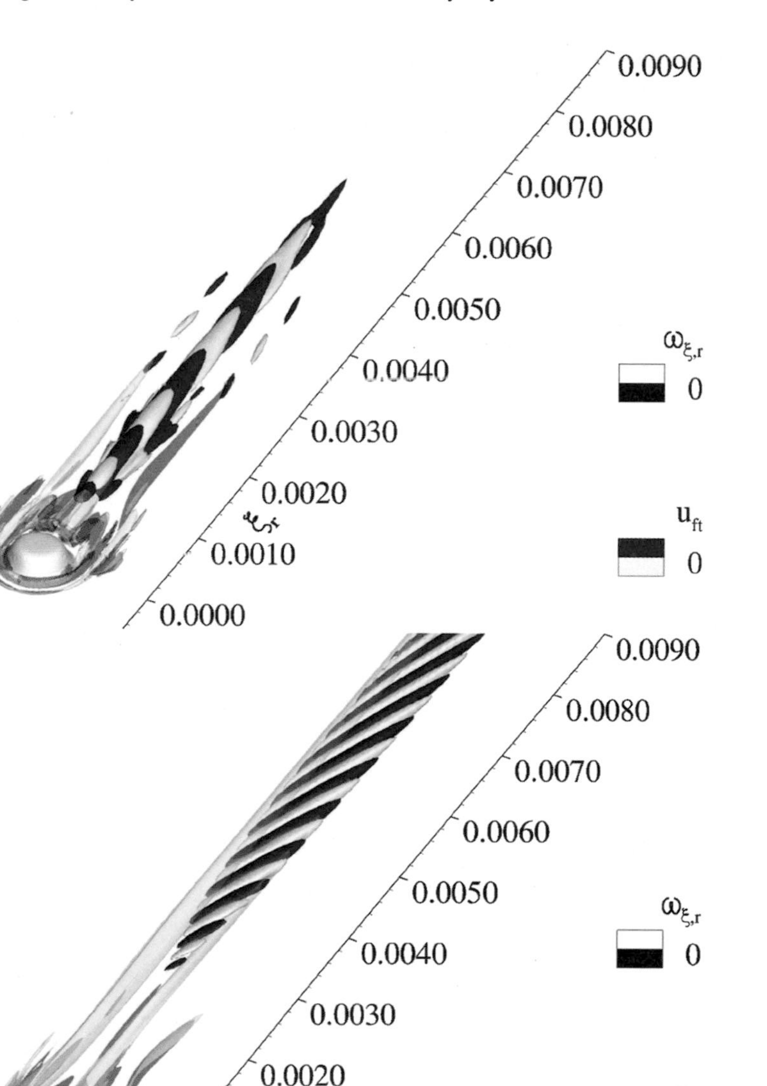

Fig. 7 Isosurfaces of $u_{ft} = u'_{s,(h)} \cdot \cos(\phi) = \pm 0.10$, respresenting a "snapshop" of the global amplitude distribution. *Upper*: 2-d base flow, $k = 2.500$, $Re_{kk} = 537$, $h = 7$. *Lower*: 3-d base flow, $k = 1.375$, $Re_{kk} = 487$, $h = 14$

found for every grid point is multiplied by the cosine of the corresponding phase, $u_{ft} = u'_{s,(h)} \cdot \cos(\phi)$, yielding a "snapshot" of the disturbance distribution at an arbitrary instant in time. Subsequently, the result is normalized by the maximum value found in the investigated domain, and isosurfaces are plottet at $u_{ft} = \pm 0.10$.

In the 2-d base flow, symmetric isosurfaces appear in two regions. At the decaying legs of the horseshoe vortex, and in the downstream centerline of the roughness, while the latter are dominant. The structures sit right on top of the inner vortex pair, and therefore in the low speed streak, indicating a connection of the instability and the low speed streak. For a clarification, however, a stability analysis based on bi-global LST with 2-d eigenfunctions in the crosscut planes is required.

The respective result for the 3-d boundary layer for $h = 14$ shows the growth of oblique structures close to the wall, originating from the downstream centerline of the roughness. The longer region of instability is clearly visible. Note that the normalization is done with a larger value than in the 2-d case (cf. Fig. 4), disabling a quantitative comparison of the onset of the growth.

3.2 Super-Effective Roughness Elements

For an investigation of directly flow tripping roughness elements, we choose $k = 3.000$, $Re_{kk} = 646$ for the 2-d boundary layer and $k = 2.000$, $Re_{kk} = 881$ for the 3-d base flow. Although the roughness Reynolds number of the latter is considerably higher, the observed effects are similar as illustrated below. Figure 8 shows the temporal Fourier results for both base flows for the same frequencies as in Fig. 4. The steady amplitudes $h = 0$ are larger for the 3-d boundary layer; behind the element, the amplitudes remain at a high level, only a slight decay is observed.

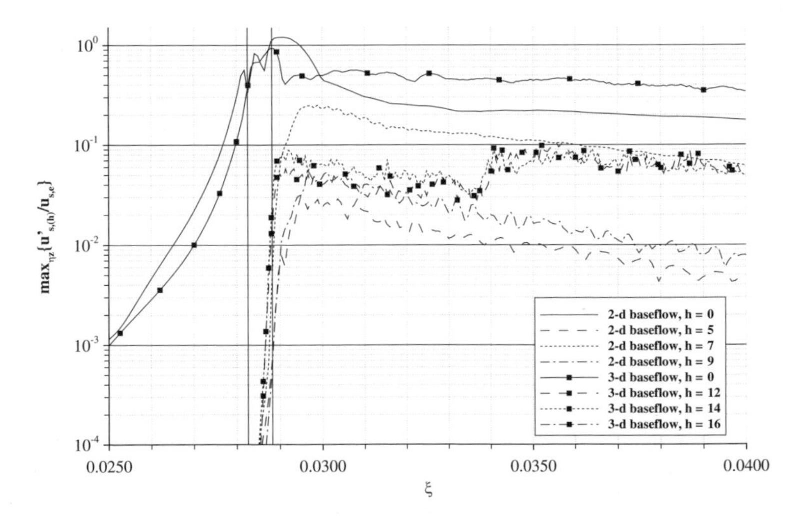

Fig. 8 Downstream development of selected steady and unsteady modal amplitudes for super-effective roughness elements in a 2-d and a 3-d boundary layer, plotted against the tangential coordinate ξ. 2-d results without symbols, 3-d results with squares

Unsteady disturbances now increase in amplitude right at the roughness position; the sharp rise is most likely due to a global instability in the recirculation zone near downstream of the roughness elements, since the rise seems too sharp for a convective growth. Also the fact that the amplification seems to take place right at the roughness indicates a global instability in the vicinity of the roughness. Note that for the super-effective roughness elements, no controlled background disturbances were introduced.

In the 2-d case, the frequency $h = 7$ again reaches the highest amplitudes, while other harmonics stay at lower levels. After the sharp rise behind the roughness, the amplitudes decay. This is due to the favorable pressure gradient, stabilizing the 2-d boundary layer and eventually leading to a relaminarization of the flow further downstream. Also note the rather low Reynolds number based on the local displacement thickness, reaching a value of $Re_{\delta1,R} \approx 170$ at the roughness position.

Unlike the 2-d base flow, the roughness in the boundary layer on the swept wing yields a flat disturbance spectrum, indicated by the fact that all shown frequencies yield similar amplitudes. At $\xi = 0.0340$, the disturbances show a second phase of growth, and then remain at approximately 5–10 %.

Figure 9 illustrates the vortex structures found behind the roughness elements under consideration. For both base flows, the roughness elements trigger turbulence. The broader disturbance spectrum in the 3-d case is manifested by smaller and more complex vortex structures.

4 Computational Aspects

4.1 General Performance Analysis for NS3D on CRAY XE6 and CRAY XC40

The direct numerical simulations for the present study were conducted with our DNS code *NS3D* on the massively parallel CRAY XE6 *Hermit* at HLRS. Parallelization is accomplished by a hybrid approach, where the computational domain is decomposed along the streamwise and the wall-normal direction into smaller subdomains and distributed over compute nodes. Communication is accomplished by means of MPI, while the spanwise direction is parallellized via the shared-memory approach OpenMP.

NS3D employs a time-accurate FD scheme to solve the non-linear Navier-Stokes equations. The time-advancing scheme first computes the temporal derivate for every grid point based on a sum of spatial derivatives, then applies a Runge-Kutta scheme to advance the flow field in time. As mentioned in Sect. 2.1, *NS3D* incorporates both 8th-order excplicit and 6th-order compact FD discretization for computation of the spatial derivatives. Despite the formally lower order, the compact scheme has proven a higher accuracy and stability of the scheme for numerious computational setups and is therefore the preferred method. For the compact

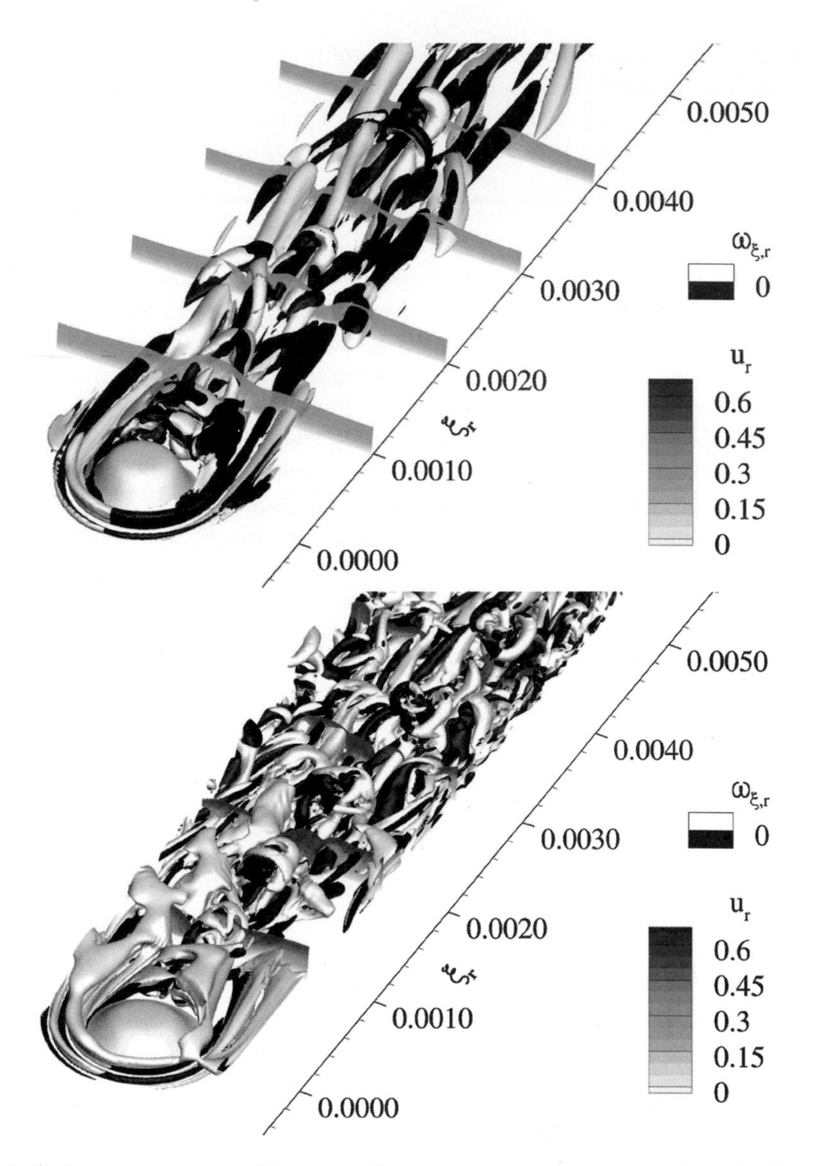

Fig. 9 Vortex systems generated by super-effective roughness elements, snapshot, visualized by isosurfaces at $\lambda_2 = -20{,}000$. *White* color indicates clockwise rotation sense, *black* color counter-clockwise rotation. The *slices* show contours of the streamwise veloctiy component u_r. *Upper*: 2-d base flow, $k = 3.000$, $Re_{kk} = 646$. *Lower*: 3-d base flow, $k = 2.000$, $Re_{kk} = 881$

FDs, tri-diagonal linear systems have to be solved at every time step in order to compute the spatial derivatives of the flow quantities. Performance optimization for parallelized simulations of the employed Thomas-algorithm is accomplised by a pipelining approach explained in Babucke [1], however with a rising number of sub-domains, the method suffers from increasing idling time, cf. Fig. 10. The

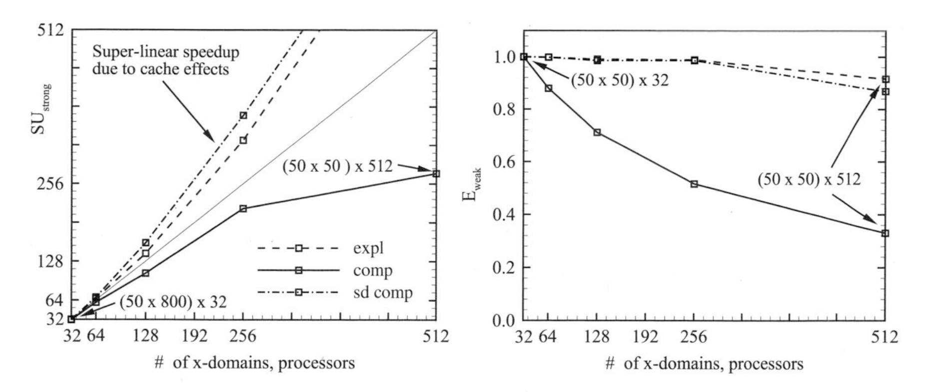

Fig. 10 Speed-up SU_{strong} for strong scaling (*left*) and parallel efficiency E_{weak} for weak scaling (*right*) versus the number of used processors. The curves are obtained with *NS3D* on CRAY XE6 at HLRS for explicit, compact and sub-domain compact finite difference discretization; the *solid line* without symbols indicates the ideal speed-up. *Numbers in brackets* indicate the number of grid points in y- and x-direction, respectively, for each sub-domain (From [6])

explicit scheme resolves the parallelization problem by eliminating the need for a tri-diagonal linear system; derivatives for every grid point can be computed independently from other points.

In order to combine the accuracy of the compact scheme with the computational efficiency of the explicit counterpart, a subdomain-compact scheme was implemented. Here, the interior of each sub-domain is treated with compact finite differences, while the global tri-diagonal system is split up by employing explicit differences near sub-domain boundaries. In fact, all derivatives in a sub-domain can be computed by compact finite differences, since the explicit derivatives are computed on overlapping ghost points and are only used as left-hand-side input for the compact scheme (cf. [7]). Thus, every sub-domain can be treated without the need to wait for neighboring processes.

A quantitative evaluation of the parallel performance of *NS3D* has been accomplished by two series of speed-up tests performed on CRAY XE6. A 2-d setup with only one point in the spanwise direction was used, thus only MPI parallelization was evalutated in the absence of OpenMP parallelization. Two types of speed-ups are of interest, weak scaling and strong scaling. For the former, up to 512 sub-domains with resolution $N_x \times N_y = 50 \times 50$ where lined up in streamwise direction. The parallel efficiency $E_{weak} = t_0/t_n$ is a measure for the dependence on the number of processes and should stay close to unity, meaning the wall-time should be only a weak function of the number of processes. The reference time t_0 was obtained from a run with 32 MPI tasks, t_n is the wall-time when using n MPI tasks. For strong scaling tests, a domain with $N_x \times N_y = 25{,}600 \times 50$ was decomposed to up to 512 x-domains, preserving the total number of points. Here, the performance is measured in terms of the speed-up $SU_{strong} = t_0/t_n$. The latter is a measure for the saving of wall-time when distributing the domain to an increasing number of processors.

Table 3 Specific CPU time per grid point and time step

	CRAY XE6 *Hermit* (μs)	CRAY XC40 *Hornet*, Installation phase 1 step 2a (μs)
Full 3-d grid transformation	47.7	12.8
Specific 3-d grid transformation	29.7	9.5

Results for all three discretization schemes are given in Fig. 10. The compact scheme shows poor performance for both the strong and the weak scaling, while the strong scaling reveals a super-linear speed-up for the explicit and the sub-domain compact scheme. This behavior can be explained with cache effects. The parallel efficiency obtained from the weak scaling test for the explicit and the sub-domain compact scheme remains at a high level up to 256 x-domains, then it decreases to approximately 0.9. Note the slightly lower value for the sub-domain compact scheme.

The performance of our DNS code has also been investigated on the new test and development system CRAY XC40 *Hornet*, Installation phase 1 step 2a. The test system features one cabinet with 164 compute nodes, each with 16 Intel SandyBridge 2.6 GHz processors. Test runs on a single domain with $N_x \times N_y \times N_z = 37 \times 47 \times 128$ with a full 3-d grid transformation (cf. Sect. 2.1) and an optimized transformation, accounting for the uniform spacing in z and the plane z-planes, $\frac{\partial \Phi}{\partial z} = f_3\left(\frac{\partial \Phi}{\partial \zeta}\right)$, were performed on both CRAY XE6 and CRAY XC40. The results for the specific CPU time per grid point and time step are summarized in Table 3. On the CRAY XE6, the optimized transformation reduces the computational costs to 62 % of the general transformation, on CRAY XC40 the simulation time still goes down to 74 %.

Comparing the absolute values for the specific CPU time, the new CRAY XC40 test and development system is 3.7 times faster than the current system CRAY XE6, with the specific grid transformation it is 3.1 times faster. These results were obtained without specific code optimizations for the new CRAY XC40.

4.2 Performance Analysis for the Present Numerical Study

The simulations for the present numerical study were performed on the CRAY XE6 using the sub-domain compact scheme and the specific 3-d grid transformation presented in Sect. 4.2. The setup for the 2-d base flow used $N_x \times N_y \times N_z = 987 \times 188 \times 128$ points and was decomposed to 84 sub-domains with a resolution of $N_x \times N_y \times N_z = 47 \times 47 \times 128$ points. Each subdomain and its respective MPI process was assigned to one compute node, and 32 OpenMP threads were used to treat the spanwise direction, summing up to a total of 2,688 CPUs. The 3-d setup was simulated on a larger domain with a total of $N_x \times N_y \times N_z = 1{,}254 \times 188 \times 128$

grid points and $N_x \times N_y \times N_z = 37 \times 47 \times 128$ per sub-domain, distributed over 136 compute nodes. The total number of processors was 4,352 CPUs, again with 32 OpenMP threads per node. The 2-d setup yielded a specific CPU time per grid point and time step of 39.3 μs, the 3-d setup 35.5 μs. We note that these numbers vary by some percent from run to run owing to the the periodic output of restart files and its dependence on the present file system loading.

5 Conclusions

A spanwise row of discrete cylindrical roughness elements was placed in a 2-d unswept and a 3-d swept-wing boundary layer with strong negative streamwise pressure gradient. The focus of the investigation was on the differences and similarities appearing in the near wake. Two heights were investigated in both cases, a sub-effective and a (super-) effective case. A Fourier analysis of the unsteady disturbances in the flow fields was performed, and the vortex systems forming around the elements were visualized.

For the sub-effective roughness elements, a recirculation zone behind the elements is identified by means of streamline visualizations. While the flow in the 2-d boundary layer is fully symmetric, the streamlines in the 3-d base flow reveal an asymmetry both in spanwise and wall-normal direction. Both flows support the amplification of unsteady background disturbances in the region downstream of the roughness. In both cases, two dominant vortex systems are identified: (i) a horseshoe vortex rolling up in front of the element, with two trailing legs. In the 3-d boundary layer, only the leg with the rotation sense pertaining to the basic crossflow direction near the wall is supported, the other leg is suppressed shortly downstream; (ii) an inner vortex pair originating from the recirculation zone: again, the 3-d base flow supports one of the counter-rotating vortices more than the other.

The main difference for the 3-d boundary layer is the triggering of steady streamwise crossflow vortices even by rather small, sub-effective roughness elements. These vortices become secondarily unstable at a certain downstream position near their saturation, depending on their inital amplitude. Thus, transition is at first continuously shifted upstream by increasing (sub-effective) roughness height. Prior to the effective height, a second but confined convectively unstable region develops downstream the roughness in the near wake.

For super-effective heights, the flow becomes turbulent right behind the roughness. For both base flows, there are indications for a global instability right behind the roughness elements. The actual value $Re_{kk,eff}$ for the roughness being effective can not be determined exactly by means of the presented data, however based on the similarities found for the sub-effective elements, the respective values for 2-d and 3-d boundary layers may be very close.

Acknowledgements This work has been supported by the European Commission through the FP7 project 'RECEPT' (Grant Agreement no. ACPO-GA-2010-265094). Further, the provision of computational resources by the Federal High Performance Computing Center Stuttgart (HLRS) within the project 'LAMTUR' is gratefully acknowledged.

References

1. Babucke, A.: Direct numerical simulation of noise-generation mechanisms in the mixing layer of a jet. Dissertation, Institut für Aerodynamik und Gasdynamik, Universität Stuttgart (2009)
2. Babucke, A., Kloker, M.J.: Accuracy analysis of the fundamental finite-difference methods on non-uniform grids. Internal report . http://www.iag.uni-stuttgart.de/people/markus.kloker/text/IAG-09_FD-nonuniformgrid_babu-klok.pdf (2009)
3. Dryden, H.L.: Review of published data on the effect of roughness on transition from laminar to turbulent flow. J. Aeronaut. Sci. Inst. Aeronaut. Sci. **20**(7), 477–482 (1953)
4. Friederich, T.A., Kloker, M.J.: Control of the secondary cross-flow instability using localized suction. J. Fluid Mech. **706**, 470–495 (2012)
5. Gaitonde, D.V., Visbal, M.R.: High-order schemes for navier-stokes equations: algorithm and implementation into FDL3DI. Technical report, DTIC Document, Air Force Research Laboratary Wright-Patterson AFB (Ohio), 1998
6. Keller, M., Kloker, M.J.: Direct numerical simulations of film cooling in a supersonic boundary-layer flow on massively-parallel supercomputers. In: Resch, M.M., Bez, W., Focht, E., Kobaysahi, H., Kovalenko, Y. (eds.) Sustained Simulation Performance 2013: Proceedings of the Joint Workshop on Sustained Simulation Performance, University of Stuttgart (HLRS) and Tohoku University, 2013, pp. 107–128. Springer, Cham (2013)
7. Keller, M., Kloker, M.J.: DNS of effusion cooling in a supersonic boundary-layer flow: influence of turbulence. In: 44th AIAA Thermophysics Conference, San Diego. AIAA Paper, pp. 2013–2897 (2013)
8. Kurz, H.B.E., Kloker, M.J.: Effects of a discrete medium-sized roughness in a laminar swept-wing boundary layer. In: Dillmann, A., et al. (eds.) New Results in Numerical and Experimental Fluid Dynamics IX. Notes on Numerical Fluid Mechanics and Multidisciplinary Design, vol. 124, pp. 173–181. Springer, Cham (2014)
9. Tani, I., Komoda, H., Komatsu, Y., Iuchi, M.: Boundary-layer transition by isolated roughness. Aeron. Res. Inst. Univ. Tokyo Rep. **375**, 129–143 (1962)

Towards Large Multi-scale Particle Simulations with Conjugate Heat Transfer on Heterogeneous Super Computers

Gonzalo Brito Gadeschi, Christoph Siewert, Andreas Lintermann, Matthias Meinke, and Wolfgang Schröder

Abstract We present numerical methods based on hierarchical Cartesian grids for the simulation of particle flows of different length scales. These include Eulerian-Lagrangian approaches for fully resolved moving particles with conjugate heat-transfer as well as one-way coupled Lagrangian particle models for large-scale particle simulations. The domain decomposition of all phases involved is performed on a joint hierarchical Cartesian grid where the individual cells can belong to one or more sub-grids discretizing different physics, such that numerical methods can operate independently on these sub-sets of the joint mesh to solve, e.g., the Navier-Stokes equations, the heat equation, or the particle motion. Due to the wide range of length scales involved, we first demonstrate the scalability of our automatic mesh generation approach. We then proceed to detail the method for fully-resolved particle simulation and the first steps towards its porting to heterogeneous supercomputers. Finally, we detail the parallelization strategy for the particle motion used by large scale one-way Lagrangian particle simulations.

1 Introduction

The study of particle-laden turbulent flows [4] is of high interest in fields as diverse as the study of water droplet growth within vapor clouds [18], the combustion process of pulverized coal [3], or the debris deposition during electrical discharge machining [12]. Still, fluid-particle flows are very hard to investigate experimentally and numerically due to the multi-scale nature of the problems. For instance, the range of scales in the rain formation process spans from the size of aerosol particles (micrometers) to the cloud length (kilometers).

Since the particle diameters are typically small compared to the fluid scales, their motion is governed by the smallest scales of turbulence requiring the use of computationally expensive direct numerical simulations [26]. Still, to accurately

G. Brito Gadeschi (✉) • C. Siewert • A. Lintermann • M. Meinke • W. Schröder
Institute of Aerodynamics, RWTH Aachen University, Aachen, Germany
e-mail: office@aia.rwth-aachen.de

W.E. Nagel et al. (eds.), *High Performance Computing in Science and Engineering '14*, DOI 10.1007/978-3-319-10810-0_21

predict, e.g., the ignition point and heat release of the coal combustion process or the energy transfer during electrical discharge machining the conjugate heat transfer between the particles and the fluid needs to be considered such that besides the smallest scales of turbulence, the local flow field around the particles themselves also needs to be resolved [18]. The required computational effort thus restricts fully resolved particle simulations in turbulent flows to $O(10^3)$ particles on modern supercomputers, which is at least two orders of magnitude less than the minimum required to gather accurate statistics [11].

Combined multi-scale approaches are a promising solution strategy to tackle multi-scale problems like fluid-particle flows. In this work, Eulerian-Lagrangian approaches for fully-resolved particle simulations involving conjugate heat transfer are developed as a tool to validate and improve Lagrangian models that can be applied for large-scale simulations in order to pave the way for simpler, even larger-scale Eulerian particle models. First, the scalable mesh generation within the framework of hierarchical Cartesian grids is introduced. Then, the hierarchical mesh is extended to handle multiple phases that can be solved by different methods independently. These techniques are demonstrated with fully-resolved Eulerian-Lagrangian particle simulations involving conjugate heat transfer between a moving body and a fluid. Afterwards, the scalability within a single heterogeneous computing node for one of the methods is shown. Finally, a Lagrangian point-particle model is introduced and utilized in large scale simulations to gather statistics about the particle motion in turbulent flows.

2 Massively Parallel Automatic Grid Generation

To model the smallest scales of turbulence in large-scale particle simulations, high grid resolutions are required. This involves the creation of unstructured grids with billions of cells; a very time consuming task if manual grid generation is used. An automatic hierarchical Cartesian mesh generator employing a Hilbert space-filling curve for optimal domain decompositioning [14] enables the creation of very large grids ($O(10^{11})$ cells) in a matter of seconds. The Hilbert ordering ensures optimal intra- and inter-node cell locality due to the self-similarity of the curve.

The scalability of the proposed approach [14] is demonstrated on the IBM BlueGene/Q system (JUQUEEN) at the Jülich Supercomputing Center (JSC). Strong-scaling results are shown in Fig. 1 for the generation of a grid consisting of $9.82 \cdot 10^9$ cells. The parallelization uses MPI and scales linearly as long as each core is in charge of generating about $3 \cdot 10^5$ cells or more. The approach is described in [14], where additionally strong and weak scalings on the CRAY XE6 system (Hermit) at the High Performance Computing Center Stuttgart (HLRS) are shown. A detailed analysis of the thermal flow in the human nasal cavity using this approach is presented in [13].

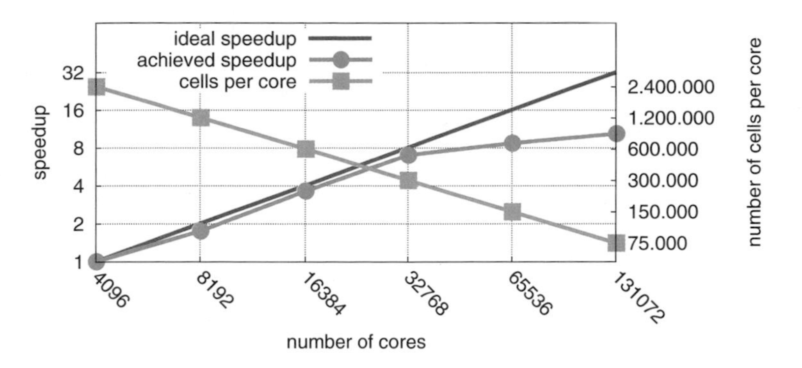

Fig. 1 Results of the strong scaling for the generation of a grid consisting of $9.82 \cdot 10^9$ cells on the JUQUEEN supercomputer at the JSC in Jülich. The ideal speed up (*red*), the actual achieved speedup (*green*), as well as the number of cells generated per core (*blue*) are shown

3 Fully-Resolved Simulation of Moving Particles Involving Conjugate Heat Transfer

For multiphysics simulations, all grids involved in the simulation are stored within a single hierarchical Cartesian data structure. Each solution method is then applied independently to subsets of cells describing different grids. The connectivity information between these grids is maintained implicitly, facilitating adaptive mesh refinement due to, e.g., moving boundaries when multiple phases are involved. The domain decomposition using a space-filling curve is performed in the joint hierarchical mesh containing all grids, resulting in optimal communication patterns between the different solution methods since adjacent cells remain topologically close to each other in memory even if they belong to different grids.

The data layout employed by our approach is illustrated in Fig. 2, where the cells of the hierarchical mesh are colored depending on the solver grid they belong to. The state required for each grid cell is then stored locally by the corresponding solution method. A bidirectional map between the cells in the hierarchical mesh and the cell data within the solvers allows to compactly store the solver state and to sort the solver data to maximize spatial and temporal memory locality, resulting in significant performance improvements on memory bound applications.

The hierarchical mesh also offers excellent complexity guarantees for computing connectivity information between the different grids, since the operations required are generally linear in the number of levels within the balanced tree, and thus *at worst* logarithmic in the number of mesh cells.

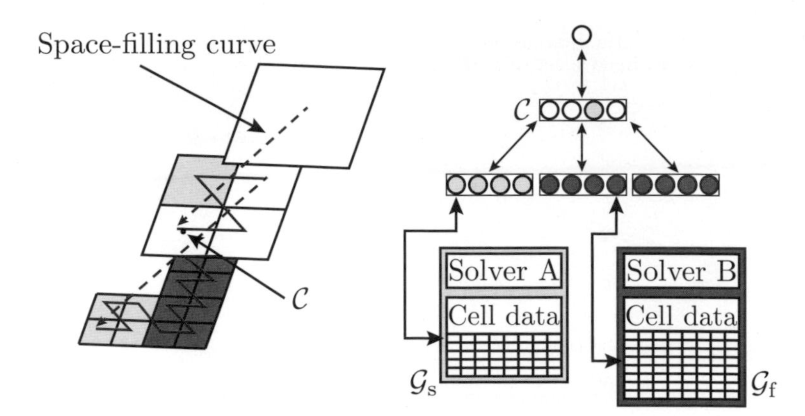

Fig. 2 (*Left*) Cells belonging to the fluid grid (*dark grey*), the solid grid (*light grey*), and empty cells belonging to no physical domain (*white*). The space-filling curve is depicted in (*red*). (*Right*) Data layout of the hierarchical Cartesian grids. The graph of the hierarchical Cartesian grid as well as the links between the grid cells and the cell data of the different solvers is shown

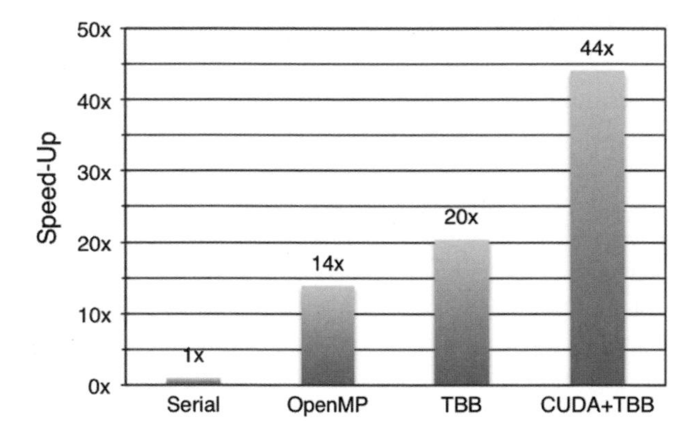

Fig. 3 Single node performance in a computing node with 2 Intel Xeon E5-2650 running at 2 Ghz and two nVidia Tesla K20x (only one Tesla card is used)

3.1 Shared Memory Parallelism

Shared memory parallelism is provided by the Thrust library [9]. This allows the development of generic numerical methods that can efficiently target single core, multi-core, and heterogeneous systems in a unified way. The single-node performance is demonstrated on a node equipped with two Intel Xeon E5-2650 running at 2 Ghz and two nVidia Tesla K20x. The performance between the OpenMP back-end (using guided scheduling), the Intel Thread Building Blocks (TBB) back-end, and nVidia CUDA back-end is compared in Fig. 3, where only one of the two available Tesla cards is used for the CUDA back-end.

The OpenMP back-end scales almost linearly with the 16 cores available. The TBB back-end is able to exploit the multi-threading capabilities of the CPUs and achieves a 1.25x performance increase over the ideal scaling for the 16 cores available. Still, offloading the whole computation to a single accelerator offers a 2x speed-up with respect to what TBB achieves for the full node. Larger speed-ups are to be expected if both accelerators were used by, e.g., creating an MPI process per CPU (and accelerator).

3.2 *Conjugate Heat Transfer Between a Moving Particle and a Laminar Flow*

A fully-resolved Euler-Lagrange simulation of a moving particle including conjugate heat transfer is performed to demonstrate the approach. The flow is assumed laminar $\mathrm{Re}_\infty = 150$ and incompressible $M = 0.1$. An oscillating movement with the amplitude $A_y = 2D$ perpendicular to the flow direction and angular frequency $\overline{\omega}_\mathrm{h} = 2/5\pi$ is forced on the particle. A temperature load of $1.25T_\infty$ is imposed on a small region at the center of the particle. Within this simulation the cells in which the coupling surface of the PDEs for the fluid and particle is located, are continuously changing with the particle movement. This update of the cells with coupling interface is straightforward to formulate, due to the connectivity information inherently stored in the joint hierarchical Cartesian mesh.

The results are shown in Fig. 4, where an instantaneous snapshot of the temperature field is plotted. Since the particle generates a von Karman vortex street, warm fluid heated by the particle is entrained into the vortices, which are then shed and transported downstream. Within the particle the temperature field shows the cooling

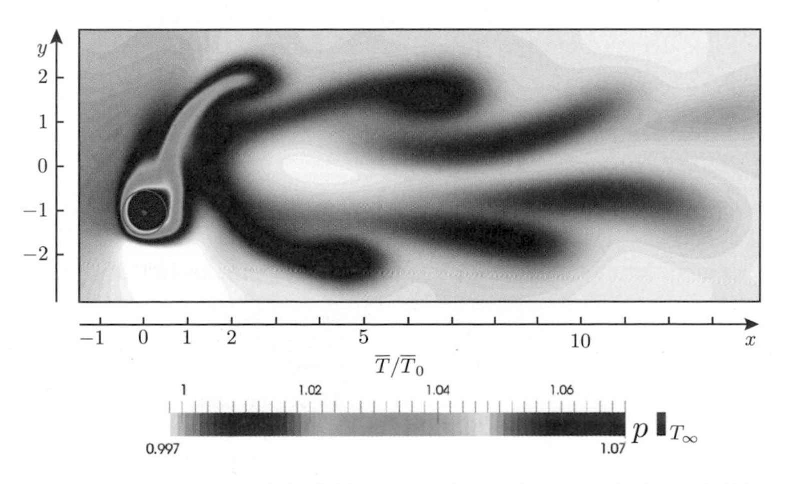

Fig. 4 Instantaneous temperature field T/T_0 over the fluid and the particle. The *solid black circle* indicates the particle surface

effect of the fluid, which is larger at the front stagnation point compared to the separated flow in the particle wake.

4 Large-Scale Simulations of Ellipsoidal Particles on Turbulent Flows

In Sect. 3 it is shown that fully resolved particle simulations can be used to gain a detailed understanding of the interaction between a moving heat-emitting particle and a viscous flow. These simulations are conducted to deduce physical coherences which can be used to improve Lagrangian point-particle models. In this section it is shown how these particle models are used to study the statistical properties of particles suspended in a turbulent flow.

4.1 Flow Solver

For small particle volume fractions the influence of the particles on the flow can be neglected [5]. This means that in principle the flow field could be computed once and reusing in subsequent particle simulations. However, the storage required to save the time dependent fluid motion necessary to advance the particles in times renders this approach unfeasible. Thus a massively parallel solver for Lagrangian particles on hierarchical Cartesian grids is coupled online with the finite-volume cut-cell method for compressible fluids developed at the Institute of Aerodynamics in Aachen [7, 8, 16, 20]. Both methods are fully parallelized on distributed memory systems using the Message Passing Interface (MPI). The results of a strong scaling experiment performed at the CRAY XE6 system of the HLRS in Stuttgart are shown in Table 1 and demonstrate the efficiency of the parallelization strategy.

4.2 Particle Solver

In the Lagrangian particle solver rotationally symmetric ellipsoidal particles are used as a first order model for general non-spherical particles [23–25].

Table 1 Strong scaling on Hermit

Cores	Cells	Cells/core	Speed-up
2,048	0.153e9	74.9e3	1.0
4,096	0.153e9	37.4e3	1.84
8,192	0.153e9	18.7e3	2.64
16,384	0.153e9	9.4e3	4.82
32,768	0.153e9	4.7e3	9.93

4.2.1 Kinematics

The position of a particle relative to an inertial coordinate system is tracked by the position of its center of mass $\mathbf{x}$ and by the rotation of its current orientation relative to its initial orientation. This rotation is tracked by quaternions $(\epsilon_1, \epsilon_2, \epsilon_3, \eta)$ [28]. Due to its velocity $\mathbf{v}$ the position of the particle changes over time:

$$\frac{d\mathbf{x}}{dt} = \mathbf{v} \,. \tag{1}$$

The particle orientation evolves due to its angular velocity $\hat{\boldsymbol{\omega}}$ measured with respect to a particle fixed coordinate system (indicated by a hat)

$$\begin{pmatrix} \frac{d\epsilon_1}{dt} \\ \frac{d\epsilon_2}{dt} \\ \frac{d\epsilon_3}{dt} \\ \frac{d\eta}{dt} \end{pmatrix} = \frac{1}{2} \begin{pmatrix} +\eta & -\epsilon_3 & +\epsilon_2 \\ +\epsilon_3 & +\eta & -\epsilon_1 \\ -\epsilon_2 & +\epsilon_1 & +\eta \\ -\epsilon_1 & -\epsilon_2 & -\epsilon_3 \end{pmatrix} \hat{\boldsymbol{\omega}} \,. \tag{2}$$

4.2.2 Dynamics

For very small and heavy particles in dilute suspensions the equation of motion is simplified since the local flow field around the ellipsoids can be approximated by Stokes flow. The shapes of these rotationally symmetric ellipsoids are characterized by the half-axes a, $b = a$, and $c = a\beta$, with β being the aspect ratio. For $\beta < 1$ and $\beta > 1$ the ellipsoids are denoted as and prolate, respectively, while ellipsoids with $\beta = 1$ are just spheres. Besides the hydrodynamical forces and torques evaluated under creeping flow conditions gravitational forces act on the particles.

These simplifications result in a particle acceleration that only depends on the gravity and the drag force [17]

$$\frac{d\mathbf{v}}{dt} = \mathbf{g} + \frac{24}{9} \frac{\beta^{2/3}}{\tau_p} \mathbf{A}^{-1} \left(\mathbf{E} \begin{pmatrix} \frac{1}{\chi_0 a^{-2} + \alpha_0} \\ \frac{1}{\chi_0 a^{-2} + \alpha_0} \\ \frac{1}{\chi_0 a^{-2} + \beta^2 \gamma_0} \end{pmatrix} \right) \mathbf{A} \left(\mathbf{u} - \mathbf{v} \right) \,, \tag{3}$$

where $\mathbf{A}$ is the rotation matrix that applies the rotation specified by the quaternion $(\epsilon_1, \epsilon_2, \epsilon_3, \eta)$, $\mathbf{E}$ the identity matrix, χ_0, α_0, and γ_0 are shape factors [6], and $\mathbf{u}$ is the fluid velocity at the center of mass. The particle response time $\tau_p = 2\rho_e r^2/(9\rho_f \nu)$ is that of an equivalent sphere at radius r with the same volume and mass as the ellipsoid. For $\beta = 1$ Eq. 3 reduces to the simplified Maxey-Riley equation [1, 2, 15, 19].

The evolution of the rotational velocity in the particle fixed coordinate system is

$$\begin{pmatrix} \frac{d\omega_{\hat{x}}}{dt} \\ \frac{d\omega_{\hat{y}}}{dt} \\ \frac{d\omega_{\hat{z}}}{dt} \end{pmatrix} = \begin{pmatrix} \omega_{\hat{y}}\omega_{\hat{z}}\frac{\beta^2-1}{1+\beta^2} \\ \omega_{\hat{z}}\omega_{\hat{x}}\frac{1-\beta^2}{1+\beta^2} \\ 0 \end{pmatrix} + \frac{40}{9}\frac{\beta^{2/3}}{\tau_p} \begin{pmatrix} \frac{1}{\alpha_0+\beta^2\gamma_0} \\ \frac{1}{\alpha_0+\beta^2\gamma_0} \\ \frac{1}{2\alpha_0} \end{pmatrix} \begin{pmatrix} \frac{1-\beta^2}{1+\beta^2}\tau_{\hat{z}\hat{y}} + \left(\zeta_{\hat{z}\hat{y}} - \omega_{\hat{x}}\right) \\ \frac{\beta^2-1}{1+\beta^2}\tau_{\hat{x}\hat{z}} + \left(\zeta_{\hat{x}\hat{z}} - \omega_{\hat{y}}\right) \\ 0 + \left(\zeta_{\hat{y}\hat{x}} - \omega_{\hat{z}}\right) \end{pmatrix}. \tag{4}$$

The rotational velocity depends only on the fluid shear stress $\hat{\boldsymbol{\tau}}$ and the fluid vorticity $\hat{\boldsymbol{\zeta}}$ but not on the gravity [10].

4.2.3 Parallelization

DNS of turbulent flows are computationally intense. The proposed parallelization strategy allows the division of the effort between a large number of processors as shown in Table 1. When the particles are simultaneously evolved in time according to the flow, the load is distributed according to the method proposed in [22]. The Lagrangian point-particle approach requires the local velocity of the carrier fluid as well as the velocity gradients to be interpolated at the particle center of mass, see Eqs. 3 and 4. This is performed using a third-order accurate method [27]. Since the interpolation is the most time consuming step of the numerical method, a predictor-corrector scheme which requires only a single interpolation per time-step is used. The update of the particle motion is performed by the process who owns the fluid cell in which the particle is located at the beginning of the time-step, ensuring the locality of the operation. However, for the collision detection the particles at the boundaries have to be transferred between the processors such that all particles pairs can be identified. The transferred particles are marked as halo particles which are only used for collision detection and are deleted again afterwards [21, 24, 25].

4.3 Ellipsoidal Settling Velocity

4.3.1 Quiescent Environment

In a quiescent environment, i.e., $\mathbf{u} = \mathbf{0}$, the gravitational force and the drag force balance each other, such that Eq. 3 can be solved for the resulting particle settling velocity

$$\mathbf{v}_t = \frac{9}{24}\frac{\tau_p}{\beta^{2/3}} g\left(\beta^2\gamma_0 - \alpha_0\right)\begin{pmatrix} \cos\theta_p \sin\theta_p \sin\phi_p \\ -\cos\theta_p \sin\theta_p \cos\phi_p \\ \cos^2\theta_p - \frac{\chi_0 a^{-2}+\alpha_0}{\beta^2\gamma_0-\alpha_0} \end{pmatrix}. \tag{5}$$

Without loss of generality, the z coordinate is chosen to point against the gravitational force. For clarity of presentation the rotation matrix **A** is written here in terms of the Euler angles ϕ_p, θ_p, and ψ_p instead of the quaternions [28].

Equation 5 shows that the settling velocity is independent of ψ due to the rotational symmetry of the ellipsoids. In contrast to spheres, which have an orientation-independent settling velocity, ellipsoids exhibit a drift, i.e., they fall in an intermediate direction in between the direction of gravity and the direction of their symmetry axis.

4.3.2 Turbulent Environment

To investigate the average settling velocity of ellipsoids in a turbulent flow a DNS is conducted. The time required for the automatic grid generation process is less than 1 min, see Sect. 2. Due to the specific layout of the Interlagos CPUs 4,096 cores are allocated such that the MPI tasks do not share a floating-point unit, i.e., only 2,048 out of the 4,096 cores allocated are used. The turbulent flow is seeded in equal parts with ellipsoids of five aspect ratios. The total number of particles is $\sim 45 \times 10^6$. The ellipsoids all have the same volume and mass as the sphere with the radius 20 μm, and the particle density is 843 larger than that of the fluid. The aspect ratios $\beta = \{0.25,\ 0.5,\ 1.0,\ 2.0,\ 4.0\}$ are considered. The total particle volume fraction is $\Phi \sim 10^{-5}$, i.e., the point-particle assumption is satisfied. Each simulation requires about $96 \cdot 4{,}096 \approx 400 \cdot 10^3$ core hours and therefore needs to be queued four times. The flow field has to reach a statistically stationary state before the particles can be seeded. Only after the particles have adapted to the flow field, i.e., are independent of their initial release conditions, statistics can be gathered.

The path line of an exemplary ellipsoid shows the influence of the turbulence on the particle motion (see Fig. 5). The turbulence significantly influences the orientation of the ellipsoid and thus via the orientation dependent drag force also the settling velocity. A turbulent settling velocity v_{turb} is obtained by averaging the settling velocity along the paths of the ellipsoids and additionally ensemble-averaged over all ellipsoids of the same type. In Fig. 6 this turbulent settling velocity is compared to the theoretical mean settling velocity v_{ODF} calculated from the turbulent orientation distribution function (ODF) and Eq. 5. Ellipsoids fall faster the higher the turbulent intensity which is characterized by the dissipation rate ϵ. This phenomenon is independent of the particle shape and results from the interaction of turbulent vortices, particle inertia, and gravity [23].

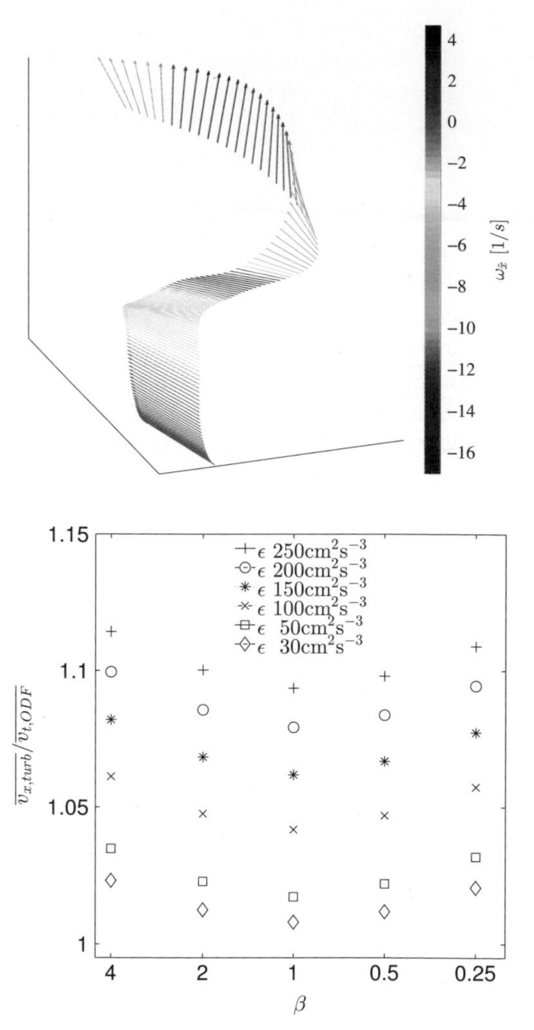

Fig. 5 Exemplary path of a oblate ellipsoid at $\beta = 0.25$ in the turbulent flow. The direction of the symmetry axis c is indicated by vectors that are colored by the rotation rate around this axis

Fig. 6 Ratio of the ensemble-averaged settling velocity v_{turb} and the theoretical mean settling velocity $v_{t,ODF}$ for ellipsoids (volume of a sphere at radius 20 μm) at five β values and the six dissipation rates ϵ

5 Conclusions

Recent developments aimed at tackling large-scale multi-scale particle-laden turbulent flows have been presented. A massively parallel automatic grid generation strategy for hierarchical Cartesian grids based on self-similar space-filling curves for domain decompositioning is shown to scale well. An extension of hierarchical Cartesian grids for tackling multi-physic problems using different numerical methods is presented and demonstrated by a fully-resolved particle simulation involving conjugate heat transfer. Furthermore, first steps into porting these strategies to heterogeneous computing nodes are shown and the performance of accelerators is compared against two common approaches to shared memory parallelism. Finally, a massively parallel Euler-Lagrange scheme for large-scale simulations of ellipsoidal particles on turbulent flows is considered. The suitability of the flow solver for massively parallel simulations has been demonstrated, the particle solver as well as its parallelization strategy are described. Then the gravitation induced settling velocity of ellipsoids in turbulent flows is compared to the settling velocity in quiescent environment. On that point, large scale simulations are conducted demonstrating that the average settling velocity of the particles increases with the turbulence intensity. This results shows that large scale simulations on supercomputers are needed to extend the knowledge about particle-fluid flows in order to develop simpler model equations.

Acknowledgements This work has been financed by the research cluster *Fuel production with renewable raw materials* (BrenaRo) at RWTH Aachen University as well as by the German Research Foundation (DFG) within the framework of the SFB/Transregio 129, SFB/Transregio 136, the collaborative research center SFB 686, and within the framework of the priority program SPP1276 METSTROEM under grant number SCHR 309/39. The support is gratefully acknowledged. The authors thank for the provided computing time at the IBM BlueGene/Q system (JUQUEEN) at the Jülich Supercomputing Center (JSC) and the CRAY XE6 system (HERMIT) at the High Performance Computing Center Stuttgart (HLRS).

References

1. Ayala, O., Rosa, B., Wang, L.-P., Grabowski, W.W.: Effects of turbulence on the geometric collision rate of sedimenting droplets. Part 1. Results from direct numerical simulation. New J. Phys. **10**, 075015 (2008)
2. Bec, J., Biferale, L., Cencini, M., Lanotte, A.S., Toschi, F.: Intermittency in the velocity distribution of heavy particles in turbulence. J. Fluid Mech. **646**, 527–536 (2010). doi:10.1017/S0022112010000029
3. Buhre, B.J.P., Elliott, L.K., Sheng, C.D., Gupta, R.P., Wall, T.F.: Oxy-fuel combustion technology for coal-fired power generation. Prog. Energy Combust. Sci. **31**(4), 283–307 (2005). doi:http://dx.doi.org/10.1016/j.pecs.2005.07.001. ISSN 0360-1285. http://www.sciencedirect.com/science/article/pii/S0360128505000225
4. Cate, A.T., Derksen, J.J., Portela, L.M., Van, H.E.A., Akker, D.: Fully resolved simulations of colliding monodisperse spheres in forced isotropic turbulence. J. Fluid Mech **519**, 233–271, 11 (2004). doi:10.1017/S0022112004001326. ISSN 1469-7645. http://journals.cambridge.org/article_S0022112004001326
5. Hagemeier, T., Hartmann, M., Thévenin, D.: Practice of vehicle soiling investigations: a review. Int. J. Multiph. Flow **37**(8), 860–875 (2011). doi:10.1016/j.ijmultiphaseflow.2011.05.002. ISSN 0301-9322
6. Happel, J., Brenner, H.: Low Reynolds Number Hydrodynamics. Prentice-Hall, Englewood Cliffs (1965)
7. Hartmann, D., Meinke, M., Schröder, W.: An adaptive multilevel multigrid formulation for Cartesian hierarchical grid methods. Comput. Fluids **37**, 1103–1125 (2008). doi:10.1016/j.compfluid.2007.06.007
8. Hartmann, D., Meinke, M., Schröder, W.: A strictly conservative Cartesian cut-cell method for compressible viscous flows on adaptive grids. Comput. Meth. Appl. Mech. Eng. **200**, 1038–1052 (2010)
9. Hoberock, J., Bell, N.: Thrust: a parallel template library (2011). Version 1.4.0.

10. Jeffery, G.B.: The motion of ellipsoidal particles immersed in a viscous fluid. Proc. R. Soc. Lond. A **102**, 161–179 (1922)
11. Kidanemariam, A.G., Chan-Braun, C., Doychev, T., Uhlmann, M.: Direct numerical simulation of horizontal open channel flow with finite-size, heavy particles at low solid volume fraction. New J. Phys. **15**(2), 025031 (2013). doi:10.1088/1367-2630/15/2/025031. http://stacks.iop.org/1367-2630/15/i=2/a=025031
12. Lei, L., Lin, G., Xuecheng, X., Wansheng, Z.: Influence of flushing on performance of EDM with bunched electrode. Int. J. Adv. Manuf. Technol. **58**(1–4), 187–194 (2012). doi:10.1007/s00170-011-3357-8. ISSN 0268-3768. http://dx.doi.org/10.1007/s00170-011-3357-8
13. Lintermann, A., Meinke, M., Schröder, W.: Fluid mechanics based classification of the respiratory efficiency of several nasal cavities. Comput. Biol. Med. **43**(11), 1833–1852 (2013). doi:10.1016/j.compbiomed.2013.09.003. ISSN 00104825
14. Lintermann, A., Schlimpert, S., Grimmen, J.H., Günther, C., Meinke, M., Schröder, W.: Massively parallel grid generation on HPC systems. Comput. Methods Appl. Mech. Eng. **277**, 131–153 (2014). doi:doi:10.1016/j.cma.2014.04.009
15. Maxey, M.R., Riley, J.J.: Equation of motion for a small rigid sphere in a nonuniform flow. Phys. Fluids **26**, 883–889 (1983)
16. Meinke, M., Schneiders, L., Günther, C., Schröder, W.: A cut-cell method for sharp moving boundaries in Cartesian grids. Comput. Fluids **85**(0), 135–142 (2013). doi:http://dx.doi.org/10.1016/j.compfluid.2012.11.010. ISSN 0045-7930. http://www.sciencedirect.com/science/article/pii/S0045793012004422. International Workshop on Future of CFD and Aerospace Sciences.
17. Oberbeck, A.: Ueber stationäre Flüssigkeitsbewegungen mit Berücksichtigung der inneren Reibung. Crelle's J. **81**, 79 (1876)
18. Pruppacher, H., Klett, J.: Microphysics of Clouds and Precipitation. Kluwer Academic, Dordrecht (1997)
19. Rosa, B., Parishani, H., Ayala, O., Grabowski, W.W., Wang, L.-P.: Kinematic and dynamic collision statistics of cloud droplets from high-resolution simulations. New J. Phys. **15**, 045010 (2013)
20. Schneiders, L., Hartmann, D., Meinke, M., Schröder, W.: An accurate moving boundary formulation in cut-cell methods. J. Comput. Phys. **235**(0), 786–809 (2013). doi:http://dx.doi.org/10.1016/j.jcp.2012.09.038
21. Siewert, C., Bórdas, R., Wacker, U., Beheng, K.D., Kunnen, R.P.J., Meinke, M., Schröder, W., Thévenin, D.: Influence of turbulence on the drop growth in warm clouds, part I: Comparison of numerically and experimentally determined collision kernels. Meteorol. Z. doi:10.1127/0941-2948/2014/0566 (2014)
22. Siewert, C., Meinke, M., Schröder, W.: Efficient coupling of an Eulerian flow solver with a Lagrangian particle solver for the investigation of particle clustering in turbulence. In: Nagel, W.E., Kröner, D.H., Resch, M.M. (eds.) High Performance Computing in Science and Engineering '13, pp. 393–404. Springer, Cham (2013). doi:10.1007/978-3-319-02165-2_27. ISBN 978-3-319-02164-5.
23. Siewert, C., Kunnen, R.P.J., Meinke, M., Schröder, W.: Orientation statistics and settling velocity of ellipsoids in decaying turbulence. Atmos. Res. **142**, 45–56 (2014). doi:10.1016/j.atmosres.2013.08.011. The 16th International Conference on Clouds and Precipitation
24. Siewert, C., Kunnen, R.P.J., Meinke, M., Schröder, W.: On the collision detection for ellipsoidal particles in turbulence. In: ASME 2014 4th Joint US-European Fluids Engineering Division Summer Meeting, Chicago, Aug 2014. ASME. Accepted for publication, Paper No. FEDSM2014-21982
25. Siewert, C., Kunnen, R.P.J., Schröder, W.: A collision mechanism for small ellipsoids settling in turbulent flows. J. Fluid Mech. **758**, 686–701 (2014). doi:10.1017/jfm.2014.554
26. Wang, L.-P., Ayala, O., Rosa, B., Grabowski, W.W.: Turbulent collision efficiency of heavy particles relevant to cloud droplets. New J. Phys. **10**, 075013 (2008). doi:10.1088/1367-2630/10/7/075013

27. Yeung, P.K., Pope, S.B.: An algorithm for tracking fluid particles in numerical simulations of homogeneous turbulence. J. Comput. Phys. **79**, 373–416 (1988)
28. Zhang, H., Ahmadi, G., Fan, F.-G., McLaughlin, J.B.: Ellipsoidal particles transport and deposition in turbulent channel flows. Int. J. Multiph. Flow **27**(6), 971–1009 (2001). doi:10.1016/S0301-9322(00)00064-1

A Parallelized Method for Direct Numerical Simulations of Rigid Particles in Multiphase Flow

Philipp Rauschenberger, Andreas Birkefeld, Martin Reitzle, Christian Meister, and Bernhard Weigand

Abstract The here described new technique allows to simulate rigid bodies in the multiphase code Free Surface 3D (FS3D). The scheme is formulated in an Eulerian framework. Multiple rigid bodies may be considered that are identified by a single Volume-of-Fluid (VOF) variable. Collisions between particles are treated with the exact solution of the excentrical, frictionless collision problem of two arbitrarily shaped particles. The phenomena of drafting, kissing and tumbling are identified when two solid spheres interact in a Newtonian fluid. This case provides a basis for a performance analysis on massively parallel architectures.

1 Introduction

A current research topic in both meteorology and aviation is the freezing of supercooled water droplets in the atmosphere. It has great impact on precipitation of atmospheric clouds and weather prediction in general. Airplane safety is a big issue when a layer of frozen water develops on an airplane wing. Direct Numerical Simulations (DNS) of multiphase flow provide an apt tool to obtain insight into the prevailing microphysical processes in both areas.

Our in-house code Free Surface 3D allows DNS of incompressible multiphase flow. It is based on the Volume-of-Fluid method. The code is entirely parallelized with the Message Passing Interface (MPI) standard. In this framework, ice is an additional phase that has the properties of a rigid body. At the same time, the liquid-solid phase change of supercooled water to ice shall be taken into account. The numerical scheme for phase change is described in [20]. The phase change makes it necessary to represent the additional solid phase by another VOF-Variable. Then, the treatment of the rigid bodies must be Eulerian.

P. Rauschenberger (✉) • A. Birkefeld • M. Reitzle • C. Meister • B. Weigand
Institute of Aerospace Thermodynamics, University of Stuttgart, Pfaffenwaldring 31, 70569 Stuttgart, Germany
e-mail: philipp.rauschenberger@itlr.uni-stuttgart.de

W.E. Nagel et al. (eds.), *High Performance Computing in Science and Engineering '14*, DOI 10.1007/978-3-319-10810-0_22

In this context, the here described technique transports the solid bodies by advecting the VOF-variable representing the solid phase. To the best knowledge of the authors, all other authors in this field treat rigid bodies as Lagrangian particles in some way. Hence, the new position of a rigid body is obtained by moving its centroid according to its velocity vector and the current time step. Its orientation is determined in the same way with knowledge of the angular velocity. A widely used method for rigid particles in flows is the Immersed-Boundary (IB) method [5,14,18,19,25]. Direct-Forcing (DF) methods [14,25,30] compute the body forces excerted on the rigid body from the boundary conditions at the interface of the rigid body and the fluid. The Arbitrary-Lagrangian-Eulerian (ALE) method [13] uses Finite Element (FE) discretisation and a boundary fitted grid. The Distributed-Lagrange-Multiplier/Fictious-Domain Method (DLM/FD) with FE-discretisation [8] extends the fluid equations over the rigid body regions and imposes rigid body motion with the Lagrange multiplier. This idea was pursued improved by other authors [4, 17]. In that paper, an FD method with explicit computation of the rigid body velocities is used rendering the Lagrange multiplier unnecessary [2,16,24,27]. Several other methods are described in the literature [7,23,29].

In this paper, we pursue the idea of the FD approach and use explicit expressions for the rigid body velocities. The particles are identified with a VOF variable and since the velocity field in the regions occupied by rigid bodies is the rigid body velocity field, the VOF variable is convected with a transport equation using the interfaces known from Piecewise Linear Interface Calculation (PLIC). In this connection, it has to be made sure that the collective velocity field in Ω is solenoidal. Hence, the method works completely within an Eulerian framework.

This paper provides the mathematical description in Sect. 2 and an outline of the numerical scheme in Sect. 3, where Sect. 3.5 gives an example for recent modifications of the applied code which influence its parallel performance. Section 4 shows two spheres following each under closely while being accelerated by gravity is taken as an example for the ability of the code to simulate rigid body motion in an incompressible flow with collisions. In the final section the performance of FS3D on the massively parallel Cray XE6 supercomputer at HLRS is analysed.

2 Mathematical Description

The governing equations for incompressible flow reduce to mass and momentum conservation. Mass conservation implies volume conservation in this case:

$$\nabla \cdot \mathbf{u} = 0. \tag{1}$$

The jump condition at the interface between fluid and solid is

$$\mathbf{u}_s \cdot \hat{\mathbf{n}}_\gamma = \mathbf{u}_l \cdot \hat{\mathbf{n}}_\gamma \tag{2}$$

and the no-slip condition between the two phases reads

$$\hat{\mathbf{n}}_\gamma \times (\mathbf{u}_s - \mathbf{u}_l) = \mathbf{0}, \tag{3}$$

where $\mathbf{u}_s$ and $\mathbf{u}_l$ denote the velocities of the solid and fluid phases respectively and $\hat{\mathbf{n}}_\gamma$ the unit vector normal to the interface. The momentum equation reads

$$\frac{\partial}{\partial t}(\rho \mathbf{u}) + \nabla \cdot (\rho \mathbf{u} \otimes \mathbf{u}) = -\nabla p - \nabla \cdot \mathbf{S} + \rho \mathbf{g}, \tag{4}$$

$$S = \mu \left(\nabla \mathbf{u} + \nabla \mathbf{u}^T \right). \tag{5}$$

Surface tension has not to be taken into account at the interface shared by the rigid region and the surrounding fluid. Hence, the respective jump condition is

$$[\mathbf{S}_s - \mathbf{I} p_s] \cdot \hat{\mathbf{n}}_\gamma = [\mathbf{S}_l - \mathbf{I} p_l] \cdot \hat{\mathbf{n}}_\gamma. \tag{6}$$

3 Numerical Method

The Navier-Stokes equations for incompressible flow are discretized with finite volumes. According to the Marker and Cell (MAC) method [10], scalar variables are stored at cell centres and velocities on cell faces. Second order accurate upwind schemes are applied to discretize the convective terms, the diffusive terms use second order accurate central difference schemes. Godunov type schemes compute the fluxes. Spatial derivatives are restricted with a Total Variation Diminishing (TVD) limiter [26] assuring that no new velocity maxima are generated.

3.1 Volume-of-Fluid Method

An additional scalar variable f identifies the rigid bodies according to the classical VOF method [12]. It is defined as

$$f(\mathbf{x}, t) = \begin{cases} 0 & \text{in the fluid,} \\ ]0, 1[& \text{at the interface,} \\ 1 & \text{in the solid.} \end{cases} \tag{7}$$

The local variables density ρ and dynamic viscosity μ are defined by the local value of f (one-field formulation):

$$\rho(\mathbf{x}, t) = \rho_s f(\mathbf{x}, t) + \rho_l (1 - f(\mathbf{x}, t)), \tag{8}$$

$$\mu(\mathbf{x}, t) = \mu_s f(\mathbf{x}, t) + \mu_l (1 - f(\mathbf{x}, t)). \tag{9}$$

The solid phase is advected by application of the transport equation:

$$\frac{\partial f}{\partial t} + \nabla \cdot (f\mathbf{u}) = 0. \tag{10}$$

FS3D makes use of piecewise linear interface reconstruction (PLIC) [21,22] for the computation of VOF fluxes. Hence, a plane separating the two phases fluid and solid is reconstructed in interface cells. This plane is orthogonal to the local normal vector $\hat{\mathbf{n}}_\gamma$ that is the negative gradient of the volume fraction f. This allows to compute the fluxes with first order accuracy.

3.2 Pressure Correction

To obtain a solenoidal velocity field, a pressure projection scheme [1] is employed (Helmholtz-Hodge decomposition). As a result of volume conservation in incompressible flow, a Poisson equation needs to be solved for pressure:

$$\nabla \cdot \left[\frac{1}{\rho(f)} \nabla p \right] = \frac{\nabla \cdot \mathbf{u}}{\Delta t} \tag{11}$$

FS3D applies a multigrid solver to the set of linear equations [22,28]. The smoother is a Red-Black Gauss-Seidel algorithm. The algorithm can be run as a V- or W-cycle scheme. In the present case the V-cycle is used.

3.3 Rigid Bodies

Up to this point, the numerical scheme treats the computational domain G including the rigid body regions $K_i(t)$ as an incompressible Newtonian fluid. Hence, an additional constraint must be applied to the latter. Based on the intermediate solenoidal velocity field $\tilde{\mathbf{u}}$, the translatory velocity $\tilde{\mathbf{U}}_i$ and the angular velocity $\tilde{\boldsymbol{\omega}}_i$ of the i-th rigid body can be computed from conservation of linear and angular momentum:

$$M\tilde{\mathbf{U}} = \int_{K_i(t)} \rho(f)\tilde{\mathbf{u}}\, dV \qquad \text{in } K_i(t), \tag{12}$$

$$\mathbf{J}\tilde{\boldsymbol{\omega}} = \int_{K_i(t)} \mathbf{r} \times \rho(f)\tilde{\mathbf{u}}\, dV \quad \text{in } K_i(t). \tag{13}$$

M is the mass and $\mathbf{J}$ the inertia tensor of the respective rigid body. The velocities in the region $K_i(t)$ then read

$$\breve{\mathbf{u}} = \tilde{\mathbf{u}} \qquad \text{in } G \setminus K_i(t), \tag{14}$$

$$\breve{\mathbf{u}} = \mathbf{u}_r = \tilde{\mathbf{U}} + \tilde{\boldsymbol{\omega}} \times \mathbf{r} \quad \text{in } K_i(t). \tag{15}$$

By this means, momentum is conserved and the no-slip condition is obtained at the interface between rigid and fluid region. Since the VOF variable f is advected with the linear transport equation (10), position and orientation of the rigid regions are obtained within the Eulerian framework. When the velocities in $K_i(t)$ are overwritten according to Eq. (15), the velocity field $\breve{\mathbf{u}}$ is not solenoidal because the velocities on the faces of interface cells do not match necessarily. A second pressure correction is conducted excluding the rigid regions and smoothing the divergence on $\partial K_i(t)$. Then, the velocity field is solenoidal while the rigid body velocities remain untouched.

The rigid regions K_i are all identified by the same VOF variable f. If more than one rigid body exists in G, they must be distinguished. For this purpose, all cell indices of a region K_i are stored in a separate list. With these index lists, each rigid body can be addressed unambiguously and computation time is saved. In order to identify the rigid bodies correctly, a minimum distance of three cells must be preserved between two bodies. At the same time, it assures that the second pressure correction has enough degrees of freedom.

At each time step, the distances between rigid bodies are roughly estimated and in case they approach one another, their exact distance and the collision normal are computed. If the respective collision distance is deceeded, the force onto each body is computed as the exact solution of the excentrical, frictionless two-body collision problem [15]. Two bodies just before the collision are depicted in Fig. 1. The first index of $\mathbf{U}$ and $\boldsymbol{\omega}$ identifies rigid body 1 or 2. The second index 0 means that these are the initial velocities before the collision. After the collision, the second index is omitted. The vector pointing from the centroid to the point of collision is denoted by $\mathbf{r}_c$. Then, the impulse I exchanged in direction of the collision normal $\hat{\mathbf{n}}_c$ reads

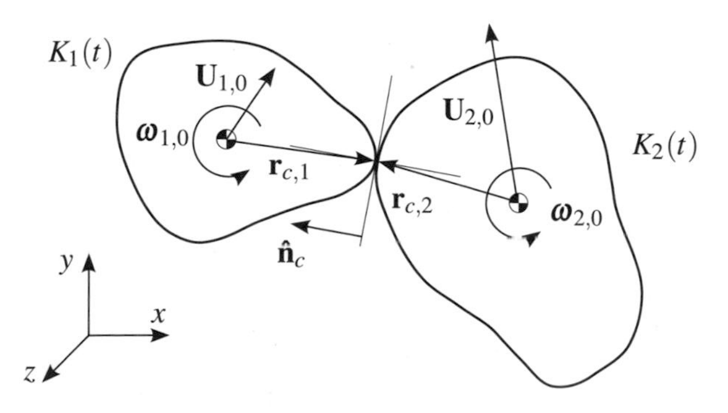

Fig. 1 Excentrical, frictionless collision of two particles

$$I = (1+\epsilon)\frac{[-\mathbf{U}_{1,0} - \boldsymbol{\omega}_{1,0} \times \mathbf{r}_{c,1} + \mathbf{U}_{2,0} + \boldsymbol{\omega}_{2,0} \times \mathbf{r}_{c,2}] \cdot \hat{\mathbf{n}}_c}{\sum_{k=1}^{2} \left[\frac{1}{M_k}\hat{\mathbf{n}}_c + \{\mathbf{J}_k^{-1} \cdot (\mathbf{r}_{c,k} \times \hat{\mathbf{n}}_c)\} \times \mathbf{r}_{c,k} \right] \cdot \hat{\mathbf{n}}_c}, \tag{16}$$

where the coefficient of restitution is $0 < \epsilon < 1$. $\epsilon = 0$ accounts for an inelastic collision and $\epsilon = 1$ for an elastic one. The velocities after collision read

$$\mathbf{U}_1 = \mathbf{U}_{1,0} + \frac{I}{M_1}\hat{\mathbf{n}}_c, \tag{17}$$

$$\mathbf{U}_2 = \mathbf{U}_{2,0} - \frac{I}{M_2}\hat{\mathbf{n}}_c, \tag{18}$$

$$\boldsymbol{\omega}_1 = \boldsymbol{\omega}_{1,0} + \mathbf{J}_1^{-1} \cdot (\mathbf{r}_{c,1} \times I\hat{\mathbf{n}}_c), \tag{19}$$

$$\boldsymbol{\omega}_2 = \boldsymbol{\omega}_{2,0} - \mathbf{J}_2^{-1} \cdot (\mathbf{r}_{c,2} \times I\hat{\mathbf{n}}_c). \tag{20}$$

3.4 Time Integration

Two time integration schemes are implemented: a first order explicit scheme and a second order Runge-Kutta scheme. Time step restriction is according to the Courant-Friedrichs-Lewy (CFL) condition [11]:

$$\alpha = \frac{u\Delta t}{h}, \tag{21}$$

where α denotes the CFL number and h the grid cell size. Acceleration of particles with no initial velocity can be accounted for with the acceleration $\mathbf{g}$. It determines the CFL limit at the first instances:

$$\alpha = \Delta t\sqrt{\frac{|\mathbf{g}|}{2h}}. \tag{22}$$

The implicit treatment of the viscous terms does not cause a time step restriction.

3.5 Numerical Modifications for Parallelization

Despite the fact that the Red-Black Gauss-Seidel based multigrid solver, which is used for the pressure correction (see Sect. 3.2), is fully parallelized using MPI as well as OpenMP, it causes up to 90 % of the computational effort in the overall numerical scheme, most of it contributed by the smoothing. Besides the computational load this also leads to an enormous communication effort since boundary conditions have to be applied after every smoothing cycle. The communication

```
subroutine Multigridcycle(l)
    if l = 1
        y^1 = CoarseGridSolver(A, b)
    else
        do n = 1 to ν_pre:   y^l = Smoother(A^l, y^l, b^l)
        b^{l-1} = Restrict(A^l y^l − b^l)
        if l - 1 = SwitchLevel:   call Collect(b^{l-1})          (*)
        if l - 1 > SwitchLevel or myRank = 0                     (*)
            y^{l-1} = 0
            call Multigridcycle(l − 1)
        end if                                                   (*)
        if l - 1 = SwitchLevel:   call Distribute(y^{l-1})       (*)
        y^l = y^l - Prolongate(y^{l-1})
        do n = 1 to ν_post:   y^l = Smoother(A^l, y^l, b^l)
    end if
end subroutine
```

Fig. 2 Schematic description of V-cycle multigrid solver with additional local-global routines (∗)

to computation ratio gets worse on the coarser grids with their exponentially decreasing grid cell numbers.

Hence it is favourable to put the domain decomposition aside at a certain level of coarsening and agglomerate the whole domain. Therefore a routine was implemented which gathers the restricted solution at the root process which then does the computations on all the coarser levels while the other processes are idling (see Fig. 2). This completely avoids MPI communication during the handling of these levels. The major drawback is the necessity to store the whole linear system of the global levels in the memory of the node with the root process, which means an increase of the memory demand on that node of $n_{cores}/32$ for the XE6 architecture. Another disadvantage is the increasing complexity of the multigrid algorithm and its data structures.

A rough estimation of the possible improvement by these changes can be given by a comparison of the number of invocations of MPI_SENDRECV since this routine is used for the whole boundary exchange within the scheme. It was done with the case presented in Sect. 4, computed for $48 \times 24 \times 24$ grid cells on 64 cores. For the initialization and one time step the number of invocations could be reduced with the new scheme from 14,625 to 3,708.

4 Drafting, Kissing and Tumbling

The interaction of particles in fluids was investigated experimentally [6] in the past. Of special interest is the case of two spheres that are placed above one another in a Newtonian fluid and are dropped. Certain characteristic phenomena may be observed. At the beginning the upper sphere falls in the wake of the lower one and

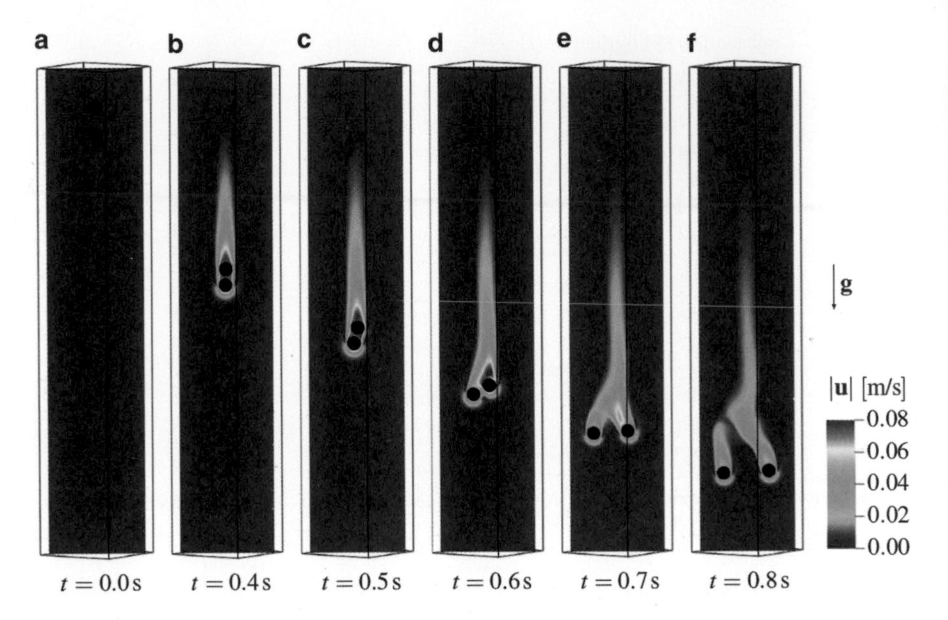

Fig. 3 Interaction of two particles in a Newtonian fluid

catches up (drafting). Then, the two spheres collide (kissing). After this, they tumble and are torn apart (tumbling).

These phenomena may also be investigated numerically. The numerical setup and the results are shown in Fig. 3. The channel has no slip walls with the boundary condition $\mathbf{u} = 0 \forall t$ including the lower wall. On the upper wall, a homogeneous Neumann boundary condition is set. This setup was investigated before [9, 24]. The spheres have a density of $1,140.0\,\mathrm{kg/m^3}$ and diameters $D = 1.66 \times 10^{-3}$ m. The fluid has the dynamic viscosity $\mu_l = 1.0 \times 10^{-3}\,\mathrm{kg/(ms)}$ and density $\rho_l = 1{,}000.0\,\mathrm{kg/m^3}$. The two spheres are initially ($t = 0\,\mathrm{s}$) at rest. Their centroids are a distance 3.4×10^{-3} m apart. The phenomena drafting, kissing and tumbling [6] are clearly identified in Fig. 3.

Figure 4 compares the vertical velocities of the two spheres to the results from literature [9, 24]. The respective velocities obtained by the three different methods appear to be quite diverse. However, the results of [9, 24] are within the same range during the whole time considered. Tumbling is the results of an instability and hence, a good agreement cannot be expected here for reason of the respective method and small disturbances in the fluid. FS3D observes higher velocities until the time of the first collision.

There are several reasons for these deviations. First, the methods and grid resolutions differ. But in particular, the expected accuracies in predicting terminal velocities vary. In the cases considered here, the relevant velocity range is in between 3×10^{-2} and 8×10^{-2} m/s. Hence, Reynolds numbers based on the sphere diameter range in between 49.8 and 132.8. According to [9], terminal velocities are underestimated and the relative deviation of the terminal velocity in a cylindrical

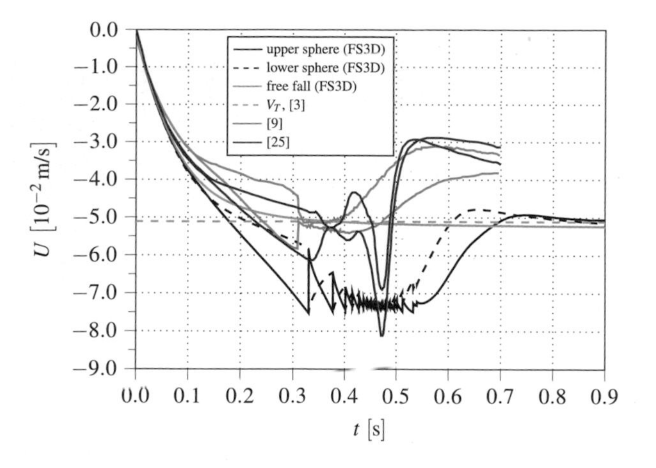

Fig. 4 Vertical velocities of two particles interacting in a Newtonian fluid

channel for $Re = 78.3$ and $D/D_{cyl.} = 0.3$ is 17 % and for $Re = 118$ and $D/D_{cyl.} = 0.4$ is 15.2 %. The terminal velocities in a rectangular channel computed by [24] in the range $11 < Re < 42$ and a ratio of diameter to side length being $D/L = 0.2$ is in between 12 % to 15 %. Terminal velocities are also underestimated. The FS3D code is validated in a theoretically unbounded region. In the range $5 < Re < 200$, terminal velocities deviate from correlations [3] only marginally (< 5 %).

According to [9] their method is of first order. They do not show a convergence analysis of free falling rigid spheres. Sharma and Patankar [24] report the algorithm to be first order with respect to the grid size according to their convergence analysis. For FS3D, the convergence index computed for the terminal velocities at $Re = 102.9$ is 1.54 at a grid refinement ratio of 2.

The terminal velocities of spheres in cylindrical channels are reported in literature. According to [3], the terminal velocity with present solid and fluid properties is $V_T = 5.12 \times 10^{-2}$ m/s for a single sphere. This serves as an estimate for the rectangular channel. The value is displayed in Fig. 4 (dashed green line) together with the terminal velocity obtained from a calculation with FS3D of the single sphere in the rectangular channel (green line). The slightly higher absolute value of the terminal velocity in the rectangular channel (2.4 %) is explained by the larger cross sectional area. It proves that FS3D predicts terminal velocities in a bounded region very accurately.

Higher velocities predicted by FS3D are explained by the above reflections. This is one of the reasons for different times for the first collision. Another one is that different collision models are applied that vary in the distance between the spheres at the instant of collision. Furthermore, multiple collisions take place during the period of kissing ($0.33\,\text{s} \leq t \leq 0.54\,\text{s}$). In contrast to the smooth velocity profiles of [9, 24], the spheres experience abrupt accelerations in FS3D and show jagged velocity profiles. After kissing and tumbling, the spheres approach the predicted terminal velocity of the single sphere.

5 Performance Analysis

All of the following performance results are obtained with the case presented in Sect. 4. The computations are performed on the CRAY XE6 supercomputer at HLRS. The baseline case, which is used for the results discussed before, is computed on a grid of $768 \times 128 \times 128$ grid cells on 384 cores. Parallelization is done purely by spatial domain decomposition with MPI data exchange. For the performance analysis the number of processes is varied for the given grid size to examine the strong scaling behaviour of the code. Furthermore, for the given baseline of 384 cores the grid size is varied. Where possible the processes are distributed such that the grid portion of each process is cubical.

For the analysis the speed-up S of parallel versions is defined by

$$S = \frac{T_2}{T_N}, \tag{23}$$

where T_i is the wall-clock time of a computation of 200 time steps on i cores. The high number of time steps allows to neglect the initialization time of the code. Memory limits do not allow a computation on one core for the chosen test case. Hence, the computation on two processes was used as reference for the speed-up. The ideal speed-up would be reached for $S = N/2$, leading to the definition of the parallel efficiency E (Table 1):

$$E = \frac{2T_2}{NT_N} = \frac{2S}{N} \tag{24}$$

The shown speed-up behaviour is not ideal, especially for high core numbers. There are several reasons that prevent a better parallel efficiency: As a solver for incompressible multiphase flows FS3D contains a collection of different numerical schemes that are applied consecutively. All of them demand one ore more invocations of point-to-point or even global MPI data exchange and thereby cause communication overhead which grows with increasing numbers of cores.

Table 1 Speed-Up of FS3D with respect to core number ($768 \times 128 \times 128$ cells)

Number of cores	2	4	6	48	384	3,072
Speed-up S	1.00	1.64	2.31	11.8	55.8	24.7
Parallel efficiency E (%)		81.8	76.8	49.1	29.1	1.61

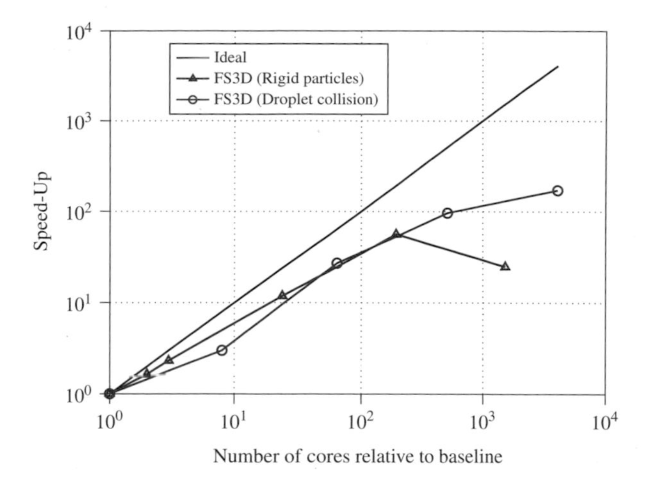

Fig. 5 Speed-Up of FS3D versus ideal Speed-Up (768 × 128 × 128 cells)

Furthermore the individual parts of the overall scheme have different serial fractions which corrupt the parallel efficiency.

Two components should be highlighted here: At first, the already mentioned multigrid solver, which not only consumes most of the overall computing time and produces a high communication load, its local-global structure also contains a noticeable serial fraction by design. For the given case, the resulting performance drawback is increased by the fact that the Poisson equation has to be solved twice every cycle when rigid bodies are present (see Sect. 3). The second part that is relevant for the here described case is the handling of the rigid particles: As mentioned before the rigid bodies have to be identified separately in the numerical grid and their particular properties (i.e. angular and linear momentum, mass and inertia tensor) must be calculated. The first one is done sequentially, leading to temporarily idling processes. The second one is done by the processes whose grid portions contain the particles which is a serious load balancing drawback in the given computation. These special issues are probably the reason for the meltdown of the parallel efficiency at higher core numbers in a comparison to a test case without rigid particles, see 'FS3D (Droplet collision)' in Fig. 5.

Table 2 shows the performance behaviour for a given number of cores and a varying grid size. It shows that an increase of the number of cells per process does not necessarily lead to a better performance as it could be expected as consequence of the better communication to computation ratio of larger grid portions. However this table also shows that the number of multigrid levels and consequently the number of smoothing calls and boundary exchanges increases. This obviously levels out the first effect at a certain point. The complex connections between the topology parameters of the multigrid scheme and their influence on convergence as well as performance make the prediction of optimum parameters very challenging. Nevertheless, for most rigid particle problems 32^3 grid cells per core have been

Table 2 Relative core time demand with respect to the number of cells with 384 cores

Number of cells	384 × 64 × 64	768 × 128 × 128	1,536 × 256 × 256
Relative number of grid cells	1/8	1	8
Relative core time demand	0.31	1.00	9.38
Coretime per grid cell [$10^{-3} S$]	21.8	8.91	10.4
Number of cells per core	16^3	32^3	64^3
Number of local coarsening levels	4	5	6
Number of global coarsening levels	2	2	2

identified as a sweet spot with regard to the computational effort needed, resulting in a typical utilization of up to 4,096 cores.

6 Conclusion

The rigid body extension to the FS3D code is presented along with the case of two spheres falling in a Newtonian fluid that is an apt case to show the ability of a code to capture the interactions and dynamics of fluids with submerged rigid bodies. The experimentally determined phenomena are captured and resulting velocity profiles are compared to results available in literature. The numerical computation also serves as a test case for the performance analysis on the massively parallel Cray XE6 high performance computer architecture. The paper provides a detailed description of the code modifications necessary for the parallelization of the FS3D inherent multigrid solver.

Acknowledgements The authors greatly acknowledge the *High Performance Computing Center Stuttgart* (HLRS) for support and supply of computational time on the Cray XE6 platform under the Grant No. FS3D/11142. The authors also greatly acknowledge financial support of this project from DFG for the collaborative research center SFB-TRR 75.

References

1. Bell, J.B., Colella, P., Glaz, H.M.: A second-order projection method for the incompressible Navier-Stokes equations. J. Comput. Phys. **85**(2), 257–283 (1989)
2. Carlson, M.: Rigid, melting and flowing fluid. PhD thesis, College of Computing Georgia Institue of Technology (2004)
3. Clift, R., Grace, J.R., Weber, M.E.: Bubbles, drops and particles. Dover Publications, Inc., Mineola (2005)

4. Diaz-Goano, C., Minev, P.D., Nandakumar, K.: A fictitious domain/finite element method for particulate flows. J. Comput. Phys. **192**, 105–123 (2003). doi:10.1016/S0021-9991(03)00349-8
5. Fadlun, E.A., Verzicco, R., Orlandi, P., Mohd-Yusof, J.: Combined immersed-boundary finite-difference methods for three-dimensional complex flow simulations. J. Comput. Phys. **161**, 35–60 (2000). doi:10.1006/jcph.2000.6484
6. Fortes, A.F., Joseph, D.D., Lundgren, T.S.: Nonlinear mechanics of fluidization of beds of spherical particles. J. Fluid Mech. **177**, 467–483 (1987). doi:10.1017/S0022112087001046
7. Gibou, F., Min, C.: Efficient symmetric positive definite second-order accurate monolithic solver for fluid/solid interactions. J. Comput. Phys. **231**, 3246–3263 (2012). doi:10.1016/j.jcp.2012.01.009
8. Glowinski, R., Pan, T.W., Hesla, T.I., Joseph, D.D.: A distributed lagrange multiplier/fictitious domain method for particulate flows. Int. J. Multiph. Flow **25**, 755–794 (1999). doi:10.1016/S0301-9322(98)00048-2
9. Glowinski, R., Pan, T.W., Hesla, T.I., Joseph, D.D., Périaux, J.: A fictitious domain approach to the direct numerical simulation of incompressible viscous flow past moving rigid bodies: application to particulate flow. J. Comput. Phys. **169**(2), 363–426 (2001). doi:10.1006/jcph.2000.6542
10. Harlow, F.H., Welch, J.E.: Numerical calculation of time-dependent viscous incompressible flow of fluid with free surface. Phys. Fluids **8**(12), 2182–2189 (1965). doi:10.1063/1.1761178
11. Hirsch, C.: Numerical computation of internal & external flows. Elsevier Butterworth-Heineman, Amsterdam (2007)
12. Hirt, C.W., Nichols, B.D.: Volume of fluid (VOF) method for the dynamics of free boundaries. J. Comput. Phys. **39**(1), 201–225 (1981). doi:10.1016/0021-9991(81)90145-5
13. Hu, H.H., Patankar, N.A., Zhu, M.Y.: Direct numerical simulations of fluid-solid systems using the arbitrary Lagrangian-Eulerian technique. J. Comput. Phys. **169**, 427–462 (2001). doi:10.1006/jcph.2000.6592
14. Kim, J., Kim, D., Choi, H.: An immersed-boundary finite-volume method for simulations of flow in complex geometries. J. Comput. Phys. **171**(1), 132–150 (2001). doi:10.1006/jcph.2001.6778
15. Lehmann, T.: Elemente der Mechanik III: Kinetik, 2nd edn. Vieweg, Braunschweig/Wiesbaden (1983)
16. Patankar, N.A.: A formulation for fast computations of rigid particulate flows. Center for Turbulence Research, Annual Research Briefs, pp. 185–196 (2001)
17. Patankar, N.A., Singh, P., Joseph, D.D., Glowinski, R., Pan, T.W.: A new formulation of the distributed lagrange multiplier/fictitious domain method for particulate flows. Int. J. Multiph. Flow **26**, 1509–1524 (2000). doi:10.1016/S0301-9322(99)00100-7
18. Peskin, C.S.: Numerical analysis of blood flow in the heart. J. Comput. Phys. **25**, 220–252 (1977). doi:10.1016/0021-9991(77)90100-0
19. Peskin, C.S.: The immersed boundary method. Acta Numer. **11**, 479–517 (2002). doi:10.1017/S0962492902000077
20. Rauschenberger, P., Criscione, A., Eisenschmidt, K., Kintea, D., Jakirlić, S., Tuković, Ž., Roisman, I.V., Weigand, B., Tropea, C.: Comparative assessment of volume-of-fluid and level-set methods by relevance to dendritic ice growth in supercooled water. J. Comput. Fluids **79**(0), 44–52 (2013). doi:10.1016/j.compfluid.2013.03.010
21. Rider, W.J., Kothe, D.B.: Reconstructing volume tracking. J. Comput. Phys. **141**(2), 112–152 (1998). doi:10.1006/jcph.1998.5906
22. Rieber, M.: Numerische Modellierung der Dynamik freier Grenzflächen in Zweiphasenströmungen. Dissertation, Universität Stuttgart (2004)
23. Sanders, J., Dolbow, J.E., Mucha, P.J., Laursen, T.A.: A new method for simulating rigid body motion in incompressible two-phase flow. Int. J. Numer. Methods Fluids **67**(6), 713–732 (2010). doi:10.1002/fld.2385
24. Sharma, N., Patankar, N.A.: A fast computation technique for the direct numerical simulation of rigid particulate flows. J. Comput. Phys. **205**, 439–457 (2005). doi:10.1016/j.jcp.2004.11.012

25. Uhlmann, M.: An immersed boundary method with direct forcing for the simulation of particulate flows. J. Comput. Phys. **209**, 448–476 (2005). doi:10.1016/j.jcp.2005.03.017
26. van Leer, B.: Towards the ultimate conservative difference scheme. V. A second-order Sequel to Godunov's method. J. Comput. Phys. **32**(1), 101–136 (1979). doi:10.1016/0021-9991(79)90145-1
27. Veeramani, C., Minev, P.D., Nandakumar, K.: A fictitious domain formulation for flows with rigid particles: a non-Lagrange multiplier version. J. Comput. Phys. **224**, 867–879 (2007). doi:10.1016/j.jcp.2006.10.028
28. Wesseling, P.: An Introduction to Multigrid Methods. Wiley, Chichester (1992)
29. Yoon, H.S., Jeon, C.H., Jung, J.H., Koo, B., Choi, C., Shin, S.C.: Simulation of two-phase flow–body interaction problems using direct forcing/fictitious domain–level set method. Int. J. Numer. Methods Fluids (2013). doi:10.1002/fld.3797
30. Yu, Z., Shao, X.: A direct-forcing fictitious domain method for particulate flows. J. Comput. Phys. **227**, 292–314 (2007). doi:10.1016/j.jcp.2007.07.027

Certain Aspects of Conditional Moment Closure for Spray Flame Modelling

S. Ukai, A. Kronenburg, and O.T. Stein

Abstract Large-eddy simulations (LES) have been coupled with a conditional moment closure (CMC) method for the improved modelling of small scale turbulence-chemistry interactions in turbulent spray flames. Partial pre-evaporation of the liquid fuel prior to exiting the injection nozzle requires a modified treatment for the boundary conditions in mixture fraction space and mixture fraction subgrid distribution and conditionally averaged subgrid dissipation need to be known. Different modelling approaches for the subgrid distribution of mixture fraction have been assessed, but the modelling of subgrid scalar dissipation that is responsible for the subgrid fuel transport from the droplet surface towards the cell filtered mean has not been forthcoming. Instead, we introduce a new conditioning method based on two sets of conditional moments conditioned on two differently defined mixture fractions: the first mixture fraction is a fully conserved scalar, the second mixture fraction is based on the fluid mass originating from the liquid fuel stream and is strictly not conserved due to the evaporation process. The two-conditional moment approach is validated by comparison with measurements from a turbulent ethanol spray flame and predicted temperature and velocity profiles could significantly be improved when compared to conventional LES-CMC modelling.

1 Introduction

The numerical simulation of turbulent spray flames needs to account for interactions of turbulence with fuel evaporation and chemistry. These interactions tend to be small scale effects and need to be modelled. One model that can account for these interactions is the conditional moment closure (CMC) method. Previous CMC computations of turbulent spray flames have used the single phase CMC equations in the RANS [17, 18] and LES [4] context. Some modifications that accounted for the presence of sprays were used by [7, 13], but Mortensen and Bilger [9] only later derived a fully consistent CMC formulation for two-phase flows. The extended

S. Ukai • A. Kronenburg (✉) • O.T. Stein
Institut für Technische Verbrennung, University of Stuttgart, Stuttgart, Germany
e-mail: kronenburg@itv.uni-stuttgart.de

W.E. Nagel et al. (eds.), *High Performance Computing in Science and Engineering '14*, DOI 10.1007/978-3-319-10810-0_23

CMC equations were applied to auto-ignition studies [3, 14] where one conditional moment can sufficiently approximate the correct solution. Ukai et al. [15] applied two-phase LES-CMC to a dilute acetone spray jet flame with pre-evaporation and discussed the impact of the additional spray terms on the conditional moments. Key issues that limited the quality of predictions were identified as (a) the correct choice of the upper boundary in mixture fraction space, (b) the accurate modelling of the sub-grid distribution of mixture fraction and (c) the modelling of the transport of fuel from the droplet surface to the surrounding gas phase. The latter issue is equivalent to the diffusion in mixture fraction space from the mixture fraction value prevalent at the surface to the LES filtered mean.

The remainder of this chapter focuses primarily on certain aspects for the closures in the CMC context. First the governing equations for the flow and mixing fields are presented. The CMC concept is introduced and issues for the modelling of the sub-grid distribution of mixture fraction and its dissipation arising from the presence of the spray are assessed. A methodology for the dynamic treatment of the upper boundary in mixture fraction space is introduced and a new two-conditional moment approach is discussed. The final modelling approach is then validated by comparison with measurements from a well defined experiment by Masri and Gounder [8].

2 Gas and Liquid Phase Formulations

LES solves spatially filtered equations with a modelling of subgrid scale effects. The Favre-filtered continuity and momentum equations read

$$\frac{\partial \bar{\rho}}{\partial t} + \frac{\partial}{\partial x_j}(\bar{\rho}\widetilde{u_j}) = \bar{\dot{\rho}}, \tag{1}$$

$$\frac{\partial}{\partial t}(\bar{\rho}\widetilde{u_i}) + \frac{\partial}{\partial x_j}(\bar{\rho}\widetilde{u_i}\widetilde{u_j}) = -\frac{\partial \bar{p}}{\partial x_i} + \frac{\partial \bar{\tau_{ij}}}{\partial x_j} - \frac{\partial \tau_{ij}^{sgs}}{\partial x_j} + \bar{\dot{F_i}}, \tag{2}$$

where ρ is density, u_i are the velocities in i direction, p is pressure and τ denotes the viscous stress tensor, $\bar{\dot{\rho}}$, $\bar{\dot{F_i}}$ are the source terms for mass and momentum that can be associated with the liquid phase, τ_{ij}^{sgs} is the subgrid stress tensor modelled by a dynamic subgrid eddy viscosity model closure.

Mixture fraction is a useful concept to analyze non-premixed flames. The mixture fraction is usually set to unity in the fuel stream, and the extrema of mixture fraction within the domain are bounded by 0.0 and 1.0. However, in turbulent spray flames with partial pre-evaporation, the maximum mixture fraction can exceed the value associated with the gaseous fuel jet at the inlet due to the evaporation process within the combustion chamber. We therefore suggest to solve transport equations for two mixture fractions simultaneously: the total mixture fraction, ξ_{tot}, and the conserved mixture fraction, ξ_{cons}. The definition of ξ_{tot} is based on the inlet conditions (a pilot flame and fuel originating from pre-evaporated droplets that lead to fuel vapour at the jet exit) *and* the fuel originating from droplet evaporation within the domain. ξ_{cons} is based on the fuel from pre-evaporation within the nozzle and the fuel elements of the pilot flame only. ξ_{cons} is not affected by the fuel evaporating from the droplets within the domain. Solving these two mixture fractions, the mixture fraction evaporated from the droplets after exiting the nozzle can be computed as $\xi_{\Delta} = \xi_{tot} - \xi_{cons}$, and it plays an important role in the following modelling approach. The maximum value of ξ_{tot} changes dynamically and is defined to be unity when the mixture is pure fuel. The maximum value of ξ_{cons} corresponds to the maximum inlet mixture fraction. The Favre-filtered scalar transport equations for the two mixture fractions are

$$\frac{\partial}{\partial t}(\bar{\rho}\tilde{\xi}_{tot}) + \frac{\partial}{\partial x_j}(\bar{\rho}\widetilde{u_j}\tilde{\xi}_{tot}) = -\frac{\partial J_{tot}}{\partial x_j} - \frac{\partial J_{tot}^{sgs}}{\partial x_j} + \bar{\dot{\xi}}, \tag{3}$$

$$\frac{\partial}{\partial t}(\bar{\rho}\tilde{\xi}_{cons}) + \frac{\partial}{\partial x_j}(\bar{\rho}\widetilde{u_j}\tilde{\xi}_{cons}) = -\frac{\partial J_{cons}}{\partial x_j} - \frac{\partial J_{cons}^{sgs}}{\partial x_j}, \tag{4}$$

where J is the diffusion flux of mixture fraction, $\bar{\dot{\xi}}$ is the source term of the mixture fraction due to evaporation within the domain, and J^{sgs} is the subgrid diffusion term. A Lagrangian particle tracking scheme is used for the liquid phase and droplets are tracked individually. Position, velocity, temperature and evaporation rate of each droplet are calculated and additional fluctuation terms are included to account for sub-grid effects [2].

3 Conditional Moment Closure

The CMC equations are formulated to solve for the conditionally averaged reactive scalars. The conditional moment of the species α is defined as $Q_\alpha(\xi, \mathbf{x}, t) \equiv \langle Y_\alpha(\mathbf{x}, t) | \xi(x, t) = \eta \rangle$, and it is often assumed that the conditional fluctuations within the flow are relatively small. However, in practice, the conditional fluctuation can be very large for spray flames, and it might be inaccurate to represent the properties by only one conditional moment. Therefore, two sets of conditional moments conditioned on ξ_{cons} and ξ_{tot} are solved simultaneously, and the solution is obtained by interpolation between the two conditional moments as presented later. The conservative form of the CMC equation for the conditional moment conditioned on ξ_{cons} is

$$\frac{\partial}{\partial t}Q_{\alpha,cons} + \frac{1}{\bar{\rho}_\eta \tilde{P}_\eta}\nabla \cdot \left[\bar{\rho}_\eta \tilde{P}_\eta \left(U_\eta Q_{\alpha,cons} - D_{t,\eta}\nabla Q_{\alpha,cons}\right)\right]$$
$$= \tilde{\omega}_{\eta,\alpha} + N_\eta \frac{\partial^2}{\partial \eta^2} Q_{\alpha,cons} + \frac{Q_{\alpha,cons}}{\bar{\rho}_\eta \tilde{P}_\eta}\nabla \cdot \left(\rho_\eta \tilde{P}_\eta U_\eta\right), \tag{5}$$

where subscript η denotes conditionally averaged quantities, N is the scalar dissipation rate, $\tilde{\omega}_\alpha$ is the chemical source term of species α, D_t is the turbulent diffusivity and P is the probability density function. The subscript *cons* indicates conditioning on the random variable, ξ_{cons}. Note that all conditioned quantities in Eq. (5) are conditioned on ξ_{cons}, but the subscripts are only shown for $Q_{\alpha,cons}$ to preserve some clarity of presentation. The CMC equation for the moments conditionally averaged on ξ_{tot} includes additional spray source terms [3, 9] as

$$\begin{aligned}\frac{\partial}{\partial t} Q_{\alpha,tot} + \frac{1}{\bar{\rho}_\eta \tilde{P}_\eta}\nabla\cdot\left[\bar{\rho}_\eta \tilde{P}_\eta\left(U_\eta Q_{\alpha,tot} - D_{t,\eta}\nabla Q_{\alpha,tot}\right)\right]\\ = \tilde{\omega}_{\eta,\alpha} + N_\eta \frac{\partial^2}{\partial \eta^2} Q_{\alpha,tot} + \frac{Q_{\alpha,tot}}{\bar{\rho}_\eta \tilde{P}_\eta}\nabla\cdot\left(\rho_\eta \tilde{P}_\eta U_\eta\right)\\ + \left[Q_{1,\alpha} - Q_{\alpha,tot} - (1-\eta)\frac{\partial}{\partial \eta} Q_{\alpha,tot}\right]\Pi_\eta,\end{aligned} \tag{6}$$

where $Q_{1,\alpha}$ denotes the composition of the liquid fuel, Π is the volumetric fuel evaporation rate. The subscript *tot* indicates conditioning on the random variable ξ_{tot} and all conditional properties in Eq. (6) are conditioned on ξ_{tot}. Corresponding equations are solved for the conditionally averaged enthalpies, Q_h.

4 Closures

The LES-filtered quantities can be obtained from the conditionally filtered quantities by integration across mixture fraction space. This implies not only modelling of the conditionally averaged quantities (Π_η, N_η, U_η and $D_{t,\eta}$) but also of the filtered density function P_η. The modelling of U_η and $D_{t,\eta}$ has been adopted from [10]. The accurate prediction of conditional scalar dissipation rates, N_η, is of primary importance for CMC simulations. Approaches developed in past studies can be mainly sorted into two categories: macroscopic and microscopic. Here, a macroscopic approach is defined as a methodology where the subgrid models are a function of filtered quantities only. In contrast, a microscopic approach is understood as an approach where a reconstruction of subgrid (or micro-sized) structures at or around the droplets is attempted.

4.1 Macroscopic Approach

In the macroscopic approach, the conditional scalar dissipation rate is constructed taking samples from the LES cells as discussed in [10]. Therefore, the modelling of the filtered subgrid scalar dissipation, $\tilde{N}_{sgs}$, plays an important role. The scalar dissipation rate is closely related to the mixture fraction variance, $\widetilde{\xi''^2}$, as often modelled as

$$\tilde{N}_{sgs} = \frac{1}{2} \frac{\mu_t}{Sc_t \Delta^2 C_\xi} \widetilde{\xi''^2}. \tag{7}$$

One of the simplest models to estimate the mixture fraction variance is an algebraic model that is based on local mixture fraction gradients as

$$\widetilde{\xi''^2} = C_\xi \Delta^2 \left(\frac{\partial \tilde{\xi}}{\partial x_j} \right)^2, \tag{8}$$

where C_ξ is usually set to equal $C_\xi = 0.09$. Another approach is to solve a transport equation of the mixture fraction variance for two-phase flow written as

$$\frac{\partial}{\partial t}(\bar{\rho}\widetilde{\xi''^2}) + \frac{\partial}{\partial x_j}(\bar{\rho}\tilde{u}_j \widetilde{\xi''^2}) = \frac{\partial}{\partial x_j}\left[\left(\bar{\rho} D + \frac{\mu_t}{Sc_t} \right) \frac{\partial \widetilde{\xi''^2}}{\partial x_j} \right] - 2\bar{\rho}\tilde{N}_{sgs} + 2\frac{\mu_t}{Sc_t}\left(\frac{\partial \tilde{\xi}}{\partial x_j} \right)^2 + \sigma_s, \tag{9}$$

where σ_s is the source term associated with the evaporation process given by

$$\sigma_s = 2\bar{\rho}(\widetilde{\xi \Pi} - \tilde{\xi}\tilde{\Pi}) + \bar{\rho}(\tilde{\xi}^2 \tilde{\Pi} - \widetilde{\xi^2 \Pi}), \tag{10}$$

and three deferent closures were suggested as listed in Table 1. Model 1 was developed by Pera et al. [11] and is fully based on the local LES quantities within the cell. Model 2 is used by Borghesi et al. [3], and it computes filtered terms by summing up the droplet evaporation given by the Lagrangian particle tracking method. Model 3 was proposed by Réveillon and Vervisch [12] and couples the subgrid scale spray evaporation with filtered quantities.

Table 1 Closures of σ_s based on the variance transport equation. $\tilde{\Pi}$ is the volume expansion rate per unit volume, $\tilde{\Pi} = \frac{1}{\bar{\rho}V} \sum_{i=1}^{N_d} \dot{m}_i$, $\dot{m}_i$ is the mass evaporation rate of i-th droplet, N_d is the number of the droplets per LES cell, V is the cell volume, and ξ_{surf} is the mixture fraction at each droplet surface

Model 1 [11]	$\sigma_s = \alpha \bar{\rho} \widetilde{\xi''^2} \left(\tilde{\Pi} / \tilde{\xi} \right)$ $\alpha = 0.5$
Model 2 [3]	$\sigma_s = 2\bar{\rho}(\widetilde{\xi \Pi} - \tilde{\xi}\tilde{\Pi}) + \bar{\rho}(\tilde{\xi}^2 \tilde{\Pi} - \widetilde{\xi^2 \Pi})$ $\widetilde{\xi \Pi} = \frac{1}{\bar{\rho}V} \sum_{i=1}^{N_d} \xi_{surf,i} \dot{m}_i$, $\widetilde{\xi^2 \Pi} = \frac{1}{\bar{\rho}V} \sum_{i=1}^{N_d} \xi^2_{surf,i} \dot{m}_i$
Model 3 [12]	$\sigma_s = 2\bar{\rho}\widetilde{\xi'' \Pi}(1 - \tilde{\xi}) - \bar{\rho}\widetilde{\xi''^2 \Pi}$ $\widetilde{\xi'' \Pi} = \frac{1}{\bar{\rho}V} \sum_{i=1}^{N_d} \sqrt{\widetilde{\xi''^2}} \dot{m}_i$, $\widetilde{\xi''^2 \Pi} = \frac{1}{\bar{\rho}V} \sum_{i=1}^{N_d} \widetilde{\xi''^2} \dot{m}_i$

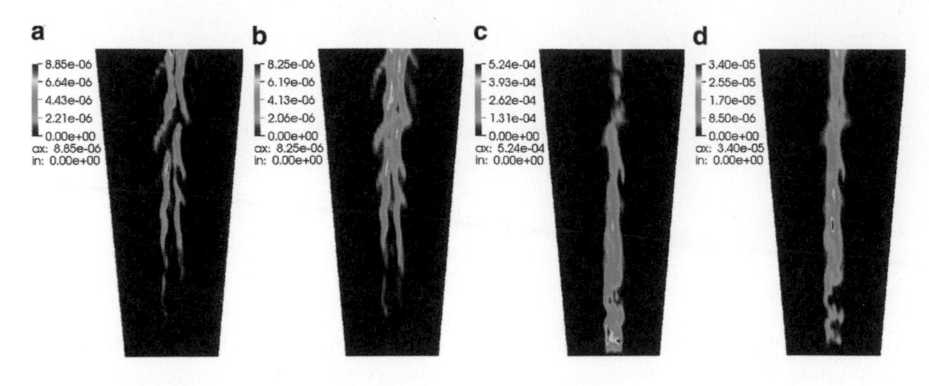

Fig. 1 Comparison of instantaneous mixture fraction variances for the different models specified in Table 1 and Eq. (8). (**a**) Algebraic model (Eq. (8)), (**b**) Model 1, (**c**) Model 2, (**c**) Model 3

These models are tested by simulating a piloted acetone spray flame (named AcF3) with partial pre-evaporation as described in [8]. The inflow gas velocity is 24 m/s and the liquid fuel mass flow rate is 150 g/min. The liquid fuel is seeded upstream of the nozzle exit, and a certain portion of the liquid fuel evaporates and appears in the gaseous phase. The velocities of the pilot and the co-flow are 11.9 and 4.5 m/s. The diameter of the jet, D, and the outer annulus diameter of the pilot and the co-flow are 10.5, 25 and 104 mm, respectively. The inlet cross-section of the computational domain is 10Dx10D and diverging towards the exit, and the domain length is taken to be 40D. The LES grid size is $90 \times 90 \times 240$, and the cells are clustered around the nozzle. Since LES subgrid models are to be investigated, the CMC grid can be rather coarse and comprises $3 \times 3 \times 40$ cells only. An acetone mechanism based on GRI-Mech 3.0 with seven additional acetone sub-steps [5] is used. The transport equations are solved with the respective closures, and results are compared to the algebraic model.

Figure 1 shows the mixture fraction variances computed by each approach, and it becomes immediately apparent that notable differences exist. Not only the magnitude of the variance is different, but also the profile shapes. For example, the algebraic model and Model 1 show small $\widetilde{\xi''^2}$ at the center of the jet near the inlet, but Models 2 and 3 predict large $\widetilde{\xi''^2}$ at the same location. Also, the maximum value in Model 2 is found to be 5×10^{-4} whereas the algebraic model and Model 1 predict peak values almost two orders of magnitude lower. This implies the variance source term estimated by Model 2 is at least two orders of magnitude larger than the shear production term (third term on the RHS in Eq. (9)). Such a large $\widetilde{\xi''^2}$ might lead to a large scalar dissipation rate (Eq. 7) that can extinguish flames.

4.2 Microscopic Approach

While the macroscopic approach provides scalar dissipation as a function of other filtered quantities, a microscopic approach attempts to obtain detailed information on the small scale structures around the (unresolved) droplet. However, there is no simple model applicable to general flow fields. We discuss two methodologies introduced in the existing literature: (1) conditioning on droplet surface quantities and (2) a single droplet stationary flow assumption.

4.2.1 Conditioning on Droplet Surface

Borghesi et al. [3] have discussed the scalar dissipation and the PDF at the spray surface. They used the relationship between liquid surface density, Σ_s, and the mixture fraction gradient for the computation of the PDF:

$$P(\xi_{surf}) \left|\langle |\nabla\xi| \,|\xi_{surf}\rangle\right| = \Sigma_s. \tag{11}$$

Since the surface density and the gradient of mixture fraction at the surface can be modelled by

$$\Sigma_s = \frac{1}{V} \sum_{i=1}^{N_{d,\xi_{surf}}} 4\pi r_i^2, \tag{12}$$

and

$$|\nabla\xi|_{\xi_{surf}} = ln[1 + B_{m,v}]Sh\frac{1-\xi_{surf}}{2r}, \tag{13}$$

where $N_{d,\xi_{surf}}$ is the number density with the specific surface mixture fraction computed from phase equilibrium, $B_{m,v}$ is the Spalding number for the vapour, r is the radius of the droplet, and Sh is the Sherwood number. The probability for the surface condition is given by

$$P(\xi_{surf}) = \frac{\Sigma_s}{|\langle\nabla\xi|\xi_{surf}\rangle|} = \frac{N_{d,\xi_{surf}}}{V} \frac{\sum_{i=1}^{N_{d,\xi_{surf}}} 4\pi r_i^2}{\sum_{i=1}^{N_{d,\xi_{surf}}} ln[1 + B_{m,v}]Sh\frac{1-\xi_{surf}}{2r}}. \tag{14}$$

Also, the conditional scalar dissipation at the droplet surface can be estimated based on the local gradient as

$$\langle N|\xi_{surf}\rangle = D\langle\nabla\xi^2|\xi_{surf}\rangle = \frac{D}{N_{d,\xi_{surf}}} \sum_{i=1}^{N_{d,\xi_{surf}}} \left(ln[1 + B_{m,v}]Sh\frac{1-\xi_{surf}}{2r}\right)^2. \tag{15}$$

The advantages of the approach are the following: the model can be valid even in a complex flow field, and it can be applied to polydispersed droplet clouds. However, the PDF and the conditional scalar dissipation rate of the mixture fraction are given *only* at the surface, but not for values between the filtered mean and at the surface ($\tilde{\xi}_{LES} < \xi < \xi_{surf}$). Therefore, it is difficult to establish a subgrid model that covers this mixture fraction region. The next subsection proposes some modelling of subgrid structure but it may be more constructive and practical to modify the boundary conditions in mixture fraction space as suggested in Sect. 4.3 below.

4.2.2 Droplet in Stationary Flow

Bilger [1] has derived an expression for conditional scalar dissipation and the PDF around a stationary, spherical, isolated droplet, leading to

$$\langle N|\eta\rangle = \frac{D}{r_s^2}\frac{[ln(1-\eta)]^4}{[ln(1+B)]^2}(1-\eta)^2, \tag{16}$$

$$P(\eta) \equiv \frac{\Sigma_\eta}{\langle|\nabla\xi||\eta\rangle} = 3f_v\frac{[ln(1+B)]^3}{[ln(1-\eta)]^4\,(1-\eta)}, \tag{17}$$

where r_s is the radius of the droplet, $B = \frac{\xi_{surf}-\xi_\infty}{1-\xi_{surf}}$, f_v is the volume fraction of the droplets and Σ_η and $\langle|\nabla\xi||\eta\rangle$ are defined as

$$\Sigma_\eta = \frac{3f_v}{r_s}\frac{[ln(1+B)^2]}{[ln(1-\eta)]^2}, \tag{18}$$

$$\langle|\nabla\xi||\eta\rangle = \frac{[ln(1-\eta)]^2}{[ln(1+B)]}\frac{(1-\eta)}{r_s}. \tag{19}$$

Note that $\frac{1}{1-\eta}$ in Eq. (17) is missing in the original paper. The advantage of Bilger's model is that it can model the PDF and N_η between the far field and ξ_{surf}. Bilger only considered far field conditions as $\xi_\infty = 0.0$, but it could be generalized for LES by setting ξ_∞ equal to the cell filtered value, $\xi_\infty = \tilde{\xi}_{LES}$. The models given in Eqs. (16)–(19) can then be rewritten as

$$\langle N|\eta\rangle = \frac{D}{r_s^2}\frac{\left[ln\left(\frac{1-\eta}{1-\tilde{\xi}_{LES}}\right)\right]^4}{[ln(1+B)]^2}(1-\eta)^2, \tag{20}$$

$$P(\eta) \equiv \frac{\Sigma_\eta}{\langle|\nabla\xi||\eta\rangle} = 3f_v\frac{[ln(1+B)]^3}{\left[ln\left(\frac{1-\eta}{1-\tilde{\xi}_{LES}}\right)\right]^4(1-\eta)}, \tag{21}$$

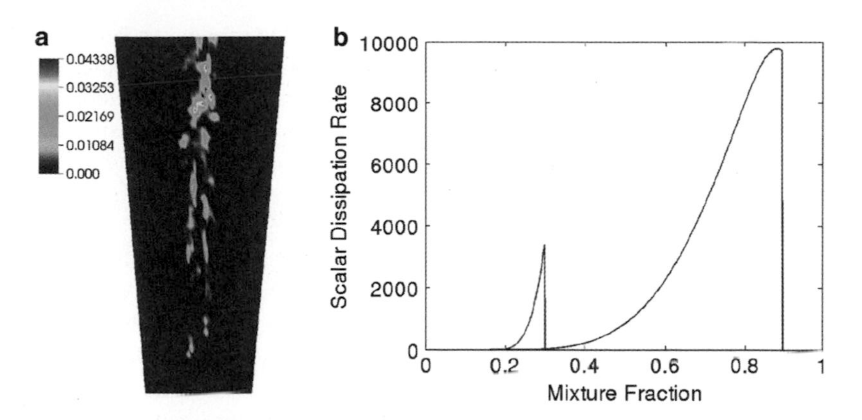

Fig. 2 Cell filtered scalar dissipation rates (*left*) in comparison with dissipation rates modelled by the microscopic model (*right*). (**a**) Contour plots of the filtered scalar dissipation rate with Model 1. (**b**) The conditional scalar dissipation rate around a droplet with 10 μm radius, for two different conditions (droplet temperatures) with $\tilde{\xi} = 0.16$ and $\xi_{surf} = 0.9$ (*blue line*) or $\xi_{surf} = 0.3$ (*red line*)

$$\Sigma_\eta = \frac{3f_v}{r_s} \frac{[ln(1+B)]^2}{\left[ln\left(\frac{1-\eta}{1-\tilde{\xi}_{LES}}\right)\right]^2}, \tag{22}$$

$$\langle|\nabla\xi||\eta\rangle = \frac{\left[ln\left(\frac{1-\eta}{1-\tilde{\xi}_{LES}}\right)\right]^2}{[ln(1+B)]} \frac{(1-\eta)}{r_s}. \tag{23}$$

The scalar dissipation rates obtained by the macroscopic and microscopic approaches are shown in Fig. 2. The macroscopic approach (here Model 1) shows $\tilde{N}_{sgs}$ of the order of 0.01 1/s whereas the microscopic model results in $\langle N|\eta\rangle \approx$ 1,000 1/s or even higher near the droplet surface. This is several orders of magnitude larger than the filtered dissipation rate and is caused by the very sharp gradients near the droplet surface. Such a large $\langle N|\eta\rangle$ implies that the time scale of vapour dissipation is very small and cannot be resolved within one CFD time step. In practical simulations, turbulence and shear flows increase the scalar dissipation rate further, so it might be adequate to assume fast mixing between $\tilde{\xi}$ and ξ_{surf}. Thus, a linear relation such as

$$Q_\alpha(\eta) = Q_\alpha\left(\tilde{\xi}_{LES}\right) + \frac{\eta - \tilde{\xi}_{LES}}{\xi_S - \tilde{\xi}_{LES}}\left[Q_\alpha(\xi_{surf}) - Q_\alpha\left(\tilde{\xi}_{LES}\right)\right] \tag{24}$$

can be assumed for the conditional moments for $\xi \in \left[\tilde{\xi}_{LES}, \xi_{surf}\right]$. However, there are still some difficulties to apply such models. One example is the estimation of the probability $P(\eta)$ in this interval.

Therefore, the remaining studies are based on the algebraic model presented in the previous subsection, and the scalar dissipation rates are estimated using a correction factor as presented in [15].

4.3 Boundary Condition in Mixture Fraction Space

The use of the macroscopic model makes us revisit the boundary conditions in mixture fraction space that are required for the discretization of the dissipation term (second RHS term in Eqs. (5) and (6)). In the acetone example flame under investigation some of the fuel is pre-evaporated leading to a rich mixture of $\xi_{jet} = 0.162$ exiting the jet, and the filtered mixture fraction can locally increase due to further evaporation of the droplets. Thus, the upper boundary within mixture fraction space changes and needs to be specified. We have seen above that fixing the boundary at the droplet surface does not lead to convincing closures due to the unknown subgrid structure and missing models for P_η and N_η for $\xi \in \left[\tilde{\xi}_{LES}, \xi_{surf}\right]$. For any other upper boundary the arising issues can easily be illustrated. Consider a CMC cell that spans from the fuel jet with pre-evaporation well into the pilot as shown in Fig. 3a. This is not an uncommon scenario due to the relatively large size of CMC cells. If there is no liquid droplet, the conditional temperature profile within the CMC cell would follow the black line as shown in Fig. 3b, and the upper bound could be set to $\xi_{UL} = \xi_{jet}$. If a single droplet moved – due to turbulent dispersion –

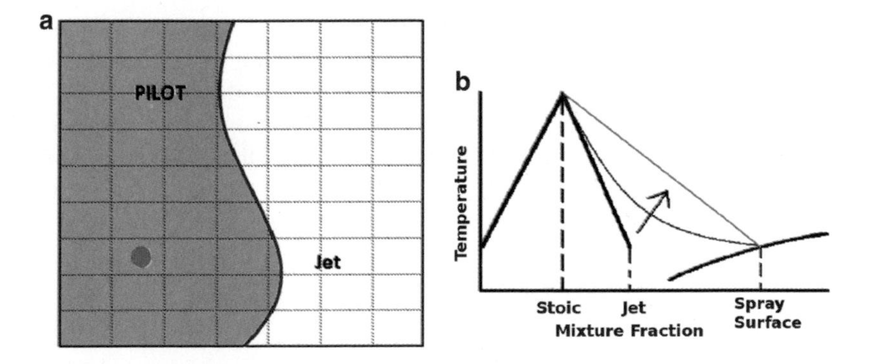

Fig. 3 Conceptual issues with mixture fraction boundaries in two-phase flows. (**a**) Representation in physical space: The large box with the *bold line* represents a CMC cell, and small boxes shown by *dotted lines* are LES cells. The left half (*red zone*) of the domain is a reacting solution originating from a piloted flame, and the right half of the domain (*white zone*) is non-reacting gas from the jet. The *blue circle* represents a droplet. (**b**) Mixture fraction space representation: The *thick solid black line* indicates the conditionally averaged temperature close to the jet exit, the *curved black line* indicates the droplet surface temperature as function of the equilibrium surface mixture fraction. The *grey* and *black dashed lines* illustrate the evolution of the conditional temperature due to diffusion in mixture fraction space (Adapted from [15])

into an LES cell located in the pilot, this LES cell will locally have a high mixture fraction equal to the surface value of the droplet, such as $\xi_{UL} = \xi_{surf}$, and a different solution of the conditional moment is likely to result as indicated by the green line in Fig. 3b. However, conventional CMC allows for one solution within a CMC cell only. If mixing between the jet and pilot was enforced, the high mixture fraction regime around the droplet would follow a non-reacting solution (the black line). If mixing between the droplet surface and the pilot mixture fraction is considered, $\xi_{UL} = \xi_{surf}$, and the temperature increases for $\eta = \xi_{jet}$ due to diffusion in mixture fraction space (green line). This implies that the non-reacting solution cannot be preserved at ξ_{jet}, even though the droplet does not physically interact with the jet as indicated in Fig. 3a. Thus, in theory, one single droplet in the pilot suffices to raise the jet temperature since the reactive-diffusive balance (the first and second RHS-terms in Eqs. (5) and (6)) is independent of mixture fraction probabilities and cannot distinguish between one droplet with low probability and a spray of pure fuel with high probability.

4.4 Selection of Upper Mixture Fraction Bound

For completeness, we will now present the method to determine ξ_{UL} from LES-filtered quantities. In a single phase flow, a non-reacting scalar mixes linearly between fuel and oxidizer in mixture fraction space. Thus, if the value of the conserved scalar, ϕ, at a certain mixture fraction is known, the value of ϕ at the upper boundary, ϕ_{UL}, is given by

$$\phi_{UL} = \frac{\xi_{UL}}{\xi}\phi, \tag{25}$$

and the slope is given by the mixing line. Using a single phase flow as an analogy, ξ_{UL} must be unmixed with the surroundings. In other words, the solution must lie on the unmixed limit which is denoted as the baseline ξ_{base}. The unmixed limit can vary, it is dependent on the amount of fuel evaporated in the LES cell and can be deduced from the two mixture fractions governed by Eqs. (3) and (4). The minimum amount of vapour emitted from the droplets within the domain can be defined as

$$\xi_{base} = \begin{cases} 0 & \text{for } \xi_{tot} < \xi_{jet} \\ \frac{\xi_{tot}-\xi_{jet}}{1-\xi_{jet}} & \text{for } \xi_{tot} \geq \xi_{jet} \end{cases}. \tag{26}$$

In other words, fluid originating from the jet and vapour originating from the liquid fuel exist along ξ_{base} for $\xi_{tot} \geq \xi_{jet}$, and mixing with the surrounding fluid (e.g. from the pilot or co-flow) has not yet occurred. These relationships are illustrated with the configuration of AcF3 in Fig. 4, and it shows the mixture fraction originating from the droplets in the domain for two different CMC cells. Since there is no mixing with

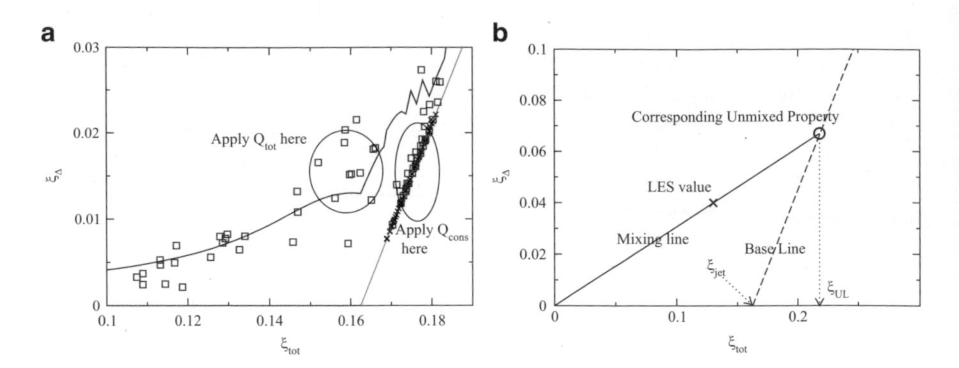

Fig. 4 Definition of a baseline mixture fraction and definition of the unmixed property with a corresponding upper bound in mixture fraction space. (**a**) Realizations of $\widetilde{\xi}_{\Delta}$ within a CMC cell that lies within the jet core (*black*) and a CMC cell at the interface of the jet core and the pilot (*red*). The *green line* is ξ_{base} (Eq. 26). The condition is based on AcF 3. (**b**) Estimation of ξ_{UL}. *Cross* – actual quantity in an LES cell ($\widetilde{\xi}_{tot}$, $\widetilde{\xi}_{\Delta}$), *circle* – (imaginary) unmixed property corresponding to the LES cell (ξ_{UL}, $\xi_{\Delta,@\xi_{UL}}$), *solid line – mixing line*, *dashed line* – ξ_{base}

the surrounding fluid, ξ_{Δ} within the jet core clusters near ξ_{base}, and the conditional fluctuation is relatively small. However, the samples from the CMC cell containing the jet core and the pilot show a wide distribution of ξ_{Δ} around $\xi_{tot} \approx 0.16$. It is obvious that a simple first order closure is not forthcoming with only one conditional moment since realizations along the baseline must represent cold (or non-reacting) solutions but those with large ξ_{Δ} can be hot and thus, conditional fluctuations will not be negligible. Thus, we have proposed a selection method of ξ_{UL} with the combination of the base mixture fraction and the mixing line as shown in Fig. 4 [16]. From the baseline mixture fraction an amount of $\xi_{\Delta} \equiv \xi_{tot} - \xi_{cons}$ at ξ_{UL} can be determined as

$$\xi_{\Delta,@\xi_{UL}} = \frac{\xi_{UL} - \xi_{jet}}{1 - \xi_{jet}}, \tag{27}$$

and the linear mixing line as in Eq. (25), (with $\phi_{UL} = \xi_{\Delta,@\xi_{UL}}$ and $\phi = \widetilde{\xi}_{\Delta}$) can be deduced,

$$\xi_{\Delta,@\xi_{UL}} = \frac{\xi_{UL}}{\widetilde{\xi}_{tot}}\widetilde{\xi}_{\Delta}. \tag{28}$$

Thus, the upper mixture boundary in mixture fraction space, ξ_{UL}, is obtained as a pseudo-unmixed material and can be written as

$$\xi_{UL} = \frac{\xi_{jet}}{1 - \frac{\widetilde{\xi}_{\Delta}}{\widetilde{\xi}_{tot}} + \frac{\widetilde{\xi}_{\Delta}}{\widetilde{\xi}_{tot}}\xi_{jet}}. \tag{29}$$

4.5 Interpolation Method

As indicated in Sect. 3 and illustrated by Fig. 3b, one conditional moment may not suffice to represent the species composition in one CMC cell. Therefore two sets of CMC equations are solved (cf. Eqs. (5) and (6)) with the appropriate boundary treatment. A suitable interpolation method between $Q_{\alpha,tot}$ and $Q_{\alpha,cons}$ needs to be established. It can be stated that $Q_{\alpha,cons}$ is the solution along ξ_{base}, and $Q_{\alpha,tot}$ is approximately the solution along the conditional mean of the evaporated fuel $\langle \xi_\Delta | \eta \rangle$, and the interpolation shall be dependent on $\tilde{\xi}_\Delta$. Thus, when $\tilde{\xi}_\Delta$ is close to ξ_{base}, the solution must be close to $Q_{\alpha,cons}$, and a large amount of $\tilde{\xi}_\Delta$ should weight the solution towards $Q_{\alpha,tot}$ as shown in Fig. 4. The weighting factor θ is given by [16]

$$Q_{\alpha,\eta} = \theta Q_{\alpha,\eta,tot} + (1-\theta) Q_{\alpha,\eta,cons}, \tag{30}$$

with

$$\theta = \begin{cases} \frac{\tilde{\xi}_\Delta - \xi_{base}}{\langle \xi_\Delta | \eta \rangle - \xi_{base}} & \text{if } \xi_{base} < \tilde{\xi}_\Delta < \langle \xi_\Delta | \eta \rangle \\ 1.0 & \text{if } \tilde{\xi}_\Delta > \langle \xi_\Delta | \eta \rangle \end{cases}. \tag{31}$$

After the weighted conditional mean is found, the LES-filtered solution can be obtained by integration across mixture fraction space using a bounded β-FDF (see e.g. [6]).

5 Results and Discussion

Here, a selection of results using the two-conditional moment closure approach is shown. A more detailed discussion of the simulation results can be found elsewhere [16]. The CMC grid size used here is $15 \times 15 \times 40$ and thus differs from the CMC grid used in Sect. 4. Computed profiles of the unconditionally averaged temperature are compared with experiments in Fig. 5. The comparisons of the two-conditional moment approach and the conventional one-moment approach used in [15] are shown for AcF3. Very low temperatures are observed near the centerline when using one moment only, since the upper mixture fraction bound was fixed at ξ_{jet} and the temperature rise in rich mixtures along the centerline cannot possibly be predicted. The two-conditional moment approach predicts the increase of centerline temperature successfully.

The computed spray axial velocity statistics are compared with measurements in Fig. 6. The conventional one-moment approach leads to lower velocities, whereas the two-conditional moment approach markedly improves the predictions. This is due to the improved temperature prediction: the increased temperature along the centerline leads to thermal expansion and increased droplet velocities.

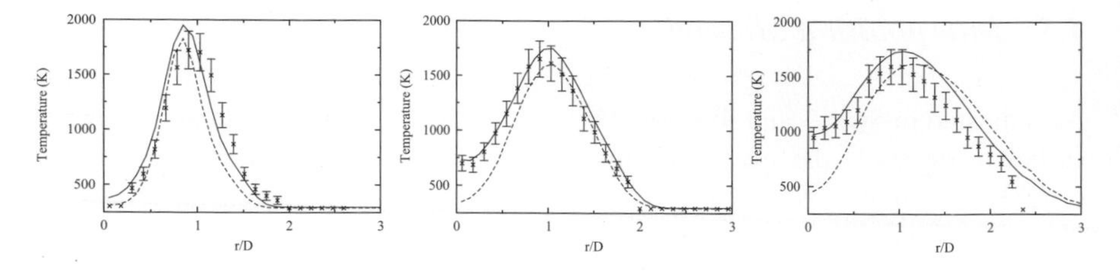

Fig. 5 Radial profiles of mean temperature at different downstream locations: z/D=10 (*left*); z/D=20 (*center*); z/D=30 (*right*). *Crosses* denote experiments [8] , the *solid line* denotes the two-conditional moment approach and the *dashed line* the conventional one-conditional moment approach [15], respectively

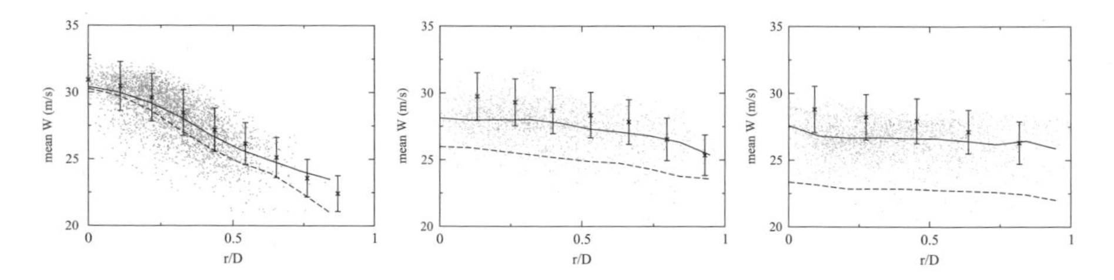

Fig. 6 Mean axial droplet velocity profiles for the diameter range 20–30 μm. The *crosses* denote experiments [8], scatter points are single realization of a simulated droplet, the *solid line* denotes the two-conditional moment approach and the *dashed line* the one-conditional moment approach [15], respectively

6 Performance

Code parallelization efficiency of the current implementation of the inhouse code BOFFIN on the CRAY XE6 (hermit) for the cases presented here is satisfactory. Most current computations have been carried out using 162 and 243 cores yielding a parallelization efficiency of above 70 %. Further test cases with approximately ten million ($144 \times 144 \times 512$) LES cells and 20,000 CMC cells have been carried out and results are presented in Fig. 7 . The scaling is satisfactory up to 972 cores. Note that the parallel scaling performance is measured here by computing the strong scaling efficiency and that the current domain decomposition does not allow an optimal use of the cluster architecture based on 32 cores per node. Weak scaling is likely to provide satisfactory scalability much beyond 2,000 cores. However, a corresponding demonstration is not attempted here since larger computational setups do not seem sensible for the relatively simple jet flame configurations under investigation. Future computations using the finer grids will require 486 and 972 cores and efficiencies are around or above 55 % for reacting flow simulations with liquid fuels. In addition, some restructuring of the code will be attempted to increase the efficiency beyond 70 % as could be obtained for LES computations for DME flames of similar size.

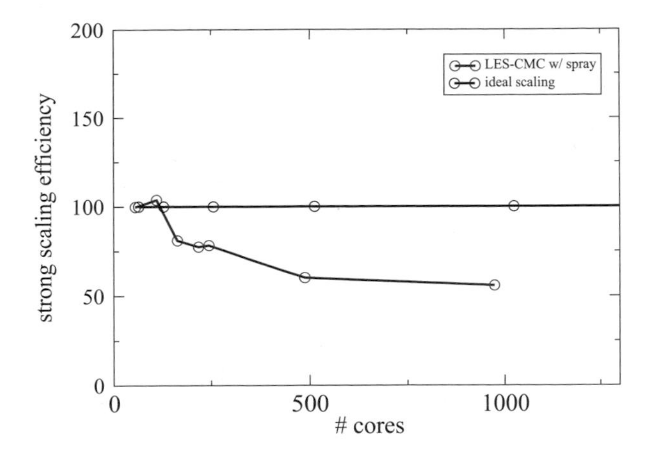

Fig. 7 Scaling based on approx. 10 m LES and 20,000 CMC cells

The presented analysis required approximately 100,000 CPU hours for one converged run. Approximately 78 % of the run time is used for the computation of the evolution of the conditionally averaged reactive species. The CPU usage for anticipated runs with finer LES and CMC meshes and different target flames and the development of a different modelling strategy based on the solution of an additional progress variable can be estimated to be around an additional 5 m CPU hours.

References

1. Bilger, R.W.: A mixture fraction framework for the theory and modeling of droplets and sprays. Combust. Flame **158**, 191–202 (2011)
2. Bini, M., Jones, W.: Large eddy simulation of an evaporating acetone spray. Int. J. Heat Fluid Flow **30**, 471–480 (2009)
3. Borghesi, G., Mastorakos, E., Devaud, C.B., Bilger, R.W.: Modeling evaporation effects in conditional moment closure for spray autoignition. Combust. Theory Model. **15**(5), 725–752 (2011)
4. Bottone, F., Kronenburg, A., Gosman, D., Marquis, A.: The numerical simulation of diesel spray combustion with LES-CMC. Flow Turbul. Combust. **89**(4), 651–673 (2012)
5. Chong, C., Hochgreb, S.: Measurements of laminar flame speeds of acetone/methane/air mixtures. Combust. Flame **158**(8), 490–500 (2011)
6. Ge, H.W., Gutheil, E.: Probability density function (pdf) simulation of turbulent spray flows. At. Sprays **16**(8), 531–542 (2006)
7. Kim, W.T., Huh, K.Y.: Numerical simulation of spray autoignition by the first-order conditional moment closure model. Proc. Comb. Inst. **29**, 569–576 (2002)
8. Masri, A.R., Gounder, J.D.: Turbulent spray flames of acetone and ethanol approaching extinction. Combust. Sci. Technol. **182**(4), 702–715 (2010)
9. Mortensen, M., Bilger, R.W.: Derivation of the conditional moment closure equations for spray combustion. Combust. Flame **156**, 62–72 (2009)

10. Navarro-Martinez, S., Kronenburg, A., di Mare, F.: Conditional moment closure for large eddy simulations. Flow Turbul. Combust. **75**, 245–274 (2005)
11. Pera, C., Réveillon, J., Vervisch, L., Domingo, P.: Modeling subgrid scale mixture fraction variance in LES of evaporating spray. Combust. Flame **146**, 635–648 (2006)
12. Réveillon, J., Vervisch, L.: Spray vaporization in nonpremixed turbulent combustion modeling: a single droplet model. Combust. Flame **121**, 75–90 (2000)
13. Rogerson, J.W., Kent, J.H., Bilger, R.W.: Conditional moment closure in a bagasse-fired boiler. Proc. Combust. Inst. **31**, 2805–2811 (2007)
14. Tyliszczak, A., Cavaliere, D., Mastorakos, E.: LES/CMC of blow-off in a liquid fueled swirl burner. Flow Turbul. Combust. **92**, 237–267 (2014)
15. Ukai, S., Kronenburg, A., Stein, O.T.: LES-CMC of a dilute acetone spray flame. Proc. Combust. Inst. **34**, 1643–1650 (2013)
16. Ukai, S., Kronenburg, A., Stein, O.T.: Simulation of dilute acetone spray flames with LES CMC using two conditional moments. Flow Turbul. Combust., **93**, 405–423 (2014)
17. Wright, Y., Margari, O.N., Boulouchos, K., De Paola, G., Mastorakos, E.: Experiments and simulations of n-heptane spray auto-ignition in a closed combustion chamber at diesel engine conditions. Flow Turbul. Combust. **84**, 49–78 (2010)
18. Wright, Y.M., DePaola, G., Boulouchos, K., Mastorakos, E.: Simulations of spray autoignition and flame establishment with two-dimensional CMC. Combust. Flame **143**, 402–419 (2005)

Comparative Numerical Investigation of a Sonic Jet in a Supersonic Turbulent Crossflow

S. Eberhardt and S. Hickel

Abstract We scrutinize new and previously published Large Eddy Simulation (LES) data for a sonic jet injected into a supersonic crossflow. Jet and crossflow consist of air, the crossflow Mach number is $M_\infty = 1.6$ and the jet to crossflow momentum ratio is $J = 1.7$. For this case, experimental measurements and rich numerical data are available in literature. Results of previous numerical studies and our LES are in excellent agreement. However, all numerical data sets show very similar deviations from the available measurements. We analyse sensitivities on uncertain flow parameters to identify the parameters that are most likely responsible for the discrepancy between the numerical and experimental data.

1 Introduction

In a scramjet combustor, efficient and fast mixing of injected fuel with the surrounding airflow is essential to enable supersonic combustion because of the extremely short residence time of the reactants in the combustion chamber. Even for fast reacting fuels like Hydrogen, the ignition delay time typically is of the same order of magnitude as the flow through time. State of the art and a widely studied way of injecting fuel into a scramjet combustor is a sonic jet of gaseous fuel that is injected perpendicularly into the supersonic main stream.

A schematic of the jet in supersonic crossflow (JISC), showing the flow from the side (a) and top (b), is provided in Fig. 1. The flow configuration consists of a wall with a circular hole through which the jet enters the supersonic crossflow. At the wall a fully turbulent supersonic boundary layer (1) has developed in front of the injection location. The jet that exits the hole penetrates through the boundary layer into the free stream where it blocks and displaces the supersonic part of the flow. This blockage causes a bow shock (2), which interacts with the boundary layer. The boundary layer separates due to the induced pressure gradient. The recirculation region thickens the boundary layer and acts like a compression corner

S. Eberhardt (✉) • S. Hickel
Institute of Aerodynamics and Fluid Mechanics, Boltzmannstr. 15, 85748 Garching, Germany
e-mail: sebastian.eberhardt@tum.de; sh@tum.de

W.E. Nagel et al. (eds.), *High Performance Computing in Science and Engineering '14*, DOI 10.1007/978-3-319-10810-0_24

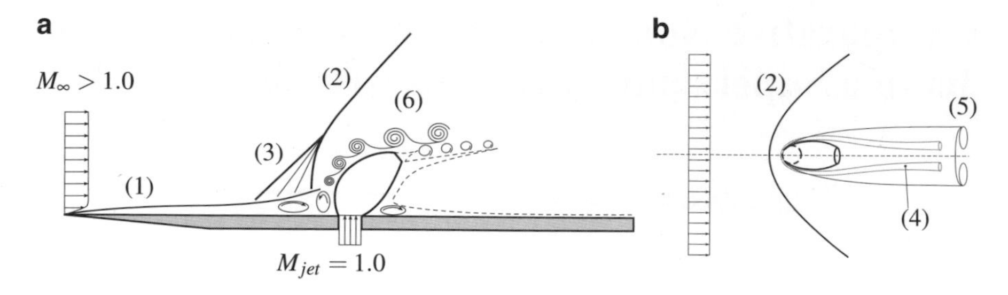

Fig. 1 Sketch of a jet in supersonic crossflow from the side (**a**) and top (**b**)

with a separation shock (3). Around the jets base close to the wall a horse shoe vortex is created (4). A second large counter-rotating vortex pair (5) rolls around the jet and is transported downstream by the free stream. An additional recirculation forms at the downstream side of the jet. The separated flow regions around the jet exit play an important role with respect to the mixing of the jet and crossflow fluids. The turbulent mixing and shear layer (6) on the upper side of the jet carries disturbances that originate from the upstream recirculation. The shear stress, due to different velocities in the two streams, dominates the near-field mixing-layer development [12]. In this case the vertical velocity of the jet at the exit is very large compared to the vertical velocity of the free-stream, which is almost zero. Kelvin-Helmholtz instability, triggered by upstream perturbations, leads to periodic vortex formation. The roll up has several effects; it pulls in fluid from both streams and thereby increases the area of the interface between the two fluids. The mixing finally occurs due to molecular diffusion, which is enhanced by the enlarged interface and concentration gradients.

A larger convective velocity increases the shear rate between the two fluids. In general this leads to a faster roll up and entrainment and consequently to faster mixing. This however produces losses. Turbulent shear layers transport momentum, vorticity and molecules of the two fluids. In a free shear layer this causes an increase in viscous dissipation of kinetic energy to thermal energy and irreversible entropy increase and total-pressure loss.

In compressible flows the effect of strong roll ups and the associated faster shear layer growth rate is counteracted by compressibility effects, which decrease the growth rate significantly while the convective Mach number rises. The near-field mixing is dominated by the macroscopic stirring of the fluids, while the far-field mixing depends on the small scale turbulence and molecular diffusion. For jet in supersonic crossflow configurations the transition from near-field to far-field mixing occurs at a downstream position of 10–20 jet diameters [12]. The simulations in this study focus on the near-field mixing.

We conducted implicit Large Eddy Simulations (LES) of an under-expanded sonic air jet in a supersonic air crossflow [2], see Sects. 2 and 3, with a crossflow Mach number of $M_\infty = 1.6$ and a jet to crossflow momentum ratio of $J = 1.7$. This particular configuration is based on the experimental work of Santiago and

Dutton [11], Everett et al. [3] and Lerberghe et al. [13]. Our initial setup (baseline) follows a previous numerical study of Kawai and Lele [8], who, for computational feasibility reasons, simulated a six times lower Reynolds number as compared to the experiments. Kawai and Lele conducted the simulation for both, a fully turbulent and a laminar boundary layer in front of the jet. As the results for the turbulent case match the experimental measurements closer, we use a fully turbulent boundary layer. We also include numerical results of Chai and Mahesh [1] in our analysis. In Sect. 5 we analyze the mentioned numerical and experimental data. We then evaluate the sensitivity of numerical predictions on uncertain flow parameters, which are the possible reasons for observed differences between the experimental and numerical results, in Sect. 6.

2 Numerical Method and Thermodynamic Model

The non-dimensionalized, compressible Navier Stokes equations are solved on an adaptive Cartesian grid. For time-integration the explicit 3rd order accurate Runge-Kutta scheme of Gottlieb and Shu [4] and a finite-volume spatial discretization are used. The subgrid-scale (SGS) turbulence model is provided by the Adaptive Local Deconvolution Method (ALDM), see Ref. [6, 7], which follows an implicit LES (ILES) approach. ALDM is implemented for Cartesian collocated grids and used to discretize the convective terms of the Navier-Stokes equations. The diffusive terms are discretized by 2nd order centered differences. A fully conservative immersed boundary technique, based on a cutcell approach [5], is employed to represent the circular geometry of the jet injection nozzle on the Cartesian grid.

Jet and crossflow fluids are modeled as perfect gas with thermodynamic properties of air, a Prandtl number of $Pr = 0.7$ and a Schmidt number of $Sc = 1.0$. The viscosity is provided by Sutherlands law with a reference state of $T_{ref} = 195.58\,\mathrm{K}$, $\mu_{ref} = 9.95 \cdot 10^{-5} Ps \cdot s$ and the Sutherland constant $S = 0.564475$. A passive scalar is added to the jets fluid to track its massfraction as it mixes with the crossflow.

3 Computational Details

All LES in this study were performed using our in-house code INCA[1] which is written in FORTRAN language. It uses a classical block-structured grid topology which is decomposed for parallelization. Three ghost-layers need to be exchanged as INCA uses discretization schemes that operate on six point stencils. For blocks

[1]http://www.inca-cfd.org

which are not part of the same process non-blocking communication according to the MPI-2.2 standard[2] is employed, otherwise values are copied directly.

To demonstrate the scaling capabilities of INCA we present a strong scaling study of INCA on Hermit. The test problem is chosen to be a compressible Taylor-green vortex. The computational domain is a cube with periodic boundary conditions in all directions. The domain is discretized by $N = 256^3$ cells with homogeneous grid spacing and the cube has an edge length of 2π on all sides. The domain is divided in equi-sized parts according to the number of cores used, starting with 32 cores up to 2,048. The number of cells per core, ghost cells per core, and ratio of ghost cells to cells is given in Table 1.

Figure 2 shows the strong scaling of INCA on Hermit. For two and three nodes (64 and 128 cores), we observe a super-linear scaling that we attribute to optimized cache usage for these processor configurations. Up to the maximum investigated number of nodes, we observe good strong scaling properties of INCA.

Table 1 Grid statistics for strong scaling of INCA

No. of cores	Cells/core	Ghost cells/core	Ghost cells/cells [%]
32	524,288	132,312	25.2
64	262,144	80,856	30.8
128	131,072	55,128	42.1
256	65,536	35,544	54.2
512	32,768	22,104	67.5
1,024	16,384	15,384	93.9
2,048	8,192	10,200	124.5

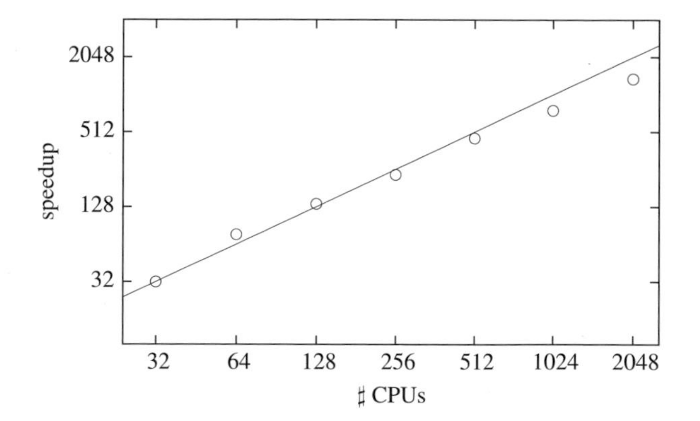

Fig. 2 Strong scaling of INCA on Hermit: speed-up on up to 64 nodes

[2]http://www.mpi-forum.org

Based on the strong scaling study, we conclude that an almost linear speed-up is possible by decomposing a computational grid into grid blocks with approximately 30,000 cells.

The computational grid in this case was divided into 784 blocks. As the blocks do not have identical cell count, but a maximum of 32,768, they were load balanced for 512 CPUs using the METIS[3] library. Runtimes for the simulations presented in this report are typically around 21 days for approximately 800,000 timesteps.

4 Flow Conditions and Numerical Setup

A sonic jet entering a supersonic crossflow of Mach number $M_\infty = 1.6$ is considered. All flow parameters are based on the experimental studies of Santiago and Dutton [11], VanLerberghe et al. [13] and Everett et al. [3]. The jet to crossflow momentum ratio is $J = \rho_j u_j^2 / \rho_{CF} u_{CF}^2 = p_j / p_{CF} = 1.7$ with ρ being the density, u the velocity and p the dynamic pressure. The subscripts j and CF denote quantities in the jet and in the crossflow, respectively. The crossflow has a stagnation temperature of $T_{CF,0} = 295\,\mathrm{K}$ and a stagnation pressure of $p_{CF,0} = 241\,\mathrm{kPa}$ as described by VanLerberghe et al. [13]. With the given Mach number this yields a static temperature and pressure of $T_\infty = 195.1\,\mathrm{K}$ and $p_\infty = 56.7\,\mathrm{kPa}$. The jet is defined by its stagnation temperature $T_{j,0} = 300\,\mathrm{K}$ and pressure $p_{j,0} = 476\,\mathrm{kPa}$. The jet exit orifice diameter measures $D = 4\,\mathrm{mm}$. The boundary layer of the cross flow is fully turbulent and has a thickness of about $\delta = 3.1$ mm a Reynolds number based on its thickness of roughly $Re_\delta = 1.1 \cdot 10^5$ at the injection location. As mentioned before, this Reynolds number is lowered in the baseline simulation to $Re_\delta = 18{,}611$.

The computational grid follows an adaptive mesh refinement (AMR) strategy with hanging nodes to adjust the mesh to different resolution requirements. Additionally a hyperbolic point distribution is applied above the wall to sufficiently resolve the boundary layer. The grid consist of about 4.8 million cells in the boundary layer region upstream of the injection and 11.8 million cells for the resolved volume around the jet. Layers of coarse buffer cells surround the resolved part of the grid to increase the domain size in order to avoid spurious effects of the boundary conditions. The resolved area measures $10.4D$ in x-direction, $3.9D$ in y-direction and $4D$ in z-direction. The resolution is based on the medium grid size of Kawai and Lele's grid convergence study [8]. In the boundary layer region it satisfies the $y^+ < 1.0$ condition, the x and z directions are resolved with $x^+ < 25.0$ and $z^+ < 12.5$. The bottom plate with the jet nozzle is modeled as an adiabatic wall in all simulations. The nozzle geometry is modeled using the immersed boundary technique [5].

[3]http://glaros.dtc.umn.edu/gkhome/views/metis

The experiment has been conducted in a test section that is large compared to the jet diameter (with a width of $19D$ and a height of $8D$, Everett et al. [3]), which means that the tunnel walls can be neglected in the numerical setup, if only the small volume around the jet is of interest. We therefore use a characteristic farfield condition for the top boundary and the side boundary conditions are symmetries. The distance of the symmetries to the resolved part of the grid is sufficiently large to ensure that shock reflections do not enter the domain of interest. At the supersonic outlet all gradients are set to zero. A constant total temperature and pressure is applied at the jet nozzle inlet. Transient turbulent inflow data for the supersonic, turbulent boundary layer is generated by a recycling-rescaling technique [9]. Samples for a statistical analysis have been collected for 157 D/u_∞ (jet diameter overflow times).

5 Baseline Case

In order to validate the baseline simulation, results are compared to the already mentioned experimental [3, 11, 13] and computational [1, 8] data. Here and in the following all quantities are non-dimensionalized by their free stream counterparts and the jet exit diameter D. Figures 3 and 4 give an overview of the observed instantaneous flow field on a wall normal slice ($z/D = 0.0$) and a wall parallel slice ($y/D = 0.005$). The four plots in Fig. 3 display the density gradient magnitude (a), the Mach number (b), the jet fluid massfraction (c) and the static temperature (d). In Fig. 4 the density gradient magnitude is displayed in (a) and the jet fluid massfraction in (b). We clearly recognize the characteristic compressible flow features, such as the primary bow shock and barrel shock with Mach disk.

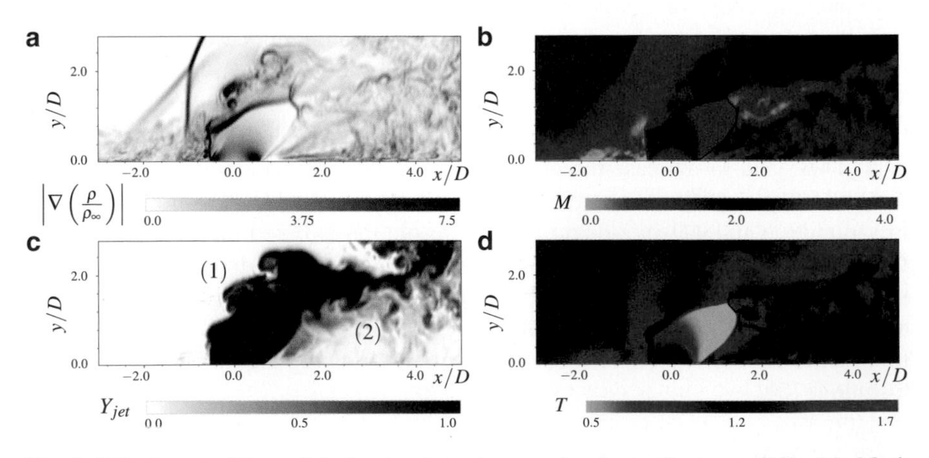

Fig. 3 2-D slice at $z/D = 0.0$ showing instantaneous density gradient magnitude (**a**), Mach number (**b**), jet fluid mass fraction (**c**) and static temperature (**d**) of a JISC

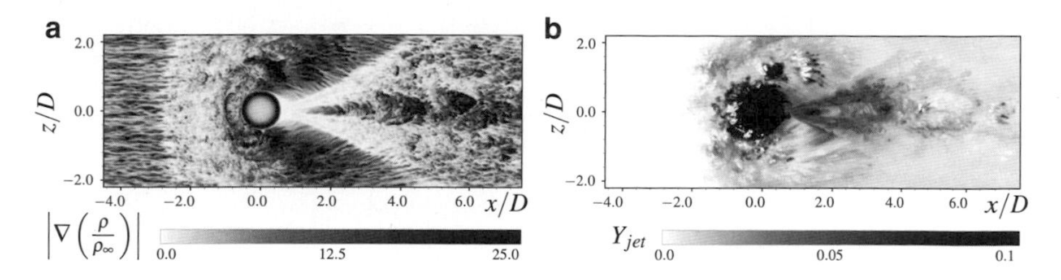

Fig. 4 2-D slice at $y/D = 0.005$ showing instantaneous density gradient magnitude (**a**) and jet fluid massfraction (**b**) of a JISC

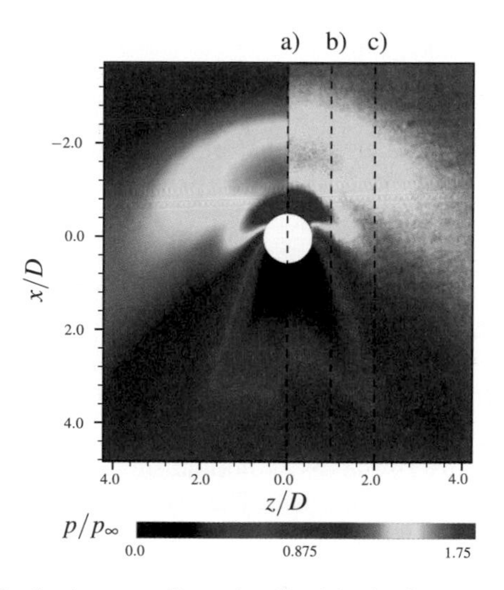

Fig. 5 Wall pressure distribution, non-dimensionalized by its freestream value p_∞. *Left*: LES simulation of this study, *right*: experimental, obtained by pressure sensitive paint taken from Ref. [3]

Turbulence structures of different length scales are resolved, e.g., large scales in the jets upstream side mixing layer ((1) in Fig. 3c), and small scales in the wake at the downstream side ((2) in Fig. 3c). The upstream recirculation and the signature of the horse shoe vortex are visible in the density gradient top view of Fig. 4. A typical quantity which can be measured comparatively easy in supersonic flows is the time averaged wall pressure. In Ref. [3] pressure measurements have been conducted with both pressure taps and pressure sensitive paint. The pressure sensitive paint yields two-dimensional time-averaged data of the wall pressure. A qualitative comparison of this two-dimensional field is displayed in Fig. 5. In front of the nozzle exit two high pressure regions are visible. The top one, which has a lower pressure value and is more spread out, originates from the separation bubble and the pressure rise across the separation shock. The second high pressure region is created by the primary bow shock and its maximum is limited by the theoretical pressure rise across a normal shock at M_∞. The upstream extension of the high pressure region

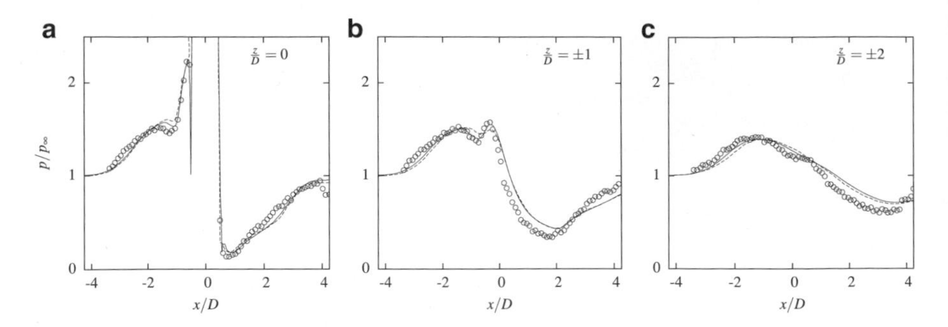

Fig. 6 Non-dimensionalized wall pressure p/p_∞ at three positions. ○ experimental Ref. [3], — LES of this study, - - - LES of Ref. [8]

of the simulation seems smaller than that of the experiment, but this is assumed to be an inaccuracy of the color scale of the paint measurement. This argument is strengthened by one-dimensional profiles extracted from the pressure-paint data, see Fig. 6. In Fig. 6 three stations are compared, the center line (a)), and lines one (b)) and two (c)) diameters off the center line, these are marked in Fig. 5 as well by the dashed lines. All three stations show very good agreement on the upstream side of the injection when compared to the measurements (○ symbols) and the reference simulation of Ref. [8] (dashed line, Ref. [1] does not provide pressure data). On the centerline the maximum of p/p_∞ reaches a value of 2.25, which is $\approx$80 % of the theoretical pressure rise across a normal shock at $M_\infty = 1.6$ and in very good agreement with the reference data. The two-dimensional low pressure region of our simulation is in excellent agreement with LES data of Kawai and Lele [8] (c.f. Fig. 10a of Ref. [8]). Both simulations, however, show almost identical small deviations from the experimental data in the low pressure region on the downstream side.

Streamwise, u, and wall-normal, v, velocity components one-dimensional profiles were measured with Laser Doppler Velocimetry (LDV). Figures 7 and 8 show these profiles at four locations, including the experimental values (○ symbols) [11], numerical reference data [8] (dashed line) and [1] (dotted line) together with the data of our simulation (solid line). Our results for the streamwise velocity are in excellent agreement with the reference LES data of Kawai and Lele [8] and in good agreement with the LES results of Chai and Mahesh [1] and the experimental data of Santiago and Dutton [10]. Best agreement between all simulations and the experiment is found for the $x/D = 4$ and $x/D = 5$ stations. For the stations closer to the injection ($x/D = 2$ and $x/D = 3$) we observe some discrepancies. First, the experiment shows higher flow speeds in direct vicinity to the wall than the simulations. Second, the location of the characteristic local minimum in the velocity profile and the value of the minimum velocity (dashed lines) differ. This indicates a lower penetration δ_p of jet fluid into the freestream. The location of the local minimum in the velocity profiles correlates with the penetration depth since the jet fluid has to be accelerated from a pure vertical motion to a horizontal one by the entrainment of the freestream.

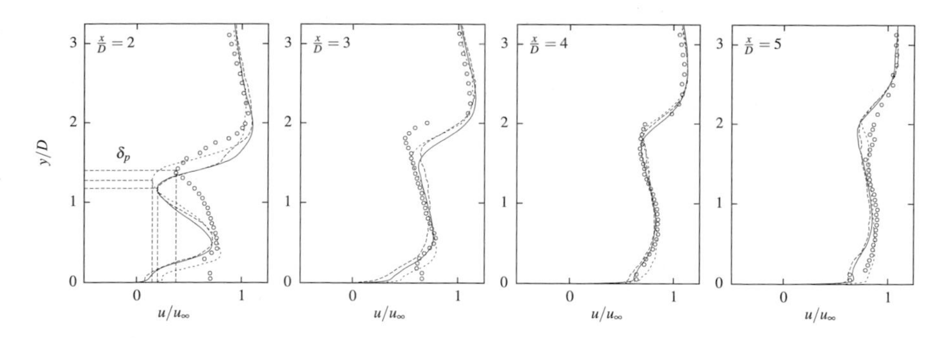

Fig. 7 1-D profiles of streamwise velocity u/u_∞ at several downstream stations on the center plane of the simulation. ∘ experimental Ref. [11], - - - LES Ref. [8], ··· LES Ref. [1], — LES this study

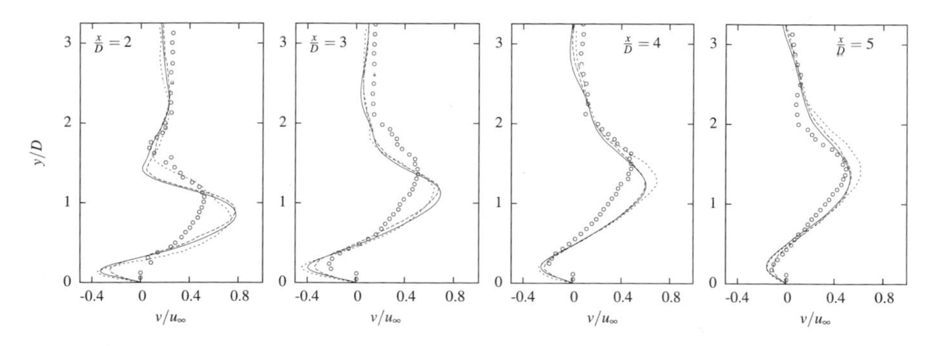

Fig. 8 1-D profiles of wall normal velocity v/u_∞ at several downstream stations on the center plane of the simulation. ∘ experimental Ref. [11], - - - LES Ref. [8], ··· LES Ref. [1], — LES this study

As the minimum velocity is subsonic, and the subsonic speed behind a shock is determined by its strength, this would indicate a stronger shock in the simulations than in the experiment. This argument is supported by comparison of the maximum Mach number reported in the experiment $M = 2.66$ [11] and the maximum Mach number of the simulations $M_{max} = 3.6$. Particle lag of the particles used for the LDV measurements may have biased the measurements towards a lower Mach number in front of the barrel shock and a higher velocity on the downstream side of the shock, as mentioned in Ref. [11].

While the streamwise velocity is lower than expected, we observe a higher than expected wall normal velocity, see Fig. 8. The penetration depth initially (at $x/D = 2$ and $x/D = 3$) is lower than observed in the experiment and agrees well with the experimental data at the downstream stations ($x/D = 4$, $x/D = 5$). This is consistent with the normal velocity being over predicted where the penetration depth is lower than in the experiment. In general profiles of the wall normal velocity show a similar tendency as the streamwise velocity profiles. All simulations deviate from the experiment but agree very well amongst each other. The difference to the experiment becomes smaller with increasing distance from the injection.

In conclusion, we see excellent agreement in the wall pressure comparison with the experimental measurements, as well as with other numerical investigations. Mean velocity field and turbulence statistics agree very well with previous simulations, which allows the conclusion that our numerical model is accurate in simulating a JISC. Results for all simulations, however, slightly deviate from the experimental data. We observe the most significant discrepancy in the penetration depth of the jet at the $x/D = 2$ and $x/D = 3$ stations.

6 Parameter Variation

In order to analyze the cause for some of the differences to the experiment, a parameter variation has been performed. First, we change the Reynolds number of the flow. This is the obvious starting point when searching for reasons for differences since this parameter has been intentionally changed for reducing the resolution requirements in the simulations, as mentioned earlier. Therefore the Reynolds number has been adjusted to twice the baseline value, then four times and in a last step to six times the baseline value, which corresponds to the experimental value.

The second influencing parameter we studied is the nozzle geometry and the inflow profile at the nozzle inlet. In all other simulations of this study a constant total pressure and total temperature are prescribed at the bottom of the jet nozzle. This resulted in a naturally developed velocity and density profile at the jet nozzle exit, which is plotted with a solid line in Fig. 9a. The dashed line is the nozzle-exit profile for the modified nozzle. This modified profile results from replacing the convergent nozzle with a short tube and prescribing a parabolic velocity profile at its bottom. The density (Fig. 9b)is adjusted in such a way that the momentum (Fig. 9c) integrated across the exit area is the same as in the original configuration. Third, we

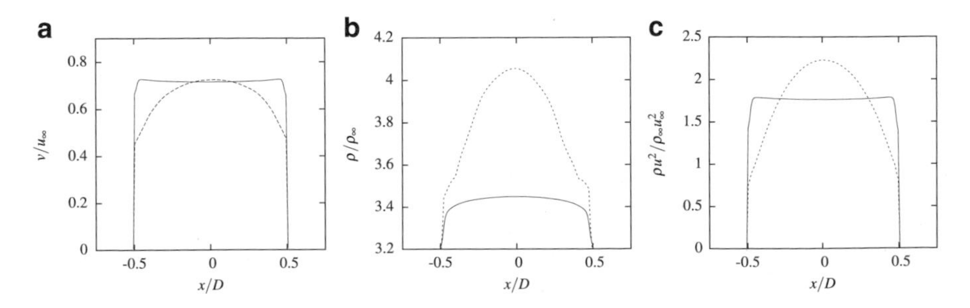

Fig. 9 Jet nozzle exit profiles. (a) vertical velocity v/u_∞, (b) density ρ/ρ_∞ and (c) momentum $\rho u^2/\rho_\infty u_\infty^2$. — standard nozzle, - - - modified nozzle

Table 2 Characteristic quantities of the boundary layer five jet diameters upstream of the injection location

(A)	(B)	(C)	(D)	(E)	(F)
$Re = 2Re_{bl}$	$Re = 4Re_{bl}$	$Re = 6Re_{bl}$	$T_0 = 0.9T_{0,bl}$	$p_0 = 1.1p_{0,bl}$	geometry

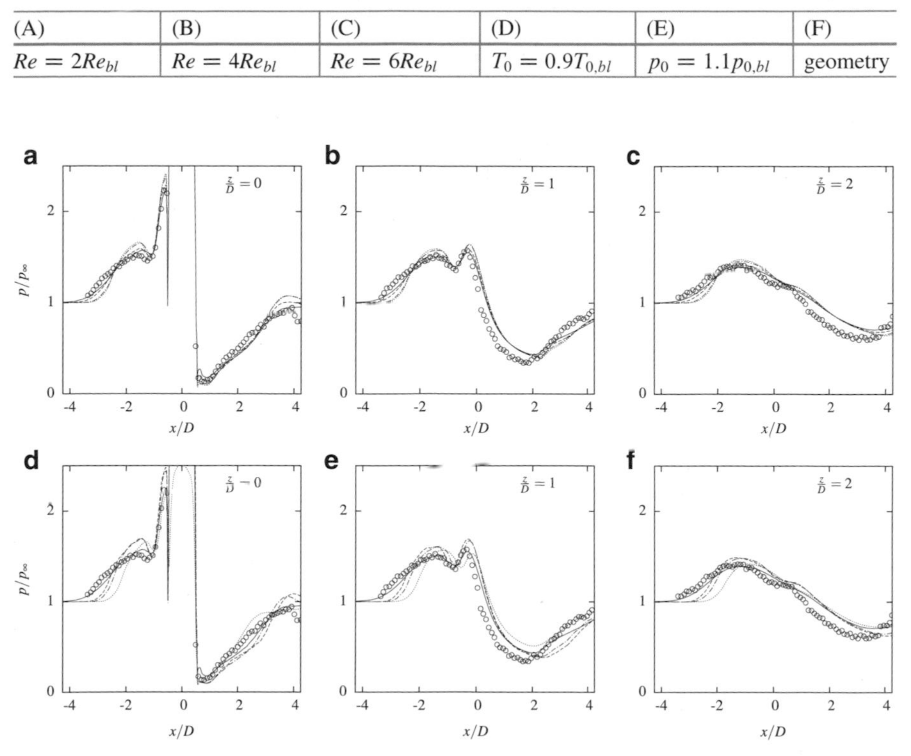

Fig. 10 Non-dimensionalized wall pressure p/p_∞ at three positions. ∘ experimental Ref. [3] in all plots. (a) to (c): — LES Re_{bl}, –– LES $2Re_{bl}$, –· LES $4Re_{bl}$, ··· LES $6Re_{bl}$. d) to f): — LES Re_{bl}, –– LES Δp_0, –· LES ΔT_0, ··· LES modified nozzle geometry

changed the total temperature and total pressure in separate simulations to slightly increase the momentum ratio. Table 2 summarizes all simulated cases with changes relative to the baseline case. The individual simulations will be referred to by the capital letters given in the table, the original setup is referred to as baseline for the remainder of the document. Investigating the wall pressure at the same locations as earlier, we can see in Fig. 10a–c that the effect of the Reynolds number on the wall pressure is relatively small. It affects mainly separated flow regions, because recirculation regions become smaller with lower viscosity. This is the reason for the curves staying at the value 1.0 for a slightly larger distance. The separation shock is steeper and thus stronger for a shorter recirculation, so the first pressure peak is slightly higher than in the baseline simulation. The simulations with changed inflow conditions are conducted with the highest Reynolds number. In the plots 10d–f we see very little influence of the changed total pressure and temperature, whereas the particular choice of the nozzle geometry has a larger effect. The modified nozzle-exit profile has less momentum at the jet edges and more momentum in the jet center.

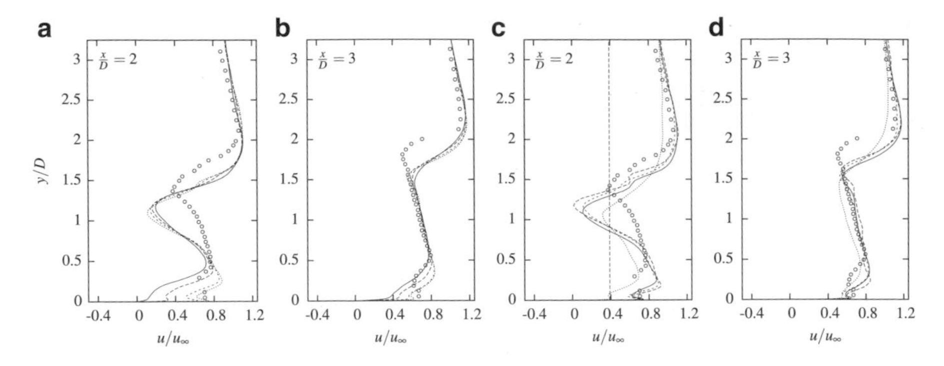

Fig. 11 1-D profiles of streamwise velocity u/u_∞ at two downstream stations on the center plane of the simulation. ∘ experimental Ref. [3] in all plots. a) and b): — LES Re_{bl}, –– LES $2Re_{bl}$, –· LES $4Re_{bl}$, ··· LES $6Re_{bl}$. c) and d): — LES Re_{bl}, –– LES Δp_0, –· LES ΔT_0, ··· LES modified nozzle geometry

As a result, there is less momentum exchange at the edges of the jet when it exits the nozzle and consequently recirculations and pressure rise are less emphasized.

Figure 11a, b show profiles of the streamwise velocity for the simulations with changed Reynolds number. Panels (c) and (d) show the effect of the jets pressure, temperature and nozzle variations. Only the two stations near the injection are plotted, since the differences diminish further downstream of the nozzle exit. In the profiles the most significant influence of the Reynolds number can be seen close to the wall. Figure 11 shows that the location and peak value of the velocity minimum is not notably affected by Reynolds number. However, the overall shape of the profiles has the tendency to agree better with the experiment.

Changing the total pressure, temperature and inflow profile has visible effect on the local velocity minimum. The location is shifted slightly upwards, as expected, because the modified boundary condition result in a slightly increased momentum ratio. The simulation with the modified nozzle geometry shows the most significant difference compared to the rest of the simulations. The location of the local velocity minimum moves closer to the wall than before but the actual minimum velocity is almost identical to the experiment, see vertical dashed line in Fig. 11c. At the $x/D = 3$ station the profile already recovers the general shape of the other simulations.

In comparison to the streamwise velocity, the vertical velocity is generally less affected by the parameters investigated here. Figure 12 displays v/u_∞ profiles at the same locations as Fig. 11. The most prominent effect can be seen at $x/D = 3$ for the inflow profile variation (F). Figure 13 shows the time averaged 2-D Mach number statistics for all four simulated Reynolds numbers. In contrast to the 1-D velocity profiles, a strong influence of the Reynolds number can clearly be seen. The maximum Mach number in the jet significantly increases with rising Reynolds number from $M_{max} = 3.6$ to $M_{max} = 4.2$. To visualize this effect, the sonic line and a $M = 3.5$ and a $M = 4.0$ isoline have been added to the plots in Fig. 13.

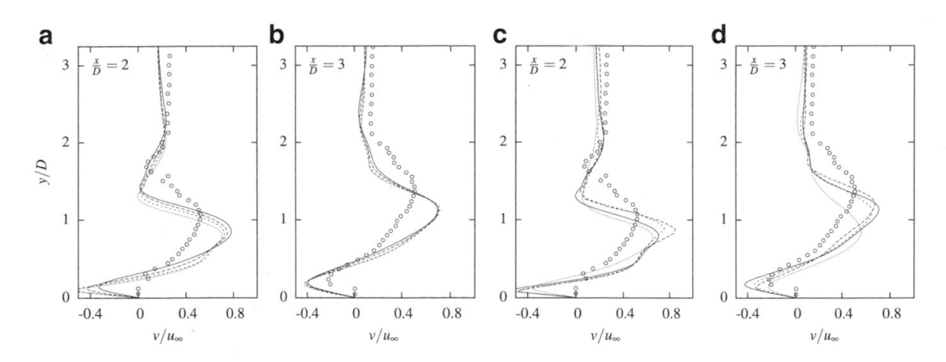

Fig. 12 1-D profiles of wall normal velocity u/u_∞ at two downstream stations on the center plane of the simulation. ∘ experimental Ref. [3] in all plots. a) and b): — LES Re_{bl}, -- LES $2Re_{bl}$, -· LES $4Re_{bl}$, ··· LES $6Re_{bl}$. c) and d): — LES Re_{bl}, -- LES Δp_0, -· LES ΔT_0, ··· LES modified nozzle geometry

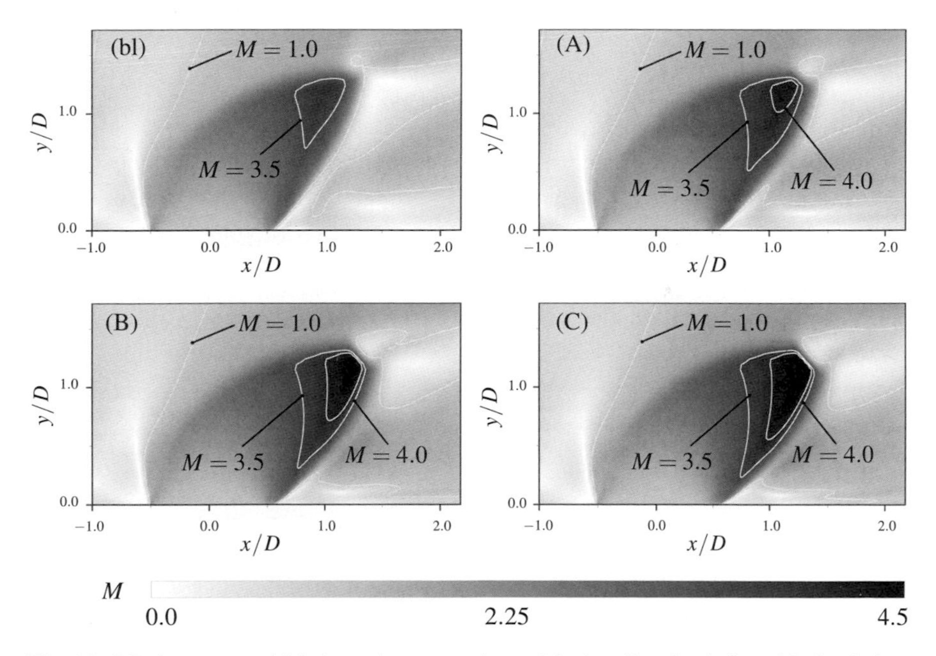

Fig. 13 2-D time averaged Mach number comparison of the baseline simulation with simulations A-C, the yellow solid line marks the sonic line

The recirculation regions shrink for the high Reynolds number cases, which results in a larger expansion angle α at the downstream side of the nozzle exit, see Fig. 14. The larger downstream expansion together with the unchanged penetration depth also increases the size of the Mach disk, which is clearly visible in Fig. 13. As the Mach disk is a normal shock in the jet flow, this also affects the subsonic region following the jet: a larger Mach disk leads to a larger subsonic region. The lowest

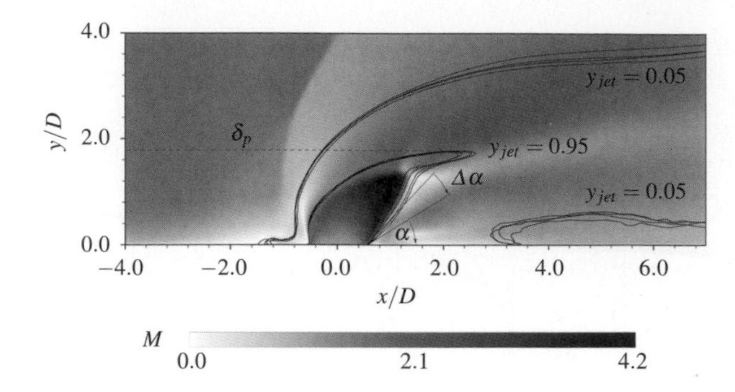

Fig. 14 2-D Mach number plot with added isolines of 5 and 95 % jet fluid mass fraction. Angle α depicts the expansion angle, $\Delta\alpha$ is the change from weakest to strongest expansion

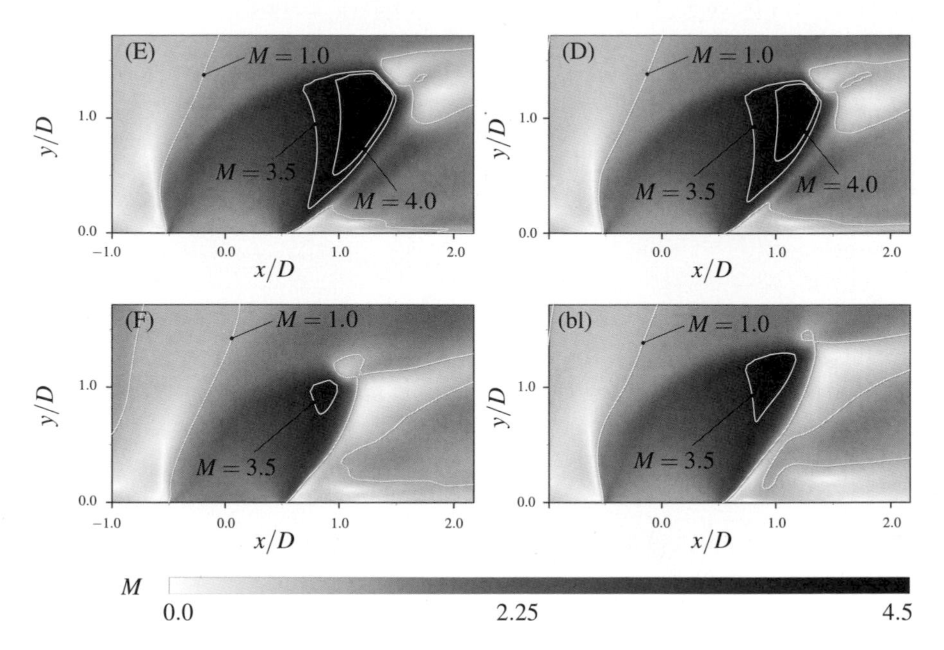

Fig. 15 2-D time averaged Mach number comparison of the baseline simulation with simulations D-F, the yellow solid line marks the sonic line

Mach number, which is also smaller than the one of the baseline simulation, yields the modified nozzle (F), see Fig. 15. This result resembles the experimental value closest of all simulations, however, the value measured in the experiment is still lower.

The penetration depth δ_p of the jet is visualized in Fig. 14 with isolines of jet fluid massfractions. The inner line depicts 95 % jet fluid, and the outer lines show the outer bound of the jet with a 5 % massfraction line.

7 Conclusion

We observed excellent agreement of three independently performed LES for a sonic jet that enters into a $M_\infty = 1.6$ supersonic crossflow (with jet to crossflow momentum ratio $J = 1.7$). Characteristic deviations from the reference experiment are recognized in all simulations, that is, all LES consistently show similar differences to the measurements.

In order to quantify uncertainties in reproducing the reference experiment by numerical simulations, we repeated the LES with different Reynolds numbers and jet inflow parameters. With the Reynolds number matching the experiment, the only obvious discrepancy of previous LES from the experimental setup had been eliminated. Nevertheless, we observed little improvement of the predicted penetration depth with increasing Reynolds number. On the other hand, Reynolds number has important effects on separated flow regions and on the maximum Mach number inside the jet in front of the Mach disk. We found that small variations in the jet inflow parameters, such as nozzle geometry and total pressure and temperature, can have a strong effect on the LES results.

Acknowledgements The support of this research within the Research Training Group "Aero-Thermodynamic Design of a Scramjet Propulsion System for Future Space Transportation Systems" 1095/2 by the Deutsche Forschungsgemeinschaft (DFG) and the High Performance Computing Center Stuttgart (HLRS) is greatly acknowledged.

References

1. Chai, X., Mahesh, K.: Simulations of high speed turbulent jets in crossflows. AIAA J. (2011)
2. Eberhardt, S., Hickel, S.: Implicit LES of a sonic jet in a supersonic crossflow. In: Direct and Large-Eddy Simulation IX. Springer (2013)
3. Everett, D.E., Woodmansee, M.A., Dutton, J.C., Morris, M.J.: Wall pressure measurements for a sonic jet injected transversely into a supersonic crossflow. J. Propuls. Power (1998)
4. Gottlieb, S., Shu, C.-W.: Total variation diminishing Runge-Kutta schemes. Math. Comput. (1998)
5. Grilli, M., Hickel, S., Hu, X.Y., Adams, N.A.: Conservative immersed interface method for compressible viscous flows with heat transfer. In: Academy Colloquium on Immersed Boundary Methods: Current Status and Future Research Directions. Amsterdam, The Netherlands (2009)
6. Hickel, S., Adams, N.A., Domaradzki, J.A.: An adaptive local deconvolution method for implicit LES. J. Comput. Phys. (2006). doi:10.1016/j.jcp.2005.08.017
7. Hickel, S., Larsson, J.: On implicit turbulence modeling for LES of compressible flows. Advances in Turbulence XII. Springer (2009). doi:10.1007/978-3-642-03085-7.209
8. Kawai, S., Lele, S.K.: Large-eddy simulation of jet mixing in supersonic crossflows. AIAA J. (2010)
9. Petrache, O., Hickel, S., Adams, N.A.: Large eddy simulations of turbulent enhancement due to forced shock motion in shock-boundary layer interaction. AIAA J. (2011)
10. Pirozzoli, S., Bernardini, M., Grasso, F.: Characterization of coherent vortical structures in a supersonic turbulent boundary layer. J. Fluid Mech. (2008)

11. Santiago,J.G., Dutton, J.C.: Velocity measurements of a jet injected into a supersonic crossflow. J. Propuls. Power (1997)
12. Segal, C.: The Scramjet Engine, Processes and Characteristics. Cambridge University Press (2009)
13. VanLerberghe, W.M., Santiago, J.G., Dutton, J.C., Lucht, R.P.: Mixing of a sonic transverse jet injected into a supersonic flow. AIAA J. (2000)

LES of Temporally Evolving Turbulent Cavitating Shear Layers

Christian Egerer, Stefan Hickel, Steffen Schmidt, and Nikolaus A. Adams

Abstract We present LES results of temporally evolving cavitating shear layers. Cavitation is modeled by a homogeneous equilibrium mixture model whereas the effect of subgrid-scale turbulence is accounted for by the Adaptive Local Deconvolution Method (ALDM). We quantitatively compare LES results with experimental data available in the literature. In terms of computational performance, we present a *strong scaling* study of our MPI-parallelized in-house FORTRAN code INCA on Cray XE6 "Hermit" at the High Performance Computeing Center Stuttgart (HLRS).

1 Introduction

Cavitation refers to the formation of vapor cavities in a liquid that undergoes phase change when subjected to low pressures. Coherent structures in turbulent shear flows play an important role in the cavitation process. High negative pressure peaks associated with turbulent eddies are the main source of cavitation in shear flows. Vice versa, cavitation can generate vorticity resulting in a complex mutual interaction, see Arndt [5] for a complete review.

Experimental studies of cavitating shear layers carried out in water tunnels have been published, for example, by O'Hern [18], Iyer and Ceccio [13], or Aeschlimann et al. [3, 4]. Questions related to cavitation and turbulence modulation by phase change have been addressed by the authors. Direct Numerical Simulation (DNS) of cavitating separated flow and cavitating shear layers based on the *incompressible* Navier-Stokes equations have been performed by Kajishima et al. [15] and Okabayashi and Kajishima [19], respectively.

By investigating a temporally evolving turbulent shear layer in water under non-cavitating and cavitating conditions, we address questions related to the mutual interaction of turbulence and cavitation, the resulting modulation of turbulence due to phase change, and its influence on properties of the shear layer. Within the

C. Egerer (✉) • S. Hickel • S. Schmidt • N.A. Adams
Lehrstuhl für Aerodynamik und Strömungsmechanik, Technische Universität München, Boltzmannstr. 15, 85748 Garching bei München, Germany
e-mail: christian.egerer@aer.mw.tum.de

W.E. Nagel et al. (eds.), *High Performance Computing in Science and Engineering '14*, DOI 10.1007/978-3-319-10810-0_25

present project, we plan to add to the rather scarce numerical data available on the aforementioned topics. In particular, effects of compressibility will be investigated.

2 Governing Equations and Numerical Method

We consider the 3-D compressible Navier-Stokes equations written in integral form for an arbitrary control volume Ω with surface $\partial\Omega$

$$\partial_t \int_{\Omega} \mathbf{U} d\Omega + \int_{\partial\Omega} [\mathbf{C}(\mathbf{U}) + \mathbf{D}(\mathbf{U})] \cdot \mathbf{n} dS = 0, \tag{1}$$

where $\mathbf{U} = [\rho, \rho u_i, \rho E]$ is the vector of conserved variables comprising density ρ, momentum densities ρu_i, and total energy $\rho E = \rho e + \rho u_k u_k/2$; $\mathbf{n}$ denotes the normal vector of the surface increment dS. The total flux across the control volume surface $\partial\Omega$ is split into its inviscid and viscous part

$$C_i(\mathbf{U}) = \begin{bmatrix} u_i \rho \\ u_i \rho u_1 + \delta_{i1} p \\ u_i \rho u_2 + \delta_{i2} p \\ u_i \rho u_3 + \delta_{i3} p \\ u_i(\rho E + p) \end{bmatrix}, \quad \text{and} \quad D_i(\mathbf{U}) = \begin{bmatrix} 0 \\ -\tau_{i1} \\ -\tau_{i2} \\ -\tau_{i3} \\ -\tau_{ik} u_k + q_i \end{bmatrix}. \tag{2}$$

p denotes the pressure,

$$\tau_{ij} = \mu \left(\frac{\partial u_i}{\partial x_j} + \frac{\partial u_j}{\partial x_i} - \frac{2}{3} \delta_{ij} \frac{\partial u_k}{\partial u_k} \right) \tag{3}$$

denotes the viscous stress tensor for a Newtonian fluid with dynamic viscosity μ, and

$$q_i = -\kappa \frac{\partial T}{\partial x_i} \tag{4}$$

denotes heat conduction according to Fourier's law, where T is the temperature and κ is thermal conductivity.

Considering an arbitrary control volume Ω, the solution is represented by the volume average of the vector of conserved variables

$$\overline{\mathbf{U}} = \frac{1}{V_\Omega} \int_{\Omega} \mathbf{U} dV; \qquad V_\Omega = \int_{\Omega} dV. \tag{5}$$

A homogeneous mixture of liquid and vapor is assumed in two phase regions, with $\alpha = \Omega_v/\Omega$ being the vapor volume fraction. Therefore, the mixture or volume averaged density and internal energy are given by the convex combination of the liquid and vapor densities and internal energies ρ_l, ρ_v, e_l, and e_v:

$$\overline{\rho} = (1-\alpha)\rho_l + \alpha\rho_v, \quad \text{and} \tag{6}$$

$$\overline{\rho e} = (1-\alpha)\rho_l e_l + \alpha\rho_v e_v. \tag{7}$$

Assuming that phase change is in thermodynamic equilibrium, i.e., infinitely fast, isentropic and in mechanical equilibrium, the vapor volume fraction can be computed as follows

$$\alpha = \begin{cases} 0 & , \overline{\rho} \geq \rho_{l,sat}(T) \\ \dfrac{\rho_{l,sat}(T) - \overline{\rho}}{\rho_{l,sat}(T) - \rho_{v,sat}(T)} & , \text{else} \end{cases}, \tag{8}$$

where $\rho_{l,sat}(T)$ and $\rho_{v,sat}(T)$ denote the temperature-dependent densities of the vapor and liquid phase at the saturation point. One major advantage of the homogeneous mixture methodology is its capability to reproduce subgrid-scale two-phase regions, e.g., small vapor bubbles, as well as fully resolved vapor structures, e.g., vapor clouds, with the underlying numerical grid. Note, however, that surface tension effects have to be neglected since no interface is reconstructed. Furthermore, no empirically calibrated constants are needed in the computation of mass transfer terms between the liquid and vapor phase. The modeling of two-phase flows using the homogeneous mixture model presented above has proven to be able to accurately describe cavitating flows. A homogeneous model has been successfully applied for the prediction of inertia controlled sheet and cloud cavitation around hydrofoils [22], in bore holes of injection devices [23], in micro channels [8, 12], as well as the prediction of erosion sensitive areas in cavitating flows [17].

The set of conservation Eq. (1) is closed by applying a modified version of Tait's equation of state in pure liquid regions:

$$\overline{p} = (p_{sat}(T) + B)\left(\frac{\overline{\rho}}{\rho_{l,sat}}\right)^N - B, \quad \text{if } \alpha = 0. \tag{9}$$

In two phase regions, the pressure is equal to the temperature-dependent saturation pressure $p_{sat}(T)$ due to the equilibrium assumption. The temperature-dependent saturation properties are approximated by the polynomials provided in Schmidt and Grigull [21]. The internal energy of liquid water and water vapor are modeled assuming temperature-independent specific heats at constant volume, $c_{v,l}$ and $c_{v,v}$, i.e.,

$$e_l = c_{v,l}(T - T_0) + e_{l,0}, \tag{10}$$

Table 1 Reference properties and fitted constants of liquid water and water vapor

Property	Value	Unit
$c_{v,l}$	4,180.0	J/kg K
$c_{v,v}$	1,410.8	J/kg K
$c_{l,0}$	617.0	J/kg K
$L_{v,0}$	2,501.3	kJ/kg K
T_0	273.15	K
N	7.1	–
B	3.06×10^8	Pa

$$e_v = c_{v,v}(T - T_0) + e_{l,0} + L_{v,0}, \tag{11}$$

where $e_{l,0}$ is the liquid reference internal energy and $L_{v,0}$ is the latent heat of evaporation at T_0. The temperature T is computed iteratively from the definition of the internal energy in two-phase regions, see Eq. (7).

We assume that the effective dynamic viscosity of the liquid-vapor mixture satisfies a quadratic law with a maximum in the two-phase region reading [6]

$$\overline{\mu}(T, \overline{\rho}) = (1 - \alpha)\left(1 + \frac{5}{2}\alpha\right)\mu_l(T, \rho_l) + \alpha\mu_v(T, \rho_v), \tag{12}$$

where μ_l and μ_v denote the dynamic viscosities for the liquid and vapor phase. A linear blending in two-phase regions is performed for the thermal conductivity:

$$\overline{\kappa}(T, \overline{\rho}) = (1 - \alpha)\kappa_l(T, \rho_l) + \kappa_v(T, \rho_v). \tag{13}$$

The dynamic viscosity and thermal conductivity of pure liquid and pure vapor are evaluated from the IAPWS correlations neglecting critical enhancement [1, 2]. The refernce properties of water and the fitted constants N and B are summarized in Table 1.

The volume-averaged conserved variables, Eq. (5), are advanced in time by a finite volume method. The inviscid flux is discretized by means of the Adaptive Local Deconvolution Method (ALDM) for compressible fluids [11, 12]. ALDM is a non-linear discretization method incorporating a subgrid-scale turbulence model. ALDM reconstructs the primitive variables $\varphi \in [\rho, u_i, p, \rho e]$ at cell faces, $\breve{\varphi}^{\pm}_{i\pm1/2}$, by combining Harten-type deconvolution polynomials, $\breve{g}^{\pm}_{kr}$, up to order three non-linearly and solution-adaptively:

$$\breve{\varphi}^{\pm}(x_{i\pm1/2}) = \sum_{k=1}^{3}\sum_{r=0}^{k-1} \omega^{\pm}_{kr}(\gamma_{kr}, \overline{\varphi})\breve{g}^{\pm}_{kr}(x_{i\pm1/2}). \tag{14}$$

The weights $\omega^{\pm}_{kr}$ of the deconvolution polynomials introduce free parameters γ_{kr} which are used to control the truncation error of ALDM. Additionally, a suitable numerical flux function comprising the physical convective flux and a secondary regularization term is used:

$$\breve{C}_{i\pm\frac{1}{2}} = C\left(\frac{\breve{\varphi}^{+}_{i\pm\frac{1}{2}} + \breve{\varphi}^{-}_{i\pm\frac{1}{2}}}{2}\right) - R\left(\varepsilon, \breve{\varphi}^{\pm}, \overline{\varphi}\right). \tag{15}$$

A physically consistent implicit subgrid-scale model is obtained by optimizing the free parameters $\{\gamma_{kr}, \varepsilon\}$ of ALDM so that the effective spectral numerical viscosity matches the eddy viscosity from the Eddy-Damped Quasi-Normal Markovian theory for isotropic turbulence [10].

The viscous flux is discretized by a linear second order accurate scheme. Time integration is performed by the explicit third order accurate Runge-Kutta method of Gottlieb and Shu [9], which is total-variation diminishing (TVD) for CFL ≤ 1 if the employed spatial discretization is TVD.

3 Strong Scaling of INCA on Hermit

Our in-house finite-volume code INCA[1] is written in FORTRAN language. Parallelization is obtained with a classical domain decomposition by a block-structured grid topology. INCA provides various discretization schemes that operate on six point stencils, such as ALDM, which was described in the previous section. Consequently, three ghost-cell layers need to be exchanged at every inter-block interface. If two neighboring blocks reside on the same process, the three ghost-cell layers are filled by copying values directly; otherwise, non-blocking communication according to the MPI-2.2 standard[2] is employed.

Since INCA uses structered Cartesian grids, complex boundaries are represented by a conservative immersed interface method. Refinement towards immersed interfaces is achieved by employing an adaptive mesh refinement (AMR) algorithm. Grids generated by the AMR algorithm can be composed of several thousand grid blocks. In this case, optimal load balancing with respect to cells per core is achieved either by an integrated load balancing algorithm which uses the number of cells per process and the inter-block connections as balancing criterion, or by utilizing the graph partitioning tool METIS[3] in a pre-processing step.

In the following, we present a strong scaling study of INCA on the Cray XE6 Supercomputer "Hermit" at the High Performance Computing Center Stuttgart (HLRS). As test problem we compute a compressible three-dimensional Taylor-Green vortex. The computational domain is a cube with edge length 2π and periodic boundaries. The domain is discretized by $N = 256^3$ cells with homogeneous grid

[1] http://www.inca-cfd.org

[2] http://www.mpi-forum.org

[3] http://glaros.dtc.umn.edu/gkhome/views/metis

Table 2 Grid statistics for strong scaling of INCA

No. of cores	Cells/core	Ghost cells/core	Ghost cells/cells (%)
32	524,288	132,312	25.2
64	262,144	80,856	30.8
128	131,072	55,128	42.1
256	65,536	35,544	54.2
512	32,768	22,104	67.5
1,024	16,384	15,384	93.9
2,048	8,192	10,200	124.5

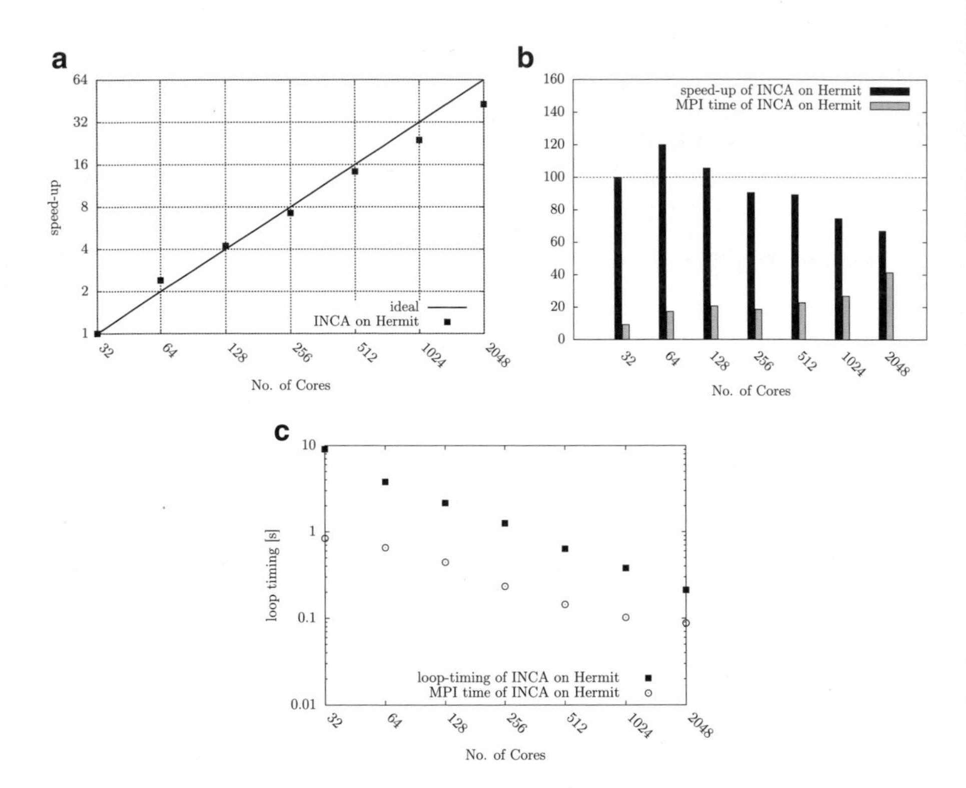

Fig. 1 Strong scaling of INCA on Hermit: (**a**) speed-up normalized to one node, (**b**) percentage of ideal scaling and percentage of MPI-time of total loop-timing, and (**c**) total wall time (loop timing) and MPI time of INCA on Hermit necessary to advance the solution one time-step

spacing in all directions. The domain is decomposed in equi-sized blocks according to the number of cores used, starting with one node up to 64 nodes, i.e., from 32 up to 2,048 cores. The number of cells per core, ghost cells per core, and ratio of ghost cells to cells is given in Table 2.

Figure 1a shows the strong scaling of INCA on Hermit, where the results have been normalized to one node (32 cores). For two and three nodes (64 and 128 cores), we observe a super-linear scaling that we attribute to optimized cache usage for these processor configurations. Up to the maximum investigated number of nodes, we observe good strong scaling properties of INCA. For 64 nodes, e.g., we still observe a parallel performance larger than 60 %, see Fig. 1b, despite having more ghost cells than cells for one grid block, cf. Table 2. The percentage of time spent for exchanging data between processors (MPI time) of the total wall time necessary to advance the solution one time step (loop timing) is less than 20 % up to 16 nodes. Absolute numbers for MPI time and loop timing are provided in Fig. 1c.

Based on the strong scaling study, we conclude that an almost linear speed-up is possible by decomposing a computational grid into grid blocks with approximately 30,000 cells.

4 Computational Setup

Two parallel streams with velocities U_1 and $U_2 = -U_1$, and velocity difference $\Delta U = U_2 - U_1$, characterize the shear layer. The initial condition for the mean velocity is

$$u_i = \begin{bmatrix} \frac{1}{2}\Delta U \tanh\left(\frac{2y}{\delta_{\omega,0}}\right) \\ 0 \\ 0 \end{bmatrix}, \tag{16}$$

where the vorticity thickness is defined as

$$\delta_\omega = \frac{\Delta U}{\left.\frac{\partial u_1}{\partial x_2}\right|_{\max}}. \tag{17}$$

Three-dimensional random velocity fluctuations with a maximum amplitude of 10 % of ΔU are superimposed on the mean flow to trigger transition. The velocity fluctuations are additionally restricted to the initial shear layer thickness by multiplying them with an exponential function, $\exp(-y^2/(2\delta_{\omega,0}))$. Constant initial density and temperature complete the initial conditions.

We consider a rectangular domain of size $L_x = 1{,}200\delta_{\theta,0}$, $L_y = 400\delta_{\theta,0}$, and $L_z = 240\delta_{\theta,0}$ in streamwise, spanwise, and cross-stream direction. The domain is discretized by $N_x \times N_y \times N_z = 768 \times 192 \times 192$ cells. The grid spacing is homogeneous in streamwise and spanwise direction, while a hyberbolic tangent function stretches the grid towards the center of the shear layer in the cross-stream direction. The domain has been decomposed into 1,024 equi-sized grid blocks and all computations were carried out on the same number of cores on Hermit (32 nodes).

Table 3 Simulation parameters of temporally evolving cavitating shear layers matching cavitation and Reynolds number of the experiment of Aeschlimann et al. [3, 4]

σ_c	ΔU	ρ_∞	p_∞	T_∞	$\delta_{\omega,0}$
–	m/s	kg/m^3	$\times 10^5$ Pa	K	$\times 10^{-3}$
1.0	40.0	998.52	8.012	293.15	0.1
0.167	40.0	998.22	1.357	293.15	0.1
0.1	40.0	998.19	0.822	293.15	0.1

The mutual interaction of turbulence and phase change is investigated by varying the cavitation number

$$\sigma_c = \frac{2(p_\infty - p_{\text{sat}})}{\rho_\infty (\Delta U)^2}. \tag{18}$$

Since we keep the characteric velocity difference constant, the variation in σ_c is obtained by setting the initial density and by imposing the far-field pressure p_∞ at the cross-stream boundaries according to Eq. (18). The far-field pressure p_∞ is either imposed by a simple pressure boundary condition (PSTAT) or by an essentially non-reflecting boundary condition (NSCBC) [16, 20]. If not otherwise noted, we use the NSCBC boundary condition. Periodicity is assumed in streamwise and spanwise directions.

In the following section, we present results for three cavitation numbers σ_c: one with ensentially no cavitation ($\sigma_c = 1.0$), and two with significant cavitation ($\sigma_c = 0.167$ and $\sigma_c = 0.1$), which have been chosen according to the experiments by Aeschlimann et al. [3, 4]. The experiments provide data of the shear layer thickness, velocity statistics, as well as vapor volume fraction by means of X-ray measurements. The setup and simulation parameters for the three cavitation numbers are specified in Table 3. The initial vorticity thickness, $\delta_{\omega,0}$, is choosen so that the Reynolds numbers, $Re = \rho_\infty \Delta U \delta_\omega / \mu_\infty$, based on the vorticity thickness of the LES and the experiment are equal at the begin of the self-similar region. Cross-stream statistics for the evolution of the shear layers until $t = 5 \times 10^{-3}$ s have been optained by averaging in streamwise and spanwise directions.

5 Results

5.1 Coherent Structures

Figure 2 shows coherent vortical structures identified by means of the λ_2-criterion [14] (left column) and the corresponding vapor cavities visualized by an iso-surface with $\alpha = 0.1$ at cavitation number $\sigma_c = 0.1$ (right column) at four instants in time, $t = [1.25, 2.5, 3.75, 5.0]$ s. Local low pressure regions associated with turbulent eddies are sources for cavitation. At $t = 3.75$ s, a pairing process

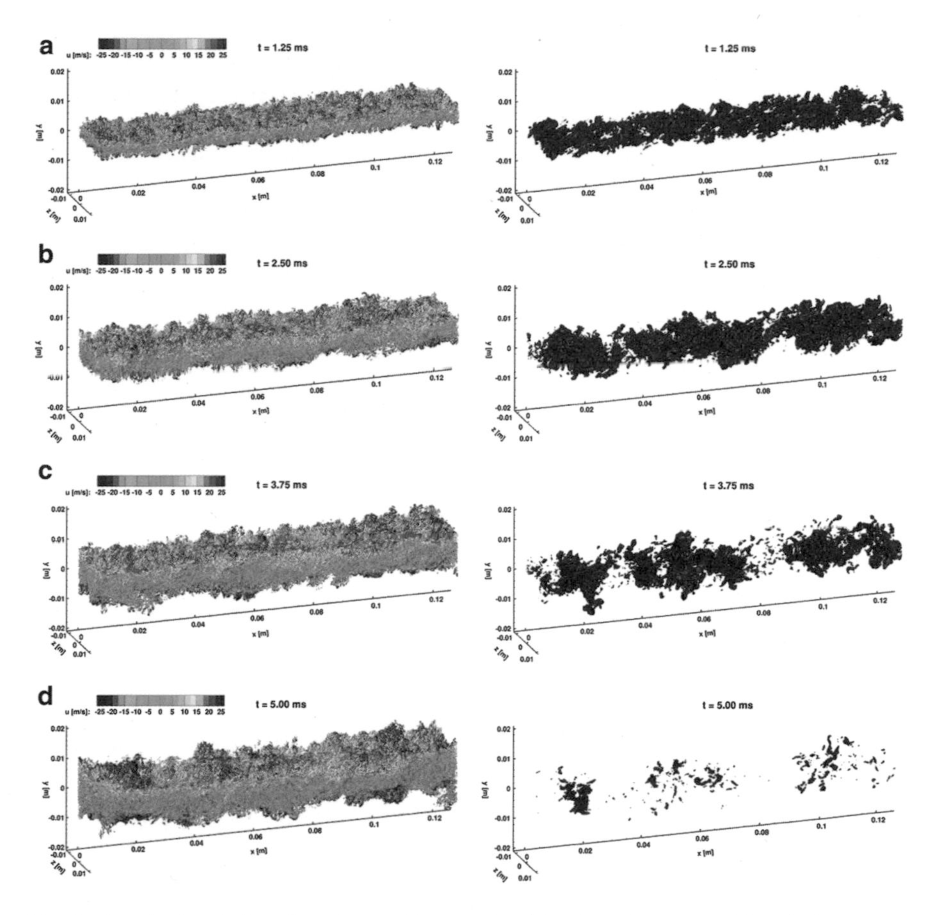

Fig. 2 Turbulent cavitating shear layer with $\sigma_c = 0.1$: (**a**) coherent structures identified by means of the λ_2-criterion, (**b**) iso-surfaces with vapor volume fraction $\alpha = 0.1$

in the turbulent shear layer leads to the formation of streamwise segregated vapor regions, cf. Fig. 2c; subsequently, recondensation occurs in the shear layer, cf. $t = 5.0\,\text{s}$ and Fig. 2d.

5.2 *Evolution of Vapor Volume Fraction*

Figure 3 shows the evolution of the maximum and integrated vapor volume fraction for the LES in comparison with the experiment. The streamwise coordinate of the experiment has been transformed to a time coordinate by

$$t = x \left.\frac{\Delta U}{U_c}\right|_{exp} \left.\frac{1}{\Delta U}\right|_{LES}, \tag{19}$$

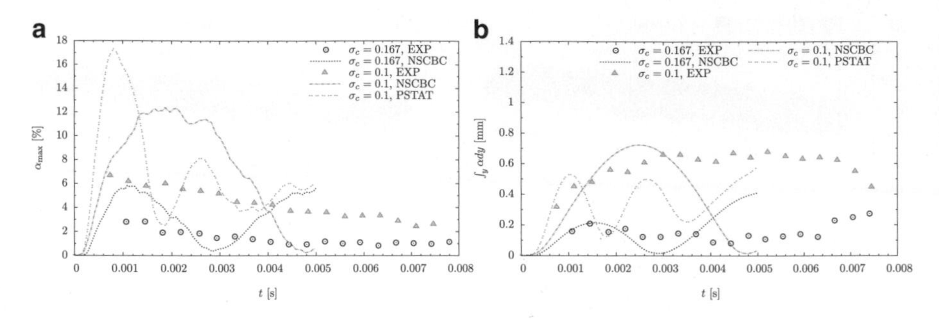

Fig. 3 Comparison between LES (*lines*) and experiment (*symbols*) of the evolution of (**a**) maximum vapor volume fraction and (**b**) integrated vapor volume fraction

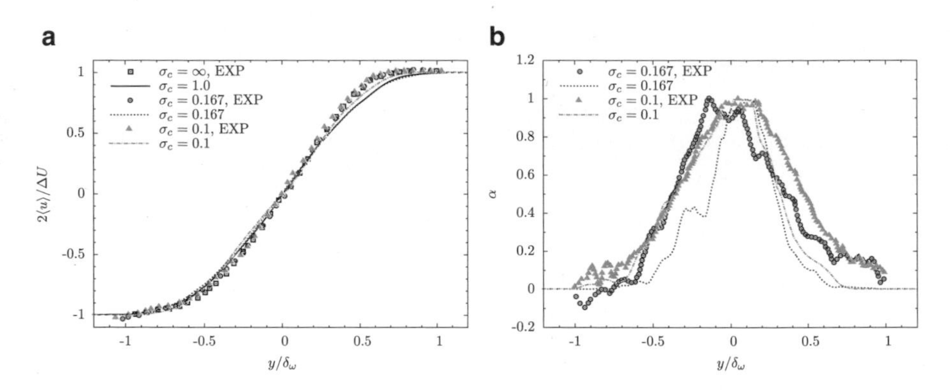

Fig. 4 Comparison between LES (*lines*) and experiment (*symbols*) of (**a**) mean streamwise velocity profiles and (**b**) vapor volume fraction profiles normalized with the maximum vapor volume fraction

where U_c is the convective velocity of the experiment. The level of the maximum vapor volume fractions, see Fig. 3a, are in good agreement with the experimental data. However, the LES results show larger variations in time than the experiment. Employing a reflective boundary condition (PSTAT) instead of a non-reflecting one (NSCBC) for case $\sigma_c = 0.1$ results in a better agreement with the experiment. The integrated vapor volume fraction, see Fig. 3b, shows the same trends between LES and experiment.

5.3 *Cross-Stream Statistics*

Figure 4a shows normalized velocity profiles in the self-similar region of the shear layers. Good agreement between LES and experiment is observed. Slight discrepancies can be seen in the high-speed stream, i.e., $y \approx 0.5$, between LES and experiment.

Although the absolute values of the maximum vapor volume fractions showed a different temporal evolution in the LES, cf. Fig. 3a, the cross-stream profiles compare well when normalized by their maximum value, see Fig. 4b. For $\sigma_c = 0.167$, the vapor volume profiles predicted by the LES appear a little narrower than the experimental ones.

6 Summary and Conclusions

The results for LES of turbulent cavitating shear layers presented above compare well with experimental data. Quantitative comparison in terms of vapor volume fraction and velocity statistics have been provided. Further studies will focus on the influence of the cross-stream boundary condition on the evolution of the vapor volume fraction since the presented results for $\sigma_c = 0.1$ indicated a connection between them. Furthermore, refined LES of spatially developing cavitating turbulent shear layers, of which preliminary results are already available [7], are currently under way for further comparison with experimental data and thus validation of our numerical method for turbulent cavitating flows.

Acknowledgements This project is supported by the German Research Foundation (DFG) under contract AD186/20-1.

References

1. Release on the IAPWS formulation 2008 for the viscosity of ordinary water substance. Technical report, The International Association for the Properties of Water and Steam (2008)
2. Release on the IAPWS formulation 2011 for the thermal conductivity of ordinary water substance. Technical report, The International Association for the Properties of Water and Steam (2011)
3. Aeschlimann, V., Barre, S., Djeridi, H.: Velocity field analysis in an experimental cavitating mixing layer. Phys. Fluids **23**(5), 055105 (2011)
4. Aeschlimann, V., Barre, S., Legoupil, S.: X-ray attenuation measurements in a cavitating mixing layer for instantaneous two-dimensional void ratio determination. Phys. Fluids **23**(5), 055101 (2011)
5. Arndt, R.E.A.: Cavitation in vortical flows. Annu. Rev. Fluid Mech. **34**, 143–175 (2002)
6. Beattie, D.R.H., Whalley, P.B.: A simple two-phase frictional pressure drop calculation method. Int. J. Multiph. Flow **8**(1), 83–87 (1982)
7. Egerer, C., Hickel, S., Schmidt, S., Adams, N.A.: LES of Turbulent Cavitating Shear Layers, chap. 24, pp. 349–359. Springer (2013)
8. Egerer, C.P., Hickel, S., Schmidt, S.J., Adams, N.A.: Large-eddy simulation of turbulent cavitating flow in a micro channel. Phys. Fluids **26**(8), 085102 (2014)
9. Gottlieb, S., Shu, C.W.: Total variation diminishing Runge-Kutta schemes. Math. Comput. **67**(221), 73–85 (1998)
10. Hickel, S., Adams, N.A., Domaradzki, J.A.: An adaptive local deconvolution method for implicit LES. J. Comput. Phys. **213**(1), 413–436 (2006)

11. Hickel, S., Larsson, J.: On implicit turbulence modeling for LES of compressible flows. In: B. Eckhardt (ed.) Advances in Turbulence XII, vol. 132, pp. 873–875. Springer Berlin Heidelberg (2009)
12. Hickel, S., Mihatsch, M., Schmidt, S.J.: Implicit large eddy simulation of cavitation in micro channel flows. In: Li, S.C. (ed.) WIMRC 3rd International Cavitation Forum. University of Warwick, Coventry (2011)
13. Iyer, C.O., Ceccio, S.L.: The influence of developed cavitation on the flow of a turbulent shear layer. Phys. Fluids **14**(10), 3414–3431 (2002)
14. Jeong, J., Hussain, F.: On the identification of a vortex. J. Fluid Mech. **285**, 69–94 (1995)
15. Kajishima, T., Ohta, T., Sakai, H., Okabayashi, K.: Influence of cavitation on turbulent separated flow. In: Friedrich, R., Adams, N.A., Eaton, J.K., Humphrey, J.A.C., Kasagi, N., Leschziner, M.A. (eds.) Fifth International Symposium on Turbulence and Shear Flow Phenomena, Munich, pp. 829–834 (2007)
16. Lodato, G., Domingo, P., Vervish, L.: Three-dimensional boundary conditions for direct and large-eddy simulation of compressible viscous flows. J. Comput. Phys. **227**(10), 5105–5143 (2008)
17. Mihatsch, M.S., Schmidt, S.J., Thalhamer, M., Adams, N.A.: Numerical Prediction of Erosive Collapse Events in Unsteady Compressible Cavitating Flows, pp. 499–510. International Center for Numerical Methods in Engineering (CIMNE), Lisbon (2011)
18. O'Hern, T.J.: An experimental investigation of turbulent shear flow cavitation. J. Fluid Mech. **215**, 365–391 (1990)
19. Okabayashi, K., Kajishima, T.: Investigation of turbulent modulation by cavitation for subgrid-scale modelling in les. In: Proceedings of the 7th International Symposium on Cavitation, Ann Arbor (2009)
20. Poinsot, T., Lele, S.K.: Boundary conditions for direct simulations of compressible viscous flows. J. Comput. Phys. **101**(1), 104–129 (1992)
21. Schmidt, E., Grigull, U.: Properties of Water and Steam in SI-Units, 4th edn. Springer, Berlin/München (1989)
22. Schmidt, S.J., Thalhamer, M., Schnerr, G.H.: Inertia controlled instability and small scale structures of sheet and cloud cavitation. In: Proceedings of the 7th International Symposium on Cavitation, Ann Arbor (2009)
23. Sezal, I.H., Schmidt, S.J., Schnerr, G.H., Thalhamer, M., Förster, M.: Shock and wave dynamics in cavitating compressible liquid flows in injection nozzles. Shock Waves **19**(1), 49–58 (2009)

Investigations of Unsteady Transonic and Supersonic Wake Flow of Generic Space Launcher Configurations Using Zonal RANS/LES and Dynamic Mode Decomposition

V. Statnikov, T. Sayadi, M. Meinke, P. Schmid, and W. Schröder

Abstract Dynamic mode decomposition (DMD) is applied to time-resolved results of zonal RANS/LES computations of three different axisymmetric generic space launcher configurations having freestream Mach numbers of 0.7 and 6.0. Two different after-body geometries consisting of an attached sting support mimicking an endless nozzle extension and an attached cylindrical after-expanding nozzle are considered. The distinguishing feature of the investigated configurations is the presence of a separation bubble in the wake region. The dynamics of the bubble and the dominant frequencies are analyzed using DMD. The results are then compared to the findings of spectral analysis and temporal filtering techniques. The transonic case displays distinct peaks in the Fourier transform spectra. We illustrate a close agreement between the frequencies of the DMD modes and the aforementioned peaks. From the shape of the DMD modes, it is deduced that most of the peaks are related to the dynamics of the separation bubble and the corresponding shear layer. The same geometry with higher Mach number shows the existence of two distinct DMD modes of axisymmetric and helical nature, respectively. The presence of the helical mode is illustrated through composite DMD of the cylindrical velocity components. Finally, when considering the free-flight configuration we were able to identify DMD modes with frequency $Sr_D \approx 1$ which results in the flapping motion of the shear layer and a mode of a lower frequency which causes the breathing of the recirculation bubble.

1 Motivation and Objectives

Although in most cases the base geometry of a space launcher is quite simple and can be generically modeled by two coaxial cylinders mimicking the main body and the attached nozzle extension, the wake flow field is determined by different intricate

V. Statnikov (✉) • T. Sayadi • M. Meinke • P. Schmid • W. Schröder
Institute of Aerodynamics, RWTH Aachen University,
Wüllnerstr. 5a, 52062 Aachen, Germany
e-mail: v.statnikov@aia.rwth-aachen.de

W.E. Nagel et al. (eds.), *High Performance Computing in Science and Engineering '14*, DOI 10.1007/978-3-319-10810-0_26

unsteady phenomena such as flow separation at the base shoulder, reattachment of the shear layer at the outer nozzle wall, interaction with the jet plume, etc. It is generally known that the base drag of axial cylindrical bodies, which is caused by the low pressure at the recirculation area at the base, constitutes up to 35 % of the overall drag as shown by Rollstin [1]. Moreover, the unsteady dynamics of the separation bubble, the reattaching shear layer and their interaction lead to significant wall pressure oscillations and, consequently, dynamic loads that may excite critical structural modes which is broadly known as a buffeting phenomenon. At supersonic speeds, the wake is dominated by shock and expansion waves, drastically changing the dynamic wake flow behavior. Moreover, at higher altitudes the nozzle usually operates at under-expanded mode with a strongly after-expanding jet plume. The resulting displacement effect on the base flow generated by a wide jet plume leads to an increase of the base pressure level and, consequently, to a reduction of the base drag. On the other hand, the periodic and stochastic base pressure oscillations become stronger and might excite vibrations of critical amplitude. In addition to the aero-elastic aspects, convection of the hot gases from the jet upstream to the base area driven by the recirculation zones can lead to confined hot spots and thermal loads of the structure.

For these reasons, an accurate prediction of the wake dynamics and an understanding of its dynamics and controllability is of vital importance for the development of efficient and safe space transportation systems. Different aspects of the wake flow have been investigated in the past experimentally and numericallly. Mathur and Dutton [2] addressed base drag reduction effects of base bleeding at supersonic conditions with $Ma_\infty = 2.46$. Janssen and Dutton [3] experimentally investigated the temporal behavior of the supersonic base flow and detected dominant frequencies around $Sr_D = 0.1$ of the wall pressure fluctuations referred to inner motion of the large structures inside the separation region. The interaction of the supersonic base flow at $Ma_\infty = 2$ and 3 with an after-expanding nozzle jet was studied by Bannink et al. [4] and by Oudheusden et al. [5] who also observed several dominant lower frequencies tracked back to dynamics of the separation bubble while higher frequencies were attributed to shed turbulent structures. In the transonic regime, the buffeting phenomenon was experimentally addressed by Deprés [6] who investigated dynamics of the wake of the main stage of Ariane 5 in ONERA's S3Ch wind tunnel and detected dominant peaks in the wall pressure signal at $Sr_D = 0.2$ and 0.6. The performed two-point correlation measurements showed that the flow is dominated by a highly coherent antisymmetric mode at the vortex shedding frequency. Meliga and Reijasse [7] extended the experimental investigations on a set-up of Ariane 5 with solid boosters and detected a shift of the dominant frequency of the wall pressure fluctuations to $Sr_D = 0.34$ assumed to be caused by an axi-symmetric mode. The wake flow has also been studied numerically using turbulence models that range from various Reynolds-averaged Navier-Stokes (RANS) models [8, 9] via detached-eddy simulations (DES) [10, 11] and large-eddy simulations (LES) [12] to direct numerical simulations (DNS) [13–15]. RANS models were found to be suitable only for prediction of the attached flow and fail to provide accurate results concerning the low pressure recirculation area behind the base. DNS is still restricted to low Reynolds numbers and a small integration domain. In contrast, LES (Meliga et al. [16]) and particularly hybrid approaches like DES (Deck and Thorigny [17]) and zonal RANS/LES (Statnikov et al. [18]) allow time-resolved computation of the dynamic wake flow field at practically relevant Reynolds-numbers and were found to be a good compromise between costs and accuracy for computation of the wake flow of space launchers.

The objective of this work is to study the numerically computed turbulent wakes of generic space launcher configurations in order to better understand the base-flow dynamics and its controllability. As shown by Statnikov et al. [18–20] for both transonic and supersonic cases, numerical results such as computed flow field topology, time-averages and fluctuations of base pressure signals as well as its spectra satisfactorily agree with the experimental data from literature and experimental investigations conducted within TRR 40 [21–23]. Using classical spectral analysis and correlation methods, several distinct frequencies were identified indicating a periodic behavior of the base-flow dynamics, e.g., $Sr_D = 0.1$ (shear layer flapping), $Sr_D = 0.2$ (von Kármán vortex shedding) as well as a number of dominant peaks at lower ($Sr_D \approx 0.05$) and considerably higher frequencies ($Sr_D \approx 0.7$ and $Sr_D \approx 1$) with not fully understood origins.

Since the wake dynamics is to a great extent dominated by a superposition of different periodic and stochastic flow phenomena, the detection of responsible coherent flow structures by using classical statistical data analysis is still a difficult task. In this respect, a combined application of Dynamic Mode Decomposition (DMD), classical statistical analysis and temporal filtering methods to the computed flow fields is a promising approach for a precise identification of significant coherent flow structures responsible for the observed unsteady behavior. As demonstrated in the paper at hand, this method allows a straightforward detection of the underlying coherent flow motion, yields a better understanding of the base-flow dynamics, and may aid in the development of efficient active and passive base-flow control mechanisms. Furthermore, DMD is applicable to any choice of flow quantity as well as to a combination of different quantities (composite DMD). Thus, composite DMD is applied to a variety of flow variables in an attempt to establish a correlation and interaction pattern between different flow quantities, e.g., pressure, velocity, and vorticity. Each DMD mode represents an optimal phase-averaged structure with a unique frequency in contrast to POD, where each mode is a multi-frequency signal. This makes DMD suitable for analyzing broadband spectra with pronounced peaks, in which the coherent flow patterns can be separated from the stochastic background fluctuations as is the case in turbulent rocket wakes.

This paper is organized as follows. In Sect. 2 the three different computational setups are presented along with the respective flow parameters. Furthermore, a brief description of the computational method is provided. Dynamic mode decomposition is briefly described in Sect. 3. The results of the current approach for the three different flow geometries are presented and discussed in Sect. 4. Finally, a summary and conclusions of the current study are given in Sect. 5.

2 Computational Setups

Two different supported wind tunnel models and one free-flight configuration generically approaching an Ariane V-like launcher are considered, Fig. 1. The wind tunnel models feature an attached sting support mimicking an endless nozzle extension, while the free-flight configuration possesses an attached cylindrical after-expanding truncated ideal contour (TIC) nozzle. The same as for the main stage of Ariane V launcher, nozzle-main body ratios of $d_{nozzle}/D_{main\,body} \approx 0.4$ and $l_{nozzle}/D_{main\,body} \approx 1.2$ are used.

Two assumed trajectory points of an Ariane V-like space launcher are defined at transonic ($Ma_\infty = 0.7$, $Re_D = 1.08 \cdot 10^6$) and supersonic ($Ma_\infty = 6.0$, $Re_D = 1.73 \cdot 10^6$) conditions that follow the corresponding wind tunnel tests carried out in the framework of the TRR 40 project.

Time-resolved numerical computations of the flow field around the generic rocket configurations are performed using a zonal RANS/LES approach [24–26]. In this approach the computational domain is split into two zones, see Fig. 2. In the zone with an attached flow, e.g., the main body and the inner nozzle flow, the turbulent flow field is predicted by solving Reynolds-averaged Navier-Stokes (RANS) equations. In the wake region a large eddy simulation (LES) is performed. Using such a hybrid approach allows an efficient time-resolved computation of the dynamic wake flow field via LES at a fraction of the costs of a pure LES. The computations are done on a structured vertex-centered multi-block grid using an in-house zonal RANS/LES finite-volume flow solver. The Navier-Stokes equations of three-dimensional unsteady compressible flow are discretized in conservative form by a mixed centered upwind AUSM (advective upstream splitting method) scheme [27] of second-order accuracy for the Euler terms and by a second-order accurate, centered approximation for the viscous terms with low numerical dissipation. The temporal integration is performed by an explicit 5-stage Runge-Kutta method with second-order accuracy. The LES formulation is based on the monotone integrated LES (MILES) approach [28] modeling the impact of the sub-grid scales by numerical dissipation. A detailed description of the fundamental LES solver is given by Meinke et al. [29] and its convincing solution quality for fully turbulent sub- and supersonic flows is e.g. discussed in Alkishriwi et al. [30]

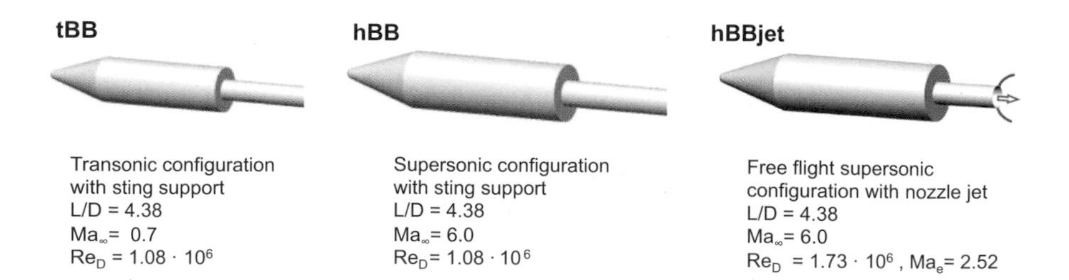

Fig. 1 Investigated generic space launcher configurations

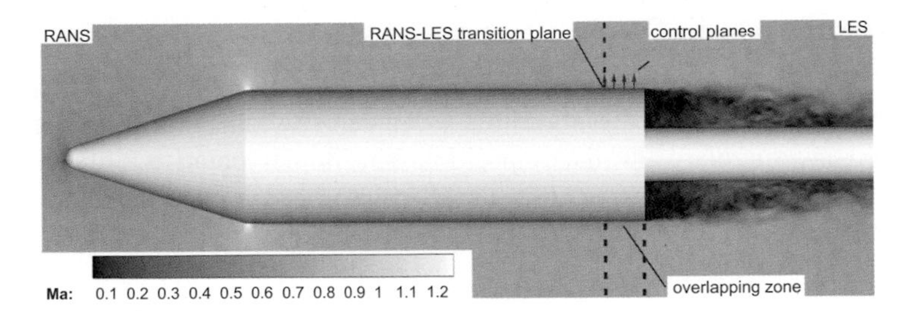

Fig. 2 Flow field decomposition into RANS and LES zones

and El-Askary et al. [31]. The RANS equations are descretized using the same overall discretization scheme. A one-equation turbulence model of Spalart and Almaras [32] is employed to close the time-averaged equations.

The transition from the RANS to the LES zone is performed by applying Refomulated Synthetic Turbulence Generation (RSTG) methods developed by Roidl et al. [24–26] that allow a reconstruction of the time-resolved turbulent fluctuations from the time-averaged upstream RANS solution. The RSTG methods are based on the synthetic eddy method (SEM) of Jarrin et al. [33] and Pamiès et al. [34] and describe turbulence as a superposition of coherent structures. These structures are generated over the LES inlet plane by superimposing virtual eddy cores which are defined in a virtual volume V_{virt} around the inlet plane that has the streamwise, wall-normal, and spanwise dimensions of the turbulent length scale l_x, the boundary-layer thickness at the inlet δ_0, and the width of the computational domain L_z. The turbulent length scales that describe the spatial properties of the synthetic structure depend on the distance from the wall and are derived from the turbulent viscosity μ_t of the upstream RANS solution and scaled with the Reynolds number and the associated convection velocity. As a result, the final velocity signal is composed of an averaged velocity component $\overline{u_i}$ which is provided from the upstream RANS solution and the normalized stochastic fluctuations u_i' which are subjected to a Cholesky decomposition to assign values of the Reynolds-stress tensor corresponding to the turbulent eddy viscosity.

To minimize the transition zone between the RANS and LES domains, controlled body forces f_i can be added to the wall-normal momentum equation at a number of control planes at different streamwise positions to enforce a match of the turbulent flow properties of the LES with the given RANS values as shown in Fig. 2. The amplitude of the force term is defined by a proportional integral controller, which controls the deviation between the target and the current profile of the reconstructed Reynolds shear stresses. A detailed description of the methods including the shape functions and length scale distributions is given in [24, 25].

3 Dynamic Mode Decomposition (DMD)

DMD is performed using the algorithm presented in [35], with a preprocessing step based on a singular value decomposition. This implementation allows for rank-deficient snapshot sequences, and it avoids ill-conditioned companion matrices [36]. In its simplest form, DMD provides the following representation of a flow field U,

$$U(x, y, z, t) = \sum_{n=1}^{N} a_n \exp(\lambda_n t)\phi_n(x, y, z), \tag{1}$$

where x, y, and z are spatial coordinates and t is time. In Eq. (1), ϕ_n's are the DMD modes, a_n's are the amplitudes and λ_n's are the frequencies of the respective modes. ϕ_n, a_n, and λ_n are complex-valued quantities. The eigenvalues of the modes can be deduced from the frequencies following the relation, $\lambda_n = log(\mu_n)/\Delta t$, where μ_n are the eigenvalues of the coupled inter-snapshot mapping and Δt is the time step between two consecutive snapshots. In DMD, the modes and frequencies are determined without the need for specifying an inner product or a norm. Compared to POD, this gives DMD the advantage of being applicable to any choice of flow quantities or spatial domains.

Once the modes and frequencies of the system are computed, we recover the complex amplitudes through a reconstruction of the original data. To solve for the magnitudes of each mode, a_n, we apply a pseudo-inverse to find the bi-orthogonal basis, ψ, of dynamic modes, ϕ such that

$$< \psi_i . \phi_j >= \int \psi_i^* \phi_j dV = \begin{cases} 0, \, i \neq j \\ 1, \, i = j. \end{cases} \tag{2}$$

While this could be done over all snapshots, in this report we choose to apply the pseudo-inverse to the first snapshot only, and then use the remaining snapshots to evaluate the closeness of fit for later times. If the dynamic modes are normalized, $|\tilde{a}_n|$ gives the relative magnitude of the nth mode. This algorithm has been applied to the direct numerical simulation (DNS) data of H-type transition on a flat-plate boundary layer [37].

4 Results

The results are discussed in three sections. In Sect. 4.1 the post-processing methodology is applied to the unsteady streamwise velocity field in the near wake of the supersonic free-flight configuration with after-expanding jet. Particular attention is paid to an interpretation of the extracted DMD modes required for understanding the results presented in the following sections. In the next section the transonic wake is considered and the existent peaks in the Fourier spectra of the pressure signal in the vicinity of the separation bubble are interpreted using DMD. In addition, the concept of composite DMD is introduced and applied to the pressure and velocity signals. Finally, in Sect. 4.3, composite DMD is performed on the cylindrical velocity components of the hypersonic wake of the same geometry as the transonic setup.

4.1 DMD Analysis of the Free-Flight Configuration

To introduce the applied post processing methodology, the wake dynamics of the free-flight configuration is analyzed. At the base shoulder, the supersonic boundary layer separates and undergoes an expansion leading to formation of a low-pressure region and subsonic recirculation zone associated with radial deflection of the shear layer towards the axis of symmetry. Farther downstream, a strongly after-expanding jet plume emanates from the nozzle leading to a displacement effect on the base flow. As a result, the main body shear layer is reflected away from the nozzle wall and no reattachment occurs. Thus, a confined subsonic cavity region is formed between the base and jet plume in the axial direction as well between the nozzle wall and shear layer in the radial direction. The resulting wake flow topology is presented in Fig. 3 showing the time-averaged axial velocity, Mach-number and streamline distribution.

The flow behind the base shoulder and just in front of the nozzle, which includes the separation bubble and the cavity, is investigated using DMD. Figure 4a shows

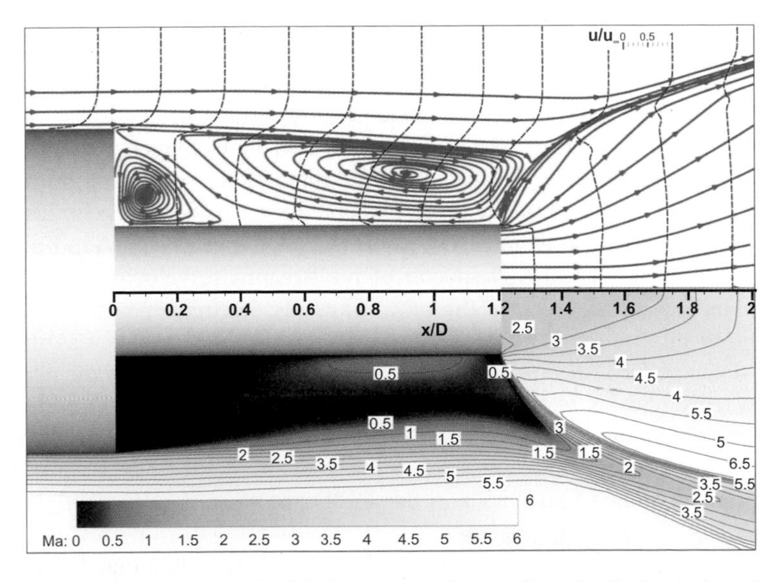

Fig. 3 Time-averaged axial velocity, Mach-number and streamlines distribution in the wake of the free-flight configuration

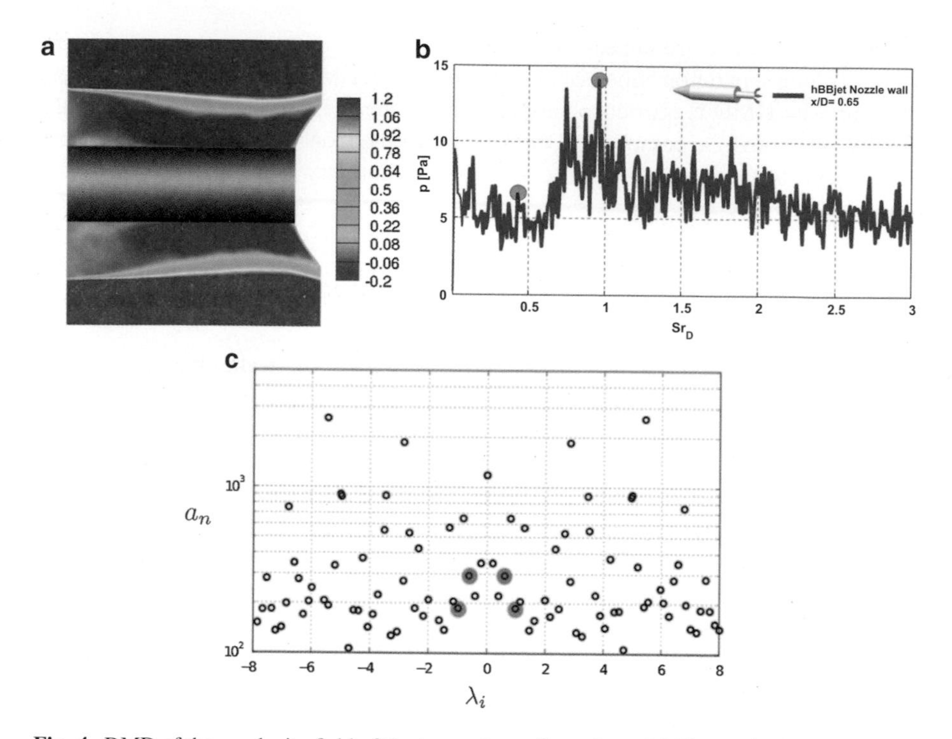

Fig. 4 DMD of the u-velocity field of the transonic configurations. (**a**) Mean velocity flow field in the $x - y$ slice of the thee-dimensional flowfield. (**b**) DFT of the pressure fluctuations. •, $Sr_D = 0.4$; •, $Sr_D = 0.98$. (**c**) DMD spectra of the u-velocity component. •, $\lambda_i = 0.4$; •, $\lambda_i = 0.98$

a cross section of the mean streamwise velocity inside this domain. The discrete Fourier transform (DFT) spectra of the pressure fluctuations at the wall can be split in two regions of low and high frequencies extending from $Sr_D = 0$ to 0.5 and $Sr_D \geq 0.5$, respectively. Inside each low-and high-frequency zones, peaks are also identified. DMD is performed on the three-dimensional data, in order to construct a global shape of the mode responsible for each frequency. In this report we concentrate on one mode in each zone of the DFT spectra shown in Fig. 4b by blue and red circles. The DMD spectra of the streamwise velocity are shown in Fig. 4c. The axis of abscissas in the plot represents the frequency of each DMD mode. From this information the modes with the same frequencies as the ones identified via the DFT spectra are chosen. These modes are shown in the plot using blue and red circles as in Fig. 4b. Figure 4c shows two frequencies in the DMD spectra corresponding to the single DFT frequency. A pair of complex conjugate modes are needed to provide the correct phase of the modes with respect to the original data. The two frequencies in the DMD spectra represent a single DMD mode.

The high-frequency DMD mode with $Sr_D = 0.98$ is shown in Fig. 5. This mode oscillates with constant frequency by construction and, to represent this oscillation,

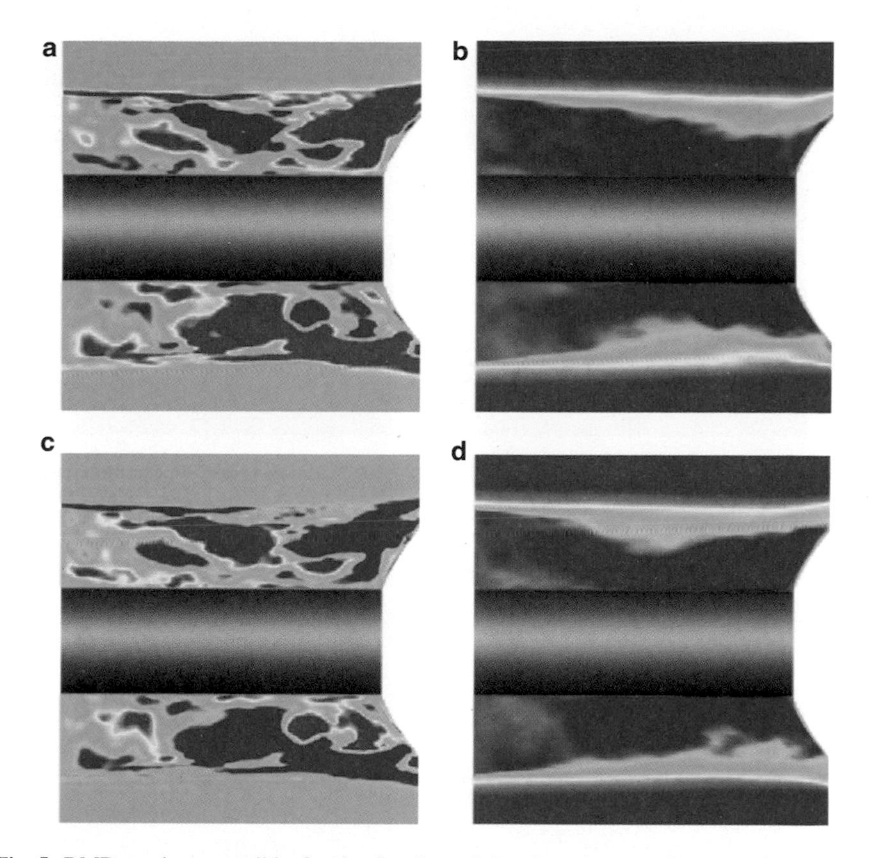

Fig. 5 DMD mode responsible for the flapping of the shear layer at frequency $\lambda_i = 0.98$. Contour levels: (*blue*), negative values; (*red*), positive values. (**a**) Mode, beginning of the cycle. (**b**) Modulation of the mean beginning of the cycle. (**c**) Mode, middle of the cycle. (**d**) Modulation of the mean middle of the cycle

the shape of the mode in the beginning and middle of each period is shown in Fig. 5a, c, respectively. The mean-flow solution shows the location of the separation bubble in a time-averaged manner. By adding the DMD mode to the time-averaged mean flow, the modulation of the mean through each mode is illustrated as shown in Fig. 5b, d for $Sr_D = 0.98$. The shape of this mode shows that the main contribution to the total mean flow is at the location of the shear layer edge. Concentrating on the upper half of the domain shown in Fig. 5a, this interaction with the mean shear is in the form of two areas of low and high velocity magnitude, which are reversed in position in the middle of the period (Fig. 5c). The lower half of the mode is 180° out of phase with the upper half. Ultimately, the combination of this mode and the mean results in the flapping of the shear layer and in this frame of reference is 180° out of phase in the upper and the lower half of the domain.

The same exercise can be performed on the mode with lower frequency. Figure 6a, c shows the mode shape in the beginning and the middle of the period. The

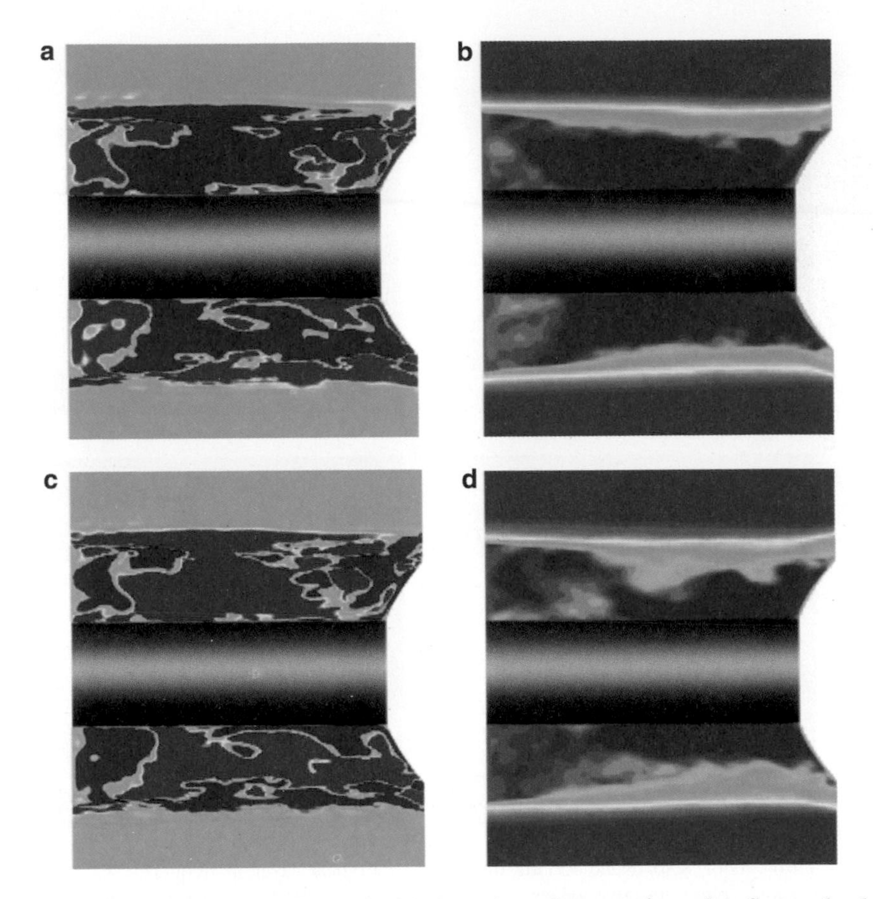

Fig. 6 DMD mode responsible for the bubble dynamics at frequency $\lambda_i = 0.4$. Contour levels: (*blue*), negative values; (*red*), positive values. (**a**) Mode, beginning of the cycle. (**b**) Modulation of the mean beginning of the cycle. (**c**) Mode, middle of the cycle. (**d**) Modulation of the mean middle of the cycle

modulation of the mean streamwise velocity through this mode is also demonstrated in Fig. 6b, d. In contrast to the previous results, there is little dependence on the transverse direction when the shape of the mode is considered. The mode also interacts mainly with the separation bubble itself. This is illustrated through Fig. 6b, d, which shows the increase and decrease in the size of the averaged separation bubble when the magnitude of the DMD mode is, respectively, negative or positive.

These two examples illustrate the advantage of the three-dimensional DMD, since the evolution of the flow is determined through the shape of the mode and not only the frequencies. We have also illustrated how modes of different frequencies and shape influence the dynamics of the flow by studying the interaction of each mode with the averaged solution.

Fig. 7 Instantaneous vortex structures by means of λ_2-contours colored by Mach-number and wall-pressure coefficient distribution in the wake of the transonic configuration

4.2 Transonic Configuration

The interaction between the shedding shear-layer vortices and the main engine's nozzle reaches its maximum intensity at transonic speeds and leads to increased mechanical loads on the nozzle structure. This effect is investigated for an axisymmetric configuration at $Ma_{\infty} = 0.7$ shown in Fig. 1. The axisymmetric backward-facing step causes a strong separation of the incoming turbulent boundary layer leading to the formation of a low-pressure recirculation region downstream of the base. As a result, the shear layer sheds from the shoulder of the cylindrical fore-body and broadens further downstream because of turbulent mixing effects. Due to the low-pressurized region at the base, the shear layer is deflected towards the sting and reattaches forming a closed recirculation zone, Fig. 7.

To gain insight into the global wake dynamics, discrete Fourier transformation (DFT) of the pressure fluctuations are performed at several axial positions on the wall of the attached sting. Figure 8a shows the resulting spectra at three different axial positions representing regions of interest inside the recirculation bubble; close to the base ($x/D = 0.2$), at the reattachment point ($x/D = 1.2$) and downstream of the separation bubble ($x/D = 2.0$). These regions show a slight variation of the detected dominant modes and their respective amplitudes. The global averaged DFT of the pressure fluctuations on the sting is presented in Fig. 8b. Through this approach, DFT spectra of the various streamwise locations are subsequently averaged to highlight the global dominant modes. Six distinct global modes of the wall-pressure fluctuations lower than $Sr_D = 1.0$ (summarized in Table 1) are identified.

In order to extract the spatial shape of the modes associated with peaked frequencies in the globally averaged DFT spectra, DMD is performed on the streamwise field signal of the three-dimensional domain surrounding the averaged separation bubble. Figure 9a shows the eigenvalues of the DMD spectra. Since the flow is turbulent and fully non-linear we expect all eigenvalues to be on the unit circle. Figure 9a shows, however, that there are some modes with eigenvalues that

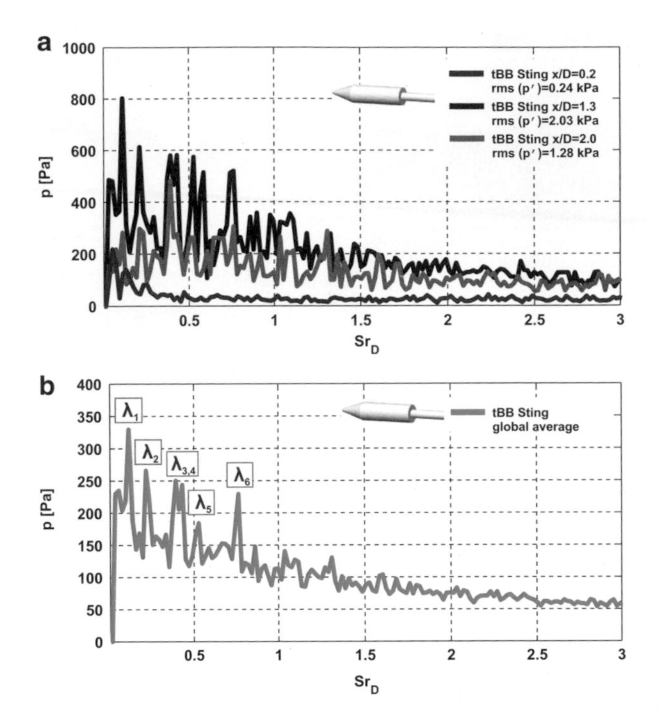

Fig. 8 DFT of the wall pressure fluctuation of the transonic configuration. (**a**) Local. (**b**) Globally averaged

lie inside the circle. These modes represent transients that have high amplitude in the beginning of the signal but are quickly damped and have little influence on the overall dynamics of the flow. As a result, in this study they are identified as transient modes and are neglected in our further analysis. Figure 9b shows the DMD spectra corresponding to the DMD eigenvalues. The transient modes can be identified by having higher magnitudes than the mean located at $\lambda_i = 0$. The amplitudes a_n's, as explained in Sect. 3, are calculated by projection on the data of the first snapshot, where these transient modes have the most contribution, which causes them to appear as artificial peaks in the DMD spectra, even though, they have a strong decay rate and are irrelevant in the global dynamics. We can therefore conclude that Fig. 8a, b each give part of the information on the significance of a single DMD mode and only by considering them together we are able to provide an accurate interpretation of the results. The low-frequency modes identified in the DFT spectra can also be extracted from DMD spectra of Fig. 9b and are represented by colored symbols. As mentioned before, each mode consists of a complex conjugate pair and can be seen in both Fig. 9a, b.

The estimated frequencies from the DMD analysis are compared to the DFT-results in Table 1. This table shows that there is good agreement between the frequencies extracted by either method. We have also analyzed the pressure signal

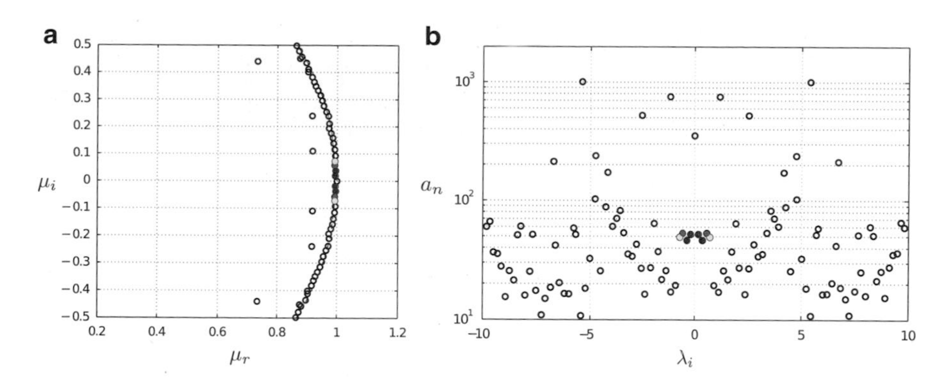

Fig. 9 DMD of the streamwise velocity of the transonic configuration. •, λ_2; •, λ_3; •, λ_5; , λ_6. (**a**) Eigenvalues, μ. (**b**) DMD spectra

Table 1 Comparison of the frequencies extracted by using DFT and DMD

Frequency	λ_2	λ_3	λ_4	λ_5	λ_6
DFT	0.21	0.39	0.43	0.57	0.76
DMD	0.19	0.37	0.46	0.59	0.72

at the wall, which is the signal used in the DFT analysis, and we have verified that the results remain unchanged. As DMD allows sub-domain analysis of the full data, we could also establish that mode λ_4 is relevant close to the step, since this frequency appears once the step is included in the DMD domain. Mode $\lambda_1 = 0.1$ is not detected using the DMD formalism. This mode could be a product of a non-linear interaction of two higher-frequency modes of frequencies 0.2 and 0.3 that is artificially magnified. This frequency can also be related to local dynamics close to the step. This is a plausible explanation, since Fig. 8a shows a decrease in the magnitude of this peak with an increase of the streamwise distance from the step. In DMD a larger three-dimensional domain is analyzed, which could cause the very local dynamics of the flow to remain undetected.

Figure 10 shows the instantaneous streamwise (u) and mean velocity (λ_0) contours in the domain of consideration for the DMD analysis. This figure also illustrates the shape of the modes with frequencies λ_2, λ_3, λ_5 and λ_6 which are related to the dynamics of the separation bubble. Mode λ_5 is more active close to the wall, whereas mode λ_3 has the most influence away from the wall. This can be deduced from the location of high- and low-magnitude velocities (red and blue regions) in each mode. Modes λ_2 and λ_6, however, show the most variation along the edge of the mean shear layer. Therefore, they are more relevant for the dynamics of the shear layer itself. This exercise illustrates how, with the help of DMD and extracting the three-dimensional modal shape for a single frequency from the fully turbulent data, the associated dynamics can be interpreted.

DMD is a method of data-driven spectral analysis; therefore, any type of data can be considered for evaluation. Furthermore, different flow parameters, for example vorticity, velocity, etc., can be combined as an input to this analysis (composite

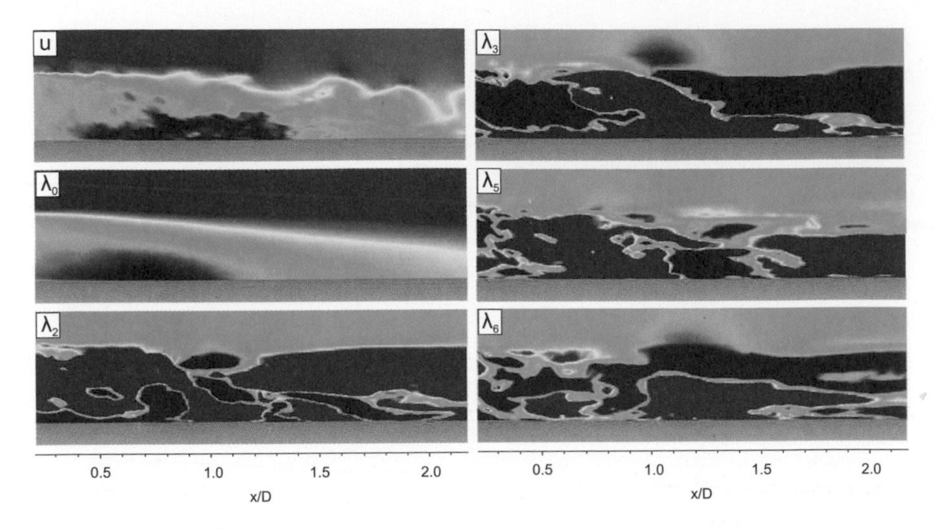

Fig. 10 Instantaneous (u) and mean velocity (λ_0) contours of the DMD domain. Mode shapes of frequencies λ_2, λ_3, λ_5 and λ_6 (*blue*: negative values; (*red*): positive values)

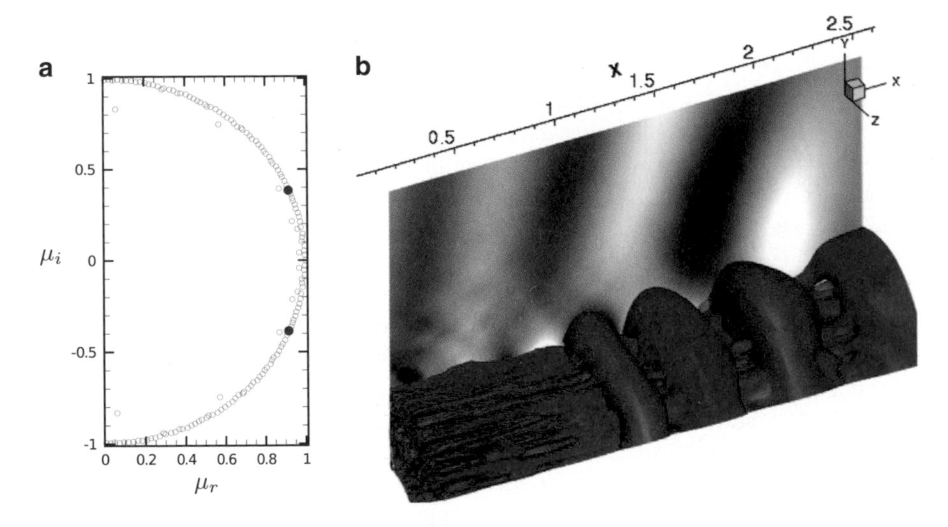

Fig. 11 Composite DMD of the pressure (*black-white* contours) and streamwise velocity (*red* and *blue* iso-surfaces). (**a**) Eigenvalues, μ. (*red*) $\lambda_i = 4$. (**b**) modal shape, $\lambda_i = 4$

DMD). Composite DMD allows the assessment of a coupling between various flow variables on a basis of their frequency. We have performed composite DMD on the pressure and streamwise velocity of the transonic data, and the resulting eigenvalue spectrum of this analysis is shown in Fig. 11a. For the purpose of this analysis we have considered a domain which extends further downstream than the original DMD domain. As the domain of interest is not centered on the separation bubble, other dynamics are also included. In particular, a mode of a higher frequency

$\lambda_i = 4$, shown in Fig. 11b, contains structures at the edge of the shear layer with a growing wavelength downstream. The pressure signal associated with this velocity also shows a trace in the freestream highlighted by the oscillating black and white pressure contours. This figure shows that the velocity, iso-surfaces of red and blue, are 180° out of phase with the pressure signal. In addition, by analysis of the evolution of this mode in time we have determined that the pressure in the freestream travels upstream while the velocity component is moving downstream. This is reminiscent of an acoustic wave in such circumstances, the proof of which needs further investigation.

4.3 Supersonic Configuration

In this setup the main body geometry and freestream conditions ($Ma_\infty = 6.0$) are identical to the supersonic configuration with an attached sting support presented in Sect. 4.1. Therefore, the flow along the main body up to the base shoulder is the same as for the supersonic free-flight configuration. The wake flow topologies, however, differ significantly from the free-flight case due to the use of an endless nozzle extension instead of a working TIC nozzle. As a result, no displacement effect of the jet is present and the resulting shear layer consequently is deflected towards the axis of symmetry and reattaches on the surface of the sting support. The time-averaged wake flow field is presented in Fig. 12 showing the Mach-number, static pressure ratio, streamlines, and axial velocity profiles.

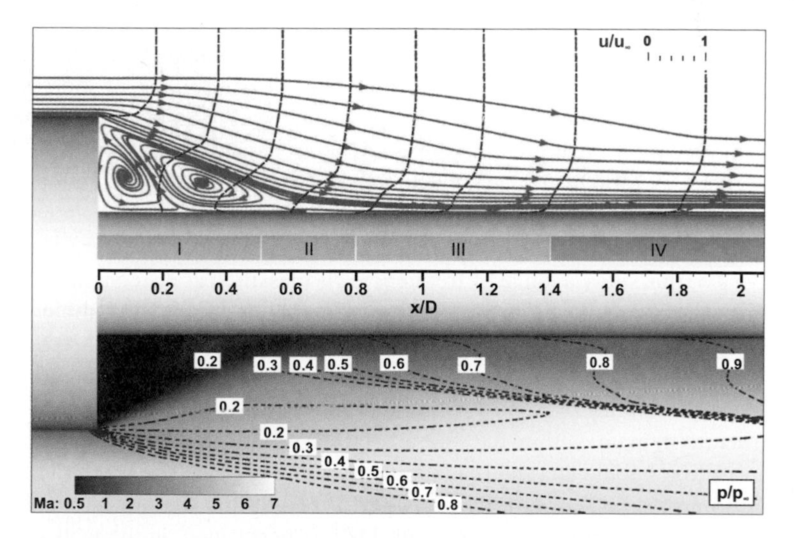

Fig. 12 Time-averaged wake flow topology of the supersonic configuration with sting. Different dynamic regions: (*i*) recirculation bubble; (*ii*) reattachment line; (*iii*) recompression waves; (*iv*) recompression shock

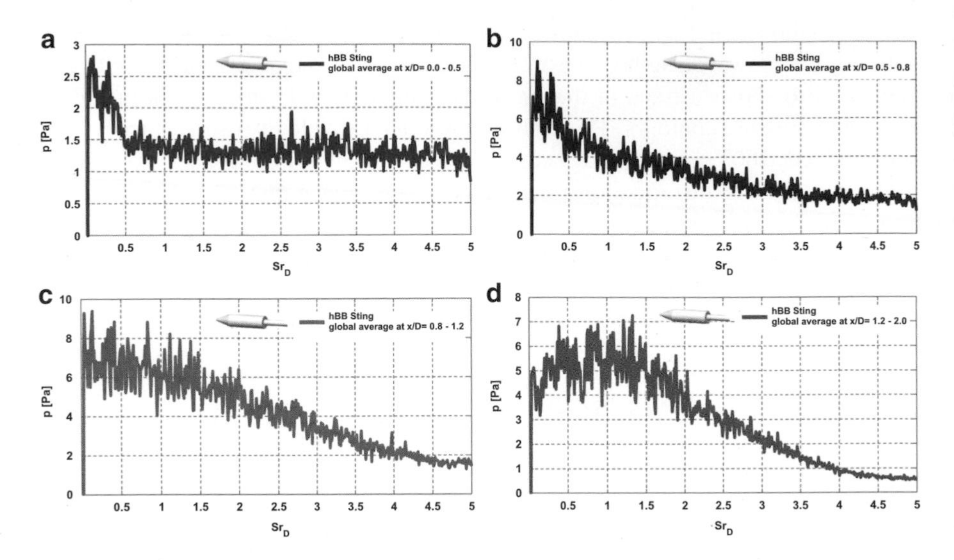

Fig. 13 DFT of the wall-pressure fluctuations for the hypersonic configuration with sting of four different dynamic regions. (**a**) Recirculation bubble (i). (**b**) Reattachment line (ii). (**c**) Recompression waves (iii). (**d**) Recompression shock (iv)

As schematically highlighted by colored rectangles in Fig. 12, the wake of the supersonic configuration can be divided into four regions; the recirculation bubble (**i**), the reattachment region (**ii**), the re-compression waves (**iii**), and the re-compression shock (**iv**). Unlike the transonic case, the dynamics and consequently the dominant frequencies at $Ma_\infty = 6.0$ differ significantly depending on each of these four regions. The spectra in the vicinity of the separation bubble show lower-frequency modes of higher amplitudes, however, the amplitude of these modes reduces as streamwise positions further downstream are considered. The spectra taken at the location of the re-compression shock show little dependance on the modal frequency. Moreover, the DFT spectra of wall-pressure fluctuations show rather dominant frequency ranges than sharp peaks of the wall-pressure fluctuations (Fig. 13). Therefore it is difficult to isolate a single frequency of interest, and for the purpose of this study a range of frequencies is classified as a frequency-domain. These frequency-domains for the region including the recirculation bubble (region (i)) are; $D_1 : 0 \leq Sr_D \leq 0.2$, $D_2 : 0.2 \leq Sr_D \leq 0.2$ and $D_3 : Sr_D > 0.5$. When applying the DMD algorithm, a frequency within the defined frequency-domain is selected to represent the dynamics within that range.

In order to remain within the scope of this report, we will concentrate on the recirculation bubble and the reattachment line; therefore, the DMD domain is restricted to regions (i) and (ii). DMD is performed on the streamwise, radial and transverse velocity components. Although DFT is performed on the wall-pressure signal, we realized that due to the presence of the shock wave inside the domain of interest the pressure has a higher value downstream than upstream. As a result the

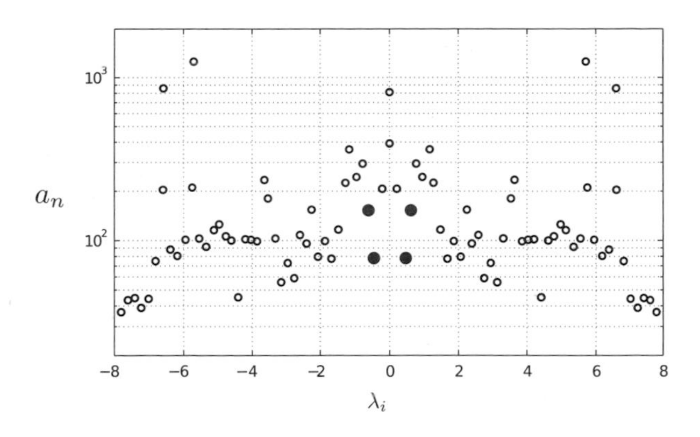

Fig. 14 DMD spectra. •, $Sr_D = 0.4$; •, $Sr_D = 0.64$

pressure signal would be more relevant for studying the dynamics of the shockwave or the downstream pressure. Since we are interested in the separation bubble located upstream of the domain, velocity components are a better choice for the DMD analysis. In addition, composite DMD allows for a complete three-dimensional picture of the dynamics in the flow and would provide information on how each velocity component is coupled to the others within a single modal frequency. We have chosen the cylindrical rather than Cartesian velocity components in order to take advantage of the axisymmetric nature of the flow. Moreover, the presence of helical modes have been reported for these types of geometries, and the transverse velocity would allow direct assessment of the existence of such modes.

Figure 14 shows the DMD spectra of the three velocity components within the subdomain of region (i) and (ii). As mentioned previously, the modes which appear with a higher amplitude than the mean are transient modes, which decay rapidly and have only a short-term influence on the dynamics of the flow, and are therefore neglected. We have focused on the two frequency-domains of D_2 and D_3. We have chosen not to include D_1 since its dynamics is of lower frequency and more snapshots are needed to properly capture these modes correctly. The chosen DMD modes in domains D_2 and D_3 are shown in Fig. 14 by the blue and red symbols.

Figure 15a shows the shape of the DMD mode corresponding to $Sr_D = 0.4$ in D_2. For each velocity component slices in the $x - y$ and $y - z$ planes are shown. The radial velocity component of this mode is symmetric across the plane of reference of the $x - y$ slice. The $y - z$ slice of this velocity component shows that symmetry is not preserved throughout the geometry. However, this mode is not a traveling mode in the transverse direction, which is illustrated in the $y - z$ slice shown in Fig. 15b, where the modulation of the transverse mean flow due to the transverse component of the mode is plotted. This component has little effect on the mean-flow value suggesting that, although the bubble is wrinkled in the transverse direction, it represents a standing wave. The same results apply when considering the modulation of the streamwise velocity. The evolution of the mean flow through this mode shows the pinching of the separation bubble when the radial component is

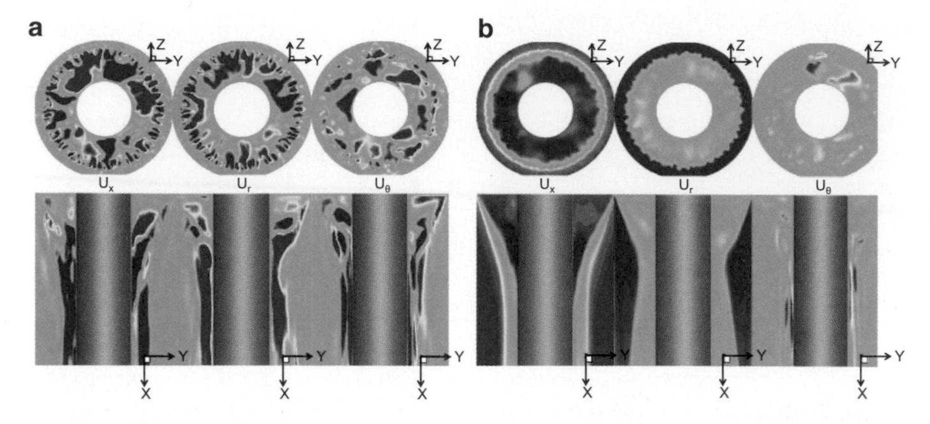

Fig. 15 DMD mode with frequency $Sr_D = 0.4$. Velocity components from *left* to *right*: U_x, streamwise velocity; U_r, radial velocity; U_θ, transverse velocity. (*Blue*), negative values; (*Red*) positive magnitude. (**a**) Velocity components of mode $\lambda_1 = 0.4$. (**b**) Velocity components the mean and mode $\lambda_1 = 0.4$

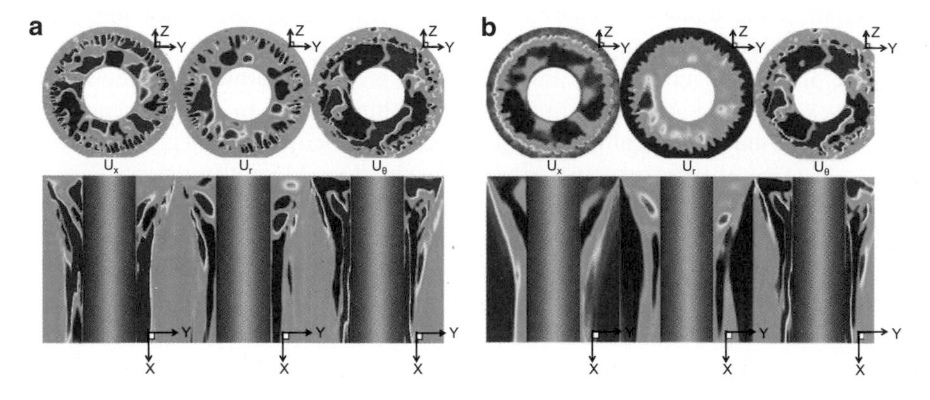

Fig. 16 DMD mode with frequency $Sr_D = 0.64$. Velocity components from *left* to *right*: U_x, streamwise velocity; U_r, radial velocity; U_θ, transverse velocity. (*Blue*), negative values; (*Red*) positive magnitude. (**a**) Velocity components of mode $\lambda_1 = 0.64$. (**b**) Velocity components the mean and mode $\lambda_1 = 0.64$

negative and the release of the bubble when the value is positive. This in turn results in the breathing of the separation bubble.

Figure 16a shows the shape of the higher-frequency DMD mode. In contrast to the lower-frequency mode this mode has a bigger signature in both the streamwise and the transverse velocity components. The mean-flow velocity is symmetric along the streamwise axis due to the presence of axi-symmetry in the mean flow. As a result, the transverse component is of negligible value, as seen in Fig. 15b. The modulated mean velocity of this higher-frequency mode, however, shows a distinct variation along the transverse direction. The evolution of this mode in time confirms this conclusion, and the mode is shown to move along the transverse direction periodically.

Modes $\lambda_1 = 0.4$ and $\lambda_2 = 0.64$ contain two completely different dynamics of the separated flow. One mode causes the breathing of the separation bubble with a small trace in the transverse direction, while, the higher-frequency mode is of a helical nature, illustrated by the modulated mean transverse velocity component.

4.4 Computational Resources

The time-resolved zonal RANS/LES computations for this study were performed on the CRAY XE6 (Hermit) at the HLR Stuttgart. The computational grids are divided into blocks which reside on a single CPU Core. Data is exchanged using the message passing interface (MPI). The good scalability of the used house-internal zonal RANS/LES flow solver is demonstrated in Fig. 17. The computational details for the three performed simulations are given in Table 2. It is evident that the total number of the mesh points, which is the first major parameter with respect of the required computational resources, is significantly larger for the transonic case than for the hypersonic ones, even though the transonic grid spans over 60° in the circumferential direction and the hypersonic domains extend over full 360°. The main reason for this fact is that at transonic speeds the pressure waves, which propagate with the speed of sound, also travel upstream. Therefore, the boundaries of the computational domain have to be positioned at a satisfactorily large distance

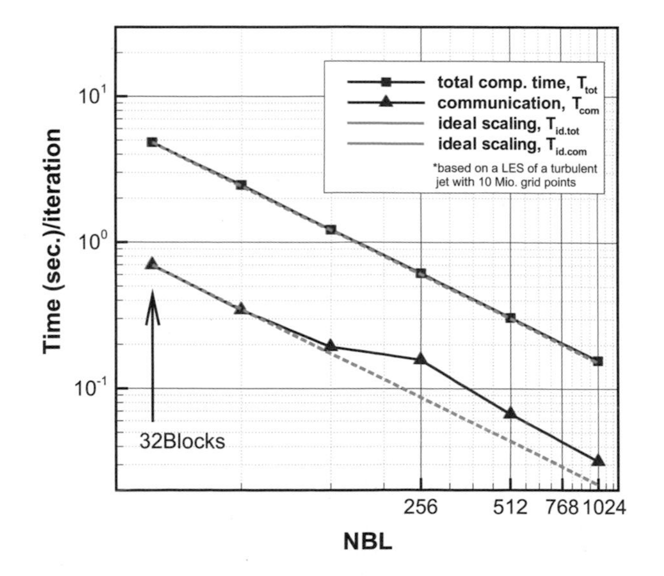

Fig. 17 Strong scaling of the used house-internal zonal RANS/LES flow solver

Table 2 Required resources for CFD computations

Configurations	Transonic w. sting	Hypersonic w. sting	Hypersonic w. jet
Total grid points	$76.1 \cdot 10^6$	$22.5 \cdot 10^6$	$35.5 \cdot 10^6$
Number of CPUs	4,096	512	1,024
Number of nodes	128	16	32
Grid points/CPU	18,557	43,946	34,667
Comp. phys. time $[t_{ref}]$	50	50	50
Comp. speed $[t_{ref}/h]$	0.5	0.1	0.13
Total. user time [h]	100	500	385
Total. core hours [CPU· h]	0.4096 Mio.	0.2560 Mio.	1.2 Mio.
Used HDD storage [TB]	2.3	0.7	1.1

Table 3 Required resources for DMD analysis

Configurations	Transonic w. sting	Hypersonic w. sting	Hypersonic w. jet
Total grid points	$76.1 \cdot 10^6$	$22.5 \cdot 10^6$	$35.5 \cdot 10^6$
Peak RAM use $[GB]$	232	73	115
HDD storage/Variable $[GB]$	187	66	93

from the launchers body. At supersonic speeds, no flow information can travel upstream and the computational domain can be made compact covering only the bow shock region around the rocket's body. On the other hand, the hypersonic cases were found to feature a smaller time-step than their transonic counterpart, which is caused by significant compressibility effects and occurrence of strong shock waves at $Ma_\infty = 6$ limiting the numerical stability.

To assess the length of the computed physical time interval and respective computational resources required for a reliable statistical analysis of the simulation data, a reference time unit t_{ref} is used, with 1 t_{ref} being the time needed by a particle moving with the freestream velocity u_∞ to cover one reference length equal to the main body diameter D of the launcher. To achieve satisfactory statistical quality, a minimum time interval of $50\,t_{ref}$ is required. Taking into account the different computational speed of each case, which depends on the total number of grid points per CPU and maximum stable time-step, we obtain the total required CPU hours summarized at the end of Table 2.

While the major requirement for the large CFD computations is the number of CPUs, the main driving parameter for the post processing of the simulation data is the virtual memory. For the performed DMD analysis, the shared node SMP of the HLR Stuttgart with 1 TB memory was used. The computational resources required for the performed DMD analysis are summarized in Table 3. The peak virtual memory use reaches up to 232 GB for the composite DMD of three velocity components. Physical storage of one post-processed variable, e.g., the axial velocity component u, requires up to 0.2 TB which linearly scales with the total number of the grid points.

5 Summary and Conclusions

The separation bubble dynamics in three different flow configurations of axisymmetric space launchers has been analyzed numerically using DMD together with a conventional discrete Fourier transform technique. The freestream Mach number has been varied from 0.7 to 6.0 consistent with the common operating points of the space launchers.

In the transonic case, DFT of the pressure fluctuations at the wall reveal distinct frequencies, which appear as peaks. We have shown that these frequencies can be extracted with close agreement using DMD. In addition, DMD provides a modal shape corresponding to each frequency. The shapes of the modes verify that the low-frequency modes are indeed related to the dynamics of the separation bubble and the reattachment line. Applying composite DMD we have been able to show the coupling between the streamwise velocity and pressure of a single high-frequency mode to be reminiscent of an acoustic wave.

The flow over the same geometry at the supersonic Mach number shows different trends in the DFT spectra. The single-frequency peaks are no longer distinguishable; instead, a region of dominant frequencies are identified, whose amplitudes decrease as we move downstream. The domain of interest encircles the separation bubble, and the DMD has been performed on this domain. Two modes within these higher and lower frequency ranges have been analyzed. The low-frequency mode is responsible for the breathing motion while the high-frequency mode produces a helical motion commonly detected in such geometries. The wave-length of this mode in the transverse direction illustrates that this mode would have been difficult to be detected without including the full geometry. Moreover, using composite DMD we have been able to establish a coupling of each velocity component with a single frequency mode.

Finally, a free-flight configuration has been considered. Although the Mach number is similar to the supersonic case, due to the presence of an after-expanding nozzle jet, the dynamics and hence the relevant frequencies change significantly. The two modes studied here display two different dynamics of the cavity flow, which develops behind the step. The DMD mode corresponding to the low-frequency range of the DFT spectra is more active close to the region of the averaged separation bubble and results in the low-frequency breathing of the bubble. In contrast, a higher-frequency mode with $Sr_D \approx 1.0$ is active in the region of the averaged shear layer edge, and in combination with the mean flow, results in the flapping motion of the shear layer.

In conclusion, although the conventional DFT offers some idea on the relevant frequencies of the flow, without a clear picture of modal shapes, the assessment of the effect of each frequency on the entire flow dynamics would be difficult. Using DMD not only provides a three-dimensional shape associated with the respective frequency, it also allows the determination of the coupling between different flow variables.

Acknowledgements The support of this research by the German Research Association (DFG) in the framework of the Sonderforschungsbereich Transregio 40 and the High Performance Computing Center Stuttgart (HLRS) is gratefully acknowledged.

References

1. Rollstin, L.R.: Measurement of in-flight base pressure on an artillery-fired projectile. AIAA Paper **87**, 2427 (1987)
2. Mathur, T., Dutton, J.: Base-bleed experiments with a cylindrical afterbody in supersonic flow. J. Spacecr. Rockets **33**, 30–37 (1996)
3. Janssen, J., Dutton, J.: Time-series analysis of supersonic base-pressure fluctuations. AIAA J. **42**, 605–613 (2004)
4. Bannink, W., Houtman, E., Bakker, P.: Base flow / Underexpanded exhaust plume interaction in a supersonic external flow. AIAA Paper **98**, 1598 (1998)
5. Scarano, F., van Oudheusden, B., Bannink, W., Bsibsi, M.: Experimental investigation of supersonic base flow plume interaction by means of particle image velocimetry. In: 5th European Symposium on Aerothermodynamics for Space Vehicles, Cologne, pp. 601–607. Number ESA SP-563 (2004)
6. Deprés, D., Reijasse, P., Dussauge, J.: Analysis of unsteadiness in afterbody transonic flows. AIAA J. **42**, 2541 (2004)
7. Meliga, P., Reijasse, P.: Unsteady transonic flow behind an axisymmetric afterbody equipped with two boosters. AIAA Paper **4564**, (2007). doi:10.2514/6.2007-4564
8. Benay, R., Servel, P.: Two-equation k-σ turbulence model: application to a supersonic base flow. AIAA J. **39**, 407–416 (2001)
9. Papp, J., Ghia, K.: Application of the RNG turbulence model to the simulation of axisymmetric supersonic separated base flows. AIAA Paper 2001-2027 (2001)
10. Forsythe, J., Hoffmann, K., Cummings, R., Squires, K.: Detached-eddy simulation with compressibility corrections applied to a supersonic axisymmetric base flow. J. Fluid Eng. **124**, 911–923 (2002)
11. Kawai, S., Fujii, K.: Computational study of a supersonic base flow using LES/RANS hybrid methodology. AIAA Paper 2004-68 (2004).
12. Fureby, C., Kupiainen, K.: Large-eddy simulation of supersonic axisymmetric baseflow. In: Turbulent Shear Flow Phenomena, TSFP3, Sendai (2003)
13. Sandberg, R., Fasel, H.: High-accuracy DNS of supersonic base flows and control of the near wake. In: Proceedings of the Users Group Conference, Nashville, pp. 96–104. IEEE Computer Society (2004)
14. Sandberg, R., Fasel, H.: Numerical investigation of transitional supersonic axisymmetric wakes. J. Fluid Mech. **563**, 1–41 (2006)
15. Sandberg, R.: Numerical investigation of turbulent supersonic axisymmetric wakes. J. Fluid Mech. **702**, 488–520 (2012)
16. Meliga, P., Sipp, D., Chomaz, J.: Elephant modes and low-frequency unsteadiness in a high Reynolds number, transonic afterbody wake. Phys. Fluids **21**, 054105 (2009)
17. Deck, S., Thorigny, P.: Unsteadiness of an axisymmetric separating-reattaching flow: numerical investigation. Phys. Fluids **19** (2007). doi:10.1063/1.2734996
18. Statnikov, V., Meiß, J.-H., Meinke, M., Schröder, W.: Investigation of the turbulent wake flow of generic launcher configurations via a zonal RANS/LES method. CEAS Space J. **5**, 75–86 (2013). doi:10.1007/s12567-013-0045-6
19. Saile, D., Gülhan, A., Henckels, A., Glatzer, C., Statnikov, V., Meinke, M.: Investigations on the turbulent wake of a generic space launcher geometry in the hypersonic flow regime. EUCASS Prog. Flight Phys. **5**, 209–234 (2013)

20. Statnikov, V., Glatzer, C., Meiß, J.-H., Meinke, M., Schröder, W.: Numerical investigation of the near wake of generic space launcher systems at transonic and supersonic flows. EUCASS Prog. Flight Phys. **5**, 191–208 (2013)
21. Saile, D., Gülhan, A., Henckels, A.: Investigations on the near-wake region of a generic space launcher geometry. AIAA 2011–2352 (2011)
22. Bitter, M., Scharnowski, S., Hain, R., Kähler, C.J.: High-repetition-rate PIV investigations on a generic rocket model in sub- and supersonic flows. Exp. Fluids **50**, 1019–1030 (2011)
23. Bitter, M., Hara, T., Hain, R., Yorita, D., Asai, K., Kähler, C.J.: Characterization of pressure dynamics in an axisymmetric separating/reattaching flow using fast-responding pressure-sensitive paint. Exp. Fluids **53**, 1737–1749 (2012)
24. Roidl, B., Meinke, M., Schröder, W.: A zonal RANS-LES method for compressible flows. Comput. Fluids **67**, 1–15 (2012)
25. Roidl, B., Meinke, M., Schröder, W.: Reformulated synthetic turbulence generation method for a zonal RANS-LES method and its application to zero-pressure gradient boundary layers. Int. J. Heat Fluid Flow **44**, 28–40 (2013).
26. Roidl, B., Meinke, M., and Schröder, W.: "Boundary layers affected by different pressure gradients investigated computationally by a zonal RANS-LES method," Int. J. Heat Fluid Flow **45**, 1–13 (2014).
27. Liou, M.S., Steffen, C.J.: A new flux splitting scheme. J. Comput. Phys. **107**, 23–39 (1993)
28. Boris, J., Grinstein, F., Oran, E., Kolbe, R.: New insights into large eddy simulation. Fluid Dyn. Res. **10**, 199–228 (1992)
29. Meinke, M., Schröder, W., Krause, E., Rister, T.: A comparison of second- and sixth-order methods for large-eddy simulations. Comput. Fluids **31**, 695–718 (2002)
30. Alkishriwi, N., Meinke, M., Schröder, W.: A large-eddy simulation method for low Mach number flows using preconditioning and multigrid. Comput. Fluids **35**, 1126–1136 (2006)
31. El-Askary, W., Schröder, W., Meinke, M.: LES of compressible wall-bounded flows. AIAA Paper 2003–3554 (2003)
32. Spalart, P., Allmaras, S.: A one-equation turbulence model for aerodynamic flows. AIAA Paper 92–0439 (1992)
33. Jarrin, N., Benhamadouche, N., Laurence, S., Prosser, D.: A synthetic-eddy-method for generating inflow conditions for large-eddy simulations. J. Heat Fluid Flow **27**, 585–593 (2006)
34. Pamiès, M., Weiss, P., Garnier, E., Deck, S., Sagaut, P.: Generation of synthetic turbulent inflow data for large eddy simulation of spatially evolving wall-bounded flows. Phys. Fluids **21**, 045103 (2009)
35. Schmid, P.J.: Dynamic mode decomposition of numerical and experimental data. J. Fluid Mech. **656**, 5–28 (2010)
36. Rowley, C.W., Mezic, I., Bagheri, S., Schlatter, P., Henningson, D.S.: Spectral analysis of nonlinear flows. J. Fluid Mech. **641**, 115–127 (2009)
37. Sayadi, T., Nichols, J.W., Schmid, P.J., Jovanovic, M.R.: Dynamic mode decomposition of h-type transition to turbulence. In: Proceedings of the Summer Program 2012, Center for Turbulence Research, Stanford, pp. 5–13 (2012)

CFD Performance Analyses of Wind Turbines Operating in Complex Environments

Pascal Weihing, Christoph Schulz, Thorsten Lutz, and Ewald Krämer

Abstract This paper presents results from CFD simulations of wind turbines performed within the project *WEALoads*. The focus of this project is devoted to the unsteady load response of wind turbines under realistic environmental conditions, as for example operation inside of a wind farm or in complex terrain which are both subject of this paper. The first case shall investigate the behavior of a wind turbine operating half in the wake of an upstream turbine, in order to derive the dominant interference effects between wind turbines. Secondly, a wind turbine shall be analyzed which is sited on a hill to elaborate the main effects arising from the interaction of the atmospheric boundary layer with the hill and finally the wind turbine. Both simulations were performed using the flow solver *FLOWer* from DLR (German Aerospace Center) and the Detached Eddy Simulations (DES) approach. Results of the flow fields are shown in terms of wake development, as well as turbulence intensity. Regarding the case of the turbine sited in complex terrain, a site assessment study has been performed, in order to find designated positions where maximum power output of the wind turbine can be expected. Finally, for both cases, blade load evaluations showed significant influence of the operating environment. For the case of the interacting turbines the load response of the shadowed turbine showed a massively asymmetric loading of the entire rotor. For the turbine located on the hill, significant augmentation of the entire load level could be observed.

1 Introduction

The demand to reduce production costs in the generation of electric energy from wind is a major objective in energy policy. For this purposes wind turbines require to be sited in regions of relatively higher wind speeds. These can be found for example offshore, where the turbines are embedded in large wind farms or onshore on hills. To develop reliable turbines which require little maintenance, it is important to accurately predict the aerodynamic loads on the blades. Particular interest is put

P. Weihing (✉) • C. Schulz • T. Lutz • E. Krämer
Institute of Aerodynamics and Gas Dynamics, Pfaffenwaldring 21, 70569 Stuttgart, Germany
e-mail: weihing@iag.uni-stuttgart.de; schulz@iag.uni-stuttgart.de

W.E. Nagel et al. (eds.), *High Performance Computing in Science and Engineering '14*, DOI 10.1007/978-3-319-10810-0_27

on the fluctuations of these loads which lead to enhanced fatigue loading of the whole turbine [4, 8]. These load fluctuations result from rapid changes in the local angle of attack due to atmospheric turbulence, which is particularly produced in wind turbine wakes or in complex terrain.

In the framework of load prediction methods the role of Computational Fluid Dynamics (CFD) is becoming increasingly important as high-performance computing capacities are steadily increasing. These render possible to simulate the unsteady behaviour of whole wind turbines operating in their environment. In contrast to the widely used engineering tools, being based on blade element momentum theory, solving the Navier-Stokes equations and directly representing all relevant wind turbine components within CFD, poses the most accurate approach to calculate unsteady aerodynamic effects as dynamic stall, wake-turbine interaction or effects due to complex terrain without the need for further modeling [9, 12, 15].

This paper presents results of simulations performed in the project*WEALoads* using the CFD code *FLOWer* [7] by *DLR (German Aerospace Center)*. The first simulation presented here is devoted to interference of two offshore wind turbines which are sited behind each other at half-wake conditions as found in the German research wind farm *Alpha Ventus*. This simulation is referred to as *CaseA*. Secondly, hereafter referred to as *CaseB*, results are shown from simulations of a turbine sited on a hill near *Schnittlingen, Baden-Württemberg, Germany*. The latter case is subdivided into three parts: *CaseB1* serves for reference and considers the wind turbine sited on flat terrain under turbulent inflow; *CaseB2* refers to a simulation of the hill in combination with turbulent inflow but without wind turbine, in order to discuss site specific aspects; finally, in *CaseB3* the wind turbine is included on top of the hill.

2 Numerical Setup and Computational Details

The computational domain of *CaseA* is depicted in the left subplot of Fig. 1. The turbines under consideration have a rotor diameter of 126 m, a hub-height of 95 m and a rated power of 5 MW and are located 845 m behind each other which

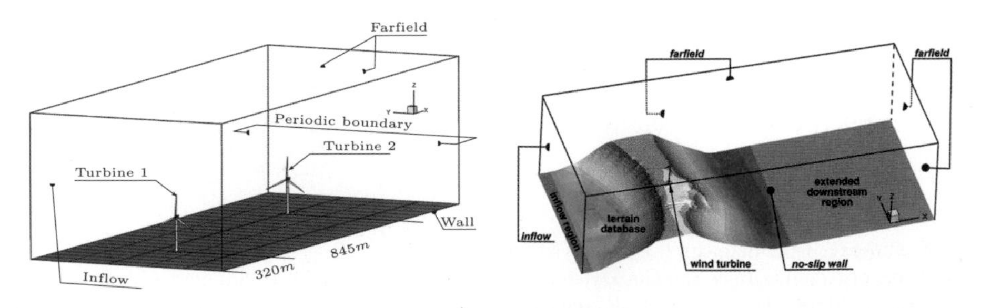

Fig. 1 Computational domain and boundary conditions (l.) *CaseA*, (r.) *CaseB3*

corresponds to 6.7 rotor diameters. The lateral offset relative to the wind direction is half a rotor diameter. The total domain spans $1{,}216 \times 560 \times 404$ m in streamwise, lateral and vertical direction. At the inlet, data of a realistic turbulent atmospheric boundary layer is prescribed which was gathered from a precursor LES simulation performed at *ForWind*, Oldenburg, Germany with the solver *PALM* [9, 10]. The averaged inflow velocity at hub height is 15.27 m/s. The outlet and top patches imply farfield conditions, whereas at the side planes periodic conditions are applied. Turning now to the setup of *CaseB3*, shown in the right subplot of Fig. 1, the domain spans $1{,}100 \times 400 \times 475$ m and is divided into three different regions: An even inflow region, the region of interest including the hill and the wind turbine as well as an extended downstream region. The boundary conditions in both grids were chosen similarly – inflow at the front plane, no-slip walls at the ground, farfield boundary conditions at the sides, the top and downstream. The turbine sited on top of the hill has a rated power of 2.4 MW, a rotor diameter of 109 m, and a hub height above ground of 95 m. The inflow dataset is that of *CaseA*, however, the mean inflow velocity is scaled to 11.01 m/s.

In both cases all the relevant turbine components like the blades, tower, hub and the nacelle are represented as full model being meshed with fully resolved boundary layers ensuring $y^+ = 1$ of the first grid point. The components are meshed separately and connected within *FLOWer* using the overset grid technique [1] as exemplarily shown in Fig. 2 for one turbine of *CaseA*. As for this case it is important to propagate the wake from the upstream turbine to the downstream one, a ring-shaped refinement mesh was introduced connecting both turbines. In the case of complex terrain, a major difficulty is the connection of the turbine at arbitrary positions, since a proper overlapping of the tower to the terrain mesh must be guaranteed. Hence, as shown in Fig. 3, it turned out to be the best to create a locally flat platform, where the cell distribution accurately matches the one of the tower.

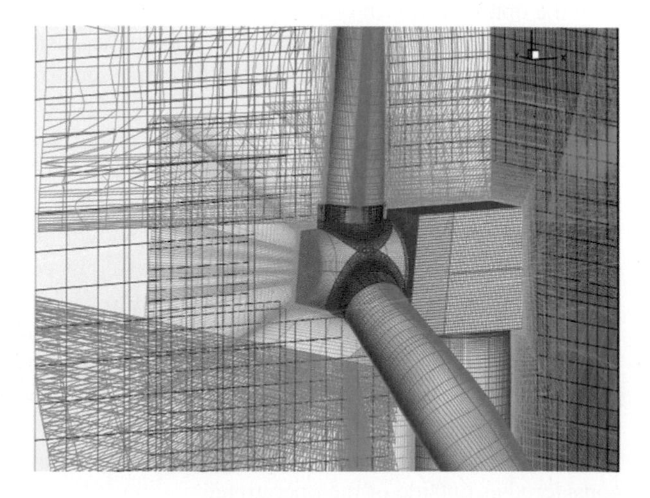

Fig. 2 Component meshes embedded with overset grid technique (*CaseA*)

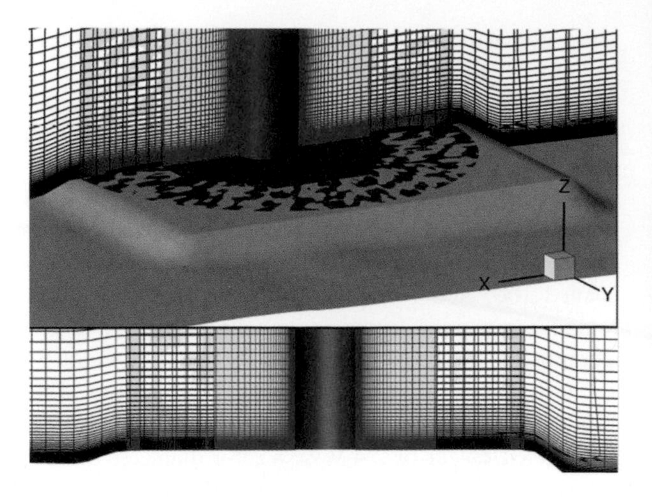

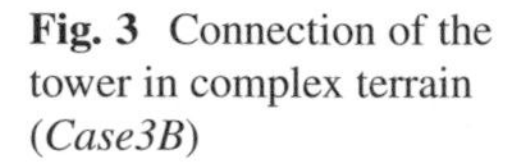
Fig. 3 Connection of the tower in complex terrain (*Case3B*)

In total, the domain of *CaseA* is discretized by approx. 110 mio. cells, *CaseB* contains approx. 68 mio. cells. Spatial discretization is conducted using the second order central differencing scheme JST [6], whereas for temporal discretization an implicit dual time stepping scheme [5] is used with a time-step equivalent to 3 ° azimuthal blade movement in *CaseA* and 2.5° in *CaseB*. Both studies use the Detached-Eddy Simulation approaches developed by Spalart [13,14] in combination with the Spalart-Allmaras turbulence model with Edwards modification [3] to adequately propagate the atmospheric turbulence through the domain. In order to obtain converged loads, 36 revolutions have been performed in both cases before data was extracted for another 12 revolutions in *CaseA* and 13 revolutions in *CaseB* (this corresponds to a periodic window in the inflow data).

All simulations have been performed at *High Performance Computing Center Stuttgart (HLRS)* with *AMD Interlagos* processors. *CaseA* used 1,700 processors, with the option to use only every second processor to increase memory and speed, whereas *CaseB* was computed on 1,024 processors using every processor. In total, *CaseA* required approx. 300,000 CPUh, compared to approx. 100,000 CPUh for *CaseB*.

In addition to the presented simulations strong scaling tests of the *FLOWer*-code have been performed, in order to evaluate the suitability regarding massively parallel computations on the Cluster *Hermit*. Previous investigations on the scaling of FLOWer on the cluster *Hermit* have been shown in [2]. In Fig. 4 is plotted the speedup vs. the number of used cores for different compiler settings on *Hermit*. The total number of blocks was kept constant to 512. The scaling test started with two cores, since the number of blocks exceeded the memory capacity of one CPU. Therefore, the speedup is related to the baseline speed of two CPUs. The curves show almost linear speedup for up to 1,024 cores with both compilers. However, there is a constant shift to the ideal speedup mainly due communication time between the blocks. The kink visible for the crayftn compiler for 16 CPUs can be considered as outside of the operational regime.

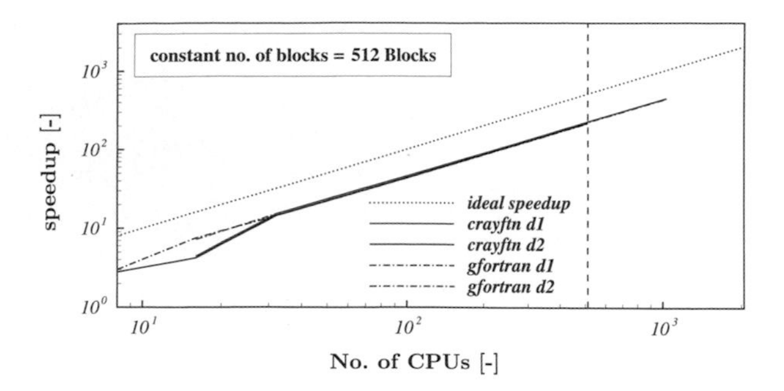

Fig. 4 Strong scaling tests for FLOWer on the cluster *Hermit*

3 Results

Due to confidential clauses the results of power and loads presented in the following are normalized by appropriate mean values. The azimuthal position of 0° refers to the upward position of the blade. When looking in streamwise direction the rotor rotates clockwise.

The instantaneous flow field of the interacting wind turbines is shown in Fig. 5. Therefore, vortex structures in the domain are visualized by means of an appropriate λ_2 iso-surface. Further, the relative velocity contours $V_{rel} = \sqrt{u^2+v^2+w^2}/U_\infty$ are included in a slice at hub-height. At the inlet the vortices clearly show the incoming atmospheric turbulence. It is to be noted, that due to numerical dissipation of the used JST scheme small scale turbulence decays when propagating downstream and therefore cannot be resolved by the selected λ_2 value. However, lower frequencies in the regime being relevant for the prediction of loads, are still adequately propagated. Regarding the transport of tip and root vortices, the introduction of wake refinement meshes turned out to be beneficial. Approximately 2.5 rotor diameters downstream of the first turbine the tip vortices start breaking down, going along with large scale motion of the entire wake. This results in significant changes in velocity and angle of attack at the shaded side of the downstream turbine, whereas the non-shaded remains still in mainly undisturbed conditions. This effect also results in an asymmetric wake of the downstream turbine.

The corresponding vortex analysis for *CaseB3* is depicted in Fig. 6. At the windward side of the hill the thin vortex layer at the slopeside indicates a speedup of almost two compared to freestream velocity at the inflow plane, whereas on the lee side the flow is retarded and massively separated, showing two and three dimensional vortex structures. Due to that highly turbulent flow pattern the tip vortices in *CaseB* decay significantly earlier compared to *CaseA*. However, it is to be noted, that no wake refinement mesh has been introduced in that case.

As the power output of a wind turbine is proportional to the cubic of the mean wind speed, the vertical profiles of the time averaged streamwise velocity component can provide an estimation on that behaviour. In *CaseA*, the profiles

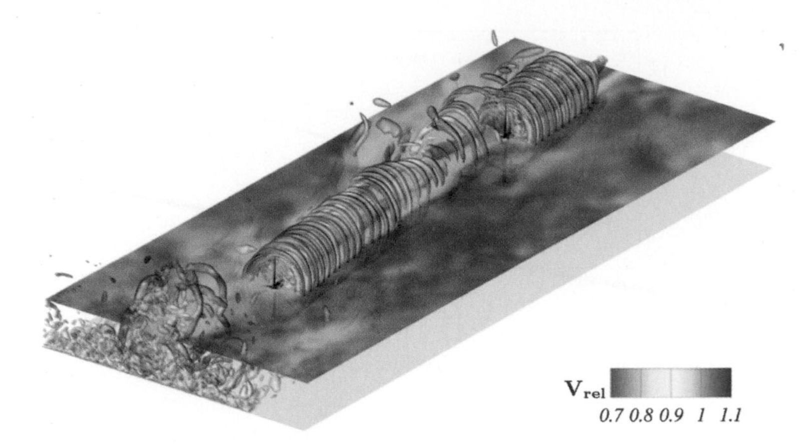

Fig. 5 Wake and vortex structures of *CaseA*: λ_2 iso-surface, Velocity contours at hub-height $V_{rel} = \sqrt{u^2+v^2+w^2}/U_\infty$

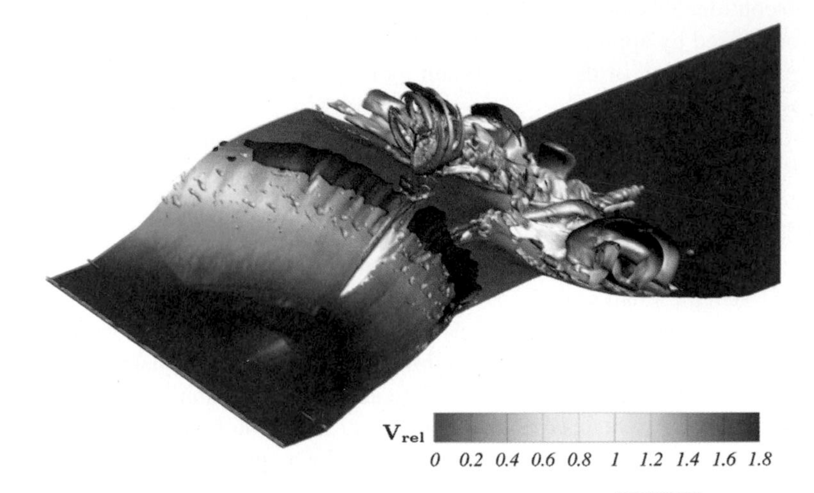

Fig. 6 Wake and vortex structures of *CaseB3*: λ_2 iso-surface, $V_{rel} = \sqrt{u^2+v^2+w^2}/U_\infty$

were extracted inline with the wind direction one half rotor diameter upstream and one rotor diameter downstream of each turbine (Fig. 7). Upstream of Turbine1, the profile is that of a slightly sheared neutral atmospheric boundary layer. In the wake of Turbine1, the velocity deficit is about 20 %. Upstream of Turbine2 the profile partly recovered an got rounder. As Turbine2 operates in the wake of upstream turbine, it is obvious that wake deficit is higher compared to that of Turbine1.

Turning to *CaseB*, the velocity profiles over flat terrain and over the hill are compared in Fig. 8. Due to flow separation at edge of the hill, streamwise velocity is reduced in a layer above the surface ($z < 25$ m), whereas a uniform speedup of $\approx$20 % in the designated rotor area gives reason to expect significant power augmentation when siting the turbine on the hill (further details in Fig. 12).

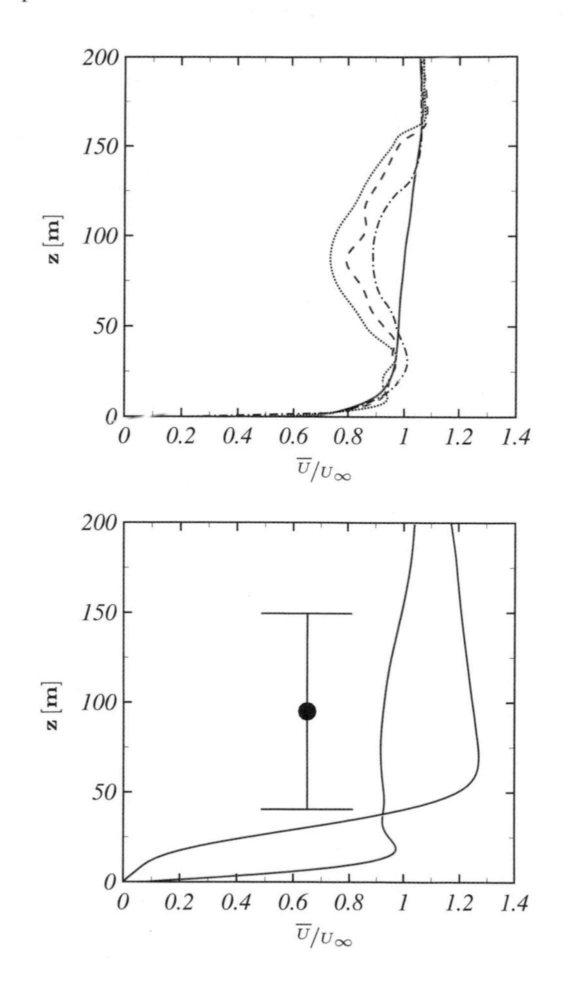

Fig. 7 Vertical time averaged streamwise velocity profiles for various downstream positions aligned in direction of each turbine (*CaseA*): $-0.5D$ Turbine1 (—), $1D$ Turbine1 (– –), $-0.5D$ Turbine2 (—), $1D$ Turbine2 (. . .)

Fig. 8 Vertical time averaged streamwise velocity profiles for flat terrain (*CaseB1*) (—) and at the designated turbine position (*CaseB2*) (—), the *black symbols* indicate the rotor

An important property correlating with power and load fluctuations of wind turbines is the turbulence intensity. Higher turbulence intensities correlate with high load and power fluctuations and overall with an increase of fatigue and a decrease of power performance [4]. The vertical profiles of turbulence intensity *Ti* are plotted in Fig. 9 for *CaseA*. Upstream of Turbine1, turbulence intensity is relatively higher in a stripe between 50 and 120 m and near the surface, whereas it decays above. Overall, the level of ambient turbulence (2–3 %) is relatively low. In the wake of the first turbine turbulence intensity increases where tip and root vortex shedding occurs. However, for the downward position of the blade the tower seems to damp fluctuations at blade passing frequency. Upstream of Turbine2 the shape is similar to the corresponding profile in front of Turbine1, showing slightly higher values in the hub region. In the wake of the donwstream turbine the turbulence intensity is higher compared to Turbine1. The reason for the lower peak is that the profile was extracted with a small lateral offset from the tower.

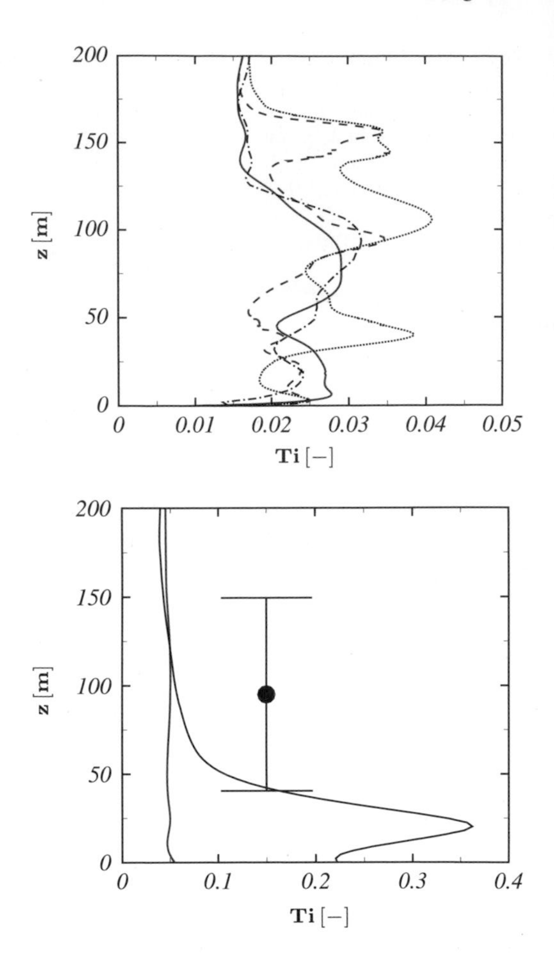

Fig. 9 Vertical *Ti* profiles for various downstream positions aligned in direction of each turbine (*CaseA*): $-0.5D$ Turbine1 (—), $1D$ Turbine1 (– –), $-0.5D$ Turbine2 (—), $1D$ Turbine2 (. . .)

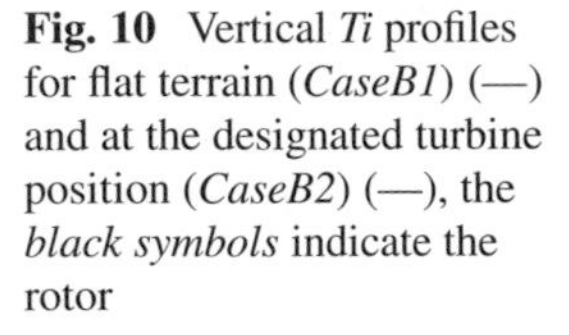

Fig. 10 Vertical *Ti* profiles for flat terrain (*CaseB1*) (—) and at the designated turbine position (*CaseB2*) (—), the *black symbols* indicate the rotor

The corresponding profiles of *CaseB* are plotted in Fig. 10 comparing turbulence intensity over flat terrain (*CaseB1*) with the values expected for the designated turbine position on the hill (*CaseB2*). In the former case the level is similar to *CaseA*, whereas on top of the hill, turbulence intensity increases by almost an order of magnitude ($z = 25$ m) and approaches the curve of *CaseB1* for $z > 100$ m. With focus on site assessment, turbulence intensity was quantified over the entire domain in a slice at hub-height ($\Delta z = h_{hub} = const.$). The contours in Fig. 11 indicate very small fluctuations over the flat terrain and in the accelerating flow on the rising hillside, moderate values over the hill and due to flow separation very high on the leeside.

Figure 12 shows an estimation of the possible turbine power output in complex terrain (*CaseB2*). The power is calculated by integration of the wind speed over tilted, circular areas representing the rotor [11]. This integration is performed for all points in a slice at hub-height. The data are normalized by the value gained under

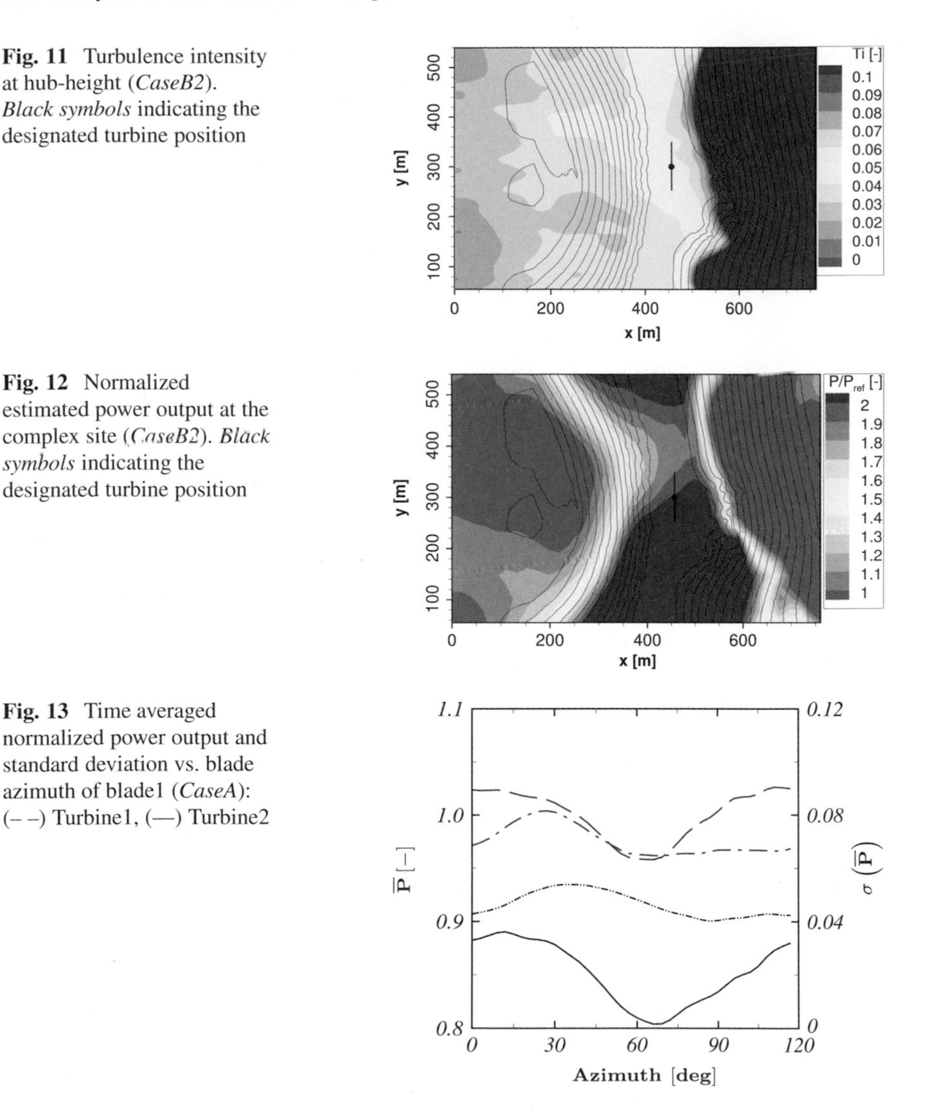

Fig. 11 Turbulence intensity at hub-height (*CaseB2*). *Black symbols* indicating the designated turbine position

Fig. 12 Normalized estimated power output at the complex site (*CaseB2*). *Black symbols* indicating the designated turbine position

Fig. 13 Time averaged normalized power output and standard deviation vs. blade azimuth of blade1 (*CaseA*): (– –) Turbine1, (—) Turbine2

uniform inflow conditions in flat terrain with a wind speed of 11.01 m/s. As shown in Fig. 12 the estimated power is seen to increase over the hill by a factor of two.

The mean power output obtained for the turbines of *CaseA* is shown in Fig. 13, plotted vs. the azimuth position of blade1. Due to 120° symmetry, the time series was split into 120° packages. Then, the time averaging and calculation of the standard deviation was performed for each azimuthal position, respectively. The data is normalized with the mean value of the upstream turbine. Both turbines show a local minimum when blade1 is at around 60° since then, the next blade is in front of the tower which experiences a drop in the aerodynamic forces. For the downstream turbine, the minimum is slightly asymmetric and shifted to 65° and the

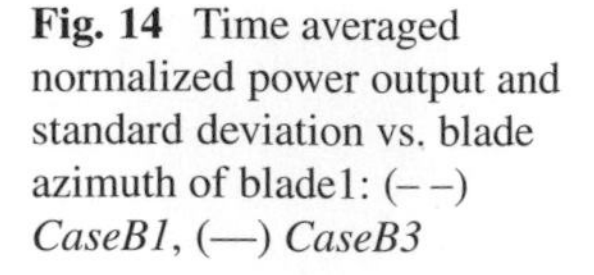
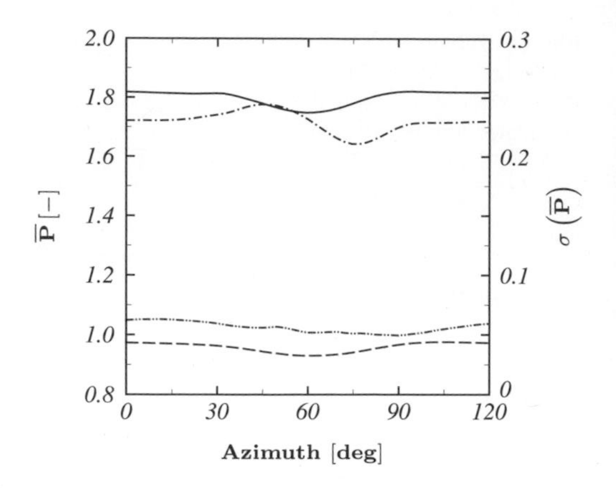

Fig. 14 Time averaged normalized power output and standard deviation vs. blade azimuth of blade1: (– –) *CaseB1*, (—) *CaseB3*

overall average is about 15 % lower than for the upstream turbine. Regarding the standard deviation, this is seen to increase before one of the blades passes the tower. It is to be noted, that for each constant azimuthal position the latter is higher for the upstream turbine.

In *CaseB* the power and forces have been normalized with manufacturer reference data for flat terrain and uniform inflow. For complex terrain (*CaseB3*), the power curves plotted in Fig. 14 very well confirm the power estimation presented in Fig. 12. When siting the turbine on the hill the power output can be increased by a factor of more than 1.8. However, due to the enhanced turbulence intensity above the hill, the turbine is exposed to significantly higher power fluctuations than in flat terrain.

Regarding the thrust behavior of the wind turbines, the axial blade force of blade1 shall be analyzed vs. its azimuthal position. In *CaseA*, the tower impact can be clearly observed for Turbine1. Furthermore, due to the tilt angle of −6° the upward moving blade is exposed to a higher angle of attack, resulting in higher axial forces compared to the downward moving blade ($\overline{F_x}$ (270°) $>$ $\overline{F_x}$ (90°)). Regarding the half-shadowed turbine, the forces at the unshaded side remain almost the same as for Turbine1, whereas the shadowed side reveals a dramatic drop of more than 30 %. Hence, it can be concluded that load reduction due to wake turbine interaction is dominant over the local tower effect. Furthermore, it should be noted that the asymmetrically loaded rotor induces yaw moments which have to be absorbed by whole transmission system and finally the grounding (Fig. 15).

For the flat terrain case (*CaseB1*) the trend is similar to that of Turbine1 in *CaseA*. In correspondence to the power output shown before, siting the turbine on the hill leads to an elevation of the load level by almost 40 %. Again, due to the very high turbulence intensity near the ground, the blade load fluctuations indicate a maximum before passing the tower. Directly in front of the tower, fluctuations are damped. The value before passing the tower is not reached, while the blade is in the layer of high turbulence intensity (Fig. 16).

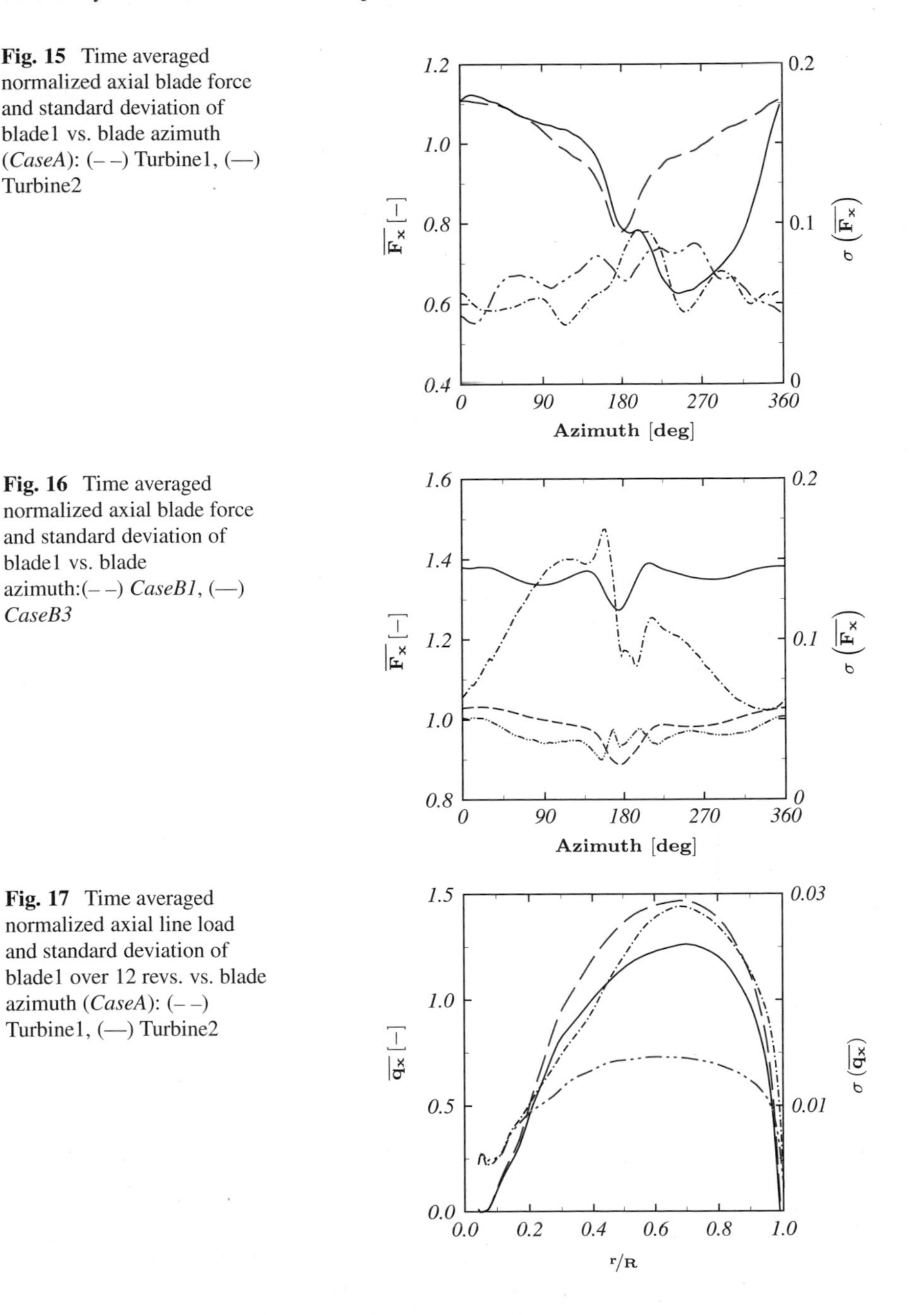

Fig. 15 Time averaged normalized axial blade force and standard deviation of blade1 vs. blade azimuth (*CaseA*): (– –) Turbine1, (—) Turbine2

Fig. 16 Time averaged normalized axial blade force and standard deviation of blade1 vs. blade azimuth:(– –) *CaseB1*, (—) *CaseB3*

Fig. 17 Time averaged normalized axial line load and standard deviation of blade1 over 12 revs. vs. blade azimuth (*CaseA*): (– –) Turbine1, (—) Turbine2

Finally, turning to the sectional blade loads, these are plotted vs. the relative rotor radius in Fig. 17 for *CaseA* and in Fig. 18 for *CaseB*. The sectional blade loads of the shadowed turbine are reduced due to the wake of the upstream turbine. Near the root the effect is small, since the blade only grazes the wake of the upstream turbine

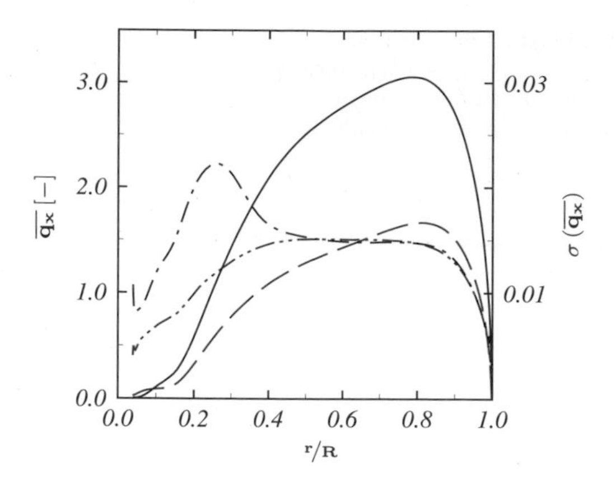

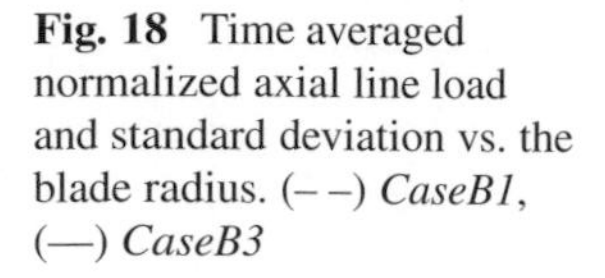

Fig. 18 Time averaged normalized axial line load and standard deviation vs. the blade radius. (– –) *CaseB1*, (—) *CaseB3*

in a small azimuthal range. In the mid portion, the blade dives more and more into the wake and the reduced velocity particularly causes a reduction in angle of attack. In the outer region, the tip vortices of the respective turbine dominate the loading. Regarding the standard deviation of the loads, the consideration of constant radial positions leads to increasing load fluctuations by a factor of more than 2.5 for the shadowed turbine, since the load significantly changes over one revolution.

For *CaseB3*, not only the level of the loads increases, but further the shape of the blade loading is altered. In the root region, where circumferential speed is low and the lift depends only little on the angle of attack (separated flow), the increasing loads mainly result from the increased axial velocity over the hill. Therefore, in that part also higher load fluctuations occur. In the outer region, the increase in relative velocity seen by the airfoils is small compared to the circumferential speed. The load increase follows primarily from the increasing angle of attack. As there, the flow is attached, the load fluctuations collapse with those of *CaseB1*.

4 Conclusions

In this paper Detached Eddy Simulations have been performed of wind turbines operating in complex environments. The first simulation aimed to investigate interference effects between wind turbines located at half-wake conditions behind each other. It could be observed that the power of the downstream turbine decreased by approximately 15 %, whereas the axial blade loads showed a significant azimuthal asymmetry and higher fluctuations over most of the radius. Secondly, a wind turbine was analyzed which was sited on a hill and comparisons were drawn to a reference case in flat terrain. Regarding the flow field, a speedup in the velocity could be observed above the hill going along with a massive increase of turbulence intensity in the surface layer on top of the hill. The speedup could almost double the power output. However, blade loads and their fluctuations increased as well.

Acknowledgements The authors gratefully acknowledge the *High Performance Computing Center Stuttgart* for providing computational resources.

References

1. Benek, J.A., Steger, J.L., Dougherty, F.C., Buning, P.G.: Chimera. A Grid-Embedding Technique. NASA, Technical Documents (1986)
2. Busch, E., Wurst, M., Keßler, M., Krämer, E.: Computational aeroacoustics with higher order methods. In: Nagel, W.E., Kröner, D.H., Resch, M.M. (eds.) High Performance Computing in Science and Engineering'12, pp. 239–253. Springer, Berlin/Heidelberg (2013)
3. Edwards, J.R., Chandra, S.: Comparison of eddy viscosity-transport turbulence models for three-dimensional, shock-separated flowfields. AIAA J. **34**(4), 756–763 (1996)
4. Hau, E.: Wind Turbines: Fundamentals, Technologies, Application, Economics. Springer, New York (2006)
5. Jameson, A.: Time dependent calculations using multigrid, with applications to unsteady flows past airfoils and wings. AIAA Pap. **1596**, 1991 (1991)
6. Jameson, A., Schmidt, W., Turkel, E., et al.: Numerical solutions of the euler equations by finite volume methods using Runge-Kutta time-stepping schemes. AIAA Pap. **1259**, 1981 (1981)
7. Kroll, N., Rossow, C.-C., Becker, K., Thiele, F.: The MEGAFLOW project. Aerosp. Sci. Technol. **4**(4), 223–237 (2000)
8. Mandell, J.F., Samborsky, D., Combs, D., Scott, M., Cairns, D.: Fatigue of composite material beam elements representative of wind turbine blade substructure. Technical report, National Renewable Energy Lab., Golden (1998)
9. Meister, K., Lutz, T., Krämer, E.: Simulation of a 5mw wind turbine in an atmospheric boundary layer. J. Phys. (To be published)
10. Raasch, S., Schroter, M.: Palm – a large-eddy simulation model performing on massively parallel computers. Meteorologische Zeitschrift **10**(5), 363–372 (2001)
11. Schulz, C.: CFD of wind turbines in complex terrain. In: IEA R&D Wind Task 11 – Topical Expert Meeting Challenges on Wind Energy Deployment in Complex Terrain, Stuttgart (2013)
12. Schulz, C., Klein, L., Weihing, P., Lutz, T., et al.: CFD studies on wind turbines in complex terrain under atmospheric inflow conditions. J. Phys. Conf. Ser. **524**, 012134 (2014). IOP Publishing
13. Spalart, P., Jou, W., Strelets, M., Allmaras, S.: Comments of feasibility of LES for wings, and on a hybrid RANS/LES approach. In: Liu, C., Liu, Z., Sakell, L. (eds.) Advances in DNS/LES. Greyden Press, Columbus, (1997)
14. Spalart, P.R., Deck, S., Shur, M., Squires, K., Strelets, M.K., Travin, A.: A new version of detached-eddy simulation, resistant to ambiguous grid densities. Theor. Comput. Fluid Dyn. **20**(3), 181–195 (2006)
15. Weihing, P., Meister, K., Schulz, C., Lutz, T., et al.: CFD simulations on interference effects between offshore wind turbines. J. Phys. Conf. Ser. **524**, 012143 (2014). IOP Publishing

Flow Simulation of Francis Turbines Using Hybrid RANS-LES Turbulence Models

Timo Krappel, Albert Ruprecht, and Stefan Riedelbauch

Abstract The operation of Francis turbines in part load condition causes high pressure fluctuations and dynamic loads in the turbine as well as high flow losses in the draft tube. Owing to the co-rotating velocity distribution at the runner blade trailing edge a low pressure zone arises in the hub region finally leading to a rotating vortex rope in the draft tube. The goal of this study is to reach a quantitatively better numerical prediction of the flow at part load and to evaluate the necessary numerical depth with respect to effort and benefit.

The focal point of the investigations is on flow simulations of part load conditions using two different kinds of hybrid RANS-LES turbulence model approaches: Scale Adaptive Simulation (SAS) and Improved Delayed Detached Eddy Simulation (IDDES).

The flow simulations are done for a high specific speed Francis turbine and a low specific speed Francis pump turbine.

The mainly dominating flow phenomenon, the rotating vortex rope, with a frequency of around one third of the runner frequency, requires long physical time and therefore long computational time using many processors.

1 Introduction

The operation of Francis pump turbines in part load turbine conditions causes high fluctuations and dynamic loads in the turbine and draft tube. At the hub of runner outlet a rotating low pressure zone arises and propagates into the draft tube cone, the rotating vortex rope phenomenon. The flow field is strongly three dimensional and unsteady, which is very challenging to predict by numerical flow simulations.

Mostly, RANS turbulence models fails in the prediction, but as the Reynolds number is very high and the turbine consists of many blades, a fully resolved LES is nowadays unfeasible. Therefore, in this study a SAS and a IDDES-type hybrid RANS-LES turbulence model is used.

T. Krappel (✉) • A. Ruprecht • S. Riedelbauch
Institute of Fluid Mechanics and Hydraulic Machinery, Pfaffenwaldring 10, 70550 Stuttgart, Germany
e-mail: timo.krappel@ihs.uni-stuttgart.de

W.E. Nagel et al. (eds.), *High Performance Computing in Science and Engineering '14*, DOI 10.1007/978-3-319-10810-0_28

Different levels of modelling effort are compared with each other and, in the case of the pump turbine, validated against measurements.

2 Numerical Method

For the flow simulation of the Francis turbine the commercial code Ansys CFX version 14.5 and for the Francis pump turbine the open source code OpenFOAM® version 1.6-ext is used. Both CFD codes are able to handle the rotation of the turbine and to couple different meshes by an interface, e.g. [1]. They are both using finite-volume method for discretisation, while the volumes of discretisation are built around the cell nodes for Ansys CFX and around the cell center for OpenFOAM.

2.1 SAS-SST Turbulence Model

The SAS approach enables the unsteady SST RANS turbulence model [10] to operate in SRS (Scale Resolving Simulation) mode [14]. This is achieved by introducing a new quantity, namely Q_{SAS}, into the ω-equation of the SST model [2,3,12], Q_{SAS} is defined as:

$$Q_{SAS} = \max\left[\rho\zeta_2\kappa S^2\left(\frac{L}{L_{vK}}\right)^2 - C\frac{2\rho k}{\sigma_\Phi}\max\left(\frac{|\nabla\omega|^2}{\omega^2}, \frac{|\nabla k|^2}{k^2}\right), 0\right], \quad (1)$$

containing the turbulent length scale L and the von Karman length scale L_{vK}. Based on the theory of Rotta [15], L_{vK} describes the second derivative of the velocity field. As the SAS model requires a lower constraint on the L_{vK} value, it is

$$L_{vK} = \max\left(\kappa S/\left|\nabla^2 U\right|, C_S\sqrt{\kappa\zeta_2/((\beta/c_\mu)-\alpha)}\,\Delta\right), \Delta = \Omega_{\mathrm{CV}}^{1/3} \quad (2)$$

The purpose of this limiter is to control damping of the finest resolved turbulent fluctuations. For the turbulent eddy viscosity of the SAS model the following relation can be derived:

$$\mu_{t,SAS} = \rho\left(\sqrt{((\beta/c_\mu)-\alpha)/\kappa\zeta_2}L_{vK}\right)^2 S \quad (3)$$

This has similar structure as the Smagorinsky LES model

$$\mu_{t,LES} = \rho\left(C_S\Delta\right)^2 S \quad (4)$$

and can therefore be used as the high wave number damping limiter for the SAS model. The limiter can prevent the SAS turbulent eddy viscosity from decreasing below the LES subgrid-scale turbulent eddy viscosity, if

$$\mu_{t,SAS} \geq \mu_{t,LES} \tag{5}$$

$\mu_{t,SAS}$ and $\mu_{t,LES}$ can be substituted again, which results in the L_{vK} limiter of equation (2).

2.2 *Theoretical Aspects of the SST-Based IDDES Turbulence Model*

In this work the focus is on the RANS-SST turbulence model [13] based IDDES approach. The length scales of the SST-based IDDES turbulence model are according to e.g. [16]:

$$l_{RANS} = \frac{\sqrt{k}}{\beta^* \omega} \qquad l_{LES} = C_{DES}\Delta \tag{6}$$

$$C_{DES} = (1 - F_1)C_{DES,k-\epsilon} + F_1 C_{DES,k-\omega} \tag{7}$$

whereas $C_{DES,k-\epsilon} = 0.58$ and $C_{DES,k-\omega} = 0.78$ are used as DES-coefficients [4], depending on the RANS-SST constant F_1. Further details of the combined DDES and WMLES length-scale and other constants can be found in e.g. [16] and [5].

The validation of the IDDES-SST turbulence model with several test cases is described in [9].

3 Francis Turbine Case

3.1 *Computational Setup*

The used turbine components are the spiral casing, stay and guide vanes, runner blades and draft tube with expansion tank (see Fig. 2). An expansion tank has been chosen as draft tube extension to replicate the test rig which is ready to operate in the near future. The simulation models dimensions are about 4 m in length and 2 m in height.

At the inlet of spiral casing steady boundary conditions are applied with a mass flow specification of the part load operating point. For temporal discretisation a second order backward Euler scheme and for spatial discretisation a bounded second order central differencing scheme [6] is used. For the turbulence quantities a bounded second order backward Euler scheme is applied for the temporal discretisation and a first order scheme for the spatial discretisation [11]. The SAS model, presented in Sect. 2.1, is applied as hybrid RANS-LES turbulence model.

Two different grids of entirely hexahedral type are evaluated for the simulation of the Francis turbine: one with approximately 16 million elements (16M) and a refined one with around 40 million elements (40M). Although the grid size is quite large, wall resolution is still away from resolving the boundary layer into the laminar region. The majority of the mesh points, about 50 %, are concentrated in the draft tube to obtain a good resolution of the complex vortex flow.

As it is essential to have a Courant-number smaller than one to resolve turbulent structures to small scales, the time steps size corresponds to 2° of runner revolution for the coarser (16M) respectively 0.5° for the finer mesh (40M).

3.2 *Global Machine Data*

First of all, global machine data like hydraulic head losses and efficiency are compared for different levels of modelling. The steady state simulation approach with SST turbulence model overestimates the total losses for the complete turbine, especially in the runner and draft tube (see Fig. 1), by more than 4 % of total head compared with the reference simulation SAS-SST 40M. By using a URANS approach, the deviation is reduced to about 1 %. As the 16M mesh already has a quite good resolution for predicting the large structures of the vortex rope, the losses are quite similar compared to the reference with 0.2 % higher losses. Generally, the deviations of hydraulic head losses are in the runner and draft tube domain, where most of transient phenomena occur. In the upstream components spiral casing and stay and guide vanes the flow is mainly steady state.

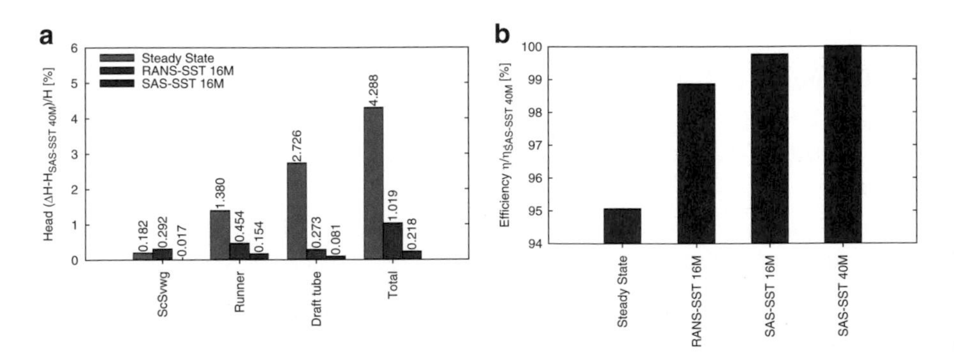

Fig. 1 Hydraulic head losses per component and relative total efficiency. (**a**) Hydraulic head losses for each simulation approach relative to the reference simulation with fine mesh SAS-SST 40M. (**b**) Relative total machine efficiency

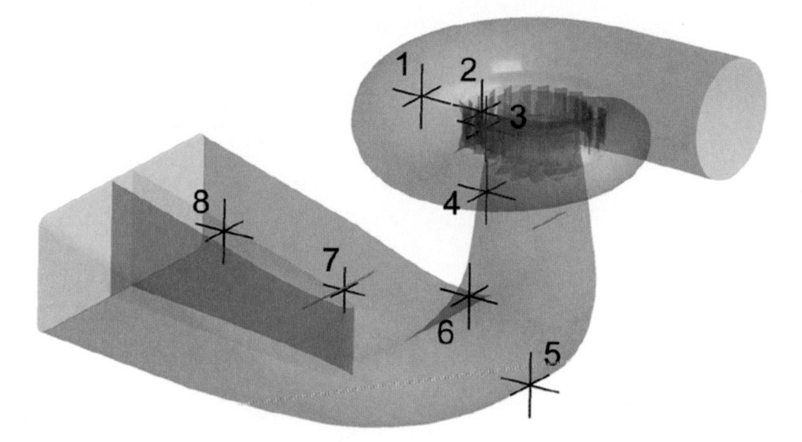

Fig. 2 Evaluation lines in the cone and diffuser (*green lines*) and monitor points at different positions in the turbine (*black crosshairs*)

3.3 *Flow Analysis*

A flow analysis is done on an evaluation line half way down the cone for all velocity vector components (see Fig. 2). The steady state simulations show large deviations from the transient simulations for all velocity components (see Fig. 3). For transient simulations the differences between RANS-SST and SAS-SST turbulence model are moderate for the coarser mesh (16M). However, the result for the finer mesh (40M) with SAS-SST model shows a larger axial velocity component without any back flow in the core region. The tangential component is co-rotating along the entire evaluation plane except a tiny region close to the rotation axis. This might be due to a better resolution of the vortex rope phenomenon at the runner outlet. For the radial component only minor differences are visible.

Velocity vector components are also displayed on a exemplary evaluation line in the right channel of the diffuser (see Fig. 2). In the diffuser a similar tendency is visible for the steady state simulation like in the cone (see Fig. 4). The meridional velocity component of the steady state simulation has the largest relative deviation from the other simulations. The transient RANS simulation has also some deviations, whereas the SAS simulations with both meshes show quite similar results. For the other velocity components in transversal direction similar tendencies are visible.

3.4 *Vortex Rope*

The contour of the vortex rope is visualised with pressure isosurfaces (see Fig. 5). Downstream of the runner in the center region of the draft tube a low pressure zone

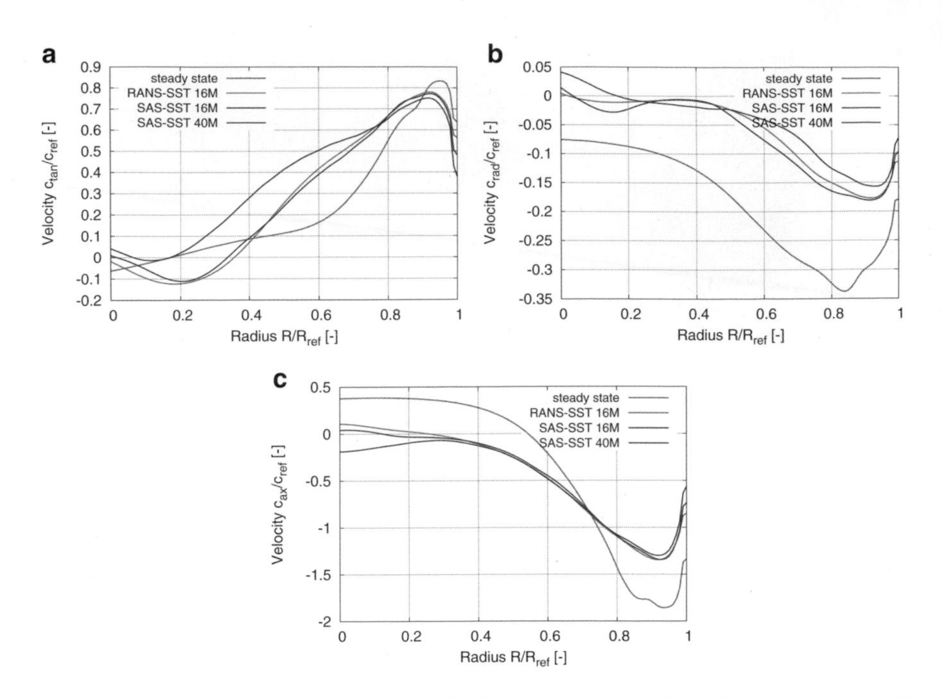

Fig. 3 Velocity vector components in the draft tube cone. (**a**) Tangential velocity component. (**b**) Radial velocity component. (**c**) Axial velocity component

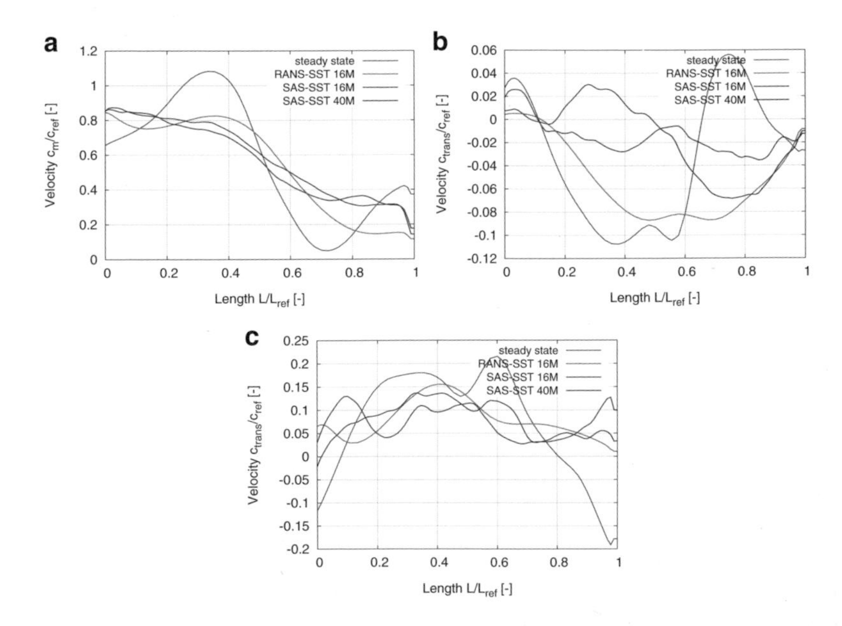

Fig. 4 Velocity vector components in the draft tube diffuser on a horizontal line in the right channel in streamwise direction. (**a**) Streamwise velocity component. (**b**) Horizontal, transversal velocity component. (**c**) Vertical, transversal velocity component

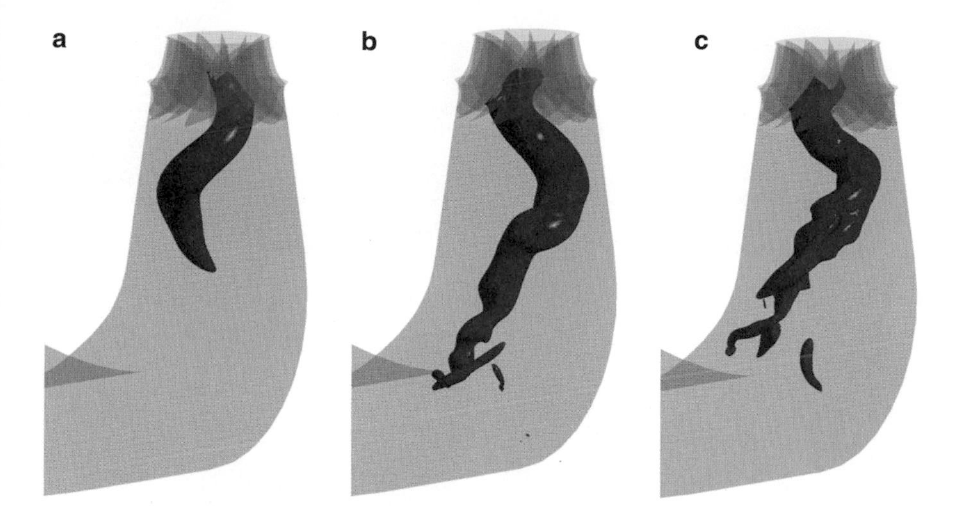

Fig. 5 Visualisation of vortex rope shape by isosurfaces of pressure. (**a**) 16M RANS-SST. (**b**) 16M SAS-SST. (**c**) 40M SAS-SST

is generated due to the velocity distribution (see Fig. 3). The result of the transient RANS-SST simulation shows a comparably short vortex rope with a quite clear elliptical shape. The vortex rope obtained by SAS-SST simulation with both mesh sizes propagates further into the cone towards elbow. With larger mesh resolution using the SAS model the elliptic shape of the vortex rope is less pronounced. The result shows more details of turbulent structures, both on half height where the vortex rope is rotating around its own axis and at the end in the elbow where it decays to turbulence. These turbulent structures are well known from LES simulations which indicates a LES-like resolution of the vortex rope for the fine resolution mesh (see Fig. 5c).

A further visualisation of the vortex rope and the turbulent structures generated by it is depicted in Fig. 6. This picture shows snapshots of the Q-isosurface of the SAS-SST 16M simulation every 300° of runner revolution between successive pictures. Within this time period of around three runner revolutions, the vortex rope is rotating once around the main rotation axis. This indicates a vortex rope rotating frequency of about $1/3$ of runner frequency. Direct after the runner only some smaller turbulent structures are visible showing the runner blades wake. In the draft tube cone the clear vortex rope shape is dominating. This structure is decaying in the elbow to smaller turbulent structures, which are propagating through the diffuser.

Pressure fluctuation amplitude in the draft tube cone is highest for transient RANS simulation and gets lower with higher modelling effort (see Fig. 7a). Also the frequency is somewhat shifted towards higher frequencies for the SAS-SST 40M simulation. If the vortex rope passes the cone wall with a clear and compact shape, a local high pressure peak is generated. In contrast, if the shape is less clear elliptic due to the resolution of more details of turbulence, that peak gets reduced. A reason

Fig. 6 Visualisation of vortex rope shape and turbulent structures by isosurfaces of velocity invariant $Q = 90$, coloured with pressure of the SAS-SST 16M simulation; snapshots for time steps after 300° of runner revolutions each

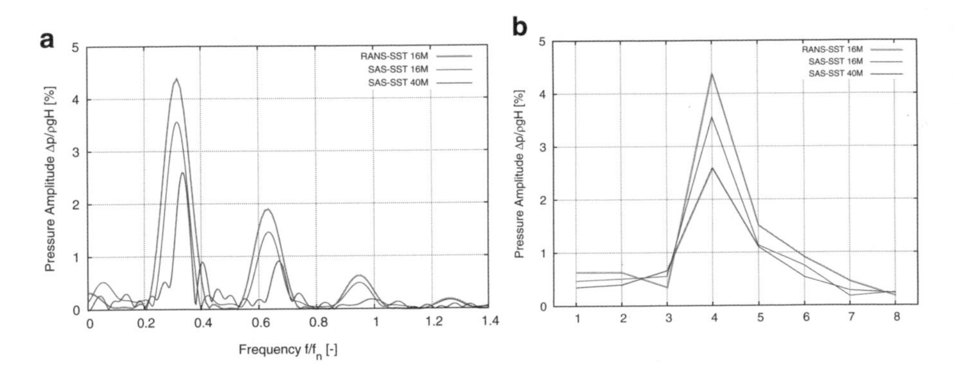

Fig. 7 Pressure amplitude at different locations in the Francis turbine; the positions for (**b**) can be found in Fig. 2. (**a**) Draft tube cone. (**b**) Different positions in the machine

for the frequency shift of the SAS-SST 40M simulation could be the higher level of tangential velocity component (see Fig. 3a).

Pressure fluctuation amplitudes are also evaluated at other positions in the machine (see Fig. 7b) as well as upstream and downstream of the runner. In the whole draft tube the trend of higher pressure amplitudes with less exact numerical modelling approaches is propagating downstream through the draft tube. Further upstream the same trend as in the draft tube is visible for spiral casing (position 1) and stay vanes (position 2).

In the vaneless space between guide vanes and runner another trend is visible: The results obtained with an exacter modelling approach have higher pressure amplitudes.

3.5 Turbulence Evaluation

As the SAS model is a hybrid RANS-LES turbulence model without a geometrical demarcation, in contrast to e.g. the IDDES approach, a variable Δ_{μ_t} is introduced to visualise and localise regions with higher turbulent eddy viscosity values than obtained by a pure LES model:

$$\Delta_{\mu_t} = (\mu_{t,SAS} - \mu_{t,LES})/\mu \tag{8}$$

In Fig. 8 the results of this variable for both mesh sizes are depicted. At the beginning of the draft tube cone only minor differences are visible. In the draft tube elbow, where the vortex rope decays, the result of the coarser mesh shows a larger region of high turbulent eddy viscosity values than the result of the finer mesh. The distinct region of the fine mesh result indicates that this mesh does not have LES-like resolution in this area. In the diffuser only smaller regions are visible, indicating a good mesh resolution in terms of activating the LES or SRS capabilities of the SAS turbulence model.

To quantify the amount of resolved turbulence in the flow simulation the turbulent kinetic energy spectra are compared with each other (see Fig. 9). As expected the RANS simulation does not represent the inertial range of turbulence as the turbulent content of the flow is modelled. In contrast to that, the simulations with the SAS model show a good represented inertial range of turbulence. The difference between the two meshes is not as clear as expected.

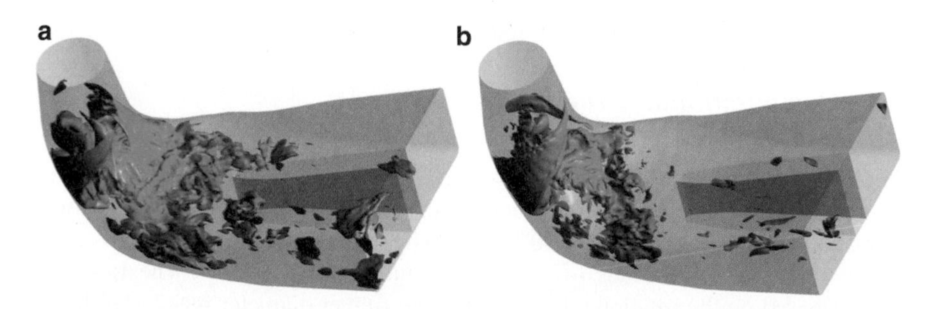

Fig. 8 Visualisation of the difference between the SAS model eddy viscosity and the Smagorinsky LES model eddy viscosity by an isosurface of $\Delta_{\mu_t} = 100$ (see Eq. 8). (**a**) 16M SAS-SST. (**b**) 40M SAS-SST

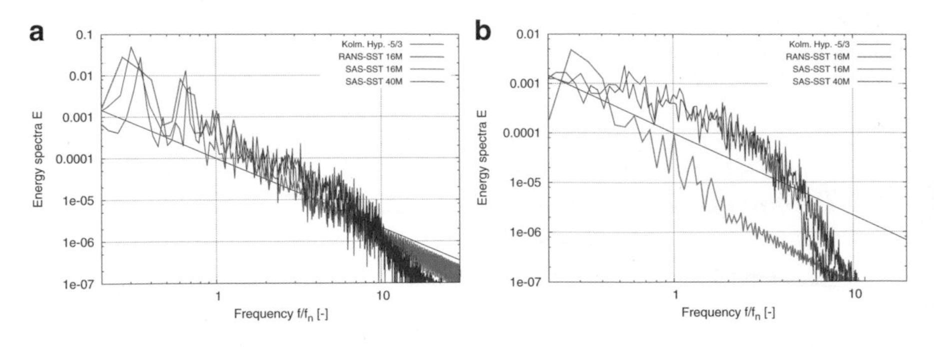

Fig. 9 Turbulent kinetic energy spectra of each one point in draft tube cone and diffuser. (**a**) Draft tube cone. (**b**) Draft tube diffuser

Fig. 10 Geometry used for flow simulations, the inflow is located top left at the spiral casing, overall height is ~4 m; *left*: complete geometry, *right*: detail

4 Francis Pump Turbine Case

4.1 Computational Setup

The Francis pump turbine geometry is depicted in Fig. 10. It contains the spiral casing, 20 stay vanes and guide vanes, 7 runner blades and the straight diffuser followed by a pipe section and a 90° elbow. The inlet diameter of the spiral casing is $d = 0.19\,\text{m}$ which leads to a Reynolds number of $5.8 \cdot 10^5$. The Reynolds number based on the rotational velocity and the runner diameter is $8.2 \cdot 10^6$ based on 1,405 rev/min. The pump turbine is operated in part load turbine conditions.

For the pump turbine case, the method of a recirculated pipe flow simulation with storage of all relevant transient flow values is chosen as inlet boundary condition for the simulation with IDDES model. Two mesh sizes are investigated: a 10 million element mesh (10M) and 20 million element mesh (20M) in total for all components. The Courant number is <1 almost for the whole computational domain.

The y^+-values in wall-normal direction are in the range of 1–2 for both meshes. The corresponding x^+-values in streamwise direction are 1,250 for the coarse and 900 for the fine mesh and the z^+-values in spanwise direction are 550 and 350.

For temporal discretisation a second order backward differencing scheme and for spatial discretisation a bounded second order central differencing scheme [6] was used and for turbulence quantities a bounded second order schemes was applied for temporal and a first order scheme for spatial discretisation.

4.2 Flow Analysis

A flow analysis is done for an evaluation line ∼0.1 m below the runner at the beginning of the straight diffuser. At this position laser measurements are available. The axial and radial components of the velocity were measured with Particle Image Velocimetry (PIV) and the circumferential component of the velocity with Laser Doppler Velocimetry (LDV).

The measurements fit quite well for the time-averaged axial velocity profiles (see Fig. 11), except for the core flow. There, the deviation from the measurements is smaller for IDDES-type turbulence model, especially for the finer mesh, compared to the RANS-type turbulence model. The tangential velocity component is somewhat overestimated for all simulations. The reason for the deviation could be that there is some difference in the geometry due to simplifications of e.g. gaps between rotating and non-rotating parts.

The fluctuations of velocity are well predicted by the hybrid RANS-LES turbulence model (see Fig. 12). This is explained by the LES content of the turbulence

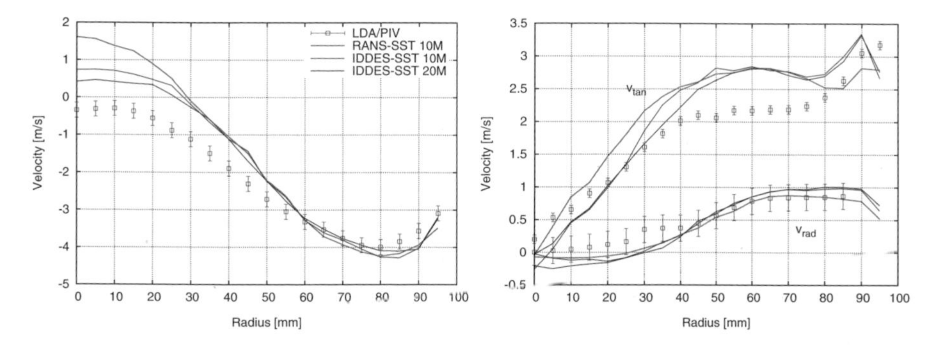

Fig. 11 Comparison between simulation results with LDA/PIV measurements of time-averaged velocity vector components; *left*: axial components, *right*: tangential and radial components; legend is the same for both figures

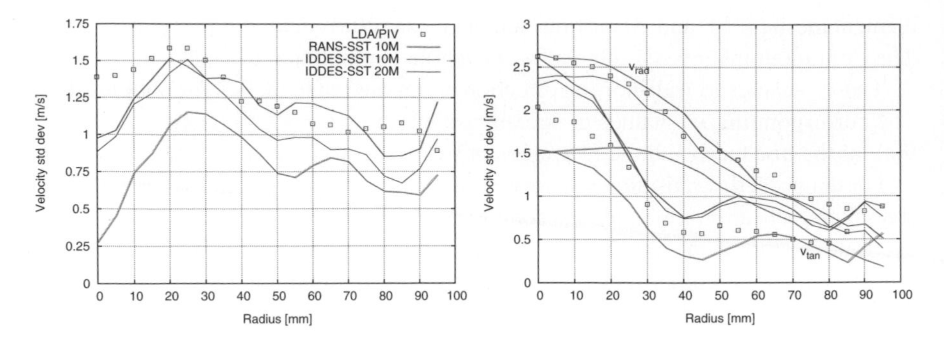

Fig. 12 Comparison between simulation results with LDA/PIV measurements of time-averaged standard deviation of velocity vector components; *left*: axial components, *right*: tangential and radial components; legend is the same for both figures

model, which can resolve the turbulent structures of the flow. The simulation with the finer mesh (20M) predicts higher fluctuation values as the coarser mesh (10M).

4.3 Vortex Rope

The vortex rope is better represented using the IDDES-SST turbulence model (see Fig. 13). The shape is quite similar compared with visualisations from measurements [7]. The result of the RANS-SST simulation shows only a small vortex rope shape. In contrast, the IDDES-SST turbulence model predicts a longer vortex rope shape with more turbulent structures. At its end the vortex rope decays to small turbulent structures.

5 Computational Resources

All flow simulations were performed on the CRAY XE6 (HERMIT) installed at the HLRS Stuttgart.

The Francis turbine flow simulation was done with Ansys CFX version 14.5 using standard MPI up to 128 cores as the parallel performance degrades for higher core numbers for this application (more details can be found in [8]). Almost 40 runner revolutions are necessary to get time-averaged velocity profile data in the whole draft tube, corresponding to almost 30,000 time steps. The simulation with 40 million elements took nearly 180 days.

In contrast to the parallel performance obtained with Ansys CFX version 14.5, the current version 15.0 seems to be more promising. Especially, as the CRAY-MPI is supported in version 15.0.

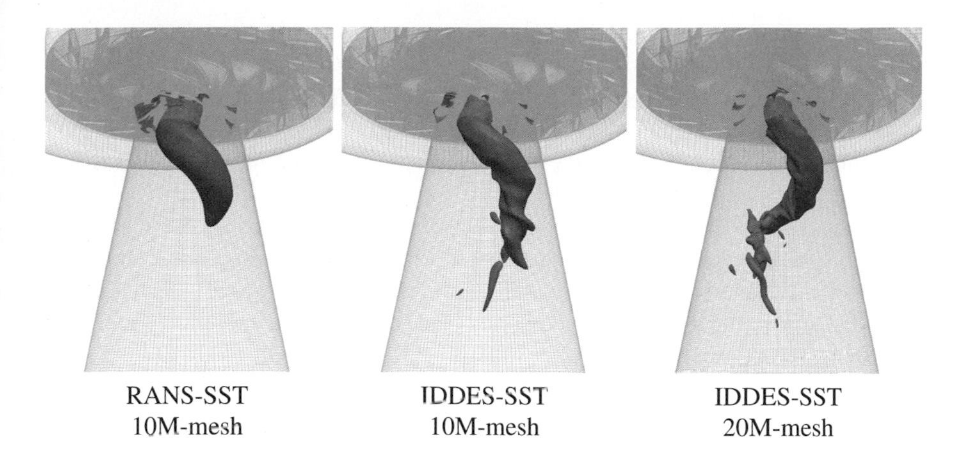

Fig. 13 Visualisation of the vortex rope shape with an iso-pressure contour

The Francis pump turbine was simulated with OpenFOAM®-1.6-ext using the optimised CRAY-MPI. The 10 million element mesh was run on 256 cores leading to around 39,000 elements per core. One time step takes around 25 s using the RANS-SST turbulence model. This leads to an overall time – for 12,600 time steps – of almost 90 h of simulation time. As the usage of the hybrid IDDES-SST turbulence model needs – compared to RANS-SST- some additional operations in the determination of the turbulence properties, the run times are around 4 % higher. The 20 million element mesh using the IDDES-SST turbulence model was run on twice as many cores as the smaller mesh, namely 512. The execution time for this comparison shows a weak scaling of around 21 % higher run time. The octree search algorithm of the mesh interface [1] could be improved leading to around 6 % of performance improvement for the 20M mesh.

6 Conclusion

Flow simulations at part load operating conditions for a Francis turbine were performed using the commercial code Ansys CFX. Different numerical approaches were compared, from steady state to transient simulations using RANS and SAS turbulence models. It could be shown that the prediction of the vortex rope phenomenon in more detail requires such a hybrid RANS-LES turbulence model with a high mesh resolution. The comparison with a LES turbulent eddy viscosity indicates that in the elbow region, where the vortex rope decays to small turbulent structures, the mesh resolution is still away from LES, even though using more than 40 million elements.

For the flow simulations, also at part load operating conditions, for a Francis pump turbine were performed using the open source code OpenFOAM®-1.6-ext.

Another type of hybrid RANS-LES turbulence model, the IDDES approach, was validated for this flow type. The results show a good agreement with measurements, especially for the fluctuations of the velocity. This indicates a better prediction of turbulent flow structures. For this test case could also be shown that a hybrid RANS-LES turbulence model predicts the vortex rope phenomenon in more detail than a RANS model.

Those types of transient flow simulations using hybrid RANS-LES turbulence models require large computational resources. Many time steps have to be simulated to get time-averaged results or enough transient data. The number of mesh nodes is large due to complex geometries with many blade channels for operation points with high Reynolds numbers.

References

1. Beaudoin, M., Jasak, H.: Development of a generalized grid interface for turbomachinery simulations with OpenFOAM. In: Open Source CFD International Conference, Milano (2008)
2. Egorov, Y., Menter, F.R.: Development and application of SST-*SAS* turbulence model in the DESIDER project. Adv. Hybrid RANS-LES Model. Not. Numer. Fluid Mech. Multidiscip. Des. **97**, 261–270 (2008)
3. Egorov, Y., Menter, F.R., Cokljat, D.: The scale-adaptive simulation method for unsteady turbulent flow predictions. Part 2: application to aerodynamic flows. J. Flow Turbul. Combust. **85**(1), 139–165 (2010)
4. Fuchs, M., Mockett, C., Gual-Skopek, M., Le Chuiton, F., Thiele, F.: Helicopter fuselage wake prediction using Detached-Eddy simulation methods in OpenFOAM®. In: 7th OpenFOAM® Workshop, Darmstadt, 25–28 June 2012
5. Gritskevich, M.S., Garbaruk, A.V., Menter, F.R.: Fine-tuning of DDES and IDDES formulations to the k-ω Shear Stress Transport model. Prog. Flight Phys. **5**, 23–42 (2013)
6. Jasak, H., Weller, H.G., Gosman, A.D.: High resolution NVD differencing scheme for arbitrarily unstructured meshes. Int. J. Numer. Methods Fluids **31**, 431–449 (1999)
7. Kirschner, O.: Experimentelle Untersuchung des Wirbelzopfes im geraden Saugrohr einer Modell-Pumpturbine. Dissertation, IHS-Mitteilungen, vol. 32 (2011)
8. Krappel, T., Ruprecht, A., Riedelbauch, S.: Flow simulation of a Francis turbine using the SAS turbulence model. In: High Performance Computing in Science and Engineering'13. Springer, Berlin (2013)
9. Krappel, T., Kuhlmann, H., Kirschner, O., Ruprecht, A., Riedelbauch, S.: Validation of an IDDES-type turbulence model and application to a Francis pump turbine flow simulation. In: 10th International ERCOFTAC Symposium on Engineering Turbulence Modelling and Measurements, Marbella, 17–19 Sept 2014 (to be published)
10. Menter, F.R.: Two-equation eddy-viscosity turbulence models for engineering applications. AIAA J. **32**(8), 269–289 (1994)
11. Menter, F.R.: Best Practice: Scale-Resolving Simulations in ANSYS CFD Version 1.0. ANSYS Germany GmbH (2012)
12. Menter, F.R., Egorov, Y.: The scale-adaptive simulation method for unsteady turbulent flow predictions. Part 1: theory and model description. J. Flow Turbul. Combust. **85**(1), 113–138 (2010)
13. Menter, F.R., Kuntz, M., Langtry, R.: Ten Years of Industrial Experience with the SST Turbulence Model. Turbulence, Heat and Mass Transfer, vol. 4, pp. 625–632. Begell House, New York (2003)

14. Menter, F.R., Schütze, J., Gritskevich, M.: Global vs. zonal approaches in hybrid RANS-LES turbulence modelling. In: Progress in Hybrid RANS-LES Modelling. Notes on Numerical Fluid Mechanics and Multidisciplinary Design, vol. 117, pp. 15–28. Springer, Berlin (2012)
15. Rotta, J.C.: Turbulente Strömungen. BG Teubner, Stuttgart (1972)
16. Shur, M.L., Spalart, P.R., Strelets, M.K., Travin, A.K.: A hybrid RANS-LES approach with delayed-DES and wall-modelled LES capabilities. Int. J. Heat Fluid Flow **29**, 1638–1649 (2008)

Enhancement and Applications of a Structural URANS Solver

Christian Stanger, Benjamin Kutz, Ulrich Kowarsch, E. Rebecca Busch, Manuel Keßler, and Ewald Krämer

Abstract The enhancement of the structured Computational Fluid Dynamics code FLOWer and three applications are presented. FLOWer has been improved by implementing a fifth order scheme for flow state reconstruction. The benefits of the improved vertex conservation are shown in a trimmed and weakly coupled main rotor simulation. Additionally, because of the reduced numeric dissipation, results of flow simulations of brownout conditions are massively improved. Finally the possibilities for acoustic evaluations are shown at the example of two different setups of Contra-Rotating Open Rotors.

1 Introduction

With the continuous increase of supercomputing power, it is not only possible to simulate setups with several million cells in acceptable time, but also the usage of higher order methods. In combination with the Chimera technique this enables detailed simulations of complex structures with moving parts like helicopters or Contra-Rotating Open Rotors (CRORs). This paper presents some results of the helicopter group of the Institute of Aerodynamics and Gas Dynamics (IAG) of the University of Stuttgart.

One goal of the group is the implementation of higher-order methods in the used Computational Fluid Dynamics (CFD) code. As already mentioned in the last HELISIM report [10], the implementation of the weighted essentially non oscillatory (WENO) method showed already good results and was further improved and validated. The consideration of several surrounding cells showed to be a disadvantage, especially in areas with high gradients. Section 3 shows a now more compact reconstruction scheme which overcomes the drawback. The focus is to achieve high performance on the HLRS cluster platform CRAY XE6 Hermit, which is shown exemplarily on a weakly coupled and trimmed main-rotor simulation.

C. Stanger (✉) • B. Kutz • U. Kowarsch • E.R. Busch • M. Keßler • E. Krämer
Institut für Aerodynamik und Gasdynamik, Universität Stuttgart, Pfaffenwaldring 21, D-70569 Stuttgart, Germany
e-mail: stanger@iag.uni-stuttgart.de

W.E. Nagel et al. (eds.), *High Performance Computing in Science and Engineering '14*, DOI 10.1007/978-3-319-10810-0_29

Two other applications of the highly parallelized code at the high performance cluster are presented in Sects. 4 and 5. One is the brownout phenomenon, which occurs when a helicopter operates in low altitudes over dust or snow covered ground, and particles are dispersed. Due to the loss of visibility this is very dangerous and can limit possible mission scenarios for a helicopter. Another research topic are CRORs. These aero-engines promise to save a lot of fuel because of their very high bypass-ratio. However due to the absence of a nacelle and the interaction of the two rotors, these engines have a high noise emission. Detailed CFD-simulations in combination with computational aeroacustics help to understand and reduce the noise generating mechanisms.

2 Numerical Method

All CFD simulations presented in this paper are performed with the structured finite volume flow solver FLOWer [11], developed by the German Aerospace Center (DLR) and enhanced by IAG. The code solves the unsteady three-dimensional Reynolds-averaged Navier-Stokes equations (URANS) for numerous time steps. For spatial discretisation a cell-centered finite volume formulation is used. Time integration is achieved by a hybrid multi-stage Runge-Kutta scheme developed by Jameson [5, 6]. Multigrid and residual smoothing are implemented as convergence accelerators. In order to enable relative movement of grids and simplify grid generation for complex structures the Chimera technique is used. A component grid is integrated into a background grid, where a hole is cut. Data is exchanged at the outer boundary of the component grid and at the hole, thus the two-way data exchange is ensured. An efficient computation at high performance clusters with many CPUs is achieved by a multi-block structure of the grid to enable parallel computing at the block level.

3 Resolving Helicopter Aerodynamics Using Higher Order Methods

In the course of last years HELISIM report [10], the implementation of the fifth order WENO scheme in the FLOWer CFD-Code was presented. The extension of the code showed significant improvements in vortex preservation at equal computational resource requirements. Due to the highly promising results this research topic was further progressed and within the last year additional methods were implemented. These improvements of the code were used to resolve complex helicopter phenomena with focus on vortex preservation. Further an overview of the methods implemented in the last year is given as well as application to computations performed on the HLRS Cray XE6 Hermit Cluster.

3.1 Compact Reconstruction Weighted Essentially Non-oscillatory Scheme

The Compact Reconstruction Weighted Essentially Non-Oscillatory scheme (CRWENO) was introduced by Ghosh in [3] as a method merging the advantage of compact reconstruction approach with the robust fifth order WENO scheme [7] which is now available in FLOWer. Both higher order schemes use cell centered values to reconstruct the fluid state of the cell boundary, which is of concern for the flux computation solved by an upwind Riemann solver. In order to achieve a higher order reconstruction, several surrounding cells are considered, providing stencils for the polynomial ansatz. Aim of the compact reconstruction is to reduce the distance of the stencils from the cell boundary in order to improve the spectral resolution of the reconstruction [3]. This is achieved by moving the stencils a half cell width closer to the considered cell boundary leading to the usage of the reconstructed cell boundary values from left sided and right sided cells. Figure 1 shows the stencil formation for the CRWENO and WENO scheme reconstruction of the left fluid state at the cell boundary $i + \frac{1}{2}$. Due to the correlation between the reconstructed values on the cell boundaries, the method requires an implicit reconstruction in space. Forming the equations for the implicit reconstruction leads to a linear system of equations (LSE) of the structure

$$\mathbf{A} \cdot f = x, \tag{1}$$

with a tridiagonal matrix $\mathbf{A}$, the vector with the unknown fluid states at the boundaries f, and the explicitly given cell centered values x. The implementation of the method in the FLOWer Code is done setting up the implicit system for every computational unit represented by a grid block. The boundary conditions of

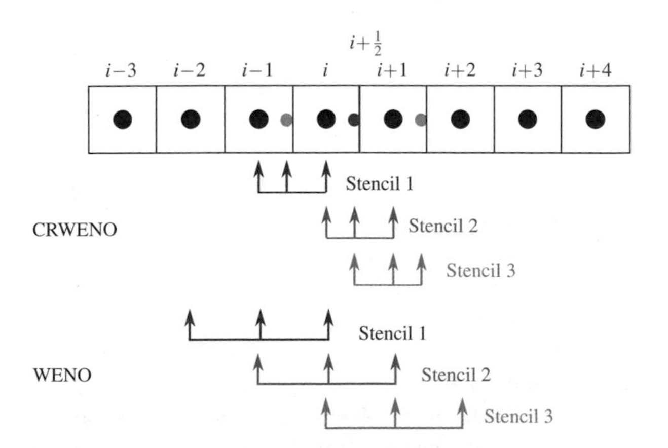

Fig. 1 Stencils for 5th order WENO and 5th order CRWENO left handed fluid state reconstruction at the cell boundary $i + \frac{1}{2}$ (•) using the cell centered values (•)

the implicit system cover the block boundaries or if applicable Chimera boundary conditions. These boundary conditions are computed using the standard WENO method. This procedure was proposed by Ghosh [3]. To solve the resulting tridiagonal LSE the Thomas algorithm is used, which is established as the most efficient numerical method for the solution of this problem. After the fluid reconstruction for the grid block, the gained fluid states are forwarded to the Riemann solver to compute the numerical flux. As for the WENO scheme, the HLLC Riemann solver according to [24] is used for the CRWENO scheme in the FLOWer code. The resulting procedure for the numerical flux computation is designated as the CRWENO method. The computational cost of the CRWENO scheme was found to be approximately two times higher than the WENO scheme. Hence an improved vortex preservation results in a higher efficiency in the overall computational performance achieving the same quality in the physical result.

3.2 CFD Helicopter Simulation Using Higher Order Methods

The following section gives a brief overview of an investigation performed on the CRAY XE6 Hermit cluster using the higher order methods introduced in the previous section. The test case shows the improvements in reproducing vortex phenomena around the helicopter.

Besides the improvement using the higher order methods to resolve aerodynamic phenomena in case of highly resolved simulations, coarse simulations for optimization and load computations also benefit from the higher order method. Therefore a setup of a standard simulation used for performance computation or blade shape optimization is used. The five bladed isolated rotor setup includes eight million grid cells and uses the Chimera technique to realize the relative movement of the rotor blades in the background mesh. Those simulations benefit in two ways from the efficiency of the higher order method, either using less grid cells for performance issues or achieving a higher aerodynamic quality of the resulting flow field with less additional computational cost. Both approaches show significant higher efficiency than using the ordinary second order methods as shown in [10]. Figure 2 shows the resulting flow field for the setup which is coupled with the Computational Structural Dynamic (CSD) code CAMRADII [8] to achieve a comparable flight state. With an additional performance requirement of 25 % in case of the WENO-HLLC and 40 % in case of the CRWENO computation a significant improvement of the rotor wake is present. The simulation took about 31,000 CPU hours for the WENO-HLLC case performed on 256 cores. Figure 2 shows the preservation of the rotor tip wakes for three rotor revolutions in case of the higher order methods until the end of the computational domain. An even more precise structure with lower dissipation effects is found in the case of the CRWENO method, compared to WENO-HLLC. This stands in contrast to the second order JST scheme which is originally used in the FLOWer code, where the blade tip vortices are preserved for less than half a rotor revolution. Especially when it comes to interaction phenomena with the rotor wake,

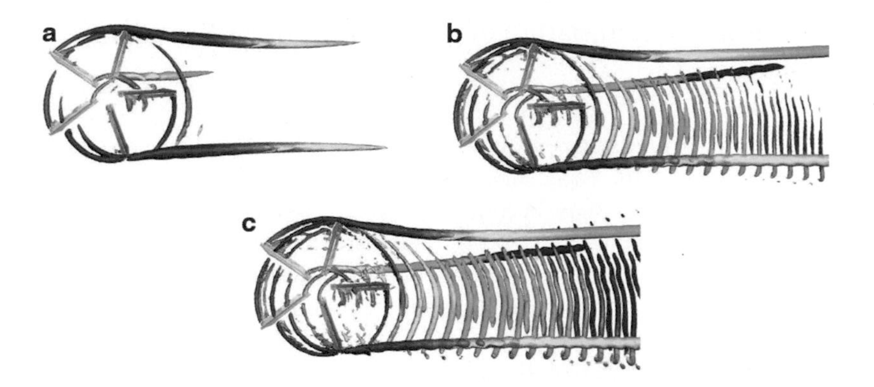

Fig. 2 Instantaneous λ_2-visualization of the vortex topology colored with the vertical position of the vortex (8th rotor revolution at $\Psi = 45^\circ$). (**a**) 2nd order JST. (**b**) 5th order WENO-HLLC. (**c**) 5th order CRWENO

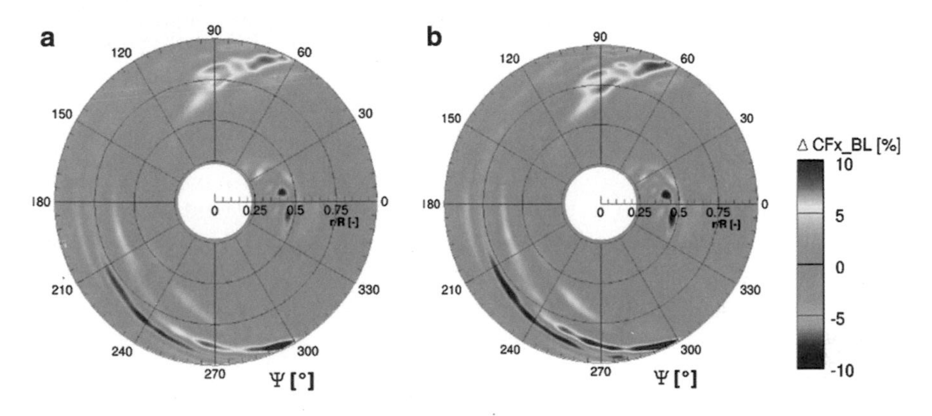

Fig. 3 Difference in rotor thrust on the rotor disc between the flux reconstruction methods. (**a**) Difference JST – WENO-HLLC. (**b**) Difference JST – CRWENO

the improved vortex conservation is a major advantage. Investigations computed on the HLRS Hermit cluster were presented at the European Rotorcraft Forum [9] showing these improvements in detail.

Besides the more precise downwash, Fig. 3 shows the differences in load distribution on the rotordisk where the higher influence of inplane vortices is found. This effect can be seen in Fig. 3a for WENO-HLLC scheme and is even stronger for CRWENO (see Fig. 3b). This improves predictions of the aerodynamic characteristics of the rotor blade and integral rotor properties. In addition, the dynamic response of the rotorblade due to airloads is more precisely captured.

4 Numerical Examination of a Model Rotor in Brownout Conditions

When a helicopter operates near the ground, the load and performance measures are changed significantly. Especially when operating near a ground surface covered with dust or snow, this may become dangerous for the flight crew or ground personnel. When the helicopter approaches the ground, the wake and trailed blade vortices are hitting the ground. Those will disperse dust right before the landing. This phenomenon is called brownout. Often the pilot looses all visual control and therefore the control of the helicopter [16]. Thus, this flight state has been subject of many studies in the past (e.g. Refs. [1, 23, 25]).

To examine the brownout phenomenon a simulation tool is desirable. At the Institute of Aerodynamics and Gasdynamics of the University of Stuttgart (IAG) the computational fluid dynamics (CFD) tool FLOWer was validated and examined in ground effect [12–14, 18]. The present results were also shown in [15].

4.1 Modeling

The flow solver FLOWer[11] is used for computation of the flow field. The brownout simulation is afterwards performed with a simulation tool developed in Ref. [4]. The particles have no impact on the fluid flow (one-way coupling). Although its validity has been questioned recently [17], this assumption has been used in many brownout simulations [21, 23].

4.1.1 Computational Fluid Dynamics

The block structured finite volume Reynolds-averaged Navier–Stokes (RANS) flow solver FLOWer employes a $k - \omega$-model for turbulence modelling. The second order central difference Jameson Schmidt Turkel (JST) method [6] and the WENO method [9] of fifth order are used for spatial discretisation. The Jameson [5] dual time stepping is used for temporal discretisation, leading to second order accuracy in space. The relative motion of the blade is realised with the Chimera technique.

A two-bladed isolated rotor is examined with a half setup with periodic boundaries. The results are compared to the experiment from [22]. Two mesh setups were compared, one with 5.9 million and a refined one with 23.3 million computational cells.

Table 1 shows an overview of the numerical test cases. Both $f = 60$ and 75 Hz rotor frequency runs were computed with seven different height-to-radius ratios, z/R. Furthermore, simulations without ground surface (OGE) were computed. The thrust was already trimmed to the experimental value.

Table 1 Test cases from [15]

f [Hz]				$z/R[-]$				
60	OGE	2.0	1.5	1.25	1.0	0.75	0.5	0.25
75	OGE	2.0	1.5	1.25	1.0	0.75	0.5	0.25

4.1.2 Brownout Simulation

The particle mass is not negligible and several forces act on the finite-size particles. Depending on the application, several effects may be neglected. The main equations of the brownout model are given in the following.

The particle motion is given by

$$\tilde{\mathbf{x}}_p = \mathbf{x}_p + s\mathbf{v}_p \tag{2}$$

$$\tilde{t} = t + s \tag{3}$$

$$\tilde{\mathbf{v}}_p = \mathbf{v}_p + s\left(\frac{\mathbf{u}(\mathbf{x}_p, t) - \mathbf{v}_p}{\tau_p} - \mathbf{g}\right), \tag{4}$$

with the particle position $\mathbf{x}_p$, speed $\mathbf{v}_p$, the step size s, the gravity constant g, the fluid velocity $\mathbf{u}$ and the particle response time τ_p. This is defined as

$$\tau_p = \frac{\rho_p d_p^2}{18\nu\rho_f} \tag{5}$$

using the particle and fluid density ρ_p, ρ_f, the kinematic viscosity ν and the particle diameter d_p.

The uplift condition used is when the friction velocity $v^\star$ is bigger than the threshold velocity $v_t^\star$ from Ref. [19]. Those are defined by

$$v^\star = \sqrt{\frac{\tau_w}{\rho_f}} \tag{6}$$

and

$$v_t^\star = A\sqrt{\frac{\rho_p}{\rho_f}\|\mathbf{g}\|d_p + \frac{\beta}{\rho_f d_p}}. \tag{7}$$

When saltating over the ground or when bombarding the ground, a particle may eject new particles. This process is modelled with a correlation of [19] which defines the number of ejected particles

$$N_s = \frac{c_s - 1.9a^2(1 - e^{-\gamma\|\mathbf{v}_p\|})}{2h^2}. \tag{8}$$

Table 2 Parameter for brownout simulation [15]

Particle density	2,650 $\mathrm{kg/m^3}$ (quarz glass)
Air density	1.2045 $\mathrm{kg/m^3}$
Kinematic viscosity	$1.532 * 10^{-5}\ \mathrm{m^2/s}$
Rotor radius R	0.085 m
Rotational frequency	60, 75 Hz
Gravity constant	9.8065 $\mathrm{m/s^2}$
Restitution e	0.6
Friction f	0.2
Coefficient A	0.1109
Coefficient β	0.0003 $\mathrm{kg/s^2}$
Coefficient c_s	0.7
Coefficient α	0.6
Coefficient h	0.08
Coefficient γ	2
Particle diameter d_p	25–45, 45–63, 25–63 μm
	Uniformly distributed

The ejecting velocity is normally distributed around the rebound velocity $\mathbf{d} = \left(d_x, d_y, d_z\right)^T$ of the impacting particle

$$d_x = v_x + \epsilon_x f(e+1)v_z \tag{9}$$

$$d_y = v_y + \epsilon_y f(e+1)v_z \tag{10}$$

$$d_z = -ev_z. \tag{11}$$

The coefficients ϵ_x and ϵ_y are the in-plane velocity components of the impacting particle. In a flat sediment bed it is $\mathbf{v}_p = \boldsymbol{\epsilon}_p$.

The selection of the particle sizes, densities and flow quantities was done according to the experimental setting in [22]. The brownout model coefficients were chosen according to the findings of [20, 21]. Table 2 from [15] gives an overview of the used parameters.

4.2 Results

In [15] the numerical results are discussed in detail. Here, the aspect of different sediment bed compositions should be reviewed.

An important influence on brownout possesses the sediment bed and its composition. Different particle sizes were simulated and the results are compared in Fig. 4. The differences are clearly visible. When only simulating small 25–63 μm (see Fig. 4a) or big 45–63 μm (Fig. 4b) particles, the dust cloud is very limited. Only the mix of big and small particles leads to a heavy dust cloud around the rotor

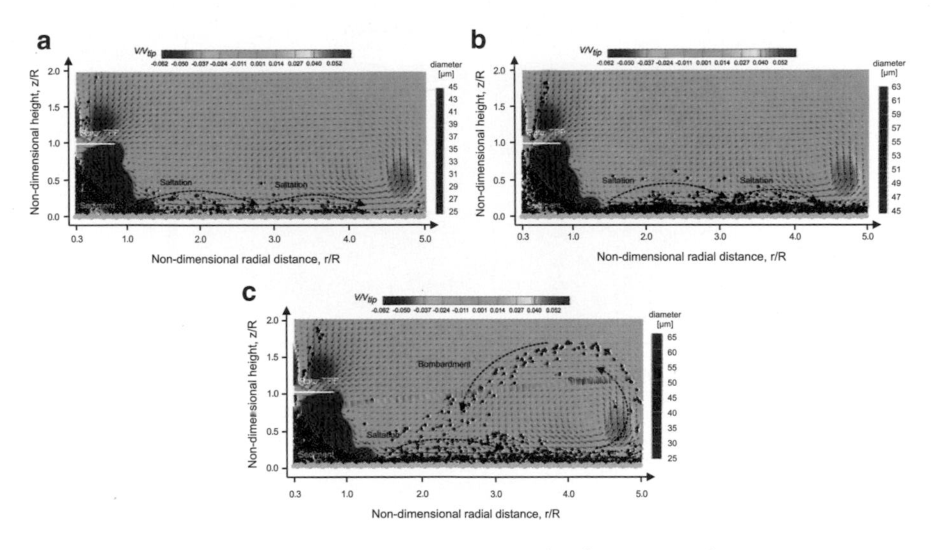

Fig. 4 WENO simulations at the high grid resolution with $f = 75\,\text{Hz}$, $z/R = 1$ with different particle sizes [15]. (**a**) $25 - 45\,\mu\text{m}$. (**b**) $45 - 63\,\mu\text{m}$. (**c**) $25 - 63\,\mu\text{m}$

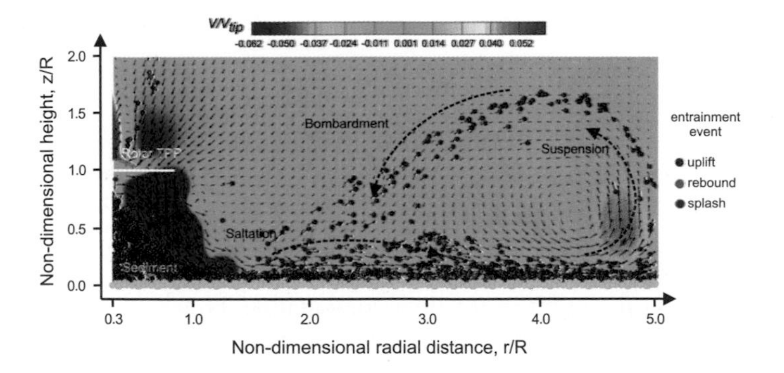

Fig. 5 WENO high resolution $f = 75\,\text{Hz}$, $z/R = 1$, particle diameter 25–63 μm

which can be seen in Fig. 4c. Especially the light particles are entrained high into the fluid flow, producing a heavy bombardment of the sediment bed in the region of $2 \le r/R \le 3$. The comparison of the three pictures leads to the finding, that the heavy particles will saltate over the ground, ejecting light particles. This fact is proved when visualising the entrainment event (see Fig. 5). Now it is obvious that especially uplifted particles are entrained high into the suspension. The rebounding heavy particles are located near the ground surface. Especially in the inner section the particles are released mostly due to uplift.

One further aspect is the interaction of the blade vortices with the ground surface see (Fig. 6). It is visible that from the position on where the vortices are hitting the ground, the cloud starts to disperse and particles are heavily entrained into the fluid flow. One may conclude that the energy of the blade vortices transferred into the

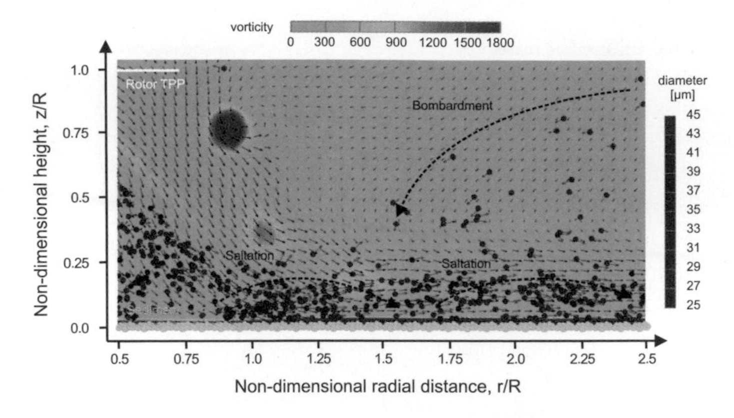

Fig. 6 $f = 75\,\text{Hz}$, $z/R = 1.0$, particle diameter 25–63 μm WENO high resolution

sediment bed is a crucial driver of brownout. Therefore, it is important to minimize the vorticity of those vortices to minimize the dust entrainment.

4.3 Conclusions

An URANS numerical examination of a rotor hovering in ground effect was performed and compared to experimental results and the literature. The JST and WENO schemes were tested, checking for the integral simulation results and the preservation of the blade tip vortices. The vortices are especially important regarding the brownout simulation.

Post processing the result of the flow computation, brownout conditions were simulated. The influence of the particle size was determined. Especially a mix of big and small particles leads to severe brownout. To prevent severe brownout, the blade tip vortices should be weakened, as those are one of the main drivers for the emission of particles.

5 Aerodynamic and Acoustic Influence of Different Blade Numbers on a Contra-Rotating Open Rotor

Two different CROR settings for a take-off and a cruise case are examined, one with 9 blades on the front rotor and 7 on the rear rotor, the other with 8 blades on each rotor. For both flight cases and CROR settings identical grids are used, which is enabled by the application of the Chimera method. The separately generated blade grids for front and rear rotor blade are inserted into a background mesh containing a cylindrical hub geometry. The blade grids are adjusted for the respective pitch angles used for take-off and cruise. With the Chimera method the modelling of the 9×7 and the 8×8 CROR is possible by simply using the corresponding number of blade grids for the assembly of front and rear rotor. For the cruise case a rotational velocity of $n = 1{,}029$ rpm with blade pitch angles of roughly $\beta = 30°$ are used for both CROR settings. The rotational velocity is set to $n = 1{,}250$ rpm with a pitch angle of $\beta = 60°$ in the take-off case. The total numerical setup consisted of roughly 50 million grid cells and was run on 800 CPUs for 8 rotor revolutions. The computational time with the 3D unsteady RANS-solver FLOWer for one rotor revolution was roughly 24 h.

5.1 Aerodynamic Results

The 9×7 and the 8×8 CROR achieve a thrust of approximately $F_x = 80$ kN for the take-off and $F_x = 19$ kN for cruise case. While the achieved thrust as well as the required torque and therefore the propulsion efficiency are comparable for both CROR settings, the flow field changes significantly for different blade numbers on front and rear rotor. Figure 7 shows the λ_2-visualization of the 9×7 and 8×8 CROR in take-off conditions. As expected the vortex distribution is depending on the number of blades and thus influences the interaction of the two rotor flow fields. While for the 9 blade front rotor the vortices are slightly weaker than for the 8 blade front rotor, the 7 blade rear rotor has stronger vortices than the 8 blade rear rotor. For the 9×7 the flow field is only periodic over one rotor revolution, whereas the 8×8 exhibits a higher periodicy, i.e. 16 per revolution. For cruise conditions the flow field is stretched in the flow direction due to the higher inflow velocity and the lower blade loading causing a smaller deflection of the flow.

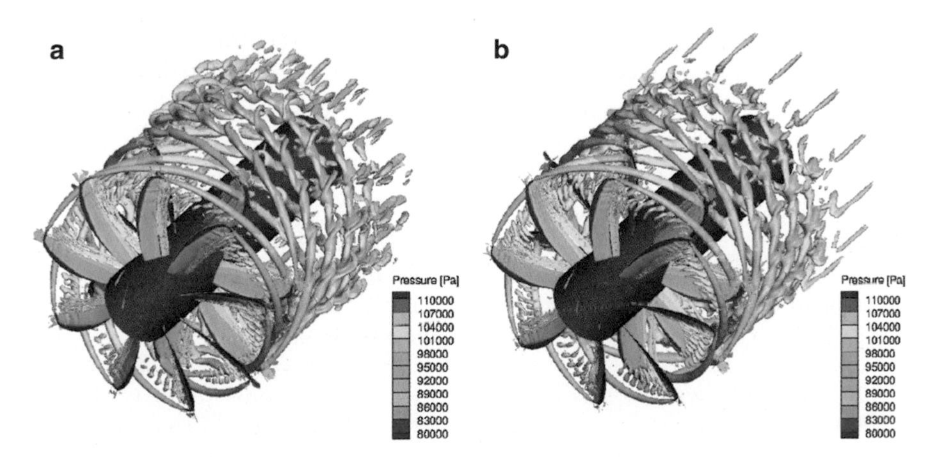

Fig. 7 λ_2-visualization of 9×7 and 8×8 CROR in take-off conditions. (**a**) 9×7. (**b**) 8×8

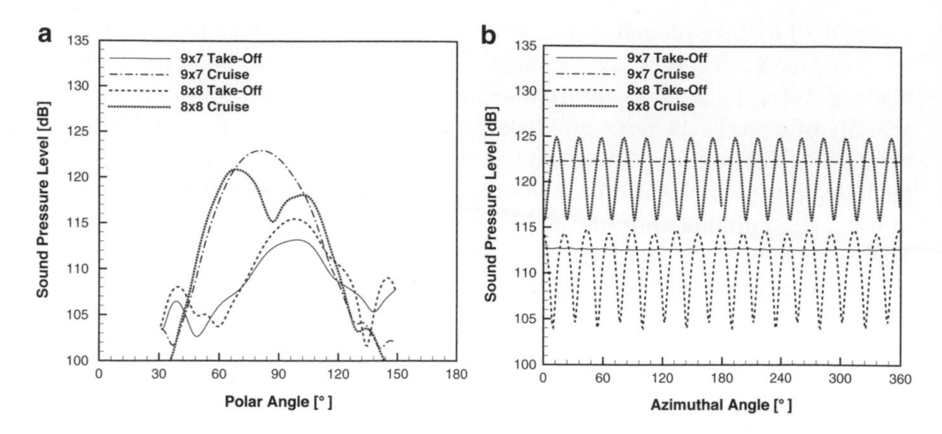

Fig. 8 Noise emission of 9 × 7 and 8 × 8-CROR in take-off and cruise conditions. (**a**) Polar directivity. (**b**) Azimuthal directivity

5.2 Acoustic Results

For the acoustic examination one rotor revolution is evaluated, with a temporal resolution of 360 time steps. The noise is transported from an acoustic evaluation surface towards a distant observer at $R = 30$ m with the Ffowcs Williams-Hawkings tool ACCO [2]. The polar directivity in Fig. 8a shows a strong variation between the two flight cases. In cruise, the sound pressure levels in proximity to the rotor plane are much higher than for the take-off case. This is caused by the dominance of single rotor tones, which have their maximum in the rotor plane and decay towards the diagonal direction. For the 9 × 7 CROR only a single peak is present, whereas the 8 × 8 CROR shows a double peak. The periodicity of the 8 × 8 leads to a periodic blade loading and thus a strong variation of the noise emitted in the azimuthal direction as Fig. 8b illustrates. The thrust is roughly constant for the 9 × 7 CROR for one rotor revolution leading to a constant emission over the azimuthal direction. The same effect can also be observed for the take-off case where the additional peaks in diagonal direction are caused by the interaction noise. The interaction noise is higher for the 8 × 8 CROR due to the simultaneous interactions on all blades. The periodic blade loading of 8 × 8 leads to a similar variation of the sound pressure levels in the rotor planes as seen for the cruise case. Overall, the 9 × 7 CROR shows a favourable noise emission compared to the 8 × 8 case as the interaction noise is lower and the thrust distribution over one rotor revolution is fairly constant. For the 8 × 8 CROR, this leads to an additional fluctuation in the single rotor noise.

6 Conclusion

The present paper shows further enhancements of the already implemented WENO scheme by using CRWENO. Compared to the previous JST scheme it shows a strong improvement in vortex conservation, which is important especially for the simulation of interaction phenomena. As example for the benefits of interaction phenomena a trimmed fluid-structure coupled simulation of a main rotor is presented. Within this simulation some performance studies have been obtained. It is shown that the additional costs for CRWENO are about +40 % compared to the previous JST scheme and +12 % compared to the WENO scheme, which is acceptable due to the better results and the higher order spatial discretization, which allows a coarser grid resolution.

Good vortex conservation is essential for the simulation of brownout conditions. It is shown that the implemented higher order methods meet the requirements concerning simulation quality and performance. Results can be achieved in an acceptable time at a massively parallel high performance computation cluster.

As much as helicopter simulations, the simulation of a CROR has high demands on the computational setup, especially if the results are to be used for an acoustic evaluation during postprocessing. Good results are achieved due to the use of fine grid resolution. Based on the better vortex conservation it is expected that the use of a WENO scheme will even improve the acoustic results of a CROR simulation, which will be examined in upcoming studies.

Acknowledgements We greatly acknowledge the provision of supercomputing time and technical support by the High Performance Computing Center Stuttgart (HLRS) for our project HELISIM.

References

1. Betz, A.: The ground effect on lifting propellers. Zeitschrift für angewandte Mathematik und Mechanik **17**(836), 1–11 (1937)
2. Busch, E.R., Wurst, M.S., Keßler, M., Krämer, E.: Computational aeroacoustics with higher order methods. In: High Performance Computing in Science and Engineering'12. Springer (2012)
3. Ghosh, D., Baeder, J.D.: Compact reconstruction schemes with weighted ENO limiting for hyperbolic conservation laws. SIAM J. Sci. Comput. **34**(3), A1678–A1706 (2012)
4. Günther, T., Kuhn, A., Kutz, B., Theisel, H.: Mass-Dependent integral curves in unsteady vector fields. Comput. Graph. Forum (Proc. EuroVis 2013) **32**(3), 211–220 (2012). http://dx.doi.org/10.1111/cgf.12108
5. Jameson, A.: Time dependent calculations using multigrid, with applications to unsteady flows past airfoils and wings. In: AIAA 10th Computational Fluid Dynamics Conference, Honolulu (1991)
6. Jameson, A., Schmidt, W., Turkel, E.: Numerical solutions of the Euler equations by finite volume methods using Runge-Kutta time-stepping schemes. In: AIAA 14th Fluid and Plasma Dynamic Conference, Palo Alto (1981)

7. Jiang, G.-S., Shu, C.-W.: Efficient implementation of weighted ENO schemes. J. Comput. Phys. **126**, 202–228 (1996)
8. Johnson, W.: CAMRAD II Comprehensive analytical model of rotorcraft aerodynamics and dynamics, 4.8 edn. Johnson Aeronautics, Palo Alto (2009)
9. Kowarsch, U., Keßler, M., Krämer, E.: High Order CFD-Simulation of the Rotor-Fuselage Interaction. European Rotorcraft Forum, Moscow (2013)
10. Kowarsch, U., Oehrle, C., Hollands, M., Keßler, M., Krämer, E.: Computation of helicopter phenomena using a higher order method. In: High Performance Computing in Science and Engineering. Springer, Stuttgart (2013)
11. Kroll, N., Eisfeld, B., Bleeke, H.: The Navier-Stokes code FLOWer. Notes on Numerical Fluid Mechanics, pp. 58–71. Vieweg, Braunschweig (1999)
12. Kutz, B.M., Kowarsch, U., Keßler, M., Krämer, E.: Numerical investigation of helicopter rotors in ground effect. In: 30th AIAA Applied Aerodynamics Conference, AIAA-2012-2913, New Orleans, 2012
13. Kutz, B.M., Keßler, M., Krämer, E.: Experimental and numerical examination of a helicopter hovering in ground effect. CEAS Aeronaut. J. **4**, 397–408 (2013)
14. Kutz, B.M., Keßler, M., Krämer, E.: Toolchain for free flight measurement and code validation purposes. In: 39th European Rotorcraft Forum, Moscow (2013)
15. Kutz, B.M., Günther, T., Rumpf, A., Kuhn, A.: Numerical examination of a model rotor in brownout conditions. In: Proceedings of the American Helicopter Society, 70th Annual Forum, AHS2014-000343, Montreal (2014)
16. Mapes, P., Kent, R., Wood, R.: DoD Helicopter Mishaps FY85-05: Findings and Recommendations. Technical report, U.S. Air Force (2008)
17. Rauleder, J., Leishman, J.G.: Particle-fluid interactions in rotor-generated vortex flows. Exp. Fluids **55**, 1–15 (2014)
18. Schmid, S., Lutz, T., Krämer, E.: Impact of ground modelling approaches on the prediction of ground effect aerodynamics. Eng. Appl. Comput. Fluid Mech. **3**(3), 419–429 (2009)
19. Shao, Y., Li, A.: Numerical modeling of saltation in the atmospheric surface layer. Bound. Layer Meteorol. **91**, 199–225 (1999)
20. Shao, Y., Lu, H.: A simple expression for wind erosion threshold friction velocity. J. Geophys. Res. **105**(D17), 22437–22443 (2000)
21. Syal, M., Govindarajan, B., Leishman, J.G.: Mesoscale sediment tracking methodology to analyze brownout cloud developments. In: Proceedings of the American Helicopter Society, 66th Annual Forum, Phoenix (2010)
22. Sydney, A., Baharani, A., Leishman, J.G.: Understanding brownout using near-wall dual-phase flow measurements. In: Proceedings of the American Helicopter Society, 67th Annual Forum, Virginia Beach (2011)
23. Thomas, S., Lakshminarayan, V.K., Kalra, T.S., Baeder, J.D.: Eulerian-Lagrangian analysis of cloud evolution using CFD coupled with a sediment tracking algorithm. In: Proceedings of the American Helicopter Society, 67th Annual Forum, Virginia Beach (2011)
24. Toro, E.F.: Riemann Solvers and Numerical Methods for Fluid Dynamics. Springer, Berlin (1997)
25. Wong, O.D., Tanner, P.E.: Photogrammetric measurements of an EH-60L brownout cloud. In: Proceedings of the American Helicopter Society, 66th Annual Forum, Phoenix (2010)

Impact of 3D Shock Control Bumps on Transonic Buffet

S. Bogdanski, P. Gansel, T. Lutz, and E. Krämer

Abstract Steady and unsteady RANS computations have been carried out to investigate the effect of three dimensional shock control bumps on the buffet behaviour of a transonic airfoil. A total of five basic bump shapes were investigated at two Mach numbers. All bumps were optimized for minimal drag at a single design point by varying the two parameters bump height and bump position. In total two design points were investigated. The first one was located at the higher Mach number at a angle of attack close to cruise conditions while the second design point was located at the lower Mach number close to the buffet boundary. Additionally, the applicability of separation as buffet indicator, the effect of streamwise vortices on the buffet alleviation and innovative bump shapes were examined. For a suitable test case the employed finite volume solver TAU shows a close-to-linear scaling up to high domain numbers beyond 10,000, which corresponds to less than 14,000 cells per partition. Additionally the differences achieved utilizing different compilers and partitioners are discussed.

1 Introduction

Two and three-dimensional shock control bumps (SCBs) have been investigated and optimized for numerous years at IAG. These SCBs expand the compression shock to a lambda-shock and reduce wave drag [1]. The main focus of the recent work has been on a robust bump design, i.e. a performance gain over a large range of lift coefficients, without performance deterioration in the design point of the baseline airfoil.

There are only few attempts to use SCBs to reduce the shock motion in transonic buffet [2]. All these attempts used steady computations for the design and evaluation of these "buffet bumps" and the criterion for the determination of the buffet onset has been the extend and shape of the separated flow rather than the unsteady behaviour [3].

S. Bogdanski • P. Gansel • T. Lutz (✉) • E. Krämer
Institute of Aerodynamics and Gas Dynamics, University of Stuttgart, Stuttgart, Germany
e-mail: bogdanski@iag.uni-stuttgart.de; lutz@iag.uni-stuttgart.de

W.E. Nagel et al. (eds.), *High Performance Computing in Science and Engineering '14*, DOI 10.1007/978-3-319-10810-0_30

It is the aim of this investigation to utilize steady and unsteady RANS methods to evaluate the impact of SCBs which are optimized for drag reduction on the buffet characteristics of a transonic airfoil. The investigation is conducted for two different Mach numbers: the design Mach number of the bumps and a smaller Mach number where the bumps are effective at a higher angle of attack, closer to the buffet onset. It can be supposed that bump design conditions closer to the buffet onset boundary will have a positive effect on the buffet characteristics. Additionally, bumps with a high design lift coefficient close to the buffet boundary have been investigated. It was found that streamwise vortices created by the bumps are beneficial for buffet alleviation and thus bumps with a forked front flank were created to

2 Airfoil and Free Stream Conditions

For all buffet computations presented in this report the pathfinder transonic airfoil with about 12 % thickness was used [4]. The Reynolds number was held constant at $Re = 20 \cdot 10^6$ and the transition was fixed at $x/c = 10\,\%$. Two Mach numbers were investigated: a higher Mach number of $Ma = 0.76$ which is more suitable for drag reduction and a lower Mach number of $Ma = 0.74$ were the baseline airfoil shows larger buffet amplitudes.

3 Numerical Methods

The simulations presented in this paper have been conducted with the unstructured CFD code "TAU" developed by the German Aerospace center (DLR) [5]. The code solves the Euler-, Navier-Stokes- or RANS-equations on grids with unstructured cells. It can be used for steady computations with various acceleration techniques as well as for unsteady computations with a second order dual time-stepping scheme. Several one-equation, two-equation and Reynolds stress turbulence models are implemented. All simulations in this paper have been carried out with the "Menter SST" model with the "scale-adaptive simulation" modification (SST-SAS) [6]. This model features a LES-like behaviour in regions where vortices can be directly resolved by the simulation. This is the case in the large separated region behind the shock during the upstream motion of the shock. In regions with attached boundary layers.

The physical time step δt was chosen to $\frac{1}{40}\frac{c}{u_{inf}}$, with c as the chord length and u_{inf} as the free stream velocity. This leads to approximately 500 time steps per buffet period and has been proven to guarantee a time step independent solution [7]. Every fifth time step, a solution is stored and used for post processing. A constant number of 100 inner iterations was used. Typically about 6,000 physical time steps had to be computed to gain three converged buffet periods which were used for post processing. For some cases with angles of attack close to the buffet boundary the simulation had to be run up to three times longer to establish a steady or periodic solution.

An implicit backward Euler scheme has been used for the inner iterations with a 3v multi grid scheme and residual smoothing. The flux discretization was realized with the second order Jameson, Schmidt, Turkel scheme and stabilized with scalar dissipation.

For post processing purposes a python tool was written which extracts slices of the unsteady surface solution. The data of each slice and time step is then processed separately and values like lift, drag, shock position and extent of the separated area are determined. These values are then integrated and time averaged. All post processing has been done with 25 slices equally spaced in spanwise direction and the tool has been verified against the results of the TAU solver.

3.1 Computational Grid

3.1.1 Grids

Hybrid grids with structured hexahedral blocks for the boundary layer treatment and unstructured tetrahedral blocks for the treatment of the inviscid regions have been created with the commercially available software "Gridgen". An additional structured block was created over the rear part of the airfoil to account for the thickened/separated boundary layer behind the shock. All grids have a spanwise extent of $0.3c$ and use periodic boundary conditions since these impose the least amount of artificial constraints on the flow solution.

Three different grid levels have been created to investigate the grid dependency of the solutions. The cell size was scaled with $\sqrt[3]{2}$ so as the overall amount of points doubles with each level. Table 1 shows the properties of each grid level and Fig. 1 shows the surface grid and one periodic boundary of the coarsest grid. The computations with the baseline airfoil have been carried out with both 2D and 3D grids, yet only minor differences occur in the solutions.

Table 1 Properties of the three grid levels

	fine	Medium	Coarse
Grid points	$4 \cdot 10^6$	$2 \cdot 10^6$	$1 \cdot 10^6$
Spanwise	121	96	76
Circumference	452	360	284
No. hexahedral layers	50	40	31
Extra prisms	10	8	6

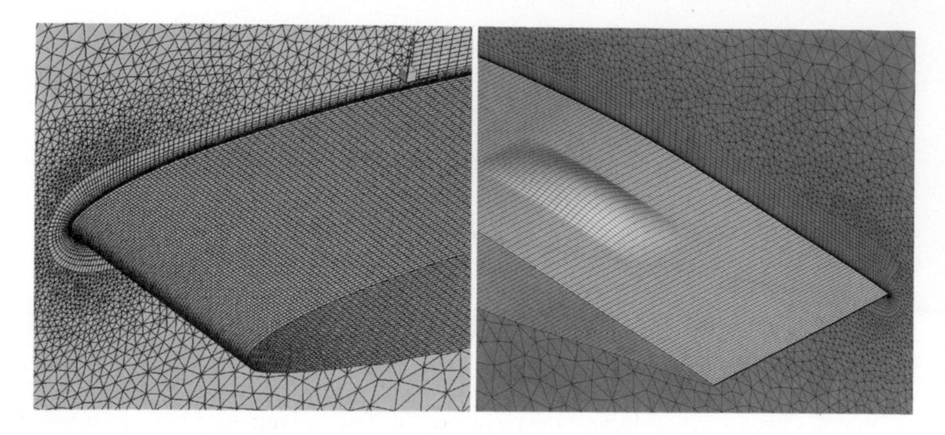

Fig. 1 Surface grid and periodic boundary, coarse mesh

3.1.2 Grid Dependency

The grid dependency has been investigated for all geometries at the higher Mach number of $Ma = 0.76$. Though the differences between the different levels are small for conditions with attached flow, large differences occurred in the buffet regime. In general the finer grids reduced the buffet amplitudes but shifted the buffet onset to smaller angles of attack. The largest differences for the lift amplitudes of about a factor of 4 occurred between the coarse and medium gird, it is therefore mandatory that fine grids are used for reliable results.

4 Results

4.1 Buffet Behaviour of Baseline Airfoil

The baseline airfoil was investigated at two different Mach numbers: $Ma = 0.76$, the design Mach number of the performance SCBs and a lower Mach number of $Ma = 0.74$, where larger buffet oscillations occur. With increasing angle of attack and shock strength the baseline airfoil develops a separation bubble at the location of the shock which is growing in streamwise direction. Once it reaches the trailing edge buffet occurs. This behaviour corresponds to "model A" according to the buffet classification of Pearcey [8]. Figure 2 (left) shows the lift polar and the lift variations $c_{l,RMS}$ of the baseline airfoil for the two Mach numbers. Steady solutions exist for all angles of attack with $\alpha \leq 4°$ for $Ma = 0.76$ and $\alpha \leq 3°$ for $Ma = 0.76$. Above these values transonic buffet occurs, with higher Mach number showing the smaller amplitudes. The evolution of lift with regard to time is shown exemplarily for one case in Fig. 2 (right).

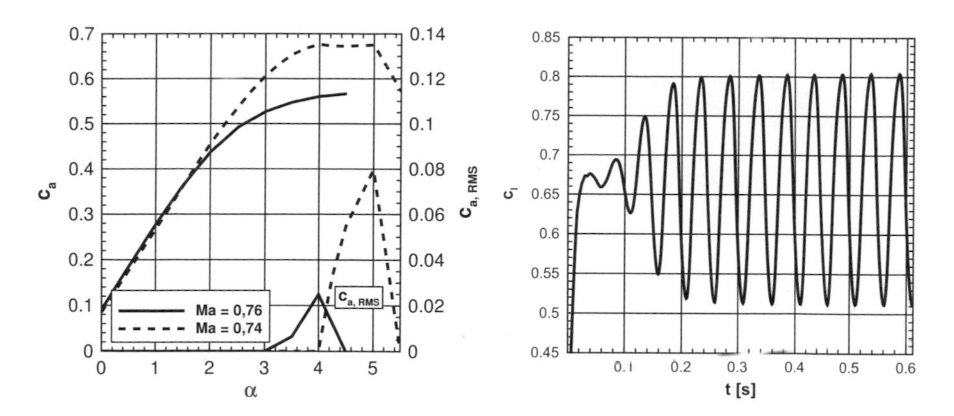

Fig. 2 Lift polar and lift variation of the baseline airfoil at the two Mach numbers, $Re = 20 \cdot 10^6$ (*left*) and evolution of c_l over time for $Ma = 0.74$ and $\alpha = 5°$ (*right*)

4.2 *Impact of Performance Bumps on Buffet Behaviour*

In this section the impact of so called performance bumps on the buffet behaviour is investigated with three different bump shapes: a smooth – hill shaped – one (HSCB), a wedge shaped one and an extended bump (see Fig. 3 (left)). These bumps were optimized purely for minimal drag at a constant angle of attack of $\alpha = 1.5°$ at the higher Mach number of $Ma = 0.76$. A lift coefficient of approximately $c_l = 0.417$ resulted. The bumps have been optimized with a downhill simplex algorithm for minimal drag by variation of the two parameters height and position. The length of the smooth and the wedge bump have been fixed at $l = 0.3c$, whereas the extended bump varies in length and reaches the trailing edge for all designs. The spanwise extent of the wedge, the smooth and the front ramp of the extended bump have been fixed at $0.1c$. When the performance of the three bumps is compared (see Fig. 3 (right)) it is evident, that the extended bumps shows the best performance at the design point. At off design conditions at lower lift coefficients the extended bump is superior to the two other bumps, too.

At the design Mach number of $Ma = 0.76$ all SCBs lower the buffet onset by an angle of attack of about $\Delta\alpha = 0.5°$ (see Fig. 4 (left)) and show larger buffet amplitudes inside the buffet regime than the baseline airfoil. However, it is not possible to determine the bump with the best/worst buffet behaviour since the buffet amplitudes vary with angle of attack. The same holds for the lower Mach number of $Ma = 0.74$. When the maximum lift coefficient is considered all SCBs show only a minor effect.

Essentially, all three SCBs that were designed for a reasonable low angle of attack / lift coefficient deteriorate the buffet behaviour at all investigated Mach numbers and regardless of the buffet criterion.

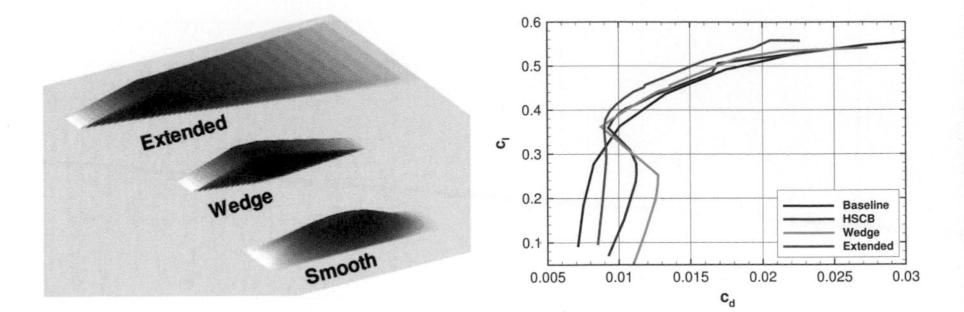

Fig. 3 Shape of the three bumps: HSCB, wedge and extended (*left*) and drag polars for the three bumps at $Ma = 0.76$ (*right*)

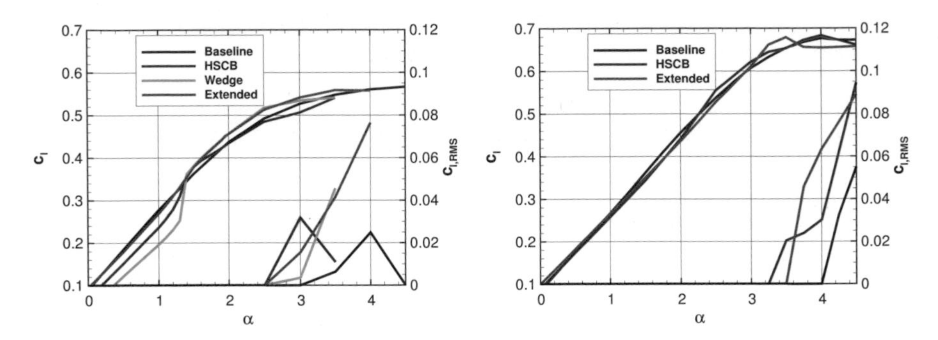

Fig. 4 Lift polar and lift variations for the three SCBs, $Ma = 0.76$ (*left*) and $Ma = 0.74$ (*right*)

4.3 Design and Analysis of Buffet Bumps

As a second investigation the impact of a large design lift coefficient on the buffet behaviour was investigated. A HSCB and an extended bump have been optimized for a high lift coefficient of $c_l = 0.65$ close to the buffet boundary at the lower Mach number of $Ma = 0.74$. The optimization aimed only at minimizing the drag coefficient at the single design point which lead to a rather "extreme" bump design. Bumps optimized for these design conditions are referred herein as buffet bumps. Figure 5 (left) shows the drag polar of the two buffet bumps and proves that both are effective at the design lift coefficient. Unfortunately the lift variations inside the buffet regime are increased by both bumps as depicted in Fig. 5 (right). The angle of attack where buffet occurs is also lower compared to the baseline airfoil. Yet the two bumps increase the maximum usable lift coefficient. At off-design conditions at lower lift coefficients both buffet bumps show a rather bad performance which is presumably prohibitive to an application on a real aircraft.

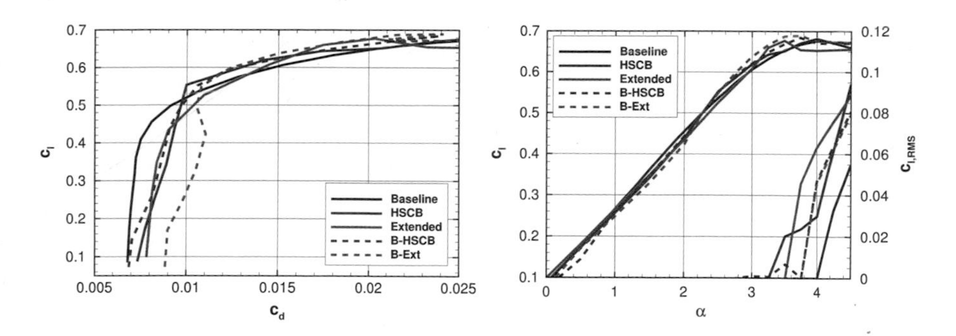

Fig. 5 Drag polar of the high cl bumps (*left*) and lift polar and variations (*right*), $Ma = 0.74$

4.4 Separation as Buffet Criterion

Separation has been used as a buffet criterion since early investigations. In the simple case of a 2D airfoil the angle of attack where the separation reaches $x/c = 90\,\%$ is usually taken as buffet onset. This procedure works for both classical buffet models: an expanding separation bubble at the shock foot or a trailing edge separation moving upstream. In the more complex case of 3D SCBs an arbitrary level separated flow on the upper surface of $20\,\%$ has been proposed (see e.g. Eastwood [3]). Yet the determination of the separated area is challenging in a 3D flow since complex flow topologies can emerge like saddle points, cross flow, vortices etc. Thus only the simplified criterion of $c_{fx} < 0$ has been employed so far. Figure 6 shows the fraction of the upper surface where separated flow (i.e. $c_{fx} < 0$) is present for for the original three bumps and the two Mach numbers. When this picture is compared to Fig. 4 it evident that the amount of separated flow does not correlate with the amount of lift oscillations. The same can be observed when comparing the separation (see Fig. 7) and the lift amplitudes (see Fig. 5 (right)) of the two buffet bumps. Thus it can be concluded that the simple separation detection criterion is not suitable for the determination of the buffet behaviour of 3D SCBs with their complex flow patterns.

4.5 3D SCBs as Vortex Generators

Due to the spanwise pressure gradients provoked by the 3D SCB geometry streamwise vortices occur behind the bumps. In the absence of separation a counter rotating vortex pair emerges behind the bump where the flow in the middle between the two vortices is directed towards the wall, herein after called common flow down. When separation appears at the rear flank of the SCB an additional pair of counter rotating vortices appears in the middle of the first pair with opposing rotational sense, common flow up. Since these vortex pairs transport high velocity fluid from the upper edge of the boundary layer towards the wall, it can be supposed that

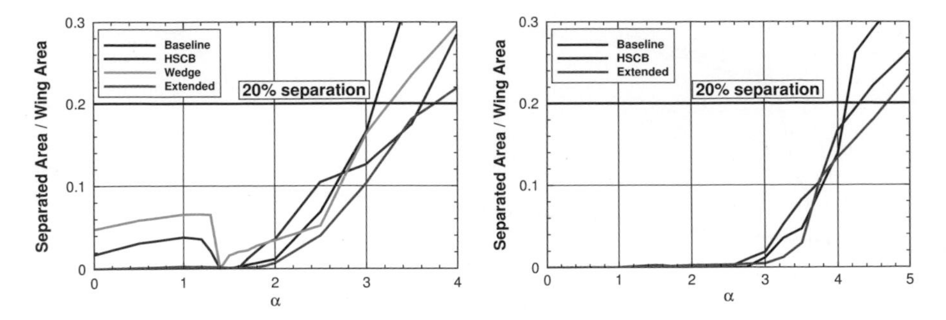

Fig. 6 Fraction of the upper surface with separated flow for the three performance bumps, $Ma = 0.76$ (*left*) and $Ma = 0.74$ (*right*)

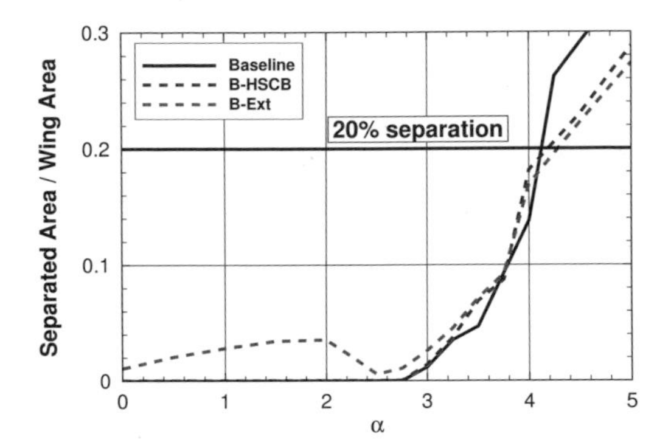

Fig. 7 Fraction of the upper surface with separated flow for the two buffet bumps, $Ma = 0.74$

they will suppress boundary layer separation and alleviate the buffet boundary. The investigation of this effect is carried out in two steps. In the first place it is tried to correlate the vortex strength behind the SCBs with local flow quantities on the surface of the SCBs and in the second place a correlation of the vortex strength with buffet behaviour is investigated.

4.5.1 Vorticity Generation and Spanwise Pressure Gradient

In this first part the driving mechanism for the vortex generation by the bumps is sought. In a first try correlations between the vortex strength and pressure gradients on the rear flank of the bumps were investigated, but with little success. The best correlation between the amount streamwise vorticity and a local flow quantity has been shown to be the one with the spanwise pressure gradient at the front flank of the bump. Figure 8 shows the correlation between these two quantities. ω_x is the maximum streamwise vorticity behind the bump and dp/dy is the maximum spanwise pressure gradient on the surface of the front flank of the bump in the middle between the beginning and the crest of the bump (see Fig. 9).

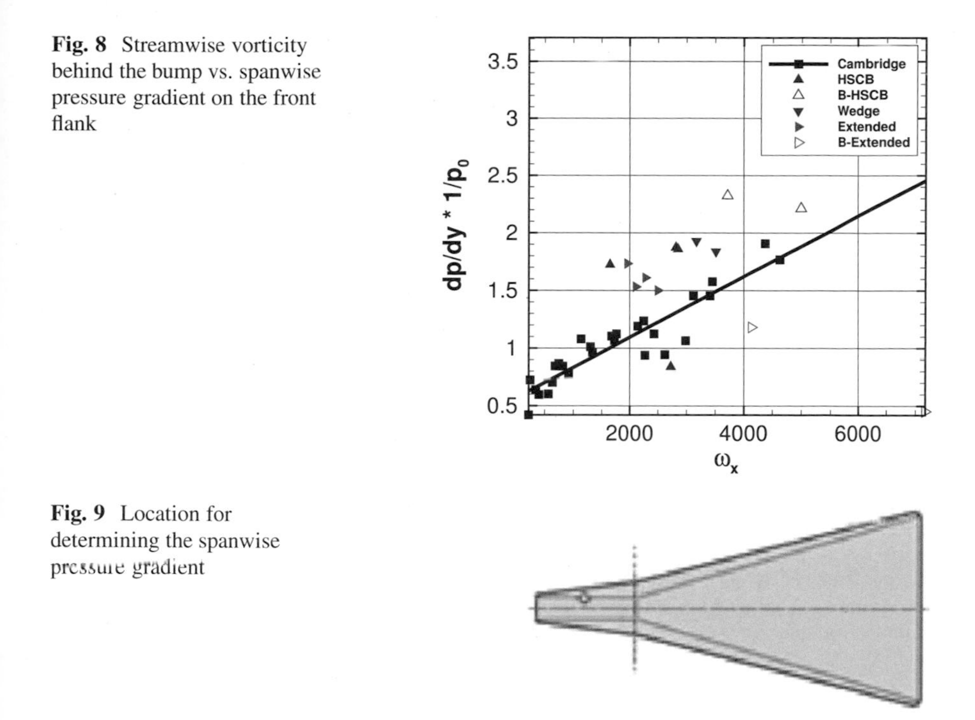

Fig. 8 Streamwise vorticity behind the bump vs. spanwise pressure gradient on the front flank

Fig. 9 Location for determining the spanwise pressure gradient

The correlation between these two quantities is better for the computations with the structured meshes and the automated post processing (black squares). The data obtained with the unstructured meshes and the manual post processing shows greater scatter (colored symbols). Three reasons can be the cause for this finding: The use of an unstructured and slightly coarser mesh which introduces more numerical dissipation, the consideration of computations with different Mach numbers and angles of attack closer to the buffet boundary and finally the determination of the maximum vorticity a fixed location of $x/c = 80\,\%$ (which is yet very close to the absolute maximum). Nevertheless the correlation could be confirmed.

4.5.2 Effect of Streamwise Vorticity on Buffet

In the second part of the investigation the effect of vortex strength behind the bumps and the buffet behaviour is investigated. As a measurement for the buffet behaviour two quantities have been used: the maximum lift coefficient that could be achieved and a quantity that accounts for the amount of lift oscillations called buffet indicator BI. Figure 10 (left) shows the correlation between the maximum lift coefficient $c_{l,max}$ and the streamwise vorticity $\omega_{x,max}$ for steady flow at four different angles of attack below the buffet onset. All three performance bumps and the two buffet bumps are included in this figure. The corresponding coefficient of determination R^2 is shown in Table 2. The values are nowhere below $R^2 = 0.7$ which means, that at least

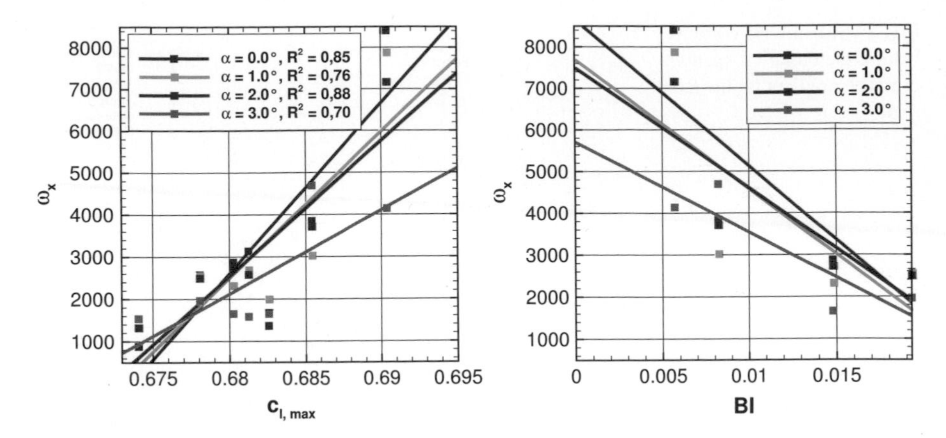

Fig. 10 Correlation between the maximum lift coefficient $c_{l,max}$ (*left*), the buffet indicator *BI* (*right*) and the maximum streamwise vorticity $\omega_{x,max}$ at $x/c = 80\,\%$

Table 2 Coefficient of determination R^2 for the correlation of $\omega_{x,max}$ and the buffet behaviour

α	$c_{l,max}$	BI
0°	0.85	0.74
1°	0.76	0.66
2°	0.88	0.78
3°	0.70	0.86

70 % of the additional lift can be attributed to the streamwise vortices. As a second measure for the buffet behaviour, the so called buffet indicator *BI* is introduced. This quantity is determined as

$$BI = \int \alpha_{low}^{\alpha_{high}} c_{l,RMS} d\alpha$$

where α_{low} is the highest angle of attack where all bumps exhibit still a steady flow solution and α_{high} is the lowest angle of attack where all bumps show lift oscillations. Essentially, the buffet indicator is a measure for the flow unsteadiness at and slightly above the buffet boundary. As depicted in Fig. 10 (right) and Table 2 the correlation is somewhat worse than the one the correlation with $c_{l,max}$. Yet, at least 66 % of the lower flow unsteadiness can be attributed to the streamwise vorticity.

4.6 Innovative Bump Shapes

The results of the previous buffet investigations lead to the conclusion that a larger streamwise vorticity is beneficial for the buffet behaviour and that the driving force for the vortex generation is the spanwise pressure gradient on the front flank. Thus it was decided to create forked bumps which exhibit two teeth or forks at the front

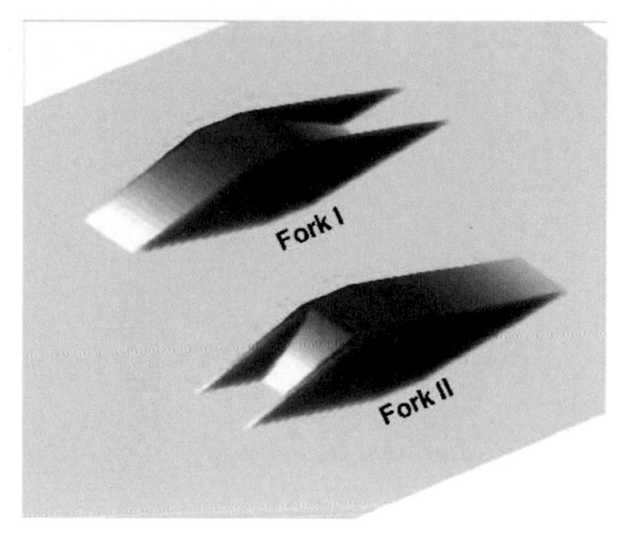

Fig. 11 Principal shape of the fork bumps

flank, pointing in the upstream direction (see. Fig. 11), referred to as "Fork2" bumps in earlier investigations. Note: The fork 1 depicted also in the picture was not investigated here due to its bad off-design behaviour. Overall, three different fork2 bumps have been created:

- A performance fork2 according to the optimization described in Sect. 4.2 (referred to as Fork2),
- A buffet fork2 according to Sect. 4.3 (referred to as B-Fork2) and
- A performance bump identical to the first one except twice as long forks (referred to as Fork2L).

Figure 12 (left) shows the drag polar of the forked bumps. It can be seen that the Fork2L shows the best off-design performance yet the Fork2 bump is slightly better at higher lift coefficients. The B-Fork2 shows a very bad off-design performance and a very small range at higher lift coefficients where it is superior to the other two fork bumps. It appears that the B-Fork2 is a rather extreme design. When the maximum lift is regarded (see Fig. 12 (right)), the Fork2 seems to be slightly better then the other two fork bumps. This is also true with respect to the lift oscillations: The Fork2 shows almost the best buffet behaviour of all bumps.

When the computations with the forked bumps are included into the two correlations presented in Sect. 4.5 the coefficient of determination drops for both to levels of about 10–20 %. This is a clear indication that the correlations are not right with respect to the forked bumps. Obviously, the front flank geometry of the forked bumps differs so strongly from that of the original bumps that no general correlation can be found.

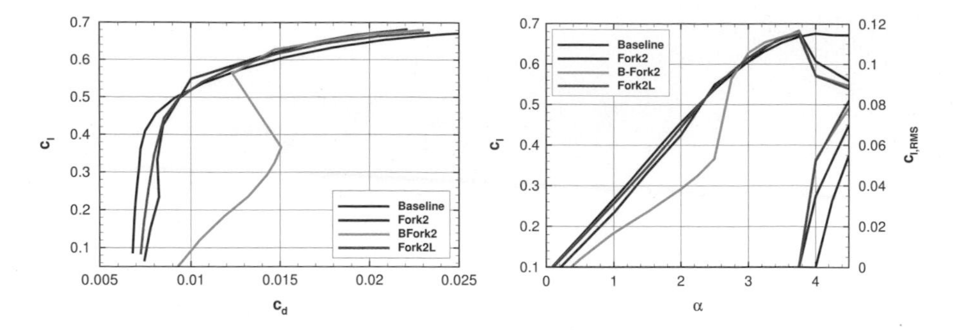

Fig. 12 Drag (*left*) and lift polar (*right*) of the forked bumps, $Ma = 0.74$

5 Computational Resources

Since the computations presented above were conducted with rather small meshes consisting of up to four million points they don't appear to be sensible for a scaling test of the TAU code. Usually, up to 64 CPU cores have been used for these computations and a whole buffet computation with 6,000 physical time steps lasted about 24 h. Whenever possible, several buffet calculations have been bundled into one job for a better queuing performance.

To have a better assess of the scaling capability of the CFD code TAU, tests with an aircraft configuration mesh with 45 million points and 137 million cells were performed. All the calculations were conducted using central differences of second order for spatial discretization and backward-Euler time integration with a LUSGS scheme. The one-equation model of Spalart and Allmaras [9] was used for turbulence modelling. With each amount of domains between 16 and 32.768 in steps of factor 2 a number of 1,000 iterations were calculated to smooth out outliers. To achieve results scalable to any iteration number, only the time difference between first and last iteration without initialization or write out time was taken into account. Every partition was simulated on an individual computation core. Because of the memory requirements of the strong scaling test case, for the simulations on 32 and 16 domains 2 nodes of 32 CPU cores each had to be allocated. That means not all available cores were used in these cases. This is also the reason why smaller amounts of domains (down to 1) were not considered.

The speed-up of the TAU code shown in Fig. 13 is very good in the tested interval and close to the ideal linear scaling up to 16.384 cores (the lowest point at 16 domains was chosen as reference). Over a wide range even a noticeable super-linear scaling occurs due to cache memory effects. However, at the last point of the curve at 32.768 cores the speed-up is lower than at 16.384, that is the walltime actually increases with the bigger number of CPUs used. Thus it would be extremely inefficient to set up a simulation with the present grid on more than around 16,000 partitions. At this limit the mean number of 8,362 cells per computation core is

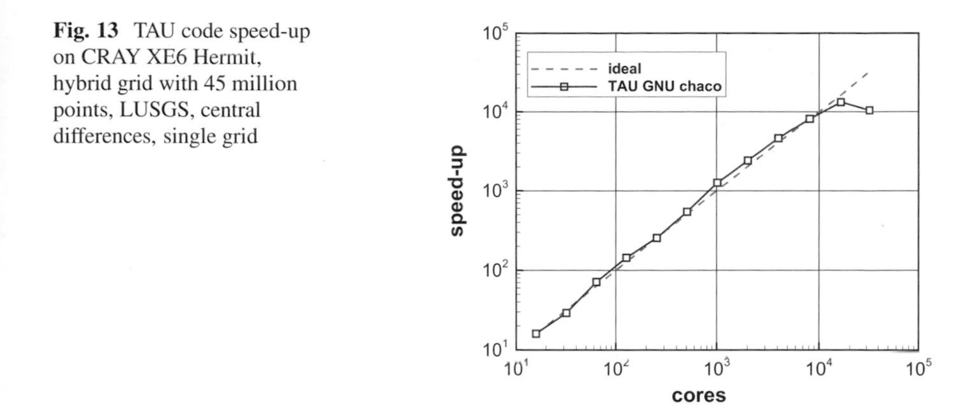

Fig. 13 TAU code speed-up on CRAY XE6 Hermit, hybrid grid with 45 million points, LUSGS, central differences, single grid

already very low. These results correspond well with previous results on similar systems [10].

To find the ideal configuration for the simulations also two different partitioning tools and four compilers were tested. Besides the native partitioner included in the TAU software package, the performance of the graph partitioner Chaco from the Sandia National Laboratories [11] in version 2.2 was investigated. Since the TAU native partitioner basically performs a bisection of the computational domain in direction of the space coordinates, the intersection surfaces are rather big in relation to the domain sizes. In contrast Chaco does the splitting in graph space and optimizes it for small bounding surfaces which results in compact clusters of cells as partitions. Although this approach seams promising to be more effective, there are a number of test points between 16 and 512 partitions, where the TAU partitioner does better (Fig. 14a). The maximal difference here is almost 8 %. In the region of 1.024–16.384 cores, which is more appropriate for a grid of the present size, the Chaco partitioning is actually advantageous. The benefit of saved wall, respectively CPU time increases continuously from 8 % at 2.048 cores to a maximum of 35 % at 16.384 cores. At 32.768 cores both tools perform comparably (bad, as mentioned above).

The following compiler versions were tested on the CRAY XE6 to compile the TAU code: Cray 8.2.3, GNU 4.8.2, Intel 14.0.2.144 and PGI 13.10-0. Except for the GNU compiler, a reduced range of partitions from 128 to 4.096 was tested. The results are plotted in Fig. 14b normalized with the run time of the GNU installation. The effect of the compiler choice is evident, while the spreading is within 10 % apart from an outlier of the Cray compiled TAU version at almost 20 %. The Cray compiler is best at some discrete points, but the GNU compiler provides more stable results over the whole range. The Intel installation is just as well at low domain numbers, but has a small offset of 2 % from 512 cores upwards. The PGI compiler builds the worst performing TAU code for all tested domain numbers which is 5–10 % slower. The shown comparisons of compilers are all done with Chaco as partitioner. The same tendencies and proportions were found with the TAU native

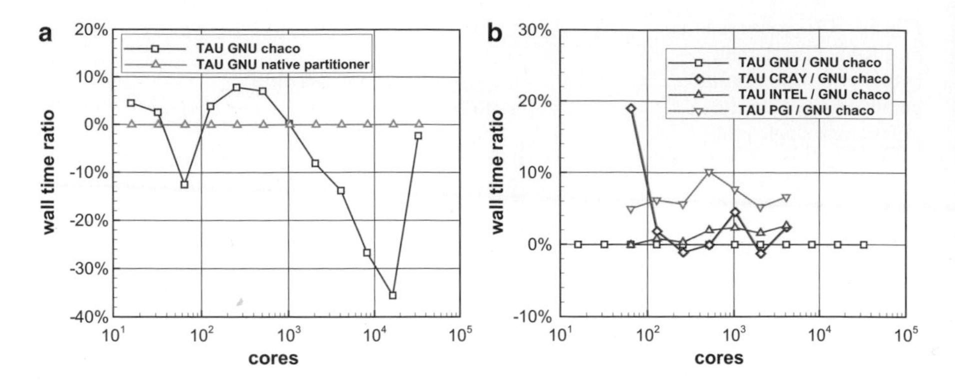

Fig. 14 Ratio of the wall time achieved (**a**) on domains split by the graph partitioner Chaco compared to the TAU native partitioning tool and (**b**) with four different compilers available on the CRAY XE6 referred to GNU

partitioner with the typical performance decrease between 512 and 16.384 domains but not plotted in the figure for the sake of clarity.

6 Conclusions

As a summary of the buffet investigations it can be concluded that:

- 3D SCBs which are purely optimized for a minimum drag at a reasonable lift coefficient far from the buffet boundary deteriorate the buffet behaviour in terms of buffet onset and buffet amplitudes.
- The shape of a 3D SCB has only a minor influence of the buffet behaviour, height and position are more crucial.
- SCBs optimized for minimum drag at high lift coefficients close to the buffet boundary of the baseline airfoil improve the maximum lift but may worsen buffet amplitudes inside the buffet regime. Yet, these type of bumps tend to show a poor off-design performance at lower lift coefficients.
- The amount of separation as determined by the simplistic $c_f < 0$ criterion is not suitable as a buffet indication for the complex flow topology encountered behind 3D SCBs.
- 3D SCBs can act as vortex generators. For "conventional" bump shapes the spanwise pressure gradient at the front flank is correlated to the vortex strength behind the bump. As a rule of thumb a 3D SCB is about half as effective in creating streamwise vorticity as a classical vortex generator.
- For SCBs with "unconventional" front flanks (i.e. the fork bumps) the correlation stated above could not be confirmed.

- The streamwise vortices created by 3D SCBs strengthen the boundary layer behind the bump and can improve the buffet behaviour in terms of maximum lift and amount of lift oscillations close to buffet onset.

In the conducted scaling tests the TAU code showed very good performance for cell amounts per core down to 8,362. At 4,181 cells per domain the scaling collapsed and the wall time even increased. In the reasonable interval of domains the graph partitioner Chaco yields better performance results than the TAU native partitioning tool. The fastest running build of the solver was achieved using the GNU compiler, compared to Cray, Intel and PGI.

Acknowledgements The research leading to these results has received funding from the European Union's Seventh Framework Programme (FP7/2007–2013) for the Clean Sky Joint Technology Initiative under grant agreement no. 271843.

References

1. Pätzold, M.: Auslegungsstudien von 3-D Shock-Control-Bumps mittels numerischer Optimierung (Design of 3D shock control bumps by numerical optimisation), PhD Thesis, Institute for Aerodynamics and Gas Dynamics, University of Stuttgart (2009)
2. Corre, C., et al.: Transonic flow control using a NS-solver and a multi-objective genetic algorithm. Fluid Mech. Appl., **73**, 297–302 (2003)
3. Eastwood, J.P., Jarrett, J.P.: Towards designing with three-dimensional bumps for lift/drag improvement and buffet alleviation. AIAA J. **50**(12), 2882–2898 (2012)
4. Streit, T., Horstmann, K.-H., Schrauf, G., Hein, S., Fey, U., Egami, Y., Perraud, J., El Din, I., Cella, U., Quest, J.: Complementary numerical and experimental data analysis of the ETW Telfona Pathfinder wing transition tests. 49th AIAA Aerospace Sciences Meeting, Orlando, AIAA Paper 2011-881 (2011)
5. Gerhold, T.: Overview of the Hybrid RANS Code TAU. In: Kroll, N., Fassbender, J.K. (eds.) MEGAFLOW – Numerical Flow Simulation for Aircraft Design. Notes on Numerical Fluid Mechanics and Multidisciplinary Design, vol. 89, pp. 81–92. Springer, Berlin/Heidelberg (2005)
6. Menter, F., Egorov, Y.: A Scale-Adaptive Simulation Model using Two-Equation Models. AIAA 2005-1095 (2005)
7. Soda, A.: Numerical Investigation of unsteady transonic shock/boundary-layer interaction for aeronautical applications. Dissertation, RWTH Aachen (2006)
8. Pearcey, H., Osborne, J.: The Interaction Between Local Effects at the Shock and Rear Separation – A Source of Significant Scale Effects in Wind-Tunnel Tests on Airfoils and Wings. AGARD, CP35 (1968)
9. Spalart, P.R., Allmaras, S.R.: A one-equation turbulence model for aerodynamic flows (1992). AIAA 92-0439
10. Gansel, P.P., et al.: Unsteady CFD simulation of the NASA common research model in low speed stall. In: High Performance Computing in Science and Engineering '13, pp. 439–453. Springer, Cham (2013)
11. Hendrickson, B., Leland, R.: The chaco user's guide, version 2.0. Technical Report SAND94-2692, Sandia National Laboratories (1994)

Simulation of a Flow in a Gas Turbine Exhaust Diffuser with Advanced Turbulence Models

S. Brouwer (née Volkmer) and D.M. Vogt

Abstract The prediction of the flow in a gas turbine exhaust diffuser of a combined cycle power plant is particularly difficult as maximum performance is obtained with highly loaded diffusers, which operate close to boundary layer separation. Computational fluid dynamics (CFD) simulations then need to cope with complex phenomena such as smooth wall separation, recirculation, reattachment and free shear layer mixing. Recent studies based on the Reynolds-Averaged-Navier-Stokes (RANS) approach demonstrate the challenge for two-equation turbulence models to predict separation and mixing of the flow correctly and identify that more accurate methods are needed. In the present study the flow in an exhaust diffuser (Reynolds number 1.5×10^6 with twice the channel height and the mean inlet velocity resulting in an inlet Mach number 0.6) is examined with unsteady RANS (URANS) simulations with the hybrid Scale Adaptive Simulation (SAS) model. The SAS model switches from URANS to a Large Eddy Simulation (LES)-like mode in unsteady flow regions to resolve various scales of detached eddies. This is achieved by decreasing the turbulent viscosity, which is generally over-predicted by common two-equation models leading to a highly dissipative behavior in terms of resolving unsteady fluctuations unless the flow instabilities are dominant. For validation, experimental data is obtained from a test rig at the Institute of Thermal Turbomachinery of the University of Stuttgart and Machinery Laboratory (ITSM).

1 Introduction

In combined cycle power plants, the diffuser connects the gas turbine and the heat recovery steam generator (HRSG) and partly converts the leaving kinetic energy at the turbine outlet into static pressure at the diffuser exit. This is achieved by a flow channel with a certain area increase, in which the subsonic flow with Mach number near 0.6 is decelerated. As the outlet pressure of the diffuser is generally fixed by the HRSG, the static pressure recovery in the diffuser leads to a decreased static pressure at the diffuser inlet or turbine outlet, respectively. This results in a higher

S. Brouwer (née Volkmer) • D.M. Vogt (✉)
Institute of Thermal Turbomachinery and Machinery Laboratory, University of Stuttgart, Pfaffenwaldring 6, 70569 Stuttgart, Germany
e-mail: damian.vogt@itsm.uni-stuttgart.de

W.E. Nagel et al. (eds.), *High Performance Computing in Science and Engineering '14*, DOI 10.1007/978-3-319-10810-0_31

enthalpy head in the turbine and therefore higher work output and efficiency of the whole cycle.

In general, the design limit of the diffuser geometry is governed by the appearance of smooth wall separation when the adverse pressure gradient in the diffuser reaches a certain limit causing the decelerated flow in the boundary layer to reverse direction. The resulting dissipation reduces both diffuser and turbine performance, and on top of this, the unsteadiness of the separated and recirculating flow can lead to increased vibration, which has to be strictly avoided. As a matter of fact, best performance is obtained with highly loaded diffusers, which are close to boundary layer separation. Apart from that, there are unsteady recirculation zones downstream of the blunt end of the hub and downstream of the struts supporting the hub and generating high losses. Diffuser flows involve complex phenomena such as smooth wall separation, recirculation, reattachment and free shear layer mixing and these are mainly controlled by the turbulence characteristics of the flow. In general there are two challenges for turbulence modelling of the above mentioned complex phenomena in Computational Fluid Dynamics (CFD) simulations. Firstly, it is necessary to predict the shear stresses in the wall shear layer to locate the correct separation point and secondly to predict the components of the Reynolds-stress tensor beyond separation in a free shear layer for the correct reattachment point and mixing of the flow.

In CFD simulations the limits for turbulence modelling of these complex phenomena with common two-equation models (currently widely used in industrial CFD simulations) and the RANS (Reynolds-Averaged-Navier-Stokes) approach have been investigated within a project in the framework of the research initiative "Kraftwerke des 21. Jahrhunderts". The trends of diffuser performance were in general captured with these models, but recent studies demonstrate the challenge for two-equation turbulence models to predict separation and mixing of the flow correctly and identify that more accurate methods are needed. As these phenomena have a large impact on the diffuser performance, the need for a detailed understanding of these was pointed out. Therefore, investigations with advanced turbulence model are the topic of the current project. A first step is the hybrid Scale-Adaptive-Simulation (SAS) model, which switches from unsteady RANS to a Large Eddy Simulation (LES) like mode in unsteady flow regions to resolve various scales of detached eddies. An adequate application of this model incorporates unsteady simulations with a 360° model as turbulence characteristics are inherently three-dimensional. This leads to an increasing computational effort and therefore high performance computing is needed.

2 State of Knowledge

The whole range of spatial and temporal scales of turbulent fluctuations is included in the Navier-Stokes equations. However, solving these equations numerically with the Direct Numerical Simulation (DNS) approach to compute all time and length scales (even the smallest), the spatial and temporal resolution must be extremely fine leading to an excessive computational effort for industrial applications featuring high Reynolds numbers. There are several approaches to reduce the computational effort by not resolving the different scales in turn, that means modelling more and more of the turbulent fluctuations. With Large Eddy Simulation (LES) the small scales are modelled and only the large scales are resolved by the simulation.

In contrast, all turbulent fluctuations are modelled by the RANS equations by a time-averaged model of the turbulence characteristics with different available turbulence models. Linear eddy-viscosity models are based on the assumption of a simple proportionality between the Reynolds stresses and the mean strain rate involving isotropic turbulence structures. The turbulent eddy viscosity can be defined by appropriate two-equation models with two differential transport equations for the surrogates of the turbulent velocity and length scales. The two-equation k-ϵ model generally predicts a high production of turbulent kinetic energy, which results in an optimistic prediction of the flow field in adverse pressure gradients e.g. in diffusers with separation (Casey and Wintergerste [1]). The advantages of the k-ω model in near-wall regions and of the k-ϵ model in free-stream regions are combined in the SST model by Menter [2]. With the SST model local changes of turbulent production in the shear layers are involved but also limited and a reduced eddy viscosity leads to a better prediction of the separation.

A basis for further CFD studies of a relatively simple planar and asymmetric diffuser geometry using the LES model is provided by Kaltenbach et al. [3]. However, the simulation of the flow is quite challenging due to the unsteady smooth wall separation in the diffuser. Nevertheless, results obtained with LES agreed very well with experimental data. These results were used in a later study by Cherry et al. [4] as reference for RANS simulation of the same diffuser flow with the k-ω-SST model. Results with LES agree better with measured data, but a strong dependency on the grid resolution was detected. Various studies of simulations with RANS equations with common two-equation turbulence models based on the linear eddy-viscosity approach have been published in the past for many two-dimensional planar and also axi-symmetric diffusers. Apsley and Leschziner [5], Chen et al. [6], Iaccarino [7] and Kluß [8] studied the application of common eddy-viscosity models for the flow in a 2D planar diffuser with separation on an inclined wall and all come to the same conclusion that the SST model predictions of the separation and reattachment achieve the best match compared to measurements.

Feldcamp and Birk [9] examined the turbulence modelling of flow in an annular diffuser and Vassiliev et al. [10] in an annular diffuser with a typical abrupt hub end, as found in real gas turbine applications. Both studies also include the struts that support the hub. In general the conclusions are similar to that of studies with planar diffusers, stating that separation is over predicted with k-ω based models. Furthermore, studies by Kluß and Stoff [11] show that the URANS approach with these common two-equation model leads to a highly dissipative behavior in terms of resolving unsteady fluctuations unless the flow instabilities are dominant (e.g. downstream of bluff bodies).

A hybrid model between the LES and RANS approach is the Detached-Eddy-Simulation (DES) which models the wall-bounded flow eddies by a RANS model and switches to LES mode in detached regions, which are determined by the ratio of RANS to LES (length of largest grid cell size) turbulent length scales. Unfortunately the DES model is not easy to apply because of required high accuracy of grid generation and grid induced separation (Menter and Egorov [12]). In the current study another hybrid model is investigated, the Scale-Adaptive-Simulation (SAS) approach. This is an improved URANS model with features comparable to LES, but not requiring exact grid information like the DES model (Menter and Egorov [13]). In contrast to the DES model, the SAS model introduced by Menter et al. [14–16] is switched on with increasing ratio of the modelled turbulent length scale to the Karman Length Scale (ratio of first to second velocity gradients). For unsteady velocities the latter is lower, leading to an additional production term in the ω-equation that increases ω and hence decreases turbulent kinetic energy and viscosity and therefore less damping of unsteady fluctuations occurs. For two-equation models, the increased local velocity gradient due to unsteadiness leads to higher production of turbulent kinetic energy and results in contrast to SAS in increased turbulent viscosity damping unsteadiness. The use of the SAS model was studied by Davidson [17] using a plane asymmetric diffuser with separation similar to that of Kaltenbach [3]. He showed that the computation of the flow is adversely affected by the SAS model in terms of worse agreement with experimental data than the SST model. The unsteady fluctuations vanish with the SAS model even in the diffuser part because the turbulent viscosity stays too high. In general, the separation is overpredicted with both models, but with the SAS model the largest separation is computed. As a consequence the results with the SST model are in better agreement with measured data than those with the SAS model. Davidson [17] stated that the reason seems to be that the solved equations are a mix of both models, RANS and LES, and that the use of the SAS model is different to what it was intended for (flows with inherent unsteadiness).

In contrast to that, Kluß and Stoff [11] investigated into the simulation of a gas turbine diffuser with a spoke wheel to produce wakes similar to that downstream of turbine blades. In this case with a cylinder flow the SAS model has a clear positive effect on the prediction of the separation zone in the diffuser even when the spoke wheel is non-rotating.

In the present study the URANS approach is applied firstly with the SST model and finally with the SAS model to the simulation of a gas turbine diffuser flow. The studies are carried out for a highly loaded sensitive diffuser test case. The diffuser has a conical annular section with a center body representing the hub in a real turbine, which is supported by five struts and which is followed by a cylindrical and conical section. In the annular section a separation zone at the casing and downstream of the abrupt hub end a recirculation zone occurs. The intention of the SAS model is generally to decrease the turbulent viscosity in detached eddy regions to provide less damping of unsteady fluctuations for better convective transport of vortices into these detached regions. This might reduce the extent of the separation in the current test case, which is overpredicted by the common RANS approach with the SST model. Upstream of the current diffuser there is no rotating component, so the unsteadiness is self-induced by the separation at the casing, the strut wakes and mainly by the flow downstream of the abrupt hub end, which is similar to the flow over a bluff body for which the SAS model was intended for [18].

3 Test Case

The results obtained from the CFD study are validated with experimental data from a test rig at ITSM, which is a scaled model (with a scale of about 1/10) of a typical gas turbine diffuser with annular and adjacent conical diffuser parts. The last turbine stage is currently not modeled but wire screens and vanes can generate turbine specific inlet flow profiles. This test rig (Fig. 1) is described in detail in Volkmer et al. [19] and in Hirschmann et al. [20]. Ambient air is drawn into the rig by a downstream compressor, whereby the mass flow rate is measured with a bell-mouth nozzle. Along the diffuser traverse pneumatic probe measurements are carried out at eight planes and in plane S0 to define the diffuser inlet conditions. The center body, which represents the hub in a real turbine, is supported by five front and rear struts. At the casing and hub wall the static wall pressure is measured.

The geometry configuration for the test case in the current work consists of an annular part with a typical conical casing followed by a cylindrical casing contour, which passes again into a conical part (Fig. 1). The inlet flow profile at plane S0 of the main flow is a circumferentially and radially uniform total pressure profile with $Ma_{Inlet} \approx 0.6$ with no swirl. The mean turbulence intensity of the main flow is around 1 %. With these flow conditions, the probe measurements in different planes indicate that the flow detaches at the casing wall in the annular section and that

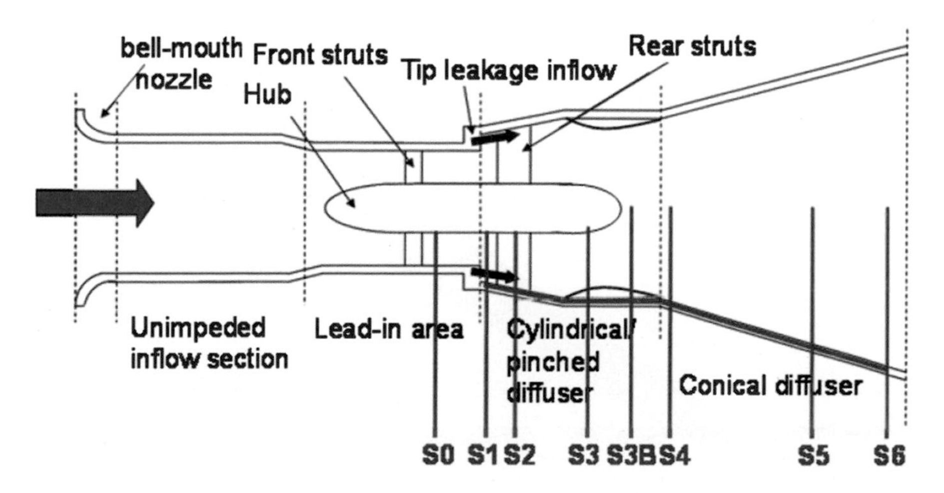

Fig. 1 Cross-section plot of the axial diffuser test rig

the flow is reattached downstream of the hub in plane S3. Downstream of the hub end there is a recirculation zone which closes between S3 and S5. The performance of the diffuser is quantified by three coefficients computed with the area-averaged static pressure $\bar{p}$ and mass-averaged total pressure $\bar{p}_{\text{tot}}$ of traverse probe data. The reference plane is the inlet plane S0. In cases with tip jet flow, its additional amount of kinetic energy is not included. The coefficients are the pressure recovery

$$C_p = \frac{\bar{p} - \bar{p}_{\text{in}}}{\bar{p}_{\text{tot,in}} - \bar{p}_{\text{in}}} \tag{1}$$

the total pressure loss coefficient

$$\zeta = \frac{\bar{p}_{\text{tot,in}} - \bar{p}_{\text{tot}}}{\bar{p}_{\text{tot,in}} - \bar{p}_{\text{in}}} \tag{2}$$

and the kinetic energy loss coefficient (leaving loss)

$$\xi = \frac{\bar{p}_{\text{tot}} - \bar{p}}{\bar{p}_{\text{tot,in}} - \bar{p}_{\text{in}}} \tag{3}$$

which together sum up to 1.

4 CFD Model

The computational domain for the CFD simulations is generally modeled assuming rotational periodicity and includes one of the five struts resulting in a 72° sector of the diffuser (Fig. 2). The inlet of the domain is set at the first measurement plane S0.

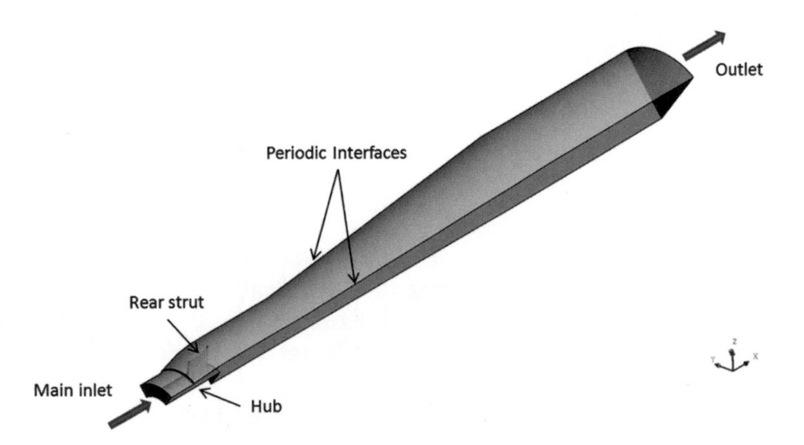

Fig. 2 Computational domain of a 72° sector of the diffuser and the boundary conditions

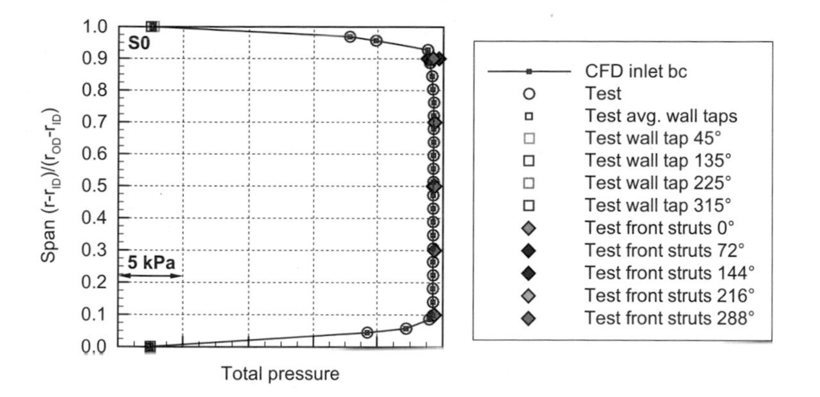

Fig. 3 Measured total pressure profiles at the inlet and the generated profile for the simulation

The short cylindrical exit part of the diffuser is extended with an artificial plenum to avoid flow reversal at the outlet. In a further step the whole diffuser is modeled as a 360° model to avoid periodic interfaces which are not reasonable as turbulence structures are not symmetric. With the RANS approach and usual two-equation models the turbulence is computed as isotropic so the simulation with a 360° model has no impact on the solution, but in case of a scale-resolving model this impact has to be investigated.

The outlet boundary condition is defined with the measured mass flow of the inlet nozzle plus the additional tip jet mass flow. The boundary conditions at the inlet are defined generally by a measured radial total pressure profile, which is extended to the measured static wall pressure towards the walls (Fig. 3). The flow direction is set axial at the inlet plane. The total temperature ($T_{tot} = 288.15$ K) and the turbulence quantities ($Tu = 0.01$ and $EVR = \mu_T/\mu = 1$) are defined as constant across the whole plane. All measured quantities are reduced to ISO-standard values. A grid sensitivity study was carried out with block-structured hexahedral meshes with about one (1M), three (3M) and five (5M) million nodes for another test case. The grid refinement was performed only in the core flow resulting in the same dimensionless distance of the first node normal to the wall for all meshes with $y^+ < 3$ ($y^+_{\text{area-avg}} < 1$). However, there is no significant impact on the prediction of the flow observed. Therefore the focus of the current work is the impact of the advection scheme, the inlet boundary conditions and finally the computational domain (72° vs. 360° model).

5 Unsteady Convergence and Time-Averaging with URANS

One issue which arises with unsteady simulations is the determination of convergence. Another one is the comparability with time-independent solutions. Firstly, a stable converged simulation state has to be achieved. For computations with rotating parts, this implies that the computed unsteady fluctuations of flow characteristics should be periodic with multiple orders of the rotational frequency. For the current study with no forced unsteadiness but with self-induced unsteadiness from separation zones or wakes, this method is not applicable. The time-dependent behavior is not periodic, but more like a saw-tooth profile, see Kaltenbach [3].

In Fig. 4 the time-dependent profile of the axial velocity at a point in plane S3 close to the hub radius (Point S3_1 in Fig. 5) and the pressure recovery close to plane S6 computed with mesh 3M and a time step of 1×10^{-4} s is shown. The computed results with the SST model are converged to a nearly constant value over time like with a steady-state solution. The computed difference in pressure recovery over time is below 0.001 percentage points (pp). Comparing the solution computed with RANS and URANS with the SST model there is no noticeable difference in the complete flow field, with a maximum difference in pressure recovery of about 0.1 pp. The turbulence characteristics of the computed flow are identical with the steady-state and transient mode, resulting in a steady-state solution although a time-dependent method is used. This behavior can be found when the eddy viscosity is computed to be too large, damping out turbulent fluctuations (Kluß [8]).

A larger impact is obvious by applying the transient SAS model with a stronger unsteady behavior of the flow field, see Figs. 4 and 5. The separations are here

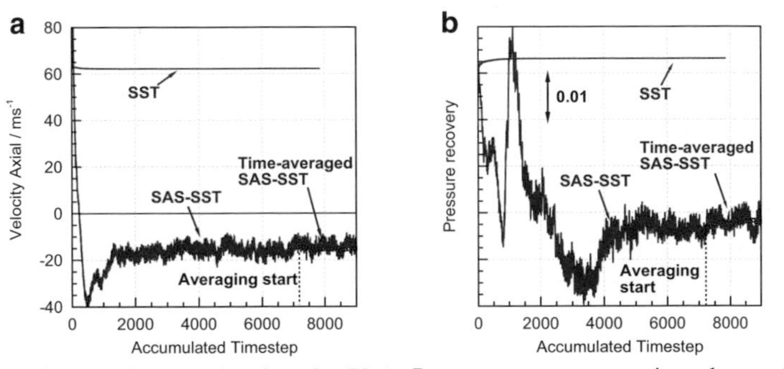

Axial velocity over time in point S3_1 Pressure recovery over time close to S6

Fig. 4 Time-dependent (**a**) local axial velocity and (**b**) pressure recovery with mesh 3M and a time step of 0.0001 s computed with the SST and the SAS model in transient mode

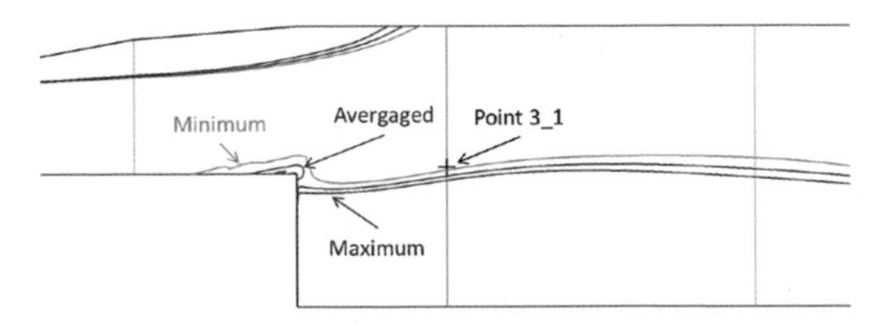

Fig. 5 Separation lines with $u = 0\,\mathrm{ms}^{-1}$ for the minimum, maximum and time-averaged separation zones computed with the SAS model in the midpath plane

unsteady which can be seen by the fluctuations in the reattachment of the flow near the casing, the onset of the separation on the hub and further downstream the closure of the recirculation. The local velocity in point S3_1 fluctuates due to this unsteady recirculation zone downstream of the hub (Fig. 4a). Furthermore, the pressure recovery close to plane S6 shows a time-dependent behavior and the variation is below 1 pp (Fig. 4b).

For evaluation and also for comparison with steady solutions, time-averaged values have to be computed. The issue is here to choose an adequate time frame for the averaging process. One method is to evaluate the time for the convective transport of unsteady wakes through the diffuser. This transport time is estimated from the mean velocity and the diffuser length to be 0.063 s. With a time step of 1×10^{-4} s this results in 630 iterations. In Fig. 4 the time-averaged values are computed starting at 7,148 iterations, so the last time-averaged value is computed with nearly 2,000 iterations. The differences are marginal and as time-averaged values are used for comparison the convergence of the unsteady simulation is clearly sufficient.

6 Comparison SST and SAS Model

The computed flow field with the SST and SAS model is compared globally by plotting the isolines of the time-averaged axial velocity equal to Zero in the midpath plane (see Fig. 6). With the SST model the steady-state and transient results in the 72° sector are similar, these are compared with the results of steady-state SST model in the full 360° model. The computed flow field are similar, so that in terms of the computational effort simulations with steady-state approach and the 72° model can be reasonably applied. In case of the SAS model this is not observed as the computed flow fields represented by the separation zone in the midpath plane differ significantly. With the 72° model the recirculation zone downstream of the hub closes downstream of plane S6, whereby with the 360° model it closes highly further upstream at plane S4.

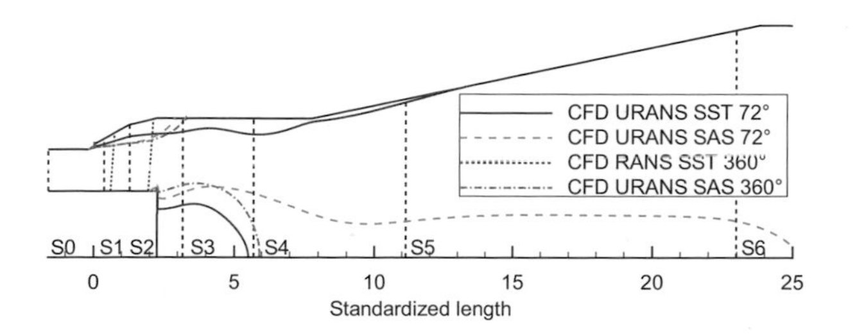

Fig. 6 Separation line with time-averaged $u = 0\,\mathrm{ms}^{-1}$ along the diffuser in the midpath plane

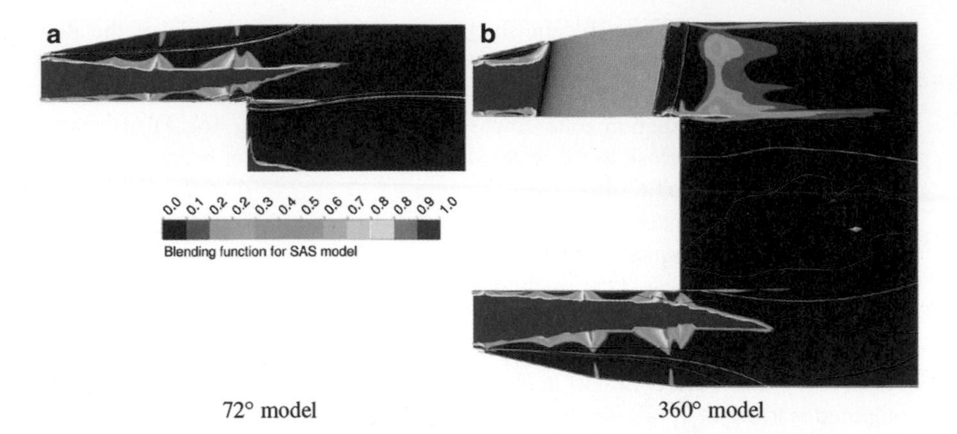

Fig. 7 Blending function for the SAS model (1 = RANS, 0 = LES) and the separation lines (avg. = *yellow*, min = *orange*, max = *red*) computed with (**a**) the 72° model and (**b**) the 360° model with the SAS model

In general, with the SAS model on the one hand the radial extent of the separation at the casing in the annular section is much larger, but on the other hand the detached flow reattaches even upstream of plane S3 compared to downstream of plane S5 with the SST model. Furthermore, the radial and axial extent of the recirculation zone downstream of the hub is much larger with the 72° model as well as with the 360° model. With the 72° model different advection schemes and inlet boundary conditions (total pressure and turbulence parameters) are applied, but all computed flow fields involve the largely computed recirculation zone downstream of the hub.

To investigate into the working mechanism of the SAS model, the regions where the RANS and SAS mode is activated can be compared by plotting the contour plots of the blending function of the SAS mode (see Fig. 7). The SAS mode is activated downstream of the onset of the separation at the casing, on the hub in the annular section and downstream of the hub. However, with the 72° model at the intersection of the hub end and the center line of the diffuser, the RANS mode is activated, whereby with the 360° model in the whole hub wake the SAS mode is activated.

The target of the SAS mode is to decreases the eddy viscosity ratio by increasing the turbulence eddy frequency (caused by an additional term in the ω-equation) to allow unsteady fluctuations (caused e.g. by a separation or bluff body) to remain. With the 72° sector and the SAS model the time-averaged EVR is already decreased by about 40 % (see Fig. 8a–c), but with the 360° model this is further decreased to 80 % lower EVR compared to the SST model. With decreasing EVR the unsteadiness of the recirculation zone is increased which can be seen by the separation lines computed with the time-averaged (blue), the minimum (orange) and maximum (red) axial velocity in Fig. 8. In Fig. 8d–f the isosurface with Invariant = Vorticity^2 − ShearStrainRate^2 = 10^5 s^{-2} visually establishes turbulent structures and obviously with the 360° model downstream of the hub the vortex shedding is much larger than with 72° model. However, with the SAS model and the 72° sector

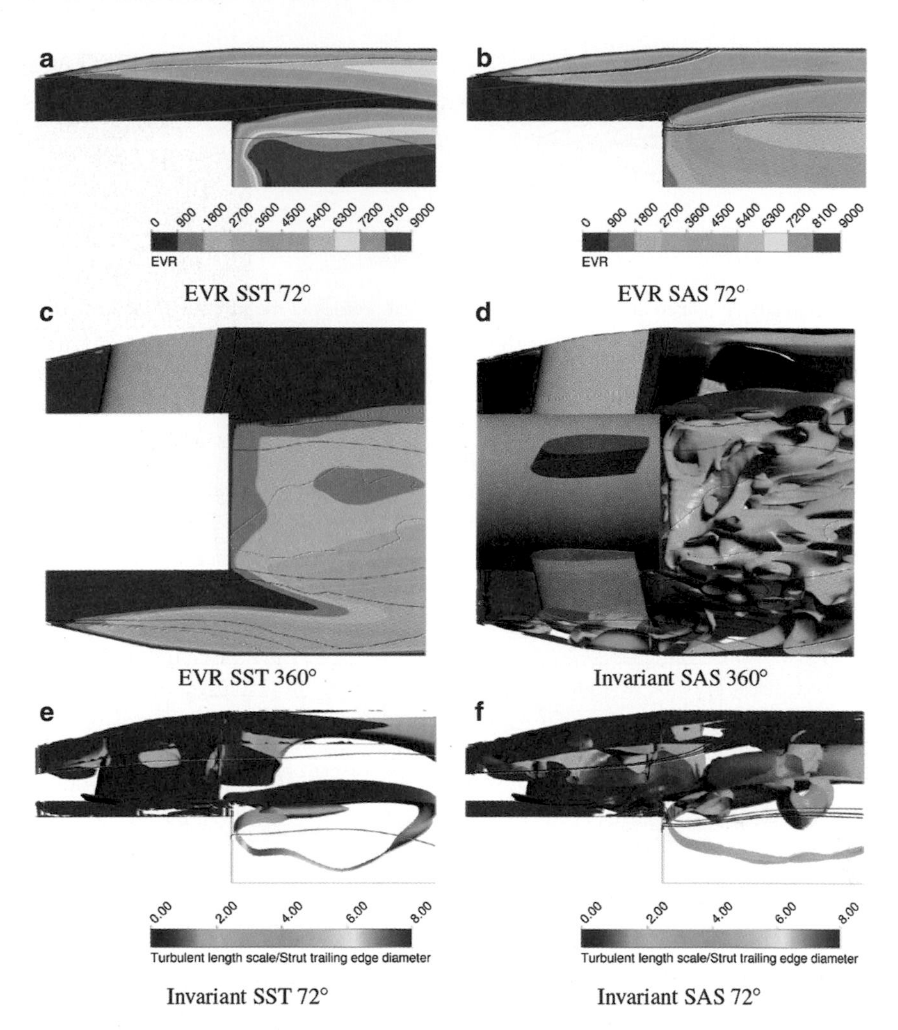

Fig. 8 (**a**)–(**c**) Time-averaged eddy viscosity ratio (EVR) and the separation lines (avg. = *blue*, min = *orange*, max = *red*) and (**d**)–(**f**) isosurfaces of the invariant = $10^5\,\mathrm{s}^{-2}$ colored by the turbulent length scale related to the strut trailing edge diameter

the turbulent structures downstream of the hub are similar to that with the SST model.

Another effect of the increased turbulence eddy frequency is the decreasing turbulence kinetic energy (see Fig. 9). With the 72° model the decreased turbulence kinetic energy has a larger effect than the unsteady fluctuating separation and this finally leads to a much larger recirculation zone downstream of the hub with a closure downstream of plane S6 (compared to S4 with the SST and 360° model).

The separation at the casing is in the annular section with the SAS model radially larger than with the SST model. Although the SAS mode is activated just

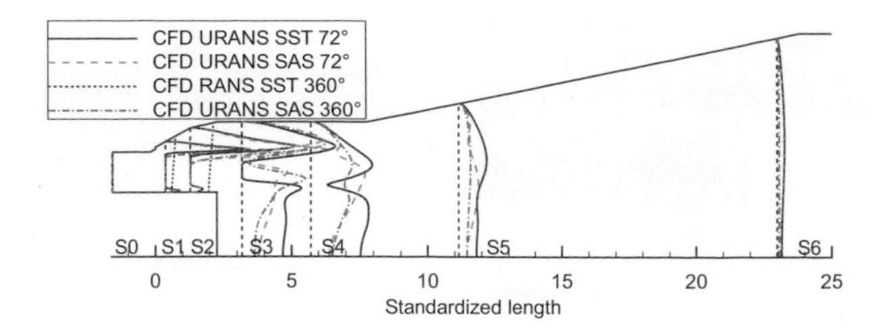

Fig. 9 Radially computed and measured time-averaged profiles of the turbulent kinetic energy at different traverses

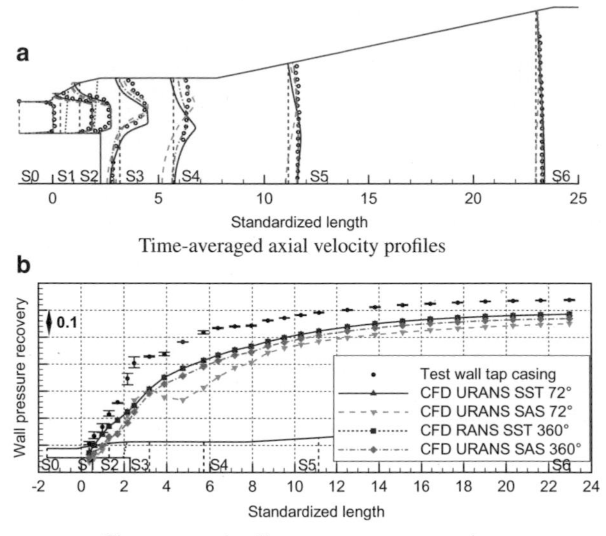

Fig. 10 (**a**) Radially computed and measured time-averaged axial velocity profiles at different traverses and (**b**) averaged measured and computed pressure recovery at casing taps

downstream of the separation onset, the unsteadiness is too low to decrease the EVR. At the onset of the separation the turbulence kinetic energy is lower with the SAS model than with the SST model and this leads to a radially larger extent of the separation. However, downstream of plane S2 the eddy viscosity ratio decreases by about 30 % and this allows the developing of unsteady fluctuations and a higher transport of high energy vortex structures (see Fig. 8d–f). This leads finally to the reattachment of the detached flow further upstream in plane S3 (compared to plane S5 with the SST model).

In Fig. 10a the radial profiles of the axial velocity are shown and compared to the measured profiles in the annular section the separation zone at the casing which is computed with the SST model is closer to the measured data than with the 72° as well as with the 360° sector and the SAS model (too large). Due to the further upstream reattachment of the casing separation with the SAS model, the profile in plane S3 computed with the SAS model matches the measured profile very well. In the further downstream planes the separation computed with the SST model leads to mismatch with measured data especially near the casing. In contrast to that, the computed profiles with the SAS model and the 360° sector are similar to measured profiles especially in plane S5 and S6. With the SAS model and the 72° sector the large extent of the recirculation zone is not detectable in the measured profiles.

According to the larger separation zone at the casing in the annular section the pressure recovery is about 4 pp lower with the SAS than with the SST model (see Fig. 10b). This difference is increased with the 72° sector to about 11 pp due to the larger blockage of the recirculation zone downstream of the hub at a standardized length of five. With the 360° sector and the SAS model the pressure recovery along the diffuser follows the same trend as with the SST model, but on a lower level (at the exit about 1.6 pp lower) due to the large separation zone in the annular section.

7 Discussion

In general, the SAS model is intended to resolve unsteady fluctuations and this is achieved by decreasing the eddy viscosity ratio. Therefore, the flow needs to generate unsteady fluctuations. In the current case the flow in the diffuser separates in the annular section at the casing and this separation does not produce enough unsteadiness in the first part of the separation and therefore the computed separation is larger with the SAS model larger than with the SST model, which is in line with the results of Davidson [17]. But towards the end of this separation the flow separation becomes unsteady and finally reattaches further upstream than with the SST model. Downstream of the abrupt hub end, there is a clear effect of the SAS model in terms of resolved unsteady fluctuations detectable. The recirculation zone is computed larger with the SAS model, but this is in good match with the measured profiles downstream of the hub. Also at the casing the velocity profiles computed with the SAS model match the measured profiles better than with the SST model due to the further upstream predicted reattachment of the separation at the casing. Nevertheless, the prediction of the separation extend at the beginning of the casing separation is too large and this results in an about 1.6 pp lower pressure recovery at the exit, although the velocity profiles downstream of the hub match the measured data very well. Furthermore, it is shown that for the application of the SAS model, a full 360° model has to be used to resolve the by nature three-dimensional anisotropic turbulent structures correctly, which drastically increases the computational effort about 40 times compared to the SST model.

Table 1 Simulation time in calendar days of simulations with the SAS model on the Hermit cluster

No. of mesh nodes in Mio.	No. of Cores	No. of nodes	Simulation time in days
3 (72°)	128	4	3
13 (360°)	192	6	12
13 (360°)	192	12	8

8 Computational Effort

The simulations were carried out on the Hermit cluster at the HLRS of the University of Stuttgart with the ANSYS CFX version 14.0. The computational effort is increased by the unsteady approach with the SAS model and with additionally 5 auxiliary (internal) loops each time step increasing the total number of iterations. Tests on the ITSM cluster demonstrate that the URANS simulation needs about 6 times more Core hours than RANS simulation for the same amount of iterations (including the 5 internal loops). The SAS model needs even about 8 times more Core hours than the RANS simulations.

To reach a stable state of convergence with the SAS model it took about up to 8,000 iterations, which is in total about 40,000 iterations including the five internal loops. The simulations are parallelized with the HPMPI method and with the 360° model (13 million nodes mesh) with 192 cores on 6 nodes (a 32 cores) the simulation took about 12 days or 55,000 core hours. This can be decreased by 30 % to 8 days using only every second core on the node due to the node architecture, that means using 192 cores on 12 nodes. With the 72° (3 million nodes mesh) model and the SAS model a simulation took about 3 days with 128 cores on 4 nodes. Clearly the need for a high performance platform is demonstrated in Table 1.

9 Conclusion

In the current work the results of time expensive transient simulations with the SAS model are investigated. However, with the SAS model the radial extent of the separation at the casing is firstly too large, but finally the flow reattaches further upstream compared to with the SST model. This prediction and the radially larger computed recirculation zone downstream of the hub match the measured velocity profiles downstream of the hub better than with the SST model. Due to the too large computed separation in the annular section the pressure recovery along the diffuser is lower with the SAS model (at the exit about 1.6 pp). Therefore, further investigations about the sensitivity of the SAS model close to the separation onset are needed. Due to the large computational effort of about 55,000 core hours high performance computing is necessary.

References

1. Casey, M., Wintergerste, T.: Best Practice Guidelines: ERCOFTAC Special Interest Group on Quality and Trust in Industrial CFD. ERCOFTAC, London (2000)
2. Menter, F.R.: Two-equation Eddy viscosity turbulence models for engineering applications. AIAA J. **32**(8), 1598–1605 (1994)
3. Kaltenbach, H.J., Fatica, M., Mittal, R., Lund, T.S., Moin, P.: Study of flow in a planar asymmetric diffuser using large eddy simulation. J. Fluid Mech. **390**, 151–185 (1999)
4. Cherry, E.M., Iaccarino, G., Elkins, C.J., Eaton, J.K.: Separated Flow in a Three-Dimensional Diffuser: Preliminary Validation. Center for Turbulence Research Annual Research Briefs, pp. 31–40 (2006)
5. Apsley, D.D., Leschziner, M.A.: Advanced turbulence modelling of separated flow in a diffuser. Flow Turbul. Combust. **63**, 81–112 (1999)
6. Chen, L., Yang, A., Dai, R., Chen, K.: Comparative study of turbulence models in separated-attached diffuser flow. In: 4th International Symposium on Fluid Machinery and Fluid Engineering, 4ISFMFE-Ch04, Beijing (2008)
7. Iaccarino, G.: Predictions of a turbulent separated flow using commercial CFD codes. J. Fluids Eng. **123**, 819–828 (2001)
8. Kluß, D.: Numerische Untersuchung des Einflusses der dreidimensionalen, instationren Turbinenabstrmung auf Nabendiffusoren. Verl. Dr. Hut, München (2010)
9. Feldcamp, G.K., Birk, A.M.: A study of modest CFD models for the design of an annular diffuser with struts for swirling flow. In: Proceedings of the ASME Turbo Expo, GT2008-50605, Berlin (2008)
10. Vassiliev, V., Irmisch, S., Florjancic, S.: CFD analysis of industrial gas turbine exhaust diffusers. In: Proceedings of the ASME Turbo Expo, 2002-GT-30597, Amsterdam (2002)
11. Kluß, D., Stoff, H.: Effect of wakes and secondary flow on re-attachment of turbine exit annular diffuser flow. J. Turbul. **131**, 041012 (2009)
12. Menter, F.R., Egorov, Y.: Turbulence Modelling of Aerodynamic Flows, International Aerospace CFD Conference, Paris (2007)
13. Menter, F.R., Egorov, Y.: Turbulence models based on the length-scale equation. In: 4th International Symposium on Turbulent Shear Flow Phenomena, Williamsburg, Paper No. TSFP4-268 (2005)
14. Menter, F.R., Kuntz, M., Bender, R.: A scale-adaptive simulation model for turbulent flow prediction. In: AIAA No. 2003-0767, Reno (2003)
15. Menter, F.R., Egorov, Y.: Revisiting the turbulent length scale equation. In: IUTAM Symposium: One Hundred Years of Boundary Layer Research, Gttingen (2004)
16. Menter, F.R., Egorov, Y.: A scale-adaptive simulation model using two-equation models. In: AIAA, No. 2005–1095, Reno (2005)
17. Davidson, L.: Evaluation of the SST-SAS Model: Channel Flow, Asymmetric Diffuser and Axi-Symmetric Hill, ECCOMAS 2006, Delft (2006)
18. ANSYS CFX: Manual ANSYS CFX-Solver Modelling Guide 14.1 (2011)
19. Volkmer, S., Kuschel, B., Hirschmann, A., Schatz, M., Casey, M., Montgomery, M.: Hub injection flow control in a turbine exhaust diffuser. In: Proceedings of ASME Turbo Expo, GT2012-69713, Copenhagen (2012)
20. Hirschmann, A., Volkmer, S., Schatz, M., Finzel, C., Casey, M., Montgomery, M.: The influence of the total pressure profile on the performance of axial gas turbine diffusers. J. Turbul. **134**, 021017 (2010)

3D Numerical Simulation of Flow with Volume Condensation in Presence of Non-condensable Gases Inside a PWR Containment

Jing Zhang and Eckart Laurien

Abstract One severe accident scenario in a Pressurized Water Reactor (PWR) is a leak in the primary circuit of a reactor resulting in hydrogen and steam injection into the containment. The steam-air-hydrogen mixture could reach the conditions for deflagration of combustion or local detonations. Because of the influence of steam condensation on the gas mixing and hydrogen stratification in the containment, the wall and volume condensation phenomena are of interest for the safety considerations. The wall condensation model is available in the CFD code ANSYS CFX. This paper presents a newly developed volume condensation model in the presence of non-condensable gases with two-phase flow for the ANSYS CFX code. The two-fluid model is applied with a continuous gas phase consisting of a steam-air-light gas mixture, and a dispersed liquid phase composed of water droplets. Both phases are modeled with separate temperatures and velocities. The motion of the droplets due to gravitational force is considered. Volume condensation is modeled as a sink of mass and source of energy at the droplet interfaces. The newly developed volume condensation model is validated with a condensation experiment TH13, which was performed in the German THAI facility within the OECD/NEA International Standard Problem (ISP-47). Finally, in order to predict the local hydrogen behavior within a real containment during a severe accident, the containment flow was simulated at the time of accident in a 'Generic Containment', which was developed based on a German PWR.

1 Introduction

1.1 Motivation

During a severe accident in PWR, e.g. Loss of Coolant Accident (LOCA), significant amounts of hydrogen can be produced by a chemical reaction between steam

J. Zhang (✉) • E. Laurien
Institute of Nuclear Technology and Energy Systems, University of Stuttgart, Pfaffenwaldring 31, 70569 Stuttgart, Germany
e-mail: jing.zhang@ike.uni-stuttgart.de; eckart.laurien@ike.uni-stuttgart.de

W.E. Nagel et al. (eds.), *High Performance Computing in Science and Engineering '14*, DOI 10.1007/978-3-319-10810-0_32

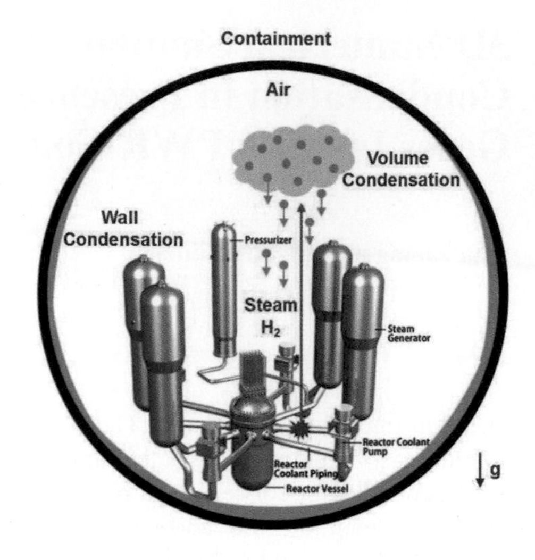

Fig. 1 Condensation inside containment

(vapor) and the zircaloy cladding ($Zr + 2H_2O \Longrightarrow ZrO_2 + 2H_2$). Through a break of surge line, steam and hydrogen can be released from the reactor coolant system in the primary loop into the containment, see Fig. 1. Condensation can mostly occur on the walls inside the containment. If the steam partial pressure within the mixture is locally higher than the steam saturation pressure, the condensation could also happen within its volume, referred to as 'volume condensation'. Because steam condensation influences the hydrogen stratification, the change of temperature and pressure, as well as the flow condition in the containment, the condensation phenomena are of interest for the safety considerations.

1.2 Status of the Literature

A main issue of the ISP-47 is the prediction of the hydrogen, steam and air distribution [1]. In order to study the flow with condensation under severe accident conditions, containment experiments with helium representing hydrogen have been performed in a German THAI facility (THAI = **T**hermal hydraulics, **H**ydrogen, **A**erosols, **I**odine) by Becker Technologies. In order to analyze the experiments, two thermal-hydraulic approaches were used:

The lumped parameter (LP) approach is often referred to as system code, e.g. COCOSYS, ASTEC, MELCOR, ECART [2, 3]. LP models have been developed, verified and used to analyze and predict transport processes within a nuclear power plant. These models are based on mass and energy budgets between control volumes. They can provide valuable information about complex flows, such as mixing, condensation and aerosol transport. However, the user influence is rather large.

Recently, the computational fluid dynamics (CFD) approach has also been used to simulate containment flows. CFD methods are based on temporally averaged mass, momentum and energy conservation equations, which appear as a set of coupled partial differential equations. Within the ISP-47 the flow inside vessels with wall and volume condensation has been modeled as a single fluid by various authors using the CFD codes CFX [4–7], Fluent [8] and GASFLOW [9]. For a two phase flow the condensation phenomena on the wall and in the volume has been simulated with one non-condensable gas of air by using ANSYS CFX [10], in which the newly developed volume condensation model was described. By using the two-fluid model, two phases, including the dispersed droplet phase, are regarded as inter-penetrating fluids and can move independently. The volume condensation model has been tested firstly in a simplified test case and then in a TH2 experiment. Their results have shown good agreement. To investigate the sensitivity of the results, the simulations were carried out with the same initial droplet volume fraction but with different droplet diameters. Because with smaller droplet diameter the calculation needs a smaller time step, for the same transient time much more computational resources are needed. The difference between the results with various droplet diameters is very small. In order to reduce the computational effort, the droplet diameter of 100 μm is enough small for the test phase.

For a steam-air-light gas mixture, which can be present in the containment during the LOCA accident, condensation in volume may have an effect on the formation of explosive mixtures. However, the volume condensation has so far not been taken into account with a two-phase flow in presence of more than one non-condensable gas.

1.3 Aim of This Study

The aim of the present work is to develop and validate a phase exchange model for volume condensation in the presence of more than one non-condensable gas for a two-phase flow using the CFD code ANSYS CFX 14.0. In order to validate the volume condensation model, a three-dimensional (3D) computational grid is used, covering one half of the TH13 vessel. In order to predict the local hydrogen behavior within a real containment during a severe accident, a 'Generic Containment' is used, which was developed based on a German PWR.

2 Computational Model

2.1 Two-Fluid Model

The volume condensation is modeled on the basis of the two-fluid model [11]. The gas phase (index G) is modeled as a continuous mixture of steam, air, and a light gas, as well as the liquid phase (index L) is treated as a dispersed phase of droplets. It is assumed that small droplets are already present as nucleation sites, i.e. heterogeneous nucleation is modeled by assuming a given droplet number density and initial droplet diameter. The volume fraction of each phase is denoted by α_k, where k $(k = L, G)$ is the phase index. Both phases may have different velocities $\overrightarrow{\overline{u}}^k$ and temperatures $\overline{T}^k$ (the overbar refers to temporal phasic averaging), thus allowing for thermal and/or mechanical non-equilibrium, e.g. wet fog. The two-fluid model describes the time-dependent mass, momentum and energy conservation for each phase in a 3D coordinate system x_m $(m = 1, 2, 3)$. The mass transport is described by the phase continuity equations:

$$\frac{\partial(\rho_k \alpha_k)}{\partial t} + \nabla(\rho_k \alpha_k \overrightarrow{\overline{u}}^k_m) = \Gamma_k, \tag{1}$$

where ρ_k is the density of each phase. The momentum balance is described by the momentum equations of each phase

$$\begin{aligned} &\frac{\partial(\rho_k \alpha_k \overline{u}^k_m)}{\partial t} + \nabla(\rho_k \alpha_k \overrightarrow{\overline{u}}^k_m \overline{u}^k_m) = \\ -&\frac{\partial(\alpha_k p)}{\partial x_m} + \nabla[(\alpha_k \overline{\overline{\tau}}^k + \overline{\overline{\tau}}^{Re,k})]_m + \overline{u}^k_m \Gamma_k + \alpha_k \rho_k g_m + M_{k,m}, \end{aligned} \tag{2}$$

in which p is the pressure, $\overline{\overline{\tau}}^k$ and $\overline{\overline{\tau}}^{Re,k}$ are the molecular stresses and the turbulent (Reynolds) stresses of each phase, g_m is the gravity acceleration and $M_{k,m}$ is the momentum exchange term. The energy equations in ANSYS CFX are expressed by enthalpy equations, using $\overline{h}^k$ for each phase

$$\frac{\partial(\rho_k \alpha_k \overline{h}^k)}{\partial t} + \nabla(\rho_k \alpha_k \overrightarrow{\overline{u}}^k \overline{h}^k) = \nabla[\alpha_k (\overline{q}^k + \overline{q}^{Re,k})] + \overline{h}^k \Gamma_k + E_k. \tag{3}$$

Here $\overline{q}^k$ and $\overline{q}^{Re,k}$ are the molecular and turbulent heat fluxes. The sum of the volume fractions α_k of both phases is 1:

$$\alpha_G + \alpha_L = 1. \tag{4}$$

Considering the mass balance, a mass that leaves the gas phase must be added to the liquid phase. The sum of mass transfer rate Γ_k is equal to 0:

$$\Gamma_G + \Gamma_L = 0. \tag{5}$$

In the real situation the droplets may have a size distribution rather than a constant size, and the number density may vary as a function of droplet sub-cooling and other nucleation parameters. In order to understand the real flow with the wet fog, the microstructure of atmospheric clouds from meteorology is shown in [12]. For example, the cloud in a small cumulus with total droplet number density of 210 cm^{-3} (solid line) has mostly droplets with diameters from 10 to 30 µm and droplet number densities between 10 and 60 cm^{-3}, as shown in Fig. 2. An implementation of such physical models into a CFD code will be possible in the near future, when experiences with a two-fluid formulation and nucleation model will be available. In the present work it is assumed that droplets have a constant diameter of 100 µm.

With constant droplet diameter the momentum exchange term results to:

$$M_{k,m} = M^D_{G,m} = -M^D_{L,m} = \alpha_L c_D \frac{3\rho_G}{4d} |\vec{u}^L - \vec{u}^G| (\vec{u}^L - \vec{u}^G). \tag{6}$$

The drag coefficient c_D is given by Schiller-Naumann correlation for a droplet:

$$c_D = \frac{24}{Re}(1 + 0.15 Re^{0.687}). \tag{7}$$

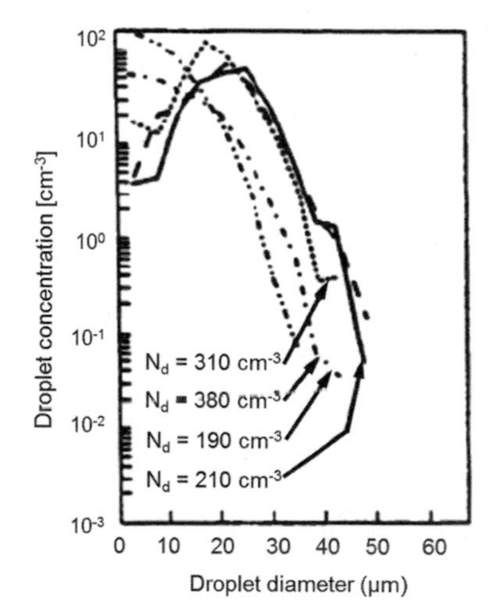

Fig. 2 Size spectra of cloud droplet in various developing small cumuli from [12]

The Reynolds number is defined as:

$$Re = \frac{\rho_G |\vec{u}^L - \vec{u}^G| d}{\mu_G}, \tag{8}$$

where ρ_G is the gas mixture density and μ_G is the dynamic viscosity of the gas mixture.

2.2 Volume Condensation Model

The volume condensation model has been developed for a two-phase flow in presence of one non-condensable gas of air in [10]. In the present work the model is extended to more than one non-condensable gas, e.g. air and a light gas. In order to understand the volume condensation process between small droplets and the surrounding gas phase, the interactions are described first between a single droplet and the steam-air-light gas mixture, as shown in Fig. 3.

The liquid droplet is assumed to be spherical with a diameter d. The liquid droplet is in thermodynamic equilibrium at its surface and has a uniform temperature $\overline{T}^L$. The gas phase at the edge of the thermal boundary layer around the droplet has a temperature $\overline{T}^G$. The partial pressure of steam directly at the interface must be equal to the saturation pressure $p_{H_2O,sat}(\overline{T}^L)$ corresponding to the droplet temperature [13]. The molar fraction at the interface $y^G_{H_2O,sat}$ is assumed to be in equilibrium and is equal to the ratio of the saturation pressure to the static pressure p

$$y^G_{H_2O,sat} = \frac{p_{H_2O,sat}(\overline{T}^L)}{p}. \tag{9}$$

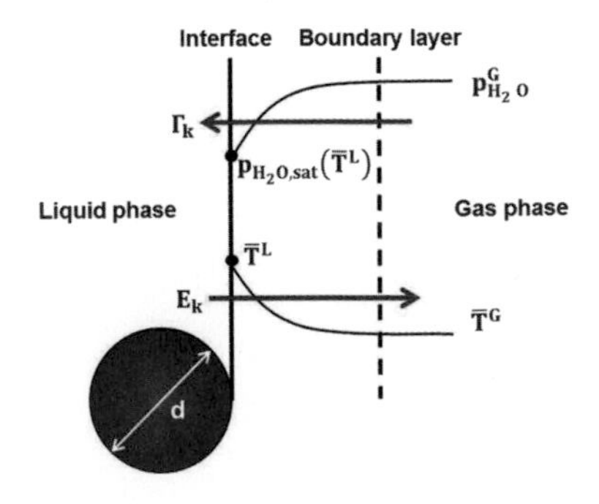

Fig. 3 Pressure and temperature profiles for a condensing droplet

The mass fraction at the interface $c^G_{H_2O,sat}$ is defined as follows:

$$c^G_{H_2O,sat} = y^G_{H_2O,sat} \frac{M_{H_2O}}{M_{G,sat}}, \tag{10}$$

where $M_{G,sat}$ is the molar mass of the saturated gas mixture with steam and non-condensable gas mixture:

$$M_{G,sat} = c^G_{H_2O,sat} M_{H_2O} + c^G_{NG,sat} M_{NG} = c^G_{H_2O,sat} M_{H_2O} + (1 - c^G_{H_2O,sat}) M_{NG}. \tag{11}$$

In the present work, the non-condensable gas mixture consists of air and a light gas (hydrogen or in the experiment helium). The molar mass of non-condensable gas mixture is given:

$$M_{NG} = c^G_{Air} M_{Air} + c^G_{LG} M_{LG} = c^G_{Air} M_{Air} + (1 - c^G_{Air}) M_{LG}. \tag{12}$$

The molar mass of water M_{H_2O} is 18 g/mol and of air M_{Air} is 29 g/mol, and of hydrogen M_{H_2} is 2 g/mol, as well as of helium M_{He} is 4 g/mol. The gradient of concentration $(c^G_{H_2O} - c^G_{H_2O,sat})$ leads to mass transport across the diffusion boundary layer. If the steam mass fraction in the gas phase is higher than under saturation conditions, the volume condensation rate is calculated from the following simple expression:

$$\Gamma_G = -\rho_G \beta A (c^G_{H_2O} - c^G_{H_2O,sat}) = -\Gamma_L. \tag{13}$$

The mass transfer coefficient β due to the flow around a sphere is given by the correlation of Ranz-Marshall [14] for the Sherwood number Sh

$$Sh = \frac{\beta d}{D_{H_2O,G}} = 2 + 0.6 Re^{\frac{1}{2}} Sc^{\frac{1}{3}}. \tag{14}$$

The diffusion coefficient of a multicomponent gas mixture $D_{H_2O,G}$ is defined with two binary diffusion coefficients $D_{H_2O,Air}$ and $D_{H_2O,LG}$, which are provided by Blanc's law in Table 1 [15]

$$D_{H_2O,G} = \left(\frac{y^G_{Air}}{D_{H_2O,Air}} + \frac{y^G_{LG}}{D_{H_2O,LG}}\right)^{-1}. \tag{15}$$

The Schmidt number is expressed as:

$$Sc = \frac{\mu_G}{\rho_G D_{H_2O,G}}. \tag{16}$$

Table 1 Diffusion coefficient [15]

Gas pair	Diffusion coefficient (cm^2/s)
$H_2O - Air$	0.292
$H_2O - He$	1.136
$H_2O - H_2$	0.927

The interface area density of all droplets is defined as follows:

$$A = \pi n d^2 = \frac{6\alpha_L}{d}. \tag{17}$$

During the volume condensation, the heat is extracted from droplets into the surrounding gas mixture. The energy source due to the release of latent heat Δh_{LG} is determined as:

$$E_G = -\Gamma_G \Delta h_{LG}, \quad E_L = 0. \tag{18}$$

The source-sink term of mass transport equations Γ_k and of energy transport equations E_k are implemented for the volume condensation model in ANSYS CFX.

3 Validation Experiment

3.1 Geometry and Boundary Conditions

To validate the volume condensation model, the TH13 experiment was chosen, which was performed by Becker Technologies. The main component of the THAI facility is a cylindrical vessel with a volume of 60 m^3, see Fig. 4. The height of the vessel is 9.2 m with a diameter of 3.16 m. An open inner cylinder with a diameter of 1.4 m and a height of 4 m is installed inside the vessel. Four equally spaced condensate trays are located in the middle of annulus and span from the inner cylinder to the vessel wall. Each tray covers 60° of the circumference. The free opening spaces enable communication between the upper and lower annulus. In the TH13 experiment cold air was initially present in the vessel and the outside wall was set to be adiabatic. The TH13 experiment consisted of four phases. During the first phase (0–2,700 s) helium was injected together with a small amount of steam through a free upward directed jet in the upper part of the annulus. During the second phase (2,700–4,700 s) the injection position was changed to the opposite side of the helium injection. Hot steam was injected upwards through a vertical nozzle into the upper annulus. During the third phase (4,700–5,700 s) hot steam was injected with a horizontal jet into the lower plenum. In the last phase (5,700–7,700 s) the facility was cooled passively without any gas injection.

Fig. 4 THAI geometry and TH13 injection positions

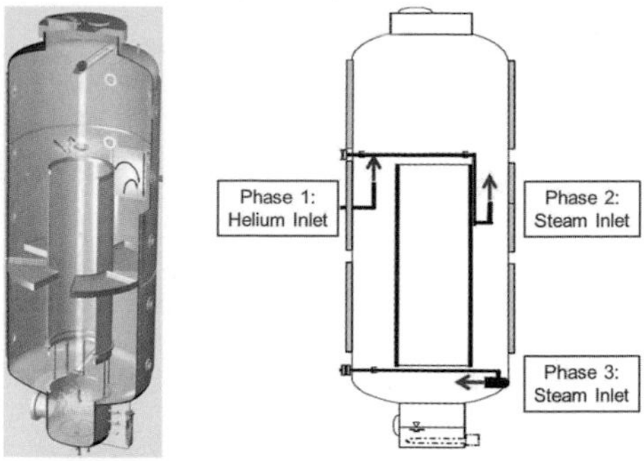

Fig. 5 3D TH13 mesh

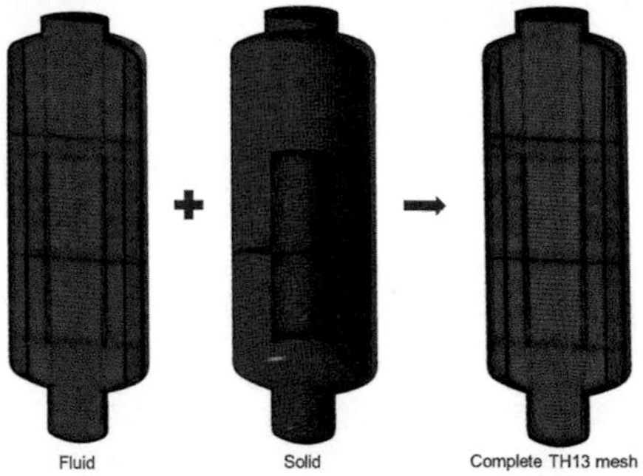

Fig. 6 Initial conditions of second phase in TH13 experiment

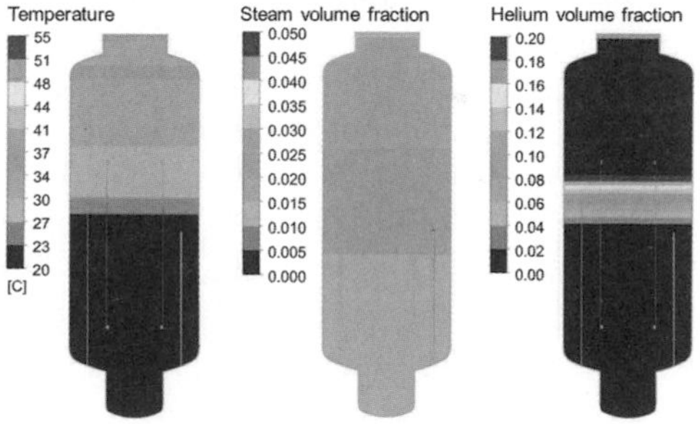

A 3D computational mesh was developed by ANSYS (Zschaeck, Private communication. ANSYS, Otterfing, 2013), covering one half of the THAI vessel. It was possible due to the assumption of symmetry relative to the vertical plane crossing the injection locations. The structured grid amounts to 1.3 million nodes, see Fig. 5.

Only the second phase of TH13 from 2,700 to 3,100 s was simulated. The values of injection temperature and mass flow rate were taken from TH13 experimental data [16]. The initial conditions (e.g. temperature, steam and helium volume fraction) were shown in Fig. 6. With lighter density steam and helium accumulate in the upper vessel.

3.2 Numerical Setup

The transient simulations of the TH13 experiment were performed using the coupled double precision solver. The Shear-Stress Transport (SST) turbulence model was used for all simulations. For the mass, momentum and energy conservation equations a second order scheme was applied. A first order scheme was chosen for the turbulence model. For the convergence root mean square (RMS) criterion was set to 5×10^{-4}.

3.3 Results

Two transient simulations of the TH13 experiment were carried out from 2,700 to 3,100 s. The first calculation was a single-fluid simulation only using the wall condensation model, which was developed by ANSYS [17]. At the domain interface on the fluid side the wall condensation model was activated, in which condensation is modeled as a sink of mass and energy, but the liquid film is not modeled. The second calculation was a two-phase simulation including the wall and volume condensation model. In this calculation droplets are present in the vessel with a constant droplet diameter of 100 μm and an initial droplet volume fraction of 10^{-5}. For the presentation of results, the positions of two monitor points (MP), which represent the conditions in the vessel, are provided in Table 2.

The right side of Fig. 7 shows the simulated helium stratification using the wall and volume condensation model in the TH13 vessel at 3,100 s. The calculated helium volume fraction is compared with the experiment at MP205, as shown in Fig. 7 left. The difference of the simulated results is very small between only using wall condensation model and with both condensation models. A significant discrepancy is observed between experimental and simulated results. The simulated results consistently show a much more intense turbulent mixing than observed in the experiment.

The right side of Fig. 8 shows the temperature distribution, which was simulated with the wall and volume condensation model, at 3,100 s. As shown in Fig. 8 left, two calculated temperatures are compared with the experiment at MP25. The hot steam injection leads to an increase of temperature. At the beginning, the fast heating of atmosphere by contact with hot steam contributes to the rapid rise in the temperature. The simulated temperature only with the wall condensation model is consistently underpredicted compared to the experiment and shows a poor

Table 2 Position of monitor points to presentation of results

Name	Vertical elevation (m)	Radius (m)	Azimuthal position (°)
MP25	7.7	0.7	300
MP205	8.7	0.65	300

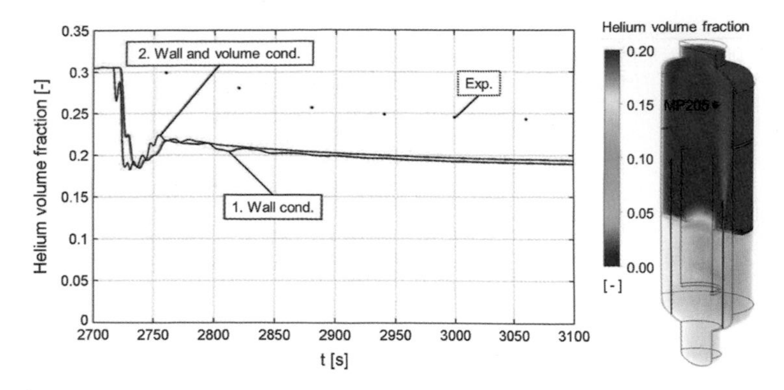

Fig. 7 *Left*: comparison of calculated helium volume fraction to TH13 experiment at MP205; *right*: helium distribution of simulation with wall and volume condensation model at 3,100 s

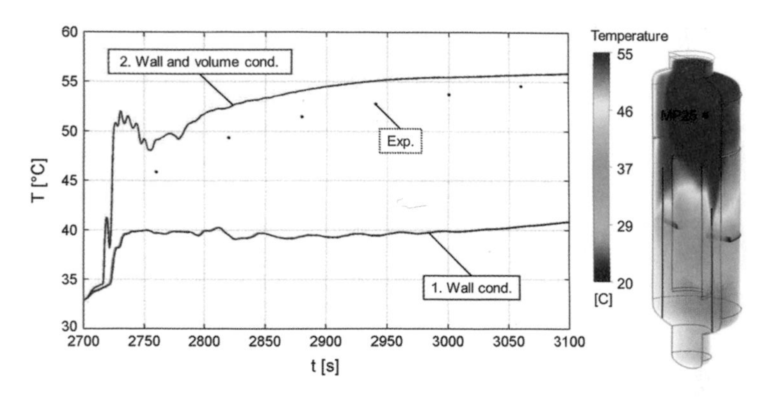

Fig. 8 *Left*: comparison of calculated temperature with TH13 experiment at MP25; *right*: temperature distribution of simulation with wall and volume condensation at 3,100 s

agreement. The calculated temperature with the wall and volume condensation model agrees well with the experiment and is higher than only with the wall condensation model. Because the volume condensation happens directly in the upper region of the vessel, the released heat leads to a rapid temperature increase.

Figure 9 shows the measured and simulated pressures in the TH13 vessel. The injection of steam leads to a continuous increase of pressure. Two simulated pressure profiles show good agreement with the measurements. Because the pressure is proportional to the temperature, the simulated pressure value with the wall and volume condensation model is also higher than only with the wall condensation model.

To understand the effect of volume condensation on the simulation with wall and volume condensation model, the integral value of volume condensation rate is compared with that of wall condensation rate, as shown in Fig. 10. After the steam injection, steam condenses mostly in the free volume of TH13 vessel. After about 2,708 s steam condensation happens more on the walls. At 3,100 s steam condenses

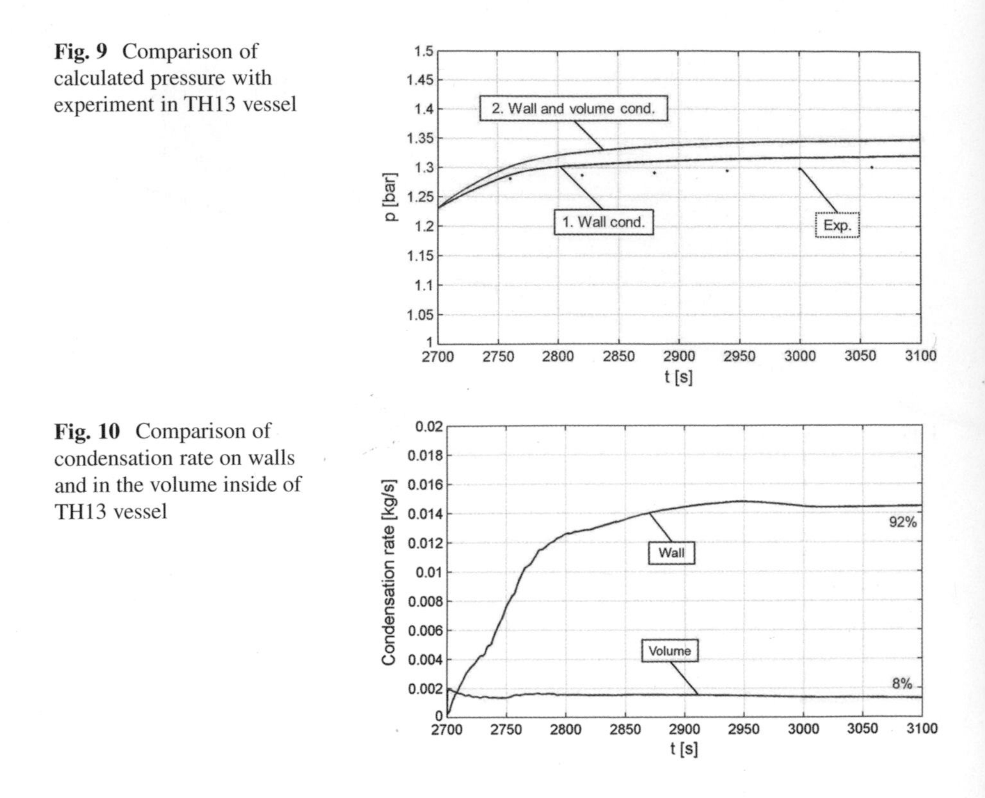

Fig. 9 Comparison of calculated pressure with experiment in TH13 vessel

Fig. 10 Comparison of condensation rate on walls and in the volume inside of TH13 vessel

92 % on the walls and 8 % in the volume. Based on the observation, the volume condensation has an influence on the complete condensation rate in this test case.

Figure 11 shows the droplet volume fraction distribution on the symmetry plane at different simulation times. The volume condensation happens mostly in the near steam jet. During the volume condensation the droplet volume fraction is increased in the vessel. The droplets flow at first with the gas stream to the dome, and then fall down into the sump due to the effect of gravity.

3.4 *Computational Performance*

All simulations were performed on the CRAY XE6 (Hermit) Cluster at the High Performance Computing Center Stuttgart (HLRS). This cluster provides 3,552 compute nodes. Each node has two sockets, each of which contains 16 cores, resulting in 32 cores per node. Data between the partitions is exchanged via Message Passing Interface (MPI). To estimate the efficiency on the CRAY XE6 (Hermit)

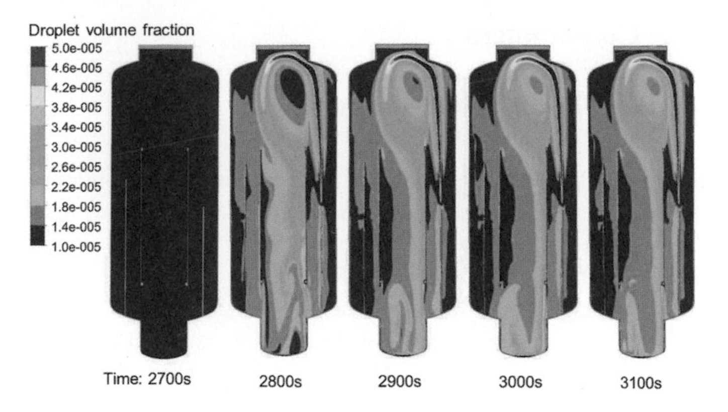

Fig. 11 Droplet volume fraction distribution at different simulation times

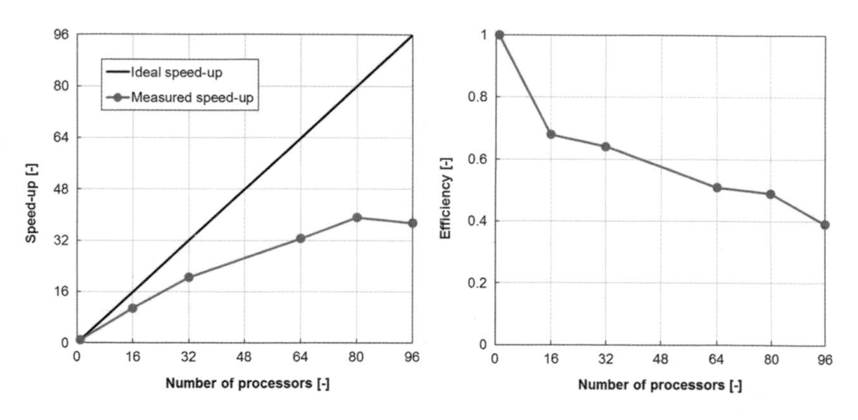

Fig. 12 Speed-up (*left*) and efficiency (*right*) with increased number of processors with TH13 experiment

Cluster, the two-phase simulations of TH13 experiment were carried out with the wall and volume condensation model for up to 96 processors. In order to analyze the performance of the parallel program ANSYS CFX, Fig. 12 shows the speed-up and efficiency of simulations with increased number of processors. The reference for the speed-up is the computation time on a single processor. If the number of processors is more than one, the measured speed-up could not reach the ideal speed-up. The efficiency for 16 processors is approximately 70 %. Parallelization on more processors makes more communications, which has further influence on the efficiency.

4 Generic Containment

4.1 Geometry and Boundary Conditions

In order to predict the local hydrogen behavior within a real containment during a severe accident, a PWR containment similar to the German design was used. A 3D geometry and its mesh were built by ANSYS (Zschaeck, Private communication. ANSYS, Otterfing, 2013) and GRS (Stewering, Private communication. Gesellschaft für Anlagen- und Reaktorsicherheit (GRS), Cologne, 2013) included main rooms, free volumes and main metallic structures, see Fig. 13. The reactor building contains a spherical steel shell, which has a volume of about 70,000 m^3. The main room was generated with hexahedral meshes. And the free volume was generated with the meshes of prisms and tetrahedrons. All computational meshes have about 10 million nodes.

A break of surge line is ca. 60 cm^2 and located at the lower part of steam generator tower as a horizontal jet. The accident started from atmospheric condition in the air filled containment with a uniform temperature of 30 °C and pressure of 1 atm. The steel wall temperature was set to 25 °C as the initial condition. In the containment the burst disks and doors will open in the case of overpressure more than 0.04 bars. Under the accident condition with a total problem time of 4,110 s, the hydrogen and steam release profiles were selected from [18]. As shown in Fig. 14, the steam injection starts at 448 s. The release profile of steam has a strong peak at 1,380 s. Hydrogen release starts at 2,040 s and occurs with much lower mass flow rate than steam. Hydrogen enters during a strong release peak at 3,470 s along with the steam injection. Steam and hydrogen are released with the same temperature, which increases strongly after injection, reaching peak at 2,780 s.

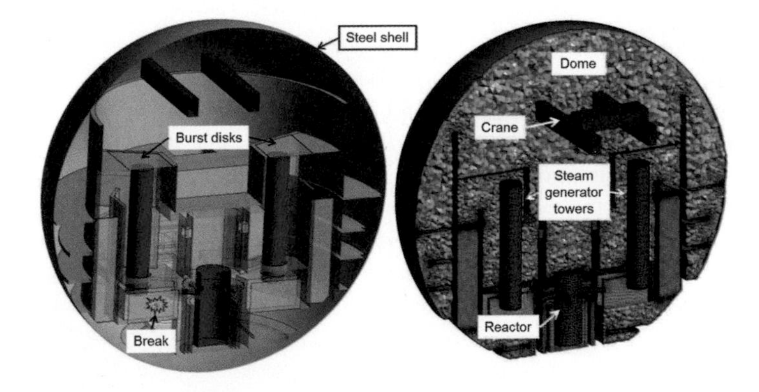

Fig. 13 Containment geometry and its mesh

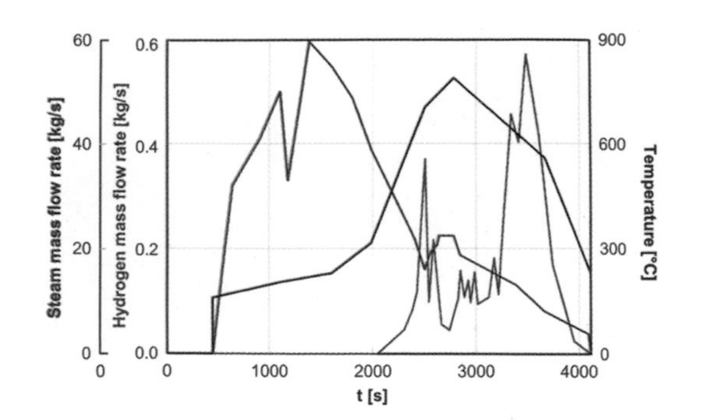

Fig. 14 Release profiles of steam and hydrogen

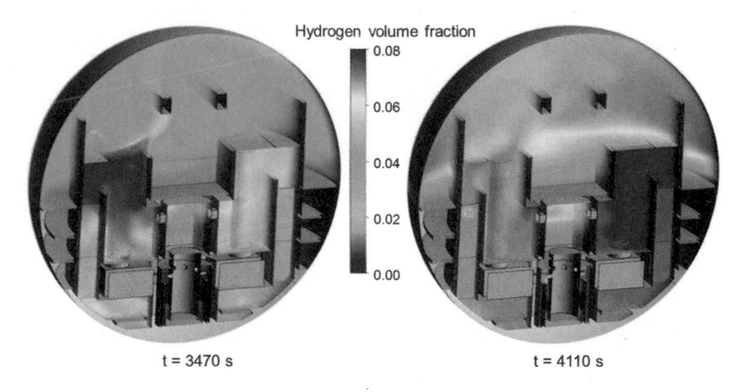

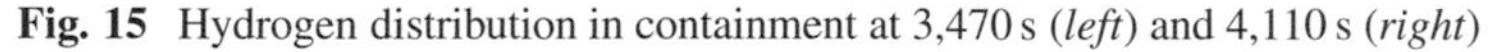

Fig. 15 Hydrogen distribution in containment at 3,470 s (*left*) and 4,110 s (*right*)

4.2 Numerical Setup

A transient simulation was performed using the coupled double precision solver. The SST turbulence model was applied for the calculation. A first order scheme was used for turbulence model equations. And a second order scheme was applied for other transport equations. The RMS criterion was set to 5×10^{-4}.

4.3 Results

In order to reduce the computational effort, a single-fluid simulation was carried out only by using the wall condensation model from 448 to 4,110 s after the time of accident. The left side of Fig. 15 shows the way of containment flow and the hydrogen distribution in the containment at 3,470 s during the strong hydrogen injection. The horizontal flow released as a jet through a break into the lower part of steam generator tower. It travels upward as a plume driven by buoyancy. Only one burst disk, which is located in the steam generator tower on the injection

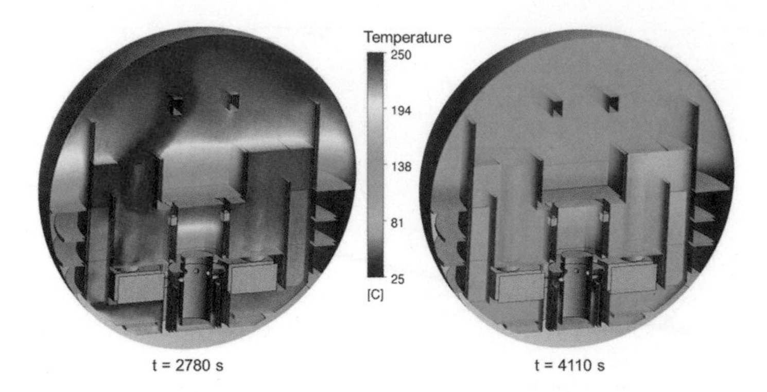

Fig. 16 Temperature distribution in containment at 2,780 s (*left*) and 4,110 s (*right*)

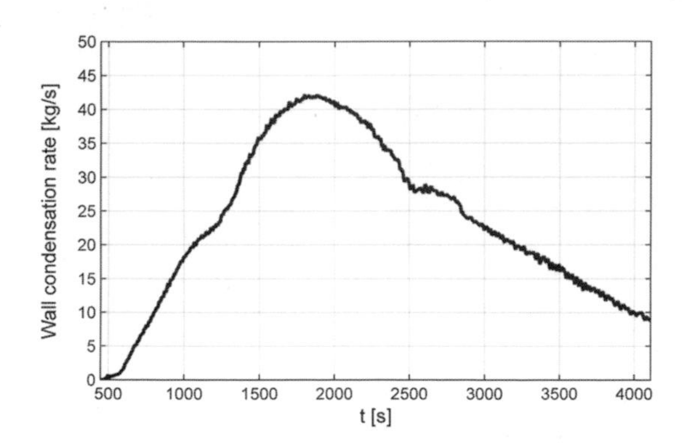

Fig. 17 Integral value of the wall condensation rate in the containment

side, is opened in the case of overpressure. Through the ruptured burst disk the flow moved towards the dome. After 3,470 s the mass flow rate of hydrogen decreased strongly. As shown in Fig. 15 right, at the end of the simulation (i.e. 4,110 s), steam condensation on the cold walls locally increases the gas density and starts a secondary downward-directed flow, and brings the hydrogen from the stratified region down into the lower part of containment. Because of the light gas of hydrogen, the hydrogen accumulates in the upper part of different rooms.

The released temperature profile has a strong peak at 2,780 s. The left hand side of Fig. 16 shows the temperature distribution in the containment at that time. The released hot gas flow of steam and hydrogen form a buoyant jet rising through the ruptured burst disk from one steam generator tower into the dome. The hot gas flow leads to a temperature increase inside the containment. After 2,780 s the temperature of injected gas decreases consistently. The right side of Fig. 16 shows a strong effect on cold walls and temperature decrease at the end of the calculation (i.e. 4,110 s).

To analyze the wall condensation effect in the containment, the integral value of the wall condensation rate in the containment was evaluated. As shown in Fig. 17, in the first about 1,800 s the integral value rises consistently, when steam is released

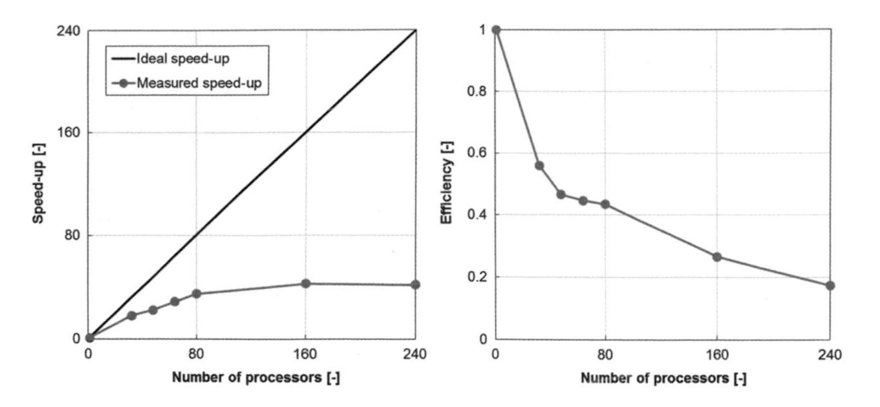

Fig. 18 Speed-up (*left*) and efficiency (*right*) with increased number of processors with 'Generic Containment'

with a large amount of mass flow rate into the containment. After that the wall condensation rate decreases gradually with the less mass flow rate of steam.

4.4 Computational Performance

Because the computational resources for ANSYS CFX are limited at the HLRS, the scalability of the ANSYS CFX code has been analyzed for up to 240 processors on the CRAY XE6 (Hermit) Cluster. The results of Fig. 18 show that the speed-up of the simulations is observable away from ideal speed-up, and the efficiency is rather poor. Using 32 processors already leads to only 56 % efficiency, which decreases faster using more than 80 processors. Based on the results of Fig. 18, the number of processors is chosen 80 for the simulation of the 'Generic Containment'.

The ratio of actual CPU time to wall clock time is evaluated. The wall clock time on the cluster is limited to 24 h. For example, a run with a total wall clock time of 22.5 h had a total CPU time of 15.7 h, so that the ratio for this run is about 0.7. It means that only 70 % of the total time is used for the actual simulation. And the rest 30 % of total wall clock time is dedicated to the communication and output.

5 Conclusion

The physical formulation of newly developed volume condensation model was presented, which simulates the dispersed droplets and a continuous gas mixture in the presence of non-condensable gases based on the two-fluid model. The thermal and mechanical non-equilibrium were considered for the temperatures and velocities between both phases. The sink term of mass and source term of energy were modeled for the volume condensation. The SST turbulence model was used for the simulations.

Then, the volume condensation model was tested in a TH13 experiment with a constant droplet diameter of 100 μm and an initial droplet volume fraction of 10^{-5}. The two-phase simulation results using the wall and volume condensation model were compared with the single-fluid simulations solely using wall condensation model, as well as with the experiment from 2,700 to 3,100 s. The transient temperature, helium volume fraction and pressure profiles with the usage of wall and volume condensation model have shown good agreement with the experiment. The droplet motion in the vessel was simulated with the volume condensation model.

Finally, a transient prediction was performed about the containment flow in a German PWR containment under the hypothetical, severe LOCA accident condition. To reduce the computational effort, a single-fluid simulation was carried out only using the wall condensation model from 448 to 4,110 s. The simulation has evaluated the time-dependent hydrogen and temperature distributions, as well as the wall condensation rate inside the containment.

Acknowledgements This work was supported by the German Federal Ministry of Economic Affairs and Energy (BMWi) on the basis of a decision by the German Bundestag, project number 1501414. The simulations were performed on the national supercomputer CRAY XE6 (Hermit) Cluster at the High Performance Computing Center Stuttgart (HLRS) under the grant number TurboCon/12843.

References

1. Allelein, H.J., et al.: International Standard Problem ISP-47 on Containment Thermal Hydraulics. NEA/CSNI, Paris (2007)
2. Allelein, H.-J., Arndt, S., Klein-Heßling, W., Schwarz, S., Spengler, C., Weber, G.: COCOSYS: status of development and validation of the German containment code system. Nucl. Eng. Des. **238**, 872–889 (2008)
3. Kelm, S., et al.: Generic Containment – A Detailed Comparison of Containment Simulations Performed on Plant Scale. International Topical Meeting on Nuclear Reactor Thermal Hydraulics (NURETH), Pisa (2013)
4. Kljenak, I., Babić, M., Mavko, B., Bajsić, I.: Modeling of containment atmosphere mixing and stratification experiment using a CFD approach. Nucl. Eng. Des. **236**, 1682–1692 (2006)
5. Babić, M., Kljenak, I., Mavko, B.: Prediction of light gas distribution in experimental containment facilities using the CFX4 code. Nucl. Eng. Des. **238**, 538–550 (2008)
6. Martín-Valdepeñas, J.M., Jiménez, M.A., Martín-Fuertes, F., Fernández, J.A.: Improvements in a CFD code for analysis of hydrogen behavior within containments. Nucl. Eng. Des. **237**, 627–647 (2007)
7. Houkema, M., Siccama, N.B., Lycklama à Nijeholt, J.A., Komen, E.M.J.: Validation of the CFX4 CFD code for containment thermal-hydraulics. Nucl. Eng. Des. **238**, 590–599 (2008)
8. Vyskocil, L., Schmid, J., Macek, J.: CFD Simulation of Air-Steam Flow with Condensation. CFD for Nuclear Reactor Safety Applications (CFD4NRS-4), Daejeon (2012)
9. Royl, P., Travis, J.R., Kim, J.: GASFLOW A Computational Fluid Dynamics Code for Gases Aerosols, and Combustion, Karlsruhe, vol. 3 (2008)

10. Zhang, J., Laurien, E.: Numerical Simulation of Flow with Volume Condensation in a Model Containment. In: Nagel, W.E., Kröner, D.H., Resch, M.M. (eds.) High Performance Computing in Science and Engineering '13: Transactions of the High Performance Computing Center, Stuttgart (HLRS) 2013, pp. 477–492. Springer, Cham, Heidelberg (2013)
11. Ishii, M., Mishima, K.: Two-fluid model and hydrodynamic constitutive relations. Nucl. Eng. Des. **82**, 107–126 (1984)
12. Pruppacher, H.R., Klett, J.D.: Microphysics of Clouds and Precipitation. Kluwer Academic, New York (2004)
13. Kolev, N.I.: Multiphase Flow Dynamics, vol. 2. Springer, Berlin (2005)
14. Ranz, W.E., Marshall, W.R.: Evaporation from drops. Chem. Eng. Prog. **48**, Part I, 141–146, Part II, 173–180 (1952)
15. Poling, B.E., Prausnitz, J.M., O'Connell, J.P.: The Properties of Gases and Liquids. McGraw-Hill, New York (2001)
16. Kanzleiter, T., et al.: Final Report: Experimental Facility and Program for the Investigation of Open Questions on Fission Product Behaviour in the Containment, ThAI Phase II. Report No. 1501272 (2007)
17. Zschaeck, G., Frank, T., Burns, A.D.: CFD Modelling and Validation of Wall Condensation in the Presence of Non-condensable Gases. CFD for Nuclear Reactor Safety Applications (CFD4NRS-4), Daejeon (2012)
18. Kelm, S., Broxtermann, P., Krajewski, S., Allelein, H.-J.: Proposal for the Generic Containment Code-to-Code Comparison – run2. Internal Research Report, Jülich (2012)

Discontinuous Galerkin for High Performance Computational Fluid Dynamics

Muhammed Atak, Andrea Beck, Thomas Bolemann, David Flad, Hannes Frank, Florian Hindenlang, and Claus-Dieter Munz

Abstract Within the scope of this report, we discuss high-order CFD simulations on the HLRS CRAY XE6 cluster. Our discontinuous Galerkin based simulation framework is tailored for high performance computing aiming at solving large scale problems. We demonstrate and analyze the HPC capabilities of the framework for an unstructured grid and present Direct Numerical Simulations and Large Eddy Simulations of compressible unsteady and turbulent flows. The investigated cases include aeroacoustic sound generation on an airfoil, high-speed compressible turbulent boundary layer flow and decaying turbulence for LES modelling development. The simulations were performed on up to 8,000 cores on the CRAY XE6 cluster exploiting the high parallel efficiency of the code.

1 Introduction

Our research project concentrates on the development of high order discretization schemes for a wide range of continuum mechanic problems with a special emphasis on fluid dynamics. One of the major interests in the group is the simulation of unsteady compressible turbulence in the context of Large Eddy Simulation (LES) and Direct Numerical Simulation (DNS). Due to the occurrence of multiple spatial and temporal scales in such problems and the resulting high demand in resolution for both, space and time, a high performance computing framework is mandatory. Therein, the main research focus lies on the class of discontinuous Galerkin (DG) schemes, which are known to have excellent scalability properties.

DG schemes may be considered as a combination of finite volume (FV) and finite element (FE) schemes. While the approximate solution is a continuous polynomial in every grid cell, discontinuities at the grid cell interfaces are allowed which enables the resolution of strong gradients. The jumps on the cell interfaces are resolved by

M. Atak (✉) • A. Beck • T. Bolemann • D. Flad • H. Frank • F. Hindenlang • C.-D. Munz
Institute of Aerodynamics and Gasdynamics, Universität Stuttgart, Pfaffenwaldring 21, 70569 Stuttgart, Germany
e-mail: atak@iag.uni-stuttgart.de; beck@iag.uni-stuttgart.de; bolemann@iag.uni-stuttgart.de; flad@iag.uni-stuttgart.de; frank@iag.uni-stuttgart.de; hindenlang@iag.uni-stuttgart.de; munz@iag.uni-stuttgart.de

W.E. Nagel et al. (eds.), *High Performance Computing in Science and Engineering '14*, DOI 10.1007/978-3-319-10810-0_33

Riemann solver techniques, already well-known from the finite volume community. Due to their interior grid cell resolution with high order polynomials, the DG schemes can use coarser grids. The main advantage of DG schemes compared to other high order schemes (finite differences, reconstructed FV) is that the high order accuracy is preserved even on distorted and irregular grids.

A very efficient variant of a discontinuous Galerkin formulation is the discontinuous Galerkin spectral element method (DG-SEM). This special variation of the DG method is based on a nodal tensor-product basis with collocated integration and interpolation points on hexahedral elements, allowing for very efficient dimension-by-dimension element-wise operations. Over the last decade, several DG codes were implemented in the research group. Our newest code iteration is called FLEXI and is based on DG-SEM. The implementation aims at high performance computing and is currently our main tool to simulate three dimensional partial differential equations (PDE) based problems.

2 Description of Methods and Algorithms

The research code framework of our group has been extended conceptually towards physical and engineering applications. This direction implies a slightly different emphasis as opposed to pure numerics research. A major premise for analysis of relevant turbulent flows is the capability of treating complex geometries and large amounts of data. Developments therefore concentrated on the existing DG-SEM code FLEXI. The code has demonstrated superior performance, accuracy, and efficiency in large scale fluid dynamics simulations on the CRAY XE6 using an unstructured data structure with a full MPI parallelization. DG-SEM is formulated on hexahedral elements where each element is coupled only to direct neighbors. Although the restriction to hexahedra may be too restrictive for some scenarios encountered in applied CFD, it allows high accuracy and efficiency and enables the accessibility to a wide range of geometries of mid-level complexity not treatable using fully structured approaches.

Besides preprocessing and the simulation itself, the analysis of turbulent flows requires increased attention on postprocessing. For complex 3D flows, typically time-averaged and statistical quantities as well as temporal or spatial spectra are employed to evaluate the flow field characteristics. A large amount of data is produced in large scale simulations and the postprocessing step itself can become very costly and time-consuming in the HPC context. In our framework, we employ a separate postprocessing unit operating in parallel and using the same data structure as the CFD code itself. Figure 1 shows the workflow of the whole DG framework comprising of the main building blocks preprocessing the mesh, the simulation code itself and postprocessing. The two latter can be executed in parallel using respective HDF5 libraries for I/O.

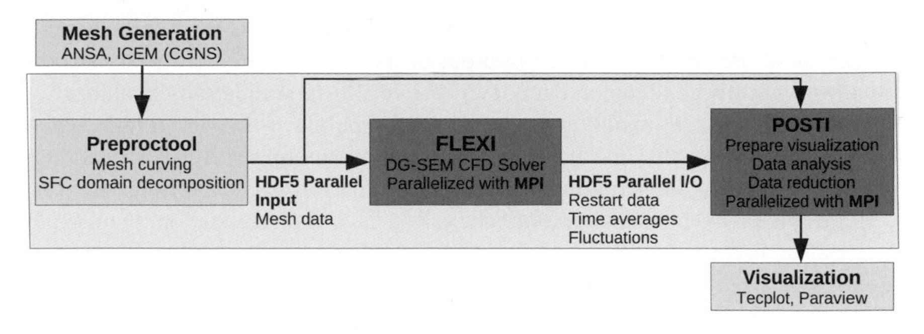

Fig. 1 Framework and workflow for a DG-SEM simulation

2.1 Pre- and Postprocessing

The meshes are generated employing existing grid generators. However, when using high order methods, wall boundary conditions need a high order representation of the wall normal. Bassi and Rebay [3] were the first to show that in the case of curved boundaries a high order DG discretization with straight-sided 2D elements yields low order accurate and even physically wrong results. To overcome this issue, they propose to use at least elements with parabolic shaped sides on the boundary. A standalone preprocessing tool, namely the *preproctool*, is able to generate curved elements at the wall boundaries to match the curved geometry, but fully relying on existing linear mesh generators. Several strategies for the element curving are at hand, using additional data provided by the grid generators, see [9] for a detailed description. In addition, the preprocessor sorts all elements along a unique space-filling curve. This leads to a contiguous one-dimensional array of elements, which is stored together with connectivity information to a parallel readable HDF5 file with data blocks for each element. This allows a fast domain decomposition during the parallel read-in of the mesh at the start of the simulation by simply splitting the one-dimensional element list in equally sized pieces for each core. The cores, in turn, read their corresponding element data in parallel. The Hilbert space-filling curve is employed for the element sorting since it has superior clustering properties [12]. The main advantage of this approach embraces that proximate three-dimensional points are also proximate on the one-dimensional curve, which is an important property for domain decomposition.

Another building block is the postprocessing. As many – from a programming point of view – very different tasks have to be addressed, the postprocessing unit is designed as a collection of relatively small tools for distinct functions. The tools rely on the same data structure as the CFD code itself and make use of the same domain decomposition technique already set up by the preprocessing unit. We created a library of common subroutines all of the tools are built upon which cover basic operations such as input and output of geometry and solution data, cell-local interpolation and integration. The concept of a common library leads to a very versatile postprocessing unit for HPC. The visualization tools perform a parallel calculation of derived quantities and output of the 3D solution and surface data to visualization software (Tecplot, Paraview). For very large simulations, visualization may be confined to a bounding box to save computing time and storage space. Furthermore for suitable meshes (e.g. with one or more structured directions), spatial averaging along coordinate axis can be performed, exploiting the very accurate quadrature-based integration inherent in the DG discretization. In unsteady or turbulent flows, time-averaged flow fields and temporal correlations such as Reynolds-stresses can be computed from the solution data. During simulation runtime, the desired quantities are accurately averaged in time. This data is written during runtime in a predefined interval, generating a collection of time-averaged solutions over short simulation time windows. In a postprocessing step, these may be merged to time-averaged solutions of arbitrary larger intervals, allowing the user to check statistical convergence by comparing time-averaged solutions of different intervals. Second order statistics (e.g. Reynolds-stresses) are treated in the same way.

2.2 *Parallelization of DG-SEM Code FLEXI*

Besides the promising fundamental efficiency of the DG-SEM scheme, its main advantages are based on its HPC capability. The DG algorithm with explicit time-stepping is inherently parallel, since all elements communicate only with their direct neighbors via solution and flux exchange. Independent of the local polynomial degree, only exchange of surface data between direct neighbors is necessary. Note that the DG operator can be split into the two building blocks, namely the volume integral – solely depending on element local DOF – and the surface integral, where neighbor information is needed. This fact helps to hide the communication latency by exploiting local element operations and further reduces the negative influence of data transfer on efficiency. It is therefore possible to send surface data while simultaneously performing volume data operations. This is realized via non-blocking send and receive commands provided by the MPI standard library [11]. To express it in a more simplistic way, the receive buffer is like a mailbox, and the receive checks the mail and would only wait, if the communication has not yet finished. Hence, the latency hiding works if the network communication is faster than the buffer operations.

The solution is represented at the interpolation nodes of the volume and the data structure separates volume and surface data. All sides except the boundaries form a pair, where each element side is either master or slave. The list of sides is sorted by the side type: The first group are boundary sides, the second includes the inner sides and the third group are sides which communicate to neighbor domains. Those groups are abbreviated as BC-sides, Inner-sides and MPI-sides. The numerical flux between element sides is unique and is computed once. This is also realized for MPI-sides, since the solution is first sent from the slave to the master side then

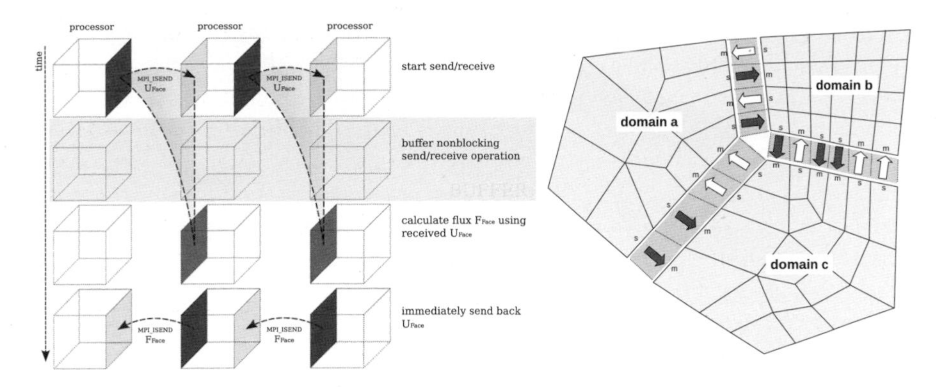

Fig. 2 Communication pattern for inter-core flux computation with non-blocking communication (*left*) and load balancing of the inter-core communication data, by assignment of master (m) and slave (s) sides (*right*)

the flux is computed on the master side and eventually sent back to the slave side, as shown in Fig. 2. Hence, except for communication, no additional operations are introduced, which is an important property of an efficiently scalable algorithm.

Surface operations at MPI-sides such as solution extrapolation and flux computation must be done before and after the communication. The domain decomposition produces domains of equal size, with an equal number of elements per core in the optimal case. However, especially on unstructured meshes, the number of elements and the number of MPI-sides per domain varies. A bad distribution of master sides would cause a load imbalance in the surface operations and the communication. Therefore, as sketched in Fig. 2, each subgroup of MPI-sides belonging to a peer-to-peer communication between two domains is splitted, so that half of the subgroup sides are master and the other slave, and a balancing of surface operations and communication is achieved. Even though not considered up to now, one could relax the location of the split from 0.5 to a factor between 0 and 1, to account for other load imbalances, such as an uneven number of elements per domain.

2.3 Parallel Performance Analysis

In this section, we investigate the scaling of the code for unstructured meshes using the example of the flow past a sphere and compare to scaling results based on cartesian meshes. The cartesian meshes are perfectly load balanced, whereas load imbalances occur for the unstructured mesh.

The unstructured mesh shown in Fig. 3 consists of 21,128 hexahedra and the elements at the sphere surface are curved using the exact point-normal approach. The domain extends $25D$ downstream and $4.5D$ upstream and circumferentially, where D is the diameter of the sphere.

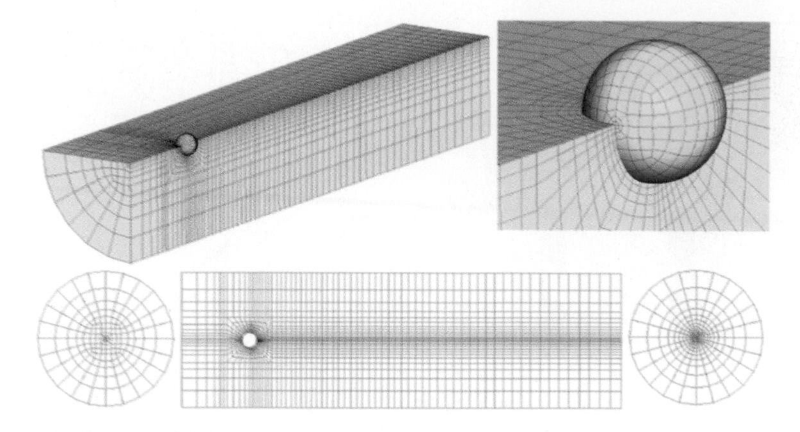

Fig. 3 Unstructured mesh of the sphere, 3D views, front view (left), slice and back view

Table 1 Statistics of the domain decomposition for 128, 1,024 and 4,096 domains, with the average, root mean square, minimum and maximum of all domains

	Elems. $n_{e/c}$	Number of sides n_s	$n_{s,inner}$	$n_{s,BC}$	$n_{s,MPI}$	$\frac{n_{s,MPI}}{n_{e/c}}$	#Neighbors
128 domains							
Average	**165.06**	**646.05**	**344.32**	**20.50**	**281.23**	**1.70**	**11.75**
Min/Max	165/166	595/711	279/401	0/159	150/374		5/23
1024 domains							
Average	**20.63**	**95.08**	**28.72**	**2.56**	**63.80**	**3.09**	**11.99**
Min/Max	20/21	84/104	21/40	0/24	34/80		5/22
4096 domains							
Average	**5.16**	**27.04**	**3.91**	**0.64**	**22.49**	**4.36**	**10.37**
Min/Max	5/6	24/35	0/9	0/8	13/34		5/20

In fact, load imbalances are unavoidable, for example if the number of elements is not dividable by the number of cores, but also from the space-filling curve domain decomposition which only balances the number of elements. Furthermore, the number of MPI neighbors, inner sides and MPI sides can vary strongly. The statistic distribution for three domain decompositions is listed in Table 1. Since the number of elements does not match the number of cores, we find an imbalance of one element between the domains. For a smaller number of elements per core, the computational imbalance created by one element increases. The domain decomposition yields an average of 11 neighbor domains, whereas the maximum number of about 20 neighbors is found for all domain decompositions. The surface to volume ratio of MPI sides to the number of elements per domain increases with the number of domains. The theoretical minimum ratio is found for a cubical domain and reads $\left.\frac{n_{s,MPI}}{n_{e/c}}\right|_{cube} = \frac{6}{\sqrt[3]{n_{e/c}}}$. For domain decompositions of $(128, 1{,}024, 4{,}096)$

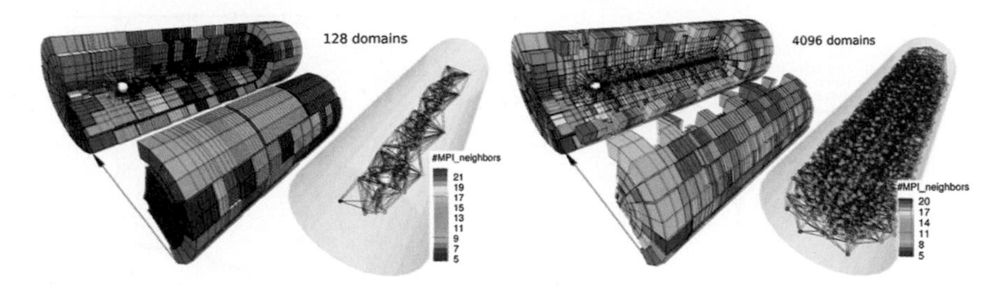

Fig. 4 Domain decomposition and communication graph for the unstructured mesh with 128 (*left*) and 4,096 (*right*) domains

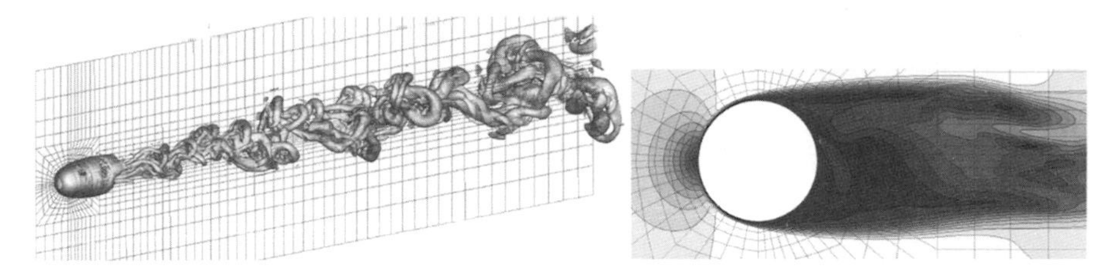

Fig. 5 Isosurfaces of $\lambda_2 = -0.001$ of the sphere flow at Re $=$ 1,000 and velocity magnitude contours (levels $[0, 1]u_\infty$) in the x-y plane

domains, the ratio equals $(1.09, 2.19, 3.47)$. The maximum ratio of 6 is reached for the limiting case of one element per domain.

We can conclude that the domain decompositions of the unstructured mesh are far from being optimal, which we illustrate in Fig. 4. Each domain is shrinked towards its barycenter by 5 % for visualization. We also plot the communication graph, showing the domain barycenter and the neighbor connections as lines.

The flow past a sphere problem (M $=$ 0.3, Re $=$ 1,000) is discretized by a polynomial degree of $N = 4$, yielding $2.64 \cdot 10^6$ degrees of freedom per conservative variable. The flow field is shown in Fig. 5. The performance analysis was conducted on the CRAY XE6 cluster for a range of 128–4,096 cores, each running for at least 10 min as a restarted simulation. We compare the strong scaling and the performance index with the perfectly load balanced cartesian meshes, for the same polynomial degree $N = 4$ in Fig. 6. The performance index (PID) is a convenient measure to compare different simulation setups: it is the computational time needed to update one degree of freedom for one time step, and is computed from the total core-hour, the total number of timesteps and degrees of freedom

$$\text{PID} = \frac{\text{Wall-clock-time \#cores}}{\text{\#DOF \#Timesteps}}.$$

It helps to compare the performance of the different runs. For example, a perfect strong scaling would result in a constant performance index.

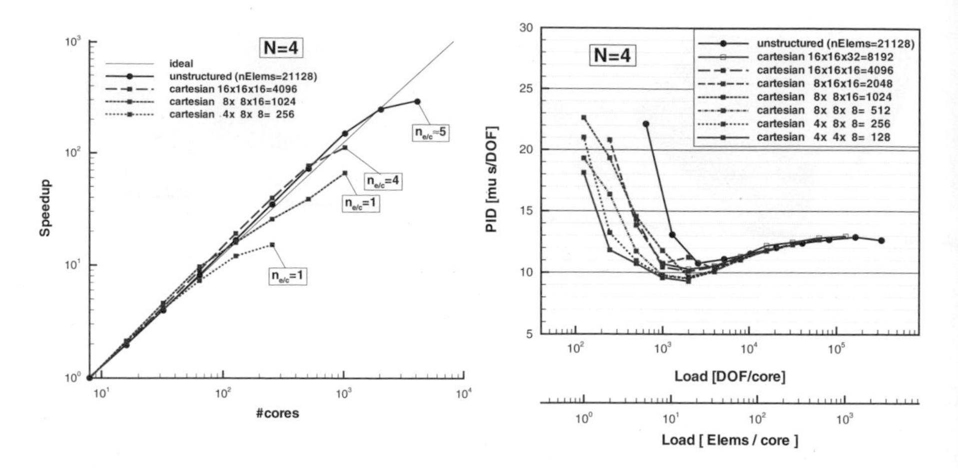

Fig. 6 Comparison of the strong scaling of the unstructured mesh for the sphere to the cartesian meshes for $N = 4$

Despite the strong imbalance caused by the unstructured mesh, an ideal strong scaling is maintained up to 10 elements per core (2,048 cores) and the scaling leaves the ideal behavior for 5 elements per core. Plotting the performance index over the load per core reveals that load imbalances slightly shift the optimum to higher loads. We conclude that the code on the CRAY XE6 cluster has the same performance either on an unstructured or a cartesian mesh, if the load per core is larger than 2,000 DOF per core.

3 Direct Simulation of Aeroacoustic Sound Generation from the Flow Around a NACA0012 Airfoil

The generation mechanisms of tonal noise are among the most important topics in aeroacoustics research. It is well known that under certain conditions, trailing edge noise from airfoils at medium Reynolds numbers exhibit dominant tonal peaks in the acoustic spectrum. The DG-SEM has been shown to be suitable for the direct simulation of aeroacoustic noise for instance in [8].

In several experimental works e.g. [1, 16] and a few papers reporting 2D direct numerical simulations [5, 10], the principal mechanism for tonal noise generation on smooth airfoils at moderate Reynolds numbers has been established and is mostly agreed. At low angles of attack, the boundary layer on the pressure side is likely to remain laminar until very close to the trailing edge. If laminar separation occurs in this region, vortex shedding with distinct frequencies around the trailing edge is possible, leading to tonal trailing edge noise. The upstream running acoustic waves generate boundary layer instability (Tollmien-Schlichting, T-S) waves in the laminar boundary layer through receptivity. Primary growth of the T-S waves along the airfoil eventually leads to vortex shedding at the same frequency which completes the so-called acoustic feedback loop. Acoustic feedback to the flow is crucial in this case, signifying the need for a direct simulation of the acoustic radiation. The numerical simulation is especially demanding as very different physical phenomena have to be captured accurately: boundary layer instability growth, noise generation at the trailing edge, upstream noise propagation and the process of acoustic receptivity.

The flow around a NACA0012 airfoil with sharp trailing edge at $\mathrm{Re} = 100,000$ and $\alpha = 0°$ following Jones and Sandberg [10] is chosen in order to demonstrate the capability of our code to reproduce complex aeroacoustic phenomena. The following results were obtained with an unstructured 2D DG-SEM code, $N = 5$ and approx. $1.47 \cdot 10^6$ degrees of freedom. The farfield boundary is at 7 chords from the airfoil and a buffer zone with grid-stretching and damping source term (sponge zone) is applied upstream of the outflow boundary. If we define the characteristic simulation time as $T^* = C/U_\infty$, with C and U_∞ being the chord of the airfoil and the freestream velocity, the computation was run for a total simulation time of $90T^*$. The last $60T^*$ were used to collect statistics. Figure 7 shows the time-averaged skin friction coefficient for pressure and suction side and a respective solution from the XFoil code [7] for reference. A comparison between two averaging intervals reveals convergence to a symmetric distribution mostly in good agreement with the XFoil solution. Here, we also see that the boundary layer separation occurs at $x \approx 0.59$. From most experimental reports, it is apparent that boundary layer separation upstream of the trailing edge is not a necessary condition but strongly increases the possibility of tonal noise generation. The inflectional velocity profiles result in strong instability growth rates and ultimately saturation and transition to 2D vortex shedding in the simulation scenario.

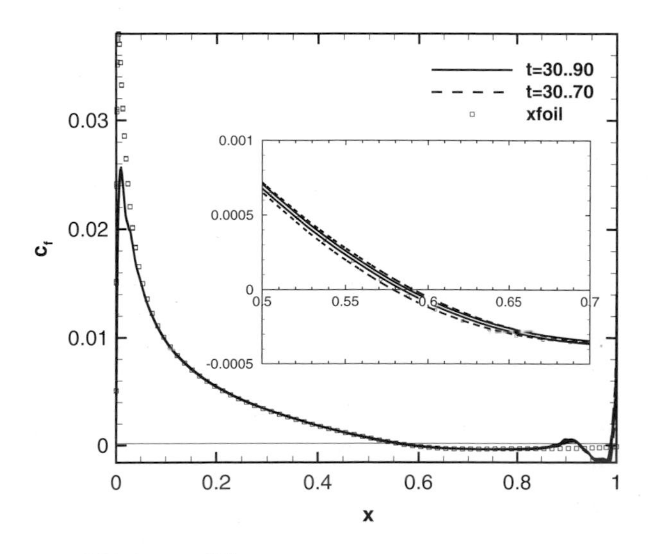

Fig. 7 Time-averaged friction coefficient c_f

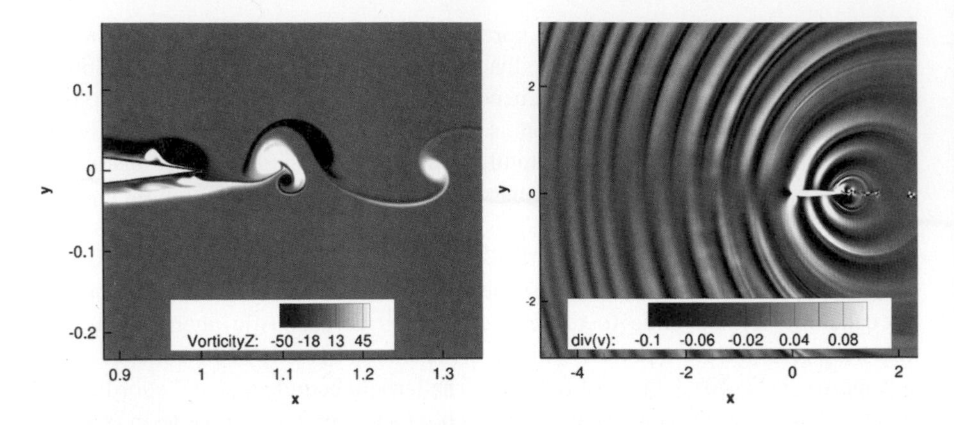

Fig. 8 *Left*: Contours of vorticity $\omega \in [-50, 50]$ at the trailing edge, *right*: Visualization by means of volume dilatation contours $\text{div}(\mathbf{v}) \in [-0.1, 0.1]$

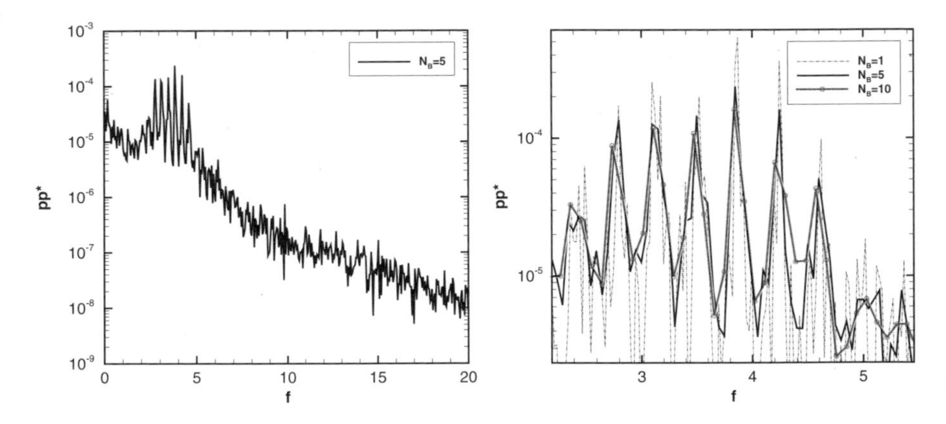

Fig. 9 Power spectral density (PSD) of the acoustic signal taken at 0.5 chords above the trailing edge

Figure 8 (left) exhibits the instantaneous flow field characterized by a contour plot of vorticity. From the separated boundary layer, vortices evolve which ultimately pass over the trailing edge generating trailing edge noise. In this symmetric setup, both sides of the airfoil contribute to this process. The acoustic field is visualized in Fig. 8 (right) by means of volume dilatation contours ($\text{div}(\mathbf{v})$), revealing the dominant dipole type noise source at the trailing edge.

A quantitative measure of the spectral content of the radiated noise is given in Fig. 9 using the power spectral density (PSD) of the pressure. The signal is recorded two chords above the trailing edge in the time interval $30T^* \leq t \leq 90T^*$, the Fourier transform of the pressure is taken separately in N_B subintervals with 50 % overlap, squared and averaged subsequently as done in [10]. This procedure resembles a statistical approach necessary due to random noise contained in the signal. In the frequency range $2 \leq f \leq 5$, at least six regularly spaced spectral

Table 2 Computational details for the NACA0012 airfoil case

N	n_{DOF}	Cores	$PID[\mu s]$	Runtime/T^*
5	$1.47 \cdot 10^6$	576	4.45	6.8 min

peaks are visible. Multiple spectral peaks are typical for the tonal noise generation mechanism described above. Figure 9 (right) demonstrates through different number of subintervals N_B that these peaks are inherent characteristics of the signal and not signal processing artefacts.

All the dominant acoustic peaks lie in the region of the most unstable T-S mode predicted by compressible linear stability analysis of the time-averaged flow-field. This supports the notion that the simulation successfully captures the acoustic feedback loop. Jones and Sandberg [10] report one dominant acoustic mode at $f \approx 3.83$, which coincides with the present most dominant peak in Fig. 9 at $f \approx 3.85$.

Table 2 summarizes the computational details of the 2D case. This simulation demonstrates the capability of the DG-SEM framework to represent very complex compressible flow effects accurately. In future analyses, extensions to more complex and turbulent 3D configurations will follow.

4 DNS of a Supersonic Turbulent Boundary Layer

The investigation of turbulent boundary layers is of fundamental importance for many engineering applications, such as for external flows around airplanes, ships and cars, or for internal flows, e.g. in turbine blades. Particularly the intake flow of scramjet-driven transportation systems are characterized by turbulent boundary layers emerging at the inlet walls and their interaction with shock waves. In this context, the turbulent boundary layer plays a decisive role which significantly determines the state of the turbulent flow that enters the combustion chamber and thus influences the efficient combustion of the fuel-air mixture. In addition to that, the boundary layer may experience intense localized heat loads which in turn could exceed the temperature limits of the materials and lead to disastrous impacts [2]. Hence, the study of turbulent boundary layers is of crucial importance not only for new cooling concepts, but also for the general functionality of such supersonic/hypersonic air-breathing propulsion systems.

The simulation of a turbulent boundary layer along the flat-plate, however, represents the base study for a canonical shock wave/boundary layer interaction (SWBLI). The simulation of a turbulent boundary layer itself can be transferred into a SWBLI by imposing a shock wave via appropriate boundary conditions. Thus, in order to lay the foundation for a SWBLI and to demonstrate the high potential of DG schemes for compressible turbulent wall-bounded flows, we will show results of a DNS of a compressible spatially-developing flat-plate turbulent boundary layer. Beyond that, the present work is also the first DNS of a compressible turbulent boundary layer with DG. The free-stream Mach number, temperature and pressure of the supersonic boundary layer are given by $M_\infty = 2.67$, $T_\infty = 568\,K$ and $p_\infty = 14{,}890\,Pa$ respectively. Air was treated as a non-reacting, calorically perfect gas with constant Prandtl number $Pr = 0.71$ and with a constant specific heat ratio $\kappa = c_p/c_v = 1.4$. Sutherland's law was used to take the temperature dependency of the dynamic viscosity μ into account. The DNS was performed with the DG-SEM code FLEXI and a polynomial degree of $N = 5$. The mesh consisted of $450 \times 100 \times 20$ elements in x-, y- and z-direction resulting in $194.4 \cdot 10^6$ DOF. The Reynolds number was set to $Re_\infty = 1.156 \cdot 10^5$ at the leading edge of the plate and isothermal no-slip conditions were applied at the wall. The wall temperature was equal to the adiabatic wall temperature $T_w = T_{ad} = 1242\,K$. To suppress any reflections from the boundaries, sponge zones were used at the outflow regions. To trigger the laminar-turbulent transition, periodic disturbances – given by amplitude and phase distributions along the inflow boundary – were added to the initial Blasius solution. In this work, we superimposed five discrete disturbances (determined by the eigenfunctions from linear stability theory) with a disturbance amplitude of $A = 0.02$. As for compressible flows the most unstable modes are oblique traveling waves, we introduced two single oblique waves with a fundamental spanwise wave number of $\gamma_0 = 21$ to rapidly reach a turbulent state.

We run the computation with 8,000 processors for a simulation time of $T = 2.3T^*$ whereas $T^* = 2L/U_\infty$ is the characteristic flowthrough time with plate length L and free-stream velocity U_∞. Figure 10 shows the instantaneous distribution of the vortices along the plate at the final simulation time visualized by the λ_2-criterion and colored by the streamwise velocity component. Here, we can clearly see how the flow stays laminar first and then breaks down into turbulence.

The downstream development of the time- and spanwise-averaged skin friction coefficient c_f is displayed in Fig. 11. Whilst the leading edge of the plate always maintains its laminar state, the flow experiences transition where a sudden increase of c_f is observable around $x_{tr} \approx 3.5$ ($Re_{x_{tr}} = 3.5 \cdot 10^5$). The resulting c_f-profile in the turbulent regime, however, agrees very well with common turbulent skin friction

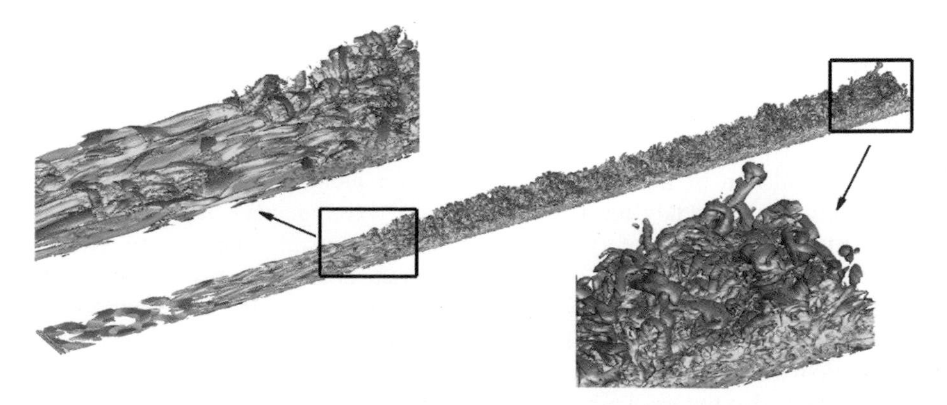

Fig. 10 λ_2-visualization of the turbulent structures along the flat-plate colored by the streamwise velocity component u

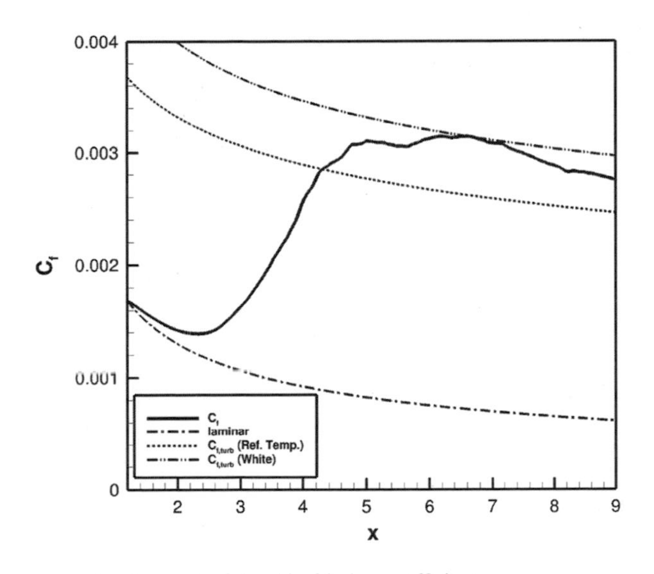

Fig. 11 Downstream development of the skin friction coefficient c_f

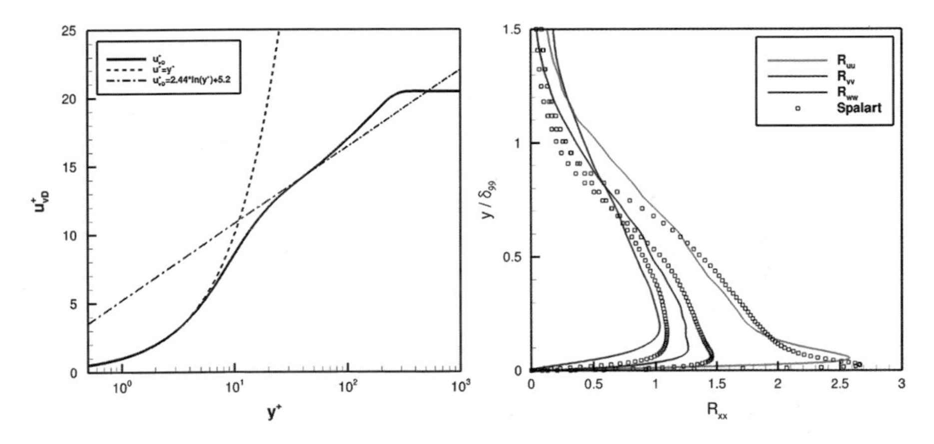

Fig. 12 *Left*: Van Driest transformed velocity profile u^+_{vD} at $Re_x = 900{,}000$, *right*: Density weighted Reynolds-stresses at $Re_\theta = 1{,}361$ compared with Spalart's incompressible boundary layer at $Re_{\theta,inc} = 1{,}410$ [15]

correlations (reference temperature concept and skin friction coefficient formula according to White [18]).

The Van Driest transformation of the streamwise velocity u accounts for compressibility effects and enables the comparison with incompressible turbulent boundary layers [18]. Figure 12 (left) illustrates the DNS results of the transformed velocity profile u^+_{vD} at the downstream position $x = 9.0$ ($Re_x = 900{,}000$) which perfectly matches the theoretical incompressible law of the wall with $u^+ = y^+$ and $u^+ = 2.44 \cdot ln(y^+) + 5.2$ respectively. Furthermore, in Fig. 12 (right) we compared

Table 3 Computational details for the compressible supersonic turbulent boundary layer case

N	n_{DOF}	Cores	PID (µs)	Runtime/T^*
5	$194.4 \cdot 10^6$	8,000	15.16	2.3

the Reynolds-stresses at $x = 9.0$ – where the momentum thickness based Reynolds number reaches values of $Re_\theta = 1{,}361$ – with Spalart's incompressible turbulent boundary layer [15]. Again we rescaled the DNS values using the density scaling to take the varying density into account. Except of the near-wall region, where slight differences can be observed, the agreement with the reference data is very good. Computational details for the simulation are given in Table 3.

The outlook of the present study is to impose an oblique shock in order to perform the first DNS of a SWBLI with a DG code. In this context, we will apply the FV subcell shock capturing which is already incorporated in our DG framework and enables the robust approximation of shock waves (Table 3).

5 High Order Large Eddy Simulation Strategy

In order to further develop strategies for high order DG methods regarding LES, a reference DNS calculation of freely decaying homogeneous isotropic turbulence (HIT) was made. The code used for the DNS is a pseudo-spectral compressible Navier-Stokes solver. It uses the 3D fast Fourier transformation library P3DFFT, which was shown to have excellent scaling properties [14]. Simulations were conducted using a fully periodic box with edge length 2π, discretized with 512 equidistant points per direction, resulting in around $134 \cdot 10^6$ DOF. De-aliasing was carried out by applying the $3/2$ rule [13], valid for incompressible flows such as the present example. The DNS was calculated on the CRAY XE6 using 9,216 cores, with a wall-clock time of 29 h for the first and 20 h for the second run described below. The initial condition was chosen to fulfill a given energy spectrum according to Ishihara et al. [19] as

$$E(k,0) = \frac{16\sqrt{2}}{\sqrt{\pi}} v_0^2 k_p^{-5} k^4 e^{-2\left(\frac{k}{k_p}\right)^2}, \tag{1}$$

where $k_p = 2$ and $v_0 \approx 2$ are the peak wavenumber and velocity respectively. The Mach number is 0.1 with respect to the peak velocity. Two runs at different Reynolds numbers were made, of which the high Reynolds number flow used the solution of the first run at $t = 2.3$ as initial state. Figure 13 shows the kinetic energy spectra for both runs.

Compared to the Kolmogorov $-5/3$ power law, the DNS spectra show the expected behavior of energy distribution over time. The DNS solutions at $t = 2.25$ and $t = 3.0$ for the low and high Reynolds number simulation respectively were used as initial state for LES runs. The current LES strategy to use polynomial de-aliasing was recently shown by Beck et al. [4] to give good results for a variety

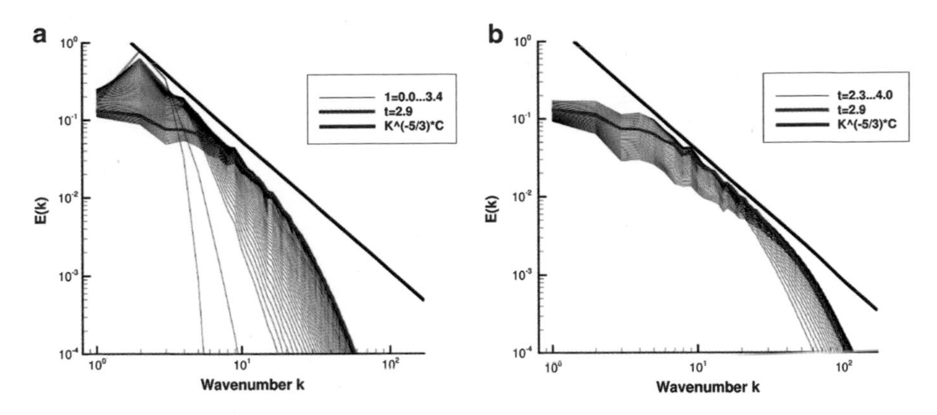

Fig. 13 Kinetic energy spectra of homogeneous isotropic turbulence for (**a**) $Re_\lambda(t = 2.9) \approx 73$ and (**b**) $Re_\lambda(t = 2.9) \approx 177$, compared to a Kolmogorov $-5/3$ power law

Table 4 Computational details for HIT LES runs using the DG-SEM code

N LES	N model	n_{DOF}	Cores	PID (µs)	Runtime
7	11	$3.73 \cdot 10^5$	216	12.7	2.3 min

of flows with moderate Reynolds numbers. For freely decaying turbulence this approach lacks a dissipation mechanism, which eventually leads to a build up of energy in the large scales. With the DG framework FLEXI, LES simulations were made on 6^3 cells with polynomial degree $N = 11$ on 216 cores, which means with one cell and 1,728 DOF per core. Our code was shown to have almost ideal scaling down to 2,000 DOF per core as discussed in Sect. 2.3. By this means a model test run took only about 2 min of wall-clock time and thus a large number of parameter tests could be performed. Considering ideal speed up would only be achieved down to 100,000 DOF per core, then one model test run would have taken about 2 h on 4 cores. This consideration clearly shows the advantage of highly efficient codes on HPC clusters for relatively small simulations, when a large number of runs is needed, an advantage often not mentioned. Details about calculation resources used are given in Table 4.

The newly developed LES model is based on cell local filtering, whenever the fluctuation energy is too high according to stability as well as physically motivated criteria. Hereinafter we refer to the model as local adaptive filtering (LAF). Similar approaches were shown to be successful for homogeneous turbulence as well as channel flow using pseudo-spectral Navier Stokes solvers, by Domaradzki et al. [17] and [6]. For underresolved simulation of turbulent flows this strategy provides a very efficient possibility of sub grid scale modeling for the DG-SEM method, making full use of the high order ansatz. It may also be seen as a possibility to make better use of the additional modes used for polynomial de-aliasing. The physical motivation of the model is to allow non-linear scale interaction and thereby drain energy from large scales, while balancing the model by filtering the small scales when needed. Figure 14 shows the kinetic energy spectra of the

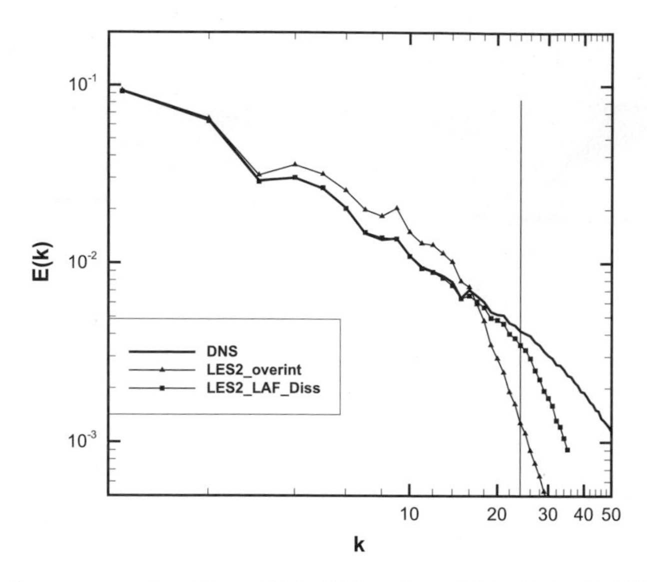

Fig. 14 Kinetic energy spectra at $t = 4.0$ for DNS, polynomial dealiasing (LES2_overint) and LAF model (LES2_LAF_Diss), the *vertical line* marks the filter wavelength

developed model (LES2_LAF_Diss) and one solution calculated with polynomial dealiasing compared to the high Reynolds number DNS result. The LES was run for approximately one large eddy turnover time. From the shown spectra the LAF model is able to reproduce the DNS results very well up to the LES resolution marked by the vertical line, where the overintegrated solution shows the expected energy build up in large scales and overdissipation in small scales.

6 Vortex Cascade

Turbulent flows are characterized by a large bandwidth of spatial and temporal scales, in which kinetic energy is transferred through a cascade from large to small vortices, where it is finally dissipated. The driving mechanism behind this cascade is the interaction of vortices, in particular, the stretching, bending and reconnection of vortices. Starting from the momentum equation of the incompressible Navier-Stokes equations, an evolution equation for the vorticity vector is obtained by applying the curl operator

$$\frac{Dw}{Dt} = (w \cdot \nabla)\, u + \nu \left(\nabla^2 w\right), \tag{2}$$

where w denotes the vorticity vector, u is the vector of the flow velocity components and ν denotes the kinematic viscosity. Thus, the material rate of change of vorticity is governed by two mechanisms: The diffusion of vorticity and the vortex stretching term $(w \cdot \nabla)\, u$. It is this term that is responsible for the increase in vorticity and the development of turbulent scales.

We have set up a test case to analyze the vortex stretching mechanisms and to investigate the turbulence cascade. Two large scale, stationary isentropic vortices of equal strength interact in a periodic domain of size $(20 \times 20 \times 20)$. The Mach number based on the average initial vortex velocity is set to $\mathrm{M} = 0.1$, while the Reynolds number based on the same velocity scale and the initial vortex diameter $(D = 2)$ is $\mathrm{Re} = 1{,}000$. The domain is discretized by 48^3 cubic elements with 8^3 inner points per cell, leading to $56 \cdot 10^6$ DOF total. We have computed this flow on $6,912$ cores, leading to a DOF per core load of about 8,000. Figure 15 shows the creation of this turbulent cascade from two single, perpendicular large scale vortices. They interact with each other through their respective fields of strain, which

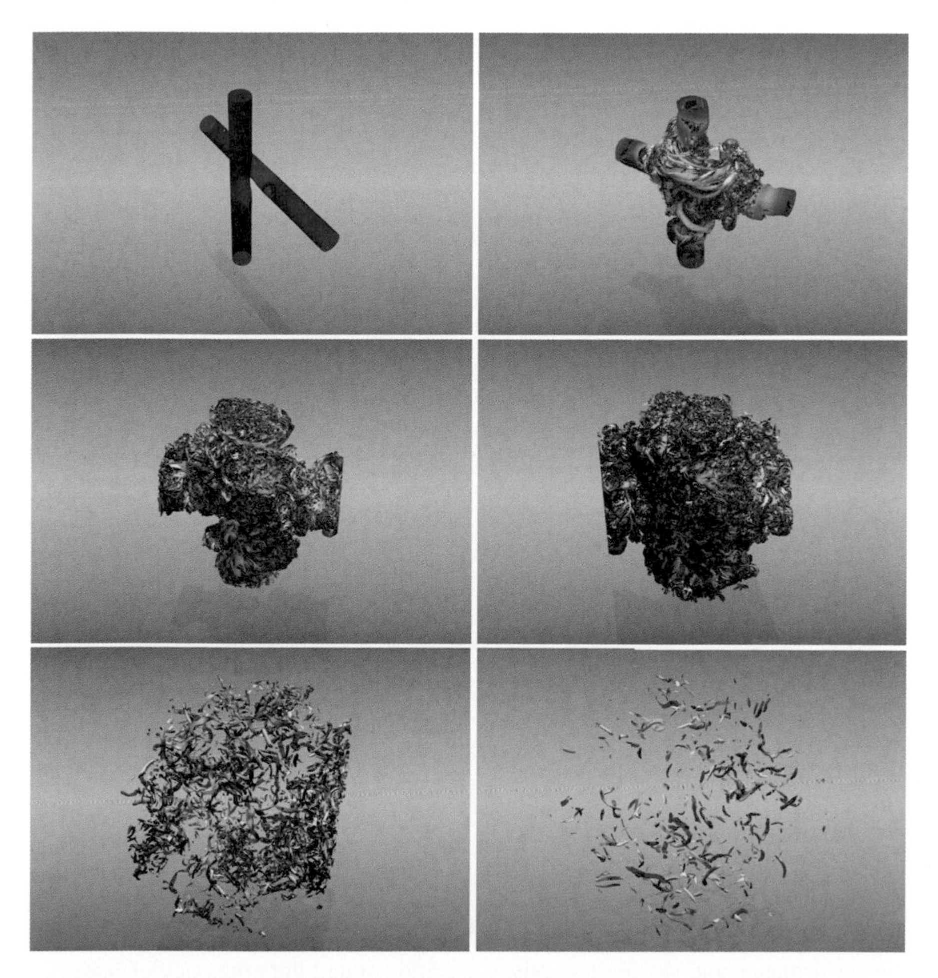

Fig. 15 Temporal development of vortical structures. Shown are $\lambda_2 = -0.001$ isocontours, colored by helicity. From upper left to lower right: $t = 0, 30, 60, 90, 250, 335$ s

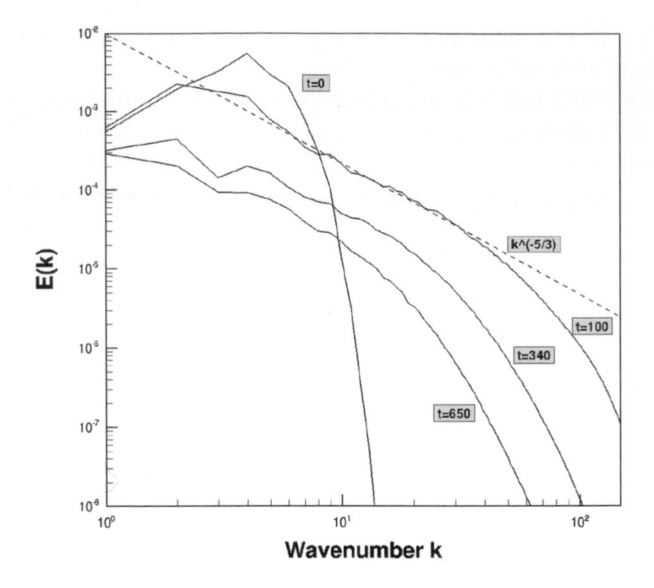

Fig. 16 Temporal development of energy spectrum for vortex cascade flow

Table 5 Computational cost for the vortex cascade computations

N	No. DOF	No. cells	Cores	$\overline{\Delta t}$	PID (μs)	Total runtime
7	$56.6 \cdot 10^6$	48^3	6912	$3.8E-3$	11.4	6,800 s

causes stretching, rotation, reconnection and squeezing through the stretching term in Eq. (2). The initial two vortices break up, reconnect and develop into a cascade of vortices, clearly demonstrating the multi-scale nature of turbulence. After the energy of the large scale vortices is transferred to the smaller scales, these slowly die out due to diffusion.

Figure 16 shows the development of the spectrum of kinetic energy $E(k)$ over time. From the initial narrow spectrum of the characteristic multiscale spectrum develops, which dissipates over time due to the lack of large scale energy input.

Table 5 summarizes the computational costs of this simulation. We have chosen the number of cores to achieve a near optimal PID, which results in a very short wall-clock time of under 2 h.

7 Summary and Outlook

In this paper we report the ongoing development and some of the applications of our highly-efficient HPC framework for compressible turbulent flows. From the performance and scaling analysis, we demonstrated the efficiency of our code for large scale parallel simulations even with unstructured grids. Load imbalance effects stemming from the domain decomposition approach were shown to be only significant for very low loads per core. The presented simulation resulted from direct numerical simulations as well as from large eddy type simulations using our DG-SEM code FLEXI. In this context, we achieved accurate results in investigations addressing aeroacoustics, decaying turbulence as well as high-speed compressible turbulent boundary layer flow. In 3D applications, we used around 5,000–10,000 cores on the CRAY XE6 cluster.

In the future, the research focus will lie on applying our HPC framework to direct numerical simulations of previously unfeasible and expensive computations to gain a deeper insight into their underlying physics. Additionally, we are interested in the further development of large eddy simulation modelling strategies using DG-SEM towards high Reynolds number flows which is accessible through the efficient use of the HLRS high performance clusters.

Acknowledgements The research presented in this paper was supported in parts by the Deutsche Forschungsgemeinschaft (DFG), amongst others within the Schwerpunktprogramm 1276: Met-Stroem and the Graduiertenkolleg 1095: Aero-Thermodynamic Design of a Scramjet Propulsion System for Future Space Transportation Systems and the research projects IDIHOM within the European Research Framework Programme, the Boysen Stiftung, Daimler AG, Audi AG and Robert Bosch GmbH. We truly appreciate the ongoing kind support by HLRS and CRAY in Stuttgart.

References

1. Arbey, H., Bataille, J.: Discrete tones of isolated airfoils. J. Fluid Mech. **134**, 33–47 (1983)
2. Babinsky, H., Harvey, J.K.: Shock Wave-Boundary-Layer Interactions. Cambridge University Press, Cambridge/New York (2011)
3. Bassi, F., Rebay, S.: High-order accurate discontinuous finite element solution of the 2D Euler equations. J. Comput. Phys. **138**(2), 251–285 (1997)
4. Beck, A.D., Bolemann, T., Flad, D., Frank, H., Gassner, G., Hindenlang, F., Munz, C.-D.: High order discontinuous galerkin spectral element methods for transitional and turbulent flow simulations. Int. J. Numer. Methods Fluids **76**(8), 522–548 (2014)
5. Desquesnes, G., Terracol, M., Sagaut, P.: Numerical investigation of the tone noise mechanism over laminar airfoils. J. Fluid Mech. **591**, 155–182 (2007)
6. Domaradzki, J.A., Loh, K.C., Yee, P.P.: Large eddy simulations using the subgrid-scale estimation model and truncated navier–stokes dynamics. Theor. Comput. Fluid Dyn. **15**(6), 421–450 (2002)
7. Drela, M.: Xfoil: an analysis and design system for low reynolds number airfoils. In: Mueller, T. (ed.) Low Reynolds Number Aerodynamics. Volume 54 of Lecture Notes in Engineering, pp. 1–12. Springer, Berlin/Heidelberg (1989)
8. Fechter, S., Hindenlang, F., Frank, H., Munz, C.-D., Gassner, G.: Discontinuous galerkin schemes for the direct numerical simulation of fluid flow and acoustics. In: Proceedings of the 33rd AIAA Aeroacoustics Conference, Colorado Springs, Colorado, USA, June 2012.
9. Hindenlang, F., Neudorfer, J., Gassner, G., Munz, C.-D.: Unstructured three-dimensional high order grids for discontinuous galerkin schemes. In: 20th AIAA Computational Fluid Dynamics Conference, Honolulu, 27–30 June 2011, number AIAA-2011-3853, 2011
10. Jones, L.E., Sandberg, R.D.: Numerical analysis of tonal airfoil self-noise and acoustic feedback-loops. J. Sound Vib. **330**(25), 6137–6152 (2011)

11. Message Passing Interface Forum: MPI: a message-passing interface standard, version 2.2. Specification, Sept 2009
12. Moon, B., Jagadish, H., Faloutsos, C., Saltz, J.: Analysis of the clustering properties of the hilbert space-filling curve. IEEE Trans. Knowl. Data Eng. **13**(1), 124–141 (2001)
13. Orszag, S.A.: On the elimination of aliasing in finite-difference schemes by filtering high-wavenumber components. J. Atmos. Sci. **28**(6), 1074–1074 (1971)
14. Pekurovsky, D.: P3dfft: a framework for parallel computations of fourier transforms in three dimensions. SIAM J. Sci. Comput. **34**(4), C192–C209 (2012)
15. Spalart, P.R.: Direct simulation of a turbulent boundary layer up to $Re_\theta = 1410$. J. Fluid Mech. **187**, 61–98 (1988)
16. Tam, C.: Discrete tones of isolated airfoils. J. Acoustical Soc. Am. **55**, 1173–1177 (1974)
17. Tantikul, T., Domaradzki, J.: Large eddy simulations using truncated navier–stokes equations with the automatic filtering criterion. J. Turbul. **11**(21), 1–24 (2010)
18. White, F.M.: Viscous Fluid Flow. McGraw-Hill, New York (1991)
19. Yamazaki, Y., Ishihara, T., Kaneda, Y.: Effects of wavenumber truncation on high-resolution direct numerical simulation of turbulence. J. Phys. Soc. Jpn. **71**(3), 777–781 (2002)

DNS/LES Studies of Turbulent Flows Based on the Cumulant Lattice Boltzmann Approach

Manfred Krafczyk, Kostyantyn Kucher, Ying Wang, and Martin Geier

Abstract In many industrial and environmental problems we encounter turbulent flows over porous surfaces which also penetrate the porous medium to different extents. Although there is a wealth of literature on macroscopic models of such phenomena which do not take the pore scale explicitly into account, these approaches typically require some additional transport coefficients to match experimentally obtained statistics for mass, momentum and energy transport across such interfaces. In this project we conduct Direct Navier-Stokes (DNS) and Large Eddy Simulation (LES) computations of turbulent flows which explicitly take into account specific pore scale geometries obtained from computer tomography imaging and do not use any explicit turbulence modeling. In this first part of the project we conducted validation studies for two canonical turbulent flows, i.e. flow around a plate and flow in a porous channel. Subsequently, we compare simulation results of turbulent flows over a porous sand and to experimental results and demonstrate the validity of our approach. Finally we discuss our approach to address evaporation processes on a pore scale which is based on a separation of time-scales. The newly developed cumulant Lattice Boltzmann scheme implemented as part of our research Code VirtualFluids shows a favorable behavior with respect to parallelization efficiency as well as to numerical stability and accuracy.

1 Kinetic Methods for CFD: The Cumulant Lattice Boltzmann Approach

The lattice Boltzmann equation (LBE) is a discrete and directly executable mathematical model used to solve various partial differential equations, most notably the Navier-Stokes equation. The basic components of the lattice Boltzmann equation are discrete local distribution functions f_q in momentum space, a uniform distribution of lattice nodes in space, an advection rule that moves the distributions according

M. Krafczyk (✉) • K. Kucher • Y. Wang • M. Geier
Institute for Computational Modeling in Civil Engineering, TU Braunschweig, Pockelsstr. 3, 38106 Braunschweig, Germany
e-mail: kraft@irmb.tu-bs.de

W.E. Nagel et al. (eds.), *High Performance Computing in Science and Engineering '14*, DOI 10.1007/978-3-319-10810-0_34

to their discrete momenta from one grid node to another, and a local collision rule that changes the magnitude of the nodal distributions in each time step. Repetition of advection and collision leads to an asymptotic convergence to a set of target equations. The target equations (here Navier-Stokes) usually leave some freedom in the design of the lattice Boltzmann equation. These include:

- *The choice of the velocity set in momentum space*, also called the lattice. The target equations impose some minimal requirements on the lattice symmetry and its degrees of freedom but these requirements are fulfilled by an infinite number of lattices. It is, however, usually desired to settle with a lattice that has as few degrees of freedom as possible and is as compact in momentum space as possible.
- *The equilibrium* is an attractor for the rearrangement of the distributions during collision. The target equations impose leading order constraints on the equilibrium. The equilibrium can be derived a posteriori from the target equation or a priori from statistical independence of different degrees of freedom.
- *The collision operator* is a set of functions acting on the local distribution function and the equilibrium to compute a new local distribution function, the so-called post-collision state. The collision is usually of relaxation type but operators differ in whether they use a single or multiple relaxation rates or in whether they use constant or non-constant relaxation rates.

The freedoms beyond the basic asymptotic requirements offered by all three aspects have a profound influence on the numerical accuracy and stability of the method. Our new LBE approach accurately and efficiently solves the weakly compressible isothermal Navier-Stokes equation even for large Reynolds numbers, i.e. turbulent flows. The choice of the microscopic velocity set is a compromise between efficiency and accuracy for an increasing magnitude of the corresponding numerical stencil reduces data locality. For solving the incompressible Navier-Stokes equation sets with 15 or 19 discrete links in three dimensions are popular. Recently evidence has been accumulated that the 19 speed model is not able to provide axis-symmetric solutions for axis-symmetric problems at moderate or high Reynolds number [7, 11, 18]. A particularly alarming observation is that the anisotropy of the solution appears to persist under grid-refinement [18]. These results put the consistency of the 15/19 speed model with the Navier-Stokes equation into question. The same studies showed that a 27 speed model can recover axis-symmetric solutions which motivates our choice for this variant for the computations presented below. The equilibrium in the lattice Boltzmann equation is usually of polynomial type and designed to fulfill the minimum requirements of the Navier-Stokes equation. Terms beyond $\mathcal{O}(Ma^2)$ are usually discarded. In this paper we follow our previous philosophy to obtain the equilibrium directly from the Maxwellian distribution and consider all terms in Mach number Ma that are supported by the velocity set [4]. Almost all lattice Boltzmann models are based on the relaxation time approximation. The pre-collision distributions are relaxed towards their equilibrium to obtain a post-collision distribution. Models differ in whether they apply a single relaxation rate or multiple relaxation rates for different observable quantities. Another distinction can be made on whether the relaxation rate(s) are constant or dependent on observable quantities. The latter is the case in various stabilization techniques like the Smagorinski model or entropic limiters. These techniques are primarily used to run simulations at high Reynolds number without loosing stability by locally increasing viscosity. Changing the relaxation rates locally usually results in non-constant transport coefficients unless the change is imposed to compensate for numerical defects.

The usage of multiple relaxation times (MRT) was introduced to maximize the number of adjustable parameters of the model motivated by the intention to tune both stability and accuracy [2]. Many of the available parameters do not influence the solution of the LBE to leading order and their exact influence on stability and accuracy might depend on the problem under consideration. Using the free parameters MRT can be both more accurate and stable than a single relaxation time (BGK) LBE or less so.

An obvious question to be asked in the design of an LBM using multiple relaxation rates is which set of observable quantities should be used. The original MRT method transforms the local distribution function into a set of raw moments by means of a linear transformation. Each moment is an observable quantity orthogonal to all the others and is allowed to relax with its own rate. The problem with this approach is that the orthogonality of the moments is not frame-invariant. Any set of moments that is orthogonal in one frame of reference is not orthogonal in any other frame of reference. In particular, due to the linear transformation, the MRT moments are only orthogonal at velocity zero. Therefore, the original MRT method has additional violations of Galilean invariance as compared to a single relaxation time model with the same velocity set and the same equilibrium function. This problem is solved with the cascaded LBM [4] that uses a non-linear transformation of the distribution function to central moments in the frame of the moving fluid. However, mutual orthogonality of the observable quantities is not sufficient to isolate their evolution. Observable quantities can also be coupled on the level of their equivalent partial differential equations. By relaxing different quantities with different rates one implicitly postulates that they evolve due to different processes. This implicit assumption was found to lead to erroneous results when relaxation rates of different moments were not optimally chosen. An ad hoc solution to this problem was proposed with the Factorized Central Moment LBM [6, 12] which is also used in the commercial flow solver XFlow [9]. The same problem was solved in a mathematically more concise way using cumulants instead of central moments as observable quantities [5]. A cumulant method for solving the Boltzmann equation was proposed by Seeger et al. [15–17]. In [15] Seeger et al. show that their cumulant ansatz and Grad's expansion from [8] are equivalent. Seeger considers the cumulant expansion in one and two dimensions and observes a difference to the moment expansion at order 6. The cumulant LBM [5], on the other hand, is so fare only derived to Navier-Stokes order. In two dimensions using a lattice with nine speeds there is no perceivable difference between cumulants and central moments. We note here that this is different in three dimensions where there are different cumulants of order 4 that differ significantly from their counterparts in two dimensions in terms of moments. The derivation of a cumulant LBM to Navier-Stokes order intrinsically requires the consideration of all three dimensions. We therefor abstain from the common praxis to first derive the method in two dimensions and than extend it to three dimension. In the following we will only consider the three dimensional case.

The original formulation of the cumulant LBM [5] was very inefficient and the model attracted little attention after its introduction. In this report we present simulations based on a revised cumulant LBM using a fast transformation from distributions to cumulants that will be described in detail in a forthcoming complementary publication.

Unlike direct discretizations of the Navier Stokes equations, the LBM equations rely on a discretization of the Boltzmann equation that is a time-dependent description of the behavior of thermodynamical ensembles:

$$f_q(x + e_q c \Delta t, t + \Delta t) = f_q(x,t) + \Omega(f_q(x,t), f_q^{eq}(x,t)), \tag{1}$$

Here, the e_q are a set of discretized particle velocities and Ω is the collision operator. The reference velocity $c = \Delta x / \Delta t$ relates the lattice spacing and the time step, such that e becomes a dimensionless quantity. The probability distribution functions f is relaxed in time towards equilibrium functions f_q^{eq}, which are often taken to be truncated versions of the Maxwell distribution, the equilibrium probability distribution for the microscopic velocities of a gas of non-interacting particles. Chapman-Enskog analysis shows that the Lattice Boltzmann equation correctly models Navier-Stokes fluid dynamics for small Mach and Knudsen numbers, i.e. when the typical hydrodynamic velocities are much lower than the speed of sound and the mean free path is small against the dimensions of the system.

2 The Research Prototype VirtualFluids and Its Parallel Performance on Hermit

The simulations were based on a new generation of the iRMB research code VirtualFluids, for which the scalability on large-scale system has been demonstrated already. The segmentation and grid-refinement is currently being revised and extends the work in [3, 14]. Despite the use of MPI-communication on distributed memory systems, the user can also opt for thread-based parallelization e.g. between the cores of a CPU node which have a common address space. VirtualFluids is an adaptive, MPI based parallel Lattice-Boltzmann flow solver which is comprised of various modules. The software framework is based on object-oriented technology and uses tree-like data structures. The flow region is divided into discrete blocks, for example on the basis of a Octrees, by means of the Block-Mesh-Refinement-Technique. Unlike conventional tree-type Finite Difference codes, the leaflets of the tree consist of these blocks (i.e. 3D matrices of grid nodes) whose size is defined as part of the setup. This additional level of granularity allows the efficient domain decomposition which occurs on the block instead of the grid node level while

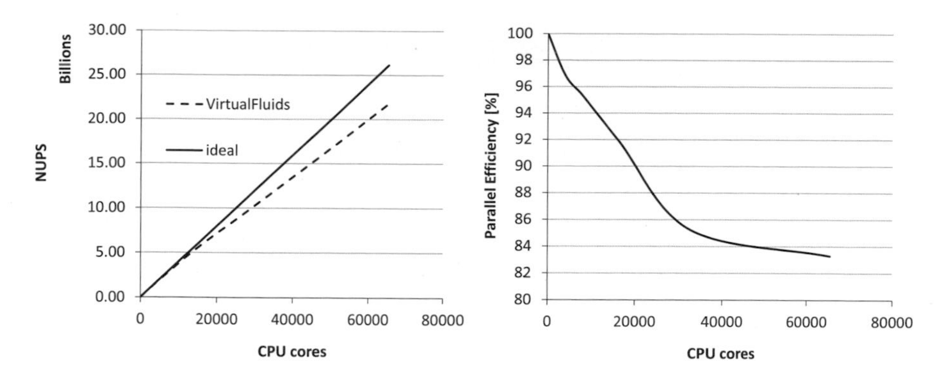

Fig. 1 Weak scaling behavior of VirtualFluids on HERMIT. The unit NUPS refers to the number of nodal updates per second, i.e. how many grid points are updated for one time step per second wall-clock time

allowing a seamless transition between uniform grids and essentially unstructured grids with exponential refinement in space and time. The scheme is based on a nested time stepping approach so that regardless of the local resolution the Courant number with respect to the microscopic velocities is always one. After a number of optimizations, the parallel scaling properties of the framework could be improved and Fig. 1 shows the weak scaling performance of VirtualFluids on the Hermit system of HLRS.

3 Benchmark 1: LES of Turbulent Flow Around a Plate

The simulation of turbulent flow around a plate is a non-trivial test case for LES simulations. This computation was based on a multi-relaxation time LBM extended by a Smagorinsky LES model. In both the experiment and the simulation we added a rigged ribbon on the upper side of the beginning of the plate which leads to rapid development of a turbulent boundary layer (see Fig. 2).

The mean flow profile over the flat plate at a distance of 75 cm from the beginning of the plate is shown in Fig. 3.

Presently we are conducting simulations of this problem with the Cumulant LBM with much higher resolution and we expect that especially the fluctuation statistics will improve as compared to the one shown in Fig. 4 as the chosen resolution will eventually allow the resolution of the boundary layer up to $y^+ = 1$. Table 1 shows parameters of the flat plate-benchmark setup.

Fig. 2 Experimental setup (courtesy of M. Mösner, LSM, TU Braunschweig) and slice representation of instantaneous horizontal velocity

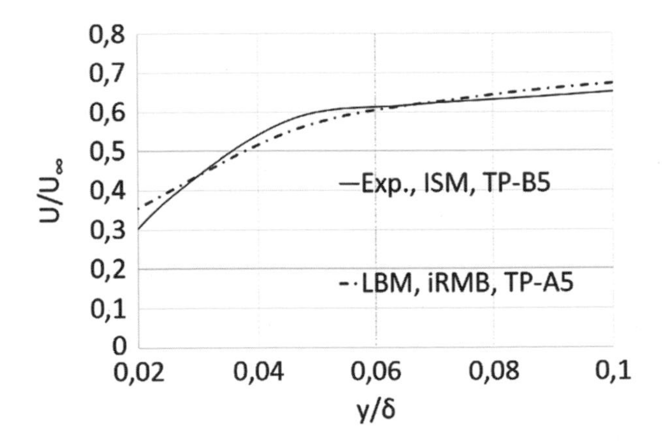

Fig. 3 Comparison of experimental and numerical time-averaged velocity profile at 75 % of plate length

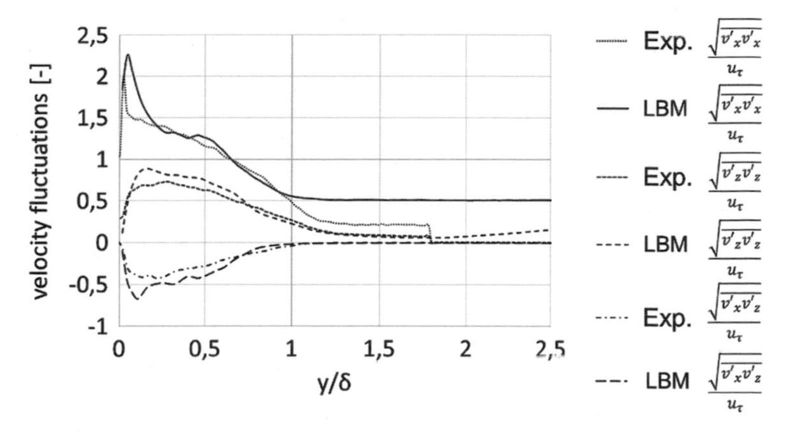

Fig. 4 Comparison of experimental and numerical data for the normalized velocity fluctuations

Table 1 Setup parameters of the plate benchmark

Parameter	Value
Inflow velocity	17 m/s
Reynolds number based on the plates thickness	11,900
Reynolds number with respect to plate length	1.13×10^6
Plate length	1 m
Domain extension	$2 \times 0.6 \times 0.0825$ m
Δx_{fine}	0.000234283 m
Δt_{fine}	3.98×10^{-7} s
Number of grid nodes	2.38×10^8
Number of degrees of freedom	6.4×10^9
Simulated time	0.48 s
Number of cores	2,048
Wall clock time	1,750 h

4 Benchmark 2: DNS/LES of Turbulent Flow in a Porous Channel

Table 2 shows parameters of the porous channel benchmark.

The boundary conditions were periodic in flow direction and a forcing was used to drive the flow with the target Reynolds number starting from an initial plug flow with constant velocity in the domain (Figs. 5 and 6).

The resolution of the upper boundary layer was chosen as to obtain a dimensionless wall distance of the first grid node of $y^+ = 1.4$.

Table 2 Setup parameters of the porous channel benchmark

Parameter	Value
Reynolds number based on domain height	13,915
Inflow velocity	0.39 m/s
Δx_{fine}	0.000182292 m
Δt_{fine}	2.217×10^{-7} s
Domain extension	$0.14 \times 0.07 \times 0.46$ m
Height of the domain	0.46 m
Number of grid nodes	1.56×10^{8}
Number of degrees of freedom	4.0×10^{9}
Simulated time	0.55 s
Number of cores	2,048
Wall clock time	1,220 h

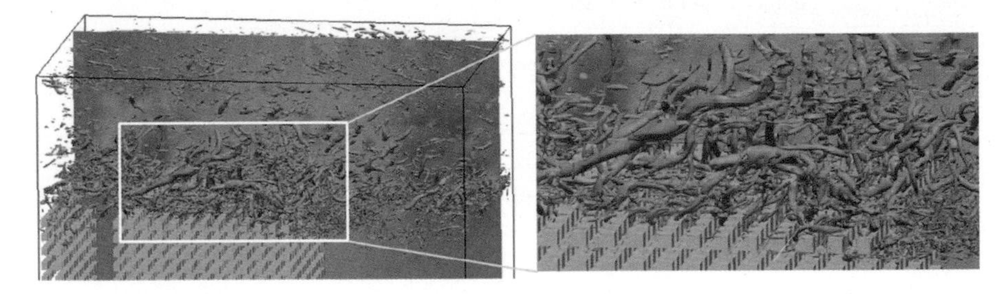

Fig. 5 Porous channel: geometric configuration and snapshot resp. *zoom* of iso-surface of Q-criterion [10]

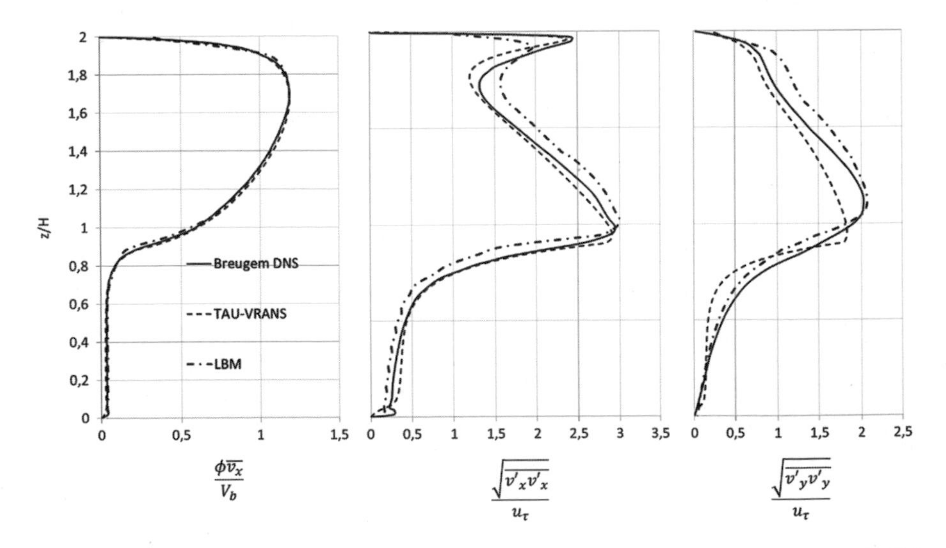

Fig. 6 Comparison of DNS [1], LBM and RANS data (courtesy of M. Mösner, LSM, TU Braunschweig) for the normalized velocity fluctuations

5 LES of Turbulent Flow Over a Porous Medium

The quantification of soil evaporation and of soil water content dynamics near the soil surface are critical in the physics of land-surface processes on many scales and are dominated by multi-component and multi-phase mass and energy fluxes between the ground and the atmosphere. Although it is widely recognized that both liquid and gaseous water movement are fundamental factors in the quantification of soil heat flux and surface evaporation, their computation has only started to be taken into account using simplified macroscopic models. As the flow field over the soil can be safely considered as turbulent, it would be natural to study the detailed transient flow dynamics by means of LES where the three-dimensional flow field is resolved down to the laminar sub-layer. Yet, this requires very fine resolved meshes allowing a grid resolution of at least one order of magnitude below the typical grain diameter of the soil under consideration. In order to gain reliable turbulence statistics, up to several hundred eddy turnover times corresponding of more than 10 s have to be simulated. Yet, the time scale of the receding saturated water front dynamics in the soil is on the order of hours. Thus we are faced with the task of solving a transient turbulent flow problem including the advection-diffusion of water vapor over the soil-atmospheric interface represented by a realistic tomographic reconstruction of a real porous medium taken from laboratory probes (see Fig. 7).

In order to validate the first step of the evaporation problem which is the TBL of the single phase flow, we computed the TBL for an unsaturated porous medium for two different grain sizes, i.e. Reynolds roughness numbers $k_s^+ = 20$ and $k_s^+ = 220$. Table 3 shows the setup parameters of turbulent boundary layer (TBL) simulation.

The boundary conditions were periodic in flow direction and a forcing was used to drive the flow with the target Reynolds number starting from an initial plug flow with constant velocity in the domain. The temporal evolution of the TBL for the fine grains is depicted in Fig. 8.

The simulation data for the resulting mean flow profiles (which includes information about the flow penetration into the porous medium) have been compared to experimental data of [13] and shows quantitative agreement for the fine grain

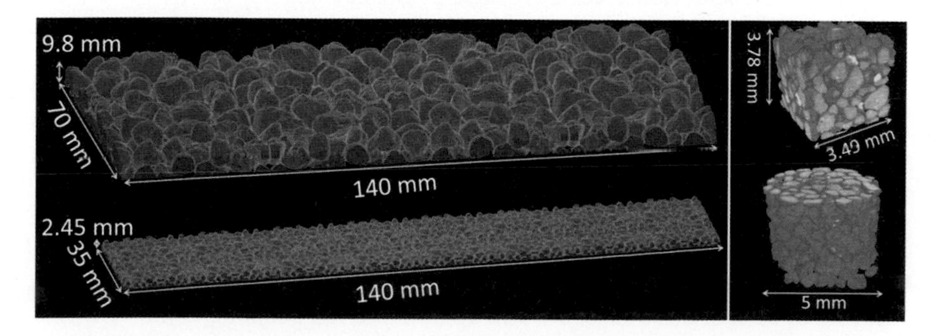

Fig. 7 Reconstructed sand probes for two different grain sizes from tomographic data

Table 3 Setup parameters of the porous medium TBL

Parameter	Value
Reynolds number based on domain height	13,915
Inflow velocity	0.39 m/s
Δx_{fine}	0.000182292 m
Δt_{fine}	2.217×10^{-7} s
Domain extension	$0.14 \times 0.07 \times 0.46$ m
Height of the domain	0.46 m
Number of grid nodes	1.56×10^{8}
Number of degrees of freedom	4.0×10^{9}
Simulated time	11.55 s
Number of cores	2,048
Wall clock time	1,220 h

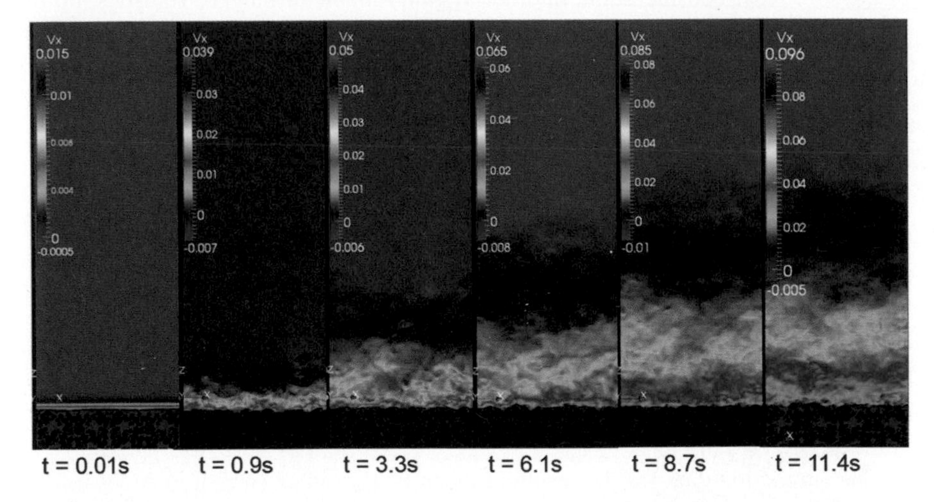

Fig. 8 Transient evolution of the turbulent boundary layer of the fine grain porous medium

setup (Fig. 9). The deviation for the coarse grain simulation is due to the fact that the Reynolds roughness number k_s^+ for the coarse grains is different from the experimental one by about a factor of 3.

Thus we demonstrated that we can conduct a quantitatively correct LES-LBM simulation of the TBL for various roughness regimes. Yet, the simulation of the full evaporation problem is hampered by the fact that the time scale of the coupled process is several days making a coupling to the LES of the TBL unfeasible. Our approach is thus to decouple the two processes by conducting simulations of evaporation without the TBL but increased diffusion coefficients for the water vapor. This will create a number of subsequent states with different water saturations, see Fig. 10. For each of these states we will conduct the full coupling to the TBL LES until the systems are in dynamic equilibrium. This in turn will allow to determine

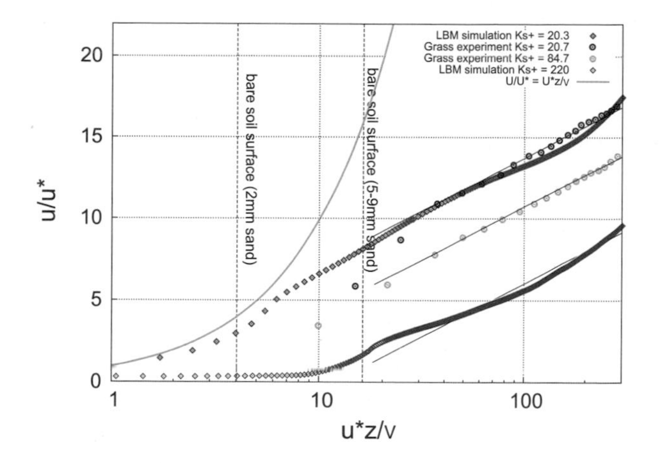

Fig. 9 Comparison of LBM simulation and experimental data [13] for two Reynolds roughness numbers

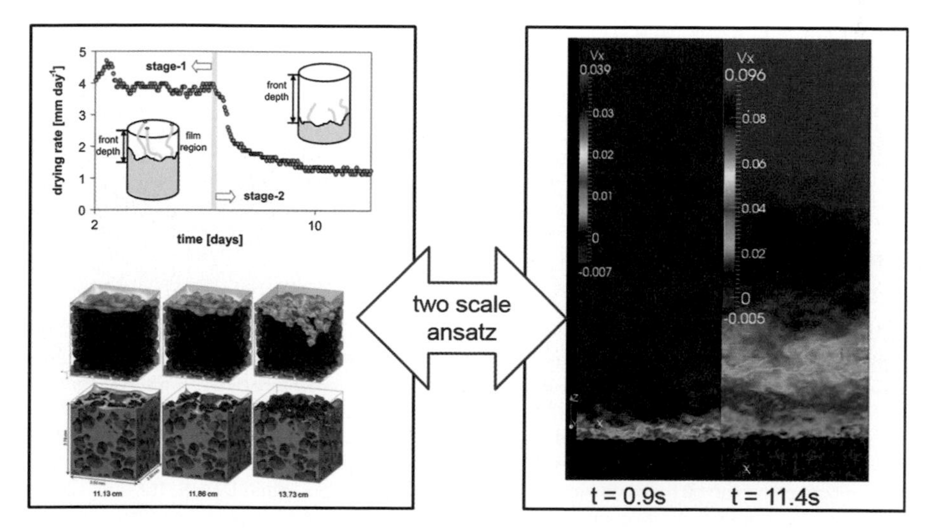

Fig. 10 Decoupling ansatz of the evaporation process and the turbulent advection

the mass flux across the soil-atmospheric interface for different saturations. These data will be interpolated and combined with pressure fluctuation statistics at the interface and will serve to calibrate RANS models for evaporation which do not take the explicit pore geometry into account and thus will be computationally much less demanding.

6 Conclusion and Outlook

The Cumulant version of the LBM which has been developed for the complex turbulent flow problems described above has been validated successfully by parallel DNS-LES computations which effectively rely on the increased stability and low numerical dispersion properties of the scheme as compared to previous LBM schemes. No explicit turbulence modeling is necessary to obtain almost quantitative agreement even for the fluctuation statistics. Although in some of the simulations, the boundary layer was resolved to $y^+ \simeq 1$, this resolution was of course not possible in the bulk flow, thus we prefer to speak of an implicit large eddy simulation approach in spite of the fact that for the cumulant LBM simulations no explicit eddy viscosity model was utilized. Future work will address improved outflow boundary conditions to reduce acoustic reflections to improve the extraction of acoustic statistics from the pressure field in combination with increased resolution on up to 50,000 cores on the Hermit system.

Acknowledgements The authors thank the Deutsche Forschungsgemeinschaft (DFG) for financial support of the Collaborative Research Center SFB 880 and the Research Training Group MUSIS FOR 1083.

The Höchstleistungsrechenzentrum Stuttgart (HLRS) is gratefully acknowledged for providing the required CPU resources.

References

1. Breugem, W.P., Boersma, B.J.: Direct numerical simulations of turbulent flow over a permeable wall using a direct and a continuum approach. Phys. Fluids **17**, 025103 (2005). doi:http://dx.doi.org/10.1063/1.1835771
2. d'Humières, D., Ginzburg, I., Krafczyk, M., Lallemand, P., Luo, L.S.: Multiple-relaxation-time lattice Boltzmann models in three dimensions. Phil. Trans. R. Soc. A **360**, 437–451 (2002)
3. Freudiger, S., Hegewald, J., Krafczyk, M.: A parallelisation concept for a multi-physics lattice Boltzmann prototype based on hierarchical grids. Prog. Comput. Fluid Dyn. **8**, 168–178 (2008)
4. Geier, M., Greiner, A., Korvink, J.G.: Cascaded digital lattice Boltzmann automata for high reynolds number flow. Phys. Rev. E **73**, 066,705 (2006). doi:10.1103/PhysRevE.73.066705
5. Geier, M., Schönherr, M., Pasquali, A., Krafczyk, M.: The cumulant Lattice Boltzmann Equation in three dimensions: theory and validation, submitted to Computers & Mathematics with Applications (CAMWA), 2014
6. Geier, M., Greiner, A., Korvink, J.G.: A factorized central moment lattice Boltzmann method. Eur. Phys. J. Spec. Top. **171**(1), 55–61 (2009). doi:10.1140/epjst/e2009-01011-1
7. Geller, S., Uphoff, S., Krafczyk, M.: Turbulent jet computations based on MRT and cascaded lattice Boltzmann models. Comput. Math. Appl. **65**(12), 1956–1966 (2013). Cited By (since 1996)1
8. Grad, H.: On the kinetic theory of rarefied gases. Commun. Pure Appl. Math. **2**(4), 331–407 (1949). doi:10.1002/cpa.3160020403
9. Holman, D.M., Brionnaud, R.M., Abiza, Z.: Solution to industry benchmark problems with the lattice-Boltzmann code xflow, http://www.xflowcfd.com/pdf/ECCOMAS2012_XFlow.pdf
10. Jeong, J., Hussain, F.: On the identification of a vortex. J. Fluid Mech. **285**, 69–94 (1995)

11. Kang, S.K., Hassan, Y.A.: The effect of lattice models within the lattice Boltzmann method in the simulation of wall-bounded turbulent flows. J. Comput. Phys. **232**(1), 100–117 (2013). doi:http://dx.doi.org/10.1016/j.jcp.2012.07.023
12. Premnath, K., Banerjee, S.: On the three-dimensional central moment lattice Boltzmann method. J. Stat. Phys. **143**(4), 747–794 (2011). doi:10.1007/s10955-011-0208-9. http://dx.doi.org/10.1007/s10955-011-0208-9
13. Schlichting, H., Gersten, K.: Grenzschicht-Theorie. Springer, Berlin (2006)
14. Schönherr, M., Kucher, K., Geier, M., Stiebler, M., Freudiger, S., Krafczyk, M.: Multi-thread implementations of the lattice Boltzmann method on non-uniform grids for cpus and gpus. Comput. Math. Appl. **61**(12), 3730–3743 (2011)
15. Seeger, S., Hoffmann, H.: The cumulant method for computational kinetic theory. Contin. Mech. Thermodyn. **12**(6), 403–421 (2000). doi:10.1007/s001610050145
16. Seeger, S., Hoffmann, K.: The cumulant method applied to a mixture of maxwell gases. Contin. Mech. Thermodyn. **16**(5), 515–515 (2004). doi:10.1007/s00161-004-0183-3
17. Seeger, S., Hoffmann, K.: The cumulant method for the space-homogeneous Boltzmann equation. Contin. Mech. Thermodyn. **17**(1), 51–60 (2005). doi:10.1007/s00161-004-0187-z
18. White, A., Chong, C.: Rotational invariance in the three-dimensional lattice Boltzmann method is dependent on the choice of lattice. J. Comput. Phys. **230**(16), 6367–6378 (2011). Cited By (since 1996)7

Highly Efficient Integrated Simulation of Electro-Membrane Processes for Desalination of Sea Water

Kannan Masilamani, Jens Zudrop, Matthias Johannink, Harald Klimach, and Sabine Roller

Abstract Electrodialysis can be used as an efficient process for sea water desalination. In this process, sea water flows through a channel with a complex spacer structure that separates the ion exchange membranes. We use a (multi-species) lattice Boltzmann method to study the hydrodynamics and ionic mass transport in this spacer filled flow channel. We present performance and scalability results of our implementation on the *Cray XE6* system *Hermit* at *HLRS* in Stuttgart. Overall, the high efficiency of our numerical solver allows us to study industrially relevant applications, ranging from a single spacer element to a laboratory-scale setup on this machine. These simulations result in a model that allows engineers to optimize the parameters in actual electrodialytic process installations.

1 Introduction

Scarcity of drinking water is a growing problem in many regions of the world. This challenges the development of new and efficient sea water desalination technologies. There are numerous sea water desalination process techniques developed to supply drinking water to mankind. However, energy and environmental impact are additional constraints, and for the mass production of drinking water an energy and cost efficient technique is required. Electrodialysis is a promising technology in this domain, which uses ion exchange membranes to separate ionic species from sea water under the influence of an electric field. For this technique, while already applied in practice, the underlying physical and chemical processes are not

K. Masilamani (✉) • J. Zudrop • H. Klimach • S. Roller
Simulationstechnik und wissenschaftliches Rechnen, University of Siegen, Hölderlinstr. 3, 57076 Siegen, Germany
e-mail: kannan.masilamani@uni-siegen.de; jens.zudrop@uni-siegen.de; harald.klimach@uni-siegen.de; sabine.roller@uni-siegen.de

M. Johannink
Aachener Verfahrenstechnik – Process Systems Engineering, RWTH Aachen University, Turmstr. 46, 52064 Aachen, Germany
e-mail: matthias.johannink@avt.rwth-aachen.de

W.E. Nagel et al. (eds.), *High Performance Computing in Science and Engineering '14*, DOI 10.1007/978-3-319-10810-0_35

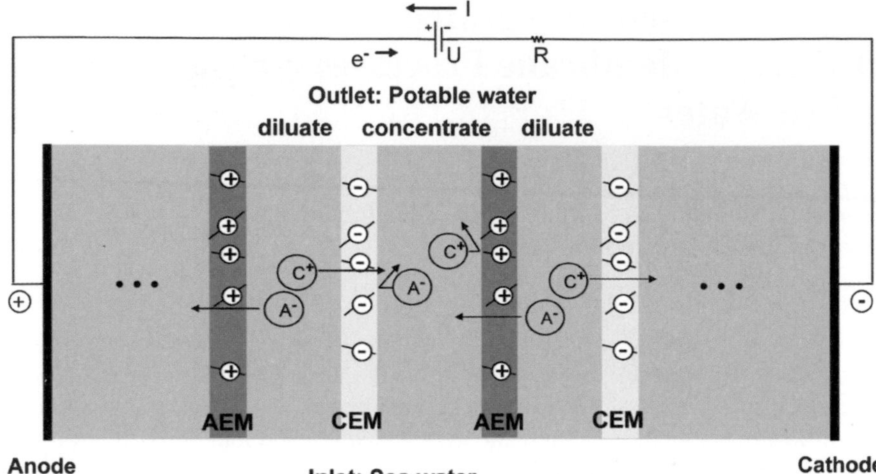

Fig. 1 Simplified structure of a desalination stack in two dimensions. A periodic arrangement of anion-exchange-membrane (AEM), diluate channel, cation-exchange-membrane (CEM) and concentrate channel is embedded between anode and cathode. After removing the ions, drinkable water can be extracted at the end of the diluate channel

understood in detail. This lack of insight constrains further optimization of devices in deployed installations. Numerical simulations on large HPC systems are used to study these underlying physical phenomena in this project.

A schematic illustration of an electrodialysis stack is shown in Fig. 1. It consists of alternatingly arranged anion (AEM) and cation exchange membranes (CEM). Anodes and cathodes are used at the ends of the stack to apply an electrical field. This electric field separates the anions and cations from the sea water by driving them in opposite directions. AEM and CEM are selective permeable membranes, they allow either anions or cations to pass through respectively. Thus, two kinds of channels are observed in the stack. One, where the ions are driven towards membranes that allow their passing, is called the diluate channel, as ions are removed from it. And the other one, where ions are driven towards membranes that block their passing is called the concentrate channel, as ions are collected by it. At the outflow of the stack, desalinated water is extracted from the diluate channels, this water is fed again into the stack several times until the desired concentration level is obtained.

Ion exchange membranes (AEM and CEM) in the stack are separated by a complex structure called spacer. This spacer structure is shown in Fig. 2. It acts as a mechanical stabilizer and also induces turbulence. The effective transport of ions through the membranes and also the total energy consumption of the stack is mainly influenced by the geometry of this spacer. A non-optimal spacer design leads to fouling and scaling effects in the flow channel which in turn results in damage of

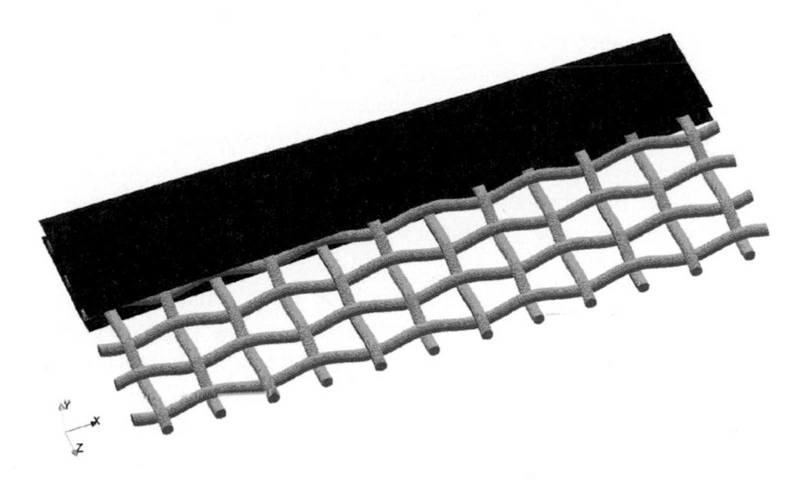

Fig. 2 Structure of a spacer used to mechanically stabilize the fluid channels of the stack

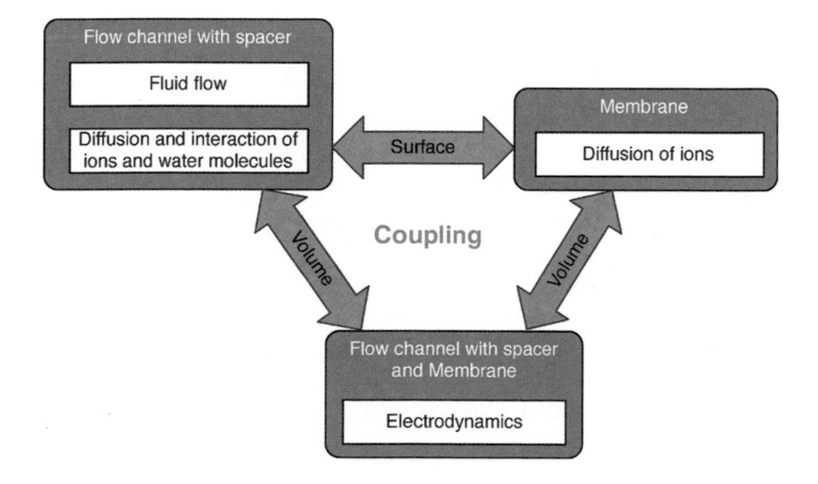

Fig. 3 Multi-physical heterogeneous system required in the simulation of the water desalination process

electrodialysis stack. Thus, the optimal design of this spacer plays a crucial role in optimization of this process.

An overview of the physical subsystems and its interactions are shown in Fig. 3. The following discussion is mainly concerned with simulations of the hydrodynamics and mass transport in the flow channel. Results will also be shown for flow overlayed by a fixed electric field, and using a simple membrane model that serves as a boundary condition to the species.

To study the physical phenomena in the spacer filled channel, a suitable numerical method is required. We chose a Lattice Boltzmann Method (LBM) for the numerical simulations, due to its advantage of easily integrating complex geometries and its high performance on large scale HPC systems [1]. For the simulations of mass transport in aqueous ionic mixtures we use a multi-species extension of the Lattice Boltzmann method, which is described in [2]. The multi-species LBM model recovers the Maxwell-Stefan equations for diffusive fluxes and the incompressible Navier-Stokes equations for the transport of ions and mixture in the macroscopic regime. This solver is embedded in our highly scalable parallel framework *APES* (Adaptable Poly-Engineering Simulator) [3], which is based on an octree data structure for the mesh [4].

In this work, we focus on the performance of our implementation of the numerical schemes on the *Cray XE6* system *Hermit*. The paper is organized as follows: In Sect. 2 we briefly describe the Lattice Boltzmann Method and the multi-species LBM for liquid mixtures. Section 3 is concerned with performance and scalability of both numerical schemes. In Sect. 4 the simulation results and some comparison to experimental data are shown. Finally, we summarize this work and highlight planned future work in Sect. 5.

2 Governing Equations and Numerical Method

In this section, we briefly review the governing equations of mass transport for an aqueous ionic mixture. For a detailed discussion we refer the reader to [2]. The mass transport is governed by a mass conservation equation for the species concentrations ($\mathbf{w}$ denotes the number density averaged velocity for the mixture)

$$\partial_t n_k + \nabla \cdot n_k \mathbf{w} = -\nabla \cdot \mathbf{J}_k, \tag{1}$$

which is closed by the Maxwell-Stefan closure relation for the diffusive fluxes $\mathbf{J}_k$ (in terms of the Maxwell-Stefan diffusivities $D_{k,l}$ and number density fraction χ_k)

$$\nabla \chi_k = \frac{1}{n} \sum_{l \neq k} \frac{1}{D_{k,l}} \left(\chi_k \mathbf{J}_l - \chi_l \mathbf{J}_k \right), \tag{2}$$

where n and n_k are total number density and species k number density respectively. The momentum balance for the mixture in the low Mach limit is given by the incompressible Navier-Stokes equation with external forcing term $\mathbf{F}$

$$\partial_t \mathbf{v} + \nabla \cdot (\mathbf{v} \otimes \mathbf{v} + pI) = \nu \nabla^2 \mathbf{v} + \mathbf{F}. \tag{3}$$

These equations are solved numerically by a multi-species lattice Boltzmann method [2]. Besides the generalizations of the lattice Boltzmann method for aqueous mixtures, the algorithm has a simple stream-collide structure for each species:

$$f_i^k(\mathbf{x} + \mathbf{e_i}\Delta t, t + \Delta t) = f_i^k(\mathbf{x}, t) - \omega^k (f_i^k(\mathbf{x}, t) - f_i^{eq,*,k}(\mathbf{x}, t)) \tag{4}$$

This makes the method presented in [2] a perfect candidate for massively parallel implementations.

3 Performance and Scalability

This section describes the performance of *Musubi* [5] on the *Cray XE6* system *Hermit* at *HLRS*. *Hermit* provides 3,552 compute nodes with two *AMD Interlagos* processors each. With 16 cores per processor this results in 32 cores per node. For our performance analysis, up to 1,024 compute nodes or 32,768 cores are used. In this analysis only *MPI* parallelism is considered. Both, intranode and internode performance are measured, i.e. performance within a single node (up to 32 cores with as many *MPI* processes) and across multiple nodes connected via the *Gemini* network. A channel, filled with a complex spacer, resembling the actual production runs is used for the performance and scaling analysis here.

The fluid flows horizontally along the filaments of the mesh between the two membranes (cf. Fig. 2). To find the optimal point of operation for our production runs, a scaling analysis was done with this complex geometry. In the following we refer to a domain covering two filaments enclosing one free space section as a single spacer element. The single spacer element has a length and width of 0.2 cm and is discretized by 660,000 cells. For the investigation of different problem sizes, multiples of this single spacer element are concatenated up to a total length of 20 cm with roughly 66 million cells. Such a channel slice has a width of 0.2 cm and periodic boundaries are assumed in this direction. Problems larger than 66 million elements are then achieved by appending multiple channel slices together. This setup covers all relevant production settings, hence all effects that influence the production performance are included in the analysis. The schematic layout of the simulation setup is illustrated in [6].

Our solver framework ensures nearly perfect balancing of the computational cells, with a difference of at most one in the cell count between any two partitions and same computational costs for each cell. However, due to the irregular domain and the large number of walls in the spacer filled channel, the communication surface of different processes might vary drastically, resulting potentially in large imbalances of communication costs. Furthermore, the computations involved in treating cells at inlet, outlet and membrane boundaries results in additional imbalances of computational load. It is worth mentioning that it is not easy to achieve load balancing with respect to both computational load and communication costs due to the space-filling curve used in our framework for cell ordering.

Two settings, the classical single-species flow and the multi-species flow consisting of water, sodium and chloride, are investigated. In both models a *D3Q19* layout with a *BGK*-like collision operator is used. Streaming and collision steps are solved at each time step for both models. However, the multi-species LBM model used in this work requires the solution of an additional cell local linear equation system, increasing the required number of floating point operations per lattice update. The single-fluid solver requires only around 169 floating point operations per cell (=lattice) update, while a simulation with 3 species requires 850 operations per update. This is a factor of 5 more in the number of operations, while the memory consumption grows only by a factor of 3 to account for all species. The operation count increases more than linearly with the number of species, as the size of the equation system to be solved increases with the number of species (hence the total operation count is proportional to the square of the number of species). Also, the amount of data to be exchanged via *MPI* increases linearly with the number of species. Expectations therefore are an increased sustained performance in the multi-species simulation, while scalability might be slightly decreased.

Hermit has a theoretical peak performance of 294.4 GFLOPS per node. With 3 species, a sustained performance of roughly 7.2 % of this peak is achieved, while with a single fluid 4.2 % are obtained on a single node. This reveals a better utilization of the *Interlagos* processors by the multi-species simulation, due to the increased number of operations per byte.

Weak scaling for a problem size of approximately 63,000 elements per node is shown in Fig. 4a. As can be seen, the weak scaling is almost perfect for single fluid as well as three species, with only a small drop in the parallel efficiency for larger counts of compute nodes.

Strong scaling for a problem size of 66.2 million cells is shown in Fig. 4b. Similarly to the weak scaling, the simulation with three species shows comparable scalability to the single fluid simulation, though both drop more significantly in this case. The performance increase in single-species simulation for 2^{10} is due to cache effects as the problem size is getting small enough to fit into cache.

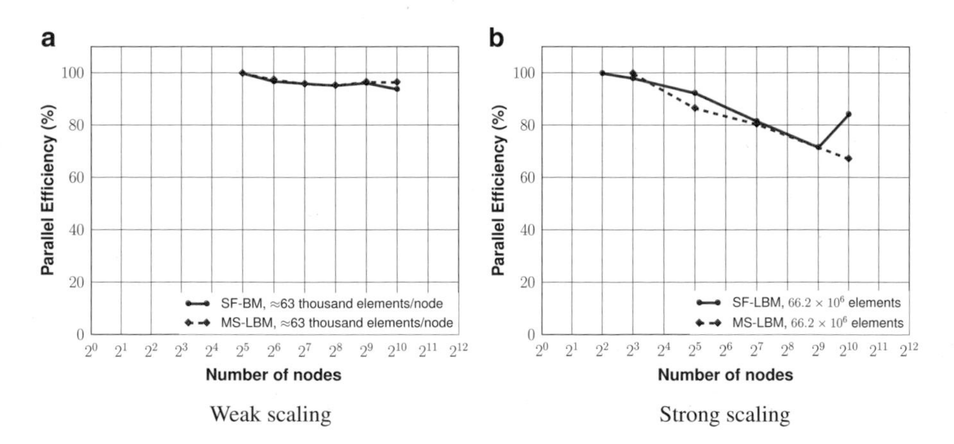

Fig. 4 Weak (**a**) and strong (**b**) scaling performance of the single-fluid and multi-species simulation with complex spacer structure

4 Simulation

In our production runs we first concentrate on the detailed simulation of pure flow through the channel. The most detailed simulations were done with 520 million cells running on 4,096 cores and simulating 0.25 s in 4 h. However, for the identification of an engineering model with few parameters that can be used to design of the overall electrodialysis process we used the following setup. The domains were discretized with around 10 million cells which were found to provide sufficient resolution to approximate the pressure for the investigated Reynolds numbers. For the execution of the simulations 1,024 cores were used, which was identified to provide a good time to solution with high performance for this problem size in our performance analysis. Each of these runs consumed around 4 h. The resulting data of a single of these run consumes 50 GB of disk space. After this analysis of the pure flow we started with the investigation of coupled physics by applying a fixed electric field perpendicular to the advective flow direction.

4.1 Comparison to Experimental Data

To validate the deployed numerical method and simulation setup, a laboratory scale woven spacer as described in Sect. 3 was simulated and the observed pressure drop compared to measured data. Figure 5 shows this comparison for varying numbers of flow channels. The shown ζ model is an engineering prediction model developed to fit the experimental data points and accounting for measure errors that are

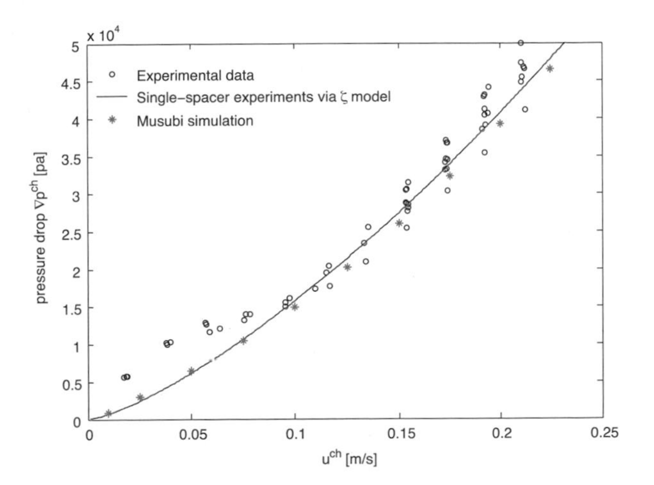

Fig. 5 Comparison of simulation results against experimental results with different spacer filled flow channels. The ζ (engineering prediction) model obtained from the experimental data is used to predict the single channel behavior

introduced due to the duct system before and after the channels. We conclude a good agreement between our numerical simulations and the experimental setup. With the simulations, a larger parameter space can be covered to develop an engineering model for the relevant design parameters related to the flow and especially the pressure drop in this electrodialysis process.

4.2 Simulations for the Pressure Drop Engineering Prediction Model

The fully systematic investigation on the influence of the spacer geometry on the performance of electrodialysis processes is still ongoing. However, for non-woven spacer, a first set of simulations was done in the course of this project. The four independent geometric design parameters are illustrated in Fig. 6. To find a model with 14 variables, a set of 216 simulations had to be performed. The study contributes to obtain profound insight into the local dynamic behavior of mass transport in electrodialysis processes. With this, it is for the first time possible to understand and optimize the process over the complete range of relevant design parameters. The engineering prediction model developed from these simulations will be published in the international Journal. Therefore, they are not presented here.

Figure 7 shows typical streamline patterns in a small section of the flow channel for two different spacer structures (woven and non-woven spacer) as observed in the flow simulations from the parameter analysis.

4.3 Multispecies Simulation

To illustrate the results of the presented multi-species Lattice Boltzmann implementation, we consider the transport of ionic species (sodium-chloride) in an aqueous solvent through a single spacer element. We impose a non-zero sodium-chloride concentration profile at the inlet of the channel and advect it downstream by imposing a mixture pressure drop along the channel. Besides the advection

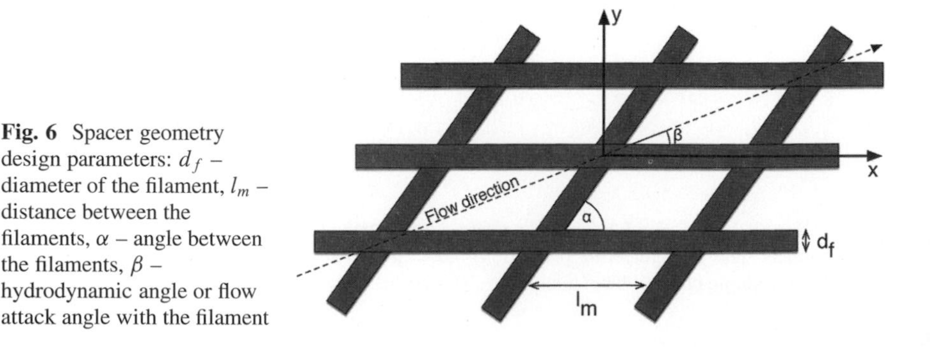

Fig. 6 Spacer geometry design parameters: d_f – diameter of the filament, l_m – distance between the filaments, α – angle between the filaments, β – hydrodynamic angle or flow attack angle with the filament

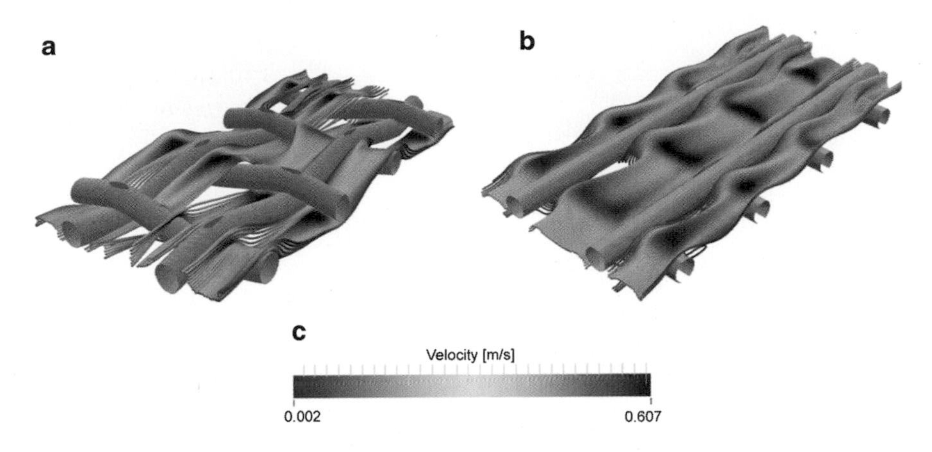

Fig. 7 Velocity distribution in spacer channel. (**a**) WoenSpacer. (**b**) nonWovenSpacer. (**c**) Color scale

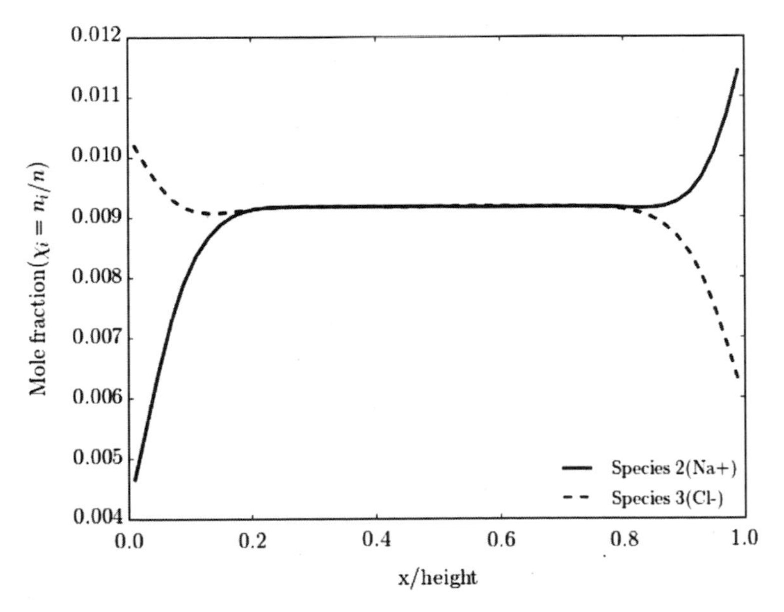

Fig. 8 Mole fraction profile of Na+ and Cl- ions along the channel height measured at the center of the channel. Ion separation is clearly seen near the membrane due to applied electric field

in downstream direction, concentration gradients lead to diffusion according to the Maxwell-Stefan relation in cross-stream-direction. Additionally we impose a fixed electric field perpendicular to the main flow direction, which drives sodium and chloride ions into opposite directions (according to their specific charges) towards the membranes as in Fig. 8. At the walls, ions are absorbed by a simple membrane model that serves as a boundary condition to the species here. The resulting concentration profiles and streamlines for various points in time are shown

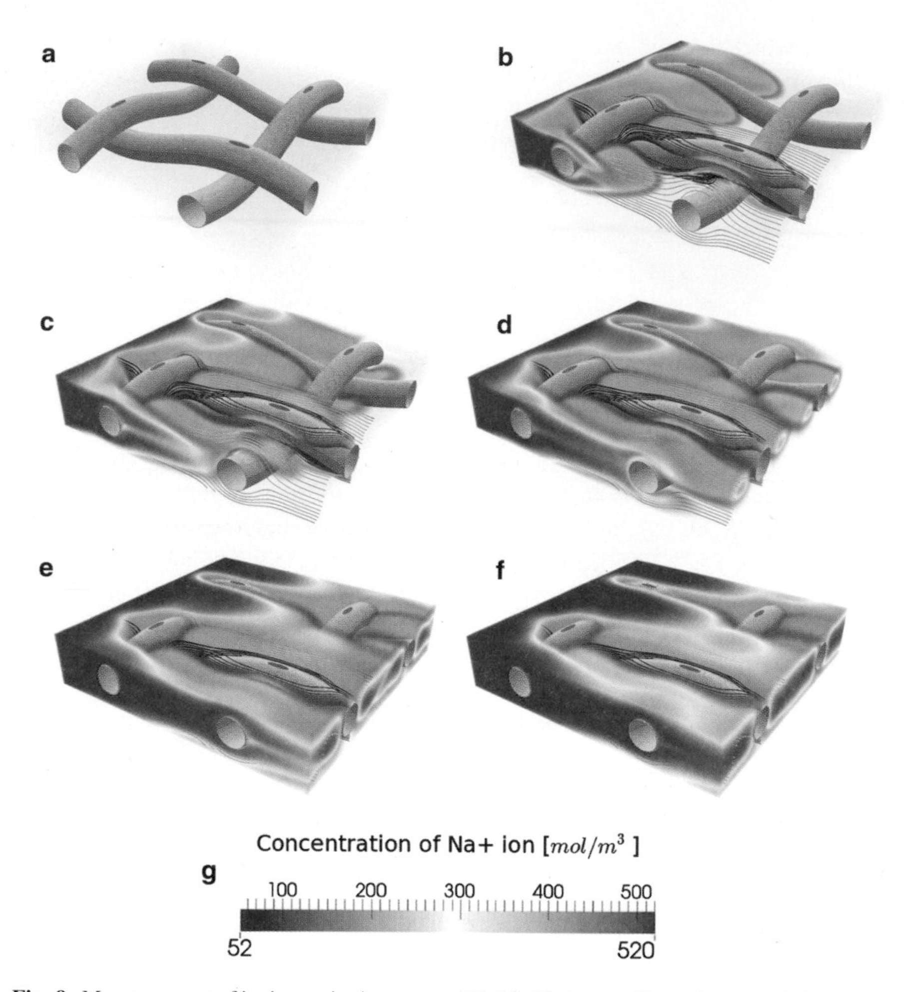

Fig. 9 Mass transport of ionic species in a spacer filled fluid channel. The volume rendering shows the advection of an ion concentration profile prescribed at the inlet, while the streamlines show the mixture velocity profile. (**a**) t = 0 s. (**b**) t = 0.5 s. (**c**) t = 0.75 s. (**d**) t = 1.0 s. (**e**) t = 1.5 s. (**f**) t = 2.0 s. (**g**) Color scale

in Fig. 9. In particular, the influence of the spacer structure on the concentration and velocity profiles can be seen.

5 Summary and Outlook

This work is concerned with the simulations for an electrodialysis process to desalinate water. We briefly describe the governing equations and physical effects to consider. The main focus in the presented work is the flow simulation in the channels, which contain a complex structure and thus a non-trivial flow field. Specifically the strong and weak scaling performance is presented for up to 1,024 compute nodes or 32,768 MPI processes. This scaling analysis was done for single species simulations as well as for simulations with three species, that is required for production runs. The scaling analysis shows that both numerical schemes scale fine on *Hermit* and large production runs where done on large process counts.

First validations of the simulated pressure drop by experiments where conducted successfully. With the help of the numerical experiments it was possible to create an engineering prediction model for few design parameters, which accurately describes the dependence of the pressure drop. Finally, the multi-species LBM is deployed to simulate transport of ionic species (sodium-chloride) in an aqueous solvent through a spacer structure under a fixed external electrical field.

In our ongoing work we focus on extending the simulations of fluid flow with overlayed electrical fields from fixed external field to a full interaction of ions from electrodynamics. To obtain stable coupled simulations results, this requires better boundary conditions with higher accuracy. Especially for liquid electrolytes, which are the major research topic in this project, further boundary conditions especially at membrane interface are under investigation. Overall, we will use the results of our research to provide an engineering prediction model, which covers all relevant design parameters of the electrodialysis process.

Acknowledgements This work was funded by the German Federal Ministry of Education and Research (Bundesministerium für Bildung und Forschung, BMBF) in the framework of the HPC software initiative in the project HISEEM. We thank *HLRS* for the granted compute time on the *Hermit* system and are grateful for their and *Cray*'s kind support.

References

1. Bernsdorf, J.: in text. Lattice-Boltzmann simulation of reactive flow through a porous media catalyst. In: Recent Progress in CFD, Japan, No. 08-07, JSAE Symposium, pp. 19–23 (2007)
2. Zudrop, J., Roller, S., Pietro, A.: Lattice Boltzmann scheme for electrolytes by an extended Maxwell-Stefan approach. Phys. Rev. E **89**, 053310 (2014). doi:10.1103/PhysRevE.89.053310, http://link.aps.org/doi/10.1103/PhysRevE.89.053310
3. Roller, S., Bernsdorf, J., Klimach, H., Hasert, M., Harlacher, D., Cakircali, M., Zimny, S., Masilamani, K., Didinger, L., Zudrop, J.: An adaptable simulation framework based on a linearized octree. In: High Performance Computing on Vector Systems: Proceedings of the High Performance Computing Center Stuttgart (2011)
4. Klimach, H., Hasert, M., Zudrop, J., Roller, S.: Distributed octree mesh infrastructure for flow simulations. In: European Congress on Computational Methods in Applied Sciences and Engineering, Vienna, pp. 1–15 (2012)
5. Hasert, M., Masilamani, K., Zimny, S., Klimach, H., Qi, J., Bernsdorf, J., Roller, S.: Complex fluid simulations with the parallel tree-based Lattice Boltzmann solver Musubi. J. Comput. Sci. **5**(5), 784–794 (2013). doi:10.1016/j.jocs.2013.11.001
6. Masilamani, K., Klimach, H., Roller S.: Highly efficient integrated simulation of electro-membrane processes for desalination of sea water. In: Nagel, W.E., Kröner, D.B., Resch, M.M. (eds.) High Performance Computing in Science and Engineering 2013, pp. 493–508. Springer, Switzerland (2013)

Mesoscale Simulations of Fluid-Fluid Interfaces

T. Krüger, S. Frijters, F. Günther, B. Kaoui, and Jens Harting

Abstract Fluid-fluid interfaces appear in numerous systems of academic and industrial interest. Their dynamics is difficult to track since they are usually deformable and of not a priori known shape. Computer simulations pose an attractive way to gain insight into the physics of interfaces. In this report we restrict ourselves to two classes of interfaces and their simulation by means of numerical schemes coupled to the lattice Boltzmann method as a solver for the hydrodynamics of the problem. These are the immersed boundary method for the simulation of vesicles and capsules and the Shan-Chen pseudopotential approach for multi-component fluids in combination with a molecular dynamics algorithm for the simulation of nanoparticle stabilized emulsions. The advantage of these algorithms is their inherent locality allowing to develop highly scalable codes which can be used to harness the computational power of the currently largest available supercomputers.

T. Krüger
Department of Applied Physics, Eindhoven University of Technology, Den Dolech 2, 5600MB Eindhoven, The Netherlands

School of Engineering, Institute for Materials and Processes, University of Edinburgh, Mayfield Road, Edinburgh EH9 3JL, Scotland, UK

S. Frijters • F. Günther
Department of Applied Physics, Eindhoven University of Technology, Den Dolech 2, 5600MB Eindhoven, The Netherlands

B. Kaoui
Department of Applied Physics, Eindhoven University of Technology, Den Dolech 2, 5600MB Eindhoven, The Netherlands

Theoretical Physics I, University of Bayreuth, Universitätsstrasse 30, 95447 Bayreuth, Germany

J. Harting (✉)
Department of Applied Physics, Eindhoven University of Technology, Den Dolech 2, 5600MB Eindhoven, The Netherlands

Institute for Computational Physics, University of Stuttgart, Allmandring 3, 70569 Stuttgart, Germany
e-mail: jens@icp.uni-stuttgart.de

W.E. Nagel et al. (eds.), *High Performance Computing in Science and Engineering '14*, DOI 10.1007/978-3-319-10810-0_36

1 Introduction

Interfaces are ubiquitous in soft matter systems and appear in various shapes and sizes. Prominent examples are capsules and all kinds of biological cells [1, 2]. In these cases, the interface is an additional material whose constitutive behavior has to be specified. Understanding the dynamics of membranes is important for disease detection by measuring mechanical properties of living cell membranes [3], targeted drug delivery [4], and predicting the viscosity of biofluids, such as blood [5]. Typical applications are lab-on-chip devices for particle identification and separation [6].

Other classes of fluid-fluid interfaces can be found in emulsions (liquid drops suspended in another liquid), foams (gas bubbles separated by thin liquid films), and liquid aerosols (liquid drops in gas) where at least two immiscible fluid phases are mixed. The interface is then defined by the common boundaries of the phases. Emulsions are of central importance for food processing (e.g., milk, salad dressings), pharmaceutics (e.g., lotions, vaccines, disinfection), and enhanced oil recovery.

Emulsions and foams are usually unstable. Drops and bubbles tend to coalesce gradually, which reduces the interfacial free energy. The traditional approach for the stabilization of emulsions is to add surfactants [7]. Surfactants are amphiphilic molecules for which it is energetically favorable to accumulate at the interface, which in turn leads to a decrease of surface tension and prevents demixing. Alternatively, emulsions can be stabilized by using colloidal particles: these particles also accumulate at the interface where they replace segments of fluid-fluid interface by fluid-particle interfaces and thus lower the interfacial free energy. However, the physical mechanism is different from that of stabilization by surfactants, and the surface tension is unchanged by the nanoparticles [8, 9]. For nanoparticle-stabilized emulsions, one distinguishes between so-called "Pickering emulsions" [10, 11] and "bijels" (bicontinuous interfacially jammed emulsion gels) [12, 13]. The former is an emulsion of discrete droplets in a continuous liquid. In the latter, both phases are continuously distributed.

Fluid-fluid interface problems are hard to solve analytically since the interface is usually deformable and its shape not known a priori. Therefore, the interface dynamics is fully coupled to that of the ambient phases. Analytical solutions are only available for academic cases. These circumstances call for numerical methods and computer simulations which can also provide access to observables not traceable in experiments such as local interface curvature or fluid and interface stresses. Therefore, computer simulations can be used to complement experiments.

In this report, we focus on capsules and emulsions in systems with dimensions comparable to the lateral extension of the interfaces. At this scale, fluid dynamics is dominated by effects due to viscosity, surface tension or elasticity, and inertia. Gravity is usually not relevant in microfluidics because the capillary length for water droplets in air is typically of the order of a few millimeters. The relative strength of the three major contributions (viscosity, surface tension, inertia) can be described by dimensionless parameters, such as the capillary number (Ca, viscous stress vs.

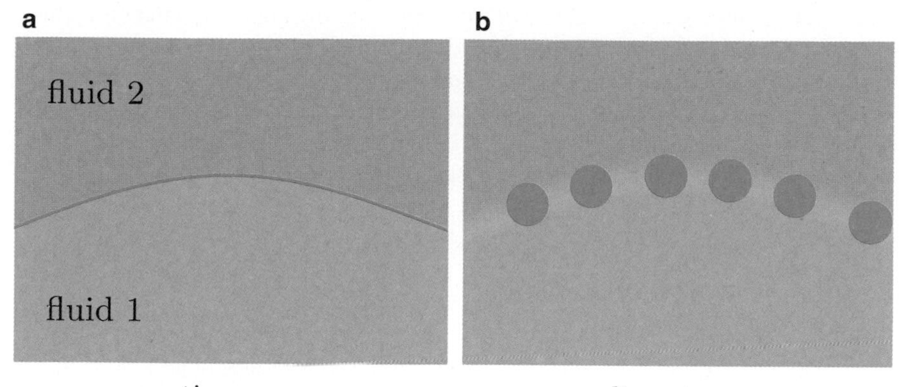

Fig. 1 Coarse-graining levels for a fluid-fluid interface (fluids 1 and 2). (**a**) the interface (*green line*) is explicitly tracked in a Lagrangian manner, and the interface properties have to be provided in form of a constitutive law. There is no direct interaction between the fluid species. The substructure of the interface is not known (continuum picture). (b) nanoparticles (*green*) are explicitly resolved (discrete picture). Their individual motions are tracked, and their geometrical and interaction properties have to be specified (e.g., shape, contact angles). The interaction of the fluid species has to be defined as well. For (**b**), the choice of the mutual interactions leads to an emergent macroscopic interface behaviour (From [1])

surface tension or elasticity), the Reynolds number (Re, inertial vs. viscous stresses), or the Weber number (We, inertial stress vs. surface tension).

In the current report, the fluid phases are treated as Newtonian fluids simulated by the lattice Boltzmann method (detailed in Sect. 2.1). Depending on the physical conditions and the desired level of coarse-graining, it may be sufficient to treat the interface as an effective two-dimensional material obeying a well-defined constitutive law or it may be necessary to resolve the substructure of the interface (cf. Fig. 1). Following Ref. [1], we consider two cases. In the continuum description for capsule membranes, the immersed boundary method is coupled to the single-component lattice Boltzmann (LB) method (Sect. 2.2). The LB Shan-Chen multi-component model is the basis for the other case where explicitly resolved solid nanoparticles interact with the fluid phases (Sect. 2.3). In Sect. 3, we present several examples (separation of red blood cells and platelets on the one hand and particle stabilized interfaces on the other hand). Finally, we conclude in section "Conclusions".

2 Numerical Approaches

The physical problem consists in capturing the dynamics of an interface separating distinct fluids. In the simplest situation, there are two different fluids separated by an interface. Mathematically, this corresponds to two domains separated by a free-moving boundary. There are two classes of numerical approaches to track the motion of an interface: (i) either *front tracking* (e.g., the boundary integral method or the immersed boundary method) in which the motion of marker points attached to the interface is tracked, or (ii) *front capturing* (e.g., the phase field method or level set methods) where a scalar field is used as an order parameter, indicating the composition of the fluid at a given point. In both approaches, the momentum of the incompressible Newtonian fluids obeys the Navier-Stokes (NS) equations.

2.1 Lattice-Boltzmann Method

Instead of solving the NS equations directly, we use a mesoscopic approach: the *lattice Boltzmann method* (LBM). The LBM has gained popularity among scientists and engineers because of its relatively straightforward implementation compared to other approaches such as the finite element method (for reviews, see [14, 15]). In the LBM, the fluid is considered as a cluster of pseudo-particles that move on a lattice under the action of external forces. To each pseudo-particle is associated a distribution function f_i, the main quantity in the LBM. It gives the probability to find a pseudo-particle at a position $\mathbf{r}$ with a velocity in direction $\mathbf{e}_i$. In the LBM, both the position and velocity spaces are discretised: Δx is the grid spacing, and $\mathbf{e}_i$ are the discretised velocity directions. The time evolution of f_i is governed by the so-called lattice Boltzmann equation,

$$f_i(\mathbf{r} + \mathbf{e}_i \Delta t, t + \Delta t) - f_i(\mathbf{r}, t) = \Omega_i, \tag{1}$$

where Δt is the discrete time step. We approximate the collision operator Ω_i by the Bhatnagar-Gross-Krook (BGK) operator, $\Omega_i = -(f_i - f_i^{\mathrm{eq}})/\tau$, which describes the relaxation of f_i towards its local equilibrium, f_i^{eq}, on a time scale τ. The relaxation time is related to the macroscopic dynamical viscosity η via $\eta = \rho c_{\mathrm{s}}^2 \frac{\Delta x^2}{\Delta t} \left(\tau - \frac{1}{2}\right)$, where $c_{\mathrm{s}} = 1/\sqrt{3}$ is the lattice speed of sound. The equilibrium distribution is given by a truncated Maxwell-Boltzmann distribution. The hydrodynamical macroscopic quantities are computed using the first and the second moments of f_i.

The LBM can also handle multi-phase and multi-component fluids and a number of corresponding extensions of the method have been published in the past [16–19]. In the Shan-Chen multi-component model [16], a system containing N miscible or immiscible fluids (with index σ) are described by N sets of distribution functions $f_{\sigma,i}$, one for each species. As a consequence, N lattice Boltzmann equations with relaxation times τ_σ have to be considered. The interaction between the fluid species is mediated via a local force density. The Shan-Chen model belongs to the class of front capturing methods. It is suitable to track interfaces for which the topology evolves in time, for example, the breakup of a droplet.

2.2 Continuum Interface Model

In this section, we present the *immersed boundary method* (IBM) as a front tracking approach [20]. Such methods are mostly employed when the interface is formed by an additional continuous material whose constitutive behaviour (e.g., elasticity, viscosity) is assumed to be known. Examples are capsules, vesicles, or biological cells.

Within the IBM, the interface is considered sharp (zero thickness) and is represented by a cluster of marker points (nodes) which constitute a moving Lagrangian mesh (cf. Fig. 2a). This mesh is immersed in a fixed Eulerian lattice representing the fluid. To consider correct dynamics, a bi-directional coupling of the lattice fluid and the moving Lagrangian mesh has to be taken into account. On the one hand, the interface is modelled as an impermeable structure obeying the no-slip condition at its surface. It is assumed that the flow field is continuous across the interface and that the interface is massless. Therefore, the interface is moving along with the ambient fluid velocity. On the other hand, a deformation of the interface generally leads to stresses reacting back onto the fluid via local forces. The stresses depend on the chosen constitutive behaviour of the interface and are not predicted within the IBM itself. The two-way coupling is accomplished in two main steps (as detailed in [20] and [21]): (i) velocity interpolation and Lagrangian node advection and (ii) force spreading (reaction).

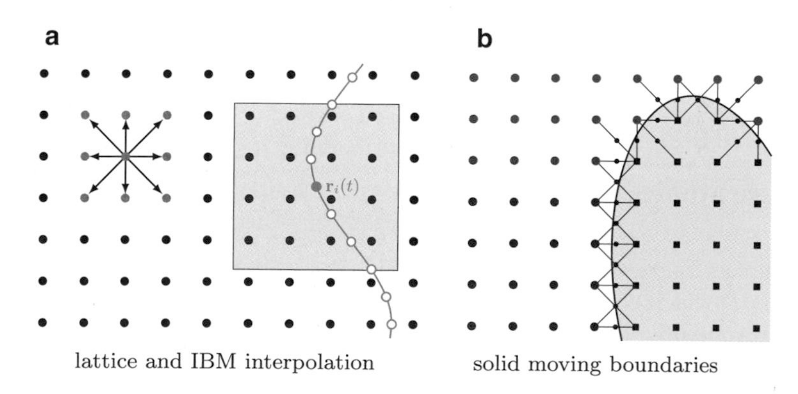

Fig. 2 Two-dimensional illustrations of the lattice Boltzmann propagation, immersed boundary interpolation, and bounce-back. (**a**) During propagation, the populations f_i move to their next neighbors (*gray lattice sites and arrows*). Within the immersed boundary method, an interface node (here: node i at position $\mathbf{r}_i(t)$) is coupled to the single-component lattice fluid. The range of the interpolation stencil (here: 4×4) is denoted by the square region. All lattice sites within this region have to be considered during velocity interpolation and force spreading. (**b**) An ellipsoidal nanoparticle is located at a fluid-fluid interface. One fluid component is indicated by *red*, the other by *blue circles*. The colour gradient illustrates the interface region. *Squares* denote lattice sites inside the particle. Populations are bounced back at points (shown as *small circles*) located half-way between neighbouring fluid and particle sites. This gives rise to an effective staircase description of the particle shape (From [1])

2.3 Discrete Interface Model

If there is no clear scale separation between immersed particles and the lateral interface extension, it may be necessary to model the particles explicitly. Here, we consider an ensemble of particles with well-defined wetting behaviour. There are several approaches to simulate a system of two immiscible fluids and particles. One example which has recently been applied by several groups is Molecular Dynamics (MD) coupled to the LBM [9, 12, 22–27]. The advantage of this combination is the possibility to resolve the particles as well as both fluids in such a way that all relevant hydrodynamical properties are included. The particles are generally assumed to be rigid and can have arbitrary shapes, where we restrict ourselves to spheres and ellipsoids.

For the fluid-particle coupling, the particles are discretised on the lattice: sites which are occupied by a particle are marked as solid. Following the approach proposed by Ladd [28], populations propagating from fluid to particle sites are bounced back in the direction they came from (cf. Fig. 2b). In this process, the populations receive additional momentum due to the motion of the particles. In the context of the Shan-Chen multi-component algorithm coupled to the particle solver, the outermost sites covered by a particle are filled with a virtual fluid corresponding to a suitable average of the surrounding unoccupied sites. This approach provides accurate dynamics of the two-component fluid near the particle surface. The wetting properties of the particle surface can be controlled by shifting the local density difference of both fluid species by a given amount $\Delta\rho$. Occasionally, when the particles move, lattice sites change from particle to fluid state, which has to be treated properly [9, 22].

3 Case Studies

The simulations presented in this report have been peformed using our simulation code *LB3D* which combines a ternary amphiphilic multicomponent lattice Boltzmann solver with a molecular dynamics module for suspended solid particles and an immersed boundary method module for deformable objects. The code has shown excellent performance and scaling behaviour in the past and was the workhorse for several previous projects on HLRS and SSC Karlsruhe HPC installations. As such, we refer the reader to the according references for technical details [1, 7, 9, 22–25, 29, 30]. In general, the simulations presented in the following case studies can be seen as building blocks for large scale simulations containing many particles, cells, or droplets. Those simulations are by definition very costly since the 3D dynamic solution of the lattice Boltzmann equation together with the interface dynamics requires generally large systems of at least 256^3 or 512^3 lattice nodes and has to run for several million timesteps. A simulation of such kind easily requires several days on several thousand cores on state of the art Tier-0 HPC resources.

3.1 Flow-Induced Separation of Red Blood Cells and Platelets

In this section, we show three-dimensional simulations of the separation of blood components as an example for the immersed-boundary-lattice-Boltzmann method (Sect. 2.2).

Efficient separation of biological cells plays a major role in present-day medical applications, for example, the detection of diseased cells or the enrichment of rare cells. While active cell sorting relies on external forces acting on tagged particles, the idea of passive sorting is to take advantage of hydrodynamic effects in combination with the cells' membrane (interface) properties. It is known that deformable and rigid particles behave differently when exposed to external flow fields. For example, in Stokes flow (Re $=$ 0), deformable capsules show a strong tendency to migrate towards the centreline of a pressure-gradient driven channel flow. Therefore, it has been proposed to use the particle deformability as intrinsic marker to separate rigid and deformable particles in specifically designed microfluidic devices. We show that the separation of red blood cells (RBCs) and platelets, which has been shown recently via experiments [31], can be simulated by means of the LBM and immersed boundary method.

The RBCs and platelets are modelled as closed two-dimensional membranes with identical fluids in the interior and exterior regions. In the present approach, a finite element method is used to compute the local surface shear deformation and area dilatation and the resulting forces [32–34]. 40 RBCs and 30 platelets are randomly distributed in the bottom 30 % of a channel segment with $50 \times 50 \times 50\,\mu\text{m}^3$ volume, cf. Fig. 3. The platelets can be considered as rigid ellipsoidal particles. The large RBC and platelet diameters are 16 and 5.5 lattice constants, respectively. The channel is bounded by two parallel walls and is periodic in the other two directions. A force f in rightward direction is used to drive the flow, resulting in a Poiseuille-

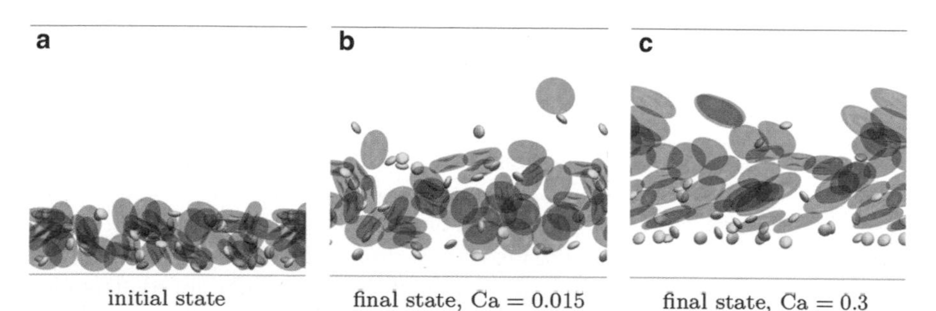

Fig. 3 Simulation snapshots of the separation of red blood cells (RBCs) and platelets. (**a**) Initially, the random suspension is located in the bottom 30 % of the channel (50 μm diameter, denoted by two black lines). RBCs (*red*) are shown with reduced opacity to reveal the platelet positions (*yellow*). Panels (**b**) and (**c**) show the lateral distribution of the cells after they have been moving downstream (rightwards) by about 20 mm on average. For rigid and tumbling RBCs, lateral separation of RBCs and platelets is not pronounced (**b**). The separation is more efficient when the RBCs are strongly deformable and are tank-treading in the vicinity of the walls (**c**) (From [1])

like velocity profile. The RBC membrane (interface) is characterised by the capillary number $\mathrm{Ca} = \eta\dot{\gamma}r/\kappa_\mathrm{S}$. Here, $\dot{\gamma}$ is the average shear rate, $r = 4\,\mu\mathrm{m}$ is the large RBC radius and κ_s is its shear elasticity which is about $5\,\mu\mathrm{N\,m^{-1}}$ for healthy RBCs. Two simulations have been run, one with the normal elasticity, leading to $\mathrm{Ca} = 0.3$ for the selected viscosity and driving force. In the other case, the RBC shear modulus has been increased by a factor of 20, and the capillary number was 0.015. This situation is typical for diseased RBCs (e.g., due to malaria or sickle cell anaemia).

The initial simulation states are shown in Fig. 3a, while the panels (b) and (c) show the state after about 20 mm downstream motion for different capillary numbers. It can be seen that RBCs and platelets can be efficiently separated when the RBCs are sufficiently deformable (Fig. 3c). The platelets marginate into the gap forming between the bottom wall and the RBC bulk. Contrarily, for the rigid RBCs, separation is visible but not pronounced (Fig. 3b). Additionally, the bulk of the RBCs moves faster towards the centreplane when the particles are more deformable, i.e., when Ca is larger. For $\mathrm{Ca} = 0.30$, RBCs near the wall are observed to tank-tread; for $\mathrm{Ca} = 0.015$, all are tumbling.

3.2 *Ellipsoidal and Spherical Particles at Fluid Interfaces*

We now consider the effect of massive nanoparticles adsorped to a fluid-fluid interface as an example of the coupled lattice Boltzmann-molecular dynamics method explained in Sect. 2.3. The first example shows the ordering of ellipsoidal particles at the interface of a spherical fluid droplet and compares it with the corresponding results of particles at a flat interface. This is followed by a case study highlighting how the presence of spherical nanoparticles affects the properties of a droplet in shear flow.

Particles adsorbed at a fluid-fluid interface can stabilise emulsions of immiscible fluids (e.g., Pickering emulsions where droplets of one fluid are immersed in another fluid and are stabilised by colloidal particles) [22–24]. These colloids are not necessarily spherical, e.g. clay particles, which have a flat shape. A simple approximation of such a particle shape is an ellipsoid. In this section, the behaviour of an ensemble of elongated ellipsoidal particles with aspect ratio $m = 2$ at a single droplet interface is discussed [35]. The results are compared to the corresponding case of a flat interface. The initial particle orientation is such that the long axis is orthogonal to the local interface (cf. Fig. 4a). As discussed in [23,24], for the case of a single prolate ellipsoidal particle at a flat interface, the equilibrium configuration is a particle orientation parallel to the interface. The simulations discussed in this section have been performed for coverage fractions $\chi = 0.153$ and $\chi = 0.305$.

Figure 4a, b show the particle laden droplet at the beginning of the simulation where all particles are perpendicular to the interface for both values of χ. For the case of $\chi = 0.305$, the particles are comparably close to each other so that capillary interactions lead to the clustering motion of particles. In Fig. 4c, at the end of the simulation, the particles have completed a rotation towards the interface

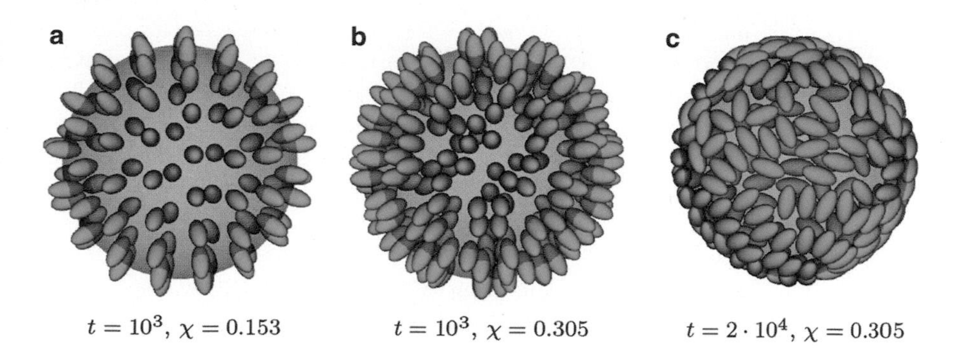

Fig. 4 Snapshots of a particle-laden droplet for different times (in time steps) and coverage fractions χ. One observes a higher global spatial order for the more dilute system in (**a**) as compared to the denser case in (**b**) (From [1])

and stabilise the largest interfacial area possible. However, their local dynamics and temporal re-ordering is a continuous interplay between capillary interactions, hydrodynamic waves due to interface deformations and particle collisions.

Inspired from liquid crystal analysis, we use the uniaxial order parameter S to characterise the orientational ordering of anisotropic particles. Figure 5a shows the time evolution of the order parameter S for particles at a droplet interface for $\chi = 0.153$ and $\chi = 0.305$ and the corresponding cases for particles at a flat interface. Initially, the order parameter assumes the value $S_{\perp} = 1$ which corresponds to perfect alignment perpendicularly to the interface. In the case of flat interfaces, the order parameter reaches the final value of $S \approx S_{\parallel} = -0.5$. This corresponds to the state where all particles are aligned with the interface plane. For particle laden droplets (cf. Fig. 4c), a value of $S \approx -0.4 > S_{\parallel}$ is found, which indicates that the particles are not entirely aligned with the interface. Furthermore, S fluctuates in time. One possible reason for this deviation of S from $S_{\parallel}$ is that the droplet does not have the exact spherical shape which is assumed for the calculation of S. The fluctuation of S can be explained by distortions of the droplet shape due to the dynamics of the particle rotations. Another difference between the curved and the flat interface is the time scale for the particle flip.

To measure the spatial short range ordering of the particles, we define a pair correlation function $g(r)$ In Fig. 5b, the time dependence of $g(r)$ is shown for a spherical interface and $\chi = 0.305$. One notices that all maxima decrease in time. This means that the local ordering decreases, particularly in the first 10^4 time steps. Afterwards, $g(r)$ changes only slightly. Figure 5c compares $g(r)$ for $\chi = 0.153$ and $\chi = 0.305$ after 10^3 time steps and can be used to explain the differences of the snapshots in Fig. 4a, b: the order is higher for a smaller concentration χ. In the case of $\chi = 0.305$, the ellipsoids are attracted by capillary forces leading to a clustering of particles. This causes disorder, which manifests itself in smaller peak amplitudes of $g(r)$.

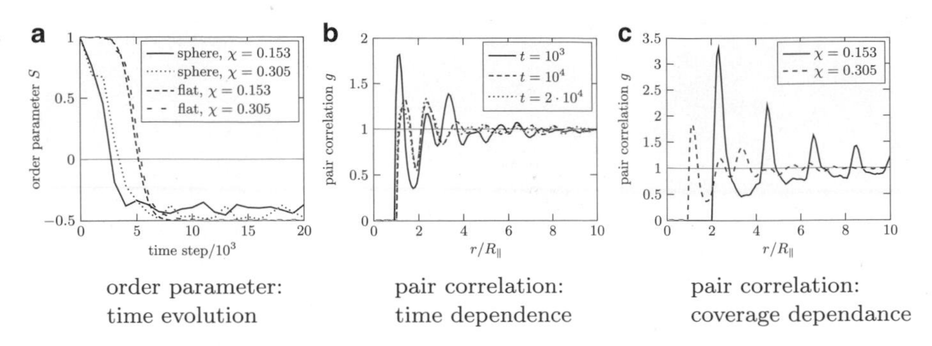

Fig. 5 (**a**) Time evolution of the uniaxial order parameter S: the order decreases faster for the case of a sphere while the coverage fraction does not play an appreciable role. (**b**) The time dependence of the pair correlation function $g(r)$ for particles adsorbed at a spherical interface ($\chi = 0.305$) is shown for different times. (**c**) Pair correlation function for the spherical interface after 10^3 time steps for different values of χ corresponding to Fig. 4a, b (From [1])

To summarize, the particle coverage χ has a marginal influence on the duration of the flipping towards the interface. However, the particles rotate faster when the interface is spherical. For increasing χ, the spatial particle ordering is reduced since the average particle distance is sufficiently small to allow for attractive capillary forces which in turn cause particle clustering.

In many industrial applications, the particle-covered droplets as discussed in the previous example are not stationary or in equilibrium, but are instead subjected to external stresses or forces. The properties of the individual droplets are then of interest as their behaviour dictates that of, for example, an emulsion formed of these droplets. In this example, we consider monodisperse neutrally wetting spherical particles. The particles are adsorped to the droplet interface which is initially spherical. External shear is realised by using Lees-Edwards boundary conditions [36]. As this shear is applied to the system, the droplet will start to deform. To quantify this deformation, for small to moderate shear rates $\dot{\gamma}$, a dimensionless deformation parameter is used: $D = (L - B)/(L + B)$ where L is the length and B is the breadth of the droplet. This parameter will be zero for a sphere and tend to unity for a very strongly elongated droplet. As the droplet loses its spherical symmetry, it will also start to exhibit an angle of its long axis with respect to the shear flow: the inclination angle θ_d.

Increasing the shear rate a droplet is subjected to will generally increase its deformation, as well as reduce its inclination angle from an initial angle of 45°: the droplet is elongated and aligns with the shear flow. However, this is only valid as long as inertia can be neglected. When inertial forces are comparable to or stronger than viscous forces, the inclination angle can first increase beyond 45°, before its eventual reduction. We now discuss the effect of increasing the particle coverage fraction χ at constant shear rates $\dot{\gamma}$. Representative snapshots of the droplets for various χ and fixed $\dot{\gamma} = 0.47 \cdot 10^{-3}$ are presented in Fig. 6a–c. The particles are not homogeneously distributed over the droplet surface when shear is applied, due to

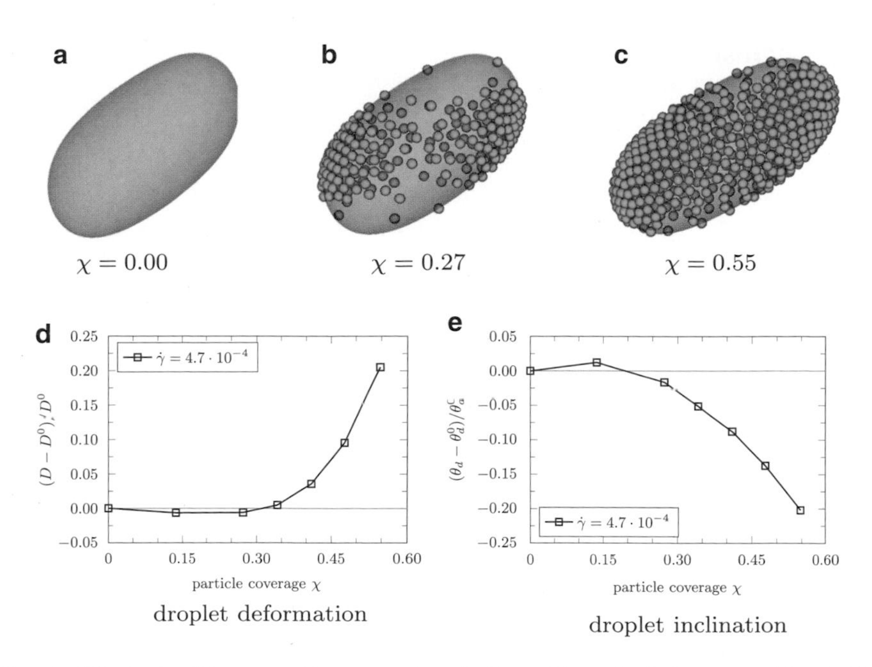

Fig. 6 Effect of adding monodisperse and neutrally wetting spherical nanoparticles to the interface of a droplet in shear flow. Representative snapshots of deformed particle-covered droplets are shown for $\dot{\gamma} = 4.7 \cdot 10^{-4}$ and (**a**) $\chi = 0$, (**b**) $\chi = 0.27$, and (**c**) $\chi = 0.55$ ([9], reproduced with permission of the Royal Society of Chemistry). The relative deformation of the droplet increases strongly for high coverage fraction, as shown in (**d**). In turn, the relative inclination angle decreases, indicating a better alignment with the shear flow, as shown in (**e**) (From [1])

an interplay between local curvature and shear velocities: high curvatures and low velocities are energetically favoured.

Figure 6d shows the change in deformation of the droplet as the particle coverage is increased. The deformation has been rescaled as $D^* = (D - D^0)/D^0$, where $D^0 \equiv D(\chi = 0)$. Due to a large decrease in free energy granted by the presence of particles at the interface, they are irreversibly adsorbed. When shear is applied, the particles are affected by this, and they start to move over the droplet surface, but cannot be swept away. As can be seen from the figure, high particle coverage fractions lead to a large increase in deformation of the droplet. The inertia of the particles plays a critical role here. The particles have to reverse direction to stay attached to the droplet, and massive particles strongly resist this change in velocity. This causes the interface to be dragged by the particles, increasing elongation from the tips. Figure 6e demonstrates the effect of particle coverage on the inclination angle of the droplet. Similar to the rescaling performed on the deformation, the inclination angle has been rescaled as $\theta_d^* = (\theta_d - \theta_d^0)/\theta_d^0$, where $\theta_d^0 \equiv \theta(\chi = 0)$. The inertial effects described above also cause the inclination angle to strongly decrease for large χ.

4 Conclusions

We presented a few of our recent applications of lattice Boltzmann based simulation methods to fluid-fluid interface problems: the immersed boundary method offers a flexible and simple approach to model deformable particles immersed in suspending fluids. As it is a front-tracking method, the interface configuration is directly known, which simplifies the force evaluation based on the interface deformation. Another advantage is that the constitutive properties of the interface can be controlled directly without tuning additional simulation parameters, and there is no need to remesh the fluid domain. Therefore, an implementation as ours can be used to simulate tens of thousands suspended deformable particles flowing in complex geometries on current supercomputers.

When the multi-component lattice Boltzmann method is coupled to a molecular dynamics algorithm for the description of suspended (colloidal) particles, particle-laden interfaces can be studied at a level where not only hydrodynamics, but also individual particles and their interactions are resolved. At the same time, efficient scaling to allow larger domains to be simulated is still provided due to the locality of the algorithm. While this method has been proven to be particularly suitable to study particle stabilised emulsions, one of its drawbacks is the diffuse interface between fluids in the Shan-Chen LBM. The particles have to be substantially larger than these diffuse interfaces in order to be able to stabilise them. Thus, even comparably simple applications as the ones provided in this report require large lattices and therefore access to supercomputing resources. However, the excellent performance and scaling behviour of the simulation method and our implementation *LB3D* has been demonstrated on several available supercomputing platforms including Hermit.

Acknowledgements Financial support is greatly acknowledged from NWO/STW (Vidi grant 10787 of J. Harting) and FOM/Shell IPP (09iPOG14 – "Detection and guidance of nanoparticles for enhanced oil recovery"). We thank the Gauss Center for Supercomputing and HLRS Stuttgart for the allocation of computating time on Hermit.

References

1. Krüger, T., Frijters, S., Günther, F., Kaoui, B., Harting, J.: Numerical simulations of complex fluid-fluid interface dynamics. Eur. Phys. J. Spec. Topics **222**, 177 (2013)
2. Pozrikidis, C. (ed.) Modeling and Simulation of Capsules and Biological Cells. Chapman & Hall/CRC Mathematical Biology and Medicine Series. Chapman & Hall/CRC, Boca Raton (2003)
3. Suresh, S., Spatz, J., Mills, J., Micoulet, A., Dao, M., Lim, C., Beil, M., Seufferlein, M.: Connections between single-cell biomechanics and human disease states: gastrointestinal cancer and malaria. Acta Biomater. **1**, 15–30 (2005)
4. Battaglia, L., Gallarate, M.: Lipid nanoparticles: state of the art, new preparation methods and challenges in drug delivery. Expert Opin. Drug Deliv. **9**(5), 497–508 (2012)

5. Fedosov, D., Pan, W., Caswell, B., Gompper, G., Karniadakis, G.: Predicting human blood viscosity in silico. Proc. Natl. Acad. Sci. **108**(29), 11772 (2011)
6. Hou, H., Bhagat, A., Chong, A., Mao, P., Tan, K., Han, J., Lim, C.: Deformability based cell margination—A simple microfluidic design for malaria-infected erythrocyte separation. Lab Chip **10**(19), 2605–2613 (2010)
7. Harting, J., Harvey, M., Chin, J., Venturoli, M., Coveney, P.V.: Large-scale lattice Boltzmann simulations of complex fluids: advances through the advent of computational grids. Phil. Trans. R. Soc. Lond. A **363**, 1895–1915 (2005)
8. Binks, B.: Particles as surfactants–similarities and differences. Cur. Opin. Colloid Interface Sci. **7**(1–2), 21–41 (2002)
9. Frijters, S., Günther, F., Harting, J.: Effects of nanoparticles and surfactant on droplets in shear flow. Soft Matter **8**(24), 6542–6556 (2012)
10. Ramsden, W.: Separation of solids in the surface-layers of solutions and 'suspensions' Proc. R. Soc. Lond. **72**, 156 (1903)
11. Pickering, S.: Emulsions. J. Chem. Soc. Trans. **91**, 2001–2021 (1907)
12. Stratford, K., Adhikari, R., Pagonabarraga, I., Desplat, J., Cates, M.: Colloidal jamming at interfaces: a route to fluid-bicontinuous gels. Science **309**(5744), 2198–2201 (2005)
13. Herzig, E., White, K., Schofield, A., Poon, W., Clegg, P.: Bicontinuous emulsions stabilized solely by colloidal particles. Nat. Mat. **6**(12), 966–971 (2007)
14. Succi, S.: The Lattice Boltzmann Equation. Oxford University Press, Oxford (2001)
15. Aidun, C., Clausen, J.: Lattice-Boltzmann method for complex flows. Ann. Rev. Fluid Mech. **42**, 439 (2010)
16. Shan, X., Chen, H.: Lattice Boltzmann model for simulating flows with multiple phases and components. Phys. Rev. E **47**, 1815 (1993)
17. Shan, X., Chen, H.: Simulation of nonideal gases and liquid-gas phase transitions by the lattice Boltzmann equation. Phys. Rev. E **49**, 2941 (1994)
18. Orlandini, E., Swift, M.R., Yeomans, J.M.: A lattice Boltzmann model of binary-fluid mixtures. Europhys. Lett. **32**, 463 (1995)
19. Dupin, M., Halliday, I., Care, C.: Multi-component lattice Boltzmann equation for mesoscale blood flow. J. Phys. A Math. Gen. **36**, 8517 (2003)
20. Peskin, C.: The immersed boundary method. Acta Numer. **11**, 479 (2002)
21. Kaoui, B., Krüger, T., Harting, J.: How does confinement affect the dynamics of viscous vesicles and red blood cells? Soft Matter **8**, 9246 (2012)
22. Jansen, F., Harting, J.: From bijels to Pickering emulsions: a lattice Boltzmann study. Phys. Rev. E **83**(4), 046707 (2011)
23. Günther, F., Janoschek, F., Frijters, S., Harting, J.: Lattice Boltzmann simulations of anisotropic particles at liquid interfaces. Comput. Fluids **80**, 184 (2013)
24. Günther, F., Frijters, S., Harting, J.: Timescales of emulsion formation caused by anisotropic particles. Soft Matter **10**, 4977 (2014)
25. Bleibel, J., Domínguez, A., Günther, F., Harting, J., Oettel, M.: Hydrodynamic interactions induce anomalous diffusion under partial confinement. Soft Matter **10**, 2945 (2014)
26. Kim, E., Stratford, K., Cates, M.: Bijels containing magnetic particles: a simulation study. Langmuir **26**(11), 7928 (2010)
27. Joshi, A., Sun, Y.: Multiphase lattice Boltzmann method for particle suspensions. Phys. Rev. E **79**, 066703 (2009)
28. Ladd, A., Verberg, R.: Lattice-Boltzmann simulations of particle-fluid suspensions. J. Stat. Phys. **104**, 1191 (2001)
29. Groen, D., Henrich, O., Janoschek, F., Coveney, P., Harting, J.: Lattice-Boltzmann methods in fluid dynamics: Turbulence and complex colloidal fluids. In: Bernd Mohr, W.F. (ed.) Jülich Blue Gene/P Extreme Scaling Workshop 2011. Jülich Supercomputing Centre, 52425 Jülich, Apr 2011. FZJ-JSC-IB-2011-02, http://www2.fz-juelich.de/jsc/docs/autoren2011/mohr1/

30. Schmieschek, S., Narváez Salazar, A., Harting, J.: Multi relaxation time lattice Boltzmann simulations of multiple component fluid flows in porous media. In: Nagel, M.R.W., Kröner, D. (eds.) High Performance Computing in Science and Engineering'12, p. 39. Springer, Heidelberg (2013)
31. Geislinger, T., Eggart, B., Braunmüller, S., Schmid, L., Franke, T.: Separation of blood cells using hydrodynamic lift. Appl. Phys. Lett. **100**, 183701 (2012)
32. Krüger, T., Varnik, F., Raabe, D.: Efficient and accurate simulations of deformable particles immersed in a fluid using a combined immersed boundary lattice Boltzmann finite element method. Comput. Math. Appl. **61**, 3485 (2011)
33. Krüger, T., Varnik, F., Raabe, D.: Particle stress in suspensions of soft objects. Phil. Trans. R. Soc. Lond. A **369**, 2414 (2011)
34. Krüger, T., Kaoui, B., Harting, J.: Interplay of inertia and deformability on rheological properties of a suspension of capsules. J. Fluid Mech. **751**, 725 (2014)
35. Ghosh, A., Harting, J., van Hecke, M., Siemens, A., Kaoui, B., Koning, V., Langner, K., Niessen, I., Rojas, J.P., Stoyanov, S., Dijkstra, M.: Structuring with anisotropic colloids. In: Proceedings of the Workshop Physics with Industry, Leiden, 17–21 Oct 2011. Stichting FOM (2011)
36. Wagner, A., Pagonabarraga, I.: Lees-Edwards boundary conditions for lattice Boltzmann. J. Stat. Phys. **107**, 521 (2002)

Part V
Transport and Climate

Christoph Kottmeier

The modelling of processes and longer-term evolutions of large natural systems like the atmosphere requires substantial efforts in HPC. Due to the non-linear interactions of dynamical and thermo-dynamical processes in the atmosphere, basically all small-scale phenomena (e.g. flow over mountains, thunderstorms, surface layer turbulence) can affect larger scale processes significantly. Consequently, there is no simple upper limit with respect to higher resolution which would allow a full realistic representation of the atmosphere. Whenever a new generation of computing facilities is available, the challenges in modelling over longer periods (climate projections), at highest resolution (process models, weather prediction, climate and ensemble techniques) result in a higher demand of HPC resources.

Currently largest efforts in the fields of transport and climate are undertaken with respect to climate reanalysis and downscaling, seamless prediction, and direct numerical simulation of atmospheric gravity waves. In the project *HRCM* ("*High Resolution Climate Modeling with the CCLM Regional Model*" – H.-J. Panitz et al., Institute for Meteorology and Climate Research, KIT Karlsruhe) climate is modelled for past periods as well as for the near and the more distant future at particular high resolution. Focus areas are the whole of Europe, with nested limited areas in mid-Europe and in Southern Germany, as well as Africa. The simulations contribute significantly to the BMBF program MIKLIP (Medium-Term Climate Prediction – Mittelfristige Klimaprognose) and the KLIMOPASS program funded by the State of Baden-Württemberg. MIKLIP is performed in close cooperation with the community on global climate modelling and is a cornerstone of the regional downscaling efforts as performed by the RLM community in Germany. Results on regionalization have been published in two recent papers by the group performing *HRCM*.

C. Kottmeier (✉)
Karlsruher Institut für Technologie (KIT), Institut für Meteorologie und Klimaforschung, Karlsruhe, Germany
e-mail: christoph.kottmeier@kit.edu

The project *WRFCLIM* (“*High-Resolution Climate Predictions and Short-Range Forecasts to Improve the Process Understanding and the Representation of Land-Surface Interactions in the WRF Model in Southwest Germany*” – K. Warrach-Sagi et al., Institute of Physics and Meteorology, University of Hohenheim) addresses both, downscaling of climate hindcasts and seamless prediction focusing on South-West Germany with the WRF-MODEL. As a thematic focus, the representation of land-surface-interaction is investigated with WRF, both with respect to longer integration: climate and short range weather forecasts.

On the other side of the spectrum with respect to the resolution of atmospheric models, the project *DINSGRAW* (*Direct Numerical Simulation of Breaking Atmospheric Gravity Waves* – S. Remmler et al., Technische Universität München, and Goethe University Frankfurt) applies direct numerical simulation of breaking atmospheric gravity waves. Currently perturbations of the prescribed gravity waves are optimized in initialization studies. A mid-term objective is to use the DNS simulations for the validation of lower order models for gravity wave breaking. Other projects in “TRANSPORT and CLIMATE” are still in a premature stage, but also show promising results. These regard the storage of CO_2 in geologic formations, a new series of simulations of the Agulhas current system, land-use cover change in Vietnam as well as the role of aerosols on weather and climate.

High Resolution Climate Modelling with the CCLM Regional Model for Europe and Africa

H.-J. Panitz, G. Schädler, M. Breil, S. Mieruch, H. Feldmann, K. Sedlmeier, N. Laube, and M. Uhlig

Abstract The High Performance Computing System (HPC) CRAY XE6 operated by HLRS is a powerful tool to study various aspects of the regional climate. Employing the regional climate model (RCM) COSMO-CLM, the research focus of IMK-TRO is on the regional atmospheric water cycle, especially on extremes, and different goals are pursued in individual research projects (MiKlip, KLIMOPASS and KLIWA). The simulation regions comprise Germany, Europe, and Africa with resolutions varying from 50 km to 2.8 km. Furthermore, different time spans are investigated. Decadal simulations are performed to assess decadal regional climate predictability. Projections of the future climate consider periods up to the end of the twenty-first century. To quantify the uncertainty of climate projections and predictions as well as the quality of the models, ensembles are built by different techniques. The Soil-Vegetion-Atmosphere-Transfer model (SVAT) VEG3D is coupled to COSMO-CLM via OASIS3-MCT to investigate the effect of soil and vegetation processes on decadal climate predictions. High resolution (2.8 km) experiments are performed for the State of Baden–Württemberg to study the potential added value and extremes. Computational capacities from 100 to 650 node–hours per simulated year (Wall Clock Time) are required for these simulations.

1 Introduction

The investigations of regional climate change include studies on how temperature and precipitation are affected in general, and specific impact studies whose results can be used for planning of adaptation and mitigation measures by authorities. Using the CRAY XE6 at the HLRS high performance computing facilities, the working group "Regional Climate and Water Cycle" of the Institute for Meteorology and Climate Research – Troposphere Research (IMK-TRO) at the Karlsruhe Institute

H.-J. Panitz (✉) • G. Schädler • M. Breil • S. Mieruch • H. Feldmann • K. Sedlmeier • N. Laube • M. Uhlig
Institut für Meteorologie und Klimaforschung Forschungsbereich Troposphäre (IMK-TRO), Karlsruher Institut für Technologie (KIT), Karlsruhe, Germany
e-mail: hans-juergen.panitz@kit.edu

W.E. Nagel et al. (eds.), *High Performance Computing in Science and Engineering '14*, DOI 10.1007/978-3-319-10810-0_37

of Technology (KIT) (www.imk-tro.kit.edu) aims to investigate the past, present, and future regional climate of Central Europe/Germany and Africa by the means of the climate version of the COSMO model (CCLM) of the German Weather Service (DWD).

The work is embedded in various national and international research programs and projects: e.g. MiKlip, KLIWA, KLIMOPASS and CORDEX.

2 The CCLM Model

The regional climate model CCLM is the climate version of the operational weather forecast model COSMO (Consortium for Small-scale Modeling) of DWD. It is a three-dimensional, non-hydrostatic, fully compressible model. The model solves prognostic equations for wind, pressure, air temperature, different phases of atmospheric water, soil temperature, and soil water content.

Further details on COSMO and its application as a RCM can be found in [1] and [2], and on the web-page of the COSMO consortium (www.cosmo-model.org).

3 Regional Climate Simulations Using the HLRS Facilities

3.1 Precipitation Statistics Derived from High Resolution Climate Simulations (KLIWA)

These studies are a continuation of the soil erosion project described in the last report [3]. In that project it was shown that high resolution has a potential to improve climate simulations, especially in terms of precipitation. Specifically, the diurnal cycle and the tails of the precipitation statistics seem to be improved, mainly for two reasons: (i) the direct simulation of convective precipitation, and (ii) the better accounting for small scale variations of orography and land use. The faster response of the (now smaller) atmospheric grid boxes might also contribute to an improvement.

Although a systematic validation and study of the potential added value is not yet completed, some first results related to extreme precipitation are presented. The statistics of extreme precipitation of COSMO-CLM simulations are compared with data derived from observations as published in the KOSTRA-Atlas (www.dwd.de/kostra). For that purpose, COSMO-CLM simulations using resolution of 2.8 km have been carried out for the period of 1971–2000. The focus of the study is on the Neckar catchment. The forcing data for these high resolution simulations are obtained from a nesting procedure, starting with a downscling of global ERA40 reanalyses data [4]. Based on these data, precipitation return values (1, 10 and 100 years) for the durations 1, 12 and 24 h have been calculated for the summer

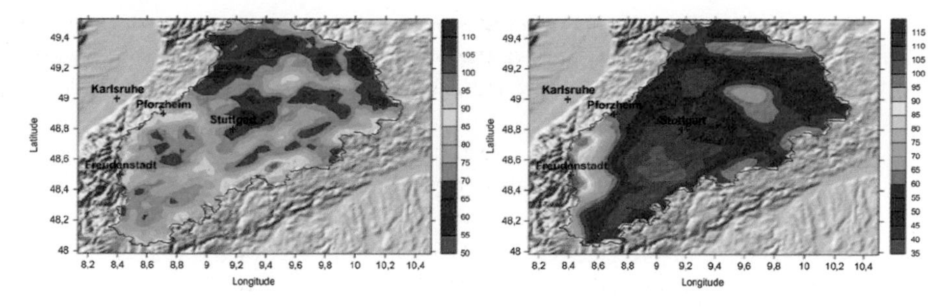

Fig. 1 Precipitation return values (return period 10 years, duration 24 h) for the summer (*left*) and winter (*right*) half years in the Neckar catchment. Note the different scales of the *colour bars*

(May to October) and winter (November to April) half years, respectively. The analysis applied the method described in [5], using the L-moment method to fit an extreme value distribution (in this case the kappa distribution), and the peak-over-threshold method to determine the return values.

As an example, Fig. 1 shows the spatial distribution of precipitation return values (period 10 years, duration 24 h) for the summer (left) and winter (right) half years. The precipitation enhancement as well as the shadowing in the lee induced by the Black Forest and the Suebian Jura can be clearly seen. In winter, the dominating large scale structures produce more homogeneous patterns than in summer, when smaller scale convection plays an important role and increases spatial variability. There is a good qualitative and quantitative agreement with the KOSTRA data and a reduced wet bias. A more thorough validation and analysis of the added value is underway.

3.2 The MiKlip Program

The German national research program MiKlip on decadal climate prediction (www.fona-miklip.de), funded by the Federal Ministry of Education and Research in Germany (BMBF), aims to establish a decadal climate prediction system that can be used operationally. The potential to predict the climate up to decadal time scales arises from studies of decadal to multidecadal variability exhibited by the climate system. One of the subsystems of climate that shows such variability over long time scales is the hydrosphere.

Much effort is put into the regionalization of global decadal predictions since climate informations are often required at a much higher spatial resolution than available from the global climate models, for example when processes such as orographic rain or mesoscale circulations are important. The MiKlip module dedicated to this regionalization of the global decadal forecasts is coordinated at IMK-TRO. Next to the coordination project of the module (REGIO_PREDICT) – which also has the task to assess the skill of the ensemble of the regional decadal hindcast experiments – two other projects aim at regional decadal predictions of the climate for Europe (project DecReg) and the West African monsoon region (project DEPARTURE), respectively.

3.2.1 Decadal Regional Predictability (DecReg)

One main objective of the DecReg project is the analysis of the feasibility and prospects of decadal regional predictions. In the framework of a joint effort we used the CRAY XE6 at the HLRS facilities for the generation of a regional 10 member hindcast ensemble with 25 km resolution using the CCLM model. The regional decadal CCLM ensemble has been initialized every 10 years, i.e. 1961, ..., 2001, and covers all of Europe.

Figure 2 shows time series of half-year temperature anomalies (long-term trend removed) at a single grid point in Poland. The CCLM ensemble mean (lilac curves) is compared with observations from E-EOBS (green curves, [6]). Thick lines represent the low pass filtered (9 year moving average) data (see also [7]). The statistically significant correlation between the filtered data (thick lines in Fig. 2) is 0.84. Further, one can see that the hit-rate is 100 % in predicting a decade with anomalies below or above zero. Thus, it seems that there is indeed a potential of regional decadal predictability.

To generalize the statements from above, we applied our analysis to the whole simulation domain shown in Fig. 3.

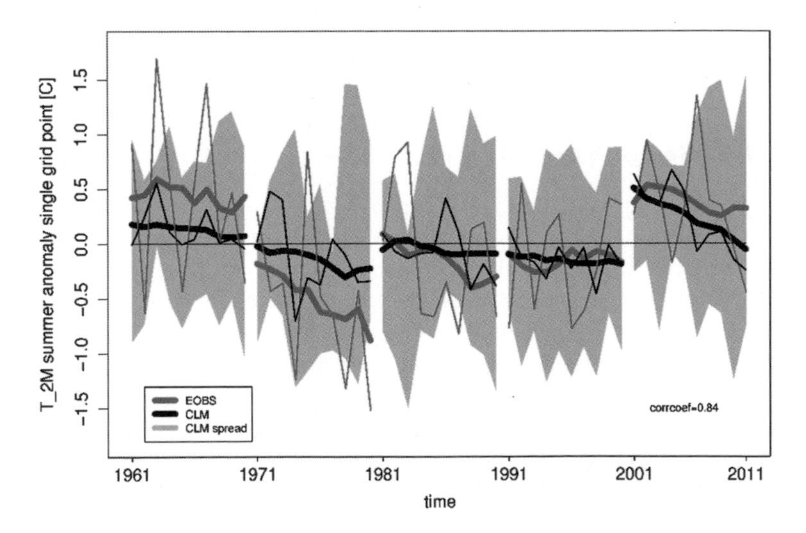

Fig. 2 E-OBS observations (*green*) and CCLM model results (*lilac*) from a grid point in Poland. The *thin lines* depict anomalies (long-term trend removed) of temperature summer half-year averages. The *shading* represents the ensemble spread. A 9 year moving average filter has been applied to the data (*thick lines*)

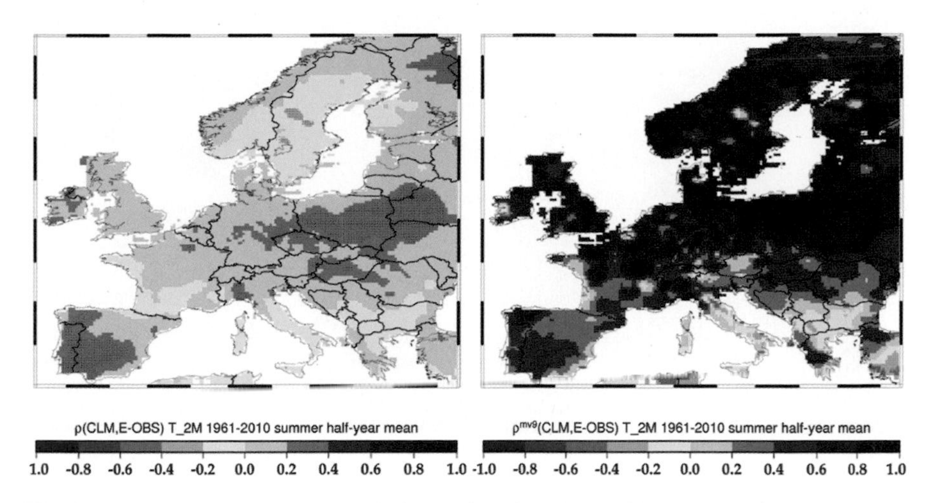

Fig. 3 Correlation between model and observations based on unfiltered data (*left*) and low pass filtered data (*right*)

The left panel shows the correlation between observational and modeled summer half-year temperatures based on unfiltered data. The right panel shows the correlations for low pass filtered data. The correlation in most parts of Europe is increased, but the temporal resolution is smeared out by the filtering. The predictive skill of unfiltered summer half-year temperature anomalies is moderate. Hence, it seems that forecasting annual values using decadal initializations is too ambitious. Temporal filtering is an option to increase the predictive skill and to capture the variability on time scales in the order of 10 years. The analysis of winter temperature and precipitation (summer and winter) can be found in [7].

An important metric to characterize a model ensemble is the reliability. Reliability is a measure of how well the ensemble spread (i.e. the model uncertainty) reproduces the "true" model error, i.e. the root mean square error (rsme) between ensemble mean and observations according to [8]:

$$R_i = \frac{\mathrm{rmse}(\mu_{t,i}, x_{t,i}) - \sqrt{\langle \sigma^2_{\mathrm{i,ens}} \rangle_t}}{\mathrm{rmse}(\mu_{t,i}, x_{t,i})},$$

where $\sqrt{\langle \sigma^2_{\mathrm{i,ens}} \rangle_t}$ represents the time averaged ensemble spread. Figure 4 shows the results for summer half-year temperature anomalies in Europe on the 25 km grid. An ensemble with $R = 0$ is called reliable, which is desirable, an ensemble with negative R is underconfident, i.e. the spread is too large, whereas a positive R indicates overconfidence, i.e. the spread is too small.

The regional decadal CCLM ensemble is reliable with $-0.2 < R < 0.2$ for the unfiltered data. Filtering in this case is not beneficial and the ensemble is

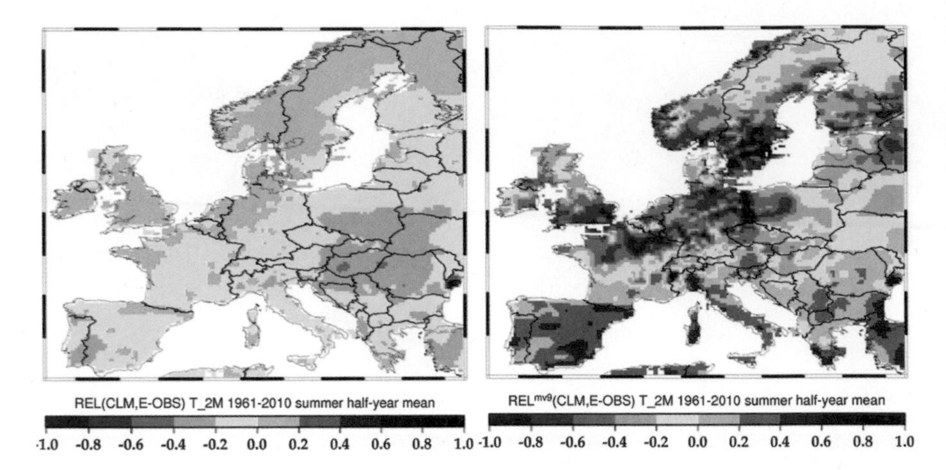

Fig. 4 Reliability of the CCLM ensemble based on unfiltered data (*left*) and filtered data (*right*)

underconfident in northern Europe and overconfident in southern Europe as shown in Fig. 4. Regarding the analysis of winter temperatures and precipitation we refer to [7].

A crucial question regarding the downscaling of decadal predictions using the regional model CCLM is whether an "added value" can be achieved with respect to the global model. We will not go too deep into the details here, but generally it seems that for climate means the predictive skill from the global driving model (MPI-ESM-LR), i.e. the correlation, is conserved by the downscaling, but not improved. The reliability, however, is improved by the regionalisation for half-year precipitation sums. The reason is probably that the regional model performs better in representing the spread of precipitation because orography and small scale processes like convection are better captured by the regional model.

3.2.2 Decadal Climate Predictability in West Africa (DEPARTURE)

The overall aim of DEPARTURE is assessing the decadal climate predictability in the West African Monsoon (WAM) region and the Atlantic region of tropical cyclogenesis. According to current knowledge, West African climate may be characterized by high decadal predictability because it is governed by processes with long-term memory such as the state of the oceans (Sea Surface Temperature, SST), land-cover and soil characteristics, greenhouse-gas and aerosol concentrations. Sensitivity studies with different lower boundary conditions, e.g. SST, and with different Soil-Vegetation-Atmosphere-Transfer (SVAT) models, which are coupled to CCLM via the OASIS MCT coupler (Sect. 3.3), allow assessing the effect of soil and vegetation processes on decadal climate predictability in the African monsoon region.

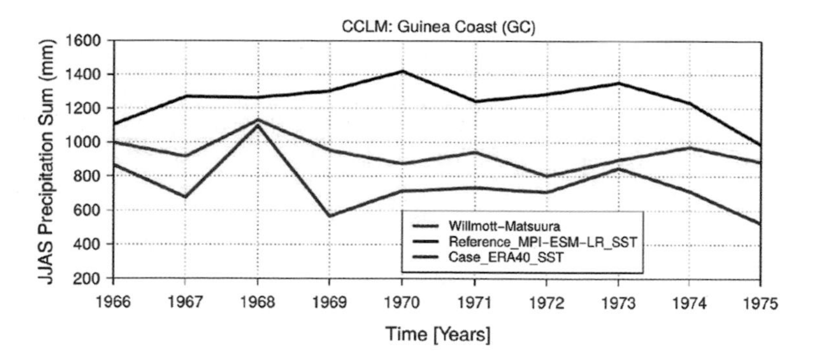

Fig. 5 Comparison of reference run using MPI-ESM-LR SST and case study using ERA40 SST with climate data (UDEL, [9]). Shown are the summer (JJAS) precipitation sums for decade 1965 (1966–1975) over the Guinea Coast region (13.5°W–10°E, 4.5°N–10°N)

Sensitivity of predictability to SST was found for precipitation in the coastal areas of the African monsoon region. When using realistic SST from ERA40 reanalyses [4] instead of strongly biased SST from global MPI-ESM-LR decadal runs, the results of CCLM simulations could be improved. This is demonstrated in Fig. 5, which shows the precipitation sum over the summer season (June to September, JJAS) of decade 1965 (1966–1975) over the Guinea Coast (GC) region (13.5°W–10°E, 4.5°N–10°N). Results of a reference simulation using the SST data provided by MPI-ESM-LR and a sensitivity run using ERA40 SST are compared with the climate data from the University of Delaware (UDEL, [9]). Whereas the reference simulation overestimates the summer precipitation considerably, on the average by more than 300 mm, the result of the sensitivity run is closer to the observation, although the observation is now underestimated (by about 190 mm on the average). The linear correlation between the climate data and the simulation results increase from −0.13 for the reference run to 0.68 when using the ERA−40 SST. Thus, this result clearly illustrates the importance of SST data, which are used as boundary conditions for the regional climate model.

In order to study the impact of soil and vegetation processes on decadal climate predictability in the African monsoon region, the SVAT TERRA-ML [10], which is implemented in CCLM, was replaced by VEG3D [11], a SVAT developed at KIT. In a preliminary sensitivity study for the year 1979, results of two ERA-Interim driven CCLM simulations, one using TERRA-ML, the other one using VEG3D, were analysed and compared with CRU (Climate Research Unit) observational data set [12]. The application of VEG3D improves the simulation results, exemplarily shown for the 2 m temperature in the Central Sahel (CS, 7°W–30°E, 10N°N–20°N) (Fig. 6). The annual cycle of temperature, resulting from the usage of VEG3D, is damped and matches the observation much better. This damping is due to an explicit vegetation layer, considered in VEG3D and not in TERRA-ML, preventing the surface of heating up too strongly during warm seasons and from cooling out during cooler seasons. Thus, already this preliminary study illustrates the huge

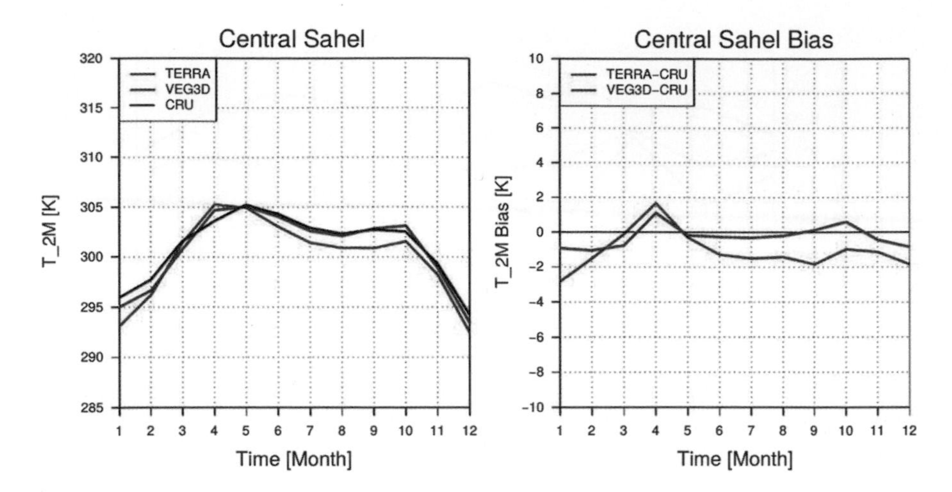

Fig. 6 Comparison of ERA-Interim driven CCLM simulations using either the SVATs TERRA-ML (*red line*) or VEG3D (*blue line*) with CRU climate data (*black line*). Shown are the 1979 annual cycles (*left*) in the Central Sahel region (7°W−30°E, 10°N−20°N) and the bias of the simulation results with respect to the CRU data (*right*)

impact of the soil-vegetation-atmosphere interaction on the climate variability of the West African Monsoon region.

3.3 Coupling of COSMO_CLM with VEG3D via OASIS3-MCT

One of the most important processes influencing climate variability is the soil-vegetation-atmosphere interaction, where the soil and vegetation characteristics control the radiation balance and the water and heat fluxes between the surface and the atmosphere. In a RCM this interaction happens between a Soil-Vegetation-Atmosphere-Transfer module (SVAT) and the atmosphere part of the model. The standard SVAT implemented in COSMO-CLM is TERRA_ML. Several studies, e.g. [13] and [14], have shown that TERRA_ML does not describe all aspects of the soil-vegetation-atmosphere interaction sufficiently. Therefore, TERRA_ML was replaced with VEG3D. Compared to TERRA_ML, VEG3D includes an explicit vegetation layer acting like one big leaf [11]. Within this vegetation layer, explicit temperature and humidity values are calculated controlling the latent and sensible heat fluxes between the surface and the atmosphere. The hydraulic conductivity in the soil is described with the parameterisation of [15]. The thermal conductivity is a function of the soil water content, described with the approach of [16].

The coupling of COSMO-CLM and VEG3D was done via the OASIS3-MCT coupler, a widely-used software in numerical climate simulations. OASIS3-MCT was developed at the Centre Européen de Recherche et de Formation Avancée en Calcul Scientifique (CERFACS) in Toulouse, France, in collaboration with the

Argonne National Laboratory, USA. The OASIS3 coupling software was interfaced with the Model Coupling Toolkit (MCT) [17]. Due to this, OASIS3-MCT allows fully parallel regridding and parallel distributed exchanges of two-dimensional coupling fields between both models via Message Passing Interface (MPI) [18]. This new coupling software is a library that only has to be linked to COSMO-CLM and VEG3D, thus, no seperate coupler executable is needed. The communication between the two models is done by including only a few selective calls to the OASIS3-MCT coupling library in the source codes of both models, controlling the different coupling steps. With the first call the coupling is initialized. To enable a regridding of the coupling fields, the grids of both models are defined with the second call. The third call describes the local field partition of each model to ensure parallel exchange between the different parallel processes. In the next steps the coupling fields are first defined and afterwards exchanged at the corresponding coupling timestep [19]. These coupling periods are defined in an external configuration file, which, in addition, includes some general coupling parameters, e.g. the number of coupled models and exchanged fields as well as specific information on each coupling field like the chosen interpolation method [18]. Compared to a subroutine coupling, changes within the model structure are substantially reduced. A further advantage of coupling via OASIS3-MCT is, that, due to numerous implemented interpolation methods, it is possible to run the SVAT in a higher resolution than the RCM. Thus, the heterogeneity of the lower boundery condition can be taken into account much better.

To couple VEG3D to COSMO-CLM via OASIS3-MCT, VEG3D had to be parallelized via MPI. Thus, in a first step, a new parallelized stand-alone version was created. Simulation results of the parallelized version are identical to the results achieved with the non-parallelized version, as shown in Fig. 7. The successfully parallelized version was then coupled to COSMO-CLM via OASIS3-MCT for further investigations on the influence of the soil-vegetation-atmosphere interaction on climate variability.

3.4 *Climate Change and Exemplary Adaptation in Baden-Württemberg (KLIMOPASS)*

Regional measures of adaptation to a changing climate require high resolution climate data in order to assess the possible changes of climate variables and statistics of practical interest, especially extremes and combinations of extremes (compound extremes, e.g. heat and drought). Furthermore, there is an increasing awareness of assessing the uncertainty of the data. For this purpose, ensemble simulations are a suitable method as has been shown, for example, in the ENSEMBLES project [20]. However, the resolution of such simulations ($\approx$25 km) is still too low for regional impact studies.

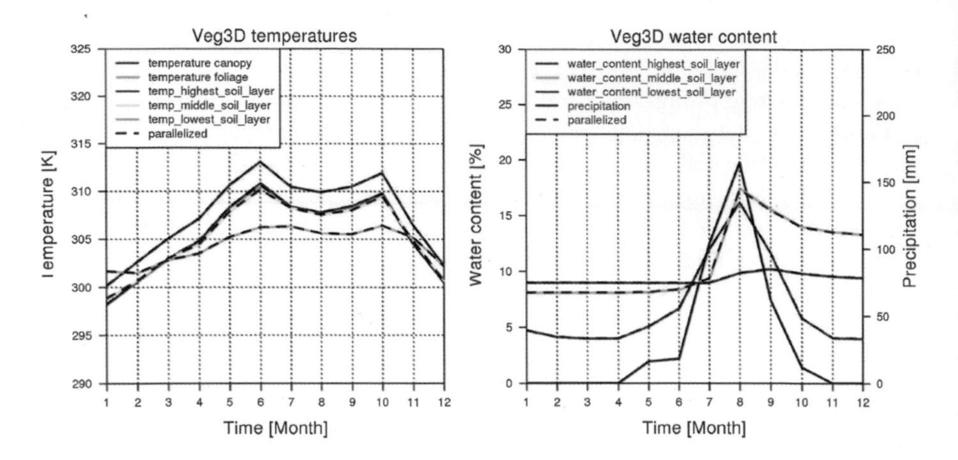

Fig. 7 Simulated soil temperatures (*left graph*) and soil water contents (*right graph*) for the non-parallelized (*coulored lines*) and the parallelized version (*dotted lines*) for different soil depths. The precipitation is illustrated in *black lines* (*right graph*)

The research program KLIMOPASS ("Klimawandel und modellhafte Anpassung in Baden-Württemberg", "Climate change and exemplary adaptation in Baden Württemberg") was initiated to investigate climate change and its impacts as well as the possibilities of adaptation measures for the state of Baden-Württemberg in Germany. The contribution of IMK-TRO to the Klimopass project is to provide an ensemble of high-resolution regional climate simulations (with a horizontal resolution of 7 km) for two 30-year time periods, one in the recent past (1971–2000) and one in the near future (2021–2050). The ensemble is analyzed with respect to the change signal of different climate variables and their statistics, especially extreme and compound extreme events. Furthermore, a quantitative estimate of the robustness of the data based on ensemble simulations is given. This information is used as input for impact studies of other projects within and outside of KLIMOPASS. Data have already been passed to different project partners dealing with topics related to climate change.

Since the last report [3], a transient run with ECHAM6 as driving model and further model runs, which downscaled different global driving models, have been finished. In addition, the simulations using the atmospheric forcing shifting method, AFS [3, 21], have been finished.

Up to now, the focus was on the analysis of univariate and bivariate temperature and precipitation extremes. One measure for precipitation extremes is the effective drought index, EDI [22], which is basically a standardised anomaly of the effective precipitation and therefore independent of linear bias and symmetric. The EDI is a local, relative measure and extremes are not necessarily record breaking events. We deduced an analogous concept for temperature, the effective heat index (EHI) as well as a compound measure for combined temperature and precipitation extremes. From the EDI/EHI timeseries, different statistics such as the number of extreme days, the number of extreme episodes with a certain length or the mean/maximum length can be derived. These statistics were calculated for a reference period (1971–2000) and the near future (2021–2050) and a climate change signal between these two time periods was deduced.

4 Computational Requirements

The finer resolution of regional models as well as the need for ensembles and scenarios require a large amount of computer resources both in computing time (up to 960 node hours for one simulation year) and data storage space (currently 2–3 TB per 10 year simulation time) depending on the number of grid-points considered, the numerical time-step necessary to receive stable numerical solutions, and the length of the simulation period. The following table (Table 1) shows the typical computational demand of the different projects. The Wall Clock Time (WCT) is given in node-hours per year (nh/y). It is scaled to one simulation year and denote the time one CRAY XE6 node needs for the simulation of 1 year assuming that all cores of the node are used. Overall, the storage requirement amounts to 2–3 TB per 10 years simulation time including the data input, output and post-processing.

5 Future Work

The CRAY XE6 at the HLRS high performance computing facilities has been essential to our group's research and will continue to be so in the future. For the project KLIMOPASS further simulations with other global climate models as driving data (e.g. the HadGEM-ES2) are planned to enlarge the already existing ensemble and thus to create a broader statistical data basis for the analysis of extremes and combinations of extremes together with their uncertainties. In addition, transient model runs for ensemble members are planned to bridge the gap between 2000 and 2021. The main objective of DecReg is now the generation of the so-called MiKlip baseline1 ensemble, downscaling the improved baseline1 ensemble of global decadal predicitions. The assessment of decadal regional predictions will be continued along with a more detailed analysis of extremes. Additional research methods like sensitivity tests in DecReg are planned to improve the predictive skill and reliability of the CCLM ensemble. To continue the work done in project KLIWA more simulations will be performed, in particular several 2.8 km simulations with different driving data from other global climate models and a climate run with a resolution of 1 km. The goal is an extensive validation with focus on extreme precipitation. Additionally, added value of higher resolution will be studied. In the DEPARTURE project further coupled simulations are planned to take advantage of the different interpolation methods implemented in the OASIS3-MCT. With the coupling software it is possible to use higher spatial resolution for the SVAT

Table 1 Computational Requirements

Research program	Project	Domain size (No. of grid-points)	Horizontal grid spacing in km	Numerical time-step (s)	No. of nodes used	WCT (nh/y)	No. of simulated years
MiKlip	DecReg	232 × 226 × 40	25	240	64	384	500
MiKlip	DEPARTURE	275 × 207 × 35	50	240	64	200	100
KLIWA	Erosion	140 × 150 × 40	2.8	25	64	650	33
KLIMOPASS	RegEns BaWü	118 × 110 × 40	50	360	64	96	105
		165 × 200 × 40	7	60	64	384	105

VEG3D to take into account the heterogenity of the surface, related e.g. to soil type and vegetation, more efficiently. Up to now CCLM and VEG3D use an identical horizontal resolution of (0.44°, about 50 km). Furthermore, an alternative SVAT, the so called Community Land Model (CLM), will be coupled to CCLM, resulting in the coupled model system CLM^2. With this new system more decadal runs will be performed within the DEPARTURE project. All planned simulations for the aformentioned projects require, large simulation times, and appreciable ensemble size. Consequently high amounts of high performance computing times and storage capacities are needed.

References

1. Baldauf, M., Seifert, A., Förstner, J., Majewski, D., Raschendorfer, M.: Operational convective-scale numerical weather prediction with the COSMO model: description and sensitivities. Mon. Weather Rev. **139**, 3887–3905 (2011)
2. Rockel, B., Will, A., Hense, A.: Regional climate modelling with COSMO-CLM (CCLM). Met. Zeitschrift, **4**(4), (2008). Special issue, ISSN 0941-2948
3. Panitz, H.-J., Fosser, G., Sasse, R., Sedlmeier, K., Mieruch, S., Breil, M., Feldmann, H., Schädler, G.: High resolution climate modeling with the CCLM regional model. In: High Performance Computing in Science and Engineering'13: Transactions of the High Performance Computing Center Stuttgart (HLRS), pp. 511–527. Springer, Berlin/Heidelberg/New York (2013)
4. Uppala, S.M., Kållberg, P.W., Simmons, A.J., Andrae, U., Da Costa Bechtold, V., Fiorino, M., Gibson, J.K., Haseler, J., Hernandez, A., Kelly, G.A., Li, X., Onogi, K., Saarinen, S., Sokka, N., Allan, R.P., Andersson, E., Arpe, K., Balmaseda, M.A., Beljaars, A.C.M., Van De Berg, L., Bidlot, J., Bormann, N., Caires, S., Chevallier, F., Dethof, A., Dragosavac, M., Fisher, M., Fuentes, M., Hagemann, S., Hólm, E ., Hoskins, B.J., Isaksen, L., Janssen, P.A.E.M., Jenne, R., Mcnally, A.P., Mahfouf, J.-F., Morcrette, J.-J., Rayner, N.A., Saunders, R.W., Simon, P., Sterl, A., Trenberth, K.E., Untch, A., Vasiljevic, D., Viterbo, P., Woollen, J.: The ERA-40 re-analysis. Q. J. R. Meteorol. Soc. **131**(612), 2961–3012 (2005)
5. Früh, B., Feldmann, H., Panitz, H.-J., Schädler, G., Jacob, D., Lorenz, Ph., Keuler, K.: Determination of precipitation return values in complex terrain and their evaluation. J. Clim. **23**(9), 2257–2274 (2010)
6. Haylock, M.R., Hofstra, N., Klein Tank, A.M.G., Klok, E.J., Jones, P.D., New, M.: A European daily high-resolution gridded data set of surface temperature and precipitation for 1950–2006, J. Geophys. Res., **113**, D20119, (2008). doi:10.1029/2008JD010201
7. Mieruch, S., Feldmann, H., Schädler, G., Lenz, C.-J., Kothe, S., Kottmeier, C.: The regional miklip decadal forecast ensemble for Europe. Geosci. Model Dev. Discuss. **6**, 5711–5745 (2013)
8. Weigel, A.P., Liniger, M.A., Appenzeller, C.: Seasonal ensemble forecasts: are recalibrated models better than multimodels. Mon. Weather Rev. **137**, 1460–1479 (2009)
9. Willmott, C.J., Matsuura, K., Legates, D.R.: Global air temperature and precipitation: regridded monthly and annual climatologies (version 2.01) (1998). Available at: http://climate.geog.udel.edu/~climate/
10. Schrodin, R., Heise, E.: The multi-layer version of the DWD soil model TERRA-ML. Technical report, COSMO Technical Report, 2, DWD, Offenbach, 2001. Available at: http://www.cosmo-model.org/content/model/documentation/techReports/default.htm
11. Schädler, G.: Triggering of atmospheric circulations by moisture inhomogeneities of the earth's surface. Bound. Layer Meteorol. **51**, 1–29 (1990)

12. Mitchell, T.D., Jones, P.D.: An improved method of constructing a database of monthly climate observations and associated high-resolution grids. Int. J. Clim. **25**(6), 693–712 (2005).
13. Kohler, M., Schädler, G., Gantner, L., Kalthoff, N., Körniger, F., Kottmeier, C.: Validation of two svat models for different periods during the west african monsoon. Met. Z. **21**, 509–524 (2012)
14. Meissner, C.: High-resolution sensitivity studies with the regional climate model COSMO-CLM. Ph.D. thesis, Falculty of Physics, University of Karlsruhe, 2008
15. van Genuchten, M.: A closed-form equation for predicting the hydraulic conductivity of unsaturated soils. Soil Sci. Soc. Am. J. **44**(5), 892–898 (1980)
16. Johansen, O.: Thermal conductivity of soils. Ph.D. thesis, University of Trondheim, Norway, 1975. (CRREL Draft Translation, 1977), ADA 044002
17. Larson, J., Jacob, R., Ong, E.: The model coupling toolkit: a new Fortran90 toolkit for building multiphysics parallel coupled models. Int. J. High Perform. Comput. Appl. **19**(3), 277–292 (2005)
18. Valcke, S., Craig, T., Coquart, L.: OASIS3-MCT User Guide, OASIS3-MCT 2.0. Technical report, TR/CMGC/13/17, CERFACS/CNRS SUC URA No 1875, Toulouse, 2012
19. Valcke, S.: The OASIS3 coupler: a European climate modelling community software. Geosci. Model Dev. **6**, 373–388 (2013)
20. van der Linden, P., Mitchell, J.F.B. (eds.): ENSEMBLES: climate change and its impacts: summary of research and results from the ENSEMBLES project. Technical report, Met Office Hadley Centre, FitzRoy Road, Exeter EX1 3PB, 2009
21. Sasse, R., Schädler, G.: Generation of regional climate ensembles using atmospheric forcing shifting. Int. J. Clim. **34**, 2205–2217, (2013). doi:10.1002/joc.3831
22. Byun, H.R., Wilhite, D.A.: Objective quantification of drought severity and duration. J. Clim. **12**(9), 2747–2756 (1999)

High-Resolution Climate Predictions and Short-Range Forecasts to Improve the Process Understanding and the Representation of Land-Surface Interactions in the WRF Model in Southwest Germany (WRFCLIM)

K. Warrach-Sagi, Hans-Stefan Bauer, T. Schwitalla, J. Milovac, O. Branch, and V. Wulfmeyer

Abstract The scope of this project is the investigation of the performance of regional climate simulations with Weather Research and Forecast (WRF) model in the frame of EURO-CORDEX (CORDEX for the European Domain: www.euro-cordex.net) at 12 km down to simulations at the convection permitting scale. Within this context the objectives are to provide high resolution climatological data to the scientific community and support the quality assessment and interpretation of the regional climate projections for Europe through the contribution to the climate model ensemble for Europe (EURO-CORDEX) for the next IPCC (International Panel of Climatic Change) report. Special attention is paid to the land-surface-vegetation-atmosphere feedback processes, namely the development of the atmospheric boundary layer. Further, the model is validated with high-resolution case studies applying advanced data assimilation to improve the process understanding over a wide range of temporal scales. This also addresses whether the model is able to reasonably represent extreme events. The current status of regional climate simulations and land-atmosphere feedback in convection permitting simulations in WRFCLIM is reported.

1 Introduction and Motivation

The application of numerical modeling for climate projections is an important task in scientific research since these projections are the most promising means to gain insight in possible future climate changes. During the last two decades, various

K. Warrach-Sagi (✉) • H-S. Bauer • T. Schwitalla • J. Milovac • O. Branch
• V. Wulfmeyer
Institute of Physics and Meteorology, University of Hohenheim, Garbenstrasse 30, 70599 Stuttgart, Germany
e-mail: Kirsten.Warrach-Sagi@uni-hohenheim.de

W.E. Nagel et al. (eds.), *High Performance Computing in Science and Engineering '14*, DOI 10.1007/978-3-319-10810-0_38

regional climate models (RCM) have been developed and applied for simulating the present and future climate of Europe at a grid resolution of 25–50 km. It was found that these models were able to reproduce the pattern of temperature distributions reasonably well but a large variability was found with respect to the simulation of precipitation. The performance of the RCMs was strongly dependent on the quality of the boundary forcing, namely if precipitation was due to large-scale synoptic events. Additionally, summertime precipitation was subject of significant systematic errors as models with coarse grid resolution have difficulties to simulate convective events. This resulted in deficiencies with respect to simulations of the spatial distribution and the diurnal cycle of precipitation. The 50-km resolution RCMs were hardly capable of simulating the statistics of extreme events such as flash floods.

Due to these reasons, RCMs with higher grid resolution of 10–20 km were developed and extensively verified. While still inaccuracies of the coarse forcing data were transferred to the results, these simulations indicated a gain from high resolution due to better resolution of orographic effects. These include an improved simulation of the spatial distribution of precipitation and wet-day frequency and extreme values of precipitation. However, three major systematic errors remained: the windward-lee effect (i.e. too much precipitation at the windward side of mountain ranges and too little precipitation on their lee side), phase errors in the diurnal cycle of precipitation and precipitation return periods. In order to provide high-resolution ensembles and comparisons of regional climate simulations in the future, the World Climate Research Program initiated the **CO**ordinated **R**egional climate **D**ownscaling **EX**periment (**CORDEX**, http://wcrp-cordex.ipsl.jussieu.fr, [1]).

The scope of this WRFCLIM project (e.g. [2, 3]) at HLRS, is to investigate in detail the performance of regional climate simulations with Weather Research and Forecast (WRF) model [4] in the frame of EURO-CORDEX1 at 12 km down to simulations at the convection permitting scale e.g. within the DFG funded Research Unit 1695 "Agricultural Landscapes under Climate Change – Processes and Feedbacks on a Regional Scale". An objective of this research unit is the development and verification of a convection-permitting regional climate model system based on WRF including an advanced representation of land-surface-vegetation-atmosphere feedback processes with emphasis on the water and energy cycling between croplands, atmospheric boundary layer and the free atmosphere.

Within this context the WRFCLIM objectives are to

- Provide high resolution climatological data to the scientific community
- Support the quality assessment and interpretation of the regional climate projections for Europe through the contribution to the climate model ensemble for Europe (EURO-CORDEX) for the next IPCC (Intergovernmental Panel of Climatic Change) report.

The main objectives of the convection permitting simulations of WRFCLIM are as follows:

- Replace the convection parameterization by the dynamical simulation of the convection chain to better resolve the processes of the specific location and actual weather situation physically.
- Gain of an improved spatial distribution and diurnal cycle of precipitation through better land-surface-atmosphere feedback simulation and a more realistic representation of orography to support the interpretation of the 12 km climate simulations for local applications e.g. in hydrological and agricultural management.

Special attention is paid to the land-surface-vegetation-atmosphere feedback processes, namely the development of the atmospheric boundary layer. Further, the model is validated with high-resolution case studies applying advanced data assimilation to improve the process understanding over a wide range of temporal scales. This also addresses whether the model is able to reasonably represent extreme events. In the following on the current status of regional climate simulations and land-atmosphere feedback in convection permitting simulations in WRFCLIM is reported. It follows a technical description of the latest simulations within WRFCLIM at HLRS between April 2013 and April 2014 and a summary of the current analysis of their results.

2 Summary of Scientific Achievements of WRFCLIM to Date

2.1 Climate Simulations

At HLRS within WRFCLIM two climate simulations with WRF were carried out for 1989–2009 to evaluate the model performance within the frame of EURO-CORDEX. The first simulation (WRF 3.1.0 on the NEC Nehalem Cluster) was evaluated by [5] with respect to precipitation and [6] with respect to soil moisture. During all seasons, this dynamical downscaling reproduced the observed spatial structure distribution of the precipitation fields. However, a wet bias was remaining. At 12 km model resolution, WRF showed typical systematic errors of RCMs in orographic terrain such as the windward-lee effect. In the convection permitting resolution (4 km) case study in summer 2007, this error vanished and the spatial precipitation patterns further improved. This result indicates the high value of regional climate simulations on the convection-permitting scale. Soil moisture depends on and affects the energy flux partitioning at the land-atmosphere interface. The latent heat flux is limited by the root zone soil moisture and solar radiation. When compared with the in situ soil moisture observations in Southern France, where a soil moisture network was available during 2 years of the study period, WRF generally reproduced the annual cycle. Over Europe the spatial patterns and temporal variability of the seasonal mean soil moisture from the WRF simulation corresponds well with two reanalyses products, while their absolute values differ significantly, especially at the regional scale. Based on these results the updated version WRF 3.3.1 was applied at Cray XE6 for EURO-CORDEX [2, 3]. The benefits from downscaling the reanalysis ERA-Interim data with WRF become especially evident in the probability distribution of daily precipitation data. Kotlarski et al. [7] and Vautard et al. [8] show that WRF 3.3.1 generally simulates the seasonal precipitation and temperature in central Europe well, namely in Germany, and fits well in the EURO-CORDEX ensemble. The evaluation within the EURO-CORDEX ensemble is still ongoing. The data is now available and used e.g. within the DFG funded Research Unit "Agricultural Landscapes under Climate Change – Processes and Feedbacks on a Regional Scale" for agricultural impact studies. However, the weakness of the 12 km resolution concerning the windward-lee effect in precipitation remains. So it is essential to proceed towards convection permitting simulations as in WRFCLIM.

2.2 Sensitivity Study on Parameterization Schemes

When setting a model for use in particular region or at particular horizontal resolution, the foremost issue is the determination of the most appropriate model configuration. Different regions experience varied conditions, and the optional setup of a model is spatially and temporally dependent. One of the most important aspects in configuring a model is to select the parameterizations to be used and therefore sensitivity tests are unavoidable part in improving a model performance. The WRF model provides multiple parameterization schemes and allows their evaluation within the same model core. Land-surface atmosphere feedback processes depend on the land surface and atmospheric boundary layer and their representation in the model. Since these are sub-grid processes, they are parameterized in WRF. The first sensitivity study of WRF to the boundary-layer and land-surface parameterizations and comparisons with Differential Absorption Lidar (Light Detection and Ranging) for September 2009 showed significant sensitivity of WRF to the choice of the land surface model and planetary boundary layer parameterizations (see e.g. [3]). The model sensitivity to the land surface model is evident not only in the lower PBL, but it extends up to the PBL entrainment zone and even up to the lower free troposphere. This study is ongoing for more experiments in spring 2013 under other weather conditions to understand the processes, their modeling and to set up a sophisticated simulation with WRF for Germany on the convection permitting scale also for climate simulations as a part of WRFCLIM.

2.3 Sensitivity Study on Land Use in a Dry Hot Climate

The introduction of large plantations into arid regions has provoked much research due to their potential for carbon sequestration, agricultural development and environmental services. Implementations of plantations on a large scale are now becoming feasible due to recent advances in desalination, wastewater and irrigation science. However, before implementation, climatic impacts of such plantations caused by changes in land surface characteristics, need to be investigated. This can be achieved via simulations within a dynamical downscaling experiment, using coupled atmospheric/land surface models (LSM) such as WRF-NOAH. Such models need to be carefully calibrated for the regional arid conditions and verified to assess confidence in simulation results. Simulations were carried out over the Israel region, for the 2012 summer season (JJA). Two model case studies were considered – WRF Impact and WRF Control. WRF Control was run with the existing land surface data where the predominant land surface in the same area is bare desert soils. WRF Impact was run with a hypothetical $10 \times 10\,\text{km}^2$ plantation introduced into the MODIS land surface model data, and this was fed with calculated optimal irrigation. For model verification, surface meteorological observations were collected from three sites in the area: a desert site and from two plantations – Jatropha and Jojoba. The first aim of the study was to configure the WRF/NOAH model for an arid region such that the model can simulate daytime surface quantities which are central to the energy balance, over a bare soil and an irrigated vegetation surface. This has to a large degree been successful in terms of magnitude and variability [9]. The comparisons show that WRF-NOAH can represent both desert and plantation land surfaces with a high degree of skill except for ground heat flux which needs adjusting. Its contribution to the energy balance is small however. Finally, the simulated plantation sensible heat fluxes ($450\,\text{Wm}^{-2}$) are around 100–$110\,\text{Wm}^{-2}$ higher than over the surrounding desert. This phenomenon is driven by high radiation and surface resistances and indicates that contrary to many studies, significant evapotranspiration is not likely if irrigation is optimized. The high peak temperatures over the plantations show that diurnal dynamics are very important. Many studies use only daily mean temperatures but this can often imply an overall irrigation cooling effect, when during the daytime the opposite may be the case.

2.4 *Quantitative Precipitation Forecast: Data Assimilation with a Rapid Update Cycle*

To prepare the model system for a European reanalysis, further experiments were made by using the WRF Rapid Update Cycle (RUC) described in [10]. The rapid update cycle was refined so that the update frequency of the model initial conditions are updated every 3 h or even less depending on the needs by the user. To improve the short-range quantitative precipitation forecast (QPF), the RUC was further enhanced that radial velocities and reflectivities from precipitation radars can be used for data assimilation in addition to other observations. The novel thing is that volume radar data from the French and German radar network can be used simultaneously. The results of [11] indicate that the application of radial velocities help to improve the location of the precipitation pattern while additional reflectivity data help to adjust the precipitation intensity.

3 WRF Simulations at HLRS From April 2013 to April 2014

This section reports the technical details (3.1) of the recent simulations on the CRAY XE6 and summarizes their first results (3.2).

3.1 Technical Description

All simulations within this period of the project were performed on the CRAY XE6 at the HLRS with the Weather and Research Forecast (WRF) model. The model is applied in Europe with focus on Germany and in Israel/Oman. WRF offers the choice of various physical parameterizations to describe subgrid processes like e.g. cloud formation and radiation transport. In WRFCLIM it is usually applied with the land surface model NOAH [12, 13] and NOAH_MP [14], the Morrison two-moment microphysics scheme [15], the YSU atmospheric boundary layer parameterization [16], the Kain-Fritsch-Eta convection scheme [17] in case of grid cells larger than 4 km and the CAM shortwave and longwave radiation schemes [18]. Experiments with the other five boundary layer parameterizations, two other radiation schemes and a new land surface model are also part of the current simulations. Within this project convection permitting WRF simulations are performed to further downscale the climate data for applications in hydrology and agriculture and an improved representation of feedback processes between the land surface and the atmosphere.

Table 1 summarizes some technical details of the WRF simulations performed from April 2013 to date (15th April 2014). Details are described in the following.

3.2 Sensitivity Study on Parameterization Schemes

Based on the results of the first sensitivity study (see Sect. 3), an enhanced sensitivity study with the most promising subset of parameterizations was performed for Germany. Five boundary layer (BL) parameterization options comprising 3 "local" (Mellor-Yamada-Janjic (MYJ); [19], Mellor-Yamada-Nakanishi-Niino, level 2.5 (MYNN 2.5); [20] quasi-normal scale elimination (QNSE); [21]) and 2 "non-local" schemes (the asymmetric convective model, version 2 (ACM2); [22] and Yonsei University (YSU); [16]), as well as to 2 land surface models (NOAH and NOAH-MP) have been performed. With these, seven experimental runs with WRF have been set up on the CRAY XE6 at HLRS for a child domain at 2-km resolution nested inside a parent domain at 6-km horizontal resolution (Fig. 1), with vertical grid consisting of 89 full sigma levels. The time step in the model has been 36 and 12 s for the parent and the child domain, respectively. The parent (child) domain has been composed of 217×194 (271×271) grid-cells. The chosen temporal resolution for the parent (child) domain of the output has been 1 h (30 min). All the simulations

Table 1 Technical details of the WRF simulations performed from April 2013 to April 2014

Simulation	No. of processors	$\Delta x, \Delta y$ (km)	Simulation period	Number of grid cells (x*y*z)	Δt (s)	No. of simulations	Storage interval	Data storage	approx. CPUh
Preparation of reanalysis on XC40 (chapter 3.1.4)	480–1,600	3	1 month	$691 \times 682 \times 57 \times 18$	6	30 min	14 TB	250,000 (on XC40)	
Spin-up (chapter 3.1.1) (2 domains)	320	6 and 2	01.06 to 16.09.2009	194*217*89, 271*271*89	36	1	6 hourly, 3 hourly	6 TB	114,133
BL-experiments (chapter 3.1.1)	320	6 and 2	21.08 to 16.09.2009	194*217*89, 271*271*89	36	7/1	6 hourly, 3 hourly	14 TB	223,253
BL-experiments part 2 (chapter 3.1.1)	320	6 and 2	08.09.2009/05.09 to 19.09.2009	194*217*89, 271*271*89	36	7/1	6 hourly, 1 hourly, 6 hourly, 3 hourly,	8 TB / 900 GB	13,493 / 8,586
CAOS (chapter 3.1.3)	640	3	26.09.2012, 27.09.2012	681*692*57	18	3	1 hourly	350 GB	1,100
Impact OMAN (chapter 3.1.2)	224	2	15.05. to 30.06.2007	220*220*89	12	3	15 min	3.5 TB	200,000
Impact Israel and Oman (chapter 3.1.2)	160	2	10.06. to 31.07.2012	200*200*89	18	4	1 hourly	1.4 TB	270,000

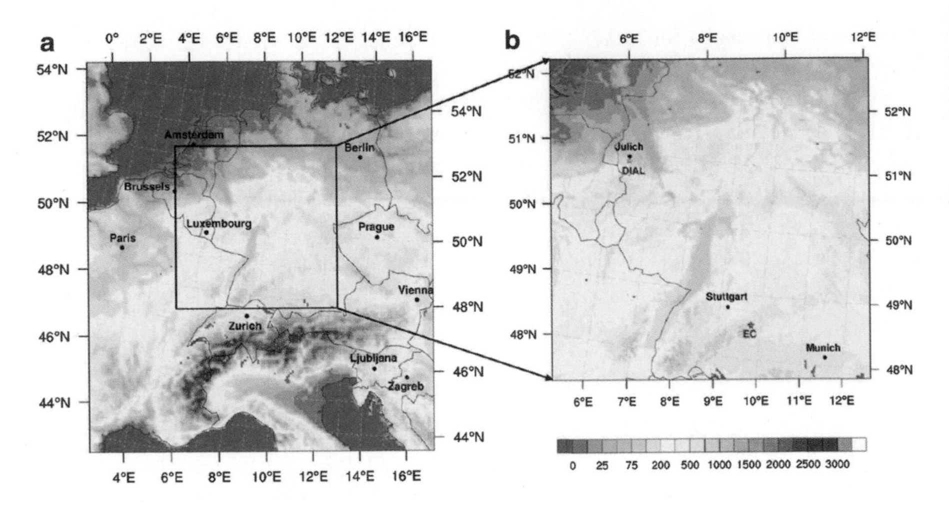

Fig. 1 Topographic representation of the parent domain (**a**) at the 6 × 6 km horizontal resolution (1164 × 1302 km), and the child domain (**b**) at the 2 × 2 km horizontal resolution (542 × 542 km). On the right panel (**b**), locations of the measurement sites used for evaluation are added (*in red*), together with the closest bigger cities (*in black*)

have been run on 320 processors (see Table 1). For the study a total of 359,465 CPU hours and 28.88 TB of storage were used.

Prior to the experimental runs, the model was run in the prognostic mode from 1st June to 21st August 2009 to spin up the soil moisture and temperature. As "prognostic model" we define that the simulation has been conducted in 30-h cycles, with each cycle starting every day at 00 UTC. While the soil state is simulated continuously, the atmospheric state is initialized with the European Centre for Medium-Range Weather Forecasts (ECMWF) operational analysis[1] data at each cycle, the first 6 h are always discarded as atmospheric spin-up. The spin-up run is completed on 21st of August at 06 UTC. From this all 7 experiments were run also in the prognostic mode from 21st August to 16th September 2009.

3.3 *Sensitivity Study on Land Use in a Dry Hot Climate*

Previous simulations for Oman [23] and Israel [9] show, that a plantation in a semi-arid desert in a coastal area may cause different impacts on the atmospheric state. To further study the land use impact in a dry hot climate on the atmospheric state. An impact study of 100 × 100 km^2 plantations has been set up for Oman for June

[1]ECMWF analysis data used in this study have been obtained from ECMWF Data Server.

2007 and (as a result) for Israel and Oman for Summer 2012 with the same model setup as in the previous Israel simulation [3, 9].

For 2007, a CONTROL simulation without the plantation using the original IGBP MODIS data set was followed by two impact studies (IMPACT_1 and IMPACT_2) with the plantation included by changing the land surface parameters in a plantation mask and using two different values for the stomatal resistance. The three simulations were started at 15th May 2007. The first two weeks until end of May were necessary to spin-up the land surface – atmosphere system of WRF. The actual simulation period of the three simulations was the whole month of June. To be able to investigate the diurnal cycle in detail, the storage interval was set to 15 min.

For the 2012 impact runs two model domains were created over Oman and Israel (see Fig. 2, black polygons). This time the horizontal dimensions were reduced from the last runs (444 × 444) to 200 × 200 (comparisons yielded only very minor differences in pressure and temperature fields). In addition to greatly reducing resource use, this means that the lateral forcings are shifted closer to the plantation whilst still allowing a reasonable margin for WRF to develop its own weather systems (50 km from boundaries to plantations – Fig. 2, red polygons).

All the simulations have been run on 160 and 224 processors respectively (see Table 1). For the study a total of approx. 470,000 CPU hours and 4.9 TB of storage were used.

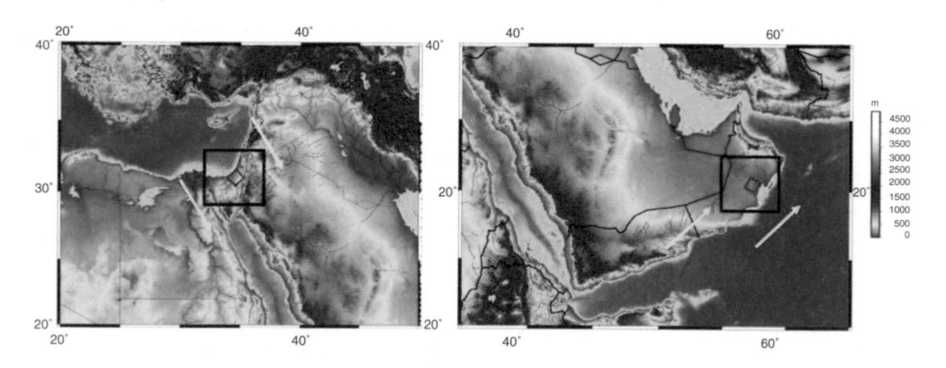

Fig. 2 Model domains marked on topographic maps of eastern Mediterranean and Arabian peninsula regions. Model domain extents are marked with a black polygon and the 100 × 100 km plantations are marked as small red polygons. The Oman domain is centered at 20.30° N, 57.58° E and Israel at 31.25° N, 34.50° E. The Oman domain is located within the tropics, 3.2° below the Tropic of Cancer (23.50° N), whilst Israel lies 7.75° above it within the subtropics. Both domains are 200 × 200 km in dimension with a 2 km horizontal grid spacing

3.4 *Quantitative Precipitation Estimation*

Within the Collaborative Research Unit "Catchments As Organized Systems" (**CAOS**) of the German Research Foundation (http://www.caos-project.de), a quantitative precipitation estimation (QPE) methodology based on numerical weather prediction was set up. For the case study simulations, a horizontal resolution of 3 km is applied in a large European domain of 681×692 grid points (see Fig. 7).

A control simulation initialized only with ECMWF analysis and one assimilation set up for a two day case study from 00Z 26th September 2012 to 00Z 28th September 2012 were performed. In the assimilation run, all available observations were used, including the French radar network, providing upstream observations of reflectivity and radial velocity.

This domain configuration and the selected physical parameterizations were tested in earlier papers [10,11] and are selected to take into account that the complete evolution of interesting synoptic situations takes place in the model domain without being cut by changes in parameterizations or the model resolution. To initialize the model as accurately as possible, a rapid update cycle for the WRF 3DVAR system was set up based on [11]. The RUC interval was set to 1 h and from each analysis a 1 h forecast was performed. Since surface precipitation is not part of the assimilation, the 1-h forecast is taken as the estimation of precipitation and shall be compared in detail and finally combined with other methods as spatial interpolation of rain gauge measurements and radar data.All the simulations have been run on 640 processors (see Table 1). For the study a total of 11,000 CPU hours and 350 GB of storage were used.

3.5 *Data Assimilation (3DVAR) with a Rapid Update Cycle (RUC)*

The developed Rapid Update Cycle (RUC) using the WRF model [3,10] was further refined and is performed on the CRAY XC40 system at a horizontal resolution of 3 km with 690×680 grid boxes in the horizontal and 57 levels in the vertical covering Central Europe (Fig. 3).

The RUC was enhanced so that the update interval is increased to 3h. In addition to conventional observations for temperature, wind, and humidity, also GPS Zenith Total Delay (ZTD) data are selected for data assimilation while satellite brightness temperatures are still not used. To further improve the model initial conditions and also QPF, a small number of sensitivity tests using Doppler radar data from France and Germany were carried out. The selected period was COPS IOP 10 (22–23 July, 2007) where intense precipitation was observed in Central France due to a strong low pressure system over the English Channel.

All the simulations have been run on 480 processors (see Table 1). For the study a total of 1,290,000 CPU hours and 14 TB of storage were used on the XC40 system.

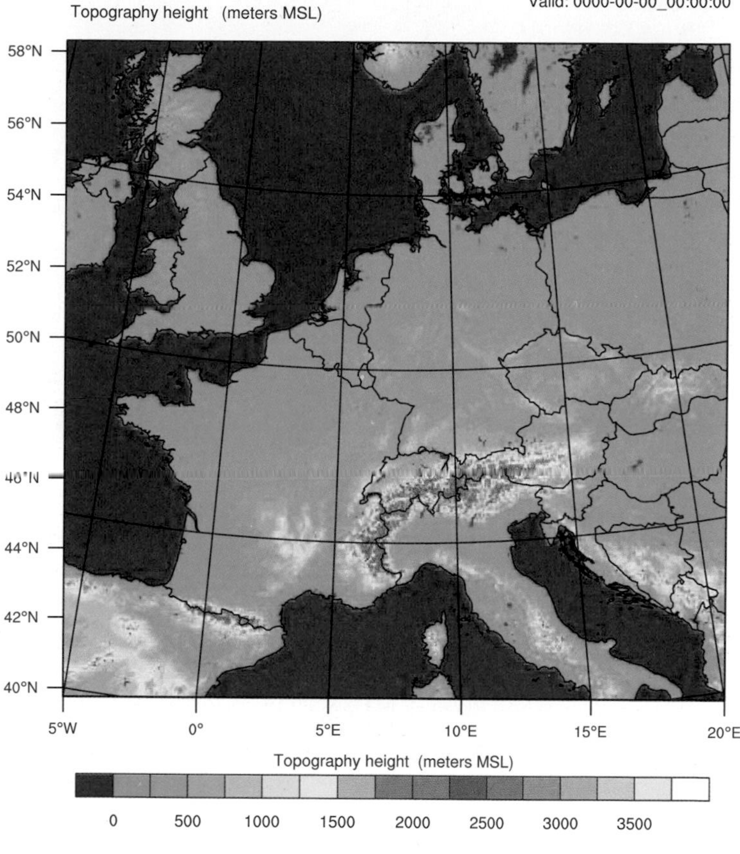

Fig. 3 Model domain for the WRF RUC

4 Preliminary Results

4.1 Sensitivity on Parameterization Schemes

In September 2009 in the study region in Western Germany absorption lidar (DIAL) measurements of absolute humidity profiles were performed during the Transregio 32 FLUXPAT campaign [24]. These vertical humidity profiles (Fig. 4) are used for a detailed study of simulated boundary layer (BL) features and processes (e.g. grid-cell averaged structure of the humidity profiles and height, entrainment and development of the convective boundary layer and the residual moist layer).

During the 8th September 2009, weather in whole central Europe was under the influence of a vast high pressure system centred in Lithuania, resulting in dry and cloudless weather conditions. The sensitivity of WRF to the land surface models (LSMs) is significantly higher than to the BL schemes (Fig. 4d–f), which is evident

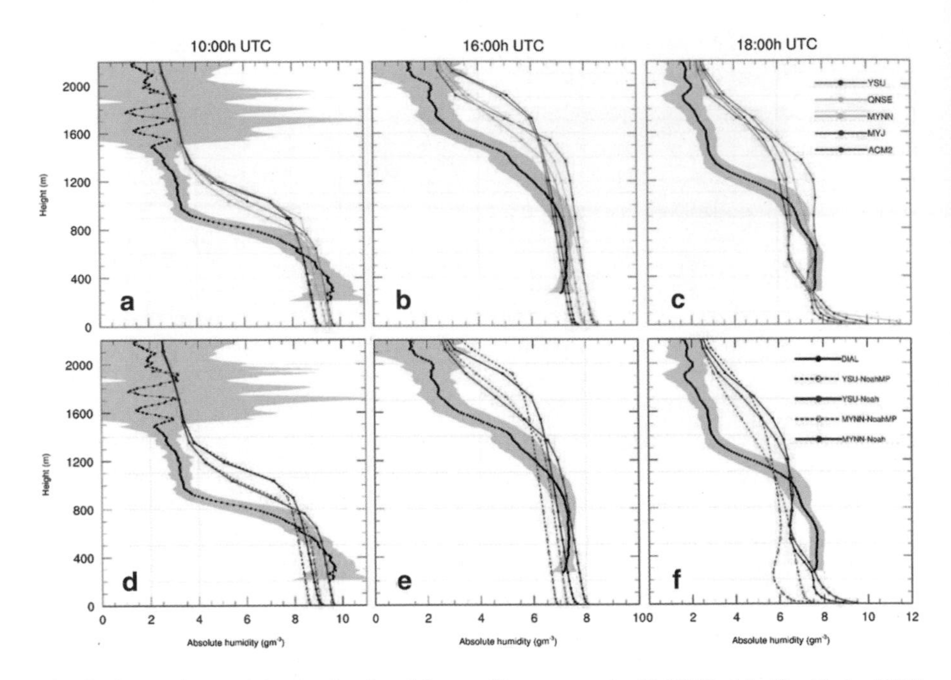

Fig. 4 Comparisons of the absolute humidity profiles measured with DIAL (*black*) with the WRF simulations on the 8th of September 2009 at 1,000 (**a**, **d**), 1,600 (**b**, **d**) and 1,800 UTC (**c**, **f**). (**a**), (**b**) and (**c**) show comparisons with the results of the WRF configured with Noah LSM and 5 BL schemes: YSU (*purple*), QNSE (*orange*), MYNN 2.5 (*green*), MYJ (*red*) and ACM2 (*blue*). (**d**), (**e**) and (**f**) denote comparisons with the WRF output configured with YSU (*blue*) and MYNN 2.5 (*red*) BL schemes in combination with Noah (*solid*) and Noah-MP (*dashed*) LSMs

not only in the lower BL, but also in the BL entrainment zone. All schemes resulted in a too high BL, accounting for too much mixing. However, the bias obtained with MYNN 2.5 is the smallest in this case study (Fig. 4a, b, d, e). Overall, the model misses to reproduce the residual layer and the start of the BL evolution (i.e. the BL transition, not shown). Furthermore, the "non-local" BL schemes, as well as NOAH-MP LSM, accounts for the drier and a bit higher convective BL than the "local" BL scheme and the NOAH LSM. While [25] reported that "nonlocal" BL schemes in general perform better than "local" BL schemes in representing the convective BL features, this cannot be supported by this case study when compared with the humidity profiles of the DIAL. This study demonstrates the great potential of DIAL measurements for evaluating the model performance and investigating the BL features. It also demonstrates the importance of such sensitivity model studies to improve process understanding and simulation. Currently the data is exploited for detailed land-surface atmosphere feedback quantification studies.

4.2 Sensitivity on Land Use in a Dry Hot Climate

Geoengineering increasingly becomes an issue in a changing climate, namely in an expected hotter and drier climate. One practice is to change land cover. One method is changing the land use through e.g. plantations. However, in a large scale style this potentially impacts the weather and climate. In Israel we have access to measurements of a real land use change experiment.

It was found that a plantation might lead to strong modification of diurnal cycles of meteorological variables such as surface fluxes, surface temperature and boundary layer depth. Variations in the estimated value for the stomatal resistance demonstrated a high model sensitivity on vegetation parameters. The observed net cooling as well as a positive impact on convection initiation, triggered by the formation of a convergence zone above the plantation under suitable initial conditions, support the assumption that afforestation in a desert region seems to be a promising mitigation option. It turned out that in terms of further assessment studies on climate engineering, diurnal dynamics need to be taken into account rather than only daily mean statistics, since they drive the evolution of processes in the boundary layer.

The June 2007 Oman simulations [26] show that a plantation might lead to strong modification of diurnal cycles of meteorological variables such as surface fluxes, surface temperature and boundary layer depth. Variations in the estimated value for the stomatal resistance demonstrated high model sensitivity on vegetation parameters. Figure 5 shows as an example the influence on the monthly averaged diurnal cycles of sensible and latent heat flux.

The amplitude of the sensible heat flux of IMPACT 1 slightly exceeds the CONTROL, whereas its amplitude in IMPACT 2 is more than 150 Wm^{-2} higher. The strong influence of the plantation and especially the changed stomatal resistance is also seen in the latent heat flux.

The observed net cooling as well as a positive impact on convection initiation, triggered by the formation of a convergence zone above the plantation under suitable initial conditions, support the assumption that afforestation in a desert region seems to be a promising mitigation option. It turned out that in terms of further assessment studies on climate engineering, diurnal dynamics need to be taken into account

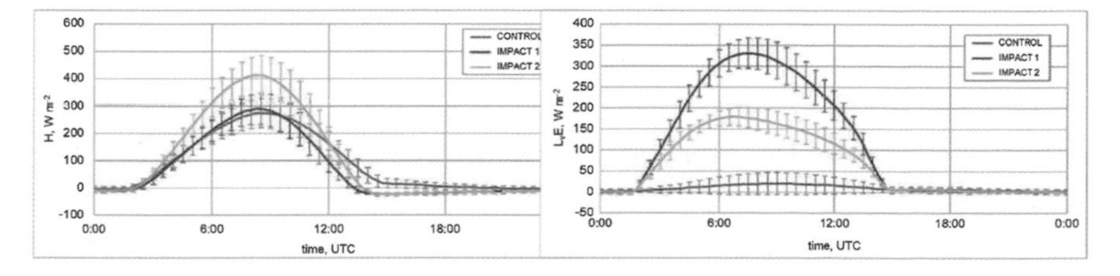

Fig. 5 Comparison of monthly averaged diurnal cycles of the sensible heat flux (H) (*left*) and latent heat flux (LvE) (*right*) between CONTROL, IMPACT 1 and IMPACT 2 over the plantation

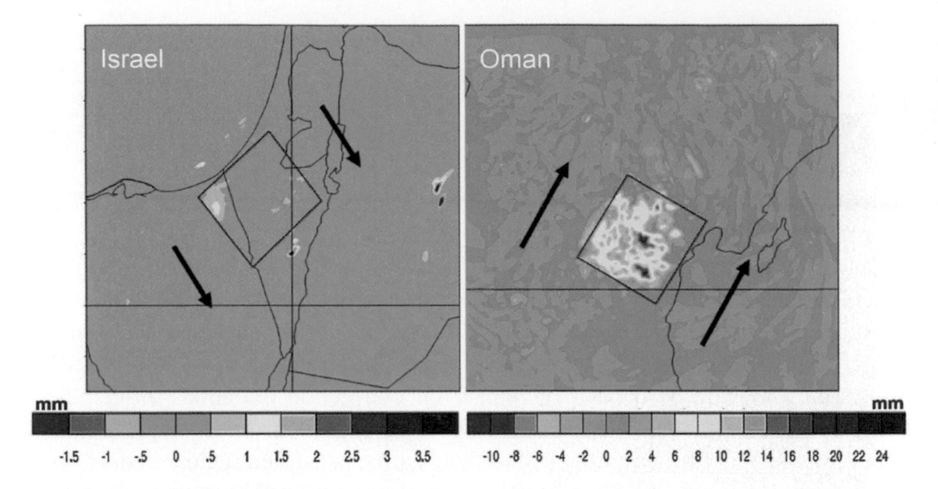

Fig. 6 Total accumulated precipitation over Israel and Oman (24.06–31.07 2012). Note the difference in scales used due to the large differences between the regional impacts. Also marked are arrows approximating the summer mean surface wind flow

rather than only daily mean statistics, since they drive the evolution of processes in the boundary layer.

Based on the different impact of the plantations in Israel and Oman, a study was set up for equally sized and shaped plantations in both countries in summer 2012. The main differences between the previous Oman simulations and these are that (a) instead of Jatropha, Simmondsia chinensis (Jojoba) is used, (b) 2012 is a non-El Niño year, unlike the previous simulations in 2007, and (c) this WRF configuration was validated with field data from a Jojoba plantation in Israel (see validation results in Sect. 2). This validation gives us reasonable confidence that the model coupling with the atmosphere is reasonable and realistic. Results – To assess the impact of the plantations on moist convection, the total precipitation over the period is plotted for Israel and Oman in Fig. 6.

A significant difference in precipitation is apparent between the two regions. In Israel, the domain precipitation increase (Impact minus Control) is only 0.019 mm. In Oman the increase is 0.37 mm. In Israel there are no patches over the plantation with values over 1.5 mm and are therefore on the verge of being considered insignificant. The results are currently under detailed investigation.

4.3 *Quantitative Precipitation Estimation*

As an example result of the QPE experiments for September 26th and 27th, Fig. 7 shows the WRF-based QPE for two time steps during the evolution of the synoptic situation. Whereas the 26th September was dominated by the passage of a frontal rain band, convection developed on the cold air on the rear side of the front on

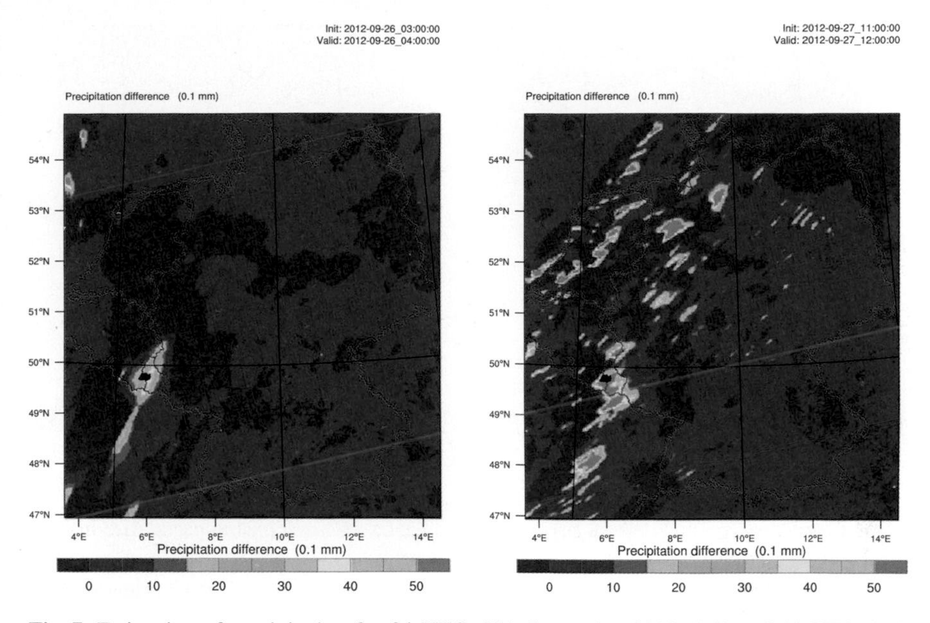

Fig. 7 Estimation of precipitation for 04 UTC, 26th September 2012 (*left*) and 12 UTC, 27th September 2012 (*right*). The region of interest for the CAOS project (the Attert catchment in Luxembourg) is outlined in *black*

the 27th. The presented region is the domain covered by the DWD RADOLAN products.

At the moment these results are compared in detail with the radar products of the German Weather Service.

4.4 Data Assimilation (3DVAR) with a Rapid Update Cycle (RUC)

The results show a positive impact on QPF when radar radial velocities are applied in addition to the other observations helping to reduce the overestimation of precipitation. If radar reflectivities are used in addition, the positive precipitation bias is significantly reduced by about 50 % (Fig. 8). Further details can be found in [11].

Recent experiments on the successor system XC40 include a soil moisture spin-up simulation of about three weeks simulation time. Also the RUC will be adjusted that the update interval can be user defined as small as 1 h. In this context, a one month reanalysis is planned by utilizing satellite radiances in addition to the conventional observations.

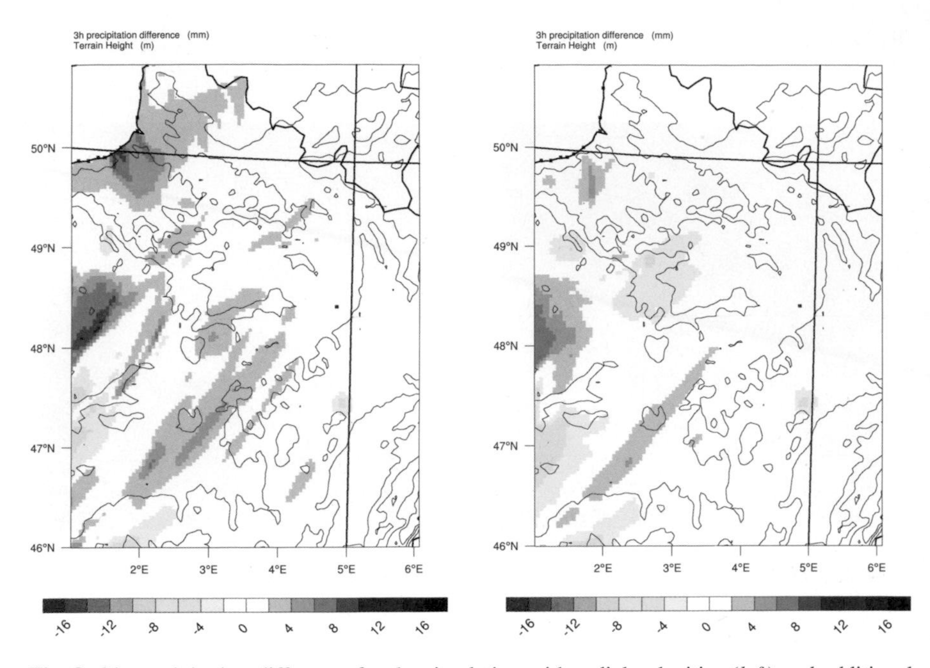

Fig. 8 3 h precipitation difference for the simulation with radial velocities (*left*) and additional reflectivities (*right*)

5 Outlook for May and June 2014

Currently a convection permitting climate simulation with WRF 3.5.1 is set up for central Europe on the CRAY XE6 for a domain of 1151×1051 grid cells and 89 vertical levels. The simulation will be started in May with more than 3,000 cores.

Additional experiments on the XC40 system will include experiments to prepare a high-resolution European reanalysis on a horizontal resolution of 3 km for at least on month period. In order to prepare the future climate simulations within the project PAK 346/RU1695 on the XC40 system, preparatory studies will be carried out to adjust the domain size, the number of cores and also to investigate the effect of using parallel NetCDF to reduce the time required for writing the model output files.

Acknowledgements This work is part of the Project PAK 346/RU 1695 funded by DFG and supported by a grant from the Ministry of Science, Research and Arts of Baden-Württemberg (AZ Zu 33-721.3-2) and the Helmholtz Centre for Environmental Research – UFZ, Leipzig (WESS project). DIAL measurements were performed within the Project Transregio 32, also funded by DFG and provided by Andreas Behrendt and Florian Späth (University of Hohenheim). Model simulations were carried out at HLRS on the Cray XE6 and the NEC Nehalem Cluster within WRFCLIM and we thank the staff for their support. WRF-based quantitative precipitation estimation is funded by DFG in the frame of the collaborative research unit "Catchments As Organized Systems" (CAOS, RU 1598).

References

1. Giorgi, F., Jones, C., Asrar, G.: Addressing climate information needs at the regional level: The CORDEX framework. Technical report WMO Bulletin 58, World Meteorological Organization (WMO) (2009)
2. Warrach-Sagi, K., Schwitalla, T., Bauer, H.S., Wulfmeyer, V.: A Regional Climate Model Simulation for EURO-CORDEX with the WRF Model. In: Resch, M. (ed.) Sustained Simulation Performance 2013, pp. 147–157. Springer, Cham (2013)
3. Warrach-Sagi, K., Bauer, H.S., Branch, O., Milovac, J., Schwitalla, T., Wulfmeyer, V.: High-Resolution Climate Predictions and Short-Range Forecasts to Improve the Process Understanding and the Representation of Land-Surface Interactions in the WRF Model in Southwest Germany (WRFCLIM). In: Nagel, W.E., Kroener, D.H., Resch, M.M. (eds.) High Performance Computing in Science and Engineering'13, pp. 529–542. Springer, Cham (2013)
4. Skamarock, W.C., Klemp, J.B., Dudhia, J., Gill, D., Barker, D.M., Duda, M.G., Huang, X.Y., Wang, W., Powers, J.G.: A description of the advanced research WRF version 3. NCAR Technical Note TN-475+STR, NCAR, P.O. Box 3000, Boulder, 80307 (2008)
5. Warrach-Sagi, K., Schwitalla, T., Wulfmeyer, V., Bauer, H.S.: Evaluation of a climate simulation in Europe based on the WRF-N OAH Model System: precipitation in Germany. Clim. Dyn. **41**, 744–755 (2013). doi:10.1007/s00382-013-1727-7
6. Greve, P., Warrach-Sagi, K., Wulfmeyer, V.: Evaluating Soil Water Content in a WRF-Noah Downscaling Experiment. J. Appl. Meteorol. Climatol. **52**, 2312–2327 (2013)
7. Kotlarski, S., Keuler, K., Christensen, O.B., Colette, A., Deque, A., Gobiet, M., Goergen, K., Jacob, D., Luethi, D., van Meijgaard, E., Nikulin, G., Schaer, C., Teichmann, C., Vautard, K., Warrach-Sagi, R., Wulfmeyer, V.: Regional climate modeling on European scales: a joint standard evaluation of the EURO-CORDEX RCM ensemble, Geosci. Model Dev.**7**, 1297-1333 (2014). doi:10.5194/gmd-7-1297-2014
8. Vautard, R., Gobiet, A., Jacob, D., Belda, M., Colette, A., Deque, M., Fernandez, J., Garcia-Diez, I.M., Goergen, K., Guettler, I., Halenka, T., Keuler, K., Kotlarski, S., Nikulin, G., Patarcic, M., Suklitsch, M., Teichmann, C., Warrach-Sagi, K., Wulfmeyer, V., Yiou, P.: The simulation of European heat waves from an ensemble of regional climate models within the EURO-CORDEX project. Clim. Dyn. **41**, 2555–2575 (2013). doi:10.1007/s00382-013-1714-z
9. Branch, O., Warrach-Sagi, K., Wulfmeyer, V., Cohen, S.: Simulation of semi-arid biomass plantations and irrigation using the WRF-NOAH model-a comparison with observations from Israel, Hydrol. Earth Syst. Sci. **18**, 1761-1783 (2014). doi:10.5194/hess-18-1761-2014, 2014.
10. Schwitalla, T., Bauer, H.S., Wulfmeyer, V., Aoshima, F.: High-resolution simulation over central Europe: Assimilation experiments with WRF 3DVAR during COPS IOP9c. Q. J. R. Meteorol. Soc.**137**, 156-175 (2011). doi:10.1002/qj.721
11. Schwitalla, T. Wulfmeyer, V.: Radar data assimilation experiments using the IPM WRF Rapid Update Cycle. Meteorol. Z. **23**, 79-102 (2014)
12. Chen, F., Dudhia, J.: Coupling an Advanced Land Surface-Hydrology Model with the Penn State-NCAR MM5 Modeling System. Part I: Model Implementation and Sensitivity. Mon. Wea. Rev. **129**, 569–585 (2001)
13. Chen, F., Dudhia, J.: Coupling an Advanced Land Surface-Hydrology Model with the Penn State-NCAR MM5 Modeling System. Part II: Preliminary Model Validation. Mon. Wea. Rev. **129**, 587–604 (2001)
14. Niu, G.Y., Yang, Z.L., Mitchell, K.E., Chen, F., Ek, M.B., Barlage, M., Kumar, A., Manning, K., Niyogi, D., Rosero, E., Tewari, M., Xia, Y.: The community Noah land surface model with multiparameterization options (Noah-MP): 1. Model description and evaluation with local-scale measurements. J. Geophys. Res. **116**, D12109 (2011). doi:10.1029/2010JD015139
15. Morrison, H., Thompson, G., Tatarskii, V.: Impact of Cloud Microphysics on the Development of Trailing Stratiform Precipitation in a Simulated Squall Line: Comparison of One- and Two-Moment Schemes. Mon. Weather Rev. **137**, 991–1007 (2009).

16. Hong, S.Y., Noh, Y., Dudhia, J.: A New Vertical Diffusion Package with an Explicit Treatment of Entrainment Processes. Mon. Weather Rev. **134**, 2318–2341 (2006).
17. Kain, J.S.: The Kain-Fritsch Convective Parameterization: An Update. J. Appl. Meteorol. **43**, 170–181 (2004).
18. Collins, W.D., Rasch, P.J., Boville, B.A., Mc Caa, J.R., Williamson, D.L., Kiehl, J.T., Briegleb, B., Bitz, C., Lin, S.J., Zhang, M., Dai, Y.: Description of the ncar community atmosphere model (cam 3.0). NCAR technical Note NCAR/TN-464+STR, 226pp. NCAR, Boulder (2004)
19. Mellor, G., Yamada, T.: A Hierarchy of Turbulence Closure Models for Planetary Boundary Layers. J. Atmos. Sci. **31**, 1791–1806 (1974).
20. Nakanishi, M., Niino, H.: Development of an Improved Turbulence Closure Model for the Atmospheric Boundary Layer. J. Meteorol. Soc. Jpn. **87**, 895–912 (2009).
21. Sukoriansky, S., Galperin, B., Perov, V.: Application of a new spectral theory of stable stratified turbulence to the atmospheric boundary layer over sea ice, Bound.-Layer Meteorol. **117**, 231–257 (2005).
22. Pleim, J.E.: A Combined Local and Nonlocal Closure Model for the Atmospheric Boundary Layer. Part I: Model Description and Testing. J. Appl. Meteorol. Clim. **46**, 1383–1395 (2007).
23. Wulfmeyer, V., Branch, O., Warrach-Sagi, K., Bauer, H.S., Schwitalla, T., Becker, K.: The Impact of Plantations on Weather and Climate in Coastal Desert Regions. J. Appl. Meteor. Climatol. **53**, 1143–1169 (2013).
24. Behrendt, A., Wulfmeyer, V., Riede, A., Wagner, G., Pal, S., Bauer, H., Späth, F.: Scanning differential absorption lidar for 3D observations of the atmospheric humidity field. In: Proceedings of the 25th International Laser Radar Conference, 5–9 July 2010, St. Petersburg, vol. 2, pp. 1187–1190. Publishing House of IAO SB RAS (2010). ISBN:978-5-94458-109-9
25. Hu, X.M., Nielsen-Gammon, J.W., Zhang, F.: Evaluation of Three Planetary Boundary Layer Schemes in the WRF Model. J. Appl. Meteorol. Climatol. **49**, 1831–1844 (2010).
26. Groner, V.: The impact of a large-scale plantation of jatropha curcas l. on convective initialization under monsoonal conditions in the coastal desert of Oman. Master thesis, University of Hohenheim (2014)

Direct Numerical Simulation of Breaking Atmospheric Gravity Waves

Sebastian Remmler, Stefan Hickel, Mark D. Fruman, and Ulrich Achatz

Abstract We present the results of fully resolved direct numerical simulations of monochromatic gravity waves breaking in the middle atmosphere. The simulations are initialized with optimal perurbations of the gives waves. Given a wavelength of 3 km, the required grid sizes range up to 3.6 billion computational cells, depending on the necessary domain size and the turbulence intensity. Our results provide an insight into the mechanics of gravity wave breaking they will be of great value for the validation of lower order methods for the prediction of wave breaking.

1 Introduction

Gravity waves are a common phenomenon in any stably stratified fluid, as the earth atmosphere. They can be excited by flow over orography [12, 18], by convection [4, 7], and by spontaneous imbalance of the mean flow in the troposphere [14, 15]. Gravity waves transport energy and momentum from the region where they are forced to the region where they are dissipated (e.g. through breaking), possibly far away from the source region. Various phenomena, such as the cold summer mesopause [8] and the quasi-biennial oscillation in the equatorial stratosphere [3], cannot be explained nor reproduced in weather and climate simulations without accounting for the effect of gravity waves. See Fritts and Alexander [5] for an overview of gravity waves in the middle atmosphere.

Since most gravity waves have a wavelength that cannot be resolved in general circulation models, the effect of gravity waves on the global circulation is usually accounted for by some parametrization based on combinations of

S. Remmler • S. Hickel (✉)
Institute of Aerodynamics and Fluid Mechanics, Technische Universität München,
85747 Garching bei, Müchen, Germany
e-mail: remmler@tum.de; sh@tum.de

M.D. Fruman • U. Achatz
Institute for Atmosphere and Environment, Goethe-University Frankfurt,
60438 Frankfurt am Main, Germany
e-mail: fruman@iau.uni-frankfurt.de; achatz@iau.uni-frankfurt.de

W.E. Nagel et al. (eds.), *High Performance Computing in Science and Engineering '14*, DOI 10.1007/978-3-319-10810-0_39

linear wave theory [11], empirical observations of time-mean energy spectra [9], and simplified treatments of the breaking process. See Kim et al. [10] and McLandress [13] for reviews of the various standard parametrization schemes.

A common weakness of most parametrization schemes is the over-simplified treatment of the wave breaking process. Improving this point requires a deeper insight into the breaking process that involves generation of small scale flow features through wave-wave interactions and through wave-turbulence interactions. Since the gravity wave length and the turbulence that eventually leads to energy dissipation into heat span a wide range of spatial and temporal scales, the breaking process is challenging both for observations and numerical simulations. Simulations must cover the breaking wave with a wavelength of some kilometres as well as the smallest turbulence scales (the Kolmogorov length η). The Kolmogorov length is on the order of millimetres in the troposphere [19] and approximately 1 m in 80 km altitude [16].

An important aspect in setting up a simulation of a gravity wave breaking event is the proper choice of the domain size and initial conditions. Since the gravity wave itself is one-dimensional, while the breaking process and the resulting turbulence is three-dimensional, proper choices have to be made for the two directions perpendicular to the wave vector. Achatz [1] and Achatz and Schmitz [2] analysed the optimal perturbations for a given monochromatic gravity wave. Fruman and Achatz [6] extended this analysis for IGWs by computing optimal secondary perturbations based on the time dependent perturbed wave. They found that the wavelength of the optimal secondary perturbation can be much smaller than the original wave. Thus a three-dimensional domain needs not to have the size of the base wavelength in all three directions. They proposed the following multi-step approach to set-up domain and initial conditions for a given monochromatic gravity wave:

1. Solution (in the form of normal modes or singular vectors) of the Boussinesq equations linearized about the basic state wave, determining the primary instability structures;
2. Nonlinear two-dimensional (in space) numerical solution of the full Boussinesq equations using the result of stage 1 as initial condition;
3. Solution in the form of singular vectors (varying in the remaining spatial direction) of the Boussinesq equations linearized about the time-dependent result of stage 2;
4. Three-dimensional DNS of the Boussinesq equations using the linear solutions from stages 1 and 3 as initial condition and their wavelengths for the size of the computational domain.

This procedure was applied to an unstable IGW by Remmler et al. [16]. In the present study we complete this work by adding two additional test cases representing different types of breaking gravity waves.

2 Governing Equations

Since the propagation angle of the studied gravity waves is very steep, it is advantageous to rotate the coordinate system with respect to the earth coordinates (x, y, z) such that one coordinate direction is aligned with the direction of wave propagation. We define the rotated Cartesian cooardinates $\boldsymbol{x} = \left(x_\parallel, y_\perp, \zeta\right)$ for a gravity wave propagating in the (x, z)-plane at an angle Θ with respect to the horizontal plane as

$$x_\parallel = (x \sin\Theta - z \cos\Theta)\cos\alpha + y \sin\alpha, \tag{1a}$$

$$y_\perp = (z \cos\Theta - x \sin\Theta)\sin\alpha + y \cos\alpha, \tag{1b}$$

$$\zeta = x \cos\Theta + z \sin\Theta. \tag{1c}$$

The corresponding velocity vector is $\boldsymbol{v} = \left(u_\parallel, v_\perp, w_\zeta\right)$ and the angle α describes the direction of the primary perturbation of the wave. We distinguish between *transverse* ($\alpha = 90°$) and *parallel* ($\alpha = 0°$) perturbations. The rotated coordinate system is sketched in Fig. 1.

The gravity waves under consideration can be investigated within the Boussinesq approximation. The corresponding equations of motion in the rotated coordinate system are

$$\nabla \cdot \boldsymbol{v} = 0, \tag{2a}$$

$$\partial_t \boldsymbol{v} + (\boldsymbol{v} \cdot \nabla)\,\boldsymbol{v} = -f\hat{\boldsymbol{e}}_z \times \boldsymbol{v} + b\hat{\boldsymbol{e}}_z - \nabla p + \nu\nabla^2\boldsymbol{v} + \mathscr{F}, \tag{2b}$$

$$\partial_t b + (\boldsymbol{v} \cdot \nabla)\, b = -N^2\hat{\boldsymbol{e}}_z \cdot \boldsymbol{v} + \mu\nabla^2 b + \mathscr{B}, \tag{2c}$$

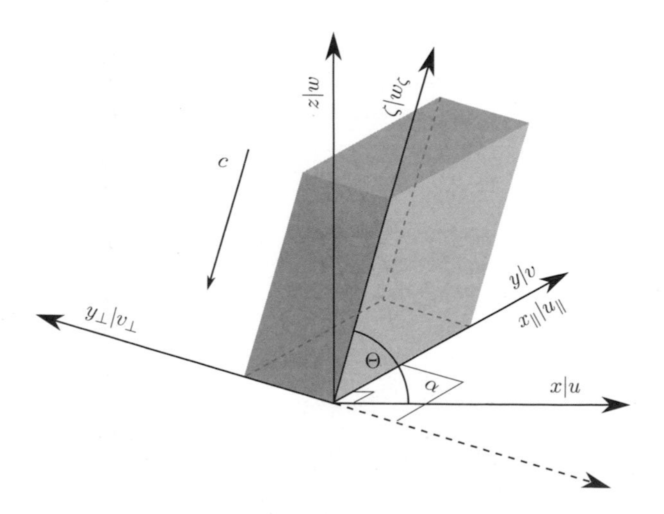

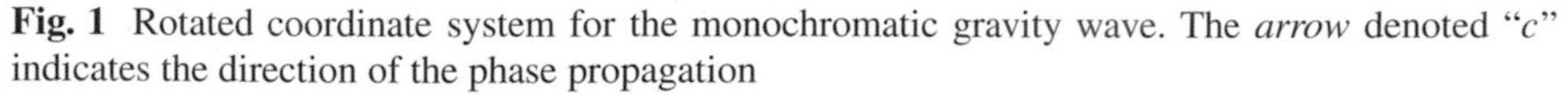
Fig. 1 Rotated coordinate system for the monochromatic gravity wave. The *arrow* denoted "*c*" indicates the direction of the phase propagation

where b is buoyancy, p is pressure normalized by a constant background density, and $\hat{\boldsymbol{e}}_z$ is the unit vector in the true vertical direction. N is the Brunt-Väisälä frequency, f is the Coriolis parameter, ν and μ are the kinematic viscosity and thermal diffusivity, respectively and $\mathscr{F}$ and $\mathscr{B}$ represent the influence of an explicit turbulence SGS parametrization.

If we look for solutions of the Boussinesq equations that resemble upward-propagating inertia-gravity waves, we find

$$\left[u_{\parallel}, v_{\perp}, w_{\zeta}, b\right] = \Re\left\{a\left[\frac{iK\omega}{km}\cos\alpha + \frac{f}{k}\sin\alpha, -\frac{iK\omega}{km}\sin\alpha + \frac{f}{k}\cos\alpha, 0, -\frac{N^2}{m}\right]e^{i\phi}\right\}, \quad (3)$$

where K is the magnitude of the wavevector, $k = K\cos\Theta$ and $m = K\sin\Theta$ are its horizontal and vertical components in the Earth frame, $\phi = K\zeta - \omega t$ is the wave phase and

$$\omega = -\sqrt{f^2\sin^2\Theta + N^2\cos^2\Theta} \quad (4)$$

is the frequency of a wave with upward group velocity. The nondimensional (complex) wave amplitude a is defined such that a wave with $|a| = 1$ is neutral with respect to static instability at its least stable point, such that waves with $|a| > 1$ are statically unstable and waves with $|a| < 1$ are statically stable. The phase speed of the wave is directed in the negative ζ-direction.

The local kinetic and available potential energy in the flow are defined as

$$E_{\mathrm{k}} = \frac{1}{2}\boldsymbol{v}^{\top}\cdot\boldsymbol{v}, \qquad E_{\mathrm{p}} = \frac{b^2}{2N^2}. \quad (5)$$

The total energy is defined as the sum of the two components: $E_{\mathrm{t}} = E_{\mathrm{k}} + E_{\mathrm{p}}$. We obtain the transport equations of these energy components by multiplying Eqs. (2b) and (2c) by $\left(\boldsymbol{v}, b/N^2\right)$ and applying the divergence constraint (2a):

$$\partial_t E_{\mathrm{k}} + \nabla\cdot(\boldsymbol{v}E_{\mathrm{k}}) = -f\boldsymbol{v}\cdot(\hat{\boldsymbol{e}}_z\times\boldsymbol{v}) - b\boldsymbol{v}\cdot\hat{\boldsymbol{e}}_z - \boldsymbol{v}\cdot\nabla p + \nu\left[\nabla^2 E_{\mathrm{k}} - (\nabla\boldsymbol{v})^2\right] + \boldsymbol{v}\cdot\mathscr{F} \quad (6a)$$

$$\partial_t E_{\mathrm{p}} + \nabla\cdot\left(\boldsymbol{v}E_{\mathrm{p}}\right) = b\boldsymbol{v}\cdot\hat{\boldsymbol{e}}_z + \mu\left[\nabla^2 E_{\mathrm{p}} + \frac{(\nabla b)^2}{N^2}\right] + \frac{b\mathscr{B}}{N^2} \quad (6b)$$

3 Numerical Method

Our flow solver INCA (www.inca-cfd.com) is a multi-purpose finite-volume solver for the solution of compressible and incompressible problems on Cartesian adaptive grids. For the present study, we used only a small part of INCA's abilities. The Boussinesq equations are discretized by a fractional step method on a staggered Cartesian mesh. Spatial discretization is based on a non-dissipative central difference scheme with 4th order accuracy for the convective terms and 2nd order central differences for the diffusive terms and the continuity equation (Poisson equation for pressure). For time advancement the explicit third-order Runge-Kutta scheme of [17] is used. The time-step is dynamically adapted to satisfy a Courant-Friedrichs-Lewy condition with $CFL \leq 1.0$. The Poisson equation is solved by a Krylov subspace solver with algebraic-multigrid preconditioning.

4 Test Cases

4.1 Physical Set-Up

We investigate three different cases of monochromatic gravity waves in an environment representative for the upper mesosphere. The parameters for all three cases are summarized in Table 1. The primary and secondary perturbations of the waves were computed based on the approach of [6].

Table 1 Basic parameters of the three breaking gravity wave cases

Case #	1	2	3
Propagation angle Θ	89.5°	89.5°	70°
Initial amplitude a_0	1.2	0.86	1.2
Phase velocity c_p	0.1 m/s	0.1 m/s	3.3 m/s
Wave period T	7.9 h	7.9 h	15.3 min
Domain size $L_{\parallel}$	4 km	2.1 km	2.9 km
Domain size $L_{\perp}$	0.4 km	0.3 km	3 km
Domain size L_{ζ}	3 km	3 km	3 km
Maximum resolution Δ	3 m	3 m	2 m
Total # of cells	172 mill.	71 mill.	3 624 mill.
Kinematic viscosity ν		$1\,m^2/s$	
Coriolis parameter f		$1.367 \cdot 10^{-4}\,s^{-1}$	
Brunt-Väisälä frequency N		$0.02\,s^{-1}$	
Gravitational acceleration g		$9.81\,m/s^2$	
Thermal diffusivity α		$1\,m^2/s$	

Case 1 is a statically unstable inertia-gravity wave with a wave period of 8 h and phase speed of 0.1 m s^{-1}. The wavelength of the leading transverse normal mode rimary perturbation) is of the order of the base wavelength, while the leading condary singular vector has a significantly shorter wavelength. The time scales of e turbulent wave breaking and of the wave propagation are similar, which makes is case especially interesting. Remmler et al. [16] pointed out that a secondary reaking event is stimulated in this cases when the most unstable part of the wave eaches the region where the primary breaking has generated a lot of turbulence before.

Case 2 is also an inertia-gravity wave with the same period and phase speed as case 1, but with a lower amplitude that is below the threshold of static instability. The wave is perturbed by the leading transverse primary singular vector ($\lambda_{\parallel} = 2.115$ km), and the leading secondary singular vector ($\lambda_{\perp} = 300$ m). Despite the wave being statically stable, the perturbations lead to a weak breaking and the generation of some turbulence. However, the duration of the breaking event is much shorter than the wave period and the overall energy loss in the wave is not much larger than the energy loss through viscous forces on the base wave in the same time.

Case 3 is a statically unstable high-frequency gravity wave with a period of 15 min and a phase speed of 3.3 m s^{-1}. It is perturbed with the leading transverse primary normal mode ($\lambda_{\parallel} = 2.929$ km) and the secondary singular vector with $\lambda_{\perp} = 3$ km. The breaking is much stronger than in cases 1 and 2 and lasts for more than one wave period. Turbulence and energy dissipation are almost uniformly distributed in the domain during the most intense phase of the breaking.

4.2 *Required Computational Resources*

For the unstable IGW, a resolution of $\Delta = 3$ m and therefore 172.8 million grid cells were required for the solution to be fully resolved. The simulation was run on the NEC SX-9 vector computer at HLRS in Stuttgart, Germany. A single node of this machine (500 GB memory, 16 vector processors with 100 GFLOP/s peak performance each) had sufficient memory to store the complete flow field. Hence we could avoid domain decomposition and relied on shared memory parallelization only. The efficient Poisson solver employs a discrete Fourier Transform in one direction in combination with a Bi-Conjugate Gradient Stabilized (BiCGSTAB) solver [20] in the plane perpendicular to the chosen direction. The Fourier Transform converts the three-dimensional problem into a set of independent two-dimensional problems, which are solved in parallel. The simulation of a flow time of 35,000 s (270,000 time steps) required a wall time of 1,100 h, which corresponds to 85.7×10^{-9} node-seconds per time step and cell.

The simulations of the stable IGW were carried out on the LOEWE cluster at CSC Frankfurt, Germany. This machine consists of nodes with two AMD Opteron 6172 CPUs (12 cores per CPU, 8.4 GFlop/s per core peak performance) and 64 GB memory. The fully resolved DNS with 71 million grid cells was decomposed into 192 blocks and simulated on eight nodes. The integration up to $t = 100$ min (38,600 time steps) took 183 h, i.e. 1.93×10^{-6} node-seconds per time step and cell.

The simulations of the unstable HGW were the most demanding and were run on the Cray XE6 cluster at HLRS Stuttgart, consisting of nodes with two AMD Opteron 6276 (Interlagos) CPUs (16 cores per CPU, 9.2 GFlop/s per core peak performance) and 32 GB memory. The fully resolved DNS with 3,624 million grid cells was decomposed into 4,096 blocks and simulated on 512 nodes using 8 processor cores per node. The integration up to $t = 46.2$ min (49,460 time steps) required a wall time of about 288 h. Hence the computational performance was 2.96×10^{-6} node-seconds per time step and cell.

5 Results

5.1 *Unstable Inertia-Gravity Wave*

The first test case is a statically unstable inertia-gravity wave with initial amplitude $a_0 = 1.2$, propagation angle 89.5° and wavelength 3 km. The wave period is 8 h and the phase speed is 0.1 ms^{-1}.

3-D DNS initialized with the IGW ($\lambda = 3$ km), the leading transverse primary normal mode ($\lambda_{\parallel} = 3.981$ km), and the leading secondary singular vector with $\lambda_{\perp} = 400$ m were run with a grid spacing Δ of about 3 m (full resolution) and 6 m (coarse) in all three directions. The amplitude for the secondary singular vector A_2, defined here as the maximum perturbation energy density divided by the maximum basic-state energy density, was 0.02. It was shown by Remmler et al. [16] that only in the fully resolved simulation was the Kolmogorov length never smaller than Δ/π but that the results of the two simulations were otherwise extremely similar (hence grid-converged). Figure 2 shows the initial buoyancy field from the full resolution simulation and a snapshot at $t = 695$ s of the buoyancy field together with the kinetic energy dissipation ϵ_k. At the instant shown, very early in the simulation, turbulence has already developed in the upper half of the wave (i.e. the less stable half) but not in the lower half. Note that the figure is plotted in the reference frame moving with the wave. The decay of the wave amplitude with time and the global mean of the total energy dissipation $\epsilon_k + \epsilon_p$ from the ensemble of 2.5-D simulations and from the 3-D DNS are shown in Fig. 3. The initial burst of turbulence is more intense in the 3-D DNS, and the wave decays more rapidly. On the other hand, in the 2.5-D simulations the initial turbulence is more sustained and the total amplitude decay greater.

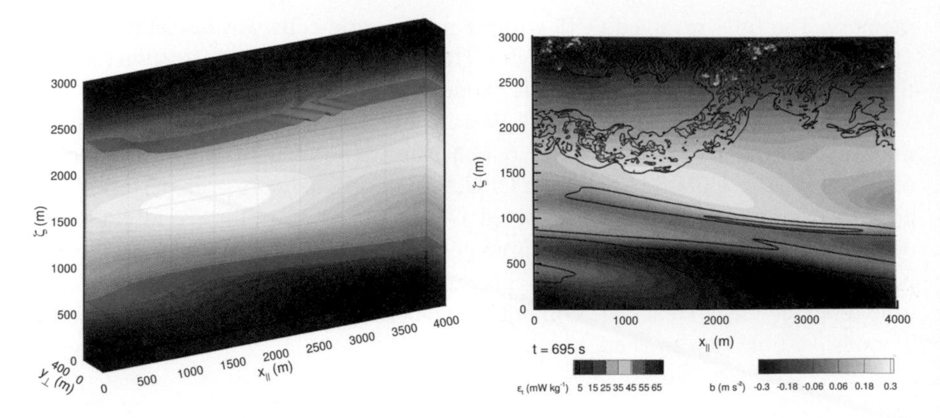

Fig. 2 Snapshots of the buoyancy field from fine 3-D DNS (1,350 × 128 × 1,000 cells) of the statically unstable IGW: *Left panel*: 3-D initial condition with an isosurface at $b = -0.02\,\mathrm{m\,s^{-2}}$ (*green colour*). *Right panel*: flow field averaged in the $y_\perp$-direction at $t = 11.6\,\mathrm{min}$ (greyscale contours: buoyancy, coloured lines: total energy dissipation)

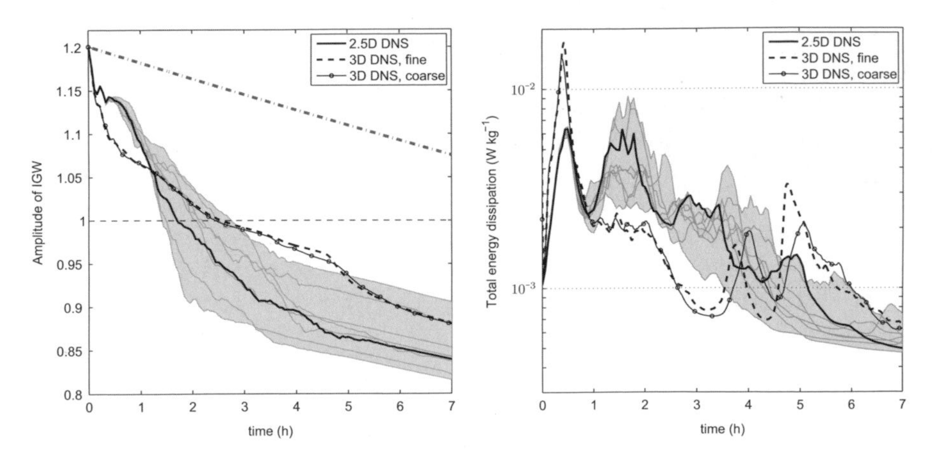

Fig. 3 Comparison of wave amplitude decay (*left*) and total energy dissipation (*right*) in 2.5-D and 3-D DNS of statically unstable IGW. *Dash-dot line* indicates amplitude decay due to laminar viscous decay of the unperturbed wave

5.2 Stable Inertia-Gravity Wave

The second case is a statically stable inertia-gravity wave, identical to the first case but with $a_0 = 0.86$, so the initial amplitude is smaller than the theshold for static stability.

The initial condition for the 3-D DNS of the stable IHW is composed of the original IGW ($\lambda = 3\,\mathrm{km}$), the leading 7.5-min transverse primary singular vector ($\lambda_\parallel = 2.115\,\mathrm{km}$), and the leading 7.5-min secondary singular vector ($\lambda_\perp = 300\,\mathrm{m}$). Simulations were run with average grid spacing $\Delta \approx 3\,\mathrm{m}$ ("fine") and $\Delta \approx 4.2\,\mathrm{m}$

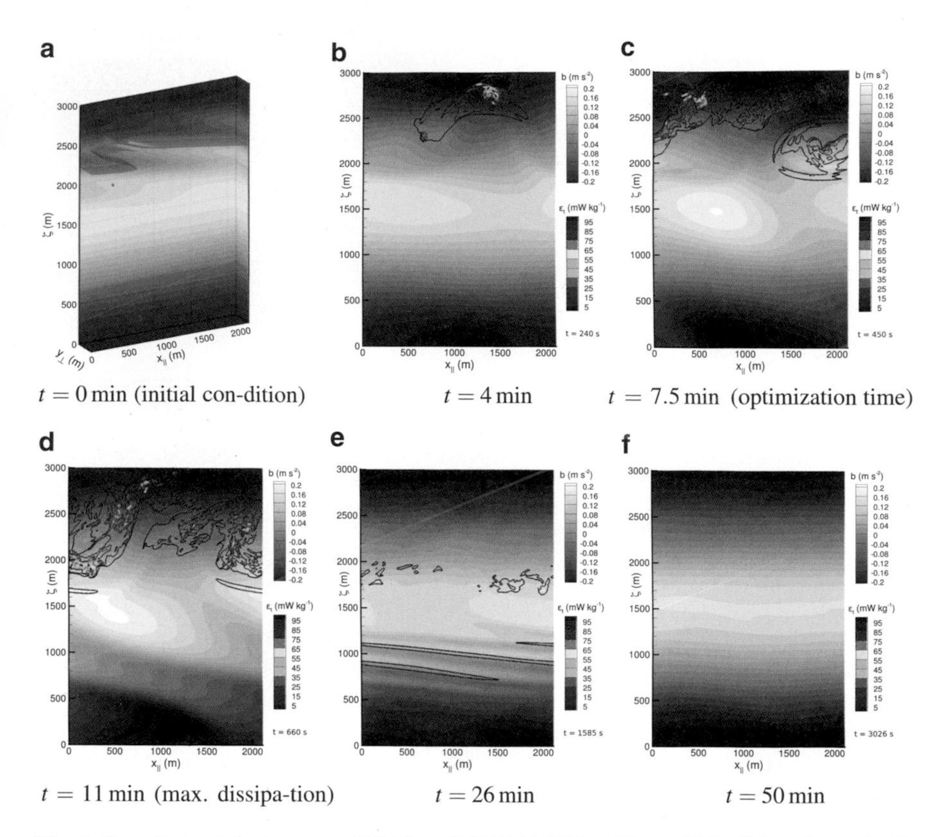

Fig. 4 Snapshots of the buoyancy field from 3-D DNS (720 × 96 × 1,024 cells) of the statically stable IGW: (**a**) 3-D initial condition with an isosurface at $b = -0.03\,\mathrm{m\,s^{-2}}$ (*green colour*). (**b**)–(**f**) Flow field averaged in the $y_\perp$-direction (greyscale contours: buoyancy, coloured lines: total energy dissipation)

("coarse"). The initial buoyancy field from the fine 3-D DNS is shown in Fig. 4. The temporal development of the flow field is visualized in Fig. 4 by contours of buoyancy and kinetic energy dissipation ϵ_k. The perturbation grows during the first minutes and generates turbulence in the least stable part of the wave. The turbulence remains confined to this region and is dissipated quickly. The peak dissipation is reached at $t = 11$ min and after 40 min the turbulence has basically vanished. During this period of turbulent decay some overturning occurs in the most stable part of the wave, similar to the case of the unstable IGW. Here, however, the overturning is too weak to create a negative vertical buoyancy gradient and breaking. It is thus simply dissipated by molecular heat transport.

Figure 5 shows the evolution of the wave amplitude and total energy dissipation from the 3-D DNS and the ensemble of 2.5-D DNS. The decay (and partial rebound) of the wave amplitude is very similar in 3-D and 2.5-D, but the onset of turbulence and the associated energy dissipation occur earlier in the 3-D simulation.

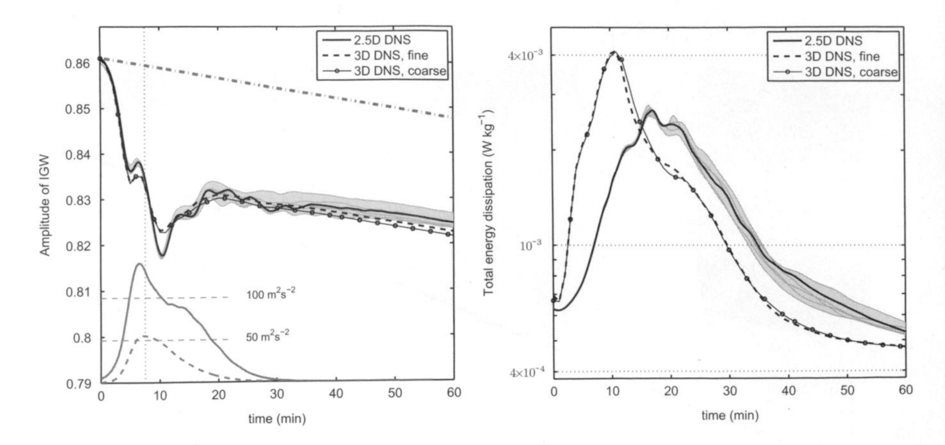

Fig. 5 As in Fig. 3 but for the statically stable IGW. The resolutions used for the 3-D DNS were $720 \times 96 \times 1{,}024$ (fine) and $512 \times 64 \times 768$ (coarse). The curves in the lower part of the left panel show the maximum (*solid line*) and mean (*dashed line*) energy in the linear 2.5-D integration initialized with the primary SV, and the vertical dotted line marks the optimization time (7.5 min). For reference, the energy density in the unperturbed IGW is 54.5 m^2s^{-2}

In the lower portion of the left panel of Fig. 5 is shown the maximum and mean perturbation energy density from a linear 2.5-D integration initialized with the primary singular vector. The mean energy in the singular vector is maximum at the optimization time and then decays. The drops in the maximum perturbation energy density from the linear integration approximately coincide in time with the rebounds of the IGW amplitude from the nonlinear simulations. The spatial distribution of the dissipation (Fig. 6) is very similar in 2.5-D and 3-D. The energy dissipation is strongly correlated with the region of $Ri < 1/4$ (bounded by the dark-grey contour), particularly in the upper half of the domain in the 2.5-D simulation.

The Kolmogorov length as a function of time from the two 3-D DNS is plotted in Fig. 7a. In the fine simulation η is always larger than Δ/π (indicated by the horizontal line), so all turbulence scales are resolved. Although η is briefly below Δ/π in the coarse simulation, the results are almost indistinguishable from the fine simulation (compare the projection and dissipation diagnostics shown in Fig. 5), so the simulations are grid-converged.

5.3 Unstable High-Frequency Gravity Wave

The third and final test case is the statically unstable ($a_0 = 1.2$) high frequency gravity wave ($\Theta = 70°$, period 15 min, phase speed 3.3 ms^{-1}).

The 3-D DNS was initialized with the original HGW ($\lambda = 3$ km), the leading transverse primary normal mode ($\lambda_{\parallel} = 2.929$ km) and the 5-min secondary singular vector with $\lambda_{\perp} = 3$ km. Three simulations were performed, with grid spacing Δ of

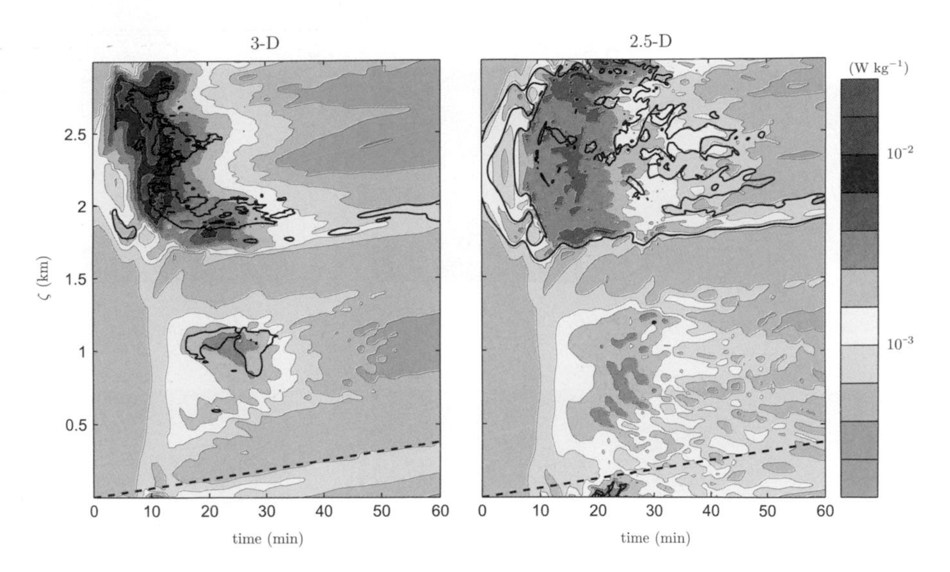

Fig. 6 Spanwise and streamwise averaged total energy dissipation from the fully resolved 3-D (*left*) and 2.5-D (*right*) DNS of the statically stable IGW. Contours equally spaced on a logarithmic (base 10) scale. *Solid light grey line* is the contour $Ri = 1/4$ and the *heavy dashed black line* represents a fixed point in the Earth-frame

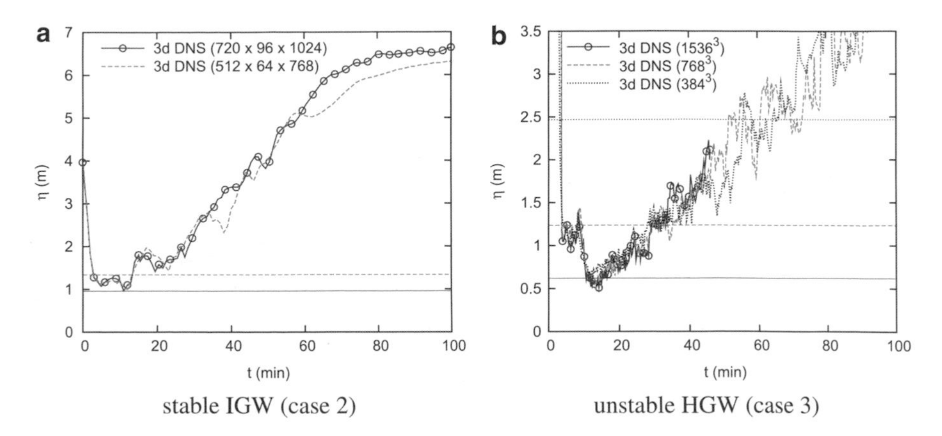

Fig. 7 Kolmogorov length in the 3-D DNS of (**a**) the stable IGW and (**b**) the unstable HGW. The threshold where the simulation is supposed to be fully resolved is indicated by a *horizontal line* for each simulation

1.9 m (fine resolution), 3.9 m (coarse 1) and 7.8 m (coarse 2). The initial buoyancy field from the fine resolution simulation is shown in Fig. 8. Comparisons of the amplitude decay and total energy dissipation are shown in Fig. 10. Both diagnostics are quite similar in 2.5-D and 3-D, although as in the previous cases the onset of turbulent dissipation occurs slightly earlier in 3-D. The spatial distribution of the energy dissipation from the medium resolution (coarse 1) run is shown in Fig. 9.

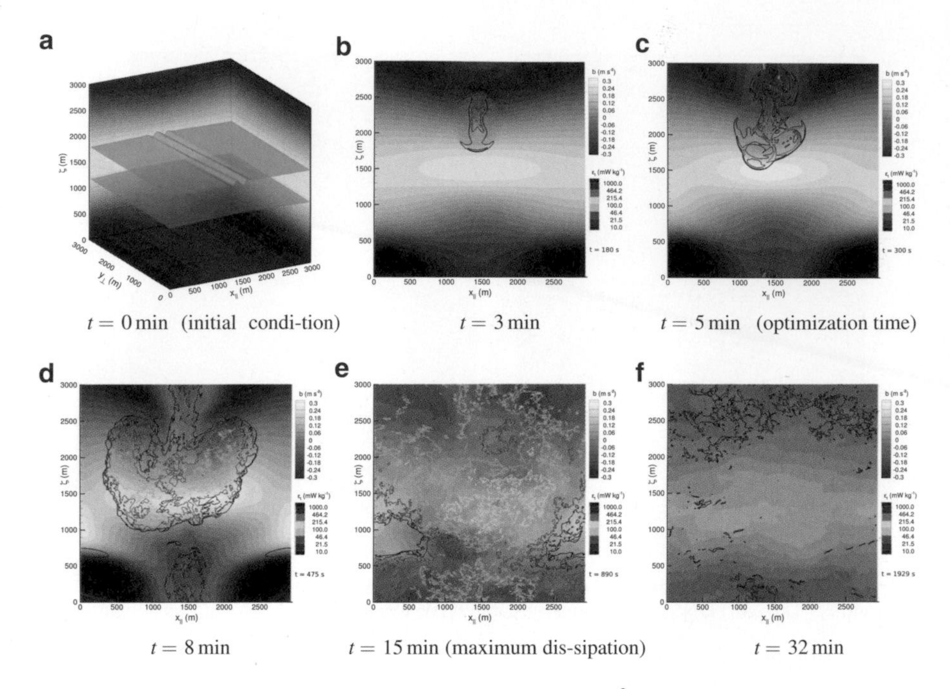

Fig. 8 Snapshots of the buoyancy field from 3-D DNS ($1{,}536^3$ cells) of the statically unstable HGW: (**a**) 3-D initial condition with an isosurface at $b = 0.2\,\mathrm{m\,s^{-2}}$ (*green colour*). (**b**)–(**f**) Flow field averaged in the $y_\perp$-direction (greyscale contours: buoyancy, coloured lines: total energy dissipation)

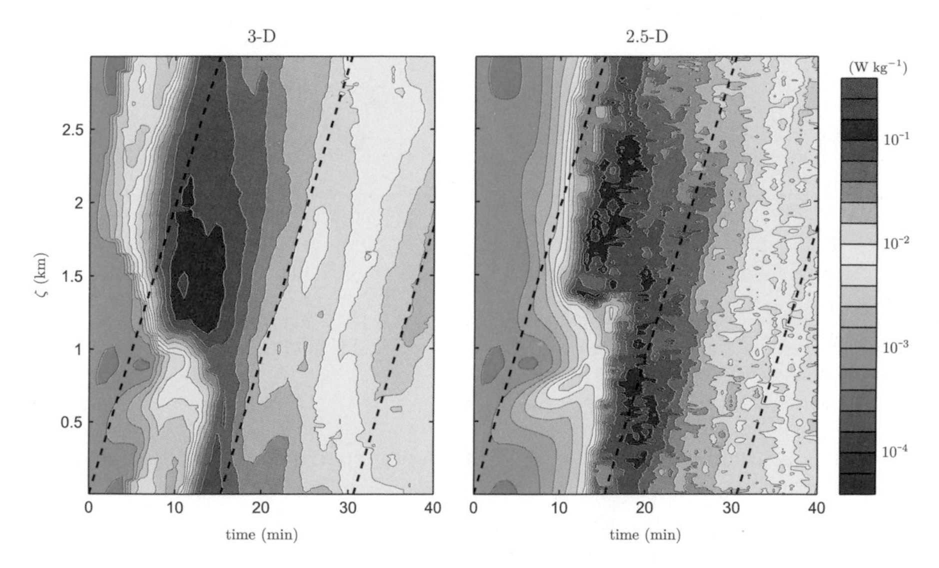

Fig. 9 As in Fig. 6 but for the unstable HGW

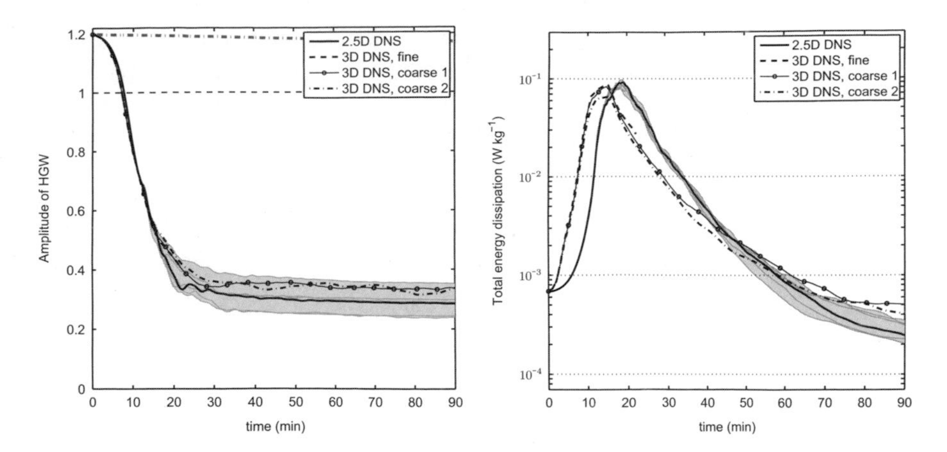

Fig. 10 As in Fig. 3 but for the statically unstable HGW. The 3-D DNS were performed with $1{,}536^3$ (fine), 768^3 (coarse 1) and 384^3 (coarse 2) gridcells

The regions of intense energy dissipation is approximately fixed in space (parallel to the heavy dashed black lines), particularly in the 2.5-D simulation.

The Kolmogorov length η from the 3-D DNS with coarse, medium and fine resolution is plotted as a function of time in Fig. 7b. In the fine simulation η is always approximately equal to or larger than Δ/π (indicated by the horizontal lines) and can hence be considered fully resolved. There is practically no difference in terms of the wave amplitude and dissipation rates (Fig. 10) in the results of the simulations with $1{,}536^3$ cells and 768^3 cells. The coarse 2 simulation with only 384^3 cells, on the other hand, has a slightly lower dissipation peak than the other two.

6 Conclusion

We reported on three-dimensional direct numerical simulations of three different cases of breaking monochromatic gravity waves in the middle atmosphere. These results provide an insight into the mechanics of gravity wave breaking in different scenarios. Furthermore, the obtained results will be of great value for the validation of lower order methods for the prediction of wave breaking, such as large-eddy simulations or extensions of linear wave propagation models.

The environmental parameters were chosen to be close the conditions found at approximately 80 km altitude. We used a wavelength of 3 km and defined (1) an unstable inertia-gravity wave, (2) a stable inertia-gravity wave and (3) an unstable high-frequency gravity wave.

The initial conditions were defined through a four-step approach which involves (1) a normal mode (NM) or singular vector (SV) analysis of the Boussinesq equations linearized about the gravity wave solution, (2) a 2.5-D nonlinear simulation of the full Boussinesq equations initialized with the inertia-gravity wave plus a leading NM or SV, (3) SV analysis on the full equations linearized about the particular time-dependent 2.5-D solution and finally (4) a 3-D DNS initialized with the inertia-gravity wave, the leading NM or SV and a leading secondary SV. The dimensions of the integration domain are equal to the wavelengths of the wave and the leading perturbations. This procedure ensures that the domain size is large enough to allow the most unstable perturbations to grow. On the other hand the domain does not need to be larger in the respective directions which saves a lot of computational resources in many cases compared to previous numerical studies.

Acknowledgements U. A. and S. H. thank Deutsche Forschungsgemeinschaft (German Research Foundation, DFG) for partial support through the MetStröm (Multiple Scales in Fluid Mechanics and Meteorology) Priority Research Program (SPP 1276), and through Grants HI 1273/1-2 and Ac71/4-2. Computational resources were provided by the HLRS Stuttgart under the grants TIGRA and DINSGRAW and by the CSC Frankfurt.

References

1. Achatz, U.: On the role of optimal perturbations in the instability of monochromatic gravity waves. Phys. Fluids **17**(9), 094107 (2005)
2. Achatz, U., Schmitz, G.: Optimal growth in inertia-gravity wave packets: energetics, long-term development, and three-dimensional structure. J. Atmos. Sci. **63**, 414–434 (2006)
3. Baldwin, M.P., Gray, L.J., Dunkerton, T.J., Hamilton, K., Haynes, P.H., Randel, W.J., Holton, J.R., Alexander, M.J., Hirota, I., Horinouchi, T., Jones, D.B.A., Kinnersley, J.S., Marquardt, C., Sato, K., Takahashi, M.: The quasi-biennial oscillation. Rev. Geophys. **39**, 179–229 (2001)
4. Chun, H.-Y., Song, M.-D., Kim, J.-W., Baik, J.-J.: Effects of gravity wave drag induced by cumulus convection on the atmospheric general circulation. J. Atmos. Sci. **58**(3), 302–319 (2001)
5. Fritts, D.C., Alexander, M.J.: Gravity wave dynamics and effects in the middle atmosphere. Rev. Geophys. **41**(1), 1–64 (2003)
6. Fruman, M.D., Achatz, U.: Secondary instabilities in breaking inertia-gravity waves. J. Atmos. Sci. **69**, 303–322 (2012)
7. Grimsdell, A.W., Alexander, M.J., May, P.T., Hoffmann, L.: Model study of waves generated by convection with direct validation via satellite. J. Atmos. Sci. **67**(5), 1617–1631 (2010)
8. Hines, C.O.: Dynamical heating of the upper atmosphere. J. Geophys. Res. **70**(1), 177–183 (1965)
9. Hines, C.O.: Doppler-spread parameterization of gravity-wave momentum deposition in the middle atmosphere. Part 1: basic formulation. J. Atmos. Sol.-Terr. Phys. **59**(4), 371–386 (1997)
10. Kim, Y.-J., Eckermann, S.D., Chun, H.-Y.: An overview of the past, present and future of gravity–wave drag parametrization for numerical climate and weather prediction models. Atmosphere-Ocean **41**(1), 65–98 (2003)
11. Lindzen, R.S.: Turbulence and stress owing to gravity wave and tidal breakdown. J. Geophys. Res. **86**, 9707–9714 (1981)
12. McFarlane, N.A.: The effect of orographically excited gravity wave drag on the general circulation of the lower stratosphere and troposphere. J. Atmos. Sci. **44**, 1775–1800 (1987)
13. McLandress, C.: On the importance of gravity waves in the middle atmosphere and their parameterization in general circulation models. J. Atmos. Sol.-Terr. Phy. **60**(14), 1357–1383 (1998)

14. O'Sullivan, D., Dunkerton, T.J.: Generation of inertia-gravity waves in a simulated life cycle of baroclinic instability. J. Atmos. Sci. **52**(21), 3695–3716 (1995)
15. Plougonven, R., Snyder, C.: Inertia gravity waves spontaneously generated by jets and fronts. Part I: different baroclinic life cycles. J. Atmos. Sci. **64**(7), 2502–2520 (2007)
16. Remmler, S., Fruman, M.D., Hickel, S.: Direct numerical simulation of a breaking inertia-gravity wave. J. Fluid Mech. **722**, 424–436 (2013)
17. Shu, C.-W.: Total-variation-diminishing time discretizations. SIAM J. Sci. Stat. Comput. **9**(6), 1073–1084 (1988)
18. Smith, R.B.: The influence of mountains on the atmosphere. Adv. Geophys. **21**, 87–230 (1979)
19. Vallis, G.K.: Atmospheric and Oceanic Fluid Dynamics. Cambridge University Press, Cambridge (2006)
20. van der Vorst, H.A.: Bi-CGSTAB: a fast and smoothly converging variant of Bi-CG for the solution of nonsymmetric linear systems. SIAM J. Sci. Stat. Comput. **13**(2), 631–644 (1992)

Part VI
Miscellaneous Topics

Wolfgang Schröder

In this chapter, contributions on geophysics, material science, computational engineering, and molecular modeling complement the research fields such as physics, solid state physics, reacting flows, computational fluid dynamics, and transport and climate which have been tackled in the previous chapters. The articles widen the field where numerical simulations represent useful tools to gain novel results and to improve the scientific knowledge. The findings evidence the close link between natural science in general and computer science in particular. Furthermore, these results provide a sound basis to develop novel scientific models. Compact mathematical descriptions will be solved by highly sophisticated and efficient algorithms using high-performance computers. This interdisciplinary collaboration between several scientific fields defines the extremely intricate numerical challenges and as such drives the progress in fundamental and applied research. The subsequent articles represent an excerpt of the projects being linked to HLRS. The computations are used to obtain some quantitative results and to corroborate physical models and to even derive new theoretical approaches. Nevertheless, to substantiate the numerical simulations, experimental investigations and analytical solutions are to be taken into account to complement the physical knowledge.

The first article by the Institute of Planetology of the University of Münster, the Institute of Planetory Research of the German Aerospace Center in Berlin, and the Department of Planetory Geodesy of the TU Berlin deals with Thermo-Chemical Mantle Convection Simulations using Gaia. Thermal and compositional convection in planetary mantles are the most dominant dynamical processes following planetary formation, being the primary mechanisms through which terrestrial planets lose their heat. In thermal convection, mantle flow is driven by density gradients caused by temperature variations. Compositional convection instead involves chemical heterogeneities caused by different mineral phases present in the interior of terrestrial

---

W. Schröder (✉)
Institute of Aerodynamics, RWTH Aachen University, Wüllnerstr. 5a, 52062 Aachen, Germany
e-mail: office@aia.rwth-aachen.de

bodies. The slow creep of silicate mantle material controls the heat transport in the interior, thus influencing surface geological structures like volcanoes, rifts and tectonic plates as well as the magnetic field generation. Two approaches exist to model thermo-chemical convection in planetary mantles. 1D models, where mantle convection is parameterized via appropriate scaling laws, or 2D-3D fully dynamical models, which solve the entire set of conservation equations for a fluid continuum. 1D models cannot predict characteristic convection patterns or lateral heterogeneities in the mantle and a direct comparison against 3D surface data is difficult. It goes without saying, that 2D-3D fully dynamical models are best suited for such a comparison. Over the years, the complexity of these models has increased remarkably such that challenges like duration of volcanism, initiation of plate-tectonics, formation of geochemical reservoirs, etc. can now be addressed.

In the contribution on Phase-Field Simulations of Large-Scale Microstructures by Integrated Parallel Algorithms of the Institute of Applied Materials, Reliability of Components and Systems of the Karlsruhe Institute of Technology and the Institute of Materials and Processes of the Hochschule Karlsruhe Technik und Wissenschaft, the extension of a phase-field model for crystal growth and solidification in complex alloy systems implemented in the software package PACE3D and the underlying parallelized algorithms are presented. The progress in three topics is discussed. First, the results of a parallel connected component labeling algorithm, being optimized to run on a large number of computing units used to simulate pore development in sintering processes, are shown. The second topic is split into a brute force study of the Read-Shockley model to get input parameters for a large multi-grain phase-field-study, such that recrystallization and abnormal grain growth can be investigated. The third topic focuses on heat transfer and fluid flow in metallic foam structures depending on the porosity as base material for new heat storage systems.

Molecular models for cyclic alkanes and ethyl acetate as well as surface tension data from molecular simulation are discussed in the next article of the Chair of Thermodynamics and Energy Technology of the University of Paderborn. Thermodynamic data for most technically interesting systems are still scarce or even unavailable despite the large experimental effort invested into their measurement over the last century. This particularly applies to mixtures containing two or more components and systems under extreme conditions. Unlike phenomenological methods, molecular modeling and simulation is based on a sound physical foundation such that it is well suited to predict such properties and processes. New molecular force field models, i.e., for four cyclic alkanes and ethyl acetate, are presented in the Paderborn contribution to widen the application area of molecular simulation. Furthermore, vapor-liquid interfaces are investigated, deepening the understanding of interfacial processes of a model fuel, consisting of acetone, nitrogen and oxygen, similar to the state of injected acetone droplets in a gaseous fluid in combustion chambers. Finally, an efficient method to generate large simulation data sets is studied with respect to the stability and accuracy using the microcanonical ensemble (NVE) instead of the canonic (NVT) ensemble.

The project "Distributed FE Analysis of Multiphase Composites for Linear and Nonlinear Material Behaviour" of Keßler, Schrader and Könke is related to the use of high performance computers. In the project the deterioration and damage in structures with heterogeneous materials is formulated and solution algorithms that are adjusted to parallel hardware architecture are developed. Due to the complex physical behaviour it is necessary to use fine grained models to capture the response at the appropriate spatial scale. This leads either to the use of CPU based or CPU-GPU based hardware parallel architectures.

The project "Parallel Performance of a Discontinuous Galerkin Spectral Element Method based PIC-DSMC Solver" of Ortwein, Binder, Copplestone, Mirza, Nizenkov, Pfeiffer, Stindl, Fasoulas, and Munz is solving Maxwell's equations with a combined Particle-in-Cell Direct Simulation Monte Carlo method. The implementation has been analyzed with respect to parallel performance for strong scaling cases, and it has been shown a significant efficiency crease in comparison to the last report, while still not reaching near scaling. The reason is due to the particle insertion, which inevitably has higher communications overhead costs for higher number processes. While parallel performance, in general, proved to be acceptable, minimum amount of elements per process has to be reached in order to be able use load balancing to efficiently treat inhomogeneous particle distributions.

The project VascCan shows that modelling, simulation, and HPC play a more and more important part in systems biology, where complex phenomena on different time and length scales have to coupled to describe the underlying systems. The project's two parts address the process of vasculogenesis and the impact of therapies on vascular tumors.

Thermo-Chemical Mantle Convection Simulations Using Gaia

Ana-Catalina Plesa, Christian Hüttig, Nicola Tosi, and Doris Breuer

Abstract Thermally and chemically driven buoyancy in planetary mantles cause the slow creep of material, which is ultimately responsible for the heat transport from the deep interior and the large-scale dynamics inside the Earth and other terrestrial planets. With the increasing computational power and the improvement of numerical methods, numerical simulations of planetary interiors have become one the principal tools for understanding the processes active during the thermo-chemical evolution of a terrestrial planet considering constraints posed by geological and geochemical surface observations delivered by various planetary missions. In the present work we present technical aspects and applications to solid-state mantle convection using our code Gaia in Cartesian/cylindrical/spherical geometry. We test the convergence of several numerical solvers that have been implemented in our code, and show the code performance on the HLRS System with up to 10,000 cores. Further we compare our results with published benchmark values.

1 Introduction

Thermal and compositional convection in planetary mantles are the most dominant dynamical processes following planetary formation, being the primary mechanisms through which terrestrial planets loose their heat. In thermal convection, mantle flow is driven by density gradients caused by temperature variations. Compositional convection instead involves chemical heterogeneities caused by different mineral

A.-C. Plesa (✉)
WWU, Institute of Planetology, Muenster and German Aerospace Center, Institute of Planetary Research, Berlin, Germany
e-mail: a_ples01@uni-muenster.de

C. Hüttig • D. Breuer
German Aerospace Center, Institute of Planetary Research, Berlin, Germany
e-mail: christian.huettig@dlr.de; doris.breuer@dlr.de

N. Tosi
Department of Planetary Geodesy, Technical University Berlin, Berlin, Germany
e-mail: nicola.tosi@tu-berlin.de

W.E. Nagel et al. (eds.), *High Performance Computing in Science and Engineering '14*, DOI 10.1007/978-3-319-10810-0_40

phases present in the interior of terrestrial bodies. The slow creep of silicate mantle material controls the heat transport in the interior, thus influencing surface geological structures like volcanoes, rifts and tectonic plates as well as the magnetic field generation.

Mantle convection manifests itself differently on the terrestrial bodies of our Solar System. The surface of the Earth, covered by seven major plates, is characterized by the generation of new oceanic crust and lithosphere occurring at spreading centers and its subsequent subduction into the mantle at convergent plate margins. On all other terrestrial bodies of the Solar System, the so-called one-plate bodies like Mercury, Venus, the Moon or Mars, convection does not involve the outermost layers. It occurs instead below an immobile lid – the stagnant lid – across which heat is transported by conduction e.g., [2].

Purely thermally-driven convection systems, where the buoyancy due to chemical components is neglected, rely on the assumption of a chemically homogeneous mantle. Thus, in order to investigate the formation and evolution of chemically distinct reservoirs, a system accounting for both thermally- and compositionally-driven convection is needed. Compositional buoyancy may lead to the long-term survival of chemical heterogeneities in planetary mantles provided that compositional gradients are strong enough e.g., [20]. Otherwise convective stirring and mixing tend to homogenize the mantle and the system behaves similarly to a purely thermally-convecting one.

Chemical heterogeneities that can strongly influence the interior dynamics have been identified for all terrestrial bodies of our Solar System and range from local to global scale. They have been inferred by seismic data, available solely for the Earth and Moon, where changes in the seismic wave velocities hint at interfaces between regions with different densities/compositions in the interior e.g., [8, 16, 22]. For other terrestrial planets, variations of surface mineral compositions observed in data sets from space missions suggest the existence of mantle geochemical reservoirs e.g., [3]. Further, laboratory analyses of meteorites/samples showing large isotopic variations also indicate chemically heterogeneous systems e.g., [6].

Data retrieved by different space missions and laboratory analyses show evidence of complex thermo-chemical processes in the interior of terrestrial bodies in our Solar System. Another "instrument" which can be employed to investigate the history of terrestrial bodies are numerical simulations of the dynamics of planetary interiors. These offer a full view of the thermo-chemical history instead of present-day snapshots as obtained from space mission data. Although the lack of plate tectonics on all other terrestrial bodies in our Solar System has maintained surface traces of events throughout the planetary history, old surfaces are sometimes buried by younger features and a time reconstruction of the past events is difficult. Even with the increasing number of meteorite samples, these are still too sparse and too few to offer a global view of the planetary thermo-chemical history. Over the years, due to the remarkable increase in computational power, numerical models have become one of the most powerful tools that can be used to understand the physical processes in the interior of planetary bodies and their implications for the thermal evolution.

When it comes to modeling thermo-chemical convection in planetary mantles, usually one can choose between two approaches: 1D models, where mantle convection is parametrized via appropriate scaling laws, or 2D-3D fully dynamical models, which solve for the entire set of conservation equations for a fluid continuum. 1D models are computationally inexpensive and are best suited to investigate large parameter ranges. However, being one dimensional, they cannot predict characteristic convection patterns or lateral heterogeneities in the mantle and a direct comparison against 3D surface data is difficult. For this, 2D-3D fully dynamical models are best suited. Over the years the complexity of these models has increased remarkably such that complex problems like duration of volcanism, initiation of plate-tectonics, formation of geochemical reservoirs, etc. can now be addressed.

Over the past years we have improved our mantle convection code Gaia and extended its application field to include complex scenarios needed to simulate interior dynamics of planetary bodies. In this study we present technical aspects of the latest version of Gaia, performance scaling on several thousands of cores and results comparison with published values for various cases.

2 Models and Methods

2.1 Physical Model

We model thermo-chemical mantle convection by solving the conservation equations of mass, momentum, energy and composition e.g., [18]. These equations are scaled using the thickness of the mantle D as length scale, the thermal diffusivity κ as time scale, the temperature drop across the mantle ΔT as temperature scale and the chemical density contrast $\Delta\rho$ as compositional scale. Assuming a Boussinesq fluid with mixed Newtonian and non-Newtonian rheology and infinite Prandtl number, as appropriate for highly viscous media with negligible inertia, the non-dimensional equations of thermo-chemical convection read e.g. [7]:

$$\nabla \cdot \mathbf{u} = 0, \tag{1}$$

$$\nabla \cdot \left[\eta(\nabla \mathbf{u} + (\nabla \mathbf{u})^T)\right] + RaT\mathbf{e}_r - Ra_C C - \nabla p = 0, \tag{2}$$

$$\frac{\partial T}{\partial t} + \mathbf{u} \cdot \nabla T - \nabla^2 T - \frac{Ra_Q}{Ra} = 0, \tag{3}$$

$$\frac{\partial C}{\partial t} + \mathbf{u} \cdot \nabla C = 0, \tag{4}$$

where η is the viscosity, p the dynamic pressure, $\mathbf{u}$ the velocity, t the time, T the temperature, C the composition that can be translated as density and $\mathbf{e_r}$ is the unit vector in radial direction. The variables in the above equations are non-dimensionalized using the relationships to physical properties presented e.g. in

[4]. In Eq. (2), Ra and Ra_C are the thermal and compositional Rayleigh numbers respectively, which are defined as:

$$Ra = \frac{\rho g \alpha \Delta T D^3}{\kappa \eta_{ref}} \quad \text{and} \quad Ra_C = \frac{\Delta \rho g D^3}{\kappa \eta_{ref}}, \tag{5}$$

where ρ is the reference density, $\Delta\rho$ the density contrast, g the gravitational acceleration, α the thermal expansivity, ΔT the temperature contrast between outer and inner boundaries, D the thickness of the mantle, κ the thermal diffusivity, and η_{ref} the reference viscosity. In Eq. (3), Ra_Q represents the Rayleigh number due to the internal heat sources and is given by:

$$Ra_Q = \frac{\rho^2 g \alpha Q_m D^5}{\kappa k \eta_{ref}}, \tag{6}$$

where Q_m is the mantle radioactive heat production and k is the thermal conductivity.

The viscosity is calculated according to the Arrhenius law for mixed diffusion and dislocation creep [13]. The non-dimensional formulation of the Arrhenius viscosity law for temperature, pressure and strain rate dependent viscosity is given by [17]:

$$\eta(\varepsilon, T, p) = \left(\frac{\varepsilon}{\varepsilon_{ref}}\right)^{\frac{1-n}{n}} \exp\left(\frac{E + pV}{n(T + T_0)} - \frac{E + p_{ref} V}{n(T_{ref} + T_0)}\right), \tag{7}$$

where E and V are the activation energy and activation volume respectively, and T_{surf} the surface temperature. The variables T_{ref}, p_{ref}, and ε_{ref} are the reference temperature, pressure and strain rate. The factor n introduces the non-linear rheology. For dislocation creep n is equal to 3.5 while for diffusion creep it is 1. Due to the strongly temperature-dependent viscosity, a stagnant lid will rapidly form on top of the convecting mantle. Additionally a pseudo-plastic approach can be used such that the stagnant lid undergoes plastic yielding if the convective stresses are high enough to overcome the imposed yield stress σ_y. An effective viscosity can thus be defined as follows:

$$\frac{1}{\eta_{eff}} = \frac{1}{\eta(\varepsilon, T, p)} + \frac{1}{\frac{\sigma_y}{2\dot{\varepsilon}}} \quad \text{with} \quad \sigma_y = \sigma_0 + z\frac{\partial \sigma}{\partial z}, \tag{8}$$

where z is the depth and $\dot{\varepsilon}$ is the second invariant of the strain rate tensor e.g., [12]. The plastic yielding introduces a plate-like behavior of the lithosphere with the surface being recycled when the lithosphere undergoes plastic failure represented by a strong reduction of the near-surface viscosity, which will necessarily allow for the surface material to flow.

2.2 Technical Realization

The Gaia code is a fluid flow solver that solves the conservation of mass, momentum and energy (Eqs. 1–3) on a fixed grid using arbitrary geometries (Fig. 1). It is used for modeling Stokes-flow with strongly varying viscosity, this scenario being often met in geophysics particularly in mantle convection simulations. However, during the last years the equations were extended to include rotation and to support flows driven by inertia (finite Prandtl number), enabling simulations with a Mach number up to 0.2. The code is based on a module system which provides the ability to add new features without having to manipulate the core functionality.

Gaia is written in C++ and does not use additional libraries [9]. The discretization of the governing equations is based on the finite-volume method and uses regular or fully irregular Voronoi grids [10, 14] with various geometries as shown in Fig. 1. Temporal discretization uses a fully implicit second-order scheme, also called an implicit Three Time Level Method [5]. In contrast to spatial discretization, the

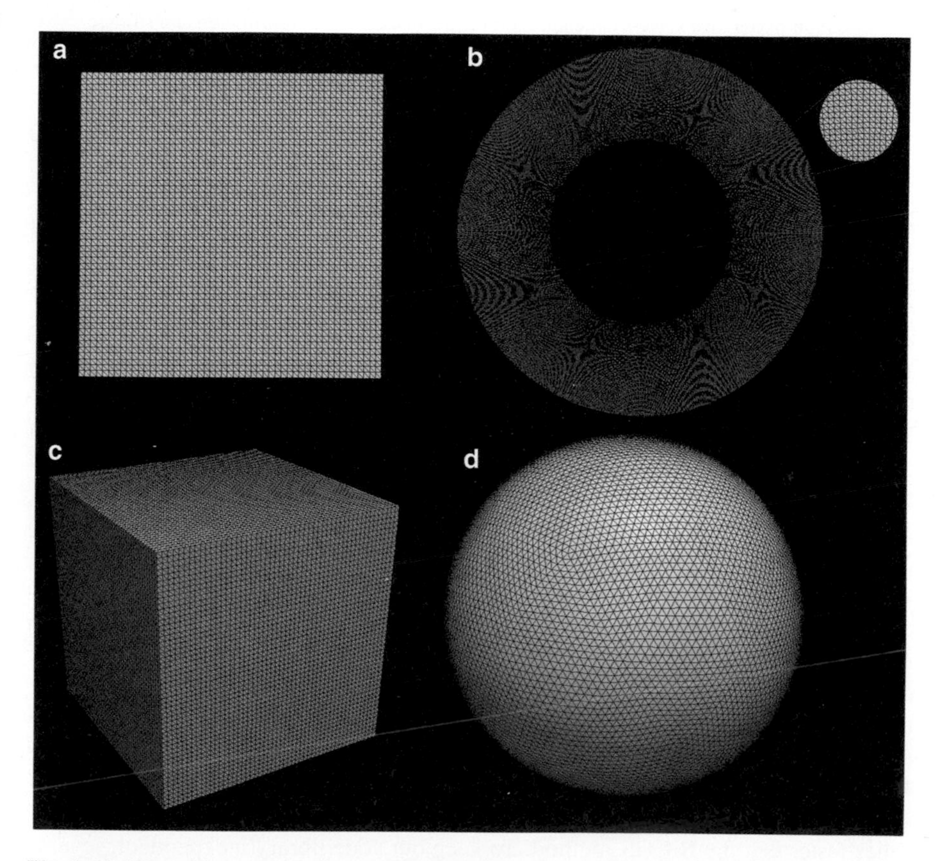

Fig. 1 Various grids used with Gaia: (**a**) 2D Cartesian box; (**b**) 2D cylinder; (**c**) 3D Cartesian box; (**d**) 3D spherical shell

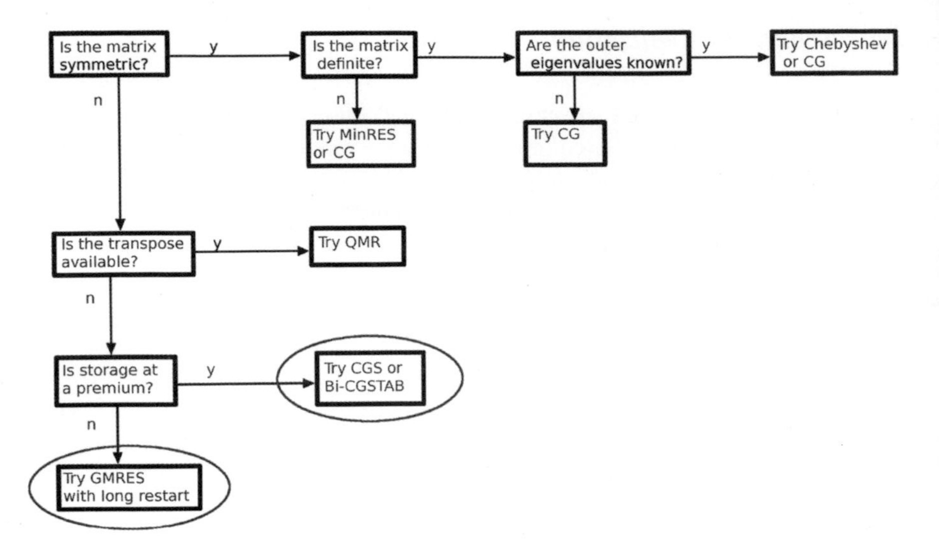

Fig. 2 Graph of iterative solvers depending on the available matrix information and storage

temporal discretization is flexible and can adapt the time step according to the problem. In order to advance a scalar field like temperature in time, one can choose between various time-stepping criteria. A time step fulfilling the Courant-Friedrichs-Lewy condition implies that no theoretical particle could cross more than the smallest cell in the simulation for the current velocity field. Another time stepping mechanism relates the next time step to the change in the velocity magnitude and thus, if a simulation reaches a steady state (no velocity change), the time steps increase considerably. Alternatively, a fixed time step can also be used.

The splitting method SIMPLE divides the original mass and momentum equations into an equation for velocity and another for pressure. Iterating between them finally yields the desired solution, but due to the strong viscous part convergence is often slow. The method was introduced to reduce the memory needed to solve the Navier-Stokes. Because available memory grew with the power of CPUs and methods like domain-decomposition, the SIMPLE process becomes obsolete as it is now possible to solve for the complete system of equations at once. Therefore, we solve for velocity and pressure simultaneously, by including the three velocity components (u, v, w) and pressure p in one single system of equations. The code uses a selective Jacobi preconditioner, which is only applied to the velocity part of the matrix. For further details, we refer the readers to [11]. To efficiently solve the system of equations for velocity and pressure, we employed PETSC to test convergence with various methods (Fig. 2).

The results in Fig. 3 show that beside BiCGS and BiCGS(l), most iterative solvers show a poor convergence or no convergence at all. In our test, BiCGS(l) was two times faster that the BiCGS method. To keep our code library independent, several

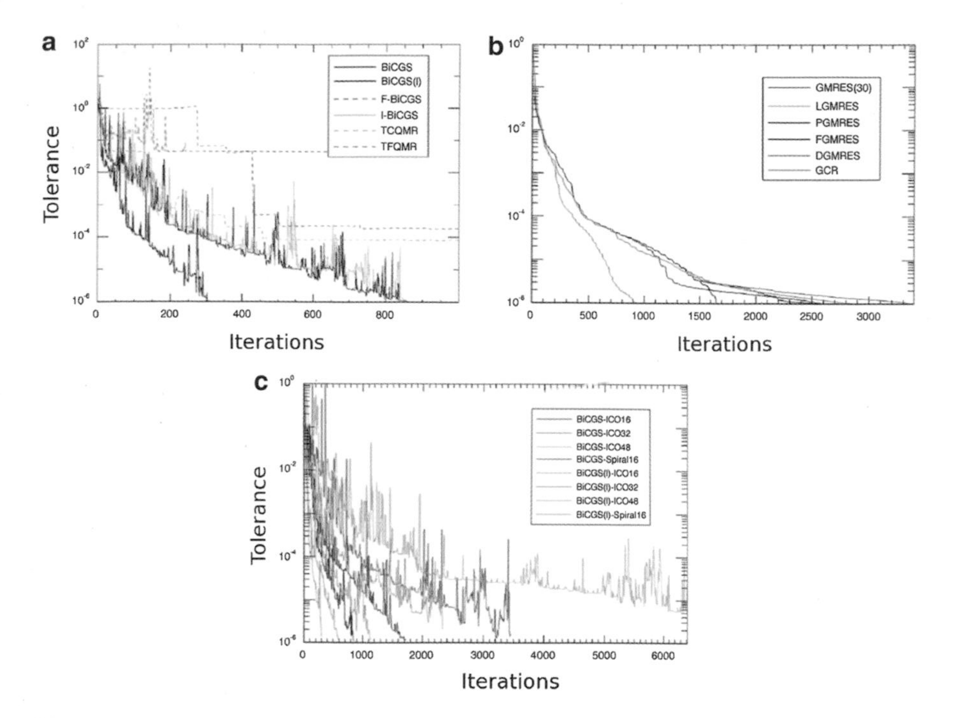

Fig. 3 Convergence tests with current PETSC library: tested was a matrix containing the coefficients from A4 case shown in the benchmark study from [23] in a 3D spherical shell (ICO16) with a viscosity contrast of 100, initial time-step from zero solution. (**a**) convergence rates for the CG and QMR methods; (**b**) convergence for the GMRES group; (**c**) size dependent scaling of the two best solvers, measured with different 3D grids. The problem sizes (unknowns) are: ICO16 = 142.000; Spiral16 = 458.400; ICO32 = 1.328.000; ICO48 = 1.792.000. A matrix multiplication grows linear with the problem size, so any factor from convergence iteration is the non-linear part

iterative solvers have been additionally implemented in Gaia. One can choose between BiCGS, BiCGS(l), Jacobi and TFQMR.

To speed up convergence for steady-state problems or to couple in non-linear effects, Gaia can employ under-relaxation for the velocity field. The method is described in [5] and computes the velocity for the current time step taking into account a fraction of the previous velocity value.

Thermo-chemical convection is modeled by using a particle in cell method (PIC) e.g. [15, 19, 21]. To account for the advection of different chemical components (Eq. 4), the motion of massless particles is first calculated from the velocity field computed on a fixed mesh. The compositional field is then interpolated from the distribution and composition of the particles. The method has the advantage over classical grid-based methods of being essentially free of numerical diffusion and to enable to naturally advect an arbitrary large number of different compositional fields by solving only one equation.

The following core features and modules are currently available in the Gaia code:

- Navier-Stokes Solver for low Mach-number flow
- Compressible flow with anelastic approximation
- Variable viscosity supported
- (In)finite Prandtl / Reynolds number (MUSCL advection)
- Co-located, cross-derivative-free finite-volume discretization
- Arbitrary irregular grid support (currently supported: 3D spherical shell, Sphere, 3D box, 2D cylinder, 2D box)
- Massively parallel (MPI), various iterative solvers, library independent.
- Fully implicit
- Module system provides additional:
  - Energy solvers
  - Different rheologies (Newtonian/ non-Newtonian / Bingham /...)
  - Particles for chemical convection
  - Melt
  - Thermal evolution (time dependent temperature boundary condition)
  - Initial conditions

3 Results and Discussion

3.1 Code Performance

The code performance was tested on the HLRS system using up to 8,192 computational cores (dual socket AMD Interlagos @ 2.3 GHz having 2 sockets per node with 16 cores each). To be able to scale with a significant amount of cores, the problem size must be chosen accordingly. For our test we used a computational grid with $\sim 8 \times 10^6$ points. The grid is a spiral grid [10] with refined inner and outer boundaries.

Table 1 shows the time in seconds needed for 100 solver steps when keeping the grid size constant ($\sim 8 \times 10^6$ points) but increasing the number of cores (strong

Table 1 Strong scaling time used to compute 100 solver steps on a $\sim 8 \times 10^6$ points grid

# Cores	Time (s)
128	175.999
256	85.789
512	44.220
1,024	24.576
2,048	14.328
4,096	9.343
8,192	7.388

Table 2 Weak scaling using various grid sizes and scaling the number of cores used accordingly, such that in all cases a problem size of ∼3,891 cells/core is used

Grid nodes	Time (s)
1.55×10^6	10.1626
3×10^6	11.5654
3.75×10^6	12.0778
8×10^6	14.3278

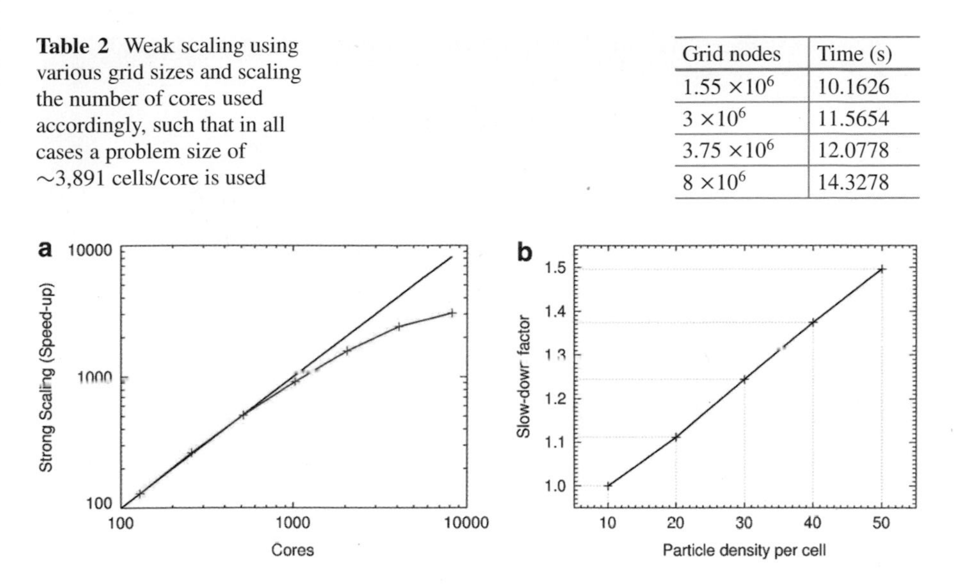

Fig. 4 (**a**) Speed-up factor when running the A4 case shown in [23] on a fully irregular grid with $\sim 8 \times 10^6$ points, on up to 8,192 computational cores; (**b**) Slow-down factor when increasing the number of particles per cell from 10 to 50. Production runs typically use 15–20 particles per cell

scaling), while Table 2 shows the time needed for 100 solver steps for various grids when keeping the problem size constant on each core (weak scaling). Figure 4a shows the strong scaling plot obtained with the data from Table 1 when running the A4 benchmark set-up published in [23], while panel b shows the slow-down when running the benchmark set-up from [21] with the PIC method and increasing the number of particles per cell.

3.2 Application: Mantle Convection in Terrestrial Planets

We have applied our code to run simulations of planetary interiors in various geometries. In Fig. 5 we show the results obtained for a purely thermal benchmark case published in [1]. The system modeled here has a constant viscosity and is heated from below (from the inner boundary in 2D cylinder and 3D sphere geometry). In all cases we use free-slip boundary conditions. The panels in Fig. 5 show the temperature distribution at steady-state together with streamlines. The initial perturbation results in a symmetric pattern with 4 upwellings in 2D cylinder and 6 upwellings in 3D spherical shell on each axis. The 2D box shows a single upwelling on the side, while the 3D box reaches two stable upwellings in the corners.

In a second suite of simulations, we have tested a purely thermal system similar to the previous one but with temperature dependent viscosity. The viscosity contrast

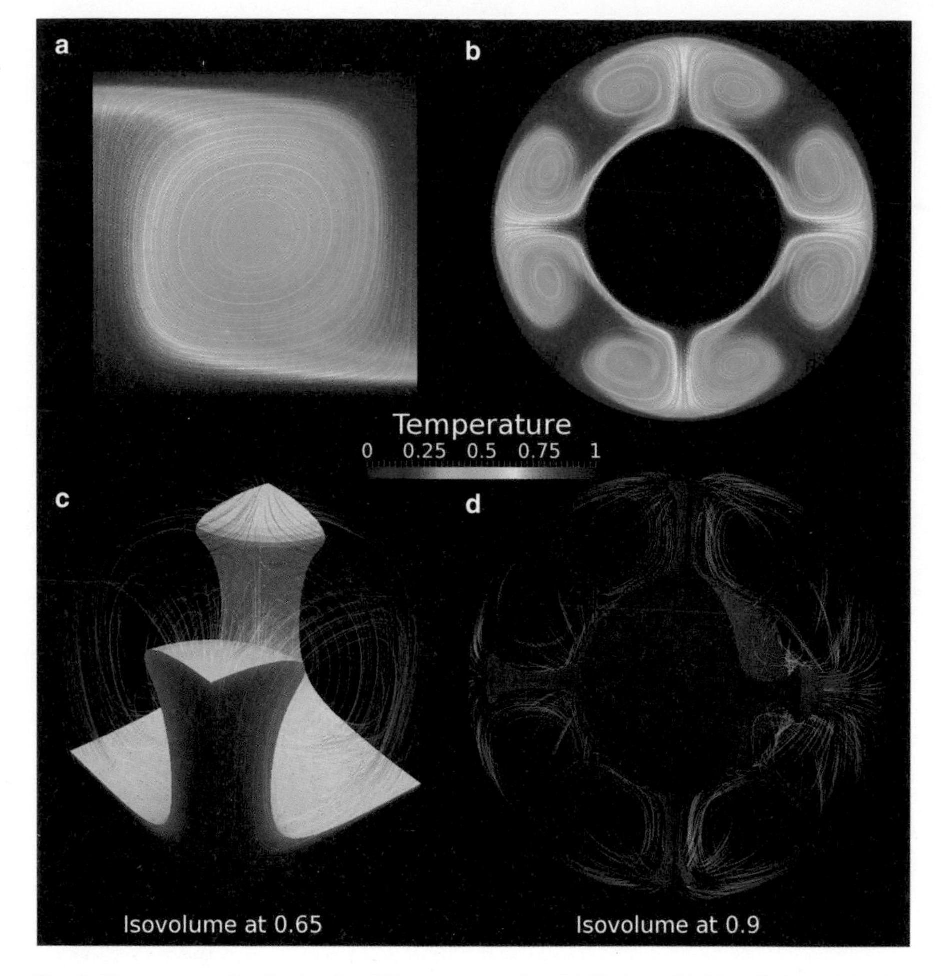

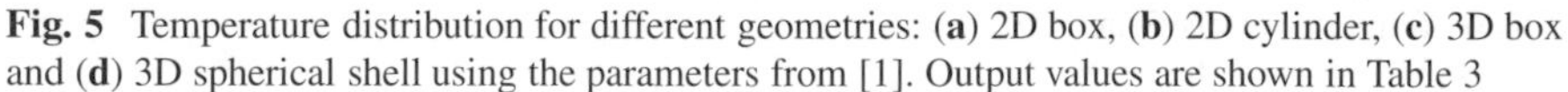

Fig. 5 Temperature distribution for different geometries: (**a**) 2D box, (**b**) 2D cylinder, (**c**) 3D box and (**d**) 3D spherical shell using the parameters from [1]. Output values are shown in Table 3

is 10^5 across the entire computational grid. In Fig. 6 we show the temperature distribution at steady-state in various geometries and in Table 4 we compare our results with values published in [23].

In this set-up, due to the temperature dependence of the viscosity, a stagnant lid establishes on top of the convective domain. Since through the stagnant lid heat is only transported by conduction, the system cannot loose heat as efficient as in the previous isoviscous case. Therefore the internal temperatures shown in Table 4 are significantly higher than the values from Table 3.

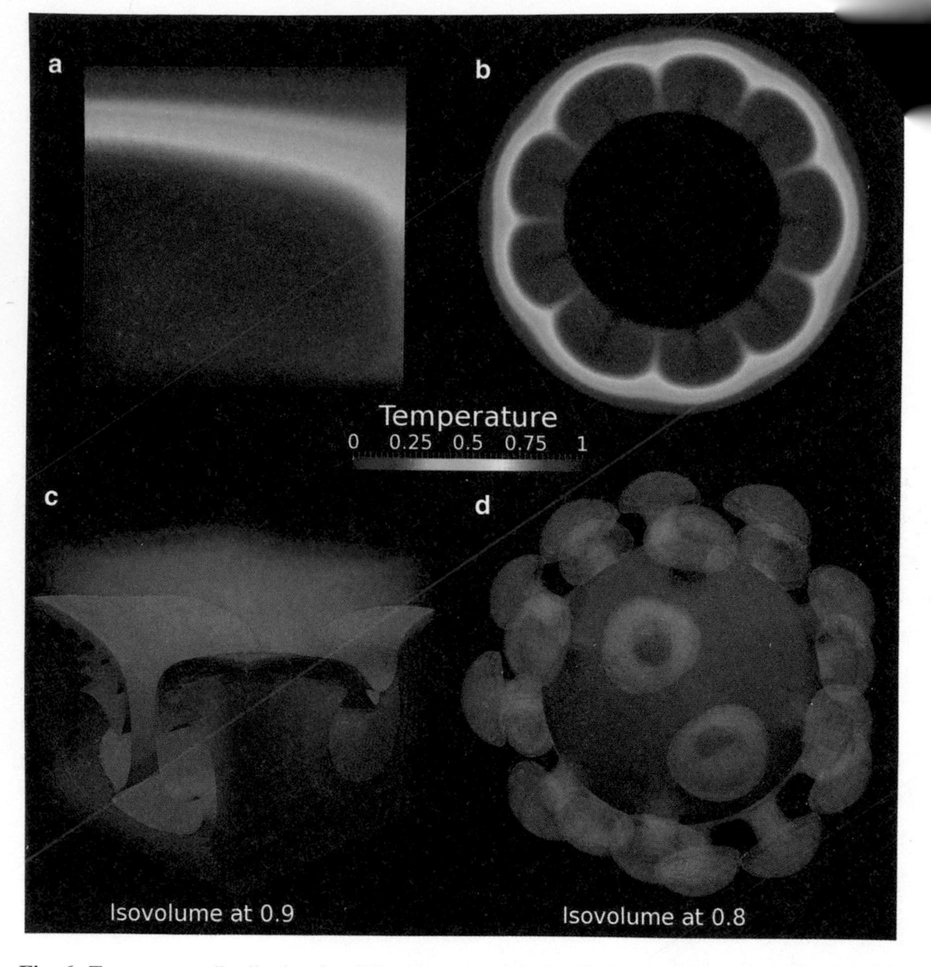

Fig. 6 Temperature distribution for different geometries: (**a**) 2D box, (**b**) 2D cylinder, (**c**) 3D box and (**d**) 3D spherical shell using the parameters from [23]. Output values are shown in Table 4

Further, we have tested and benchmarked our PIC method against published results from [21]. To be able to simulate thermo-chemical convection in planetary mantles, the code should be able to efficiently perform the advection of a large number of particles in all geometries in parallel, on various amount of CPUs. For the Rayleigh-Taylor instability test presented in Fig. 7, we used a number of 15 particles per cell in all cases to track the compositional field. Particles are being created or destroyed to maintain approximately the same number in each cell. This also guarantees the balance of the problem on the CPUs used. The system modeled

3 Output values for the benchmark case from [1]. *Nu* stands for Nusselt number, v_{rms} stands ...oot mean square velocity and $|T|$ for mean temperature. The 2D box geometry shows a good ...eement with the published results. We show all other cases for reference. Here different output ...alues are expected due to the different geometries

Code	Grid	*Nu*	v_{rms}	$\|T\|$
Bl 2D box	24 × 24	4.888	42.879	–
Gaia 2D box	50 × 50	4.873	42.826	0.5
Gaia 2D cyl	64 × 584	4.664	44.091	0.365
Gaia 3D box	50 × 50 × 50	4.704	34.806	0.5
Gaia 3D sph	50 × 10,242	3.917	35.686	0.187

Table 4 Output values for the benchmark case from [23]. *Nu* stands for Nusselt number, v_{rms} stands for root mean square velocity and $|T|$ for mean temperature. The results in 3D spherical geometry are compared with the values obtained with the CitcomS code [23]. All other cases are only shown for reference. Different output values are due to the different geometries

Code	Grid	*Nu*	v_{rms}	$\|T\|$
CitcomS 3D sph	12 × 32^3	2.735	50.21	0.5039
Gaia 3D sph	32 × 10,242	2.786	50.490	0.504
Gaia 3D box	50 × 50 × 50	2.731	113.420	0.745
Gaia 2D box	50 × 50	2.399	95.955	0.726
Gaia 2D cyl	32 × 325	2.535	75.330	0.643

here contains of a light component initially located at the bottom of the domain (composition = 0) and a heavy component initially at the top (composition = 1). A perturbation has been applied on the boundary between the two components to initialize instabilities. To enable secondary instabilities to develop, we use no-slip boundary conditions at the outer and inner boundary. In the first part of all simulations, the light material rises to the surface, while the denser component sinks toward the inner boundary. However, due to the no-slip boundary, heavy and light material sticks to the outer and inner boundary respectively. This causes the formation of secondary droplets/diapirs.

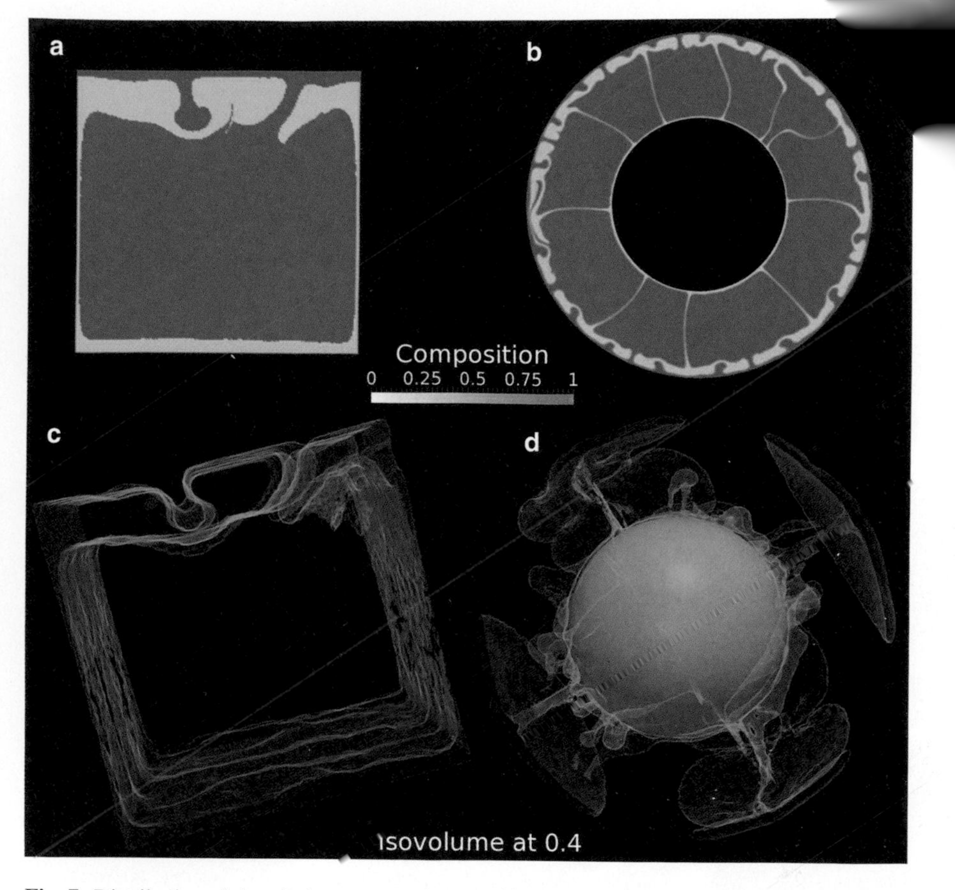

Fig. 7 Distribution of chemical component (composition) during the reorganization of fluid layers for different geometries: (**a**) 2D box, (**b**) 2D cylinder, (**c**) 3D box and (**d**) 3D spherical shell using the parameters from [21]

4 Conclusions and Outlook

In this work we have used the Gaia code with different geometries to perform numerical simulations of mantle convection considering both thermally and chemically driven buoyancy. The benchmark tests presented in previous section show a good agreement with published results. The ability of the code to use various grids makes possible the treatment of various problems with different geometries. Since additional libraries are not needed, Gaia can be easily ported on various systems.

Code performance has been tested on the HLRS system and shows a good scaling for up to 8,192 computational cores. Moreover, a suite of numerical solvers from the current PETSC library has been tested for a typical mantle convection setup. The results show fast convergence for the BICGS group, of which BICGS(l), BICGS and TFQMR have been implemented in Gaia. The code is one of the few world-wide that can use various geometries, high viscosity contrasts (up to 15 ...rs of magnitude system wide) as appropriate for mantle convection problems ...d Lagrangian particles in 3D spherical geometry to track active compositional ...ields.

With Gaia, we are able to model complex processes in the interior of planetary bodies and take into account both thermal and compositional buoyancy fields as main agents to drive mantle convection. We are currently investigating the consequences of active compositional fields in case of Mars, Mercury and the Moon, as this may explain the variations in the surface mineralogical maps or in seismic data available from various space missions. Massive parallel computing resources offered by high-performance computational centers like the HLRS facility will allow us to tackle complex fluid dynamical problems of planetary interiors and improve our knowledge about the physical processes active in the mantle of terrestrial bodies.

Acknowledgements This research has been supported by the Helmholtz Association through the research alliance "Planetary Evolution and Life", by the Deutsche Forschungs Gemeinschaft (grant TO 704/1-1), by the Interuniversity Attraction Poles Programme initiated by the Belgian Science Policy Office through the Planet Topers alliance, and by the High Performance Computing Center Stuttgart (HLRS) through the project "Mantle Thermal and Compositional Simulations (MATHECO)".

References

1. Blankenbach, B., Busse, F., Christensen, U., Cserepes, L., Gunkel, D., Hansen, U., Harder, H., Jarvis, G., Koch, M., Marquart, G., Moore, D., Olson, P., Schmeling, H., Schnaubelt, T.: A benchmark comparison for mantle convection codes. Geophys. J. Int. **98**, 23–38 (1989)
2. Breuer, D., Moore, W.B.: Dynamics and thermal history of the terrestrial planets, the Moon, and Io. Treatise Geophys. **10**, 299–348 (2007)
3. Charlier, B., Grove, T.L., Zuber, M.T.: Phase equilibria of ultramafic compositionson Mercury and the origin of the compositional dichotomy. EPSL **363**, 50–60 (2013)
4. Christensen, U.: Convection with pressure- and temperature-dependent non-Newtonian rheology. Geophys. J. R. Astron. Soc. **77**, 343–384 (1984)
5. Ferziger, J.H., Perić, M.: Computational Methods for Fluid Dynamics, 2nd rev. edn. Springer, Berlin/New York (1999). ISBN:3-540-65373-2
6. Foley, C.N., Wadhwa, M., Borg, L.E., Janney, P.E., Hines, R., Grove, T.L.: The early differentiation history of Mars from ^{182}W–^{142}Nd isotope systematics in the SNC meteorites. Geochim. Cosmochim. Acta **69**, 4557–4571 (2005)
7. Grasset, O., Parmentier, E.M.: Thermal convection in a volumetrically heated, infinite Prandtl number fluid with strongly temperature-dependent viscosity: implications for planetary thermal evolution. J. Geophys. Res. **103**, 18171–18181 (1998)
8. Hofmann, A.W.: Mantle geochemistry: the message from oceanic volcanism. Nature **385**(6613), 219–229 (1997)
9. Hüttig, C., Stemmer, K.: Finite volume discretization for dynamic viscosities on Voronoi grids. Phys. Earth Planet. Inter. (2008). doi:10.1016/j.pepi.2008.07.007
10. Hüttig, C., Stemmer, K.: The spiral grid: a new approach to discretize the sphere and its application to mantle convection. Geochem. Geophys. Geosyst. Q02018 **9** (2008). doi:10.1029/2007GC001581

11. Hüttig, C., Tosi, N., Moore, W.B.: An improved formulation of the incompressible Nav Stokes equations with variable viscosity. Phys. Earth Planet. Inter. **220**, 11–18 (201 doi:10.1016/j.pepi.2013.04.002
12. Ismail-Zadeh, A., Tackley, P.J.: Computational Methods for Geodynamics. Cambridge University 368 Press, Cambridge/New York (2010)
13. Karato, S., Paterson, M.S., Fitz Gerald, J.D.: Rheology of synthetic olivine aggregates: influence of grain size and water. J. Geophys. Res. **91**, 8151–8176 (1986)
14. Plesa, A.-C.: Mantle convection in a 2D spherical shell. In: Rückemann, C.-P., Christmann, W., Saini, S., Pankowska, M. (eds.) Proceedings of the First International Conference on Advanced Communications and Computation (INFOCOMP 2011), 23–29 Oct 2011, Barcelona, pp. 167–172. ISBN:978-1-61208-161-8. Retrieved on 3 Nov 2011. From http://www.thinkmind.org/download.php?articleid=infocomp_2011_2_10_10002
15. Plesa, A.-C., Tosi, N., Hüttig, C.: Thermo-chemical convection in planetary mantles: advection methods and magma ocean overturn simulations. In: Rückemann, C.-P. (ed.) Integrated Information and Computing Systems for Natural, Spatial, and Social Sciences. Information Science Reference, Hershey (2013)
16. Ritsema, J., van Heijst, H.J., Woodhouse, J.H.: Global transition zone tomography. J. Geophys. Res. Solid Earth **109**(B2) (2004). doi:10.1029/2003jb002610
17. Roberts, J.H., Zhong, S.: Degree-1 convection in the Martian mantle and the origin of the hemispheric dichotomy. J. Geophys. Res. E Planet. **111**, E06013 (2006)
18. Schubert, G., Turcotte, D.L., Olson, P.: Mantle Convection in the Earth and Planets. Cambridge University Press, Cambridge/New York (2001)
19. Tackley, P.J., King, S.D.: Testing the tracer ratio method for modeling active compositional fields in mantle convection simulations. Geochem. Geophys. Geosyst. **4**(4), 8302 (2003). doi:10.1029/2001GC000214
20. Tosi, N., Plesa, A.-C., Breuer, D.: Overturn and evolution of a crystallized magma ocean: a numerical parameter study for Mars. J. Geophys. Res. Planet. **118**, 1–17 (2013). doi:10.1002/jgre.20109
21. van Keken, P.E., King, S.D., Schmeling, H., Christensen, U.R., Neumeister, D., Doin, M.-P.: A comparison of methods for the modeling of thermochemical convection. J. Geophys. Res. **102**, 22477–22495 (1997)
22. Weber, R.C., Lin, P.Y., Garnero, E.J., Williams, Q., Lognonne, P.: Seismic detection of the Lunar core. Science **331**, 309–312 (2011). doi:10.1126/science.1199375
23. Zhong, S., McNamara, A.K., Tan, E., Moresi, L., Gurnis, M.: A benchmark study on mantle convection in a 3-D spherical shell using CitcomS. Geochem. Geophys. Geosyst. **9**, Q10017 (2008). doi:10.1029/2008GC002048

Phase-Field Simulations of Large-Scale Microstructures by Integrated Parallel Algorithms

Johannes Hötzer, Marcus Jainta, Alexander Vondrous, Jörg Ettrich, Anastasia August, Daniel Stubenvoll, Mathias Reichardt, Michael Selzer, and Britta Nestler

Abstract In this report, we present specific model extensions of the phase-field method [17] implemented in the software framework *PACE3D* and summarize the underlying parallelized algorithms. Three different applications of microstructure evolution processes are illustrated and the need of large representative volume elements and of efficient parallelization. Within a first application, a parallel connected component labeling algorithm, is optimized for a large number of computing units, and is used for the simulation of pore development in the sintering processes. The second topic discusses a brute force study of the Read-Shockley model, to investigate recrystallization and abnormal grain growth in anisotropic polycrystalline material systems. The third topic is focussed on heat transfer and fluid flow in metallic foam structures depending on the porosity as base material for new heat storage systems.

1 A Distributed 3D Connected Component Labeling Algorithm

In material science it is of interest to know the topology of the involved microstructure, to determine characteristic quantities of the material and to compare simulations of a statistically based representative volume element with experiments. Connected component labeling algorithms (CCL) allow to describe the topology graph and to detect changes therein. The first connected component labeling

J. Hötzer (✉) • M. Jainta • A. Vondrous • J. Ettrich • A. August • D. Stubenvoll • M. Reichardt • M. Selzer • B. Nestler
Institute of Applied Materials, Reliability of Components and Systems (IAM-ZBS), Karlsruhe Institute of Technology (KIT), Haid-und-Neu-Str. 7, 76131 Karlsruhe, Germany

Institute of Materials and Processes, Hochschule Karlsruhe Technik und Wirtschaft, Moltkestr. 30, Karlsruhe, Germany
e-mail: johannes.hoetzer@kit.edu; alexander@vondrous.de; joerg.ettrich@hs-karlsruhe.de; marcus.jainta@kit.edu; anastasia.august2@kit.edu; stubenvoll.daniel@gmx.de; mathias.reichardt@hs-karlsruhe.de; michael.selzer@hs-karlsruhe.de; britta.nestler@kit.edu

W.E. Nagel et al. (eds.), *High Performance Computing in Science and Engineering '14*, DOI 10.1007/978-3-319-10810-0_41

.hm [25] was published in the year 1966 by Rosenfeld and Pfaltz using a scan method. Based on this, various algorithms where developed [2] to use .arallel computers [4, 6] and handle large images [19]. There are improved variants [1] as well, that can treat three dimensional images [22]. We want to focus on the application of CCL, during runtime of a simulation, which requires an algorithm, that is fast and can be parallelized on a large number of computing units.

The physical process considers the sintering of ceramic materials. Here we start with granular ceramic particles with a density of 40–70 %. Due to a heat process below the melting temperature, the particles in the so-called green body connect. This reduces the size, resulting in a higher density, and a rigid body is formed. During the sintering process, pores evolve in the material, which can merge or separate by the movement of grain boundaries[26]. Due to pinning effects, this phenomenon influences the shrinking rate significantly [15, 21, 27].

We employ a phase-field model [17] to describe both, the ceramic particles, as well as the embedded gas-filled pores, using independent order parameters. In the event of merging or separation of pores, the pressure changes therein. The pressure balance between two merging pores occurs on a much smaller timescale, compared to the whole sintering process. Therefore, we describe the different pores as connectivity components. Both, the resulting pressure according to the ideal gas law as well as the phase evolution of pores are computed during runtime.

To perform large three-dimensional simulations, domain decomposition techniques on distributed systems, based on the Message-Passing Interface (MPI) are used. This allows to divide the calculation domain in each dimension into subdomains (three-dimensional domain decomposition) which are computed by one unit [29], [16, P. 595]. To use component labeling in this environment, we introduce a parallel, three-dimensional, connected-component-labeling algorithm to get a global and unique identification of the pores. The algorithm consists of two successive parts. In the first part, connected components are labeled independently in each subdomain. Secondly, the relations between the connected components in the neighboring subdomains are created and afterwards simplified according to an accumulated global graph. To further reduce the computation time for the procedure of relabeling the connected components, we use a mapping layer and a separate memory manager. For the validation of the algorithm, we utilize a simulation series for different settings on the Cray Hermit in the HLRS in Stuttgart to show the runtime and the scaling behavior with more than 12,000 computing units.

1.1 The Algorithm

The parallel distributed three-dimensional connected component labeling algorithm (PCCL) allows to detect and track components (e.g. pores) in our simulation domain. The algorithm needs a single pass for labeling, using an additional field to preserve the original data and a mapping layer to quickly relabel already

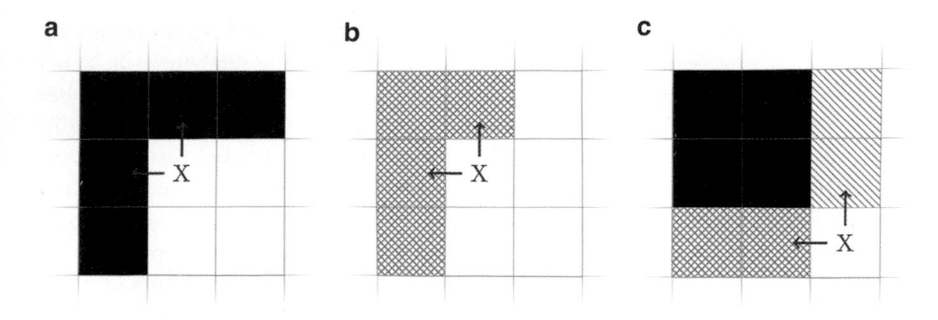

Fig. 1 Illustration of the labeling field to classify the cells in the domain. The x refers to the current cell and the arrows show the previous neighbors that have to be checked. (**a**) The cell to be checked is located next to non-matching or boundary cells. (**b**) The cell to checked is located next to matching cells with the same label. (**c**) The cell to be checked is located next to cells with different labels

found connectivity components. After labeling the connected components for each subdomain, the global dependencies are resolved in a second step.

In the first part, the algorithm iterates from the left top front corner across the three dimensional subdomain to the opposite corner. For each cell, a given user-defined condition is tested, such as the order parameter or the color of the pixel. If the cell matches the condition, it is marked as part of a component in the labeling field. Otherwise it will be marked as a non matching cell (black) in the labeling field. The matching cell is further considered and the labels of the previous neighbors are compared. The comparing algorithm has to distinguish three cases, as shown in Fig. 1a–c. In the first case (Fig. 1a), the cell matching the condition is located on the subdomain boundary or the predecessor cells are marked as non-matching cells. In this case, a new component is generated. If all marked predecessor cells are marked with the same label (Fig. 1b), the matched cell is integrated as a member of this component. If the matched cell has marked predecessor cells with different labels (Fig. 1c), the components still belong together. The label of one of the components is selected and the cell as well as the other components are marked with this label. Instead of relabeling all marked cells of this components in the labeling field, we use a mapping layer. This mapping layer holds the corresponding label of each component and the cell in the labeling field references to them. Therefore the relabeling can be done by switching the references in the mapping layer following order $\mathcal{O}(1)$. The construction of the mapping layer allows to store further information such as the cell count or minimal/maximal values and their positions.

In the second part of the PCCL algorithm, the distributed local components in each subdomain have to be transformed to global unique components. To get the global components, a graph is composed, in which all local components in the subdomains, that are connected to a neighboring subdomain, are stored. First we make sure that all components over all subdomains have a unique labeling number. To ensure this, the offset of the subdomains and a counter is defined to generate the unique labels. This graph can be used to find all connected components by a breadth-first search or depth-first search algorithms and by relabeling the local components. This is realized by searching the lowest component label number in the graph to which the local component in the subdomain is connected. To create the graph, two communication steps are needed. In the first step, the neighboring subdomains exchange the components directly at the boundary to the next subdomains to create the edges of the graph. For each subdomain, the local components can be mapped to those of the adjacent subdomain at the corresponding side. In the second step, these local edges are distributed to all computing units by collective communications. After this, all subdomains/computing units can create the total graph using the edges, and can test to which global components their local components are connected. With the global graph it is possible to assign a unique label to each component.

In order to improve the time required to create the graph, a separate memory manager is implemented for the graph nodes. This allows to redistribute a node that is already allocated, the next time the graph is generated.

1.2 Setup

In order to measure the runtime and scaling behavior, fillings as shown in the Fig. 2a–c are chosen to examine the performance of the PCCL algorithm. Two kinds of spiral fillings are considered to get many connected components as well as to ensure that they are connected over the subdomains. The filling consists of one connected spiral through the total domain and the other filling with many independent spirals, to increase the number of globally connected components. As a more realistic case, a H-tree and a real green body particle distribution are studied. For all cases, a domain of 500^3 cells is initialized. In addition, a 100^3 cell domain is evaluated for the spiral and the H-tree fillings to compare the scaling depending on the domain size. This leads to seven different settings. To investigate the strong

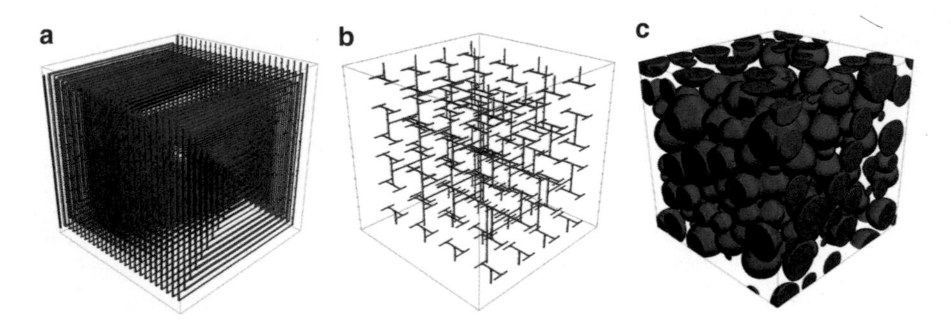

Fig. 2 Fillings to measure the runtime depending on the number of independent components. (**a**) Spiral filling. (**b**) H-tree filling. (**c**) Realistic green body particles

scaling behavior, the number of computing units p is increased proportional to x· order to decompose the domain equally in all dimensions, starting from $x = 2 \ldots 2$. as well as by an additional management unit for each of the settings. Furthermore, each of the seven settings is labeled 500 times.

1.3 Results and Discussion

In the following section, we present the runtime and strong scaling results of the distributed component labeling algorithm on the Cray XE6 Hermit. We show the runtimes of the different parts of algorithm representative for all settings.

Figure 3 displays the total runtime behavior for the seven different setups. In all settings, the PCCL first scales with the increase of the computing units. Depending on the initial filling, this changes by further increasing the computing units, which then results in an increase of the runtime. However, for more than 3,000 computing units, the algorithm requires nearly constant runtime, independent of the number of integrated computing units.

The speedup in the two spiral settings shrinks from a theoretical value of 125 to 15–34, in the range of 1,000–5,000 computing units, and to 11–18 for

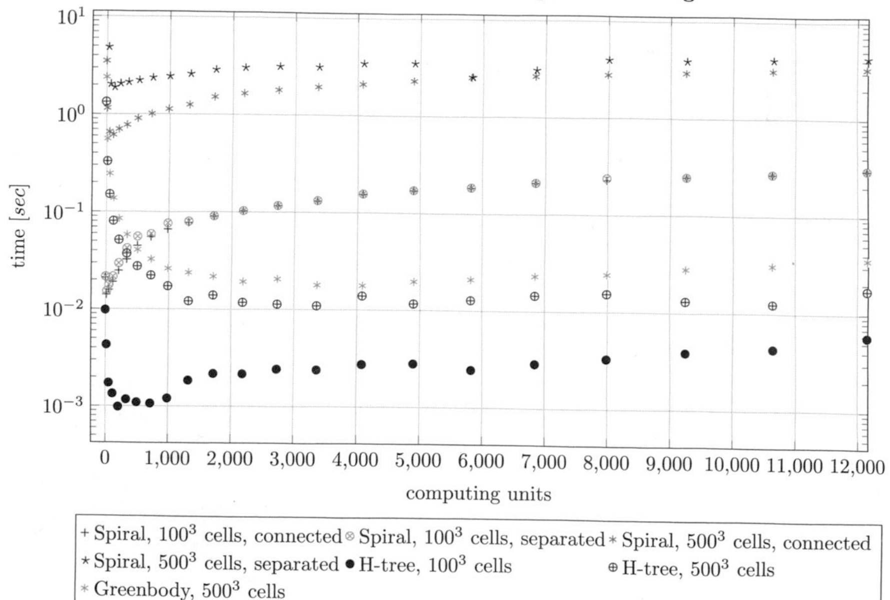

Fig. 3 Average runtime for the distributed 3D connected component labeling for the seven different setups

en more computing units. The two different spiral settings withdraw nearly the same runtime for the 100^3 cell domain. For the 500^3 cell domain, the influence of the different fillings on the runtime is obvious. This runtime differences can be explained by the number of local components resulting from the filling and the domain decomposition, and therefore the longer communication time as well as graph creation, search and destruction time. The influence of the filling, and therefore the number of individual components can also be seen between the 100^3 cell domains of the spiral and the H-tree. For the 500^3 cell domain, the connected component labeling for 12,168 computing units is in the fastest setting (H-tree) 190 times faster than the slowest (separated spiral) and, in the 100^3 domain, it is still 51 times faster.

In all settings, the runtime of the connected component labeling does not significantly change after 2000–3000 computing units. Due to the relatively slow change of the order parameter in the simulation, the PCCL only needs to run 1–100 simulation time steps. Compared to the runtime of a realistic simulation in which the wall clock time needs about 10 ms–2 s for a simulation time step , the PCCL is, with 0.035 s fast enough, even for 12,168 computing units. Figure 4 shows the scaling behavior of the present algorithm. For the H-tree and green body setting in the 500^3 cell domain, the algorithm scales up to 4,000 computing units and then slowly drops. In the other settings with more components or smaller domains, the

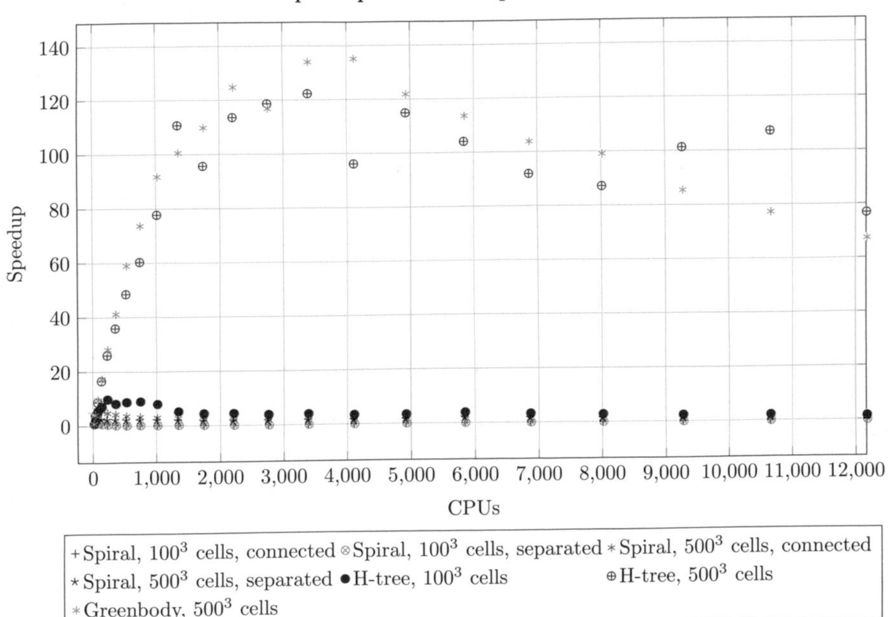

Fig. 4 Scaling behavior for the distributed 3D connected component labeling for the seven different setups

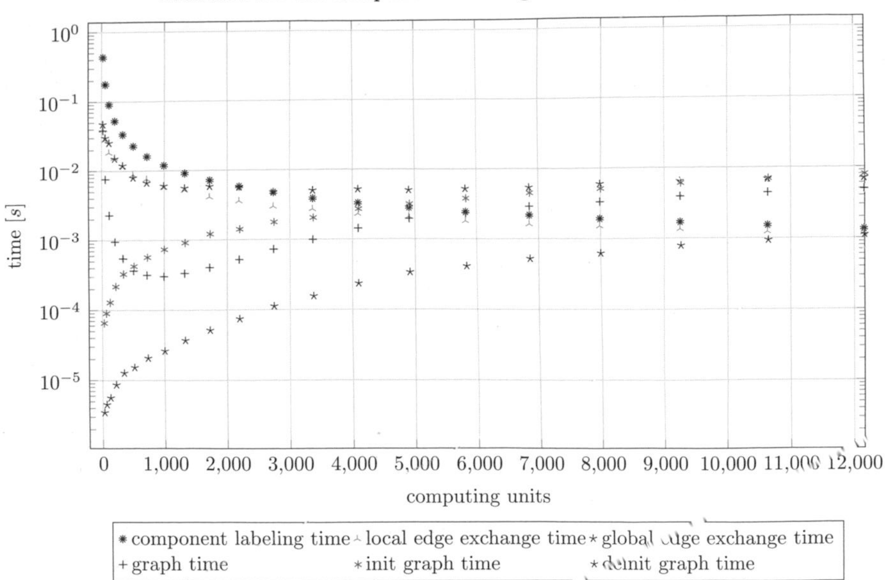

Fig. 5 Computing times of the different parts of the connected component labeling algorithm for the green body filling evaluated as functions of the computing units

algorithm scales up to 200 computing units and then drops to a nearly constant level. Figure 4 illustrates that the number of components has a significant influence on the scalability. This is not a perfect scaling behavior, but, considering that the runtime compared to a calculation time step of the simulation is equally fast and that the algorithm is not required to run in every time step, it is a suitable solution to trace the topology during the simulation.

Figure 5 discusses the runtimes of the different parts of the algorithm for the selected example of the green body setting depending on different numbers of computing units. The resulting tendency is representative for all other settings. With the increase of the computing units, the runtime of the connected component labeling algorithm decreases in each of the subdomains, as the size of subdomains decreases. The edge exchange time between the neighboring subdomains decreases, because of the shrinking subdomain size, which also leads to less data communication. With the increase of the computing units, the number of subdomains increases as well, which results in more edges and a larger graph. This leads to more data for the collective communication, as well as more involved computing units. The increase of computing units and hence the larger graph results in a longer initialization and deinitialization time. For less than 1,000 computing units, the search time for all labels in a subdomain decreases in the graph. This correlation can be explained by a decreasing number of labeled components in the subdomain and by a reduced number of graph searches, that are required to map the local labels to the globally unique ones. After around 1,000 computing units, the behavior changes to an increase of the graph search time. This is due to the fact that the graph reaches a critical size accompanied by a longer time for a single breadth first search. But the number of local components does not decrease that fast anymore to compensate the search time.

1.4 Outlook

We have introduced a parallel connected component labeling algorithm running on three-dimensional decomposed distributed memory systems. The measured speedups demonstrate up to 3,000 computing units a good scaling behavior for the realistic cases. The runtime of 0.035 s using 12,168 computing units is also an acceptable value, considering the time for one simulation time step. To further optimize the runtime, we propose to decrease the runtime of the second part of the algorithm, in which the graph is exchanged and the local components are relabeled. This can be done by exchanging only graph updates, instead of exchanging the total graph, creating and recycling it. This allows to reduce the time for searching within the graph, because the subdomains can directly decide, whether the update changes the graph and the component labels subsequently.

Further, the described PCCL enables a broad field of topology structure detections during the simulation. We have already adopted the algorithm to stop the simulation in the event of a crack separating a grain. The algorithm generates new nuclei in a metal liquid at the coldest spots using additional traced data during the PCCL run. A broad range of applications for the PCCL algorithm is related to parallel flood filling.

2 Brute Force Study of a Read–Shockley Type Model

The brute force method is employed to simulate grain structure evolution and to recover the relations of neighboring grains proposed by Elsey et al. and Kazaryan et al. [5, 13]. In the following, the growth rate distributions for the Read–Shockley surface energy model in two dimensions (2D) with regular 3- to 8-sided grains are computed. For a detailed investigation, we refer to [30].

A well known approach to describe the surface energy of grains in solid metals at the mesoscopic scale is the Read-Shockley model [20]. It is favored because of its simplicity. A full description of the surface energy requires five parameters [23], whereas the Read-Shockley model in its original form consists of one equation to describe the surface energy containing only one parameter. It is common to add a surface mobility in form of a sigmoidal function, which depends on the same parameter as the surface energy. Humphreys [11] describes both equations and applies them to identify configurations for abnormal grain growth. The

Read–Shockley formulation in the following study to describe the surface ten. γ_{ij} between the grains i and j is given by

$$\gamma_{ij}(\Theta_{ij}) = \begin{cases} \frac{\Theta_{ij}}{\Theta_{\max}} \cdot (1 - \ln(\frac{\Theta_{ij}}{\Theta_{\max}})) & \text{if } \Theta_{ij} < \Theta_{\max} \\ 1 & \text{otherwise} \end{cases} \tag{1}$$

and the respective equation for the surface mobility τ_{ij} reads

$$\tau_{ij}(\Theta_{ij}) = \frac{1}{1 + \frac{0.9}{1+\exp(-2(\Theta_{ij}-\Theta_{\max}))}}. \tag{2}$$

The misorientation angle Θ_{ij} between two crystal lattices is the input parameter to determine surface energy and mobility. To describe large angle grain boundaries with misorientation angles $\geq 15°$, we set $\Theta_{\max}$ to $15°$.

A setup of a regular n-sided grain consists of the central grain and n neighbors. Each grain has a cubic crystal lattice, which is used to compute the misorientation between two grains, in order to determine the energy of the interface. Figure 6 depicts a 7-sided grain with its neighbors and crystal lattices. If everything is set, the evolution of the system is computed to determine the growth rate of the central grain.

The Read-Shockley model is incorporated to compute the growth rate of a regular n-sided grain in 2D with the phase-field method, based on Nestler and Garcke [18] and with the vertex model, based on Kawasaki et al. [12]. Phase-field simulations are performed to validate the vertex simulations, because the computational efficiency of the vertex model requires some restrictions on the physical representation of the grain boundaries. Curvature is the major restriction and not considered by the

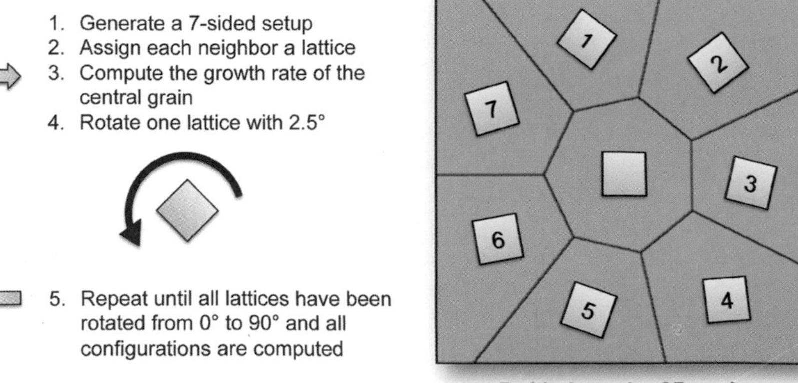

Fig. 6 Brute force scan of a regular 7-sided grain in two dimensions with a step width of 2.5°. Each grain has its own crystal lattice orientation, which is indicated by a *square*

plified vertex model. On the other hand, phase-field simulations have a higher mputational demand than vertex simulations. In order to compute billions of configurations, it is necessary to use a fast method with a small error.

For the validation of the vertex simulations, 100 configurations with random lattice orientations are generated for three to eight sided grains. The 700 growth rates are computed with the phase-field and the vertex model to correlate the results. A short time step is computed, such that no grain boundaries or triple junctions vanish. The Pearson product-moment correlation coefficient is about 0.96, which indicates a linear correlation that is high enough to conduct the study with billions of configurations.

The aim is to perform a high resolution scan of the parameter space with the vertex method. The parameter space consists of the number of neighbors n and the crystal lattice orientations of each neighbor. Because of the cubic symmetry, each lattice is rotated from $-45°$ to $45°$ ($90°$ range) to cover the parameter space. Figure 6 depicts the procedure to scan the parameter space of a 7-sided grain with a step width of 2.5°.

The evolution of large scale grain structures and the determination of neighborhood-relationships demonstrate the need for fast methods and concurrent processing. To compute more than 95 billion $(\frac{90}{2.5}+1)^7$ configurations with 1 second per configuration needs more than 3,000 years on a single processor. To obtain the results with the vertex method within reasonable time, we can decompose the parameter space because no dependencies between each configuration exists. The growth rates are stored for each process until all configurations are computed, where the message passing interface is used to sum the partial histograms of all processes into one histogram. More than 268 billion configurations are computed in total. Table 1 shows the number of configuration, the runtime and the number of CPUs.

Table 1 Setup of the vertex simulation runs on the HERMIT supercomputer in Stuttgart. The more neighbors are investigated, the bigger the step width because of the exponential growth of the configuration count. An estimation for more than eight neighbors and 4,096 CPUs takes the increasing computation time per neighbor into account

n	Step width	Configurations	CPUs	Seconds	$\frac{Configurations}{Second \cdot CPU}$
3	0.10°	731,432,701	1,024	40	17,857
4	0.25°	16,983,563,041	1,024	1,177	14,091
5	0.75°	25,937,424,601	2,048	1,084	11,683
6	1.50°	51,520,374,361	2,048	2,533	9,931
7	2.50°	94,931,877,133	2,048	4,864	9,530
8	4.00°	78,310,985,281	2,048	4,628	8,262
Estimation					
9	5.80°	91,503,485,996	4,096	3,522	6,343
10	7.90°	85,419,299,412	4,096	4,714	4,424
11	10.20°	82,213,482,971	4,096	8,012	2,505
12	12.60°	84,980,990,339	4,096	35,387	586

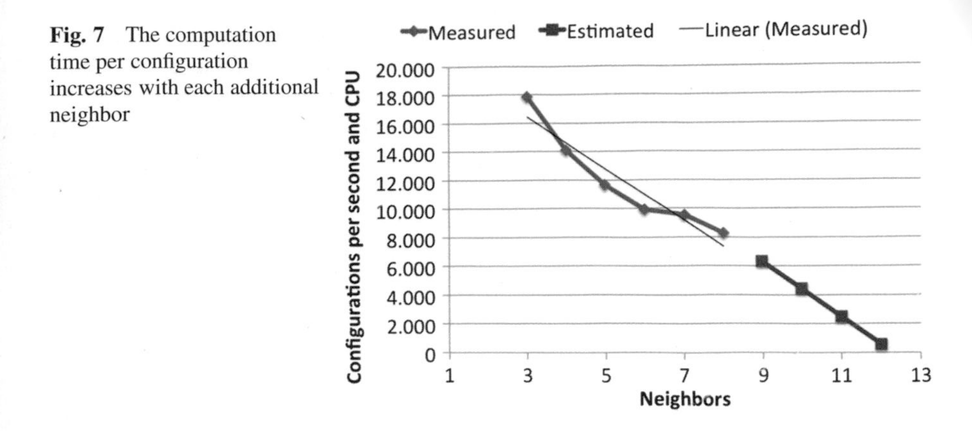

Fig. 7 The computation time per configuration increases with each additional neighbor

The last column contains the number of configurations each CPU can compute in 1 s. The second part of Table 1 is an estimation for simulations with more than eight neighbors and about 80 billion configurations for each neighbor count.

Figure 7 depicts the influence of the neighbor count on the runtime with a prediction that the behavior of more than eight neighbors requires the distribution of the computation on more CPUs.

The simulations allow the conclusion that the Read-Shockley model mainly behaves isotropically, but there exist single configurations for which the Neumann-Mullins relations are violated. Another finding is, that the standard deviation reduces for an increasing neighbor count. Such a high resolution scan of the parameter space to identify model properties is not possible without the use of high performance computing resources.

3 Recrystallization

To simulate the three-dimensional (3D) recrystallization process in polycrystalline grain structures under the influence of an external load is challenging, because of the large domain size, which is necessary to extract technologically important parameters such as the orientation distribution and the recrystallized volume fraction.

The goal is to simulate cold rolling (deformation) and annealing (recrystallization) at the end of the sheet metal production chain. An experimental investigation of the grain structure and orientation distribution (Schreijäg and Mönig, Experimental grain structure investigation of dc04 steel, private communication, 2013) is the base to parameterize and validate the simulations. A crystal plasticity finite element simulation (Bienger and Helm, Crystal plasticity investigation of a 1,000 grains microstructures. Private communication, 2013) is used as the starting conditions to simulate static recrystallization with the phase-field method.

The deformation induced stored energy, the orientation distribution and the grain structure of the finite element simulation is transferred to serve as initial grain configuration and local driving force to simulate recrystallization with the phase-field method. To incorporate the deformation induced stored energy as driving force for nucleation, the phase-field model of Nestler and Garcke [18] is extended by an appropriate formulation of the bulk free energy function. To compute a large 3D finite differences domain and many phases, we apply the message passing interface standard and 3D domain decomposition as presented by Vondrous et al. [29] and a local order parameter reduction scheme as described by Gruber et al., Kim et al. and Vedantam et al. [10, 14, 28]. The cell updates are computed with explicit Euler scheme, such that the code relates to the stencil code category of the Berkley classification [3].

The simulation domain contains $1{,}900\times660\times214$ cells and requires in total about 20 GB of main memory for the order parameter field and the stored energy field. 195,375 order parameters are introduced to represent the different orientational variants of the grain structure with the Read–Shockley surface energy model. A simple triangular matrix to store all parameter combinations requires over 142 GB of memory, such that only active parameters are computed at runtime and cached for a faster access. The domain is decomposed into parts of $25 \times 25 \times 25$ cells for 18,469 parallel processes to compute the simulation task within the 24 h restriction of the batch system.

We obtain the evolution of the grain structure until the fully recrystallized state is reached. Figure 8 depicts successive images of the microstructure during recrystallization.

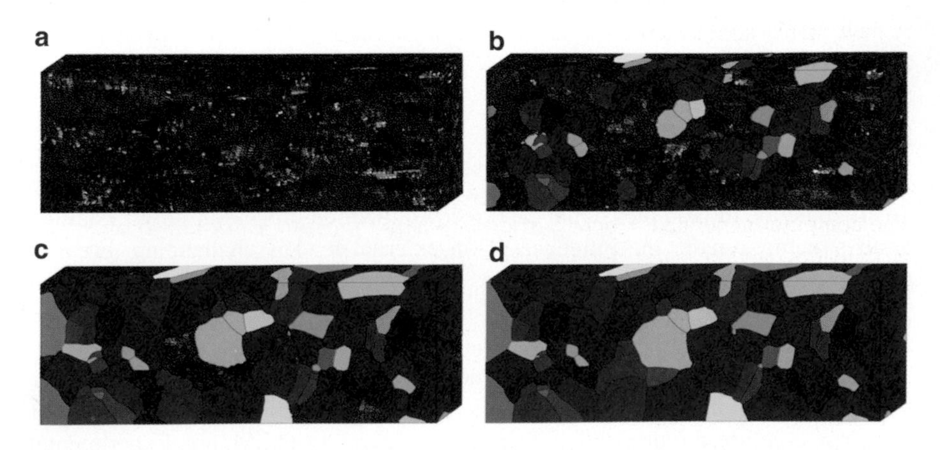

Fig. 8 Phase-field simulation of the grain structure during recrystallization in the coloring of inverse pole figures (**a**–**d**). The starting condition (**a**) consists of a subgrain like structure

4 Fluid Flow and Heat Transfer in Cellular Solids

Regarding the difficulties one has to face measuring fluid flow and heat transfer in cellular solids at the pore scale level, the benefits of employing a modern numerical simulation framework become obvious.

To the authors best knowledge, there are only few publications on the coupled fluid flow and heat transfer in open cell metal foams. Thus, attention will be drawn to the combination of an automated method for generation of cellular solids within the context of an interface capturing phase-field approach, incorporating fluid flow and heat transfer. The method applied to open cell metal foams, we employ a new algorithm developed in [24], that is capable of synthetically creating three dimensional cellular solids, foam or fabric like structures [8].

While the phase-field method originates in the simulation of solidification, crystal-growth or phase-transition problems, we utilize the method for the creation and mapping of complex pore-scale structures. Fluid flow and heat transfer is computed across a diffuse interface whereas the application does not tread any phase dynamics, i.e. the phase field is only used to distinguish between different phases (fluid|solid) which do not evolve in time. However, there are plenty of applications, topologies and configurations which can hardly be mastered by body fitted, interface tracking methods, if ever. Especially in such situations, the phase-field approach promises to be a comfortable, feasible and high potential method.

In terms of the mesoscopic phase-field approach, we employ the lattice Boltzmann method [7] for incompressible and quasi-incompressible flows, which operates on equal footage with classical Navier-Stokes solvers. Within the framework of the phase-field method, the temperature is solved by means of the energy equation. A new, stable and accurate formalism, the so called *segmented tensorial mobility approach* is established, which allows for the solution of the three dimensional time dependent temperature field [9]. Quality and accuracy of the actual implementation of fluid-flow and heat-transfer solvers is confirmed by several validations [7,9].

Besides of validation studies, the comparison with experimental measurements of fluid-flow and heat transfer in open cell metal foams shows excellent agreement. The computer generated structure of a real foam sample used for the computation of fluid-flow and heat-transfer and comparison with experimental reference values is shown in Fig. 9a, whereas the distribution of the phase-field order parameter and visualisation of the interface region is given in Fig. 9b. The overall domain is discretised with about 24×10^6 cells. With respect to the cylindrical shape and in order to save computational resources and simulation time, one quarter of the real sample is modelled, assuming a horizontal and vertical plane of symmetry in streamwise direction. Applying the measured values of pore diameter and edge width on heuristically generated most compact packing of spheres, the porosity of the generated structure matches the real sample with a residual deviation of about 0.001 %. The structure is generated from a set of input parameters in a fully automatic way without any user interaction. The fluid flow through the cellular structure is evaluated in terms of the pressure drop and Nusselt number.

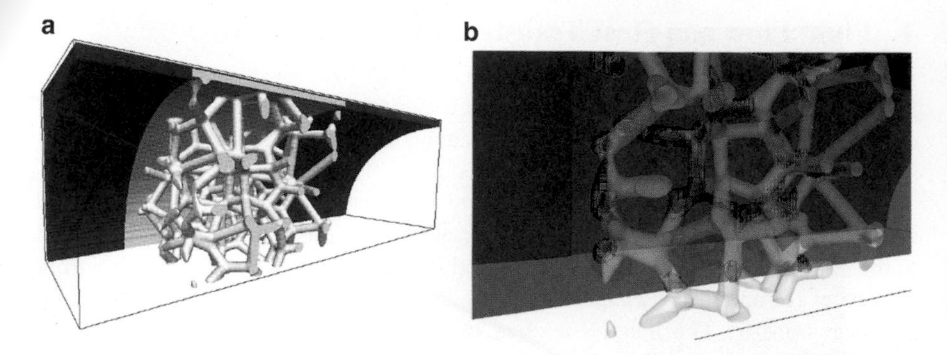

Fig. 9 (**a**) Outline of a computer generated quarter of a real open cell metal foam sample for the computation of fluid flow and heat transfer (overall resolution is 200 × 200 × 600 cells) and the comparison with experimental reference values. The foam structure is connected to a cylindrical shroud (*light-gray*), subject to a constant temperature for the heat transfer calculations, whereas the cylindrical inlet and outlet region (*dark-gray*) is subject to adiabatic boundary conditions. (**b**) Distribution of the order parameter on a streamwise cross section, where the cells belonging to the diffuse interface are highlighted

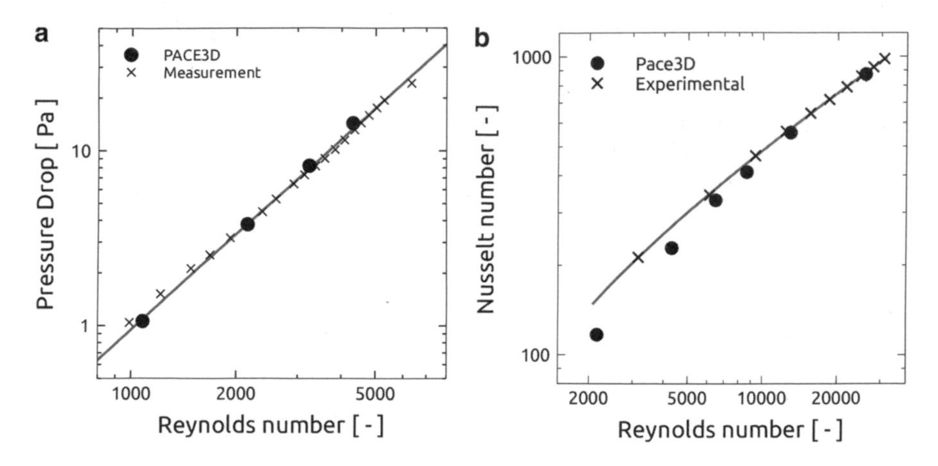

Fig. 10 Comparison of numerical and experimental values for (**a**) pressure drop and (**b**) Nusselt-number for different Reynolds numbers

The simulation results are compared with respective experimental measurements for different inflow conditions represented as different Reynolds numbers. The comparison is displayed in Fig. 10a, b. Figure 11 gives an impression of the level of details provided by the simulation, by an exemplary visualisation of streamlines in the wake of the foam edges at the outlet of the sample.

In the course of the outlined work, fluid-flow and heat transfer in open cell metal foam is successfully simulated, and the results are in very good agreement with experimental reference values.

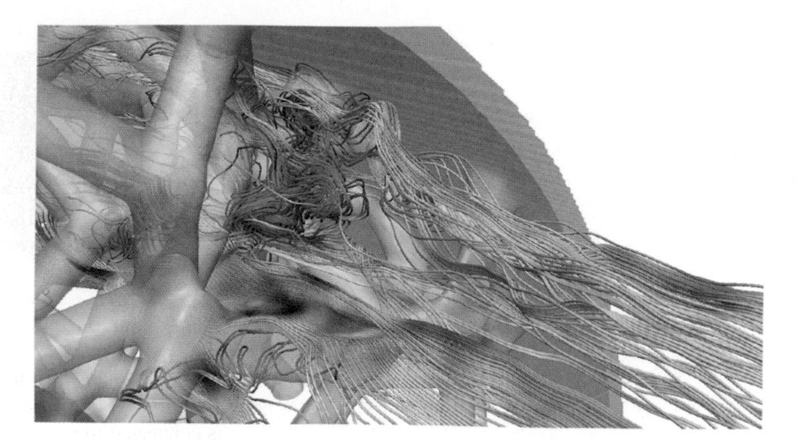

Fig. 11 Visualisation of the flow inside the foam structure with three-dimensional streamlines at a Reynolds number of about 26'000. Besides straight flow paths, detached flow is identified in the wake of a foam edge. Streamlines are coloured with velocity magnitude

References

1. AbuBaker, A., Qahwaji, R., Ipson, S., Saleh, M.: One scan connected component labeling technique. In: 2007 IEEE International Conference on Signal Processing and Communications, Dubai, pp. 1283–1286. IEEE (2007)
2. Alnuweiti, H.M., Prasanna, V.K.: Parallel architectures and algorithms for image component labeling. IEEE Trans. Pattern Anal. Mach. Intell. **14**(10), 1014–1034 (1992)
3. Asanovic, K., Bodik, R., Catanzaro, B.C., Gebis, J.J., Husbands, P., Keutzer, K., Patterson, D.A., Plishker, W.L., Shalf, J., Williams, S.W., Yelick, K.A.: The landscape of parallel computing research: a view from berkeley. Technical report UCB/EECS-2006-183, EECS Department, University of California, Cheju Island, Dec 2006
4. Choudhary, A., Thakur, R.: Evaluation of connected component labeling algorithms on shared and distributed memory multiprocessors. In: Proceedings Sixth International Parallel Processing Symposium, Beverly Hills, 1992
5. Elsey, M., Esedoglu, S., Smereka, P.: Simulations of anisotropic grain growth: efficient algorithms and misorientation distributions. Acta Mater. **61**(6), 2033–2043 (2013)
6. Ercan, M.F., Fung, Y.-F., Fung, Y.-F. Connected component labeling on a one dimensional DSP array. In: Proceedings of IEEE. IEEE Region 10 Conference. TENCON 99. 'Multimedia Technology for Asia-Pacific Information Infrastructure' (Cat. No.99CH37030), Silla Cheju, Cheju Island vol. 2, 1999
7. Ettrich, J., Nestler, B.: A lattice Boltzmann model combined with a phase-field concept for fluid dynamics in complex cellular solids. J. Comput. Phys. page submitted (2014)
8. Ettrich, J., Nestler, B., Rölle, M.: A combined mrt lattice-boltzmann and phase-field method for fluid flow and heat transfer simulations in cellular solids. In: ECCOMAS 2012 – European Congress on Computational Methods in Applied Sciences and Engineering, Vienna, pp. 8461–8472, 2012
9. Ettrich, J., Choudhury, A., Tschukin, O., Schoof, E., August, A., Nestler, B.: Modelling of transient heat conduction with diffuse interface methods. In: Modelling and Simulation in Materials Science and Engineering. Submitted, 2014
10. Gruber, J., Ma, N., Wang, Y., Rollett, A.D., Rohrer, G.S.: Sparse data structure and algorithm for the phase field method. Model. Simul. Mater. Sci. Eng. **14**(7), 1189–1195 (2006)

11. Humphreys, F.J.: A unified theory of recovery, recrystallization and grain growth, based on the stability and growth of cellular microstructures–I. The basic model. Acta Mater. **45**(10), 4231–4240 (1997)
12. Kawasaki, K., Nagai, T., Nakashima, K.: Vertex models for two-dimensional grain growth. Philos. Mag. Part B **60**(3), 399–421 (1989)
13. Kazaryan, A., Patton, B.R., Dregia, S.A., Wang, Y.: On the theory of grain growth in systems with anisotropic boundary mobility. Acta Mater. **50**(3), 499–510 (2002)
14. Kim, S.G., Kim, D.I., Kim, W.T., Park, Y.B.: Computer simulations of two-dimensional and three-dimensional ideal grain growth. Phys. Rev. E Stat. Nonlinear Soft Matter Phys. **74**(6 Pt 1), 061605 (2006)
15. Moelans, N.: Phase-field simulations of grain growth in materials containing second-phase particles. PhD thesis, Katholieke Universiteit Leuven Faculteit Ingenieurs Wetenschappen Department Metaalkunde en Toegepaste Materiaalkunde, 2006
16. Nagel, W.E., Kröner, D.B., Resch, M.M. (eds.) High Performance Computing in Science and Engineering '11: Transactions of the High Performance Computing Center, Stuttgart (HLRS) 2011, 2012 edn. Springer (2012)
17. Nestler, B.: A 3d parallel simulator for crystal growth and solidification in complex alloy systems. J. Cryst. Growth **275**(1–2), e273–e278 (2005). In: Proceedings of the 14th International Conference on Crystal Growth and the 12th International Conference on Vapor Growth and Epitaxy.
18. Nestler, B., Garcke, H., Stinner, B.: Multicomponent alloy solidification: phase-field modeling and simulations. Phys. Rev. E **71**, 041609 (2005)
19. Netzel, P., Stepinski, T.F.: Connected components labeling for giga-cell multi-categorical rasters. Comput. Geosci. **59**, 24–30 (2013)
20. Read, W., Shockley, W.: Dislocation models of crystal grain boundaries. Phys. Rev. **78**(3), 275–289 (1950)
21. Rödel, J., Glaeser, A.M.: Anisotropy of grain growth in alumina. J. Am. Ceram. Soc. **73**(11), 3292–3301 (1990)
22. Rodríguez, J.E., Cruz, I., Vergés, E., Ayala, D.: A connected-component-labeling-based approach to virtual porosimetry. Graph. Model., **73**(5), 296–310 (2011)
23. Rohrer, G.S.: Grain boundary energy anisotropy: a review. J. Mater. Sc. **46**(18), 5881–5895 (2011)
24. Rölle, M.: Füllalgorithmen zur Generierung schaumartiger Strukturen in 3D. Bachelor's thesis, Hochschule Karlsruhe – Technik und Wirtschaft, 2010
25. Rosenfeld, A., Pfaltz, J.L.: Sequential operations in digital picture processing. J. ACM **13**(4), 471–494 (1966)
26. Suk-Kang, J.L.: Sintering: densification, grain growth and microstructure, Butterworth-Heinemann, vol. 11, 2004. ISBN: 978-0-7506-6385-4
27. Tikare, V., Miodownik, M.A., Holm, E.A.: Three-dimensional simulation of grain growth in the presence of mobile pores. J. Am. Ceram. Soc. **84**(6), 1379–1385 (2001)
28. Vedantam, S., Patnaik, B.S.V.: Efficient numerical algorithm for multiphase field simulations. Phys. Rev. E Stat. Nonlinear Soft Matter Phys. **73**(1 Pt 2), 016703 (2006)
29. Vondrous, A., Selzer, M., Hötzer, J., Nestler, B.: Parallel computing for phase-field models. Int. J. High Perform. Comput. Appl. (2013)
30. Vondrous, A., Reichardt, M., Nestler, B.: Growth rate distributions for regular two-dimensional grains with Read-Shockley grain boundary energy. Model. Simul. Mater. Sci. Eng. **22**(2), 025014 (2014)

Molecular Models for Cyclic Alkanes and Ethyl Acetate As Well As Surface Tension Data from Molecular Simulation

Stefan Eckelsbach, Tatjana Janzen, Andreas Köster, Svetlana Miroshnichenko, Yonny Mauricio Muñoz-Muñoz, and Jadran Vrabec

1 Introduction

Thermodynamic data for most technically interesting systems are still scarce or even unavailable despite the large experimental effort that was invested over the last century into their measurement. This particularly applies to mixtures containing two or more components and systems under extreme conditions. In contrast to phenomenological methods, molecular modeling and simulation is based on a sound physical foundation and is therefore well suited for the prediction of such properties and processes.

In this work new molecular force field models, i.e. for four cyclic alkanes and ethyl acetate, are presented to widen the application area of molecular simulation.

In the second part of this work vapor-liquid interfaces were investigated, deepening the understanding of interfacial processes of a model fuel, consisting of acetone, nitrogen and oxygen, similar to the state of injected acetone droplets in a gaseous fluid in combustion chambers.

In the third part an efficient method for the generation of large simulation data sets is studied with respect to the stability and accuracy, using the microcanonical (*NVE*) ensemble instead of the canonic (*NVT*) ensemble.

S. Eckelsbach • T. Janzen • A. Köster • S. Miroshnichenko • Y.M. Muñoz-Muñoz • J. Vrabec (✉)
Lehrstuhl für Thermodynamik und Energietechnik (ThEt), Universität Paderborn, Warburger Str. 100, 33098 Paderborn, Germany
e-mail: jadran.vrabec@upb.de.

W.E. Nagel et al. (eds.), *High Performance Computing in Science and Engineering '14*, DOI 10.1007/978-3-319-10810-0_42

2 Optimization of Potential Models for Cyclic Alkanes from Cyclopropane to Cyclohexane

Cyclic hydrocarbons are substances that are very common in the chemical industry, having one or more rings in their molecular structure. Cyclic alkanes with a single ring are part of this class of compounds and can be represented with the general molecular formula C_nH_{2n}. These compounds, that are primarily found in crude oil, have single bonds between the carbon atoms, i.e. their carbon atoms are completely saturated.

In the petrochemical industry, for example, cyclohexane is produced from the hydrogenation of benzene, albeit the inverse reaction, i.e. the dehydrogenation of cyclohexane to yield benzene is more important. An improvement of the understanding of the nature of all involved physical forces of the cyclic alkanes may lead to various technical applications, including chemical equilibrium studies and vapor-liquid equilibrium (VLE) data extraction for both pure substances and mixtures. Furthermore, both temperature and pressure can be obtained using molecular simulation to study the development of hydrates, which is important since their formation e.g. may cause obstructions of pipes.

Moreover, it is believed that with accurate potential models it will be possible to obtain hybrid models which might decrease experimental effort and cost. Scarce experimental data are available for cyclobutane, such that molecular simulation can be a source to provide pseudo-experimental data and can contribute to extend existing models. Thus the motivation is to develop accurate molecular force field models for these substances, which constitutes the principal objective of this work.

The parameterization route introduced in prior work of our group [1] was adapted here, to obtain accurate intermolecular potential models from cyclopropane (C_3H_6) to cyclohexane (C_6H_{12}). To obtain such potential models, the cycloalkane molecules were assumed to be rigid, i.e. without internal degrees of freedom, formed by Lennard-Jones (LJ) sites bonded in a ring shaped structure with internal angles computed by quantum mechanical calculations.

The geometric structure of all molecules was obtained with the open source code for computational chemistry calculations GAMESS(US) [2], using the Hartree-Fock (HF) method with a relatively small (6-31 G) basis set. Thereby all atomic positions were determined for each hydrogen and carbon atom in the molecule. Cyclopropane is an essentially planar molecule, characterized only by bond angles and bond lengths. This is in contrast to cyclobutane, cyclopentane and cyclohexane, which require the specification of dihedral angles due to their three-dimensional structure.

A reduction of computational cost can be achieved by employing the united atom approximation, whereby the methylene group is represented by a single methylene site. The LJ force centre for each methylene site was initially located at the position of the carbon atom, as determined by the HF method.

Subsequently, the methylene site-site distance was gradually modified while keeping the bond and dihedral angles invariant, molecular simulation runs were performed to compute VLE data for the pure substances using the *ms*2 code [3] and the LJ parameters reported by [1]. The procedure was repeated for all molecules studied in this work.

The simulation results were compared to experimental data for the VLE properties, the critical point and second virial coefficient from the literature. Whenever reasonable results were obtained for VLE properties, such as saturated liquid density, vapor pressure and enthalpy of vaporization, the methylene site-site distances were fixed and the models were then subject to further optimization.

The LJ parameters were optimized using the reduced unit method by Merker et al. [4], which constitutes a computationally effective procedure, since no additional molecular simulation runs are necessary.

Figure 1 shows a snapshot for each molecular structure obtained in this work. Every geometry contains the internal angles as computed by quantum mechanical calculations, but the site-site distances were enlarged. The cyclopropane structure is planar, while the cyclobutane and cyclopentane are three-dimensional structures. For the cyclohexane case, the chair configuration is chosen.

Figure 2 shows the molecular simulation results for the VLE of each fluid studied in this work, including the critical point. The continuous lines are the data from the best reference equations of state that are available for cyclopentane and cyclohexane. For the cyclopropane and cyclobutane cases, the continuous lines are data from typical correlations that are used in the literature. Moreover, experimental data are shown for each fluid. Note that for the cyclobutane case, the experimental data are very scarce, but molecular simulation results have a similar performance as the DIPPR [5] correlations.

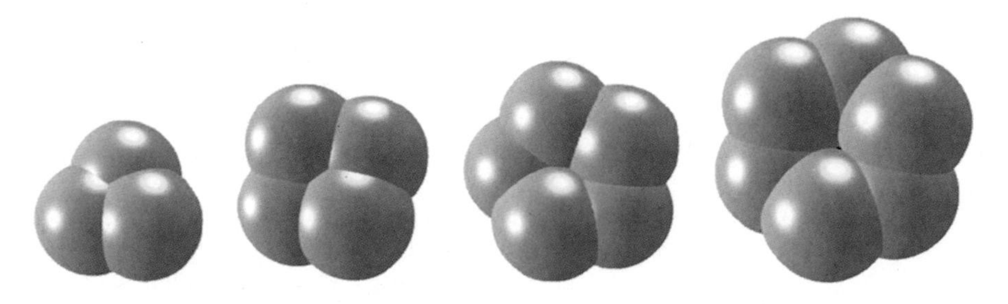

Fig. 1 Snapshots of cyclopropane (C_3H_6), cyclobutane (C_4H_8), cyclopentane (C_5H_{10}) and cyclohexane (C_6H_{12}) obtained in this work (left to right)

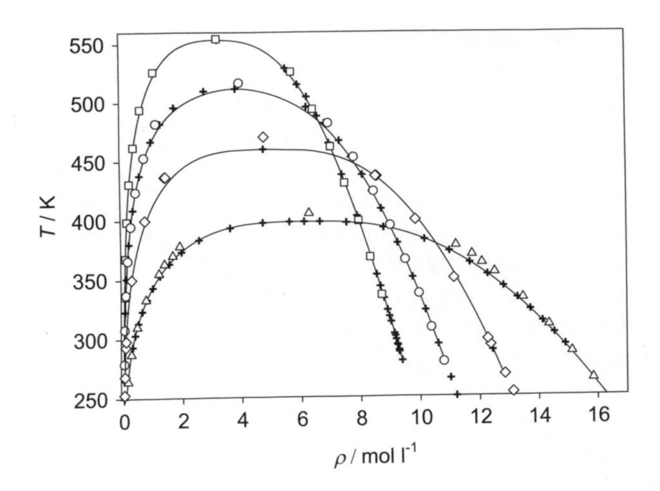

Fig. 2 Vapor-liquid envelopes of the cycloalkanes studied in this work. *Triangles* represent C_3H_6, *diamonds* C_4H_8, *circles* C_5H_{10} and *squares* C_6H_{12}. The cross symbol corresponds to experimental data from the literature

3 Ethyl Acetate

Ethyl acetate is the organic compound with the formula CH_3–COO–CH_2–CH_3. It is produced by the esterification reaction of ethanol and acetic acid and is synthesized on a large scale for use as a solvent. VLE data are necessary for the design and optimization of separation processes, therefore, a new molecular model for ethyl acetate is proposed here.

3.1 Molecular Model Class

The present molecular model includes three groups of potential parameters. These are the geometric parameters, specifying the positions of different interaction sites, the electrostatic parameters, defining the polar interactions in terms of point charges, dipoles or quadrupoles, and the dispersive and repulsive parameters, determining the attraction by London forces and the repulsion by electronic orbital overlaps. Here, the LJ 12-6 potential [6, 7] was used to describe the dispersive and repulsive interactions. The total intermolecular interaction energy thus writes as

$$U = \sum_{i=1}^{N-1} \sum_{j=i+1}^{N} \left\{ \sum_{a=1}^{S_i^{\mathrm{LJ}}} \sum_{b=1}^{S_j^{\mathrm{LJ}}} 4\varepsilon_{ijab} \left[\left(\frac{\sigma_{ijab}}{r_{ijab}} \right)^{12} - \left(\frac{\sigma_{ijab}}{r_{ijab}} \right)^{6} \right] + \right.$$

$$\sum_{c=1}^{S_i^{\mathrm{e}}} \sum_{d=1}^{S_j^{\mathrm{e}}} \frac{1}{4\pi\epsilon_0} \left[\frac{q_{ic}q_{jd}}{r_{ijcd}} + \frac{q_{ic}\mu_{jd} + \mu_{ic}q_{jd}}{r_{ijcd}^2} \cdot f_1(\boldsymbol{\omega}_i, \boldsymbol{\omega}_j) + \frac{q_{ic}Q_{jd} + Q_{ic}q_{jd}}{r_{ijcd}^3} \cdot f_2(\boldsymbol{\omega}_i, \boldsymbol{\omega}_j) + \right.$$

$$\left. \left. \frac{\mu_{ic}\mu_{jd}}{r_{ijcd}^3} \cdot f_3(\boldsymbol{\omega}_i, \boldsymbol{\omega}_j) + \frac{\mu_{ic}Q_{jd} + Q_{ic}\mu_{jd}}{r_{ijcd}^4} \cdot f_4(\boldsymbol{\omega}_i, \boldsymbol{\omega}_j) + \frac{Q_{ic}Q_{jd}}{r_{ijcd}^5} \cdot f_5(\boldsymbol{\omega}_i, \boldsymbol{\omega}_j) \right] \right\}, \qquad (1)$$

where r_{ijab}, ε_{ijab}, σ_{ijab} are the distance, the LJ energy parameter and the LJ size parameter, respectively, for the pair-wise interaction between LJ site a on molecule i and LJ site b on molecule j. The permittivity of the vacuum is ϵ_0, whereas q_{ic}, μ_{ic} and Q_{ic} denote the point charge magnitude, the dipole moment and the quadrupole moment of the electrostatic interaction site c on molecule i and so forth. The expressions $f_x(\boldsymbol{\omega}_i, \boldsymbol{\omega}_j)$ stand for the dependence of the electrostatic interactions on the orientations $\boldsymbol{\omega}_i$ and $\boldsymbol{\omega}_j$ of the molecules i and j [8, 9]. Finally, the summation limits N, S_x^{LJ} and S_x^{e} denote the number of molecules, the number of LJ sites and the number of electrostatic sites, respectively. The atomic charges were estimated here by the Mulliken method [10].

Lorentz-Berthelot combining rules were used to determine cross-parameters for Lennard-Jones interactions between sites of different type [11, 12]:

$$\sigma_{ijab} = \frac{\sigma_{iiaa} + \sigma_{jjbb}}{2}, \qquad (2)$$

and

$$\varepsilon_{ijab} = \sqrt{\varepsilon_{iiaa}\varepsilon_{jjbb}}. \qquad (3)$$

3.2 *Molecular Model of Ethyl Acetate*

A new molecular model was developed for ethyl acetate based on quantum chemical (QC) information on molecular geometry and electrostatics. In a first step, the geometric data of the molecules, i.e. bond lengths, angles and dihedrals, were determined by QC calculations. Therefore, a geometry optimization was performed via an energy minimization using the GAMESS(US) package [2]. The HF level of theory was applied with a relatively small (6-31 G) basis set. Intermolecular electrostatic interactions mainly occur due to the static polarity of single molecules that can well be obtained by QC. Here, the Møller-Plesset 2 level of theory was used

. 3 Snapshot of ethyl
etate

Table 1 Parameters of the molecular model for ethyl acetate developed in this work. Lennard-Jones interaction sites are denoted by the modeled atomic groups. Electrostatic interaction sites are denoted by point charge. Coordinates are given with respect to the center of mass in a principal axes system

	x	y	z	σ	ϵ/k_B	q
Interaction site	Å	Å	Å	Å	K	e
CH_3	−2.273	1.172	0	3.7227	74.139	
CH_2	−1.312	0.018	0	3.9212	34.800	
O	0.000	0.589	0	3.0278	59.765	
C	1.101	−0.203	0	3.8716	31.018	
O	2.188	0.321	0	2.7796	41.609	
CH_3	0.897	−1.690	0	3.7227	74.139	
Point charge (CH_2)	−1.312	0.018	0			0.306
Point charge (O)	0.000	0.589	0			−0.474
Point charge (C)	1.101	−0.203	0			0.593
Point charge (O)	2.188	0.321	0			−0.415
Point charge (CH_3)	0.897	−1.690	0			−0.010

that considers electron correlation in combination with the polarizable 6-31G basis set. Figure 3 shows the devised molecular model.

The new model for ethyl acetate is based on the work of Kamath et al. [13], which is improved by including geometry and electrostatics from QM calculations. The initial model adopted the LJ parameters by Kamath et al. which were then adjusted to experimental VLE data, i.e. saturated liquid density, vapor pressure and enthalpy of vaporization, until a desired quality was reached.

The present ethyl acetate model consists of six LJ sites (one for the methylene bridge, one for each oxygen atom and one for each methyl group). The electrostatic interactions were modeled by five point charges. The parameters of the LJ sites were adjusted to experimental saturated liquid density and vapor pressure. The full parameter set of the new ethyl acetate model is listed in Table 1.

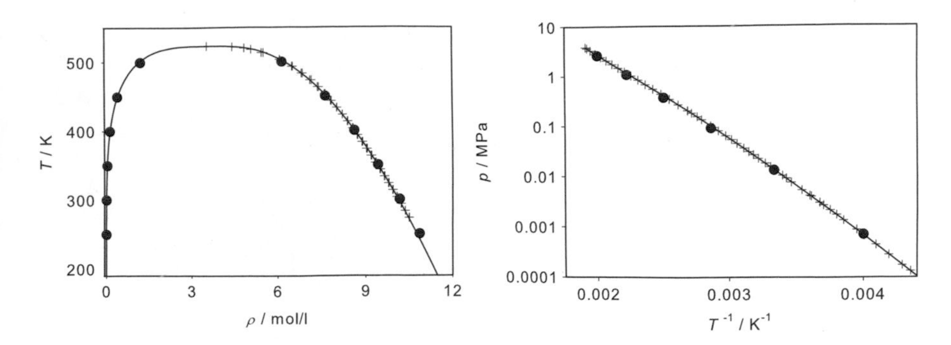

Fig. 4 Simulation results (●), experimental data [15] (+) and DIPPR correlation of experimental data [14] (—) of ethyl acetate. *Left*: saturated densities, *right*: vapor pressure. The statistical uncertainties of the present simulation data are within symbol size

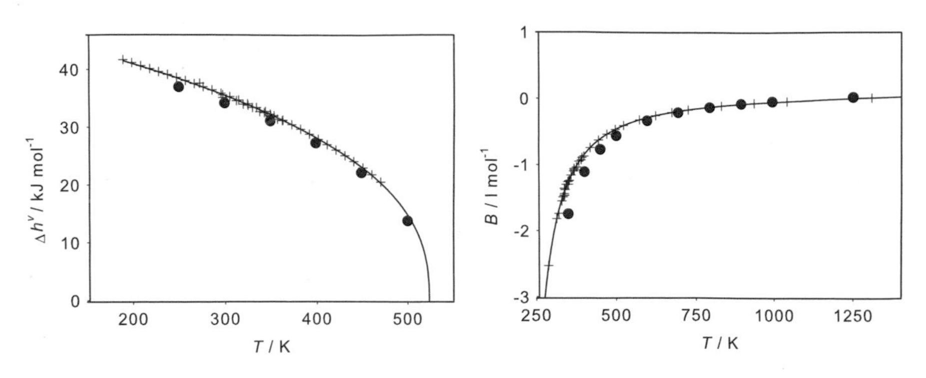

Fig. 5 Simulation results (●), experimental data [15] (+) and DIPPR correlation of experimental data [15] (—) of ethyl acetate. *Left*: enthalpy of vaporisation, *right*: second virial coefficient. The statistical uncertainties of the present simulation data are within symbol size

The pure substance VLE simulation results are shown in Figs. 4 and 5, where they are compared to experimental data and the DIPPR correlations [14]. The ethyl acetate model shows mean unsigned deviations to experimental data of 0.1 % for the saturated liquid density, 4.6 % for the vapor pressure and 4.3 % for the enthalpy of vaporization over the whole temperature range from the triple point to the critical point.

For pure ethyl acetate, predicted data on the second virial coefficient is available from DIPPR. Figure 5 shows the simulation results compared to a correlation of predicted data taken from DIPPR [14]. The present second virial coefficient agrees well with the DIPPR correlation at high temperatures, however at low temperatures, some noticeable deviations are present yielding a mean error of about 0.5 l/mol.

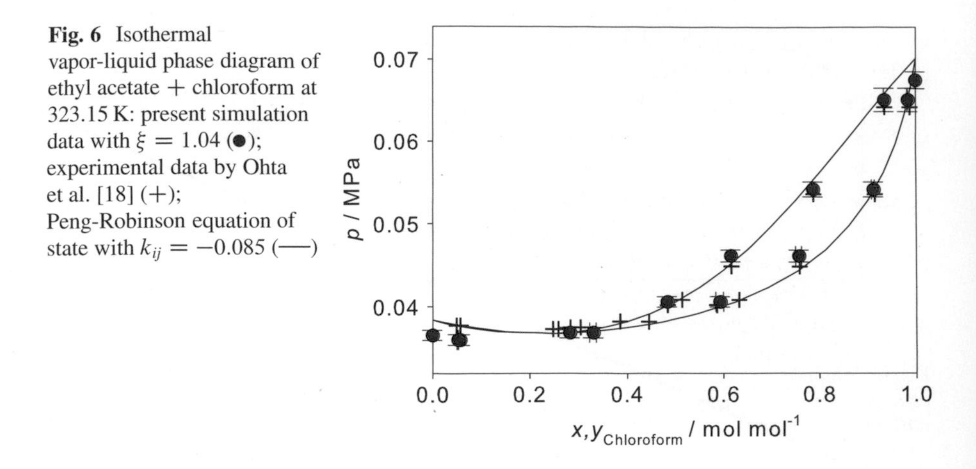

Fig. 6 Isothermal vapor-liquid phase diagram of ethyl acetate + chloroform at 323.15 K: present simulation data with $\xi = 1.04$ (●); experimental data by Ohta et al. [18] (+); Peng-Robinson equation of state with $k_{ij} = -0.085$ (—)

3.3 *Binary Vapor-Liquid Equilibria*

Based on the devised molecular model for ethyl acetate, VLE data were predicted for the binary system ethyl acetate + chloroform. The molecular model for chloroform was taken from Stoll et al. [16] and is of the two-center LJ plus point dipole type. The vapor-liquid phase behavior of the binary mixture was predicted by molecular simulation and compared to experimental data as well as the Peng-Robinson equation of state [17].

Figure 6 shows the isothermal VLE of ethyl acetate + chloroform at 323.15 K from experiment [18], simulation and Peng-Robinson equation of state. The experimental vapor pressure at a liquid mole fraction of $x_{\text{chloroform}} = 0.5$ mol/mol was taken to adjust the binary parameter of the molecular model ($\xi = 1.04$) and of the Peng-Robinson equation of state ($k_{ij} = -0.085$). It can be seen that the results obtained by molecular simulation agree well with the experimental results and the Peng-Robinson equation of state.

4 Interfacial Properties of Binary Mixtures

Interfacial properties are of interest for processes containing two or more phases, which is basically the default case for complex technical and natural systems, including separation processes, boiling and any system containing droplets. Many of these processes occur under extreme conditions of temperature or pressure, e.g. in case of flash boiling in combustion chambers. Such processes are actively being used although striking gaps remain in their essential understanding. At the same time, thermodynamic data for most technically interesting systems are still scarce or even unavailable despite the large experimental effort that was invested into their measurement. This particularly applies to mixtures of two or more components and systems under extreme conditions.

In this work, interfacial properties of the binary mixtures nitrogen + oxygen and nitrogen + acetone were investigated. These mixtures serve as a model fuel, yielding interfacial data and deepen the general understanding of the interfacial processes for the injection of acetone droplets in a gaseous fluid, consisting of nitrogen and oxygen. For this purpose, direct simulations of VLE were carried out, with particular attention to the interface region.

First, density and composition data of the bulk phases were required to set up the simulations. These data were obtained via the molecular simulation tool *ms*2 [3]. Molecular force field models, adjusted to the mixtures with an additional binary interaction parameter, from previous works of our group were used [19, 20]. They lead to very good agreement with reference data even for temperatures near the critical point, typically having only a 2 % deviation to the critical density ρ_c [20].

Further simulations were carried out with the highly parallel molecular dynamics code *ls*1 [21]. Due to the inhomogeneity of the simulated systems, the long range correction of Janeček was used [22]. After an equilibration period, the interface region was formed, which gave the possibility for further investigation. The equilibrated states of the interface region are shown exemplarily in Fig. 7 for three different compositions at a temperature of 400 K.

The surface tension can be calculated following the Irving-Kirkwood approach [23]

$$\gamma = \frac{1}{2A}\left(2\Pi_{zz} - \left(\Pi_{xx} + \Pi_{yy}\right)\right), \tag{4}$$

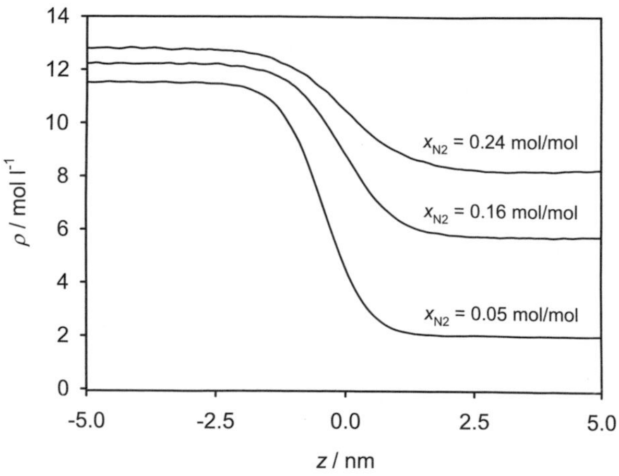

Fig. 7 Profile of the total density over the length of the simulation volume. The temperature is constant at 400 K. Three mixtures containing nitrogen and acetone are shown for three different compositions: the liquid mole fractions are $x_{N2} = 0.05$ mol/mol, $x_{N2} = 0.16$ mol/mol and $x_{N2} = 0.24$ mol/mol

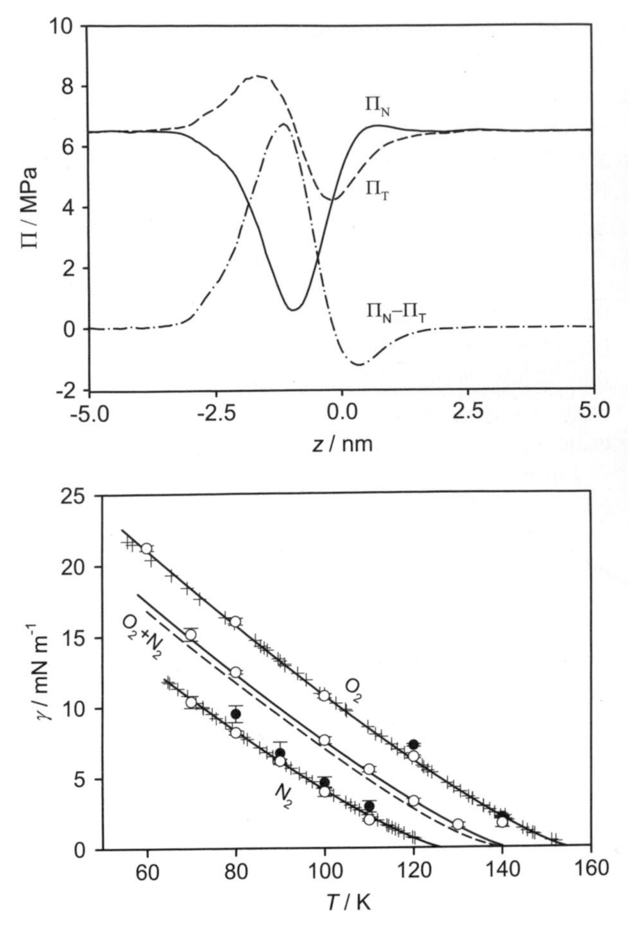

Fig. 8 Profile of the virial tensor over the length of the simulation volume. The virial tensor in normal direction Π_{N}, in tangential direction Π_{T} and the difference of these are shown. The mixture consists of nitrogen and acetone with a liquid mole fraction of $x_{\mathrm{N2}} = 0.05$ mol/mol and a temperature of 400 K

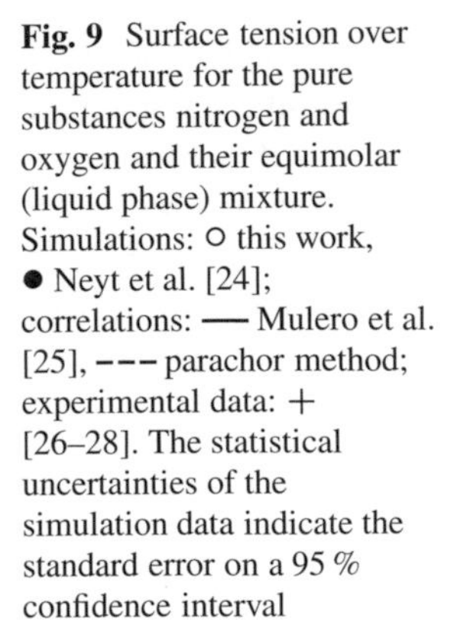

Fig. 9 Surface tension over temperature for the pure substances nitrogen and oxygen and their equimolar (liquid phase) mixture. Simulations: ○ this work, ● Neyt et al. [24]; correlations: — Mulero et al. [25], --- parachor method; experimental data: + [26–28]. The statistical uncertainties of the simulation data indicate the standard error on a 95 % confidence interval

where A is the area of the interface and $\Pi_{\alpha\beta}$ is an element of the virial tensor, which is defined as

$$\Pi_{\alpha\beta} = \left\langle \frac{1}{2} \sum_{i=1}^{N} \sum_{j=1}^{N} r_{ij}^{\alpha} f_{ij}^{\beta} \right\rangle. \tag{5}$$

The indices α and β represent the x-, y- or z-directions of the distance vector $\mathbf{r}_{ij}$ and the force vector $\mathbf{f}_{ij}$, in each case between molecules i and j. The virial tensor in normal ($\Pi_{\mathrm{N}} = \Pi_{\mathrm{zz}}$) and tangential direction ($\Pi_{\mathrm{T}} = \left(\Pi_{\mathrm{xx}} + \Pi_{\mathrm{yy}}\right)/2$) as well as the difference between these are exemplarily shown in Fig. 8 for one state point.

The simulated surface tension was compared to experimental data, which are relatively rare and only available for the pure substances. Thus the prediction of the surface tension based on the parachor [29] was used for comparison with the results for the binary mixtures. In Fig. 9 the simulation results for the pure substances

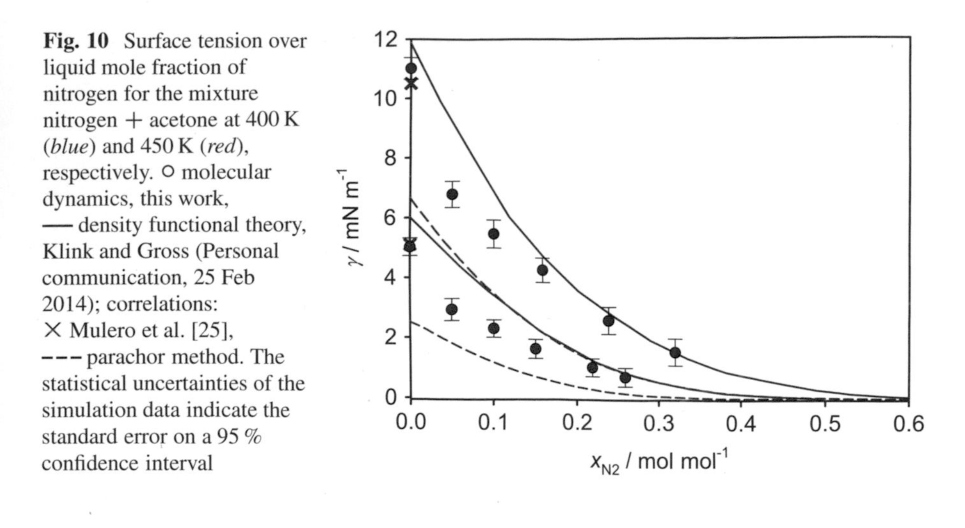

Fig. 10 Surface tension over liquid mole fraction of nitrogen for the mixture nitrogen + acetone at 400 K (*blue*) and 450 K (*red*), respectively. ○ molecular dynamics, this work, — density functional theory, Klink and Gross (Personal communication, 25 Feb 2014); correlations: × Mulero et al. [25], - - - parachor method. The statistical uncertainties of the simulation data indicate the standard error on a 95 % confidence interval

nitrogen and oxygen as well as their mixture are given. The mixture was set up in an equimolar state for the liquid phase. The simulation data for the pure substances are in good agreement with the surface tension correlation by Mulero et al. [25], which is based on experimental data, and the experimental data itself. In general, the direct simulations using the Monte Carlo scheme by Neyt et al. [24] show a slight overestimation in comparison with the reference data. The simulation results of the mixture nitrogen + oxygen also show a good agreement with the correlation [25]. The prediction following the parachor method is based on data calculated with the equations of state by Schmidt and Wagner [30] and Span et al. [31] and yields results, that are a little bit lower in the whole considered temperature range, but are relatively close to the reference.

In Fig. 10 the simulation results for the surface tension of the mixture nitrogen + acetone are plotted over the liquid mole fraction at 400 and 450 K. For higher mole fractions near the critical point, the simulated data concur with the results of Klink and Gross (Personal communication, 25 Feb 2014), determined by the density functional theory method. For smaller mole fractions, the simulation results are somewhat lower, until they match the correlation by Mulero et al. [25] for pure acetone within the estimated uncertainties of the present simulations. For this state point the data by Klink et al. overestimate the surface tension in comparison to the correlation of experimental data. Here the parachor method was based on the Peng-Robinson equation of state with the Huron-Vidal mixing rule, parameterized by Windmann et al. [20]. Its results underestimate the surface tension significantly for both investigated temperatures. The parachor method strongly depends on the input data and is thus susceptible to their quality, which, of course, is better for the highly accurate and more complex equations of state by Schmidt and Wagner [30] and Span et al. [31].

5 Thermodynamic Data by Molecular Simulations in the Microcanonical Ensemble

The importance of a solid base of thermodynamic data for the design and optimization of technological processes is crucial [32]. It was shown that molecular force field models allow for the generation of large and widespread data sets in a very short amount of time, due to the use of the molecular simulation framework proposed by Lustig [33], which gives access to any A^r_{mn} $(m, n > 0)$ in the *NVT* ensemble for a specific state point through

$$\frac{\partial^{m+n}(F/(RT))}{\partial^m(1/T)\partial^n(N/V)} \equiv A_{mn} = A^i_{mn} + A^r_{mn}, \quad (6)$$

wherein F is the Helmholtz energy, R the gas constant, T the temperature, N the number of molecules and V the volume. The ideal part A^i_{mn}, which is needed to calculate any derivative of the Helmholtz energy, can be determined with other methods [33].

In order to extend this procedure, the present work deals with the usage of the microcanonical ensemble to generate such data sets. In the *NVE* ensemble, the number of particles N, the volume V and the total energy E of the system are kept constant. Although E is not directly accessible through experimental study, there are some advantages to using the *NVE* ensemble. Compared to other statistical mechanical ensembles, it is isolated from its environment in terms of energy or mass exchange and can therefore be considered as more stable and less susceptible to fluctuations. This implies that there is no need to regulate temperature through an external thermostat and/or pressure through an external barostat in molecular dynamics simulations, because E is conserved naturally through Newton's equations of motion. Moreover, the *NVE* ensemble links mechanics and thermodynamics in a very direct way [34].

In order to validate the implementation of the *NVE* ensemble into the simulation tool *ms*2 [3], tests were carried out for the substance ethylene oxide using the molecular force field model by Eckl et al. [35]. As can be seen in Fig. 11, several simulations in the homogeneous fluid region were carried out using the *NVT* and *NVE* ensemble, respectively. It should be noted, that *NVT* ensemble simulations using this framework were compared to highly accurate equations of state in preceding work yielding excellent results. The overall agreement between the different ensemble types can be considered as quite good. Only in the vicinity of the critical point and for higher order derivatives, some deviations of the *NVE* ensemble simulation data were found. Contrary to the expectation of more stable molecular simulations in the *NVE* ensemble, the statistical uncertainties indicated by the error bars are quite similar. Therefore, further investigation is required.

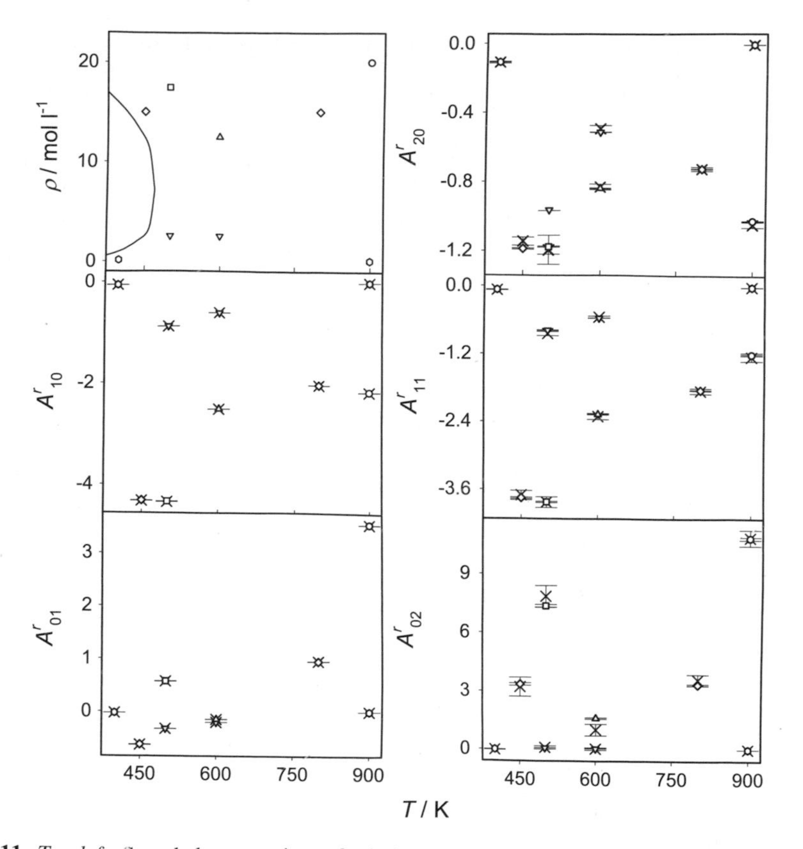

Fig. 11 *Top left*: Sampled state points of ethylene oxide in the temperature vs. density plane. *Remaining figures*: A^r_{mn} values as a function of temperature from Monte Carlo molecular simulation for ethylene oxide [35]. The *open symbols* denote various isochores in the *NVT* ensemble. The *cross symbols* represent the simulations which were carried out in the *NVE* ensemble

6 Conclusion

Five new molecular models were developed and optimized in this work. They were validated against experimental data for the VLE properties, the critical point and second virial coefficient, showing a good coincidence with the reference data.

Interfacial simulations of a model fuel were carried out, yielding the surface tension for two binary mixtures, which agree well with the experimental reference data and fill gaps where such data were unavailable.

The expected accuracy gain by using the microcanonical ensemble was not observed in the present simulations using the molecular simulation framework proposed by Lustig [33]. Here, further investigations are required.

Acknowledgements We gratefully acknowledge support by Deutsche Forschungsgemeinschaft. This work was carried out under the auspices of the Boltzmann-Zuse Society (BZS) of Computational Molecular Engineering. The simulations were performed on the Cray XE6 (Hermit) at the High Performance Computing Center Stuttgart (HLRS).

References

1. Eckl, B., Vrabec, J., Hasse, H.: Set of molecular models based on quantum mechanical ab initio calculations and thermodynamic data. J. Phys. Chem. B **112**, 12710–12721 (2008)
2. Schmidt, M.W., Baldridge, K.K., Boatz, J.A., Elbert, S.T., Gordon, M.S., Jensen, J.H., Koseki, S., Matsunaga, N., Nguyen, K.A., Su, S., Windus, T.L., Dupuis, M., Montgomery, J.A.: General atomic and molecular electronic structure system. J. Comput. Chem. **14**, 1347–1363 (1993)
3. Deublein, S., Eckl, B., Stoll, J., Lishchuk, S.V., Guevara-Carrion, G., Glass, C.W., Merker, T., Bernreuther, M., Hasse, H., Vrabec, J.: ms2: A molecular simulation tool for thermodynamic properties. Comput. Phys. Commun. **182**, 2350–2367 (2011)
4. Merker, T., Vrabec, J., Hasse, H.: Engineering Molecular Models: Efficient parameterization procedure and cyclohexanol as case study. Soft Mater. **10**, 3–25 (2012)
5. Rowley, R.L., Wilding, W.V., Oscarson, J.L., Yang, Y., Zundel, N.A., Daubert, T.E., Danner, R.P.: DIPPR data compilation of pure compound properties. Design Institute for Physical Properties, AIChE, New York (2006)
6. Jones, J.E.: On the determination of molecular fields. I. from the variation of the viscosity of a gas with temperature. Proc. R. Soc. **106A**, 441–462 (1924)
7. Jones, J.E.: On the determination of molecular fields. II. from the equation of state of a gas. Proc. R. Soc. **106A**, 463–477 (1924)
8. Allen, M.P., Tildesley, D.J.: Computer Simulations of Liquids. Oxford University Press, Oxford (1987)
9. Gray, C.G., Gubbins, K.E.: Theory of Molecular Fluids. Fundamentals, vol. 1. Clarendon Press, Oxford (1984)
10. Mulliken, R.S.: Criteria for the construction of good self–consistent–field molecular orbital wave functions, and the significance of LCAO–MO population analysis. J. Chem. Phys. **36**, 3428–3439 (2004)
11. Lorentz, H.A.: Über die Anwendung des Satzes vom Virial in der kinetischen Theorie der Gase. Ann. Phys. **12**, 127–136 (1881)
12. Berthelot, D.: Sur le Mélange des Gaz. Compt. Rend. Ac. Sc. **126**, 1703–1706 (1898)
13. Kamath, G., Robinson, J., Potoff, J.J.: Application of TraPPE–UA force field for determination of vapor–liquid equilibria of carboxylate esters. Fluid Phase Equilibr. **240**, 46–55 (2006)
14. Rowley, R., Wilding, W.V., Oscarson, J.L., Yang, Y., Zundel, N.A., Daubert, T.E., Danner, R.P.: DIPPR information and data evaluation manager for the design institute for physical properties. Version 5.0.2. Design Institute for Physical Properties, AIChE, New York (2011)
15. Dortmund Data Bank. Version 6.3.0.384. DDBST, Oldenburg (2010)
16. Stoll, J., Vrabec, J., Hasse, H.: A set of molecular models for carbon monoxide and halogenated hydrocarbons. J. Chem. Phys. **119**, 11396–11407 (2003)
17. Peng, D.-Y., Robinson, D.B.: A new two–constant equation of state. Ind. Eng. Chem. Fund. **15**, 59–64 (1976)
18. Ohta, T., Asano, H., Nagata, I.: Thermodynamic study of complex–formation in 4 binary–liquid mixtures containing chloroform. Fluid Phase Equilibr. **4**, 105–114 (1980)
19. Vrabec, J., Stoll, J., Hasse, H.: A Set of molecular models for symmetric quadrupolar fluids. J. Phys. Chem. B **105**, 12126–12133 (2001)
20. Windmann, T., Linnemann, M., Vrabec, J.: Fluid phase behavior of Nitrogen + Acetone and Oxygen + Acetone by molecular simulation, experiment and the Peng–Robinson equation of state. J. Chem. Eng. Data **59**, 28–38 (2014)

21. Niethammer, C., Becker, M., Bernreuther, M., Buchholz, M., Eckhardt, W., Heinecke, A., Werth, S., Bungartz, H.-J., Glass, C.W., Hasse, H., Vrabec, J., Horsch, M.: ls1 mardyn: The massively parallel molecular dynamics code for large systems. J. Chem. Theory Comput. **10**, 4455–4464 (2014)
22. Janeček, J.: Long Range Corrections in Inhomogeneous Simulations. J. Phys. Chem. B **110**, 6264–6269 (2006)
23. Irving, J.H., Kirkwood, J.G.: The Statistical Mechanical Theory of Transport Processes. IV. The Equations of Hydrodynamics. J. Chem. Phys. **18**, 817–829 (1950)
24. Neyt, J.-C., Wender, A., Lachet, V., Malfreyt, P.: Prediction of the Temperature Dependence of the Surface Tension Of SO2, N2, O2, and Ar by Monte Carlo Molecular Simulations. J. Phys. Chem. B **115**, 9421–9430 (2011)
25. Mulero, A., Cachadiña, I., Parra, M.I.: Recommended correlations for the surface tension of common fluids. J. Phys. Chem. Ref. Data **41** 043105–1–043105–13 (2012)
26. Blagoi, Y.P., Kireev, V.A., Lobko, M.P., Pashkov, V.V.: Surface tension of Krypton, Methane, Deuteromethane and Oxygen. Ukr. Fiz. Zh. **15**, 427–432 (1970)
27. Ostromoukhov, V.B., Ostronov, M.G.: Surface Tension of Liquid Solutions $O_2 - N_2$ at 54–77 K. Zh. Fiz. Khim. **68**, 39–43 (1994)
28. Baidakov, V.G., Khvostov, K.V., Muratov, G.N.: Surface-tension of Nitrogen, Oxygen and Methane in a wide temperature-range. Zh. Fiz. Khim. **56**, 814–817 (1982)
29. Sun, Y., Shekunov, B.Y.: Surface tension of ethanol in supercritical CO_2. J. Supercrit. Fluid. **27**, 73–83 (2003)
30. Schmidt, R., Wagner, W.: A new form of the equation of state for pure substances and its application to oxygen. Fluid Phase Equilibr. **19**, 175–200 (1985)
31. Span, R., Lemmon, E.W., Jacobsen, R.T., Wagner, W., Yokozeki, A.: A Reference Equation of State for the Thermodynamic Properties of Nitrogen for Temperatures from 63.151 to 1000 K and Pressures to 2200 MPa. J. Phys. Chem. Ref. Data **29**, 1361–1433 (2000)
32. Eckelsbach, S., Miroshnichenko, S., Rutkai, G., Vrabec, J.: Surface tension, large scale thermodynamic data generation and vapor-liquid equilibria of real compounds. In: Nagel, W.E., Kröner, D.H., Resch, M.M. (eds.) High Performance Computing in Science and Engineering'13, pp. 635–646. Springer, Berlin (2013)
33. Lustig, R.: Statistical analogues for fundamental equation of state derivatives. Mol. Phys. **110**, 3041–3052 (2012)
34. Lustig, R.: Microcanonical Monte Carlo simulation of thermodynamic properties. J. Chem. Phys. **109**, 8816–8828 (1998)
35. Eckl, B., Vrabec, J., Hasse, H.: On the application of force fields for predicting a wide variety of properties: Ethylene oxide as an example. Fluid Phase Equilibr. **274**, 16–26 (2008)

Distributed FE Analysis of Multiphase Composites for Linear and Nonlinear Material Behaviour

Andrea Keßler, Kai Schrader, and Carsten Könke

Abstract Modern and efficient adaptive multiscale models can be applied for the prognosis of complex material behavior including deterioration and damage effects of heterogeneous materials. Therewith the physical effects of damage initiation and crack propagation can be captured on the appropriate spatial scales and parameter identification for the different material constituents can be performed much easier than by applying phenomenological material models for composite materials on macroscale. As a consequence of application of these models, the number of degrees of freedom and therewith the necessary computing effort increases substantially. This statement holds for the necessary main memory as well as for the computing power to solve the underlying linear and nonlinear equation systems. The project dcmamc have been investigating partition/substructuring methods and efficient parallel algorithms to solve very large linear and nonlinear equation systems. The algorithms were implemented, tested and adapted to the HPC computing framework at the HLRS Stuttgart, using several hundred CPU nodes. A memory-efficient iterative and parallelized equation solver combined with a special preconditioning technique for solving the underlying equation system was modified and adapted in order to be applied in a mixed CPU and GPU based hardware environment. Additionally a saw-tooth algorithm has been adapted to take into account nonlinear material behavior in a sequential linear manner. Therewith the material nonlinear problem is treated as a sequential solution of purely linear problems, avoiding all drawbacks with convergence problems in classical incremental-iterative solution techniques. In return a substantially increased number of linear solution steps has to be accepted.

A. Keßler • K. Schrader • C. Könke (✉)
Institute of Structural Mechanics, Bauhaus-University Weimar, Marienstrasse 15, D-99423 Weimar, Germany
e-mail: andrea.keszler@uni-weimar.de; kai.schrader@uni-weimar.de; carsten.koenke@uni-weimar.de

W.E. Nagel et al. (eds.), *High Performance Computing in Science and Engineering '14*, DOI 10.1007/978-3-319-10810-0_43

1 Introduction

Modern trends in engineering, such as construction of lightweight structures in civil engineering, automotive or aerospace industry, lifetime extension programs and damage tolerant design concepts, are relying on sophisticated material models, taking into account the initiation and accumulation of damage over the expected lifetime of the structure [2, 8]. Therefore, up-to-date material models in engineering applications are integrating modern approaches from material science via multiscale methods. Especially for heterogeneous materials, these multiscale approaches allow a detailed insight into the material physics on appropriate scales [3, 9].

A major drawback of multiscale methods is the tremendous increase in degrees of freedom (d.o.f.s.) in the resulting equation systems when studying models on meso- or microscale. In case of problems including material nonlinear effects, the numerical effort is increased by the repeated sequential solution of the linearized equation system in an incremental-iterative solution strategy. The key idea of the approach presented in this paper is the application of an iterative distributed solver technique for the solution of the approximated Navier partial equations which are arising from these computational elasticity or inelasticity multiscale problems.

An alternative approach for solving the material nonlinear problem is applied, a sequential saw-tooth softening method [4]. In this approach the nonlinear problem is solved by sequentially applied solutions of the linearized problem and an incremental change in the stiffness matrix in each linear step due to material softening effects.

The equation system is solved in parallel with an iterative equation solver based on the conjugate gradient method (CG), and accelerated by an efficient preconditioning technique. The parallelization technique is based on a standard nonoverlapping domain decomposition method for FE problems combined with the elastic-inleastic domain split and is using the message passing interface standard (MPI) [6]. Performance investigations applying purely CPU based hardware architectures and combined CPUGPU architectures are performed and will be discussed.

Hence, the implementation takes different parallel hardware architectures into account, as well as the application in high performance computing centers. For that reason the developed algorithms for the linearized step have been evaluated considering the hybrid CPU-GPU NEC Nehalem cluster or the new petaflop system CRAY XE6 at the high performance computing center Stuttgart (HLRS).

2 Continuum Damage Material Law by Sequential Saw-Tooth-Softening (STS)

The material nonlinearity is described by a continuum damage approach within a Finite Element Formulation, which is shown in detail in [6]. Here the material degradation due to accumulating damage effects is described by modifying the material equation in each material point. The stress-strain law has to take into account the linear path up to onset of failure and a linear or nonlinear softening branch.

If the stress state in an element described by the J_2 invariant exceeds the flowrule (e.g. von-Mises)

$$0 = \sqrt{\frac{2}{3}J_2} - f_t, \tag{1}$$

with f_t as the tensile strength, the Young's modulus E of this element will be reduced in the following way

$$E_i = \frac{E}{a_i} \, for \, i = 1, 2, \ldots \, N, \tag{2}$$

with a_i as fixed reduction factor of the current stage of the tensile stress-strain softening curve. For each actual step E_i will change from the previous sequential linear analysis step according to

$$E_i = \frac{1}{a} E_{i-1}. \tag{3}$$

Additionally, the tensile strength for the corresponding critical element is also updated during each global load step. Assuming a bilinear stress-strain curve, as e.g. illustrated in Fig. 1, the notation for the update at a critical step i is based on the following criterion considering a linear softening slope: Within the constant scalar tangent modulus D the current tensile strength $f_{t,I}$ is described as the distance of the actual (total) strain state ε_i to the ultimate strain ε_u (representing the smeared crack in the element in final situation). The tensile strength is sequentially reduced to zero, if a smeared crack starts to initiate. By this, it follows

$$f_{t,i} = D(\epsilon_u - \varepsilon_i). \tag{4}$$

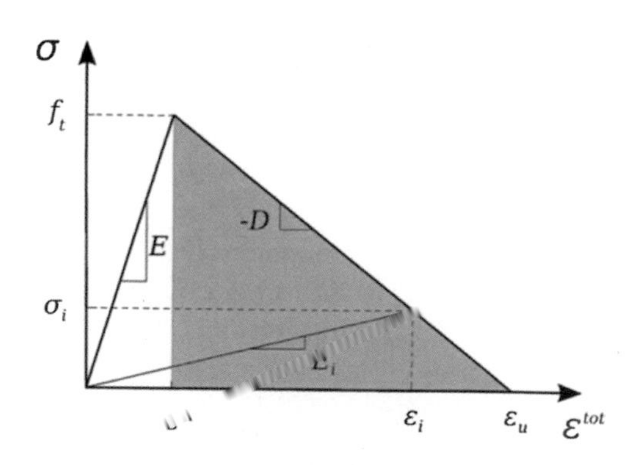

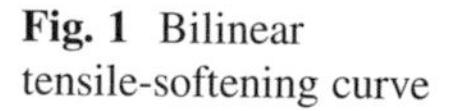
Fig. 1 Bilinear tensile-softening curve

The current total strain ε_i can then be determined considering the actual Young's modulus E_i with

$$\varepsilon_i = \frac{f_{t,i}}{E_i}. \tag{5}$$

Introducing Eq. (5) into Eq. (4) and after some transformation, we obtain

$$f_{t,i} = E_i \varepsilon_u (E_i + D)^{-1} D. \tag{6}$$

For a constant softening modulus, D remains constant during the saw-tooth procedure. Following this assumption a closed form can be presented by modifying Eq. (4). The total inelastic strain is taken as the difference between the ultimate strain and the elastic strain at the peak elastic stress

$$f_{t,i} = D(\epsilon_u - \varepsilon_{el}). \tag{7}$$

For the elastic peak stress $f_{t,I}$ (eq. to the tensile stiffness) it follows,

$$f_{t,i} = E\, \varepsilon_{el}. \tag{8}$$

Introducing Eq. (8) into Eq. (7) we obtain

$$D = f_t(\varepsilon_u - f_t E^{-1})^{-1}. \tag{9}$$

Therewith, Eqs. (6) and (9) deliver the exact result for the update of $f_{t,I}$ and for the scalar softening modulus D (as recommended in [5]). Here, for linear tensile-softening the tensile strength $f_{t,I}$ is updated with

$$f_{t,i} = \varepsilon_u E_i \frac{D}{E_i + D} \quad \text{with} \quad D = \frac{f_t}{\epsilon_u - \frac{f_t}{E}}. \tag{10}$$

Due to loss of ellipticity in the softening branch of the material response, the smeared crack approach needs to be regularized. One possible regularization method is the crack band width, predefining the width of the smeared crack by a variable h. The crack band width h may be set to the corresponding average element size, in 3D, as e.g. proposed by [4]

$$h = (V_i)^{\frac{1}{3}}. \tag{11}$$

Moreover, the design parameter h describes the dimension of the element size, which indicates the crack band width. Therewith and with the fracture energy G_f, the following ultimate strain at final failure is obtained

$$\varepsilon_u = \frac{2G_f}{f_t h}. \tag{12}$$

3 Parallel Iterative Solution Techniques for the Saw-Tooth-Softening (STS) Model: Scalable Sequential Linear Analysis

3.1 Elastic-Inelastic Decomposition

The main effort of a parallelized numerical computation model of a sequential linear analysis results in repeated solving and post-processing procedures, e.g. when extracting and evaluating the global stress and strain tensors to find the most critical element. This can efficiently be executed in parallel for all subdomains. Consequently, the sequential stiffness update of one critical element per iteration step will have a relatively low impact on the computational time compared to the overall computing time required within one SLA step. The elastic-inelastic domain split allows to reduce the numerical effort by applying the SLA only for the inelastic domains of the model, which is considering a load-balanced partitioning.

Thereby, one important task is to communicate the id of the critical element found at the subdomain n of the decomposed damage zone to the initial subdomain m where the FE data of this element are stored. Table 1 shows the algorithm of the scalable SLA technique [6].

Table 1 Algorithm for scalable SLA technique [6]

1. INIT: $\boldsymbol{f}^{(j)} = \boldsymbol{f}$ and $E_{lim} = 10^{-6}$
2. Distributed solve $\boldsymbol{K}^{(j)}\boldsymbol{u}^{(j)} = \boldsymbol{f}^{(j)}$
3. Extract strains and stresses $\boldsymbol{\sigma}^{(j)}$ and $\boldsymbol{\varepsilon}^{(j)}$ per subdomain j
4. Evaluate $\sigma_{cr}^{(j)} = \max f(\boldsymbol{\sigma})^{(j)}$ for all elements of subdomain j
5. IF i=N: SET $E_i^{(k)} = E_{lim}$ and $f_{t,i}^{(k)} = 0.01 f_t$ GOTO 10.
6. ELSE GOTO 7.
7. Get global critical stress $\tilde{\sigma}_{cr} = \max \sigma_{cr,k}^{(j)}$ close to $f_{t,i}^{(k)}$ for element k of subdomain j
8. Reduce $E_i^{(k)} = b_i E_{i-1}^{(k)}$
9. Update tensile strength $f_{t,i}^{(k)} = \varepsilon_u E_i^{(k)} D (E_i^{(k)} + D)^{-1}$ of critical element k
10. Compute element stiffness decrement $\Delta \boldsymbol{K}_i^{(k)}$ of critical element k and current SLA step i
11. Update global stiffness matrix $\boldsymbol{K}^{(j)}$ of subdomain j which includes the critical element k
12. Update global nodal forces $\tilde{\boldsymbol{f}} = \frac{f_{t,i}^{(k)}}{\tilde{\sigma}_{cr}} \boldsymbol{f}^{(j)}$ with $\boldsymbol{f}^{(j)} = \tilde{\boldsymbol{f}}$
13. GOTO 2.

4 Numerical Results

4.1 CRAY XE6 Cluster: 3D Notched Beam

Two three-dimensional notched beam examples are taken from the literature: [1] with numerical results presented in the following and [10] which is considered for further optimizations, were investigated to verify the numerical model in respect to the proposed parallelization and solver technique combined with the saw-tooth softening model. In Fig. 2 the dimensions of the 3D notched beam are given considering a thickness of $t = 160$ mm. The table in Fig. 2 contains the used material properties: E as Young's modulus, ν as Poisson's ratio, f_t as the tensile strength and G_f as the Mode I fracture energy. Considering these parameters, the initial SLA values in respect to the saw-tooth tensile-softening diagram are computed for each subdomain before starting the SLA. All relevant values are given in Table 2. The hybrid decomposition (as illustrated in Fig. 3) takes into account the initial load-balanced METIS partitioning for the distributed solver as well as a separation between potential damage zone corresponding to the inelastic domain in the region around the notch of the beam. The SLA is subsequently only applied in the inelastic region of the problem domain. The remaining region is characterized by linear-elastic material behavior.

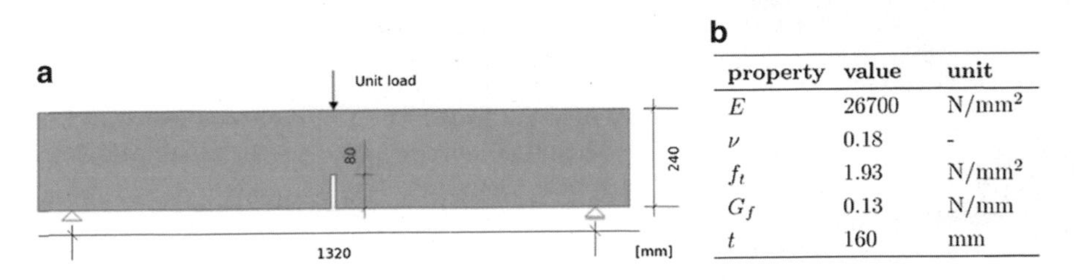

property	value	unit
E	26700	N/mm^2
ν	0.18	-
f_t	1.93	N/mm^2
G_f	0.13	N/mm
t	160	mm

Fig. 2 Geometry of notched beam and material parameters [6]. (**a**) Dimensions of the 3D notched beam. (**b**) Material properties

Table 2 Initial parameters algorithm for scalable SLA technique [6]

SLA parameter	Description	Value	Unit
N	Number of saw-teeth	20	–
a^{-1}	Reduction factor	0.5	N/mm^2
E	Limit Young's modules	0.03	N/mm^2
D	Damage modulus	29.25	N/mm^2
ε_u	Ultimate strain	0.06	–
h	Crack band width/element size	2.04	mm
g_{f,A_1}	STS diagram fracture energy	0.04	N/mm^2
g_{f,A_2}	Specific fracture energy	0.06	N/mm^2
k	Regularization term	1.17	–

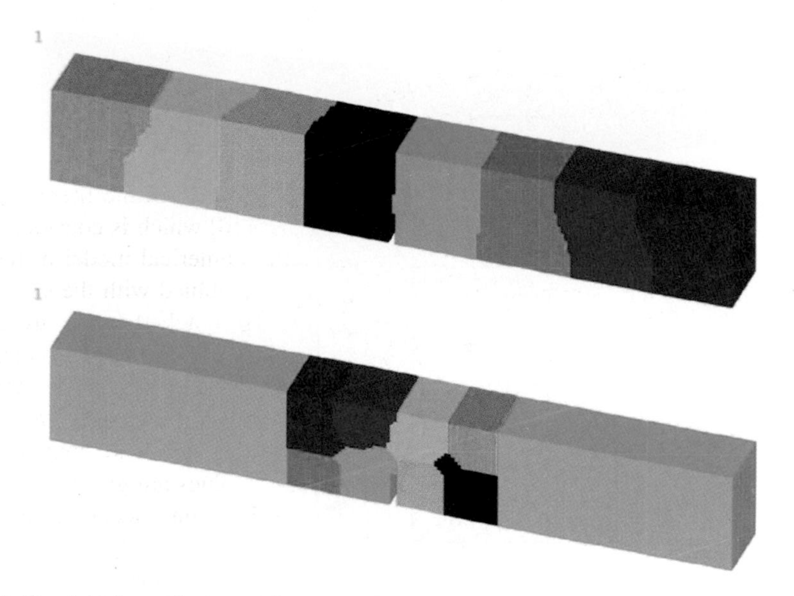

Fig. 3 *Top*: initial partitioning of the notched beam via METIS, *bottom*: hybrid elastic-inelastic domain partitioning with linear regions on the left and right boundary regions and inelastic regions around the central notch [6]

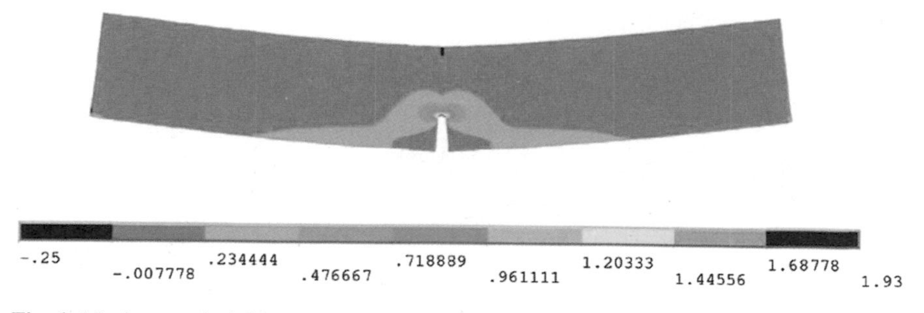

Fig. 4 Maximum principle stress at the first load step [6]

Figure 4 shows the deformed structure and the distribution of the maximum principle stress with a clearly visible hot spot around the notch. As a result of the SLA, the smeared crack is then propagating in a straight line, starting from the notch and propagating towards the applied force load in the beam center. The final crack situation at the end of the simulation after 10,000 SLA cycles is shown in Fig. 5. In this figure all fully degenerated elements reaching the ultimate limit strain ε_u have been removed from the visualization.

Moreover Fig. 6 shows the applied saw-tooth diagram describing the linear softening post-peak slope (left) and the load-displacement curve extracted during the SLA cycles with a corresponding control point in the middle of the beam (right).

Fig. 5 Final crack situation at the end of the simulation after 10,000 SLA cycles [6]

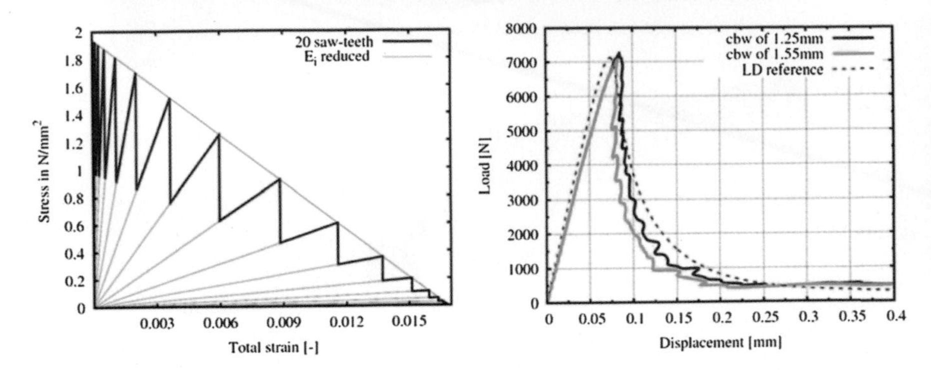

Fig. 6 *Left*: saw-tooth diagram; *right*: load-displacement curve with softening for two different crack band width parameter h [6, 7]

Table 3 Overall computing time for analysis of notched beam problem [6]

Hardware	CRAY XE6 cluster
CPU architecture	Multi-core AMD Interlagos
ccNUMA nodes	16
Number of d.o.f.s	390,000
Average SLA time	1.4 s
Number of SLA cycles	10,000
Overall (elapsed) SLA time	4 h 13 min

Some computational aspects regarding the numerical simulation considering the CRAY XE6 cluster are given in Table 3.

The load displacement curve in Fig. 6 demonstrates that there is an influence of the chosen parameter for the crack band width h on the response and that with decreasing parameter h the result approaches toward the reference solution, given by an incremental iterative load-displacement method.

5 Conclusion

Applying sequential linear solution steps and stepwise reduction of stiffness due to accumulating damage is an alternative approach which allows studying the damage evolution and crack propagation in models with large numbers of degrees of freedom. The hybrid domain decomposition technique, which separates betwe linear elastic and inelastic regions, allows to restrict the application of the SLA to those regions, in which a material nonlinear response is expected. In the presented example of a notched beam problem, the definition of the linear-elastic and nonlinear regions has been done a priori, but can be extended to an adaptive definition of elastic and inelastic regions associated to the load and damage process. Therewith the numerical effort for the SLA can be even more reduced to those local regions in which damage accumulation is taking place.

Acknowledgements This research has been supported by the German Research Foundation (DFG), which is gratefully acknowledged by the authors.

References

1. Hannawald, J.: Fracture of hollow clay units simulated by sequentially linear analysis. In: Brameshuber, W. (eds.) Modeling of Heterogeneous Materials (HetMat). Proceedings of the International RILEM Conference on Material Science (MatSci). RILEM Publications S.A.R.L. II (2010)
2. Koenke, C., Eckardt, S., Haefner, S., Luther, T., Unger, J.F.: Multiscale simulation methods in damage prediction of brittle and ductile materials. Int. J. Multiscale Eng. **8**, 17–36 (2010)
3. Raabe, D.: Continuum Scale Simulation of Engineering Materials: Fundamentals, Microstructures, Process Applications. Wiley-VCH, Weinheim (2004)
4. Rots, J.G., Invernizzi, S.: Regularized sequentially linear saw-tooth softening model. J. Numer. Anal. Methods Geomech. **28**, 821–856 (2004)
5. Rots, J.G., Belletti, B., Invernizzi, S.: Event-by-event strategies for modelling concrete structures. In: Meschke, G., de Borst, R., Mang, H., Bicanic, N. (eds.) Computational Modelling of Concrete Structures. Proceedings of EURO-C 2006, pp. 667–677. Taylor and Francis Group, London (2006)
6. Schrader, K.: Hybrid 3D simulation methods for the damage analysis of multiphase composites. Dissertation am Institut für Strukturmechanik der Bauhaus-Universität Weimar (2012)
7. Schrader, K., Könke, C.: Hybrid computing for large-scale heterogeneous 3D microstructures. Int. J. Multiscale Eng. **9**(4), 365–377 (2011)
8. Schrader, K., Könke, C.: Distributed computing for the nonlinear analysis of multiphase composites. Adv. Eng. Softw. **62–63**, 20–32 (2013)
9. Tadmor, E.B., Miller, R.E.: Modeling Materials: Continuum, Atomistic and Multiscale Techniques. Cambridge University Press, Cambridge (2011)
10. Voormeeren, L.: Appendix of the master thesis: SLA for notched beams in 2D and 3D. Section of Structural Mechanics, Faculty of Civil Enginnering, Delft University of Technology, Delft (2011)

Parallel Performance of a Discontinuous Galerkin Spectral Element Method Based PIC-DSMC Solver

P. Ortwein, T. Binder, S. Copplestone, A. Mirza, P. Nizenkov, M. Pfeiffer, T. Stindl, S. Fasoulas, and C.-D. Munz

Abstract Particle based methods are required to simulate rarefied, reactive plasma flows. A combined Particle-in-Cell Direct Simulation Monte Carlo method is used here, allowing the modelling of electromagnetic interactions and collision processes. The electromagnetic field solver of the Particle-in-Cell method has been improved by switching to a discontinuous Galerkin spectral element method. The method offers a high parallelization efficiency, which is demonstrated in this paper. In addition, the parallel performances of the complete Particle-in-Cell module and the Direct Simulation Monte Carlo module are presented.

1 Introduction

The simulation of reactive plasma flows, as encountered in planetary re-entry and electric space propulsion, pushes computational simulation efforts to the limit. The corresponding rarefied flow is characterized by non-continuum (high Knudsen numbers) and non-equilibrium conditions, to which the approach described by the Navier-Stokes or magnetohydrodynamic equations is not applicable anymore. Consequently, a more fundamental approach is required, which includes modelling the Boltzmann equation and solving Maxwell's equations. The Boltzmann equation is given by

$$\left(\frac{\partial}{\partial t} + \mathbf{v}\frac{\partial}{\partial \mathbf{x}} + \frac{\mathbf{F}}{m}\frac{\partial}{\partial \mathbf{v}}\right) f = \left(\frac{\partial f}{\partial t}\right)_{\text{coll}}, \tag{1}$$

P. Ortwein • S. Copplestone • C.-D. Munz
Institute of Aerodynamics and Gas Dynamics (IAG), University of Stuttgart, 70569 Stuttgart, Germany
e-mail: munz@iag.uni-stuttgart.de

T. Binder • A. Mirza • P. Nizenkov • M. Pfeiffer (✉) • T. Stindl • S. Fasoulas
Institute of Space Systems (IRS), University of Stuttgart, 70569 Stuttgart, Germany
e-mail: fasoulas@iag.uni-stuttgart.de

W.E. Nagel et al. (eds.), *High Performance Computing in Science and Engineering '14*, DOI 10.1007/978-3-319-10810-0_44

where $f = f(\mathbf{x}, \mathbf{v}, t)$ is the particle distribution function at a point $(\mathbf{x}, \mathbf{v})$ in six-dimensional phase space at the time t. The right-hand side $(\partial f/\partial t)_{\mathrm{coll}}$ describes intermolecular collisions between particles and is often referred to as the Boltzmann collision integral. Since no analytical solution of the Boltzmann equation exists (short of simple applications), particle methods such as Particle-in-Cell (PIC [7]) and Direct Simulation Monte Carlo (DSMC [2]) are used to find an approximate solution. These methods employ a certain number of particles N_p to approximate the distribution function as

$$f(\mathbf{x}, \mathbf{v}, t) = \sum_{k=1}^{N_p} w_k \delta(\mathbf{x} - \mathbf{x}_k(t))\delta(\mathbf{v} - \mathbf{v}_k(t)),$$

where a particle k represents a number of real molecules w_k with the position $\mathbf{x}_k$ and the velocity $\mathbf{v}_k$ at the time t. The particle weight factor w_k enables the simulation of a large number of particles without excessive computational efforts.

The combined PIC-DSMC code "PICLas" has been developed at IAG (Institute of Aerodynamics and Gas Dynamics) and IRS (Institute of Space Systems) in recent years and allows the simulation of both electromagnetic forces and collisional interactions in plasma flows. An overview of the PIC and DSMC methods is given in Sects. 2 and 3, respectively. Since the last report [10], which focused on the parallel DSMC implementation, a new field solver of the PIC module was developed. An update regarding the scalability of the coupled code is given in Sect. 3, in addition to simulations with the new field solver. Finally, a brief conclusion and outlook are given in Sect. 8.

2 Particle-in-Cell

A schematic of the different computations that comprise a time step of the explicit PIC solver is shown in Fig. 1. After the particles are located with respect to the grid, the charge and current densities they generate are interpolated, i.e., deposited, onto the computational grid using methods of varying order and computational costs [11]. The densities on the grid are then source terms for the Maxwell solver, which computes the electric and magnetic fields on the grid and is described in detail below. The fields are subsequently interpolated from the grid to the particle positions. The Lorentz force acting on the particles due to these fields is computed by

$$\mathbf{F}_L = Q(\mathbf{E} + \mathbf{v} \times \mathbf{B}), \tag{2}$$

where Q, $\mathbf{E}$ and $\mathbf{B}$ are the particle charge, the electric and the magnetic field, respectively. The particles are then moved according to the relativistic equations of motion

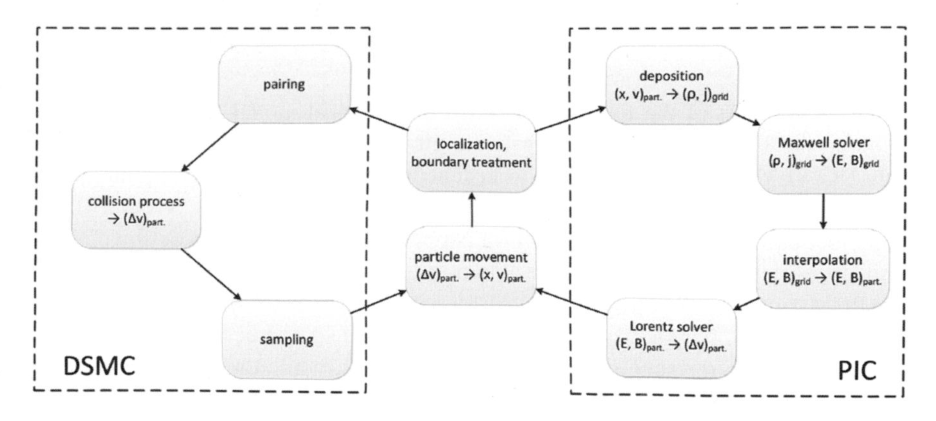

Fig. 1 Flow chart of PIC-DSMC simulation procedures performed during each time step

$$\frac{\mathrm{d}\mathbf{x}}{\mathrm{d}t} = \mathbf{v}, \qquad \frac{\mathrm{d}\mathbf{p}}{\mathrm{d}t} = \mathbf{F}, \tag{3}$$

with the relativistic momentum $\mathbf{p} = m_0\gamma\mathbf{v}$, the particle mass m_0, and the Lorentz factor $\gamma = (1 - |\mathbf{v}|^2/c^2)^{-1/2}$. In the case of PIC, $\mathbf{F}$ is comprised of only the Lorentz force $\mathbf{F}_L$. In addition to these computed fields, external fields can be defined by data file or homogeneously distributed functions in the domain. Time integration is done via an explicit fourth order low storage Runge-Kutta scheme [4].

To compute the electric and magnetic fields, Maxwell's equations are extended by Lagrangian multipliers Ψ and Φ for divergence correction purposes and the Purely Hyperbolic Maxwell (PHM) system [9]

$$\frac{\partial \mathbf{E}}{\partial t} = c^2\,\nabla \times \mathbf{B} - \chi\, c^2\,\nabla\Psi - \frac{\mathbf{j}}{\varepsilon_0} \tag{4}$$

$$\frac{\partial \mathbf{B}}{\partial t} = -\,\nabla \times \mathbf{E} - \chi\,\nabla\Phi \tag{5}$$

$$\frac{\partial \Phi}{\partial t} = -\,\chi\, c^2\,\nabla \cdot \mathbf{B} \tag{6}$$

$$\frac{\partial \Psi}{\partial t} = -\,\chi\,\nabla \cdot \mathbf{E} + \chi\,\frac{\varrho}{\varepsilon_0} \tag{7}$$

is derived, with χc being the velocity of the Lagrange multiplier propagation by which divergence errors are transported out of the domain. The PHM system is solved using a Discontinuous Galerkin – Spectral Element Method (DG-SEM), which will be presented in detail in the following section.

2.1 Field Solver

Previous versions of the field solver relied on a hybrid modal-nodal DG method to solve the PHM system, allowing for arbitrary element shapes. In contrast, the new DG-SEM version is restricted to hexahedral elements, but is computationally more efficient and characterized by low dissipation and dispersion errors [5]. The implementation strategy of DG-SEM for Navier-Stokes equations is outlined in [6]. This approach is slightly altered to obtain optimum parallel performance to solve the PHM system. To derive the DG-SEM scheme, the PHM system is written in flux formulation. For strictly hyperbolic systems, the spatial derivatives of the flux vectors switch to the state vector $\mathbf{U} = (\mathbf{E}, \mathbf{B}, \Phi, \Psi)^\intercal$, while the matrix $\mathbb{K}$ contains the constant flux coefficients of (4)–(7)

$$\frac{\partial \mathbf{U}}{\partial t} + \sum_{\mathrm{d}=1,3} \mathbb{K}_d \frac{\partial \mathbf{U}}{\partial \mathbf{x}_\mathrm{d}} = \mathbf{S}. \tag{8}$$

Here, $\mathbf{S}$ is the source vector representing the charge and current densities of the particles. Equation (8) is mapped into a unit reference space, introducing the Jacobian J of the transformation. Next, the equation is multiplied by a test function φ. The product is integrated over the reference element $\mathscr{E}$. An integration by part results in a shift of the spatial derivative from the state vector to the test function

$$\int_{\mathscr{E}} J \frac{\partial \mathbf{U}}{\partial t} \varphi d\xi + \sum_{\mathrm{d}=1,3} \int_{\mathscr{E}} \mathscr{K}_d \mathbf{U} \frac{\partial \varphi}{\partial \xi_\mathrm{d}} d\xi_\mathrm{d} - \int_{\partial \mathscr{E}} (\mathscr{F}\mathbf{n}) \varphi dS = \int_{\mathscr{E}} J \mathbf{S} \varphi d\xi. \tag{9}$$

The test function is a polynomial basis, which is constructed using a tensor product basis consisting of one-dimensional Legendre polynomials. Gaussian quadrature is applied to numerically solve the integrals over the reference space. The choice between Legendre-Gauss or Legendre-Gauss-Lobatto points is based on the following attributes: Gauss-Lobatto points are characterized by points located in the interior and at the element surface. The integration accuracy is $2N - 1$, where N is the polynomial order of the one-dimensional basis function. In contrast, Gauss points are only located inside of the volume. This requires an extrapolation of the interior states to the element interfaces to compute the surface integral. However, the resulting scheme is of higher integration accuracy, namely $2N + 1$. Eventually, the integration points are used as interpolation points and the test and interpolation functions are identical. Choosing a set of integration points alters the constructed numerical scheme. In this paper, the parallel performance of both methods for solving a strictly hyperbolic system is investigated.

3 Direct Simulation Monte Carlo

The Direct Simulation Monte Carlo method (DSMC), which was first introduced by Bird [2], is a highly efficient method for the simulation of rarefied, non-equilibrium, neutral gas flows. In order to approximate the full non-linear Boltzmann collision integral for binary collisions of mono- and diatomic gases, the gas flow is directly simulated by particles, which represent a much larger number of real gas particles. Figure 1 contains the scheme of a DSMC time step. In contrast to the PIC, the calculation of the particle velocity does not depend on an external force, so that the velocities of particles are computed in each collision directly; otherwise they are kept constant. Instead of solving the relativistic equations of motion (3), the particles are moved within the computational domain according to

$$\mathbf{x}_{n+1} = \mathbf{x}_n + \mathbf{v}_n \Delta t.$$

These particles can encounter boundaries of the flow field, where they leave the computational domain in the case of an open boundary or they will be reflected, if the boundary represents a reflecting wall. The reflection can be treated as diffuse and fully specular. Afterwards, the particles are localized in the grid cells, where they have the possibility to collide among each other. The particles are paired and the collision probability of each pair is calculated on the basis of Bird's NTC scheme [2], which is modified by the natural sample size method introduced by Baganoff and McDonald [1]. Here, the collision probability of two particles a and b is given by

$$P_{ab} = \frac{N_{p,a} N_{p,b}}{1 + \delta_{ab}} w_k \frac{\Delta t}{V_c S_{ab}} (\sigma_{ab} g_{ab}), \tag{10}$$

where N_p denotes the number of simulated particles of the species, V_c the cell volume, σ the collision cross section, g the relative velocity of the collision pair and δ_{ab} the Kronecker delta. The decision on whether a collision occurs depends on a random number R. If R is lower than the corresponding probability, the particles collide and energy and momentum exchanges and optionally relaxation processes of internal degrees of freedom as well as chemical reactions can occur. After particle interactions are treated, macroscopic values of the flow field such as temperature, density or flow velocity are determined by sampling the microscopic particle properties (velocity, location, inner energies) in each cell.

4 Investigation of Parallel Performance

The parallelization of the PIC-DSMC code is based on a domain decomposition to apply a message passing interface (MPI) standard. Therefore particle data as well as numerical flux data have to be sent through the MPI domain borders to the relevant neighbouring MPI domains. In this section, the parallel performance of the code is shown. The performance of the DSMC module is investigated and compared to last year's report[10]. The performance of the new field solver employing the DG-SEM method is analyzed with respect to different integration points. Finally, the performance of the complete PIC module, i.e., including particles, is shown. All parallel tests were conducted on the CRAY XE6 (Hermit) system of the HLRS (High Performance Computing Center Stuttgart).

4.1 *Direct Simulation Monte Carlo Solver*

In the last HRLS report, the scaling performance of only the DSMC solver was studied. For this, the three dimensional simulation of a re-entry with 336,268 cells was used as a test case. It was shown that the limiting factor for the speedup with high numbers of MPI processes was the communications in the particle emission routine. The particle emission routine uses one or all available processors for the particle emission depending on the number of particles to be emitted per process. The old version of this routine, which was used in the simulations of the last HLRS report, does not make a distinction according to the used algorithm, depending on the number of emitting processes. This means that the emission of the particles using one or all processes was performed in the same way. In the new method, the emission by using only one process was optimized. Therefore, the scaling of simulations with a high number of processes but relative small number of emitted particles is improved. Figure 2 compares the scaling result of the re-entry simulation

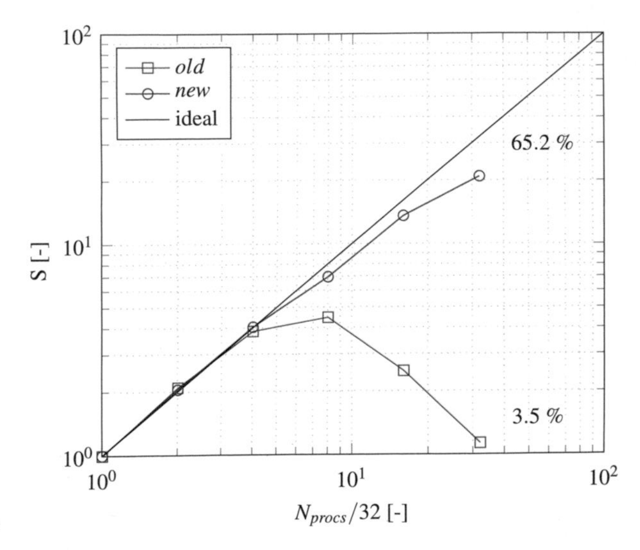

Fig. 2 Comparison of old and new scaling performance

using the new and the old particle emission routine. The new routine scales better and it is also shown that further improvements beside the particle emission are necessary.

4.2 *Qualification of Gauss and Gauss-Lobatto Integration Points for Parallel Efficiency*

The parallel performance of the DG-SEM solver is investigated by simulating an electric dipole, which is incorporated into Maxwell's equations via a source term that is handled with the same shape functions [11] as used for charged particles. A cubical domain is considered that consists of 1,331 equally distributed unstructured hexahedral elements, each containing $(N+1)^3$ degrees of freedom (DOF) for the DG-SEM scheme and is decomposed using space filling curves. The speedup is calculated by

$$S = 32\frac{t_{32}}{t}, \tag{11}$$

where t_{32} is the reference time for a simulation run on a single compute node consisting of 32 processors. The scaling capability is depicted in Fig. 3, where the number of processors is increased until reaching the limit of one processor per element. Strong scaling sets in for higher polynomial degrees, as can be seen for $N = 7$, where caching effects lead to a speedup of over 100 %. Two types of integration points are concerned, Gauss and Gauss-Lobatto, where the latter shows a slightly better scalability. The influence of the boundaries, however, has more

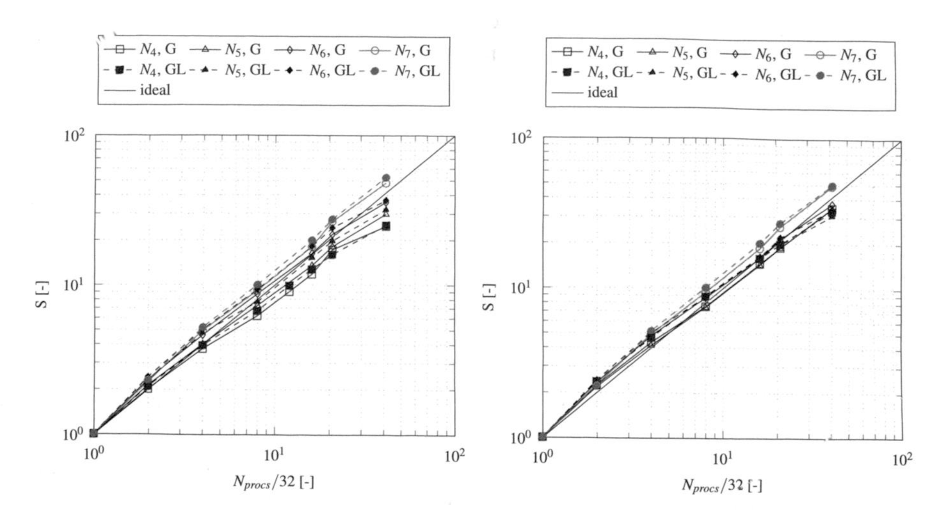

Fig. 3 Scaling performance for open (*left*) and periodic (*right*) boundaries. Comparison of different polynomial degrees N and two integration points, Gauss (G) and Gauss-Lobatto (GL)

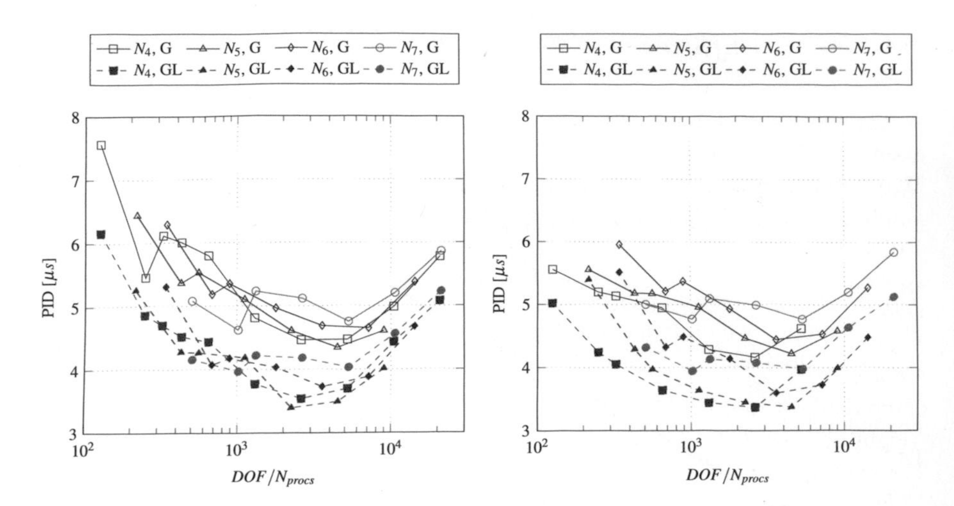

Fig. 4 PID for open (*left*) and periodic (*right*) boundaries. Comparison of different polynomial degrees N and two integration points, Gauss (G) and Gauss-Lobatto (GL)

impact, where the choice of periodic boundaries leads to a bunching of the lower polynomial degrees that scale only weakly.

The computational load on each processor can be compared using the performance index (PID) given by

$$PID = \frac{t_{wall} N_{procs}}{N_{steps} N_{DOF}}, \tag{12}$$

which is calculated by the total wall-clock time t_{wall}, the number of processors N_{procs}, the number of time steps N_{steps} and the number of degrees of freedom N_{DOF}, respectively. It measures the time that is needed to update the state at one DOF in the domain. The PID for each setup is compared in Fig. 4 and it highly depends on the number of DOF that each processors has to handle. The PID is reduced by increasing the number of processors for a given problem size due to caching effects, until a minimum is reached in the region of a few thousand DOF per processor. Then, the PID increases again for lower polynomial degrees when the number of cells per processor reduces to one, which is due to the increasing MPI communications overhead. However, for higher polynomial degrees, the PID remains nearly constant, which leads to the super-scaling characteristic. When using Gauss-Lobatto points, a lower PID level
can be observed for both open and periodic boundaries, which is due to the absent face extrapolation that is needed when using Gauss points to calculate the surface fluxes. In case of periodic boundaries, the PID distribution is less bunched and the rise of the PID for decreasing processor load is less steep than for open boundaries, which improves the scaling ability of lower polynomial degrees, as mentioned above.

4.3 Parallel Efficiency of the Particle-in-Cell Solver

A common validation of the PIC solver is the simulation of a plasma frequency [3], [8]. A periodic domain is simulated with the size of $[\pi; \frac{\pi}{2}; \frac{\pi}{2}]$ meter. Two different polynomial degrees are investigated: $N = 4$ and $N = 9$. The overall number of DOFs is kept constant, i.e., the number of elements is altered. Here, 1,024 elements are used for the higher polynomial degree, limiting the number of applied processes to 1,024. Based on the results of the previous section, Legendre-Gauss points are used due to their higher integration accuracy. The plasma frequency setup is built by initially distributing electrons sinusoidally along the x-axis. Protons are used as a neutralization background and are distributed homogeneously in the computational domain. A total of 65,536 particles are simulated with a particle weight of $w = 10^{10}$. A polynomial shape function is used for deposition, cf. [8]. The parameters are set to $\alpha = 4$ and $r_{\text{cut}} = 0.098$.

Figure 5 depicts the results of the strong scaling analysis. The scaling is similar for both computations, indicating that the parallel performance depends on the number of grid DOFs and particles of each processor. The PIC solver does not depict a super-linear scaling. The corresponding efficiency is plotted in the right-hand graph of Fig. 5. It continually drops for both polynomial degrees for up to 1,024 cores but remains above 80 %. A higher polynomial degree results in a slightly more efficient parallelization.

A problem of the PIC simulation is the particle communications overhead. Decreasing the size of the load per core by increasing the number of processes increases this communications overhead due to increased domain borders and the deposition volume of a particle stretching over multiple MPI domains. Additionally, while the simulation of one element per core is reasonable for the pure DG-SEM solver, a particle distribution is neither steady nor homogeneous throughout a simulation. Therefore, a parallelization down to one element per process cannot

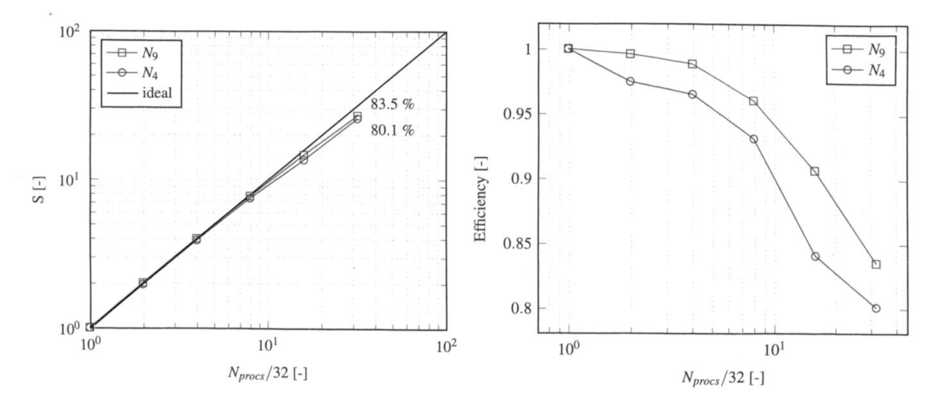

Fig. 5 Strong scaling (*left*) and parallel efficiency (*right*) for a polynomial degree of $N = 4$ and $N = 9$

possibly be 100 % efficient. Simulating at least several elements per core allows a better load balancing between different cores. Assuming each core is loaded with at least four elements, the found parallel efficiency is over 95 % for the higher polynomial degree. This demonstrates that the PIC solver benefits from the parallel performance of high-order DG-SEM schemes and generally scales reasonably well for sufficiently large computational cases.

5 Summary and Conclusions

A PIC-DSMC code has been investigated regarding parallel performance for strong scaling cases. The DSMC module shows a significant efficiency increase in comparison to the last report, while still not reaching linear scaling due to the particle insertion, which inevitably has higher communications overhead costs for higher number of processes. The electromagnetic field solver of the PIC module has been replaced by a DG-SEM based implementation and shows excellent parallel performance down to a single element per process. Scaling and efficiency using Gauss-Lobatto integration points have been compared to Gauss integration points with different polynomial degrees with overall similar results. In general, higher order polynomials are better suited for increased parallel performance. Finally, the complete PIC module, including field solver and particle treatment, has been investigated. The main result is that while parallel performance, in general, proved acceptable, a minimum amount of elements per process has to be reached in order to be able to use load balancing to efficiently treat inhomogeneous particle distributions.

Acknowledgements We gratefully acknowledge the Deutsche Forschungsgemeinschaft (DFG) for funding within the project "Kinetic Algorithms for the Maxwell-Boltzmann System and the Simulation of Magnetospheric Propulsion Systems". T. Stindl thanks the Landesgraduiertenförderung Baden-Württemberg and the Erich-Becker-Stiftung, Germany, for their previous financial support. Computational resources have been provided by the Bundes-Höchstleistungsrechenzentrum Stuttgart (HLRS).

References

1. Baganoff, D., McDonald, J.D.: A collision selection rule for a particle simulation method suited to vector computers. Phys. Fluids A **2**, 1248–1259 (1990)
2. Bird, G.: Molecular Gas Dynamics and the Direct Simulation of Gas Flows. Oxford University Press, Oxford (1994)
3. Birdsall, C.K., Langdon, A.B.: Plasma Physics via Computer Simulation. Hilger, Bristol (1991)
4. Carpenter, M.H., Kennedy, C.A.: Fourth-order 2N-storage Runge-Kutta schemes. NASA Technical Memorandum **109112**, 1–26 (1994)
5. Gassner, G., Kopriva, D.A.: A comparison of the dispersion and dissipation errors of Gauss and Gauss-Lobatto discontinuous Galerkin spectral element methods. SIAM J. Sci. Comput. **33**(5), 2560–2579 (2011). doi:10.1137/100807211

6. Hindenlang, F., Gassner, G.J., Altmann, C., Beck, A., Staudenmaier, M., Munz, C.D.: Explici discontinuous Galerkin methods for unsteady problems. Comput. Fluids **61**, 69–93 (2012)
7. Hockney, R.W., Eastwood, J.W.: Computer Simulation Using Particles. McGraw-Hill, Inc., New York (1988)
8. Jacobs, G.B., Hesthaven, J.: High-order nodal discontinuous Galerkin particle-in-cell method on unstructured grids. J. Comput. Phys. **214**, 96–121 (2006)
9. Munz, C.D., Schneider, R., Voß, U.: A finite-volume particle-in-cell method for the numerical simulation of devices in pulsed power technology. Surv. Math. Ind. **8**, 243–257 (1999)
10. Pfeiffer, M., Mirza, A., Stindl, T., Ortwein, P., Fasoulas, S., Munz, C.D.: Investigation of a parallel direct simulation monte carlo implementation. In: Poster at 16th HLRS Results and Review Workshop, Stuttgart (2013)
11. Stindl, T., Neudorfer, J., Stock, A., Auweter-Kurtz, M., Munz, C., Roller, S., Schneider, R.: Comparison of coupling techniques in a high-order discontinuous Galerkin based particle in cell solver. J. Phys. D: Appl. Phys. **44**, 194,004 (2011)

The VasCan Project: Simulation of Vascular Tumour Growth and Angiogenesis Using a Multiscale Model

Bernd-Simon Dengel, Holger Perfahl, Simona Galliani, and Matthias Reuss

Abstract In the VasCan project we use a multiscale model [1, 2] to simulate angiogenesis and vascular tumour growth. The model is based on a cellular automaton approach and couples intracellular processes, active cell movement, cell-cell interaction, extracellular diffusion, and a dynamically evolving vascular network. In order to obtain a better understanding of both our model parameters and the interaction between the model modules, we conducted a sensitivity analysis. The simulation studies are controlled via a MatLab/Java programme that generates parameter files, starts the simulations on the HLRS and finally organises the transfer and storage of the simulation results. The visualisation of the large data sets was also automated. The goal is to improve our model understanding and the predictive potential for future tumour growth and therapy simulations.

1 Introduction

Vascular tumour growth can be studied with different modelling methodologies according to the scientific questions that should be answered. The mathematical models range from systems of ordinary differential equations, to partial differential equations and agent-based approaches. While ordinary and partial differential equation models deal with spatially averaged quantities like tumour size and volume fractions, agent-based models can discretely resolve a cell population. The behaviour of the cells (which can also be called "agents") can be given by equations

B.-S. Dengel (✉) • H. Perfahl
Stuttgart Research Center Systems Biology (SRCSB), University of Stuttgart, Stuttgart, Germany

Institute of Biochemical Engineering (IBVT), University of Stuttgart, Stuttgart, Germany

S. Galliani
Stuttgart Research Center Systems Biology (SRCSB), University of Stuttgart, Stuttgart, Germany

Department of Hydromechanics and Modelling of Hydrosystem (IWS), University of Stuttgart, Stuttgart, Germany

M. Reuss
Stuttgart Research Center Systems Biology (SRCSB), University of Stuttgart, Stuttgart, Germany

W.E. Nagel et al. (eds.), *High Performance Computing in Science and Engineering '14*, DOI 10.1007/978-3-319-10810-0_45

and rules. Individual-based models usually utilise Monte-Carlo approaches to reproduce the stochastic nature of biological processes.

Agent-based angiogenesis and vascular tumour growth models provide a detailed insight into the evolution of a growing tumour. Stochastic and deterministic processes on various temporal and spatial scales are included and coupled. Intracellular models describe the progression through the cell cycle, metabolic or signalling pathways. These processes are usually influenced by extracellular factors like nutrients, growth factors, drugs, mechanical stress, etc. Cells are able to move in the simulation domain and interact with other cells. Growth factors stimulate the vascular system to build new sprouts that migrate through the simulation domain by a biased random walk up to growth-factor gradients. Vascular sprouts can anastomose to other sprouts or the already existing vascular network and build new perfused vessels. Within the vascular network, pressures and flows are calculated, and the radii of the vessel segments evolve due to growth rules. Such a model leads to a detailed insight in the interaction between the growing tumour and the induced vascular response.

In contrast to parallel implementations of e.g. CFD- or FEM-programmes- where one programme is run on several processors–there is no parallel implementation of our multiscale simulation framework. Therefore, the HLRS was used to perform parameter studies and generate a large number of stochastic realisations.

Our Monte-Carlo model has about 50 parameters and one simulation takes 2–3 h on a normal desktop PC.[1] The stochastic nature of our model together with the large number of parameters results in the following problems:

- **Large number of simulations for statistics.**
  In order to compare simulation results for different parameter sets we need a large number of realisations with different random seeds (for the initialisation of our random number generator) to draw statistically significant conclusions.
- **Large number of parameters.**
  The number of parameters and potential combinations of them result in a large parameter space. In this study we started by always varying one parameter and analysing the influence on the tumour structure.

Each simulation on the HLRS took 1–2 h and we usually run 50 independent simulations in parallel, the generated output data-size of those simulations was ~100 GB.

Before being able to use the HLRS efficiently we had to do the following changes:

- **Adaptation of our multiscale simulation framework to the CrayXE6 platform.**

[1] Intel Core i7 930@2.80 GHz

The original make-files could not be used and had to be adapted to the CrayXE6 environment. The latest implementation of our multiscale programme provides the possibility of compile our framework on the CrayXE6 platform.
- **Implementation of a simulation control framework.**
  To handle the large number of simulations efficiently a simulation control framework was implemented in MatLab/Java. The control framework automates parameter variation, automatically runs simulations and downloads and organises the simulation results.
- **Implementation of an automated visualisation framework.**
  To efficiently analyse the simulation data, a framework was implemented in which the visualisation of simulation results could be highly automated.

The following chapter gives an overview of the mathematical model and the implementation. First, the mathematical multi-scale model is described with the different included temporal and spatial scales. After the specification of the vascular tumour growth model and the underlying computational algorithm, a description of the multi-parallel simulation control framework that organises and controls the simulations on thc HLRS is given. In the results section we present first simulations outcomes. Finally, we give a conclusion and an outlook on future model extensions.

2 Multiscale Model and Computational Algorithm

The following section gives an overview of the computational model structure. The multiscale model considers four different layers: the sub-cellular, cellular, diffusible and vascular layer that describes the process of angiogenesis and vascular tumour growth. Those layers interact with each other according to predefined rules. Figure 1 shows a schematic representation of those interactions.

The sub-cellular level is implemented using a system of ordinary differential equations and describes biological processes like VEGF[2] and p53[3] production and the cell-cycle progression. On the cellular level, a cellular automaton model is used to describe cell-cell interactions and cell movement. A biased random walk is applied to describe the active movement of vessel sprouts. In the diffusible layer quasi-steady partial differential equations are used to compute the distribution of oxygen and VEGF. Finally, a flow calculation for the vascular system is executed. The vascular system itself evolves in time according to intra- and extravascular stimuli. Figure 2 illustrates the computational algorithm. The columns present the characteristic length scales while the vertical axis indicates the progression in time.

[2]Vascular Endothelial Growth Factor

[3]p53 protein, tumour suppressor gene

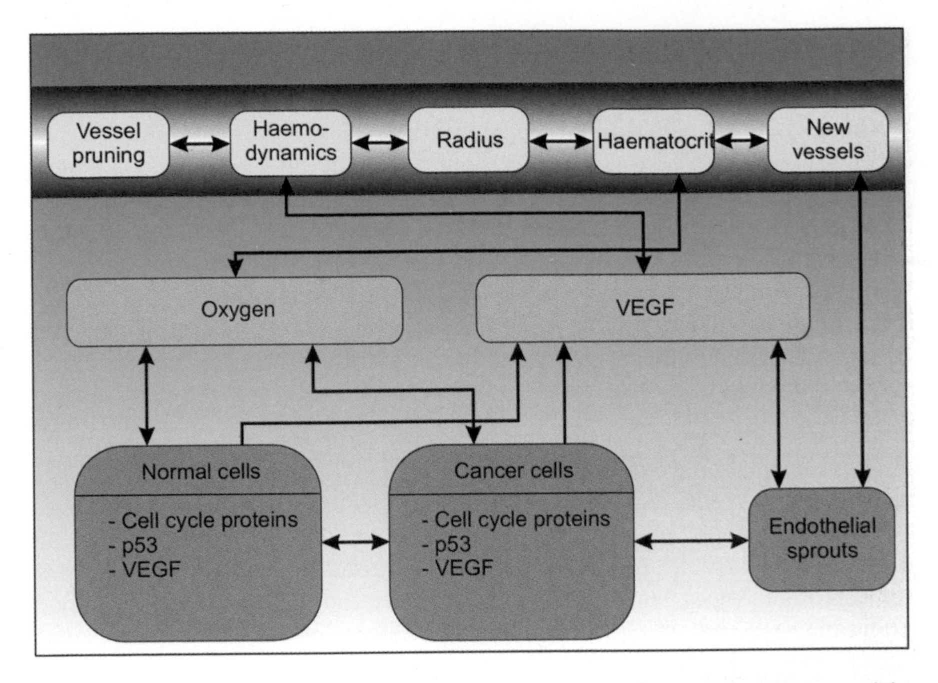

Fig. 1 Interaction diagramme. This figure shows the interaction between the different modules of the multiscale model. Normal and cancer cells with their intra-cellular cell-cycle and p53/VEGF models interact with each other. Both cell types consume oxygen and start to produce VEGF under hypoxic conditions. VEGF stimulates the vascular network to build new endothelial sprouts that move in the simulation domain by a biased random-walk. In case that endothelial sprouts connect to other sprouts or to the existing vascular network, new perfused vessels are generated. The vascular system delivers oxygen and removes VEGF from our simulation domain. Vessels adapt their radius according to several stimuli and underperfused vessels are removed from the vascular system

The simulation program is implemented in C/C++. Ordinary differential equations are solved with the CVODE[4] library. To solve the linear systems of equations for the flow calculation and the discretised partial differential equations we used a SuperLU solver[5] (see [3]). The parameters are set in the simulation framework by a parameter file that defines the simulation. Therefore, we can use the same executable to run all simulations without recompiling the code.

A detailed description of the mathematical model can be found in [1, 2].

[4] http://computation.llnl.gov/casc/sundials/description/description.html

[5] http://crd-legacy.lbl.gov/~xiaoye/SuperLU/

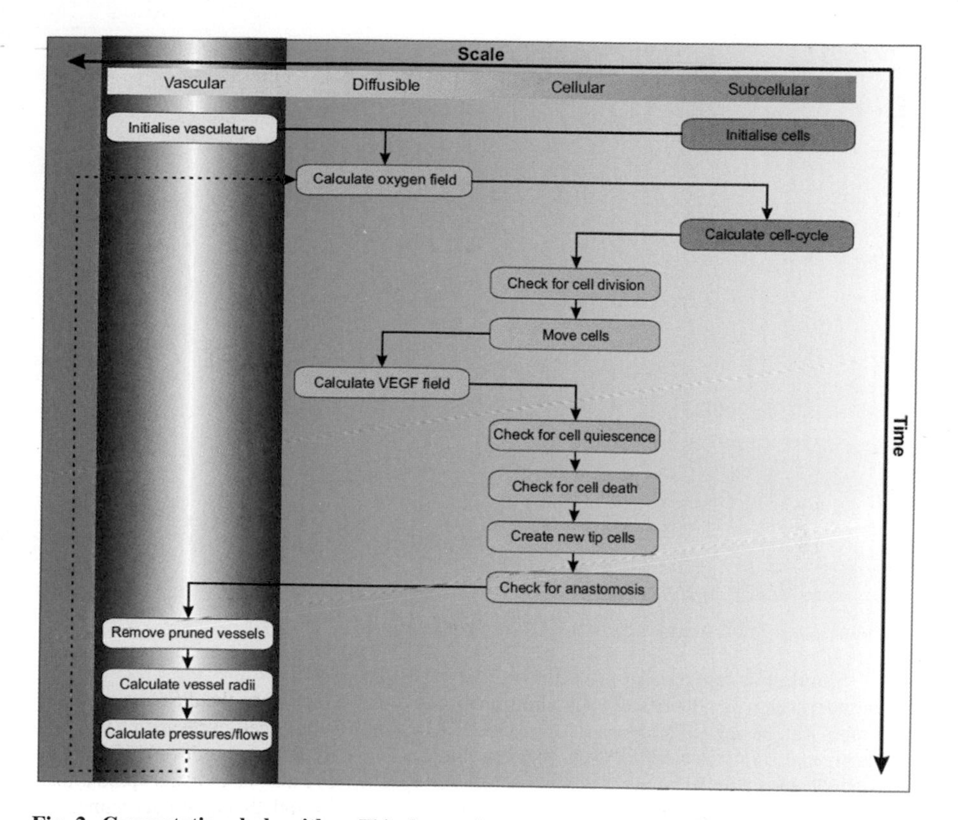

Fig. 2 Computational algorithm. This figure gives an overview of the computational algorithm and its progression in time and progression through different model layers (vascular, diffusible, cellular and sub-cellular) while it is proceeding in time. The simulations start with the initialisation of the vasculature and the cells. Then the oxygen concentration is calculated, followed by an update of the cell-cycle. Afterwards, we check for cell division, move the cells and calculate the VEGF field. Then it is checked if cells become hypoxic, death cells are removed and new tip cells generated. New tip-tip or tip-vessel connections anastomose and underperfused vessels are removed. Finally the vessel radii are updated and the current pressures and flows in the vascular system are calculated. Then the algorithm starts again with the update of the oxygen field

3 Multi-parallel Simulation Control and Visualisation Framework

To run our multiscale program on the CrayXE6 environment we had to first adapt the external libraries (CVODE and SuperLU) necessary to compute our simulation. We run a large number of simulations using different parameter sets and random seeds. The parameter files and job scripts are generated using MatLab/Java (see Fig. 3 for the workflow). For the purpose of parameter studies we applied a multi parallel application approach in which we run independent simulations with different parameters and random seeds.

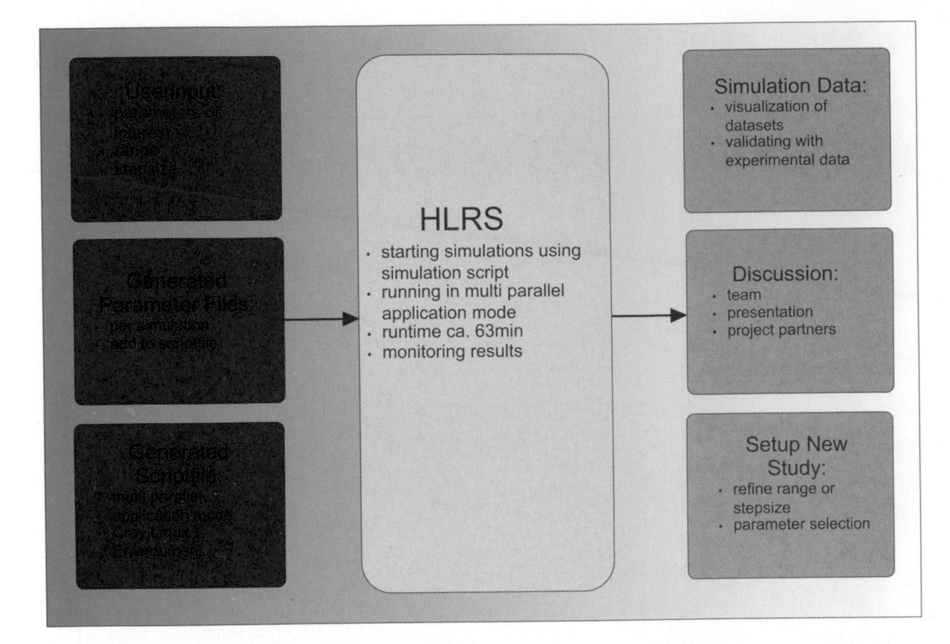

Fig. 3 Simulation control and workflow on the HLRS. This figure shows the workflow how our multi-instance simulations are run and organised on the HLRS. Parameters of interest and their range can be defined at the user input. Our MatLab/Java control programme then generates parameter and script files and uploads them to the HLRS. There the simulations are started in multi-application parallel mode. Simulation results are downloaded, stored and organised for visualisation and data analysis

For the simulations we define a template parameter file where all parameters are specified. To automate the process of parameter file generation we used a MatLab/Java programme that writes a parameter files based on the parameter names and values passed as arguments. A suitable submission script for the resource management system "TORQUE" incorporating all parameter files is also generated. The different parameter files were generated with an automate process that was set with a Matlab/Java programme. Once all the parameter and the script files have been generated, they are uploaded to the workspace and submitted. We run our simulations in parallel multi-instance mode with typically $\sim$50 instances per analysis cycle. The simulation process can be monitored during the computation by downloading intermediate data. Each simulation produces data in the range of $\sim$2 GB per instance and, for each instance, the running time takes about 45–100 min. A visualisation script was implemented to batch the visualisation of user defined simulation steps for all parallel simulations.

4 Simulations on the HLRS

An example of a typical tumour growth simulation is shown in Fig. 4. We start with a simulation domain, containing 18 straight vessel segments and then let the system evolve until a steady-state is reached. At steady-state the normal tissue is properly

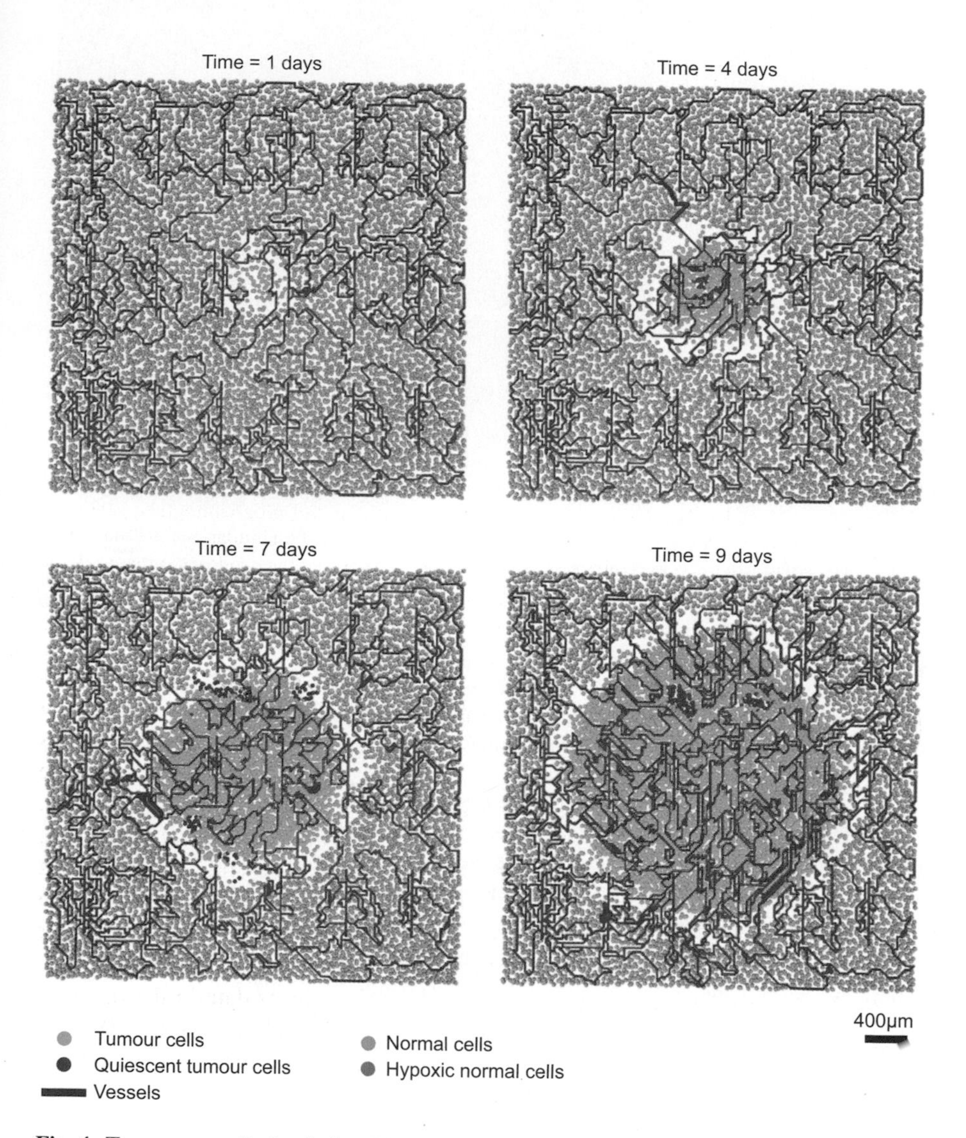

Fig. 4 Tumour growth simulation. This figure shows the evolution of a tumour within a normally vascularised healthy tissue. It can be observed that the tumour grows radially and induces an angiogenic response of the host vasculature. Normal cells are coloured in *pink*, cancer cells in *light green*, quiescent cancer cells in *dark green* and the vasculature in *red*

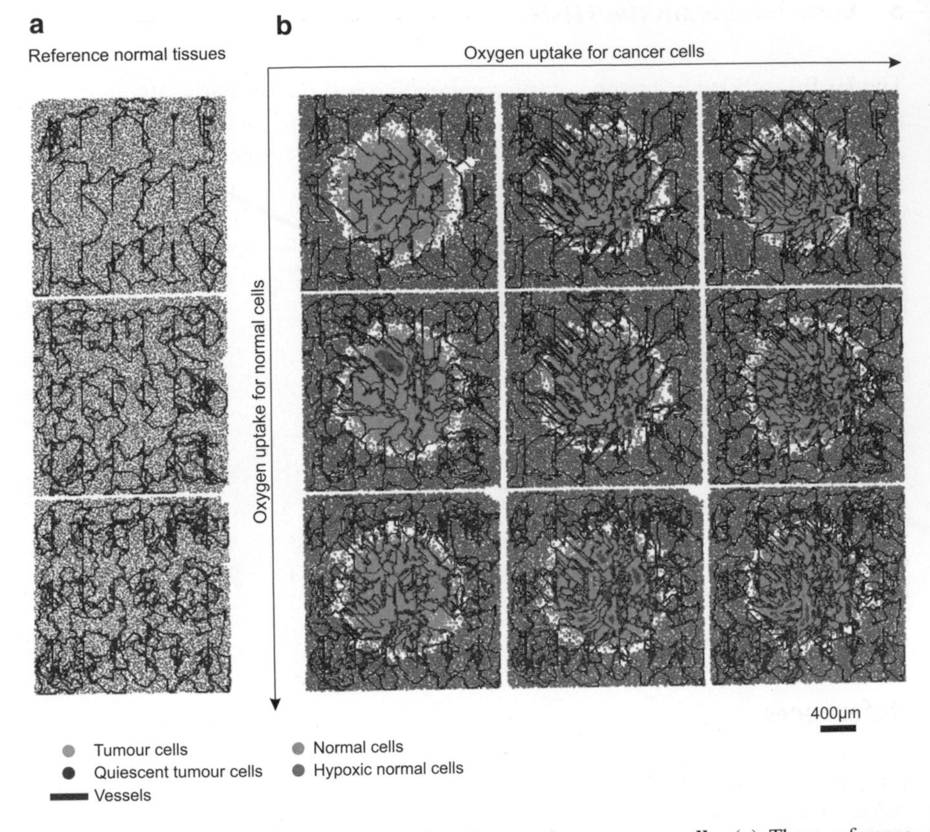

Fig. 5 Sensitivity analysis: oxygen uptake of normal vs. cancer cells. (**a**) Three reference tissues with different oxygen uptakes are generated, representing different vascularised tissues. (**b**) The oxygen uptake of normal and cancer cells are varied against each other. This approach was used to model different aggressive tumour in different vascularised host tissues

nourished and no VEGF is secreted. This normal tissue builds the initial condition for our tumour growth simulation, where we implant a small tumour in the centre of the domain and study its evolution. Then we can observe radial tumour growth and an angiogenic response of the vasculature.

Figure 5 gives an example of a parameter study in which we visualised tumours where we varied the oxygen uptake rate of normal and cancer cells. This study shows how differently aggressive tumours grow in differently vascularised host environments.

5 Conclusion and Outlook

The HLRS enabled us to run a large number of angiogenesis and vascular tumour growth simulations with an already established multiscale simulation framework [1, 2]. To organise the production of parameter files, to start multi-parallel simulations, and to transfer and organise the simulation results we implemented a simulation control framework. The framework (implemented in MatLab/Java) automatically generates parameter files for a given parameter-space. Then the programme starts parallel simulations on the HLRS and transfers the simulation results to our local harddrives. The framework is implemented in such a way that it can use the analysis of results to refine and change the parameter space for the next simulations. This provides us a loop for future model validation with experimental data. The performance gain by using the HLRS basically results from the ability to run a large number of independent simulations at the same time.

At the moment there is no parallel implementation of the vascular tumour growth model that would enable us to compute large scale simulations on the length-scale of single organs. The parallelisation of the whole multiscale programme could be a future project. There detailed patient specific data will be used to predict individual patients tumour evolution in personalised medicine.

References

1. Perfahl, H., Byrne, H.M., Chen, T., Estrella, V., Alarcón, T., Lapin, A., Gatenby, R.A., Gillies, R.J., Lloyd, M.C., Maini, P.K., Reuss, M., Owen, M.R.: Multiscale modelling of vascular tumor growth in 3D: the roles of domain size and boundary conditions
2. Owen, M.R., Stamper, I.J., Muthana, M., Richardson, G.W., Dobson, J., Lewis, C.E., Byrne, H.M.: Mathematical Modeling Predicts Synergistic Antitumor Effects of Combining a Macrophage-Based, Hypoxia- Targeted Gene Therapy with Chemotherapy
3. Supernodal LU – Li, X.S., Demmel, J., Gilbert, J., Grigori, L., Shao, M., Yamazaki, I.: http://crd-legacy.lbl.gov/~xiaoye/SuperLU/

Printing: Ten Brink, Meppel, The Netherlands
Binding: Ten Brink, Meppel, The Netherlands